Ferocactus wislizenii.

THE CACTACEAE

DESCRIPTIONS AND ILLUSTRATIONS OF PLANTS OF THE CACTUS FAMILY

BY

N. L. BRITTON AND J. N. ROSE

VOLUME III

THE CARNEGIE INSTITUTION OF WASHINGTON
WASHINGTON, 1922

CARNEGIE INSTITUTION OF WASHINGTON
PUBLICATION No. 248, VOLUME III

PRESS OF GIBSON BROTHERS
WASHINGTON

CONTENTS.

III

ILLUSTRATIONS.

PLATES.

IV

TEXT-FIGURES.

TEXT-FIGURES—continued.

TEXT-FIGURES—continued.

THE CACTACEAE

Descriptions and Illustrations of Plants of the Cactus
Family

DESCRIPTIONS AND ILLUSTRATIONS OF PLANTS OF THE CACTUS FAMILY.

Tribe 3. CEREEAE.

Subtribe 3. ECHINOCEREANAE.

Mostly low, simple or cespitose, terrestrial cacti, the stems 1-jointed or rarely few-jointed, ribbed; areoles borne on the ribs and spiniferous or rarely spineless; flowers always solitary at lateral* areoles, funnelform to campanulate; perianth-segments few to many; fruit smooth or spiny, with few exceptions fleshy and indehiscent or splitting on one side; seeds mostly black.

We recognize 6 genera, all South American except *Echinocereus*.

This subtribe, while somewhat uniform in its low, usually one-jointed stems, shows great variability in its flowers. Both *Chamaecereus* and *Austrocactus* are taken from *Cereus* of previous authors; *Echinocereus* has often been considered as a subgenus of *Cereus*. *Echinopsis* has usually been treated as a distinct genus related to *Cereus*, Bentham and Hooker, however, treating it as a subgenus of *Cereus*; in our opinion, it approaches *Trichocereus* in its flowers, but in habit resembles various genera in the *Echinocactanae*. *Lobivia* is segregated from *Echinopsis*. *Rebutia* has sometimes been recognized as a genus, but its species have usually been referred to *Echinocactus* or *Echinopsis*. The subtribe is nearest the *Cereanae*, but is also related to the *Echinocactanae*.

KEY TO GENERA.

```
Ovary and fruit bearing clusters of spines at areoles.
    Stigma-lobes always green; spines all straight..............................1. Echinocereus (p. 3)
    Stigma-lobes red; some spines hooked......................................2. Austrocactus (p. 44)
Ovary and fruit not spiny.
    Spines on tubercles as in Coryphantha; plants globular......................3. Rebutia (p. 45)
    Spines on ribs.
        Plants very small, creeping, forming low clumps.........................4. Chamaecereus (p. 48)
        Plants mostly large, solitary or cespitose.
            Flower short-funnelform to campanulate; tube short....................5. Lobivia (p. 49)
            Flower long-funnelform; tube elongated..............................6. Echinopsis (p. 60)
```

1. ECHINOCEREUS Engelmann in Wislizenus, Mem. Tour North. Mex. 91. 1848.

Plants always low, perennial, erect or prostrate, sometimes pendent over rocks and cliffs, single or cespitose, globular to cylindric, prostrate or pendent if elongated; spines of flowering and sterile areoles similar; flowers usually large, but in some species small, diurnal, but in some species not closing at night; perianth campanulate to short-funnelform, scarlet, crimson, purple or rarely yellow, the tube and ovary always spiny; stigma-lobes always green; fruit more or less colored, thin-skinned, often edible, spiny, the spines easily detached when mature; seeds black, tuberculate.

Type species: *Echinocereus viridiflorus* Engelmann.

The first part of the generic name is from ἐχῖνος hedgehog, doubtless given on account of the spiny fruit in which *Echinocereus* is conspicuously different from the true *Cereus*.

We recognize 60 species. Professor Schumann admitted 39 species in his monograph of 1898, while more than 190 species and varieties have been proposed by other authors.

The genus is confined to the western United States and Mexico. It extends as far east as central Texas, Oklahoma, and Kansas, north to Wyoming and Utah, west to the deserts of southern California, the Pacific coast, and islands of lower California, and south to the City of Mexico.

Echinocereus has often been combined with *Cereus*. Engelmann, as well as Berger, treated it as a subgenus of *Cereus*, but Schumann gave it generic rank. As we understand the genus, it is not close to *Cereus* proper, but is much nearer to some of the other genera.

Echinocereus baileyi is described as producing flowers from the young growth and appearing terminal; this habit has been observed in other species, but is inconstant.

3

In habit it simulates the South American genus, *Echinopsis*, while in flowers and fruits it comes near *Erdisia*, *Bergerocactus*, and *Wilcoxia*. While all the species are low in habit there is great variation in the manner and form of growth. Some are solitary; others grow in flat masses, and others in large rounded mounds. The flowers, while always having a spiny ovary and flower-tube and green stigma-lobes, have considerable variation in the shape and color of perianth-segments and in duration. The flower-buds as well as the young shoots are deep-seated in their origin and do not appear just at the areoles as in most cacti and hence must break through the epidermis when they develop. A somewhat similar result is produced in the flowering of some of the species of *Rhipsalis*. *Echinocereus* has been selected as the state flower of New Mexico.

Most plants of *Echinocereus* do not flower frequently in greenhouse cultivation.

The species are not readily grouped into series; our classification of them is largely artificial, taking flower-color as a more important character than it probably is in nature.

KEY TO SPECIES.

A. Flowers large, usually conspicuous, rarely only 2 to 3 cm. long.
  B. Stems covered with long weak bristles or hairs, resembling a small plant of *Cephalocereus senilis*.. 1. *E. delaetii*
  BB. Stems variously covered with spines or rarely spineless, never like the above.
    C. Flowers scarlet to salmon-colored, opening once, but lasting for several days.
      Stems usually weak, often trailing, or at least becoming prostrate; ribs nearly continuous.
        Flowers rosy red... 2. *E. scheeri*
        Flowers orange-red to salmon-colored.
          Flowers 8 to 11 cm. long; wool from areoles on flower-tube long.
            Flowers 8 to 10 cm. long; radial spines 9 or fewer...................... 3. *E. salm-dyckianus*
            Flowers 11 cm. long; radial spines 10 to 12......................... 4. *E. huitcholensis*
          Flowers 6 cm. long or less; wool from areoles on flowers shorter than subtending scale... 5. *E. pensilis*
      Stems usually erect and stout; ribs more or less tubercled.
        Plants forming large mounds, sometimes with 500 to 800 joints; spines white, long and flexuous... 6. *E. mojavensis*
        Plants in much smaller clusters; spines brownish or grayish, not long and flexuous.
          Plant body with 12 to 14 ribs.. 7. *E. leeanus*
          Plant body with 5 to 11 ribs (in one species 12).
            Ribs 5 to 8.. 8. *E. triglochidiatus*
            Ribs 9 to 12.
              Axils of flower-scales filled with long cobwebby hairs.
                Flowers 5 to 6 cm. long; spines yellowish at first..................... 9. *E. polyacanthus*
                Flowers 3 cm. long; spines reddish at first........................ 10. *E. pacificus*
              Axils of flower-scales bearing short hairs.
                Stems elongated and thinner than in *E. octacanthus*................. 11. *E. acifer*
              Stems short and thicker than in *E. acifer*.
                Stems pure green when old; central spine 1....................... 12. *E. octacanthus*
                Stems bluish green; central spines several.
                  Central spines 6; petals acutish........................... 13. *E. neo-mexicanus*
                  Central spines mostly 4, sometimes 3 or 5.
                    Central spines more or less angled, somewhat curved........... 14. *E. conoideus*
                    Central spines terete, straight.
                      Central spines white or straw-colored..................... 15. *E. coccineus*
                      Central spines gray to pinkish........................... 16. *E. rosei*
    CC. Flowers broad, rotate to campanulate, opening in sunlight, closing at night, usually purple, sometimes yellow or greenish yellow, rarely pink or nearly white, unknown in *E. standleyi*.
      D. Flowers yellow or greenish white.
        Ribs not strongly tubercled.
          Plants densely cespitose.. 17. *E. maritimus*
          Plants usually solitary.
            Ribs very stout.
              Ribs 5 to 8; spines on flower-tube and ovary short.................. 18. *E. subinermis*
              Ribs 8 or 9; spines on flower-tube and ovary acicular................ 19. *E. luteus*
            Ribs low, usually hidden by the spines.
              Flowers small, 2.5 cm. long or less.
                Areoles circular....................................... 20. *E. chloranthus*
                Areoles elliptic....................................... 21. *E. viridiflorus*
              Flowers large, 5 to 10 cm. long.
                Flowers greenish white................................. 22. *E. grandis*
                Flowers yellow-red.

KEY TO SPECIES—continued.

Central spines in more than 1 row............................ 23. E. dasyacanthus
Central spines in a vertical row............................. 24. E. ctenoides
Ribs strongly tubercled... 25. E. papillosus
DD. Flowers purple.
E. Stems weak, slender, and creeping.
Stems 2 cm. thick or less.
Areoles distant; spines not interlocking.
Perianth-segments narrowly oblong or linear-oblanceolate........... 26. E. blanckii
Perianth-segments oblong-erose.............................. 27. E. pentalophus
Areoles approximate; spines densely interlocking..................... 28. E. sciurus
Stems 3 to 4 cm. thick.. 29. E. cinerascens
EE. Stems stout, usually erect or ascending.
F. Areoles elliptic to circular, closely set, often with pectinate spines.
Areoles elliptic; spines pectinate.
Central spine often very long.
Central spine dark... 30. E. adustus
Central spine white.. 31. E. standleyi
Central spine, if present, short.
Spines of ovary and tube of flower slender and weak, the surrounding
hairs long and cobwebby.
Spines variegated..................................... 32. E. perbellus
Spines of one color.
Spines strongly pectinate and appressed..................... 33. E. reichenbachii
Spines not strongly pectinate, more or less porrect............. 34. E. baileyi
Spines of ovary and tube of flower short and stout, the surrounding
hairs short.
Central spine none.
Stems cylindric....................................... 35. E. rigidissimus
Stems globular....................................... 36. E. weinbergii
Central spines present................................... 37. E. pectinatus
Areoles circular; spines not pectinate.
Central spines brown, much longer than white radials.............. 38. E. fitchii
Central spines not longer than radials.
Areoles about 5 mm. apart; spines densely interlocking........... 39. E. scopulorum
Areoles about 1 cm. apart; spines scarcely interlocking............ 40. E. roetteri
FF. Areoles nearly circular, not so closely set; spines never pectinate.
Ovary strongly tuberculate.................................... 41. E. chlorophthalmus
Ovary not strongly tuberculate.
Flowers small, 2.5 to 5 cm. long.
Plants strongly angled; flower pinkish...................... 42. E. knippelianus
Plants not strongly angled; flower purple.
Central spines none.
Spines 3 to 5; flower-tube and ovary without long wool from
the areoles..................................... 43. E. pulchellus
Spines 6 to 8; flower-tube and ovary bearing long cobwebby
wool from the areoles .... 44. E. amoenus
Central spines 1 or more.
Central spine 1....................................... 45. E. palmeri
Central spines several, much elongated, dagger-like......... 46. E. brandegeei
Flowers large, 6 to 12 cm. long.
Central spines none................................... 47. E. hempelii
Central spines present.
Central spine solitary, rarely 2.
Spines red at base................................. 48. E. merkeri
Spines not red at base.
Plants stout, erect.............................. 49. E. fendleri
Plants weak, becoming prostrate.................. 50. E. enneacanthus
Central spines several.
Spines never white.
Spines yellowish brown to red.
Spines short, usually stout, 10 mm. long.............. 51. E. lloydii
Spines, at least some of them, very long.............. 52. E. engelmannii
Spines bluish to blackish........................... 53. E. sarissophorus
Spines usually white or straw-colored.
Ribs 7 to 9....................................... 54. E. dubius
Ribs 11 to 13.
Flowers campanulate........................... 55. E. conglomeratus
Flowers short-funnelform...................... 56. E. stramineus
AA. Flowers minute, 1.2 cm. long or less... 57. E. barthelowanus
AAA. Published species not grouped... { 58. E. mamillatus
{ 59. E. ehrenbergii
{ 60. E. longisetus

1. Echinocereus delaetii Gürke, Monatsschr. Kakteenk. **19**: 131. 1909.

> *Cephalocereus delaetii* Gürke, Monatsschr. Kakteenk. **19**: 116. 1909.

Low, 1 to 2 dm. high, densely cespitose, completely hidden by the long, white, curled hairs; ribs indistinct; areoles closely set, bearing 15 or more white reflexed hairs 8 to 10 cm. long and a few stiff reddish bristles; flowers appearing near top of plant; perianth-segments pink, oblanceolate, acute; stigma-lobes about 12; ovary covered with clusters of long, white, bristly spines; fruit not seen.

Type locality: Not cited.

Distribution: Known only from Sierra de la Paila, north of Parras, Mexico.

This is the most remarkable species in the genus; in aspect it resembles small plants of *Cephalocereus senilis*, and owing to this resemblance it was first described as a *Cephalocereus*. Its flowers, however, are so different from those of that genus that as soon as they were seen the plant was at once transferred to *Echinocereus*.

The plant is now largely imported into Europe and can be obtained from many dealers; it was named in honor of Frantz de Laet, a Belgian cactus dealer, who had imported many plants from Mexico through Dr. C. A. Purpus and other collectors.

Illustrations: Monatsschr. Kakteenk. **19**: 119, as *Cephalocereus delaetii*; Monatsschr. Kakteenk. **22**: 73; Rev. Hort. Belge **40**: after 184.

Text-figure 1 is from a photograph of the plant received from M. de Laet.

2. Echinocereus scheeri (Salm-Dyck) Rümpler in Förster, Handb. Cact. ed. 2. 801. 1885.

> *Cereus scheeri* Salm-Dyck, Cact. Hort. Dyck. 1849. 190. 1850.

Cespitose; stems procumbent, prostrate or ascending, decidedly narrowed towards the tip, 10 to 22 cm. long, yellowish green; ribs 8 to 10, rather low, not at all sinuate, somewhat spiraled; spines 7 to 12, acicular, white with brown or blackish tips; flowers 12 cm. long, rose-red to crimson, with an elongated tube; perianth-segments oblanceolate, acute; fruit not known.

Type locality: Near Chihuahua.

Distribution: Chihuahua, Mexico.

The species was named for Frederick Scheer (1792–1868), who described the cacti for Seemann in the Botany of the *Herald.*

Fɪɢ. ɪ.—Echinocereus delaetii.

Lemaire used this name as early as 1868 (Les Cactées 57), but did not formally transfer or describe it, and it is not published or even mentioned by Lemaire in Manuel de l'Amateur de Cactus (1845) as stated in Blühende Kakteen under plate 14. It seems to be a distinct species, related to *E. salm-dyckianus*, but with differently colored flowers and shorter spines. The variety *E. scheeri nigrispinus* was used by Scheer in Botany of the *Herald* 291.

The type of this species seems to have been lost; it was collected by John Potts, a mining engineer, at one time stationed in Chihuahua, Mexico. We know the species only from description and illustrations.

Echinocereus scheeri vars. *major* and *minor* (Monatsschr. Kakteenk. **15**: 175. 1905) and var. *robustior* (Monatsschr. Kakteenk. **15**: 161. 1905) are only garden forms.

Illustrations: Curtis's Bot. Mag. **132**: pl. 8096, as *Cereus scheeri;* Blühende Kakteen **1**: pl. 14; Schumann, Gesamtb. Kakteen f. 48.

Text-figure 2 is from a part of the second illustration above cited.

3. **Echinocereus salm-dyckianus** Scheer in Seemann, Bot. Herald 291. 1856.

Cereus salm-dyckianus Hemsley, Biol. Centr. Amer. Bot. 1: 545. 1880.
Echinocereus salmianus Rümpler in Förster, Handb. Cact. ed. 2. 809. 1885.
Cereus salmianus Weber, Dict. Hort. Bois 279. 1894.

Cespitose; stems more or less decumbent, 2 to 4 cm. in diameter, elongated, yellowish green; ribs 7 to 9, low, more or less sinuate; radial spines 8 or 9, acicular, yellowish, about 1 cm. long; central spine solitary, porrect, a little longer than the radials; flowers orange-colored, 8 to 10 cm. long, narrow, the tube elongated, the areoles of the flower-tube and ovary bearing white bristly spines and cobwebby hairs; perianth-segments oblanceolate to spatulate; filaments dark red; style longer than the stamens; fruit not seen.

Type locality: Near Chihuahua.
Distribution: States of Chihuahua and Durango, Mexico.

This species is in cultivation in Europe, and Dr. Rose saw it in flower at La Mortola in 1912; it was also collected by Dr. E. Palmer in Durango in 1906 (No. 205).

FIG. 2.—Echinocereus scheeri.   FIG. 3.—Echinocereus salm-dyckianus.

We have not been able to see the type specimen and it is probably not in existence.

The specific name commemorates Joseph Franz Salm-Reifferschid-Dyck (1773–1861), author of several important cactus treatises. He was the most distinguished cactologist of his time and possessed at his estate at Düsseldorf, Germany, one of the largest cactus collections ever brought together. Unfortunately, after his death the collection was permitted to disintegrate and most of his types were lost or thrown away.

Hybrids between *E. salm-dyckianus* and *Heliocereus speciosus* and with *Epiphyllum* species are reported.

Illustrations: Blühende Kakteen 1: pl. 29; Monatsschr. Kakteenk. 3: 129; Wildeman, Icon. Select. 6: pl. 202.

Text-figure 3 shows a part of the first illustration above cited.

4. Echinocereus huitcholensis (Weber) Gürke, Monatsschr. Kakteenk. **16**: 23. 1906.

Cereus huitcholensis Weber, Bull. Mus. Hist. Nat. Paris **10**: 383. 1904.

Plants 4 to 6 cm. long, 2 to 4 cm. in diameter; radial spines 10 to 12; central spine usually solitary; flowers 11 cm. long or less, narrow, with a pronounced tube; color of perianth-segments uncertain but perhaps orange, as in *E. salm-dyckianus;* spines on ovary and tube weak, acicular; areoles of flower-tube bearing long cobwebby hairs.

Type locality: Sierra de Nayarit, Jalisco, Mexico.

Distribution: Known only from the type locality.

Weber described it as a *Cereus*, but without seeing flowers or fruit, basing it on the collection of L. Diguet of 1900. Three sheets of this collection are in the herbarium of the Museum of Paris, and with them are a flower, immature fruit, and two plants of another species, perhaps undescribed.

5. Echinocereus pensilis (K. Brandegee) J. A. Purpus, Monatsschr. Kakteenk. **18**: 5. 1908.

Cereus pensilis K. Brandegee, Zoe **5**: 192. 1904.

More or less cespitose; the stems often erect, 30 cm. high or, when growing on cliffs, hanging, and then nearly 2 meters long, 3 to 4 cm. in diameter; ribs 8 to 10, low; areoles about 10 mm. apart; spines needle-like, at first yellow, becoming reddish gray, the longest not over 2 cm. long; radial spines about 8; central spine 1; flowers orange-red, narrow, 5 to 6 cm. long; areoles on ovary and tube bearing short, yellow or white wool and chestnut-colored bristly spines; fruit globular, 1.5 to 2 cm. in diameter; seeds black, rugose, very oblique at base.

Type locality: Sierra de la Laguna, Lower California.

Distribution: High mountains of the Cape region of Lower California.

This species is unlike most of the known Lower Californian species in that it grows in the high mountains of the Cape region and is in fact more closely related to the species of the mountains of the United States and Mexico than to any of its near neighbors. It is a beautiful plant; Dr. Rose saw it in flower in Darmstadt in June 1912, where it was grown from a hanging basket.

The type specimen is in the Brandegee Herbarium at the University of California, but a duplicate and a photograph of the type are preserved in the United States National Herbarium.

Illustration: Monatsschr. Kakteenk. **18**: 3.

Fig. 4.—Echinocereus mojavensis.

6. Echinocereus mojavensis* (Engelmann and Bigelow) Rümpler in Förster, Handb. Cact. ed. 2. 803. 1885.

Cereus mojavensis Engelmann and Bigelow, Proc. Amer. Acad. **3**: 281. 1856.
Cereus bigelovii Engelmann, Pac. R. Rep. **4**: pl. 4, f. 8. 1856.

Cespitose, growing in massive clumps, often forming mounds, with hundreds of stems (500 to 800 have been recorded); stems globose to oblong, 5 to 20 cm. long, pale green; ribs 8 to 13, 5 to 6 mm. high, but becoming indistinct on old parts of stem, somewhat undulate; areoles circular, about 1 cm. apart; spines all white, or in age gray; radial spines about 10, acicular, spreading, curved, 1 to 2.5 cm. long; central spine subulate, porrect or somewhat spreading, often weak, 3 to 5 cm. long; flowers rather narrow, 5 to 7 cm. long, crimson; perianth-segments broad, obtuse or even retuse; areoles on ovary with white felt and short acicular spines; fruit oblong, 2.5 to 3 cm. long.

Type locality: On the Mojave River in California.

Distribution: Southeastern California to Nevada and Utah, western Arizona, and reported from northwestern Mexico.

---

* This species was named for the Mojave Desert, California, where it was first found. The specific name is sometimes incorrectly spelled *mohavensis.* Munz and Johnston report that the flowers are "pale scarlet tinged with nopal-red."

Engelmann compares this species with *E. fendleri* and in that relationship most writers have since treated it, but in its habit and shape of flower it suggests a closer relationship to the scarlet-flowered species such as *E. polyacanthus.*

Cereus bigelovii was doubtless the first name applied to this plant, but for some reason it was afterward changed, although not in the case of the legend for the first illustration cited below.

Illustrations: Pac. R. Rep. **4**: pl. 4, f. 8, as *Cereus bigelovii;* Curtis's Bot. Mag. **126**: pl. 7705, as *Cereus mojavensis.*

Text-figure 4 is from a photograph obtained through S. B. Parish, taken by Dr. P. A. Munz near Pinos Wells, southern edge of the Mojave Desert, altitude 1,335 meters.

7. Echinocereus leeanus (Hooker) Lemaire in Förster, Handb. Cact. ed. 2. 828. 1885.

> *Cereus leeanus* Hooker in Curtis's Bot. Mag. **75**: pl. 4417. 1849.
> *Echinocereus multicostatus* Cels in Förster, Handb. Cact. ed. 2. 834. 1885.
> *Echinocereus leeanus multicostatus* Schumann, Gesamtb. Kakteen 289. 1898.

Plant erect, about 3 dm. high, 1 dm. thick at base, tapering gradually toward the top, simple so far as known; ribs 12 to 14, acute, bearing rather closely set areoles; spines about 12, acicular, very unequal in length, the central and longest about 2.5 cm. long; flowers brick-red, 5 to 6 cm. long; inner perianth-segments somewhat spreading, spatulate to obovate, 3 cm. long, acute; filaments elongated, quite as long as the style.

Type locality: Northern Mexico.

Distribution: Mexico, but range undetermined.

The only herbarium specimens so-named which we have seen are two sheets in the herbarium of the Berlin Botanical Garden, representing two flowers which are scarlet, slender, 9 cm. long, with pale brownish spine-clusters intermixed with cobwebby white hairs. We have studied a plant sent from the Berlin Botanical Garden to the New York Botanical Garden which has not flowered.

This species differs from its relatives in having more numerous ribs.

The type specimen was presented to the Royal Botanic Gardens at Kew about 1842 by Mr. James Lee, owner of the Commercial Gardens at Hammersmith, near London, for whom it was named, and it is said to have come to him from France. *Echinocereus pleiogonus* and *E. multicostatus*, which Schumann refers here as synonyms, were described about the same time from specimens introduced into France by Cels. It is not at all unlikely that all three had a common origin. *Cereus multicostatus* Cels (Schumann, Gesamtb. Kakteen 288. 1898) is only a catalogue name.

FIG. 5.—Echinocereus leeanus.

A small specimen of this species obtained from Berlin in 1914 has been grown at the New York Botanical Garden, presumably correctly identified. The young areoles are white-woolly; the spines are acicular, the outer ones white, the central ones with brown tips.

Illustrations: Curtis's Bot. Mag. **75**: pl. 4417; Gard. Mag. Bot. **2**: pl. facing 81, as *Cereus leeanus;* Schumann, Gesamtb. Kakteen f. 49.

Text-figure 5 is from a photograph of the first illustration above cited.

8. Echinocereus triglochidiatus Engelmann in Wislizenus, Mem. Tour North. Mex. 93. 1848.

> *Cereus triglochidiatus* Engelmann in Gray, Pl. Fendl. 50. 1849.
> *Cereus gonacanthus* Engelmann and Bigelow, Proc. Amer. Acad. 3: 283. 1856.
> *Cereus paucispinus* Engelmann, Proc. Amer. Acad. 3: 285. 1856.
> *Cereus hexaedrus* Engelmann and Bigelow, Proc. Amer. Acad. 3: 285. 1856.
> *Echinocereus paucispinus* Rümpler in Förster, Handb. Cact. ed. 2. 794. 1885.
> *Echinocereus gonacanthus* Rümpler in Förster, Handb. Cact. ed. 2. 806. 1885.
> *Echinocereus hexaedrus* Rümpler in Förster, Handb. Cact. ed. 2. 807. 1885.
> *Echinocereus paucispinus triglochidiatus* Schumann, Gesamtb. Kakteen 281. 1898.
> *Echinocereus paucispinus gonacanthus* Schumann, Gesamtb. Kakteen 281. 1898.
> *Echinocereus paucispinus hexaedrus* Schumann, Gesamtb. Kakteen 281. 1898.

Always cespitose, with few or many simple stems, these 2 to 6 dm. long, 5 to 8 cm. in diameter, deep green, erect or spreading, 5 to 8-ribbed; spines 3 to 8, various, nearly terete to strongly angled, when young reddish to yellow, but gray in age, usually spreading, often all radial, 3 cm. long or less; flowers scarlet, 5 to 7 cm. long; perianth-segments oblong, obtuse, 3 cm. long; areoles on the flower-tube and ovary few, white-felted, the subtending scales small and red; spines on ovary and flower-tube few, red and white; fruit at first spiny, but in age smooth, bright red, 3 cm. in diameter; seeds 1.6 mm. in diameter or less.

FIGS. 6 and 7.—Echinocereus triglochidiatus.

Type locality: Wolf Creek, New Mexico.

Distribution: Western Texas, New Mexico, and Colorado.

The species here described is very variable as to habit and number and kind of spines, and has generally been separated into three or four species; Schumann treated it as a single species with three varieties; we have recognized only a single species, but it is possible that *Echinocereus paucispinus* which has nearly terete spines should be restored for certain plants in Texas.

Echinocereus monacanthus Heese (Gartenflora 53: 215. f. 32, with wrong legend) may belong here. It is a small one-jointed plant, 10 cm. long, with 7 ribs, and bears but a single spine at an areole. The flowers and fruit are unknown. The plant is said to be a native of Mexico and Texas.

Illustrations: Pac. R. Rep. 4: pl. 5, f. 2, 3, as *Cereus gonacanthus;* Pac. R. Rep. 4: pl. 5, f. 1, as *C. hexaedrus;* Cact. Mex. Bound. pl. 56, as *C. paucispinus;* Schelle, Handb. Kakteenk. 137. f. 66; Blühende Kakteen 3: pl. 124; Rümpler, Sukkulenten 139. f. 74; Förster, Handb. Cact. ed. 2. 793. f. 102, as *Echinocereus paucispinus;* N. M. Agr. Exp. Sta. Bull. 78: pl. 16, as *E. gonacanthus;* Pac. R. Rep. 4: pl. 4, f. 6, 7, as *Cereus triglochidiatus;* Cact. Journ. 2: 18, as *Echinocereus paucispinus flavispinus.*

Text-figure 6 is from a photograph of a plant sent to the New York Botanical Garden by Dr. Rose in 1913, from near Las Vegas, New Mexico; text-figure 7 shows a flowering plant sent to Washington from near Kerrville, Texas, by Mr. B. Mackensen in 1912.

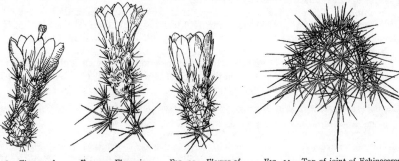

FIG. 8.—Flower of Echinocereus polyacanthus. ×o.6.

FIG. 9.—Flowering branch of Echinocereus pacificus. ×o.6.

FIG. 10.—Flower of Echinocereus neomexicanus. ×o.6.

FIG. 11.—Top of joint of Echinocereus conoideus.

9. Echinocereus polyacanthus Engelmann in Wislizenus, Mem. Tour North. Mex. 104. 1848.

Cereus polyacanthus Engelmann in Gray, Pl. Fendl. 50. 1849.

Cespitose, forming clumps of 20 to 50 stems, pale green but often tinged with red; ribs usually 10, low; areoles approximate; spines gray when old, at first pale yellow, becoming more or less purplish; radial spines about 12; centrals 4, straight, elongated; flowers crimson, 6 cm. long; spines on ovary and flower-tube yellow, intermixed with cobwebby wool; fruit and seeds unknown.

Type locality: Cosihuiriachi, Chihuahua.

Distribution: Chihuahua and Durango, Mexico, to western New Mexico and southeastern Arizona.

Echinocereus polyacanthus was described by Dr. Engelmann in 1848, based upon specimens collected by Dr. A. Wislizenus at Cosihuiriachi, a small mining town west of the city of Chihuahua.   The next year Dr. Engelmann transferred it along with the other species of his genus, *Echinocereus*, to *Cereus*, and in 1859, in his report on the Cactaceae of the Mexican Boundary, redescribed and illustrated the species; the specimens used by him for this report, however, were largely from Texas and New Mexico, and this additional material represents a quite distinct species.   In order to prove this point Dr. Rose in 1908 visited Cosihuiriachi, the type locality, and collected living and herbarium specimens which were found to be specifically distinct from the so-called *E. polyacanthus* from the

FIG. 12.—Echinocereus polyacanthus.

El Paso region which now bears the name *E. rosei* Wooton and Standley.   The habit of the two species is similar, but the armament is somewhat different and the flowers of the true *Echinocereus polyacanthus* produce an abundance of wool in the axils of the scales which is lacking in the other species.   The distribution of this species is much more restricted than has usually been given for it; it has been reported from Texas to California and as far south as La Paz, Lower California.   The plant illustrated on plate 66 of Blühende Kakteen as this species must be referred elsewhere.

Illustrations: Tribune Hort. **4**: pl. 139, as *Echinocereus polyacanthus* var.; Förster, Handb. Cact. ed. 2. f. 101; (?) Monatsschr. Kakteenk. **15**: 41; (?) **17**: 169; Schelle, Handb. Kakteenk. 139. f. 67; Shreve, Veg. Des. Mt. Range pl. 24; Bull. Torr. Club **35**: 83. f. 1; Cact. Journ. **1**: 89.

Figure 8 shows a flower of an herbarium specimen collected by Dr. Palmer near Madera, Chihuahua, in 1908; figure 12 is from a photograph of a plant collected by Professor Lloyd in the Santa Catalina Mountains, Arizona.

10. Echinocereus pacificus (Engelmann).

> Cereus phoeniceus pacificus Engelmann, West Amer. Sci. **2**: 46.   1886.
> Cereus pacificus Coulter, Contr. U. S. Nat. Herb. **3**: 397.   1896.

Cespitose, growing in clumps 30 to 60 cm. in diameter, sometimes containing 100 stems, these 15 to 25 cm. long, 5 to 6 cm. in diameter; ribs 10 to 12, obtuse; spines gray, with a reddish tinge; radial spines 10 to 12, 5 to 10 mm. long; central spines 4 or 5, the longest sometimes 25 mm. long; flowers deep red, rather small, about 3 cm. long; areoles on ovary and flower-tube bearing long tawny wool and reddish-brown bristly spines; fruit spiny.

Type locality: Todos Santos Bay, Lower California.

Distribution: Northern Lower California, recorded, apparently erroneously, from farther south.

Although the type is from the coastal hills we are inclined to refer here Mr. Brandegee's plant from the San Pedro Martir, collected May 5, 1893; the specimen shows flowers and a spine-cluster.

Mr. Brandegee's plant from Comondu Cliffs, also referred here by Coulter, may belong elsewhere; it is without flowers, however, and we are uncertain of its relationship.   The spines are long and acicular and Mr. Brandegee's notes state that the stems are not dense but sometimes hang from the rocks.

Figure 9 shows a small flowering branch of an herbarium specimen collected by C. R. Orcutt in northern Lower California in 1883.

11. Echinocereus acifer (Otto) Lemaire in Förster, Handb. Cact. ed. 2. 798.   1885.

> Cereus acifer Otto in Salm-Dyck, Cact. Hort. Dyck. 1849. 189.   1850.
> ? Echinopsis valida densa Regel, Gartenflora **1**: 295.   1852.
> Echinocereus acifer tenuispinus Jacobi in Förster, Handb. Cact. ed. 2. 798.   1885.
> Echinocereus acifer brevispinulus Jacobi in Förster, Handb. Cact. ed. 2. 798.   1885.
> Echinocereus durangensis Rümpler in Förster, Handb. Cact. ed. 2. 799.   1885.
> Echinocereus durangensis nigrispinus Rümpler in Förster, Handb. Cact. ed. 2. 800.   1885.
> Echinocereus durangensis rufispinus Rümpler in Förster, Handb. Cact. ed. 2. 800.   1885.
> Echinocereus acifer trichacanthus Hildmann, Monatsschr. Kakteenk. **1**: 44.   1891.
> Echinocereus acifer durangensis Schumann, Gesamtb. Kakteen 287.   1898.
> Echinocereus acifer diversispinus Schumann, Gesamtb. Kakteen 287.   1898.

Cespitose, glossy green, erect; ribs 10, strongly tubercled; radial spines 5 to 10, 10 to 16 mm. long, pale brownish, bulbose and purplish at base; centrals 4 (Schumann says 1), stout, purplish brown, the three upper erect, the lower and stouter one subdeflexed; flowers scarlet.

Type locality: Not cited.

Distribution: Durango and Coahuila, according to Professor Schumann.

Professor Schumann recognized three varieties, based chiefly on the differences in the spines.

We have studied a small plant secured from the Berlin Botanical Garden.

The illustration in Blühende Kakteen cited below shows a plant with almost continuous ribs and one stout central spine.   It presumably represents a different species.

Echinocereus acifer was mentioned by Lemaire in 1868 (Les Cactées 57), but the name was not published at that place.

Illustrations: Monatsschr. Kakteenk. 1: opp. 44, as *Echinocereus acifer trichacanthus;* Förster, Handb. Cact. ed. 2. 637. f. 85; (?) Gartenflora 1: pl. 29, as *Echinopsis valida densa;* (?) Blühende Kakteen 2: pl. 106; Gartenwelt 9: 410; Gard. Chron. III. 36: 245. f. 100.

12. Echinocereus octacanthus (Mühlenpfordt).

Echinopsis octacantha Mühlenpfordt, Allg. Gartenz. 16: 19. 1848.
Cereus roemeri Engelmann in Gray, Pl. Fendl. 50. 1849. Not Mühlenpfordt, 1848.
Echinocereus roemeri Rümpler in Förster, Handb. Cact. ed. 2. 792. 1885.
Cereus octacanthus Coulter, Contr. U. S. Nat. Herb. 3: 395. 1896.

Cespitose, with many simple joints; joints ovoid, yellowish green, 7 to 10 cm. long, 5 to 7 cm. in diameter; ribs 7 to 9, obtuse, somewhat tubercled; areoles when young white-woolly, in age naked, 8 to 16 mm. apart; spines rigid, grayish brown; radial spines 7 or 8, 10 to 24 mm. long; central spine solitary, stouter than the radials, porrect, 2 to 3 cm. long; flowers red, 5 cm. long, remaining open for several days; fruit unknown.

Type locality: Northern Texas.

Distribution: Known to us definitely only from northwestern Texas, but reported by Coulter from New Mexico and Utah.

M. Cary in 1907 collected at Dolores, Colorado, a plant which comes nearer this species than anything which we have seen, except a plant from Marathon, Texas, which has the armament and flowers called for by the original description. The plant referred to in the following illustration in the Garden is said to have come from northern California but this is undoubtedly an error.

Illustrations: Hort. Franc. II. 7: pl. 22; Garden 13: 291, as *Cereus roemeri.*

13. Echinocereus neo-mexicanus Standley, Bull. Torr. Club 35: 87. 1908.

Cespitose, but with only a few stout simple joints, 18 to 25 cm. long, 7 cm. in diameter, obtuse, glaucous-green; ribs 11 or 12, obtuse, low, somewhat tuberculate; areoles 1 to 1.5 cm. apart; spines slender, subulate, somewhat spreading; radial spines 13 to 16, the longest only 1.5 cm. long, white or nearly so; central spines 6, the lowest one yellowish to almost white, the others reddish, sometimes 4 cm. long; flowers abundant, appearing near the top or along the sides of the plant, 5 cm. long, narrow and not spreading at the mouth, bright scarlet; perianth-segments acute, firm in texture, 2 cm. long, 6 mm. broad; stamens about half as long as the style; stigma-lobes 7; ovary and flower-tube bearing clusters of bristly spines; fruit not known.

Type locality: Mesa west of the Organ Mountains, New Mexico.

Distribution: Known only from the type locality.

Illustrations: Bull. Torr. Club 35: f. 3, 4, 5.

Figure 10 is copied from the illustration above cited.

14. Echinocereus conoideus (Engelmann and Bigelow) Rümpler in Förster, Handb. Cact. ed. 2. 807. 1885.

Cereus conoideus Engelmann and Bigelow, Proc. Amer. Acad. 3: 284. 1856.
Echinocereus phoeniceus conoideus Schumann, Gesamtb. Kakteen 283. 1898.

Plants cespitose; joints somewhat conic at apex; ribs 9 to 11; radial spines 10 to 12, slender, rigid; central spines 2.5 to 8 cm. long, generally 5 cm. long; flowers 6 cm. long, scarlet, slender; ovary and flower-tube spiny.

Type locality: On the Upper Pecos, New Mexico.

Distribution: Southeastern New Mexico and western Texas.

The species is closely related to *Echinocereus coccineus,* perhaps not specifically distinct from it.

Coulter takes for this species the name *Cereus roemeri* Mühlenpfordt (Allg. Gartenz. 16: 19. 1848), which we refer to *E. coccineus.*

Illustrations: Pac. R. Rep. **4**: pl. 4, f. 4, 5, as *Cereus conoideus;* N. Mex. Agr. Exp. Sta. Bull. **78**: pl. [17]; Bull. Torr. Club **35**: 85. f. 2.

Figure 11 is copied from the first illustration above cited.

15. Echinocereus coccineus Engelmann in Wislizenus, Mem. Tour North. Mex. 94. 1848.

Cereus roemeri Mühlenpfordt, Allg. Gartenz. **16**: 19. 1848.
Cereus coccineus Engelmann in Gray, Pl. Fendl. 50, 51. 1849. Not Salm-Dyck, 1828.
? Echinopsis valida densa Regel, Gartenflora **1**: 295. 1852.
Cereus mojavensis zuniensis Engelmann, Proc. Amer. Acad. **3**: 281. 1856.
Cereus phoeniceus Engelmann, Proc. Amer. Acad. **3**: 284. 1856.
Echinocereus phoeniceus Rümpler in Förster, Handb. Cact. ed. 2. 788. 1885.
Echinocereus phoeniceus albispinus Rümpler in Förster, Handb. Cact. ed. 2. 789. 1885.
Echinocereus phoeniceus longispinus Rümpler in Förster, Handb. Cact. ed. 2. 789. 1885.
Echinocereus phoeniceus rufispinus Rümpler in Förster, Handb. Cact. ed. 2. 789. 1885.
Echinocereus krausei De Smet in Förster, Handb. Cact. ed. 2. 789. 1885.
Echinocereus mojavensis zuniensis Rümpler in Förster, Handb. Cact. ed. 2. 803. 1885.
Echinocereus phoeniceus inermis Schumann, Monatsschr. Kakteenk. **6**: 150. 1896.
Echinocereus roemeri Rydberg, Bull. Torr. Club **33**: 146. 1906. Not Rümpler, 1885.

Usually densely cespitose, often forming large mounds a meter in diameter, containing sometimes 200 simple stems, these 2 dm. high or less, 3 to 5 cm. in diameter; ribs 8 to 11, somewhat tubercled; radial spines acicular, 8 to 12, 1 to 2 cm. long, usually white; central spines several, longer and stouter than the radials, usually yellowish or whitish but in some specimens reddish or blackish; flowers crimson, 5 to 7 cm. long; perianth-segments broad, obtuse or retuse; areoles on flower and ovary felted and bearing short white bristly spines.

Type locality: About Santa Fé, New Mexico.

Distribution: New Mexico and Arizona to Utah and Colorado.

Schumann describes and figures a plant entirely without spines, but whether it is common or not we do not know. This was published as a variety, *E. phoeniceus inermis* (Monatsschr. Kakteenk. **6**: 150. 1896), but it often passes as *E. inermis,* although never described as such. Some years ago such a plant was sent to Washington from Utah by Mr. M. E. Jones and is still growing in the Cactus House, but it has not since flowered.

Coulter has combined this species with *Mammillaria aggregata* Engelmann (Emory, Mil. Reconn. 157. f. 1. 1848) taking both up as *Cereus aggregatus* Coulter (Contr. U. S. Nat. Herb. **3**: 396. 1896), but we do not believe that they are the same; Rydberg has used the name *Echinocereus aggregatus* (Bull. Torr. Club **33**:146. 1906) for this plant.

Echinopsis valida densa Regel (Gartenflora **1**: pl. 29. 1852; also Förster, Handb. Cact. ed. 2. f. 85) is referred by Schumann to both *Echinocereus acifer* and *E. phoeniceus* (Gesamtb. Kakteen 239, 283). To us it suggests *E. fendleri,* although it has differently colored flowers.

Illustrations: ? Gartenflora **1**: pl. 29, as *Echinopsis valida densa;* Gartenwelt **1**: 85, 89; **4**: 159, as *Cereus phoeniceus;* Gartenwelt **1**: 89, as *Cereus phoeniceus inermis;* Gartenwelt **4**: 159, as *Echinocereus phoeniceus inermis;* Curtis's Bot. Mag. **110**: pl. 6774, as *Cereus pauci-spinus* (fide Schumann); Pac. R. Rep. **4**: pl. 4, f. 9, as *Cereus bigelovii zuniensis;* Gartenwelt **4**: 157, as *Echinocereus phoeniceus;* Monatsschr. Kakteenk. **6**: 151, as *Echinocactus phoeniceus inermis* (through typographical error); N. Mex. Agr. Exp. Sta. Bull. **78**: pl. [18].

Plate 11, figure 1, shows a flowering plant collected by E. A. Goldman in Arizona; figure 2 shows an open flower.

16. Echinocereus rosei Wooton and Standley, Contr. U. S. Nat. Herb. **19**: 457. 1915.

Cespitose, forming small compact clumps, the stems 1 to 2 dm. long, 5 to 8 cm. in diameter, sometimes as many as 40; ribs 8 to 11, obtuse; areoles rather closely set; spines pinkish to brownish gray; radial spines about 10, spreading; centrals 4, 4 to 6 cm. long; flowers 4 to 6 cm. long, scarlet; inner perianth-segments broad, obtuse; spines on ovary and flower-tube brownish or yellowish, intermixed with short hairs; fruit spiny.

Type locality: Agricultural College, New Mexico.

Distribution: In mountains and dry hills and sometimes on the mesas of southern New Mexico, western Texas, and adjacent parts of northern Mexico.

f. E. Eaton del.

A.Hoen & Co

1. Top of flowering plant of *Echinocereus coccineus*.
2. Flower of same.
3. Flowering plant of *Echinocereus chloranthus*.
4. Flowering plant of *Echinocereus viridiflorus*.
5. Flowering plant of *Echinocereus maritimus*.
(All natural size.)

This species passes generally as *E. polyacanthus* Engelmann, a Mexican species with which it was confused in the Report on the Cactaceae of the Mexican Boundary, but as stated by Mr. Standley that species "is amply separated by the presence of long, white wool in the areoles of the ovary and fruit."

Illustrations: Cact. Mex. Bound. pl. 54, 55, as *Cereus polyacanthus.*

Figure 13 is from a photograph of a plant obtained on the Sierra Blanca, Texas, by Rose, Standley, and Russell in 1910, which afterward flowered in the cactus house of the U. S. Department of Agriculture.

FIG. 13.—Echinocereus rosei.　　　　　FIG. 14.—Echinocereus maritimus.

17. Echinocereus maritimus (Jones) Schumann, Gesamtb. Kakteen 273. 1898.

> *Cereus maritimus* Jones, Amer. Nat. **17:** 973. 1883.
> *Cereus flaviflorus* Engelmann in Coulter, Contr. U. S. Nat. Herb. **3:** 391. 1896.
> *Echinocereus flaviflorus* Schumann, Gesamtb. Kakteen 274. 1898.

Decidedly cespitose, often forming clumps 60 to 90 cm. broad and 30 cm. high, sometimes containing 200 joints; individual joints globose to short-cylindric, 5 to 16 cm. long; ribs 8 to 10; areoles 10 to 12 mm. apart; radial spines about 10, spreading; central spines 4, stout and angled, 2.5 to 3.5 cm. long; flowers small, including the ovary 3 to 4 cm. long, arising from near the top of the plant, light yellow; inner perianth-segments oblanceolate, rounded at apex; ovary not very spiny; fruit not seen.

Type locality: Ensenada, Lower California.

Distribution: West coast of Lower California.

This is a low, coastal species, perhaps extending all along the west coast of central Lower California. It was first found by Marcus E. Jones at Ensenada and was recently collected at the same locality by Ivan M. Johnston, April 7, 1921 (No. 3007). Dr. Rose found it in abundance about San Bartolomé Bay and introduced a great quantity into cultivation. It has frequently flowered both in the New York Botanical Garden and at Washington. A specimen of the original collection is preserved in the U. S. National Museum.

The name *Cereus glomeratus* Engelmann, unpublished, is cited by Orcutt (West Amer. Sci. **13:** 28. 1902) as a synonym of *Echinocereus maritimus.* The name is also used by Schumann (Gesamtb. Kakteen 274. 1898), but it is not found in any of Engelmann's works.

Plants collected by Mr. C. R. Orcutt in Lower California have been referred as *Echinocereus orcuttii* (Kew Bull. Misc. Inf. **1921:** 36), without description.

Illustration: Cact. Journ. **2:** 123.

Plate 11, figure 5, shows a plant in flower. Figure 14 is from a photograph of a plant of the same collection brought to Washington from San Bartolomé Bay, Lower California, in 1911 by Dr. Rose (No. 16189).

18. Echinocereus subinermis Salm-Dyck in Seemann, Bot. Herald 291. 1856.

Cereus subinermis Hemsley, Biol. Contr. Amer. Bot. 1: 546. 1880.

At first simple, 10 to 12 cm. high, afterwards a little branching at base, when young pale green, afterwards bluish and finally darker green, erect; ribs 5 to 8, broad, somewhat sinuate; spines all radial, small, conic, 1 to 2 mm. long, yellow, 3 or 4, deciduous; flowers large, 5 to 7 cm. long, yellow; perianth-segments oblanceolate, acute; spines of areoles on ovary and flower-tube short, white; fruit not known.

Type locality: Near Chihuahua, Mexico.
Distribution: Northern Mexico.

This species was introduced into Europe in 1845. It recently flowered in Germany. We have studied a plant sent from Berlin to the New York Botanical Garden, in 1902, which died before blooming. This plant is the least armed of the genus.

Illustrations: Blühende Kakteen 1: pl. 3; Monatsschr. Kakteenk. 26: 99.

Figure 15 is copied from the first illustration cited above.

19. Echinocereus luteus Britton and Rose, Contr. U. S. Nat. Herb. 16: 239. 1913.

Stem short to elongated,* sometimes branching near base, bluish green, more or less purplish, 8 or 9-ribbed; ribs rather thin, barely undulate, rounded; areoles small, 10 to 12 mm. apart; spines small, the radials 6 to 8, unequal, 2 to 8 mm. long, widely spreading, white with darker tips; central spine single, porrect; flowers on each rib appearing near top of plant and from second or third areole; flower-buds acute, reddish, covered with long, brownish bristles; areoles on ovary and flower-tube bearing white wool and light-colored spines with dark tips; flowers pale yellow, delicately sweet-scented, 7 cm. long, including the ovary; outer perianth-segments streaked with red; inner perianth-segments lemon-yellow, oblanceolate, acute; filaments light yellow.

Type locality: Above Alamos, Sonora, Mexico.
Distribution: Western Mexico.
Illustration: Contr. U. S. Nat. Herb. 16: pl. 67.

FIG. 15.—Echinocereus subinermis.

Figure 16 is from a photograph of the type specimen.

20. Echinocereus chloranthus (Engelmann) Rümpler in Förster, Handb. Cact. ed. 2. 814. 1885.

Cereus chloranthus Engelmann, Proc. Amer. Acad. 3: 278. 1856.

Cylindric, usually simple, 8 to 15 cm. long, 5 to 7 cm. in diameter; ribs about 13, often nearly hidden by the densely set spines; areoles nearly circular; radial spines several, spreading; centrals 3 or 4, not angled, in a vertical row, one much more elongated than the others, 2 to 3 cm. long; flowers yellowish green, 2 cm. long; fruit small, nearly globular, 5 to 10 cm. long, dark purplish red, covered with small bristly spines; seeds black, dull, pitted, the hilum nearly basal, round.

Type locality: About El Paso, Texas.
Distribution: Western Texas, southeastern New Mexico, and northern Mexico.

This species is somewhat like *Echinocereus viridiflorus*, having similar small flowers. It is usually more elongated, with longer central spines and with the flowers appearing lower down on the plant, generally below the middle.

*Señor Ortega has sent us an unusual specimen, 2 dm. high, from Mazatlan (exact locality not given).

Illustrations: Cact. Mex. Bound. pl. 37, 38; Amer. Gard. 11:473, as *Cereus chloranthus;* Cact. Journ. **2:** 19; Cycl. Amer. Hort. Bailey **2:** f. 747; Engler and Prantl, Pflanzenfam. **3**⁶ᵃ: f. 57, D; Stand. Cycl. Hort. Bailey **2:** f. 1375; Schelle, Handb. Kakteenk. 128. f. 59; Förster, Handb. Cact. ed. 2. 815. f. 107.

Plate 11, figure 3, shows a flowering plant sent by Dr. Rose to the New York Botanical Garden in 1913 from the east side of the Franklin Mountains near El Paso, Texas.

21. **Echinocereus viridiflorus** Engelmann in Wislizenus, Mem. Tour North. Mex. 91. 1848.

> *Cereus viridiflorus* Engelmann in Gray, Pl. Fendl. 50. 1849.
> *Cereus viridiflorus cylindricus* Engelmann, Proc. Amer. Acad. **3:** 278. 1856.
> *Echinocactus viridiflorus* Pritzel, Icon. Bot. Index **2:** 113. 1866.
> *Echinocereus viridiflorus cylindricus* Rümpler in Förster, Handb. Cact. ed. **2:** 812. 1885.
> *Echinocereus strausianus* Haage jr. in Quehl, Monatsschr. Kakteenk. **10:** 70. 1890.
> *Cereus viridiflorus tubulosus* Coulter, Contr. U. S. Nat. Herb. **3:** 383. 1896.
> *Echinocereus viridiflorus tubulosus* Heller, Cat. N. Amer. Pl. ed. 2. 8. 1900.

Plants small, nearly globular, but sometimes cylindric and 20 cm. high, simple, or more or less cespitose; ribs 14, low; areoles elongated; spines white, dark brown or variegated, usually arranged in circular bands of light and dark about the plant; radial spines about 16, appressed; centrals, when present, 2 or 3, arranged in a perpendicular row, often elongated and then 2 cm. long; flowers greenish, 2 to 2.5 cm. long; perianth-segments obtuse; fruit 10 to 12 mm. long; seeds 1 to 1.2 mm. long.

FIG. 16.—Echinocereus luteus.　　　FIG. 17.—Echinocereus viridiflorus.

Type locality: Prairies about Wolf Creek, New Mexico.

Distribution: Southern Wyoming to eastern New Mexico, western Kansas, western Texas, and South Dakota.

This species is very common on the plains of the West. It is usually deeply seated, with the low top hidden in the grass, so that it is not easily seen. It is widely distributed, rather variable in its habit and spines, but is easily distinguished from all the other species of this genus. It is frequently introduced into our collections, but lasts only a few years. It extends farther north than any other species of the genus and was one of the first to be collected in the United States, having been found by Dr. Wislizenus in 1846. It is known in Wyoming as green-flowered petaya (M. Cary).

Echinocereus viridiflorus var. *gracilispinus* (Rümpler in Förster, Handb. Cact. ed. 2. 814. 1885) and var. *major* (Monatsschr. Kakteenk. **16:** 142. 1906) are simply garden forms.

Echinocereus labouretii Förster (Handb. Cact. ed. 2. 811. 1885) is given as a synonym of this species. *Echinocereus labouretianus* Lemaire (Cactées 57. 1868) is also to be referred here. ·

Illustrations: Cact. Mex. Bound. pl. 36; Gartenwelt 1: 89; Curtis's Bot. Mag. 125: pl. 7688, as *Cereus viridiflorus;* Förster, Handb. Cact. ed. 2. 813. f. 106; Cact. Journ. 2: 19; Britton and Brown, Illustr. Fl. 2: 460. f. 2522; ed. 2. 2: 569. f. 2981; Monatsschr. Kakteenk. 15: 57; Schelle, Handb. Kakteenk. 129. f. 60; Gartenwelt 4: 159; Blanc, Cacti 59. No. 842.

Plate 11, figure 4, shows a flowering plant sent to the New York Botanical Garden by Dr. Rose from Syracuse, Kansas, in 1912. Figure 17 is from a photograph of a plant from Colorado Springs collected by F. W. Homan in 1912.

22. Echinocereus grandis sp. nov.

Stems usually single or in small clusters, sub-cylindric, 1 to 4 dm. high, 8 to 12 cm. in diameter; ribs 21 to 25, low; areoles large, longer than broad, about 1 cm. apart; spines dull white or cream-colored, rather short and stiff, the radials 15 to 25, the centrals 8 to 12, often in 2 rows; flower 5 to 6 cm. long, unusually narrow, with a short limb; ovary and flower-tube densely clothed with clusters of pale, straw-colored spines intermixed with white hairs; outer perianth-segments white, with a green medial line; inner perianth-segments narrow, 1.5 cm. long, white with green bases; filaments green; style white; stigma-lobes green; fruit densely spiny.

FIG. 18.—Echinocereus grandis.   FIG. 19.—Echinocereus dasyacanthus.

Collected on San Esteban Island in the Gulf of California, April 13, 1911, by J. N. Rose (No. 16823); on San Lorenzo Island by Ivan M. Johnston in 1921 (Nos. 3541, 4198); and on Nolasco Island by Mr. Johnston (No. 3137).

This plant was very common in a dry creek bed and in an adjoining valley as well as on the low hills on the Island of San Esteban, which Dr. Rose visited in 1911. Many fine plants were collected. One flowered in the New York Botanical Garden in 1912; three plants were grown in the Cactus House, U. S. Department of Agriculture, in 1913. of one of which a photograph was taken.

Plate 111, figure 3, shows a flowering plant sent by Dr. Rose to the New York Botanical Garden from the type locality in 1911. Figure 18 is from a photograph of one of the plants which bloomed in Washington.

23. **Echinocereus dasyacanthus** Engelmann in Wislizenus, Mem. Tour North. Mex. 100. 1848.

Cereus dasyacanthus Engelmann in Gray, Pl. Fendl. 50. 1849.
Cereus dasyacanthus neo-mexicanus Coulter, Contr. U. S. Nat. Herb. 3: 384. 1896.
Echinocereus spinosissimus Walton, Cact. Journ. 2: 162. 1899.
Echinocereus rubescens Dams, Monatsschr. Kakteenk. 15: 92. 1905.

Plants usually simple, cylindric, 1 to 3 dm. high, very spiny; ribs 15 to 21, low, 2 to 3 cm. high; areoles approximate, 3 to 5 mm. apart, short-elliptic; radial spines 16 to 24, more or less spreading, 1.5 cm. long or less, at first pinkish but gray in age; central spines 3 to 8, a little stouter than the radials, never in a single row; flowers from near the apex, often very large, often 10 cm. long, yellowish, or drying reddish; outer perianth-segments linear-oblong, 4 to 5 cm. long, acute; inner perianth-segments oblong, 5 cm. long; ovary very spiny; fruit nearly globular, 2.5 to 3.5 cm. in diameter, purplish, edible.

Type locality: El Paso, Texas.
Distribution: Western Texas, southern New Mexico, and northern Chihuahua. It has been reported from Arizona, but doubtless wrongly.
Echinocereus papillosus rubescens (Monatsschr. Kakteenk. 15: 92. 1905) was only a garden name for *E. rubescens*.
Echinocereus degandii (Monatsschr. Kakteenk. 5: 123. 1895), only a catalogue name from Rebut, is here referred by Schumann in his monograph.
This is undoubtedly the plant which Walton calls the "true *E. spinosissimus*" (Cact. Journ. 2: 162. 1899), although we do not find the name referred to elsewhere.
Illustrations: Cact. Mex. Bound. pl. 39, 40, 41, f. 1, 2; Gard. Chron. III. 32: 252; West Amer. Sci. 13: 10, as *Cereus dasyacanthus;* Monatsschr. Kakteenk. 15: 89, as *Echinocereus rubescens;* Blühende Kakteen 2: pl. 81; Förster, Handb. Cact. ed. 2. f. 110; Gartenwelt 7: 290; Schelle, Handb. Kakteenk. 130. f. 61; Cact. Journ. 1: 89; 2: 19.
Figure 19 is from a photograph of a flowering plant collected by Elmer Stearns at Juarez, Mexico, in 1906.

24. **Echinocereus ctenoides** (Engelmann) Rümpler in Förster, Handb. Cact. ed. 2. 819. 1885.

Cereus ctenoides Engelmann, Proc. Amer. Acad. 3: 279. 1856.

So far as known simple, cylindric, elongated, 10 to 40 cm. long, 8 to 10 cm. in diameter, decidedly banded with pink and gray as in the rainbow cactus; ribs 15 to 17, low; areoles crowded together, short-elliptic; radial spines often as many as 20, not spreading but standing out at an angle to the ribs; central spines 8 to 10, arranged in a single row or sometimes a little irregular; flowers up to 10 cm. long, about as wide as long when fully expanded, bright to reddish yellow; filaments yellow; style white; ovary and fruit very spiny.

Type locality: Eagle Pass, Texas.
Distribution: Southern Texas and Chihuahua.
This species is near *Echinocereus dasyacanthus;* it differs somewhat in its spines and it has a more southern range. It may not be specifically distinct.
Echinocactus ctenoides (Index Kewensis Suppl. 1. 476) is a mistake for *Echinocereus ctenoides.*
Illustrations: Cact. Mex. Bound. pl. 42*; Dict. Gard. Nicholson 4: 511, f. 7; Suppl. 217. f. 229; Watson, Cact. Cult. 73. f. 20, as *Cereus ctenoides;* Förster, Handb. Cact. ed. 2. 820. f. 109.
Figure 20 is a copy of the first illustration above cited.

25. **Echinocereus papillosus** Linke in Förster, Handb. Cact. ed. 2. 783. 1885.

Echinocereus texensis Rünge, Monatsschr. Kakteenk. 4: 61. 1894. Not Jacobi, 1856.
Echinocereus ruengei Schumann, Monatsschr. Kakteenk. 5: 124. 1895.
Cereus papillosus Berger, Rep. Mo. Bot. Gard. 16: 80. 1905.

---

* All the additional illustrations cited here are copied from this plate.

More or less cespitose, rather dark green, decumbent or ascending, 5 to 30 cm. long, 2 to 4 cm. in diameter; ribs 6 to 10, prominent, strongly tubercled; radial spines acicular, spreading, about 7, white to yellowish, 1 cm. long or less; central spine solitary, acicular, porrect, 12 mm. long or more; flowers large, 10 to 12 cm. broad, yellow with a reddish center, with rather few perianth-segments 4 to 6 cm. long, oblong-spatulate, acuminate, more or less serrate; scales on ovary red, spreading; fruit not known.

Type locality: Not cited.
Distribution: Western Texas.

Although this species is supposed to come from the vicinity of San Antonio, Texas, no specimens are known to us from that place but we have an herbarium specimen collected by Miss Mary B. Croft at San Diego, Texas. It is one of the few species in the genus with yellow flowers and ought easily to be distinguished from other Texan species.

Illustration: Blühende Kakteen 2: pl. 115.

FIG. 20.—Echinocereus ctenoides.   FIG. 20a.—Echinocereus pentalophus.

26. **Echinocereus blanckii*** (Poselger) Palmer, Rev. Hort. **36**: 92. 1865.

    Cereus blanckii Poselger, Allg. Gartenz. **21**: 134. 1853.
    Cereus berlandieri Engelmann, Proc. Amer. Acad. **3**: 286. 1856.
    Echinocereus poselgerianus Linke, Allg. Gartenz. **25**: 239. 1857.
    Echinocereus berlandieri Rümpler in Förster, Handb. Cact. ed. 2. 776. 1885.
    Echinocereus leonensis Mathsson, Monatsschr. Kakteenk. **1**: 66. 1891.
    Cereus leonensis Orcutt, West Amer. Sci. **13**: 27. 1902.
    Cereus poselgerianus Berger, Rep. Mo. Bot. Gard. **16**: 80. 1905.

*The specific name is often spelled *blankii*.

Procumbent; joints slender, 3 to 15 cm. long, 2 to 2.5 cm. in diameter; ribs 5 to 7, strongly tuberculate, or when turgid scarcely tubercled; areoles 1 to 1.5 cm. apart; radial spines 6 to 8, 8 to 10 mm. long, white; central spine solitary, 10 to 50 mm. long, brownish to black; flowers purple, 5 to 8 cm. long; perianth-segments narrow, oblanceolate, acute.

Type locality: Near Camargo, state of Tamaulipas, Mexico.

Distribution: Northeastern Mexico and southern Texas.

Dr. Rose in 1912 examined specimens labeled *E. leonensis* in the collection of Mr. Haage jr. at Erfurt which seemed to be *E. blanckii.*

Echinocereus flaviflorus Hildmann (Schumann, Gesamtb. Kakteen 264. 1898), unpublished, is referred to *E. leonensis,* its flowers not being yellow as the name would imply.

This species is named for P. A. Blanck, a pharmacist in Berlin and a friend of H. Poselger. It is closely related to *E. pentalophus.*

Echinocactus leonensis (Schumann, Monatsschr. Kakteenk. **5:** 76) seems to have been intended for *Echinocereus leonensis* Mathsson.

Fig. 21.—Echinocereus blanckii.

Illustrations: Dict. Gard. Nicholson **4:** 511. f. 5; Suppl. 217. f. 227, as *Cereus blankii;* Blühende Kakteen **1:** pl. 37; Cact. Journ. **1:** 135; **2:** 19; Förster, Handb. Cact. ed. 2. f. 97; Monatsschr. Kakteenk. **7:** 154; Ann. Rep. Smiths. Inst. **1908:** pl. 3, f. 2; Schelle, Handb. Kakteenk. 122. f. 55, as *Echinocereus berlandieri;* Cact. Mex. Bound. pl. 58; Dict. Gard. Nicholson **4:** 510. f. 4; Suppl. 216. f. 226, as *Cereus berlandieri;* Monatsschr. Kakteenk. **1:** opp. 66, as *Echinocereus leonensis;* Blühende Kakteen **2:** pl. 70; Förster, Handb. Cact. ed. 2. f. 98; Rev. Hort. **36:** following 92.

Plate III, figure 4, shows a flowering plant sent to the New York Botanical Garden from the Berlin Botanical Garden in 1902. Figure 21 is from a photograph taken by Robert Runyon at Reynosa, Mexico, July 8, 1921.

27. Echinocereus pentalophus (De Candolle) Rümpler in Förster, Handb. Cact. ed. 2. 774. 1885.

> *Cereus pentalophus* De Candolle, Mém. Mus. Hist. Nat. Paris **17:** 117. 1828.
> *Cereus pentalophus simplex* De Candolle, Mém. Mus. Hist. Nat. Paris **17:** 117. 1828.
> *Cereus pentalophus subarticulatus* De Candolle, Mém. Mus. Hist. Nat. Paris **17:** 117. 1828.
> *Cereus pentalophus radicans* De Candolle, Mém. Mus. Hist. Nat. Paris **17:** 117. 1828.
> *Cereus propinquus* De Candolle in Salm-Dyck, Allg. Gartenz. **1:** 366. 1833.

Cereus procumbens Engelmann in Gray, Pl. Fendl. 50. 1849.
Cereus pentalophus leptacanthus Salm-Dyck, Cact. Hort. Dyck. 1849. 42.   1850.
Echinocereus procumbens Rümpler in Förster, Handb. Cact. ed. 2. 781.   1885.
Echinocereus leptacanthus Schumann, Gesamtb. Kakteen 260.   1898.

Procumbent, with ascending branches, deep green; ribs 4 to 6, somewhat undulate, bearing low tubercles; radial spines 4 or 5, very short, white with brown tips; central spine 1, rarely wanting; flowers reddish violet, large, 7 to 12 cm. long; perianth-segments broad, rounded at apex; stamens borne on the lower half of throat for a distance of about 12 mm.; tube proper not much broader than the style, purple within, 8 mm. long; filaments short; style a little longer than the filaments; scales on the ovary and flower-tube bearing long cobwebby hairs and brownish spines; style stiff, 3.5 cm. long.

Type locality: Mexico.
Distribution: Eastern Mexico and southern Texas.

This is an attractive species and does fairly well in greenhouse cultivation, usually producing its beautiful flowers very early in the spring; its growth is much modified by indoor treatment, where the spines, especially, are changed.

Echinocereus procumbens longispinus (Monatsschr. Kakteenk. 12: 135. 1902) is only a form with very long spines. The wild plants in this and the following species have longer spines than the cultivated ones.

Echinocereus procumbens has usually been recognized as a distinct species but we believe we are justified in referring it as above.

In 1837 when Pfeiffer redescribed the varieties of this species he added *Cereus propinquus* De Candolle (*Echinocereus propinquus* Monatsschr. Kakteenk. 5: 124. 1895), as a synonym of variety *simplex,* and *C. leptacanthus* De Candolle, as a synonym of variety *subarticulatus,* but we do not find that De Candolle himself ever published these names.

This species was figured in Curtis's Botanical Magazine in 1839, about 10 years after its introduction from Mexico by Thomas Coulter. Although we have no definite information on this point, it is not unlikely that it was made from a part of the original stock. As the type of the species is lost, we have assumed that this illustration is typical. The species was taken up by Schumann, and by all writers since, under the much later name, *Echinocereus leptacanthus* Schumann; this is the name used by Pfeiffer, but as a synonym as mentioned above.

Illustrations: Dict. Hort. Bois 280. f. 198; Deutsches Mag. Gart. Blumen. 1868: pl. opp. 8, as *Cereus pentalophus;* Dict. Gard. Nicholson 4: 512. f. 9; Suppl. 218. f. 232; Rev. Hort. 36: opp. 171, as *Cereus leptacanthus;* Curtis's Bot. Mag. 65: pl. 3651, as *Cereus pentalophus subarticulatus;* Blühende Kakteen 1: pl. 15; Förster, Handb. Cact. ed. 2. 785. f. 100; Schelle, Handb. Kakteenk. 125. f. 57, as *Echinocereus leptacanthus;* Curtis's Bot. Mag. 117: pl. 7205; Dict. Gard. Nicholson 4: 513. f. 12; Suppl. f. 234; Cact. Mex. Bound. pl. 59, f. 1 to 11, as *Cereus procumbens;* Cact. Journ. 1: 109, 136, 164; 2: 173; Förster, Handb. Cact. ed. 2. 782. f. 99; Rümpler, Sukkulenten 136. f. 72; Schelle, Handb. Kakteenk. 124. f. 56, as *Echinocereus procumbens.*

Plate III, figure 1, shows a flowering joint of a plant sent by Dr. Rose to the New York Botanical Garden in 1913. Figure 20*a* is copied from the first illustration above cited.

28. Echinocereus sciurus (K. Brandegee).

Cereus sciurus K. Brandegee, Zoe 5: 192.   1904.

Densely cespitose, with many individuals forming clumps sometimes 60 cm. broad; stems slender, often 20 cm. long, often nearly hidden by the many spines; ribs 12 to 17, low, divided into numerous tubercles 5 to 6 mm. apart; areoles small, approximate, circular, at first woolly, becoming naked; radial spines 15 to 18, sometimes 15 mm. long, slender, pale except the brownish tips; centrals usually several, shorter than the radials; flower-buds covered with numerous slender brown-tipped spines; flowers described as 7 cm. long, about 9 cm. broad when fully open; inner perianth-segments in 2 to 4 rows, bright magenta; stamens numerous, with greenish filaments; pistil green with obtuse stigma-lobes; seeds 1 mm. long, tuberculate.

M. E. Eaton del.

1. Top of flowering plant of *Echinocereus pentalophus*.
2. Flowering plant of *Echinocereus fitchii*.
3. Top of flowering plant of *Echinocereus grandis*.
4. Flowering plant of *Echinocereus blanckii*.
(All natural size.)

Type locality: Hills near San José del Cabo, Lower California.

Distribution: Southern end of Lower California.

This species was first collected by Mr. T. S. Brandegee near San José del Cabo, Lower California, in April 1897, and described by Mrs. Brandegee in 1904 as a new species of *Cereus* of the subgenus *Echinocereus*.

In 1911 Dr. Rose re-collected it in some abundance at the type locality, and living plants were grown in the cactus collections in New York, Washington, and St. Louis. So far as we are aware it is not offered in the trade and is rare in living or herbarium collections. It has been collected in recent years also by Dr. C. A. Purpus and the name *Echinocereus sciurus* was incidentally used in referring to his collection (Monatsschr. Kakteenk. 14: 130. 1904).

Plate IV, figure 1, shows a plant collected by Dr. Rose at San José del Cabo, Lower California, in 1911.

29. Echinocereus cinerascens (De Candolle) Rümpler in Förster, Handb. Cact. ed. 2. 786. 1885.

Cereus cinerascens De Candolle, Mém. Mus. Hist. Nat. Paris 17: 116.   1828.
Cereus cinerascens crassior De Candolle, Mém. Mus. Hist. Nat. Paris 17: 116.   1828.
Cereus cinerascens tenuior De Candolle, Mém. Mus. Hist. Nat. Paris 17: 116.   1828.
Cereus deppei Salm-Dyck, Hort. Dyck. 338.   1834.
Cereus cirrhiferus Labouret, Monogr. Cact. 311.   1853.
Echinocereus cirrhiferus Rümpler in Förster, Handb. Cact. ed. 2. 778.   1885.
Echinocereus glycimorphus Rümpler in Förster, Handb. Cact. ed. 2. 800.   1885.
Cereus glycimorphus Orcutt, Seed Pl. Co. Cat. Cact. 5.   1903.

Growing in patches 6 to 12 dm. broad, branching at base, the stems ascending to about 3 dm.; ribs about 12, not very prominent, obtuse; areoles rather scattered, orbicular; spines white or pale, straight, rough, 1.5 to 2 cm. long; radials about 10; centrals 3 or 4; flowers, including ovary, 6 to 8 cm. long, the tube very short; scales on ovary and tube small, acute, their axils crowded with short white wool and 6 to 8 long white bristles; inner perianth-segments, when dry, deep purple, 3 to 4 cm. long, obtuse; stamens short; fruit not seen.

Type locality: Mexico.

Distribution: Central Mexico.

Cereus aciniformis (Pfeiffer, Enum. Cact. 101. 1837) is only a garden name supposed to be the same as *Echinocereus cinerascens* var. *crassior* (Rümpler in Förster, Handb. Cact. ed. 2. 787. 1885).

Echinocereus deppei, unpublished, belongs here according to Schumann (Monatsschr. Kakteenk. 5: 123. 1895). *Echinocereus cirrhiferus monstrosus* is an abnormal form.

Echinocereus glycimorphus was described from a sterile plant of unknown origin, obtained of F. A. Haage jr. of Erfurt; it was redescribed by Schumann, who cites definitely Mathsson's plant from Hidalgo between Ixmiquilpan and Cardonal, but whether this latter plant is the type or not is uncertain. Schumann made for it a subseries *Oleosi* of which it is the only species.

We feel justified in reducing *E. glycimorphus* to *E. cinerascens;* we have living plants of both from the Berlin Botanical Garden and they must represent essentially the same species, while the differences pointed out by Schumann seem trivial. Not only have we had *Echinocereus cinerascens* from various authentic sources, but Dr. Rose has repeatedly obtained it from the Valley of Mexico and adjacent regions. The plant is of wide distribution and has been reported from farther south than any of the other species of this genus.

Echinocereus undulatus Hildmann (Schumann, Gesamtb. Kakteen 261. 1898) is only a catalogue name for it.

Illustration: Monatsschr. Kakteenk. 14: 137.

30. Echinocereus adustus Engelmann in Wislizenus, Mem. Tour North. Mex. 104.   1848.

Echinocereus rufispinus Engelmann in Wislizenus, Mem. Tour North. Mex. 104.   1848.
Echinocereus radians Engelmann in Wislizenus, Mem. Tour North. Mex. 105.   1848.
Cereus adustus Engelmann in Gray, Pl. Fendl. 50.   1849.
Cereus rufispinus Engelmann in Gray, Pl. Fendl. 50.   1849.

Cereus pectinatus armatus Poselger, Allg. Gartenz. **21:** 134. 1853.
Cereus pectinatus spinosus Coulter, Contr. U. S. Nat. Herb. **3:** 387. 1896.
Cereus adustus radians Coulter, Contr. U. S. Nat. Herb. **3:** 387. 1896.
Echinocereus pectinatus adustus Schumann, Gesamtb. Kakteen 271. 1898.
Echinocereus pectinatus armatus Schumann, Gesamtb. Kakteen 271. 1898.
Echinocereus pectinatus rufispinus Schumann, Gesamtb. Kakteen 272. 1898.

Simple, short-cylindric, often only 4 to 6 cm. high; ribs 13 to 15; areoles closely set, elliptic; radial spines 16 to 20, appressed-pectinate, pale; the central spines wanting or solitary, sometimes elongated and porrect; flowers purplish, 3 to 4 cm. long; inner perianth-segments narrow; ovary and calyx-tube covered with clusters of short brown spines and long wool.

Type locality: Cosihuiriachi, Chihauhua.
Distribution: Mountains near type locality.

Figure 22 is from a photograph of a plant collected by Dr. Rose at the type locality in 1908.

Fig. 22.—Echinocereus adustus.   Fig. 23.—Echinocereus standleyi.

31. Echinocereus standleyi sp. nov.

Nearly globular or short-cylindric, 4 to 5 cm. in diameter; ribs 12; areoles elongated, closely set; radial spines about 16, stoutish, whitish but yellow at base; central spine one, similar to but much larger and stouter than the radials, 2 to 2.5 cm. long, porrect.

Collected by Mrs. S. L. Pattison in the Sacramento Mountains, New Mexico, and obtained from her by Mr. Paul C. Standley in 1906.

It is a little known species, resembling *Echinocereus adustus* and *E. viridiflorus*, but with different spines; neither flower nor fruit has been obtained.

Figure 23 is from a photograph of the type specimen, preserved in the U. S. National Herbarium.

32. Echinocereus perbellus sp. nov.

Stem either simple or clustered, 5 to 10 cm. high; ribs 15, low and broad; distance between the areoles about equal to the length of the areoles themselves; areoles elongated; spines all radials, 12 to 15, spreading but not widely, 5 to 7 mm. long, pale brown to reddish or nearly white below; flowers purple, 4 to 6 cm. long; perianth-segments broad, oblong to oblanceolate, acuminate, nearly 4 cm. long; areoles on flower-tube very woolly as well as spiny.

Collected by Rose and Standley at Big Springs, Texas, February 23, 1910 (No. 12215).
This is a very beautiful species which flowers abundantly in cultivation. If heretofore collected, it has doubtless passed as the next species to which it is related. Rose and Standley, who discovered it wild in 1910, also found it in cultivation in Texas.

Figure 24 is from a photograph of the type specimen.

33. **Echinocereus reichenbachii*** (Terscheck) Haage jr., Index Kewensis **2**:813. 1893.

> *Echinocactus reichenbachii* Terscheck in Walpers, Repert. Bot. **2**: 320. 1843.
> *Cereus caespitosus* Engelmann, Bost. Journ. Nat. Hist. **5**: 247. 1845.
> *Echinopsis pectinata reichenbachiana* Salm-Dyck, Cact. Hort. Dyck. 1844. 26. 1845.
> *Echinocereus caespitosus* Engelmann in Wislizenus, Mem. Tour North. Mex. 110. 1848.
> *Cereus caespitosus castaneus* Engelmann, Bost. Journ. Nat. Hist. **6**: 203. 1850.
> *Cereus reichenbachianus* Labouret, Monogr. Cact. 318. 1853.
> *Cereus reichenbachianus castaneus* Labouret, Monogr. Cact. 319. 1853.
> *Cereus caespitosus minor* Engelmann, Proc. Amer. Acad. **3**: 280. 1856.
> *Cereus caespitosus major* Engelmann, Proc. Amer. Acad. **3**: 280. 1856.
> *Echinocereus texensis* Jacobi, Allg. Gartenz. **24**: 110. 1856.
> *Mammillaria caespitosa* A. Gray, First Lessons in Botany 96. 1857.
> *Echinocereus rotatus* Linke, Wochenschr. Gärtn. Pflanz. **1**: 85. 1858.
> *Echinocereus caespitosus castaneus* Rümpler in Förster, Handb. Cact. ed. 2. 811. 1885.
> *Echinocereus caespitosus major* Rümpler in Förster, Handb. Cact. ed. 2. 811. 1885.
> *Echinocereus pectinatus caespitosus* Schumann, Gesamtb. Kakteen 272. 1898.

FIG. 24.—Echinocereus perbellus.

More or less cespitose; stems simple, globose to short-cylindric, 2.5 to 20 cm. long, 5 to 9 cm in diameter; ribs 12 to 19; areoles approximate, elliptic; spines 20 to 30, white to brown, but usually those of each individual plant of one color, pectinate, interlocking, 5 to 8 mm. long, spreading, more or less recurved; centrals 1 or 2, like the radials, or often wanting; flowers fragrant, rather variable as to size, often 6 to 7 cm. long and fully as broad, opening during the day, always closing at night and sometimes opening the second day, light purple, often reflexed; perianth-segments narrow, the margin more or less erose; filaments pinkish; fruit ovoid, about 1 cm. long; seeds black, nearly globose, 1.2 to 1.4 mm. in diameter.

Type locality: Mexico.

Distribution: Texas and northern Mexico; recorded from western Kansas.

The plant grows in a limestone country, usually among rocks.

Brandegee in 1876 reported *Cereus caespitosus castaneus* from the mesas of Saint Charles, south of Pueblo, Colorado, but we have seen no specimens. The species is not credited to Colorado

FIG. 25.—Echinocereus reichenbachii.

in recent manuals. We have seen specimens from as far south as Saltillo, Mexico (Runyon, 1921).

Cereus concolor Schott (Pac. R. Rep. **4**: Errata and Notes 11. 1856) is referred here by Coulter. The original description indicates a very different plant and it is surprising

---

* According to Walpers, the specific name is *reichenbachii*, but Labouret, when he transferred it to *Cereus*, changed it to *reichenbachianus* and this spelling is used in the Index Kewensis where the plant is taken up under *Echinocereus*. There the binomial is credited to Engelmann.

Cereus pectinatus armatus Poselger, Allg. Gartenz. **21**: 134. 1853.
Cereus pectinatus spinosus Coulter, Contr. U. S. Nat. Herb. **3**: 387. 1896.
Cereus adustus radians Coulter, Contr. U. S. Nat. Herb. **3**: 387. 1896.
Echinocereus pectinatus adustus Schumann, Gesamtb. Kakteen 271. 1898.
Echinocereus pectinatus armatus Schumann, Gesamtb. Kakteen 271. 1898.
Echinocereus pectinatus rufispinus Schumann, Gesamtb. Kakteen 272. 1898.

Simple, short-cylindric, often only 4 to 6 cm. high; ribs 13 to 15; areoles closely set, elliptic; radial spines 16 to 20, appressed-pectinate, pale; the central spines wanting or solitary, sometimes elongated and porrect; flowers purplish, 3 to 4 cm. long; inner perianth-segments narrow; ovary and calyx-tube covered with clusters of short brown spines and long wool.

Type locality: Cosihuiriachi, Chihauhua.
Distribution: Mountains near type locality.

Figure 22 is from a photograph of a plant collected by Dr. Rose at the type locality in 1908.

FIG. 22.—Echinocereus adustus.   FIG. 23.—Echinocereus standleyi.

31. Echinocereus standleyi sp. nov.

Nearly globular or short-cylindric, 4 to 5 cm. in diameter; ribs 12; areoles elongated, closely set; radial spines about 16, stoutish, whitish but yellow at base; central spine one, similar to but much larger and stouter than the radials, 2 to 2.5 cm. long, porrect.

Collected by Mrs. S. L. Pattison in the Sacramento Mountains, New Mexico, and obtained from her by Mr. Paul C. Standley in 1906.

It is a little known species, resembling *Echinocereus adustus* and *E. viridiflorus*, but with different spines; neither flower nor fruit has been obtained.

Figure 23 is from a photograph of the type specimen, preserved in the U. S. National Herbarium.

32. Echinocereus perbellus sp. nov.

Stem either simple or clustered, 5 to 10 cm. high; ribs 15, low and broad; distance between the areoles about equal to the length of the areoles themselves; areoles elongated; spines all radials, 12 to 15, spreading but not widely, 5 to 7 mm. long, pale brown to reddish or nearly white below; flowers purple, 4 to 6 cm. long; perianth-segments broad, oblong to oblanceolate, acuminate, nearly 4 cm. long; areoles on flower-tube very woolly as well as spiny.

Collected by Rose and Standley at Big Springs, Texas, February 23, 1910 (No. 12215).

This is a very beautiful species which flowers abundantly in cultivation. If heretofore collected, it has doubtless passed as the next species to which it is related. Rose and Standley, who discovered it wild in 1910, also found it in cultivation in Texas.

Figure 24 is from a photograph of the type specimen.

33. Echinocereus reichenbachii* (Terscheck) Haage jr., Index Kewensis **2:** 813. 1893.

FIG. 24.—Echinocereus perbellus.

> *Echinocactus reichenbachii* Terscheck in Walpers. Repert. Bot. **2:** 320. 1843.
> *Cereus caespitosus* Engelmann, Bost. Journ. Nat. Hist. **5:** 247. 1845.
> *Echinopsis pectinata reichenbachiana* Salm-Dyck, Cact. Hort. Dyck. 1844. 26. 1845.
> *Echinocereus caespitosus* Engelmann in Wislizenus, Mem. Tour North. Mex. 110. 1848.
> *Cereus caespitosus castaneus* Engelmann, Bost. Journ. Nat. Hist. **6:** 203. 1850.
> *Cereus reichenbachianus* Labouret, Monogr. Cact. 318. 1853.
> *Cereus reichenbachianus castaneus* Labouret, Monogr. Cact. 319. 1853.
> *Cereus caespitosus minor* Engelmann, Proc. Amer. Acad. **3:** 280. 1856.
> *Cereus caespitosus major* Engelmann, Proc. Amer. Acad. **3:** 280. 1856.
> *Echinocereus texensis* Jacobi, Allg. Gartenz. **24:** 110. 1856.
> *Mammillaria caespitosa* A. Gray, First Lessons in Botany 96. 1857.
> *Echinocereus rotatus* Linke, Wochenschr. Gärtn. Pflanz. **1:** 85. 1858.
> *Echinocereus caespitosus castaneus* Rümpler in Förster, Handb. Cact. ed. 2. 811. 1885.
> *Echinocereus caespitosus major* Rümpler in Förster, Handb. Cact. ed. 2. 811. 1885.
> *Echinocereus pectinatus caespitosus* Schumann, Gesamtb. Kakteen 272. 1898.

More or less cespitose; stems simple, globose to short-cylindric, 2.5 to 20 cm. long, 5 to 9 cm in diameter; ribs 12 to 19; areoles approximate, elliptic; spines 20 to 30, white to brown, but usually those of each individual plant of one color, pectinate, interlocking, 5 to 8 mm. long, spreading, more or less recurved; centrals 1 or 2, like the radials, or often wanting; flowers fragrant, rather variable as to size, often 6 to 7 cm. long and fully as broad, opening during the day, always closing at night and sometimes opening the second day, light purple, often reflexed; perianth-segments narrow, the margin more or less erose; filaments pinkish; fruit ovoid, about 1 cm. long; seeds black, nearly globose, 1.2 to 1.4 mm. in diameter.

FIG. 25.—Echinocereus reichenbachii.

Type locality: Mexico.

Distribution: Texas and northern Mexico; recorded from western Kansas.

The plant grows in a limestone country, usually among rocks.

Brandegee in 1876 reported *Cereus caespitosus castaneus* from the mesas of Saint Charles, south of Pueblo, Colorado, but we have seen no specimens. The species is not credited to Colorado in recent manuals. We have seen specimens from as far south as Saltillo, Mexico (Runyon, 1921).

Cereus concolor Schott (Pac. R. Rep. **4:** Errata and Notes 11. 1856) is referred here by Coulter. The original description indicates a very different plant and it is surprising

---

* According to Walpers, the specific name is *reichenbachii*, but Labouret, when he transferred it to *Cereus*, changed it to *reichenbachianus* and this spelling is used in the Index Kewensis where the plant is taken up under *Echinocereus*. Th ere the binomial is credited to Engelmann.

that it has not been re-collected.   It was collected at Escondido Springs, near the Pecos, Texas.   Schott points out how it differs from *Echinocereus caespitosus* in the following words:

"In *C. caespitosus* the flower-buds are clothed with a dense grayish wool and bear beautiful flowers 2 inches in diameter and 2 inches in length.   In *Cereus concolor* the flower-buds are perfectly naked, small, campanulate blossoms with yellowish sanguineus petals perfectly like the spines in color, 0.5 inches in diameter and 0.8 inches in length."

Echinopsis reichenbachiana Pfeiffer (Förster, Handb. Cact. 365. 1846) was used only as a synonym.

Echinocereus pectinatus castaneus (Monatsschr. Kakteenk. 1: 144. 1891), unpublished, doubtless belongs here.

Illustrations: Gray, First Lessons Bot. 96; Gray, Struct. Bot. ed. 5. 421. f. 838; ed. 6. 170. f. 317, as *Mammillaria caespitosa*;* West Amer. Sci. **7:** 238; Dict. Gard. Nicholson **4:** 511. f. 6; Suppl. 217. f. 228; Cact. Mex. Bound. pl. 43, 44; Deutsche Gärt. Zeit. **5:** 209; Watson, Cact. Cult. f. 19; Curtis's Bot. Mag. **109:** pl. 6669; Gartenflora **29:** 52, as *Cereus caespitosus;* Gartenflora **30:** 413; Garten-Zeitung **3:** 16. f. 7; Engler and Prantl, Pflanzenfam. **3**[6a]: f. 56, F; Förster, Handb. Cact. ed. 2. f. 105, 138; Cact. Journ. **1:** 107, 135; Britton and Brown, Illustr. Fl. **2:** 461. f. 2523; ed. 2. **2:** 559. f. 2982; Rümpler, Sukkulenten 140. f. 75; Ann. Rep. Smiths. Inst. **1908:** pl. 4, f. 6, as *Echinocereus caespitosus;* Monatsschr. Kakteenk. **15:** 171; Floralia **42:** 369, as *Echinocereus pectinatus caespitosus.*

Figure 26 is copied from plate 43 of the Mexican Boundary Survey, above cited; figure 25 is from a photograph furnished by Robert Runyon of a plant collected near Saltillo, Mexico.

34. Echinocereus baileyi Rose, Contr. U. S. Nat. Herb. **12:** 403. 1909.

Plant body cylindric, about 10 cm. high; ribs 15, straight or sometimes spiral; areoles elongated, separated from the adjacent ones by a space of about their own length; radial spines at first white, when mature brownish or yellowish, about 16, somewhat spreading, those at the top and base of the areole smaller; central spines none; areoles when young clothed with dense white wool, this nearly or quite wanting in age; flowers from the youngest growth appearing terminal; perianth widely spreading, 6 cm. broad or more; inner segments light purple, oblong to spatulate-oblong, the broad apex toothed or erose, the terminal teeth tapering into a slender awn; filaments short, yellow; style stout, longer than the filaments; stigma-lobes 10, obtuse; areoles of the ovary bearing 10 to 12 slender spines intermixed with cobwebby wool, the spines whitish or the central ones brownish; areoles of the tube crowning an elongated tubercle, not so closely set, bearing spines subtended by minute leaves.

Type locality: Wichita Mountains, Oklahoma.
Distribution: Mountains of Oklahoma.

This very interesting species was collected in August 1906 by Mr. Vernon Bailey, for whom it was named, in the Wichita Mountains, Oklahoma.   The following August it flowered and then died.   Until recently we supposed that this was the only collection known but, while restudying the genus, we find that a plant sent by a Mr. Merkel from Oklahoma flowered in July 1908.   We have since endeavored to collect specimens, but without success until we were reading the second proof.   On Major E. A. Goldman's return from Oklahoma in August 1921 he informed us that *Echinocereus baileyi* was very common in the Wichita National Forest near Cache and he arranged with the forest super-

---

* On inquiring of Miss Mary A. Day regarding these references we received the following reply under date of June 15, 1921:
"The name *Mammillaria caespitosa* used by Dr. Gray in his Structural Botany, edition 5, 1858, appears one year earlier in his First Lessons in Botany and Vegetable Physiology, page 96, 1857.   This is the earliest reference I find for it.   In the foot-note at the bottom of the page where this name is given, Dr. Gray himself has crossed out *Mammillaria caespitosa* and written in *Cereus caespitosus.*   He has also crossed off the words 'Upper Missouri,' and written in 'Texas.'   This would indicate that Dr. Gray himself considered the name *Mammillaria caespitosa*, or the figure of it in his First Lessons, and Structural Botany, the same as Engelmann's *Cereus caespitosus* of Texas."

.visor, Mr. Frank Rush, to have living plants sent on to Washington. These arrived in November. Besides several single plants there was a large clump, 3 dm. in diameter, consisting of 25 branches.

Illustrations: Contr. U. S. Nat. Herb. **12:** pl. 56, 57.

Figure 27 is a copy of the first illustration above cited.

FIG. 26.—Echinocereus reichenbachii.　　　　FIG. 27.—Echinocereus baileyi.

35. Echinocereus rigidissimus (Engelmann) Rose, Contr. U. S. Nat. Herb. **12:** 293. 1909.

Cereus pectinatus rigidissimus Engelmann, Proc. Amer. Acad. **3:** 279. 1856.
Echinocereus pectinatus rigidissimus Rümpler in Förster, Handb. Cact. ed. 2. 818. 1885.
Echinocereus pectinatus robustus Bauer, Gartenflora **39:** 513. 1890.

Plants simple, erect, rigid, short-cylindric, 1 to 2 dm. high, 4 to 10 cm. in diameter, usually hidden by the closely set interlocking spines; ribs numerous, 18 to 22, low; areoles approximate, elliptic, 5 to 6 mm. long; radial spines about 16, gray to reddish brown, arranged in horizontal bands, pectinate, rigid, 15 mm. long or less, often recurved; central spines none; flowers purple, 6 to 7 cm. long, fully as broad when expanded; perianth-segments oblong, 3 to 4 cm. long, acute; stamens numerous, shorter than the style; areoles on ovary somewhat floccose, very spiny; fruit globular, 3 cm. in diameter, very spiny; seeds black, tuberculate, 1.5 mm. in diameter.

Type locality: Sonora.

Distribution: Southeastern Arizona and northern Sonora.

This species is a great favorite in collections, although it does not last long. Its varicolored spines arranged in bands have given it the appropriate name of rainbow cactus,

while in Mexico it is called cabeza del viego. It has often been regarded as a variety of
E. reichenbachii, but it is abundantly distinct. It is known in the trade under various
names, among which are *Cereus candicans* and *Echinocereus candicans*, a name which
belongs properly to a very different plant from Argentina, *Cereus rigidissimus*, *C. robustior*,
and *Echinocereus robustior*, but none of which has been formally published. Here also
belong the names *Echinocereus rigidispinus*, *E. pectinatus robustior* (Monatsschr. Kakteenk.
7: 95. 1897), and perhaps *E. pectinatus candicans* (Monatsschr. Kakeenk. 3: 111. 1893).

Figs. 28 and 29.—Echinocereus rigidissimus.

The largest specimen which we have seen was collected by Dr. J. W. Gidley near
Benson, Arizona, in 1921. This plant was fully 2 dm. tall and 1 dm. in diameter. The
spines were nearly all red, separated by very narrow bands of white ones giving the plant a
brilliant and striking appearance.

Illustrations: Cycl. Amer. Hort. Bailey 2: 519. f. 748; De Laet, Cat. Gén. f. 36; Schelle,
Handb. Kakteenk. 133. f. 63, as *Echinocereus pectinatus rigidissimus;* Gartenflora 39:
pl. 1331, as *E. pectinatus robustus;* Cact. Journ. 1: pl. for September; 2: 18; (?) Balt. Cact.
Journ. 2: 88; West Amer. Sci. 7: 236, as *Echinocereus candicans;* Stand. Cycl. Hort. Bailey
2: f. 1376.

Figure 28 is from a photograph of a plant sent from Sonora, Mexico, by Carl Lum-
holtz in 1909; figures 29 and 30 are from photographs taken by F. E. Lloyd at the Desert
Laboratory of the Carnegie Institution, Tucson, Arizona; figure 31 is from a photograph
of a plant in its natural habitat, taken by Dr. MacDougal at Calabasas, Arizona, in 1908.

36. **Echinocereus weinbergii** Weingart, Monatsschr. Kakteenk. **22:** 83. 1912.

Very stout, usually simple, at first globose, becoming conical, at least in cultivation, 13 cm. in diameter; ribs 15, acute, more or less undulate; areoles elliptic, approximate; radial spines 9 to 12, pectinate, 3 to 12 mm. long, at first white or rose but in age yellowish; central spines none; flowers diurnal, 3.6 cm. broad, rose-colored; inner perianth-segments in several series, 1.5 to 3 cm. long, 4 to 5 mm. broad, lanceolate, acuminate; fruit not known.

FIGS. 30 and 31.—Echinocereus rigidissimus.

Type locality: Not cited.

Distribution: Doubtless Mexico, but known only from garden plants.

This is one of the stoutest plants of the genus known to us. It was named in honor of Frank Weinberg, a cactus dealer.

Illustration: Monatsschr. Kakteenk. **24:** 105.

Figure 32 is from a photograph contributed by Mr. Frank Weinberg.

37. **Echinocereus pectinatus** (Scheidweiler) Engelmann in Wislizenus, Mem. Tour North. Mex. 109. 1848.

> *Echinocactus pectinatus* Scheidweiler, Bull. Acad. Sci. Brux. **5:** 492. 1838.
> *Echinocactus pectiniferus* Lemaire, Cact. Gen. Nov. Sp. 25. 1839.
> *Echinocactus pectiniferus laevior* Monville in Lemaire, Cact. Gen. Nov. Sp. 26. 1839.
> *Echinopsis pectinata* Fennel, Allg. Gartenz. **11:** 282. 1843.
> *Cereus pectinatus* Engelmann in Gray, Pl. Fendl. 50. 1849.
> *Cereus pectiniferus* Labouret, Monogr. Cact. 320. 1853.
> *Echinocereus pectinatus chrysacanthus* Schumann, Gesamtb. Kakteen 272. 1898.

FIG. 32.—Echinocereus weinbergii.

Plants simple, erect, cylindric, 1 to 1.5 dm. long, 3 to 6 cm. in diameter, almost hidden by the many short interlocking spines; ribs 20 to 22, usually straight; areoles approximate, but not touching one another, elliptic, 3 mm. long; radial spines about 30, pectinate, usually much less than 10 mm. long, white or rose-colored, the colors more or less in bands about the plant; central spines several, more or less porrect; flowers purplish, 6 to 8 cm. long; areoles on ovary and flower-tube felted, very spiny; fruit spiny, becoming naked, 2 to 3 cm. in diameter.

Type locality: Near Villa del Pennasco, central Mexico.

Distribution: Central Mexico.

This species was first collected by Galeotti who sent a collection to Belgium from the states of San Luis Potosí and Guanajuato, Mexico. The type station, Villa del Pennasco, we have not located. It was soon after figured by Lemaire (Icon. Cact. pl. 14 or 15) and

Pfeiffer (Abbild. Beschr. Cact. **2**: pl. 10), very likely from the type collection. These illustrations are not very good, especially as to the areoles. In 1845 it was again described and illustrated, this time in Curtis's Botanical Magazine, plate 4190, from a specimen sent by a Mr. Staines from San Luis Potosí. This is from the region of Galeotti's type. We refer here Lloyd's No. 4 from Zacatecas.

Cereus pectinatus laevior Salm-Dyck (Cact. Hort. Dyck. 1849. 43. 1850; *Echinocereus pectinatus laevior*, Monatsschr. Kakteenk. Index 56. 1912) is only a name to be referred here. *Echinocereus pectinatus cristatus* is an abnormal form of no taxonomic importance. A very unusual illustration of it appeared in Floralia **42**: 372. This variety may or may not belong to this species. *Echinopsis pectinata laevior* Monville (Förster, Handb. Cact. 365. 1846) belongs here.

Illustrations: Curtis's Bot. Mag. **71**: pl. 4190; Lemaire, Icon. Cact.* pl. 14 or 15; Loudon, Encycl. Pl. ed. 3, 1377. f. 19371; Fl. Serr. **2**: July, pl. 7, as *Echinocactus pectiniferus;* Pfeiffer, Abbild. Beschr. Cact. **2**: pl. 10, as *Echinopsis pectinata;* Cact. Journ. **2**: 18; Förster, Handb. Cact. ed. 2. f. 108; Rümpler, Sukkulenten 141. f. 76; Ann. Rep. Smiths. Inst. **1908**: pl. 2, f. 6; Schelle, Handb. Kakteenk, 132. f. 62.

Figure 33 is copied from the first illustration cited above.

FIG. 33.—Echinocereus pectinatus.

38. Echinocereus fitchii sp. nov.

Plants short-cylindric or somewhat narrowed above, 8 to 10 cm. long, 4 to 5 cm. in diameter; ribs 10 to 12, low, rounded; areoles 4 to 6 mm. apart, small, circular; radial spines about 20, white, spreading, 4 to 6 mm. long; central spines 4 to 6, slightly spreading, 12 mm. long or less, acicular, brownish, but sometimes white at base; flowers 6 to 7 cm. long, pink; perianth-segments, oblanceolate, widely spreading, acute, serrate on the margin; ovary 2.5 cm. long, bearing numerous areoles, these spiny and with cobwebby hairs.

Living specimens were collected by Dr. Rose near Laredo, Texas, in 1913 (No. 18037) which flowered in the New York Botanical Garden, April 10, 1914.

This plant is named for Mr. William R. Fitch who accompanied Dr. Rose on collecting trips to the West Indies and to western Texas in 1913.

Plate III, figure 2, is of the type plant cited above.

39. Echinocereus scopulorum sp. nov.

Stems single, cylindric, 10 to 40 cm. long, nearly hidden by the closely set spines; ribs 13 or more, low, somewhat tuberculate; areoles circular, devoid of wool (at least in areoles of the previous year); spines highly colored, pinkish or brownish with blackish tips, in age, however, gray and stouter; radials somewhat spreading; centrals 3 to 6, similar to the radials; flower-buds or some of them appearing near the top of the plant, developing very slowly; flowers with a delicate rose perfume, widely spreading when fully expanded, 9 cm. broad; tube 2 cm. long, broadly funnelform, bearing greenish tubercles; inner perianth-segments 4 cm. long, rose or purplish rose in color, much paler on the outside, sometimes nearly white, rather loose and usually only in about 2 rows, oblanceolate to spatulate, erosely dentate, acute; stamens greenish; style stout, much longer than the stamens; stigma-lobes linear, 12.

---

* See Britton and Rose, Cactaceae **2**: 6. 1920.

Collected near Guaymas, Mexico, March 10, 1910, by Rose, Standley, and Russell (No. 12570, type), and by Ivan M. Johnston, April 14, 1921 (No. 3103). It also was found as far south as Topolabampo, Sinaloa, March 23, 1910, by Rose, Standley, and Russell (No. 13349) and at San Pedro Bay, Sonora (No. 4291), and at San Carlos Bay, Sonora (No. 4344), by Mr. Johnston in 1921.

It is related to *E. reichenbachii*, but is very distinct from it.

Mr. Johnston's No. 3103 flowered in Washington, July 22, 1921.

Figure 34 is from a photograph made in Washington from a living plant collected by Rose, Standley, and Russell at Topolabampo, Mexico.

40. **Echinocereus roetteri** (Engelmann) Rümpler in Förster, Handb. Cact. ed. 2. 829. 1885.

> *Cereus dasyacanthus minor* Engelmann, Proc. Amer. Acad. 3: 279. 1856
> *Cereus roetteri* Engelmann, Proc. Amer. Acad. 3: 345. 1856.
> *Echinocereus kunzei* Gürke, Monatsschr. Kakteenk. 17: 103. 1907.

Cespitose, or perhaps sometimes simple and occasionally budding above, 1 to 2.5 dm. high; ribs 13, straight, more or less undulate; areoles circular, or a little longer than broad, about 1 cm. apart; radial spines 15 to 17, acicular, about 1 cm. long, white or purplish; central spines 1 to 5, not in a single row, a little stouter but scarcely longer than the radials; flowers appearing below the top of the plant, large, 6 to 7 cm. long, perhaps even broader than long, light purple; outer perianth-segments greenish yellow; inner perianth-segments oblanceolate, acute, 3 to 4 cm. long; ovary and fruit spiny.

FIG. 34.—Echinocereus scopulorum.

Type locality: Near El Paso, Texas.

Distribution: Southwestern Texas; Chihuahua, near El Paso, and southeastern New Mexico.

Echinocereus kunzei which we have referred here as a synonym is usually stated to be from Arizona. It was doubtless sent out from Phoenix, Arizona, where Dr. Kunze lived, but we have a specimen in the U. S. National Herbarium labeled "southern New Mexico" in Dr. Kunze's handwriting. The illustration which Dr. Kunze uses (Price List of Cactaceae, 1913) suggests *Echinocereus viridiflorus*.

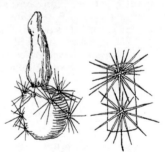

FIGS. 35 and 36.—Flower and spine-clusters of Echinocereus roetteri. ×0.9.

According to Engelmann it is similar to *E. dasyacanthus* from which it is distinguished by its fewer ribs, stouter spines, purple flowers, smaller fruit, and larger seed.

This species was named for Paulus Roetter, the artist, who made the cactus drawings for the Mexican Boundary Survey.

Illustrations: Cact. Mex. Bound. pl. 41, f. 3 to 5, as *Cereus roetteri;* Blühende Kakteen 3: pl. 128, as *Echinocereus kunzei*.

Figures 35 and 36 are drawn from a co-type herbarium specimen collected by Charles Wright in New Mexico, 1851–1852.

41. Echinocereus chlorophthalmus (Hooker) Britton and Rose, Contr. U. S. Nat. Herb. **16**: 242. 1913.

  Echinocactus chlorophthalmus Hooker in Curtis's Bot. Mag. **74**: pl. 4373. 1848.

  Cespitose, nearly globose, glaucous-green; ribs 10 to 12, somewhat tuberculate; areoles circular; radial spines 7 to 10, slender, needle-like, 12 to 18 mm. long, spreading; central spine one, stouter than the radials, the central as well as the radials pale brown but reddish at base when young; inner perianth-segments spatulate, acute, somewhat serrate towards the tip, glossy above, purple, whitish at base; stigma-lobes bright green; ovary and fruit spiny.

FIG. 37.—Echinocereus chlorophthalmus.   FIG. 38.—Echinocereus knippelianus.

Type locality: Real del Monte, Mexico.

Distribution: Known only from the type locality.

This species, although described as an *Echinocactus*, is undoubtedly an *Echinocereus*, but it is not near *Echinocereus conglomeratus* as Schumann suggests.

In 1905 Dr. Rose visited Real del Monte, the type locality, where he collected the flowers of an *Echinocereus* (No. 8730) which correspond very well to the cited illustrations.

Illustrations: Curtis's Bot. Mag. **74**: pl. 4373; Loudon, Encycl. Pl. ed. 3. 1377. f. 19374, as *Echinocactus chlorophthalmus*.

Figure 37 is copied from the first illustration above cited.

42. Echinocereus knippelianus Liebner, Monatsschr. Kakteenk. **5**: 170. 1895.

  Echinocereus liebnerianus Carp,* Balt. Cact. Journ. **2**: 262. 1896.†
  Echinocereus inermis Haage jr., Monatsschr. Kakteenk. **8**: 130. 1898.
  Cereus knippelianus Orcutt, West Amer. Sci. **13**: 27. 1902.

---

  *The authority for this name was given as Carp or as an abbreviation, Carp., which suggested that it might be an abbreviation for Carpenter, but as there was no cactus authority of this name the explanation seemed unsatisfactory.   It was known that various short articles appeared in the Baltimore Cactus Journal under this name with a California address.   This led us to write to C. R. Orcutt and then to Ernest Braunton, both of whom have long been in touch with horticultural interests in California.   From them we obtained the following information:
  Carp's real name was Daniel R. Crane.   At one time he was a dealer of cacti and advertised freely under the name of the California Cactus Company, Soldiers' Home, Los Angeles County, California.   His pen name, he explained, was a reversal of the first syllable of practical, which is of doubtful significance.   Crane served in the Union Army and was somewhat erratic in his later years.   He died about 1901.
  †This reference is taken from Schumann (Gesamtb. Kakteen 252. 1897) who cites Carp as the authority for this binomial.   The original publication of *E. liebnerianus* does not refer to Carp but to Liebner.   It occurs in the following letter of K. Schumann to the editor of the Baltimore Cactus Journal (**2**: 262. 1896):
  "The cactus found by McDowell and pictured twice in the November number of the Baltimore Cactus Journal is described in the November number of the Monatsschrift für Kakteenkunde and named by Mr. C. Liebner, *Echinocereus liebnerianus.*"

At first simple, stout, a little higher than broad, about 10 cm. high, but in cultivation elongated, 20 cm. high or more, branching, very deep green, becoming turgid and flabby; ribs 5 to 7, more prominent towards the top of the plant, sometimes strongly tuberculate, at other times only slightly sinuate; areoles minute, white-felted, 5 to 6 mm. apart; spines 1 to 3, weak, 3 to 6 mm. long, yellow; flowers pinkish, 2.5 to 3 cm. long; perianth-segments spreading, oblanceolate, acute; style cream-colored; fruit not known.

Type locality: Not cited.

Distribution: Mexico, but range unknown.

The origin of this species is unknown but it is supposed to have come from Mexico. It is not uncommon in European collections and Dr. Rose studied it in Berlin in 1912. At one time we had it in our collection but it has since disappeared; otherwise the plant is known to us only from the descriptions and illustrations.

This species is doubtless named for Karl Knippel, a well-known dealer in cacti.

Illustrations: Schumann, Gesamtb. Kakteen f. 47; Monatsschr. Kakteenk. **5:** 170; Blühende Kakteen **1:** pl. 12; Schelle, Handb. Kakteenk. 120. f. 54; Balt. Cact. Journ. **2:** 215, 228. f. 3; Kirtcht, Kakteen Zimmergarten 57.

Figure 38 is copied from the third illustration above cited.

43. Echinocereus pulchellus (Martius) Schumann in Engler and Prantl, Pflanzenfam. 3⁶ᵃ: 185. 1894.

    Echinocactus pulchellus Martius, Nov. Act. Nat. Cur. **16:** 342. 1828.
    Cereus pulchellus Pfeiffer, Enum. Cact. 74. 1837.
    Echinonyctanthus pulchellus Lemaire, Cact. Gen. Nov. Sp. 85. 1839.
    Echinopsis pulchella Zuccarini in Förster, Handb. Cact. 363. 1846.

Stems obovate-cylindric, 5 to 7 cm. high, simple, glaucous; ribs 12, obtuse, more or less divided into tubercles; spines 3 to 5, short, straight, deciduous, yellowish; flowers rosy-white, about 4 cm. broad; inner perianth-segments lanceolate, acuminate.

FIG. 39.—Echinocereus pulchellus.   FIG. 40.—Echinocereus amoenus.

Type locality: Pachuca, Mexico, *fide* Pfeiffer.

Distribution: Probably central Mexico.

This species is known to us only from descriptions and illustrations, but it seems quite distinct.

Illustrations: Nov. Act. Nat. Cur. **16:** pl. 23, f. 2, as *Echinocactus pulchellus;* Blühende Kakteen **1:** pl. 33; Monatsschr. Kakteenk. **26:** 177.

Figure 39 is copied from the first illustration above cited.

44. Echinocereus amoenus (Dietrich) Schumann in Engler and Prantl, Pflanzenfam. 3⁶ᵃ: 185. 1894.

    Echinopsis amoena Dietrich, Allg. Gartenz. **12:** 187. 1844.
    Echinopsis pulchella amoena Förster, Handb. Cact. 364. 1846.
    Cereus amoenus Hemsley, Biol. Centr. Amer. Bot. **1:** 540. 1880.
    Echinocereus pulchellus amoenus Schumann, Gesamtb. Kakteen 253. 1897.

Plants low, almost buried in the ground; ribs usually 13, low, somewhat tuberculate; young areoles bearing 6 to 8 rather stout, short, spreading spines; old areoles spineless; flowers about 5 cm. broad, magenta-colored; inner perianth-segments spatulate, with an ovate acute tip; filaments rose-colored; areoles of the ovary and flower-tube bearing brown spines and cobwebby wool.

Type locality: Mexico.

Distribution: San Luis Potosí, Mexico.

This plant has recently been introduced into Europe in great quantities. It is rather inconspicuous, but has very pretty flowers.

Echinopsis pulchella rosea (Labouret, Monogr. Cact. 292. 1853) was given as a synonym of the species.

Illustration: Monatsschr. Kakteenk. **3**: 171. f. 4, as *Echinopsis amoena.*

Figure 40 is from a photograph of a plant sent from San Luis Potosí, Mexico, by Mrs. Irene Vera.

45. Echinocereus palmeri sp. nov.

Plants small, 5 to 8 cm. high, 2 to 3 cm. in diameter; areoles closely set, round; radial spines 12 to 15, spreading, slender, brown-tipped; central spine one, porrect, 15 to 20 mm. long, brown to blackish; flower 3.5 cm. long, purple; areoles on the ovary bearing a cluster of brown spines and white wool.

Collected by Dr. E. Palmer on a small hill near Chihuahua City, April 1908 (No. 121). Only three specimens were seen, of which one was in flower.

FIG. 41.—Echinocereus brandegeei.　　FIG. 42.—Echinocereus hempelii.

46. Echinocereus brandegeei (Coulter) Schumann, Gesamtb. Kakteen 290. 1898.

Cereus brandegeei Coulter, Contr. U. S. Nat. Herb. **3**: 389. 1896.
Cereus sanborgianus Coulter, Contr. U. S. Nat. Herb. **3**: 391. 1896.
Echinocereus sanborgianus Schumann, Gesamtb. Kakteen 274. 1898.

Always growing in clumps; joints sometimes one meter long or more, 5 cm. in diameter, but usually much narrowed toward the base; ribs strongly tubercled; areoles circular; spines at first light yellow tinged with red, in age dark gray; radial spines about 12, spreading, acicular; central spines usually 4, very much stouter, more or less flattened, erect or porrect, the lowest one decidedly so, sometimes 8 cm. long; flowers purplish, about 5 cm. long; areoles on ovary and tube closely set, filled with pale acicular spines and long white wool; fruit globular, 3 cm. in diameter, spiny; seeds black, tuberculately roughened.

Type locality: El Campo Allemand, Lower California.

Distribution: Very common on the low hills along the coast of southern Lower California and adjacent islands.

The species is named for Townsend S. Brandegee, a well-known botanical collector and writer.

Illustration: Contr. U. S. Nat. Herb. 16: pl. 124.

Figure 41 is from a photograph of a plant collected by Dr. Rose at the head of Concepción Bay, Lower California, in 1911 (No. 16672).

47. Echinocereus hempelii Fobe, Monatsschr. Kakteenk. 7: 187. 1897.

Plant, so far as known, simple, erect, 1.5 dm. long or more, 6 to 7 cm. in diameter, dark green; ribs 10, strongly tuberculate; radial spines 6, spreading, white with brown tips, acicular, 1 cm. long or less; central spines none; flowers from near the top of plant, rather large, 6 to 8 cm. broad, violet; inner perianth-segments few, about 14, loosely arranged, oblong, 3 cm. long, strongly toothed above; style longer than the stamens; ovary bearing conspicuous red scales, spiny; fruit not known.

Type locality: Mexico.

Distribution: Known only from cultivated plants.

In 1912 Dr. Rose studied this plant in Berlin and thought it might be a form of *E. fendleri* but it has since been illustrated in color and shows some striking differences, as, for instance, its lack of central spines, the strongly tubercled ribs and the very loosely arranged perianth-segments.

This species was named for George Hempel (1847–1904) who collected in Mexico and South America.

Illustrations: Monatsschr. Kakteenk. 7: 185; Blühende Kakteen 3: pl. 142.

Figure 42 is copied from the second illustration above cited.

48. Echinocereus merkeri Hildmann in Schumann, Gesamtb. Kakteen 277. 1898.

    Cereus merkeri Berger, Rep. Mo. Bot. Gard. 16: 81. 1905.

Cespitose; joints erect, 12 to 15 cm. in diameter, light green; ribs 8 or 9, sinuate; radial spines 6 to 9, white, shining; central spines 1 or rarely 2, often yellowish, larger than the radials, red at base; flowers purple, about 6 cm. long; inner perianth-segments short-oblong, 3 cm. long, rounded at apex, sometimes mucronate; scales on ovary 2 to 3 cm. long, ovate, acuminate, bearing 2 to 5 long spiny bristles in their axils.

Type locality: Not cited.

Distribution: Durango to Coahuila and San Luis Potosí, Mexico.

In the original description several localities in Durango and Coahuila are assigned for this species and it is possible that some other species was confused with it.

To this species we refer Palmer's herbarium specimens from Saltillo, Mexico, 1905 (No. 510), and C. A. Purpus's specimen from northern Mexico; the latter we have living also, and it is unlike any other plant in our collections.

Echinocereus jacobyi (Schumann, Gesamtb. Kakteen 278. 1898), undescribed, belongs here.

Figure 43 is from a photograph of a plant collected by Dr. C. A. Purpus in northern Mexico.

49. Echinocereus fendleri (Engelmann) Rümpler in Förster, Handb. Cact. ed. 2. 801. 1885.

    Cereus fendleri Engelmann in Gray, Pl. Fendl. 50. 1849.
    Cereus fendleri pauperculus Engelmann in Gray, Pl. Fendl. 51. 1849.

Cespitose; stems about 8, ascending or erect, 1 to 3 dm. long, 5 to 7.5 cm. in diameter; ribs rather prominent, 9 to 12, somewhat undulate; spines very variable as to color, length, and form; radial spines 5 to 10, more or less spreading, 1 to 2 cm. long, acicular to subulate; central spine solitary, usually porrect, 4 cm. long or less, dark colored, often black-bulbose at base; flowers borne at the upper part of the plant, often very large, 10 cm. broad when fully expanded, but sometimes smaller, deep purple; inner perianth-segments spatulate, 3 to 4 cm. long, acute, the margin sometimes serrulate; filaments purple, very short, 1 cm. long or less; style very pale; ovary deep green, its areoles bearing white felt and white bristly spines; fruit ovoid, 2.5 to 3 cm. long, purplish, edible; seeds 1.4 mm. long.

Type locality: Near Santa Fé, New Mexico.

Distribution: Texas to Utah, Arizona, and northern Sonora and Chihuahua, Mexico.

The species shows considerable variation in armament and in the size of the flowers and, except in its erect habit, much resembles the next following species.

This species was named for August Fendler (1813–1883) who collected extensively in New Mexico and Venezuela.

Related to *Echinocereus fendleri* but growing at higher elevations is a plant obtained by D. T. MacDougal and Forrest Shreve from the eastern side of the Santa Catalina Mountains in March 1921 and again at Oracle, Arizona, May 6, 1921. This plant grows singly or in clumps with 13 to 16 low ribs and short spines. The central spines are from 1 to 4. More detailed field studies may prove this to be a distinct species. Figure 45 may represent this form.

FIG. 43.—Echinocereus merkeri.

FIG. 44.—Echinocereus fendleri.

Echinocereus hildmannii Arendt (Monatsschr. Kakteenk. 1: 146. pl. 11. 1891) should be compared with *E. fendleri.*

Illustrations: Curtis's Bot. Mag. 106: pl. 6533; Cact. Mex. Bound. pl. 51 to 53; Gartenflora 32: 341, as *Cereus fendleri;* Förster, Handb. Cact. ed. 2. f. 104; Plant World 11¹⁰: f. 1; Rümpler, Sukkulenten 137. f. 73; Schelle, Handb. Kakteenk. 134. f. 64; Blühende Kakteen 3: pl. 143; Floralia 42: 369.

Plate IV, figure 3, shows a flowering plant sent by W. H. Long to the New York Botanical Garden from Albuquerque, New Mexico, in 1915. Figure 44 is from a photograph taken by Dr. MacDougal in the Tucson Mountains, Arizona, in 1908; figure 45 is from a photograph of a plant collected by Dr. Rose near Benson, Arizona, in 1908.

50. Echinocereus enneacanthus Engelmann in Wislizenus, Mem. Tour North. Mex. 112. 1848.

> *Cereus enneacanthus* Engelmann, Pl. Fendl. 50. 1849.
> *Echinocereus carnosus* Rümpler in Förster, Handb. Cact. ed. 2. 796. 1885.
> *Echinocereus enneacanthus carnosus* Quehl, Monatsschr. Kakteenk. 18: 114. 1908.

Cespitose, with many stems, often forming clumps one meter in diameter or more; joints often elongated, prostrate, 5 to 7 cm. in diameter; ribs 7 or 8, prominent, more or less tuberculate, some-

M. E. Eaton del.

A. Hoen & Co.

1. Top of flowering plant of *Echinocereus sciurus*.
2. Flowering plant of *Lobivia cinnabarina*.
3. Flowering plant of *Echinocereus fendleri*.
4. Top of flowering plant of *Echinocereus lloydii*.
5. Flowering plant of *Rebutia minuscula*.
(All natural size.)

what flabby, dull green; areoles 2.5 cm. apart; radial spines unequal, usually less than 12 mm. long, acicular, at first yellowish, becoming brownish; central spine solitary, usually elongated, nearly terete, 3 to 5 cm. long; flower purple, 7.5 cm. broad; perianth-segments nearly oblong; style cream-colored, a little longer than the stamens; fruit globular, juicy, edible.

Type locality: Near San Pablo, south of Chihuahua, Mexico.

Distribution: Northern Mexico, New Mexico, and southern Texas.

There has always been more or less uncertainty about this species. Engelmann, who described the species in 1848, based it on Wislizenus's specimen which came from near San Pablo, Chihuahua. In the Cactaceae of the Mexican Boundary Report, Engelmann again describes the plant and illustrates it. His illustrations, however, represent two species. We have defined the species in the same way that Dr. Engelmann did, for it will require further field studies along the border of Texas and Mexico to determine its exact limits; a second species may be confused with it.

FIG. 45.—Echinocereus fendleri.   FIG. 46.—Echinocereus lloydii.

The type specimen consists of four flowers only. Wislizenus also collected two herbarium specimens of the stem of some other *Echinocereus* which were probably used by Engelmann in drawing up his original description. These, however, come from a different locality, Parras, and seem to represent a different species.

On account of the delicious strawberry-like flavor of the fruit this plant is known as the strawberry cactus throughout southern Texas, where the fruit is much used for jams. According to Robert Runyon, it is also called the cob cactus about Brownsville, Texas, because of the cob-like shape of its branches.

Illustrations: Cact. Mex. Bound. pl. 48, f. 2 to 4; pl. 49; Dict. Gard. Nicholson 4: 512. f. 8; Suppl. 217. f. 230; West Amer. Sci. 13: 11, as *Cereus enneacanthus;* Bull. Univ. Texas 82: pl. 3, f. 2, as *Cereus longispinus* (?); Cact. Journ. 1: 135; (?)2: 19; Schelle, Handb. Kakteenk. 127. f. 58; Förster, Handb. Cact. ed. 2. 795. f. 103.

Figure 49 is copied from the first illustration above cited.

51. Echinocereus lloydii sp. nov.

Stems in clusters of 6 or more, very stout, 20 to 25 cm. high, 10 cm. in diameter, bright green; ribs 11, about 3 cm. apart, nearly straight; areoles 15 mm. apart, rather large, circular, somewhat woolly when young; spines rather short, about 10 mm. long, wine-colored, paler at base; radial spines

14; centrals 4 to 6, nearly porrect; flowers large, 8 cm. long, reddish purple; areoles on the ovary bearing clusters of reddish spines; stigma-lobes numerous; perianth-segments narrowly obovate, obtuse or obtusish.

Collected near Tuna Springs, Texas, in 1909, by F. E. Lloyd, for whom it is named.

Plate IV, figure 4, shows the top of one of the type plants in flower at the New York Botanical Garden. Figure 46 is from a photograph of one of the type plants taken in Washington.

52. Echinocereus engelmannii (Parry) Rümpler in Förster, Handb. Cact. ed. 2. 805. 1885.

Cereus engelmannii Parry in Engelmann, Amer. Jour. Sci. II. 14: 338. 1852.
Cereus engelmannii variegatus Engelmann and Bigelow, Proc. Amer. Acad. 3: 283. 1856.
Cereus engelmannii chrysocentrus Engelmann and Bigelow, Proc. Amer. Acad. 3: 283. 1856.
Echinocereus engelmannii chrysocentrus Rümpler in Förster, Handb. Cact. ed. 2. 806. 1885.
Echinocereus engelmannii variegatus Rümpler in Förster, Handb. Cact. ed. 2. 806. 1885.

Cespitose, forming large clumps; joints erect or ascending, cylindric, 1 to 3 dm. long, 5 to 6 cm. in diameter; ribs 11 to 14, low, obtuse; areoles large, nearly circular; radial spines about 10, appressed, stiff, about 1 cm. long; central spines 5 or 6, very stout, more or less curved and twisted, terete or somewhat flattened, sometimes 7 cm. long, yellowish to brown, more or less variegated; flowers somewhat variable in size, 5 to 8 cm. long, and even broader when fully expanded, purple; perianth-segments oblong, 3 to 4 cm. long, acuminate; scales on ovary 3 to 5 mm. long, acuminate; areoles felted and bearing stout bristles; fruit ovoid to oblong, spiny, about 3 cm. long; seeds black, nearly globular, or a little oblique, 1.5 mm. in diameter or less, tuberculate.

Type locality: Mountains about San Felipe, southern California.

Distribution: California, Nevada, Utah, Arizona, Sonora, and Lower California.

The three varieties, *albispinus* Cels, *fulvispinus* Cels, and *pfersdorffii* Heyder, mentioned by Schumann (Gesamtb. Kakteen 276. 1898) are simply forms named from color differences in the spines.

Two other varieties have been mentioned but are unimportant: *robustior* Hildmann and *versicolor* (Monatsschr. Kakteenk. 4: 194. 1894).

Some Arizona and Sonora specimens have more slender and lighter-colored spines than

FIG. 47.—Echinocereus sarissophorus.

is typical, and on these the variety *chrysocentrus* was based. Additional field observations may show this to be a distinct species; the spines closely resemble those of *E. stramineus*.

Illustrations: Journ. N. Y. Bot. Gard. 6: 93, as *Echinocereus;* Cact. Mex. Bound. pl. 57; Gartenflora 33: pl. 1174a, as *Cereus engelmannii;* Pac. R. Rep. 4: pl. 5, f. 8 to 10, as *Cereus engelmannii chrysocentrus;* Pac. R. Rep. 4: pl. 5, f. 4 to 7, as *Cereus engelmannii variegatus;* Cact. Journ. 2: 132, as *Echinocactus engelmannii;* Cact. Journ. 1: pl. for September; 2: 146; Monatsschr. Kakteenk. 16: 151; Schelle, Handb. Kakteenk. 134. f. 65; Journ. N. Y. Bot. Gard. 6: f. 23; Contr. U. S. Nat. Herb. 16: pl. 8, 9.

Plate V, figure 1, shows the top of a plant, which flowered at the New York Botanical Garden, sent by S. B. Parish from southern California in 1915.

53. Echinocereus sarissophorus sp. nov.

Cespitose; stems short, thick, pale green, about 10 cm. thick; ribs 9; radial spines 7 to 10, slender; centrals several, 5 to 8 cm. long, often bluish, somewhat angled; flowers purplish, 7 to 8 cm. long; inner perianth-segments broad; areoles on ovary and flower-tube bearing short, white wool and 3 to 5 long pale bristle-like spines; fruit globular, 2 to 3 cm. in diameter, covered with clusters of deciduous spines; seeds black.

Collected near Saltillo, Coahuila, April 1898, by Dr. E. Palmer (No. 100).

This species is common in Coahuila and Chihuahua, Mexico, having been repeatedly collected by Dr. Palmer and others. It is characterized by its stout, stubby habit and by its very long, usually stiff, often bluish spines.

Figure 47 is from a photograph of a plant collected by Dr. E. Palmer in Mexico in 1908.

54. Echinocereus dubius (Engelmann) Rümpler in Förster, Handb. Cact. ed. 2. 787. 1885.

Cereus dubius Engelmann, Proc. Amer. Acad. 3: 282. 1856.

Somewhat cespitose; stems 12 to 20 cm. long, pale green, of a soft flabby texture, 7 to 9-ribbed; ribs broad; spines white; radial spines 5 to 8, 12 to 30 cm. long; centrals 1 to 4, 3.5 to 7.5 cm. long, angled, often curved; flowers pale purple, 6 cm. long or more, with rather few and narrow perianth-segments; scales on flower-tube bearing 1 to 3 white bristles in their axils; fruit very spiny, 2.5 to 3 cm. long; seeds covered with confluent tubercles.

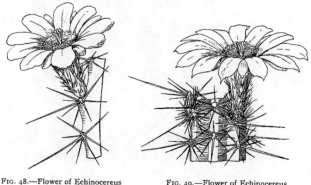

FIG. 48.—Flower of Echinocereus dubius. FIG. 49.—Flower of Echinocereus enneacanthus.

Type locality: Sandy bottoms of the Rio Grande at El Paso.

Distribution: Southeastern Texas, perhaps confined to the El Paso region.

This is said by Engelmann to be near *Echinocereus stramineus* and *E. enneacanthus.* The former, however, grows in the mountains and must be quite distinct. It is given a wide range by Schumann, who doubtless has included specimens of one or more related species. We know it only from co-type herbarium specimens of the Mexican Boundary Survey.

Illustrations: Cact. Mex. Bound. pl. 50, as *Cereus dubius;* Ann. Rep. Smiths. Inst. 1908: pl. 9, f. 3.

Figure 48 is copied from a part of the first illustration above cited.

55. Echinocereus* conglomeratus Förster, Gartenflora 39: 465. 1890.

Cereus conglomeratus Berger, Rep. Mo. Bot. Gard. 16: 81. 1905.

Cespitose, forming large clumps; joints simple, often half covered in the ground, 1 to 2 dm. long; ribs 11 to 13, slightly undulate; areoles 1 to 1.5 cm. apart, small, circular, slightly felted; spines white to brownish; radial spines acicular, 1.5 to 2.5 cm. long, spreading; central spines several, elongated, often 7 cm. long, very flexible; flowers 6 to 7 cm. long, broad and open, purplish; perianth-segments broad, 2 cm. long; spines on ovary and flower long, white, more or less curved; fruit globular, 3 cm. in diameter, somewhat acid, edible; seeds numerous.

*The generic name for this species was given in Gartenflora as *Echinocactus* in error.

Type locality: Rinconada, near Monterey, Mexico.

Distribution: Mountains in the states of Nuevo Leon, Coahuila, and Zacatecas, Mexico.

This species has usually been confused with *E. stramineus* but it has smaller, more open flowers, and it has a more southern range.

The plant is called alicoche; the fruit, which is edible, is known as pitahaya.

Illustrations: Blanc, Cacti 56, No. 736; Karsten and Schenck, Vegetationsbilder 2: pl. 19, b; 20, d; 22, a; 24.

Figure 50 is from a photograph of a plant collected by Dr. E. Palmer, near Saltillo, Mexico, and contributed by Dr. William E. Safford.

FIG. 50.—Echinocereus conglomeratus.

56. Echinocereus stramineus (Engelmann) Rümpler in Förster, Handb. Cact. ed. 2. 797.   1885.

    Cereus stramineus Engelmann, Proc. Amer. Acad. **3**: 282.   1856.

Plants grouped in masses forming immense mounds 1 to 2 meters in diameter and 3 to 10 dm. high; joints 12 to 25 cm. long, 3 to 7 cm. in diameter; ribs about 13, almost hidden by the long spines; spines at first brownish to straw-colored, in age nearly white; radial spines 7 to 14, 2 to 3 cm. long, spreading; central spines 3 or 4, 5 to 9 cm. long; flowers purple, 8 to 12 cm. long; perianth-segments oblong, 3 to 4 cm. long, rounded at apex; spines from the axils of scales on ovary and flower-tube, 2 to 5, short, white; fruit nearly globular, 3 to 4 cm. in diameter, red, spiny at first, becoming glabrous, edible; seeds 1.5 mm. in diameter, somewhat oblique.

Type locality: Mountain slopes, El Paso, Texas.

Distribution: Western Texas, southern New Mexico, and northern Chihuahua.

This species has often been given a much wider range than is here assigned to it, as it has been confused with other related species.

The plant is found only on dry mountains or hills, where it makes very large mounds; one of these observed by Dr. Rose in New Mexico was 15 dm. broad and 9 dm. high at the center, and was estimated to contain 400 to 500 joints.

Echinocereus bolansis Rünge (Monatsschr. Kakteenk. **5**: 123. 1895) was never published, but was referred here by Schumann.

The plants figured as *Echinocereus stramineus* in the Cactus Journal (**1**: 136; **2**: 19) do not seem to belong here, but to *E. fendleri.*

Illustrations: Cact. Mex. Bound. pl. 46 to 48, f. 1, as *Cereus stramineus;* Möllers Deutsche Gärt. Zeit. **25**: 482. f. 14; (?) Cact. Journ. **1**: 136; **2**: 19.

Figure 51 is from a photograph of a plant collected by Dr. Rose on the Sierra Blanca, Texas, in 1913; figure 52 shows a flower and figure 53 a fruit copied from the above cited illustrations in the Cactaceae of the Mexican Boundary Survey.

57. Echinocereus barthelowanus sp. nov.

Plants cespitose, forming large clusters; stems cylindric, 1 to 2 dm. long, 4 to 5 cm. in diameter; ribs about 10, somewhat tuberculate below, but completely hidden by the stout numerous spines; areoles approximate, 2 to 5 mm. apart, white-felted when young; spines numerous, acicular, sometimes 7 cm. long, pinkish when quite young, afterward white or yellow with brown or blackish tips, in age becoming gray; flowers only 10 to 12 mm. long; perianth-segments oblong, 3 to 4 mm. long, ovary minute, strongly tubercled, hidden under the mass of spines; spine-clusters on ovary

FIG. 51.—Echinocereus stramineus.

with 6 to 12 white or pinkish tipped spines, half as long as the flower.

Collected by J. N. Rose on the mesa, near Santa Maria Bay, Lower California, March 18, 1911 (No. 16278). Here we would refer also plants collected by C. R. Orcutt near the same locality in 1917.

The species is named for Captain Benjamin Barthelow, in whose company Dr. Rose collected the plant while making a cruise in the Lower California waters on the U. S. Steamer *Albatross* in 1911.

58. Echinocereus mamillatus (Engelmann).

Cereus mamillatus Engelmann in Coulter, Contr. U. S. Nat. Herb. 3: 405. 1896.

Cespitose; stems ascending, 2 to 3 dm. long, cylindric, 3.5 to 6 cm. in diameter; ribs 20 to 25, sometimes oblique, strongly tuberculate; spines white or pinkish; radial spines 10 to 25, acicular, 3 to 12 mm. long; central spines 3 or 4, much stouter than the radials, 1 to 2.5 cm. long; flowers and fruit unknown.

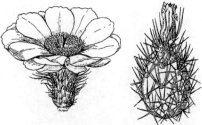

FIGS. 52 and 53.—Flower and fruit of Echinocereus stramineus. ×0.25.

Type locality: Mountain sides south of Mulege, Lower California.
Distribution: Southern Lower California.
Illustration: Schelle, Handb. Kakteenk. 97. f. 36, as *Cereus mamillatus.*

Figure 54 is from a photograph of a plant sent to the New York Botanical Garden from the Missouri Botanical Garden, in 1904.

59. Echinocereus ehrenbergii (Pfeiffer) Rümpler in Förster, Handb. Cact. ed. 2. 775. 1885.
Cereus ehrenbergii Pfeiffer, Allg. Gartenz. 8: 282. 1840.

Cespitose, 2 dm. high; joints often procumbent, pale or leaf-green; ribs 6, obtuse, sinuate, areoles remote, 2 cm. apart, white-felted; radial spines 8 to 10, slender, white; central spines 3 or 4, yellowish at base; flowers not known.

Type locality: Not cited.
Distribution: Central Mexico.

A living plant was obtained for the New York Botanical Garden from M. Simon of St. Ouen, Paris, in 1901, but it died before flowering.

This species was named for Karl Ehrenberg who spent ten years in Mexico, where he made large collections of cacti.

E. ehrenbergii cristatus (Förster, Handb. Cact. ed. 2. 776. 1885) is a garden monstrosity.

60. Echinocereus longisetus (Engelmann) Rümpler in Förster, Handb. Cact. ed. 2. 822.   1885.

Cereus longisetus Engelmann, Proc. Amer. Acad. 3: 280. 1856.

Plants simple or nearly so, cylindric, 15 to 25 cm. long, 5 to 7.5 cm. in diameter; ribs 11 to 14, somewhat tubercled; areoles circular; spines slender, elongated, white; radial spines 18 to 20, spreading, the lower 10 to 15 mm. long, much longer than the upper; central spines 5 to 7, very unequal, the lower elongated, 2.5 to 5.5 cm. long, deflexed; flowers said to be red.

FIG. 54.—Echinocereus mamillatus.

FIG. 55 a.—Echinocereus sp.

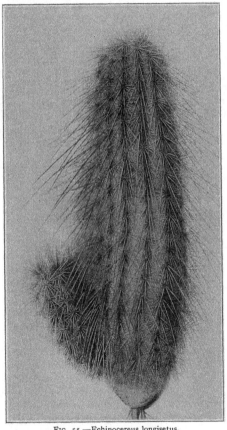

FIG. 55.—Echinocereus longisetus.

Type locality: Santa Rosa, south of the Rio Grande in Coahuila.

Distribution: Coahuila.

This plant, although collected as long ago as 1853 and well illustrated, has not been rediscovered; neither flowers nor fruit were collected and these are still desiderata; when obtained the relationship of this evidently distinct species can be determined.

Illustration: Cact. Mex. Bound. pl. 45, as *Cereus longisetus.*

Figure 55 is a copy of the illustration above cited.

ECHINOCEREUS sp.

A cylindric white-spined plant of this genus is illustrated by a photograph of Professor Lloyd's No. 27, collected by him in the Quijotoa Mountains, Arizona, in August 1906; neither flowers nor fruits were seen.

Figure 55 a is reproduced from this photograph.

ECHINOCEREUS sp.

In March 1910, in the mountain above Alamos, Sonora, a specimen of this genus was collected by Rose, Standley, and Russell, which we have not been able to identify (No. 13123); living plants have been under observation since, but no flowers have yet been produced. It may be described as follows:

Densely cespitose, usually 1, sometimes 2-jointed, dark green; ribs 7 or 8, low but distinct, the margin nearly straight; areoles approximate, 2 to 3 mm. apart; areoles minute, circular, white-felted; spines acicular, at first yellowish, often with brownish tips but soon whitish, less than 1 cm. long; radial spines about 10; central spines 3 or 4; flowers and fruit unknown.

It resembles *E. scheeri* but has slenderer joints and more delicate spines, and is of more western range.

ECHINOCEREUS PLEIOGONUS (Labouret) Croucher, Garden 13: 290. 1878.

    Cereus pleiogonus Labouret, Monogr. Cact. 317. 1853.

Short-cylindric, 5 to 13 cm. high; ribs 9 or 10; spines stiff, yellow, reflexed, 8 to 12, about 1 cm. long; flowers pinkish red, as long or longer than the plant itself; inner perianth-segments serrate.

It was introduced into France by M. Cels but its native country is not known. The illustrations cited below seem to represent some species of *Echinocereus* but we are not able to identify them. For further remarks on this species see *Echinocereus leeanus*, p. 9.

Illustrations: Dict. Gard. Nicholson 1: 299. f. 409; Watson, Cact. Cult. 82. f. 26, as *Cereus pleiogonus;* Garden 13: 291.

UNCERTAIN OR UNDESCRIBED SPECIES.

The following names have not been published or the plants have been so briefly or poorly described that they have not been identified:

CEREUS MACRACANTHUS Linke (Allg. Gartenz. 25: 239. 1857) is said to be related to *Cereus eburneus*, but Schumann thinks it is an *Echinocereus*.

ECHINOCEREUS BARCENA Rebut (Monatsschr. Kakteenk. 6: 127. 1896) is a garden name, but has never been described.

ECHINOCEREUS BICOLOR Galeotti (Wiener Illustr. Gartenz. 83. 1893, *fide* Index Kewensis Suppl. 1: 149) is not found at the place or in the work cited above. The name intended was probably *Echinocactus bicolor* Galeotti.

ECHINOCEREUS BOLIVIENSIS Poselger (Schumann, Gesamtb. Kakteen 290. 1898) does not belong to this genus.

ECHINOCEREUS CLAVIFORMIS (Haage, Preis-Verz. Cact. 22. *fide* Index Kewensis) is, so far as we know, unpublished. It is doubtless based on *Cereus (Echinocereus) claviformis* Regel and Klein (Ind. Sem. Hort. Petrop. 46. 1860).

ECHINOCEREUS GALTIERI (Monatsschr. Kakteenk. 5: 124. 1895) is only a garden name.

ECHINOCEREUS GRAHAMII (Monatsschr. Kakteenk. 20: 47. 1910) is doubtless intended for *Mammillaria grahamii.*

ECHINOCEREUS HAVERMANSII Rebut (Schumann, Gesamtb. Kakteen 290. 1898 and Monatsschr. Kakteenk. 17: 64. 1907) is only a name.

ECHINOCEREUS MALIBRANII Rebut seems to be only a catalogue name.

ECHINOCEREUS MAMILLOSUS (Rümpler in Förster, Handb. Cact. ed. 2. 787. 1885), undescribed, is supposed to be a hybrid.

ECHINOCEREUS SCHLINI, is a horticultural name, but supposed to be a misspelling of *E. scheeri.*

ECHINOCEREUS THURBERI (Monatsschr. Kakteenk. 3: 153. 1893) may have been intended for *Cereus thurberi.*

ECHINOCEREUS THWAITESII (Schumann, Gesamtb. Kakteen 290. 1898) is only a name.

ECHINOCEREUS TROCKYI is advertised for sale by A. V. Fric at 20 to 40 marks per plant (Monatsschr. Kakteenk. **28**: No. 8. 1918).

The following have been listed in the American Cyclopedia of Horticulture as being unidentifiable: *Echinocereus polycephalus*, *E. paucupina*, *E. uspenskii*, *E. uehri*, and *E. sanguineus*.

Echinocereus dahliaeflorus, with an illustration, appeared in a reputable Garden Magazine (Möllers Deutsche Gärt. Zeit. **15**: 148. 1900), issued on April 1st. The April Fool joke is so cleverly concealed that the editor deceived himself, for he carefully indexed the name at the end of the year. The name is to be ignored.

Echinocereus princeps Förster, *E. persolutus* Förster (Hamb. Gartenz. **17**: 163. 1861), and *E. raphicephalus* Förster (Hamb. Gartenz. **17**: 164. 1861) were described without flowers and we can not decide their generic alliance. The second species came from Peru and can be excluded from this genus.

Echinocereus penicilliformis Linke (Wochenschr. Gärtn. Pflanz. **1**: 85. 1858) we do not know, but since it comes from Bolivia it can be excluded from the genus *Echinocereus*.

2. AUSTROCACTUS gen. nov.

Plants low, ribbed, the areoles borne on the tuberculate ribs; spines in two series, the centrals hooked; flowers diurnal(?), borne at the upper part of areoles near the top of the plant, large, pinkish yellow, with a short, but rather definite tube and campanulate limb; perianth-segments aristate-acuminate; style as long as the stamens; stigma-lobes red to purplish; ovary and flower-tube very spiny or bristly; fruit spiny; seeds dull, flattened, reticulated.

Type species: *Cereus bertinii* Cels.

Only one species is known, inhabiting Patagonia.

The generic name is from *auster*, south, and cactus, referring to the habitat of the plant.

This genus seems to be the South American representative of the North American genus *Echinocereus*. It is like it in habit, ribs, flowers, and fruit, but its stigma-lobes are red, not green, and the central spines are always hooked. It has nothing to do with *Eulychnia* to which Berger referred it.

FIG. 56.—Austrocactus bertinii.

1. Austrocactus bertinii* (Cels).

Cereus bertini Cels, Hort. Franc. II. **5**: 251. 1863.

Simple or perhaps sometimes cespitose, olive-green, 15 to 40 cm. high; ribs 10 to 12, prominent; areoles circular, yellow-felted; radial spines about 15, acicular, straight, spreading, about 1 cm. long; central spines 4, very slender, strongly hooked, brownish to blackish, the lower and longest up to 3 cm. long; flowers when fully expanded reaching 10 cm. in breadth, about 6 cm. long; outer perianth-segments about 30, pinkish to brown, spiny tipped; inner perianth-segments about 20, pinkish yellow, oblong, long-acuminate, 4 cm. long, 1 cm. broad; stamens in two definite series; style thickish, red, longer than the stamens; stigma-lobes 16, linear; ovary and flower-tube with numerous areoles, these with clusters of 5 or 6 bristles or acicular spines.

Type locality: On the coast of Argentina, latitude 45° 30′ South.

Distribution: Southern Argentina† (Patagonia).

---

*The original spelling of the specific name was *bertini*.

†Schumann states that it comes from Paraguay while the Index Kewensis says Chile.

This species was discovered in 1855 by E. Cels, a brother of F. Cels, at one time a cactus dealer, who first described the plant; it was again collected at the type locality by Captain Bertin in 1861, for whom it was named. The first plants obtained did not live, but those of the second collection lived and flowered. Since then no plants have been reported, although the region in which it grows must have been frequently visited by collectors. Dr. Spegazzini, who knows Argentina well, was surprised to learn that such a plant was reported from southern Argentina. The illustration of *Cereus bertinii* certainly seems to represent a quite distinct genus. Our attention was first called to this species by the discovery of the illustration, cited below, by Dr. Rose, in an old book stall on the banks of the Seine in Paris in 1912.

Although described as *Cereus*, Cels calls attention to its relationship to *Echinocereus* and states that it should form a separate group. It has been cited as *E. bertinii* by Schelle (Handb. Kakteenk. 96. 1907).

Schumann (Gesamtb. Kakteen 163. 1897) established a series in *Cereus* called *Ancistracanthi* for this species, which he seems to have abandoned as a series name, and the species itself is omitted from his Keys published in 1903.

Illustration: Hort. Franc. II. **7**: pl. 14, as *Cereus bertinii*.

Figure 56 is copied from the illustration above cited.

3. REBUTIA Schumann, Monatsschr. Kakteenk. **5**: 102. 1895.

Plants small, globose to short-cylindric, single or cespitose, tuberculate, not ribbed, resembling a small *Coryphantha;* flower diurnal, arising from old tubercles, at the base or side of the plant, small, red or orange, with a slender, somewhat curved funnelform tube and a spreading or campanulate limb; scales on ovary small, naked or hairy in their axils, withering and persistent on the fruit; fruit small, red, not spiny.

Type species: *Rebutia minuscula* Schumann.

Five species, all South American, are here described. These have been referred heretofore either to *Echinocactus* or *Echinopsis*, or to both. They differ from *Echinocactus* in their lateral flowers borne at old areoles as well as in the structure of flowers, fruit, and plant-body. They are like *Echinopsis* in having lateral flowers, but otherwise very unlike any of the species of that genus. The plant-body in shape, size, and tubercles suggests some species of *Coryphantha*.

We know so little of the plants that we are not able to describe them very accurately and have depended largely upon descriptions and illustrations.

The genus was named by Schumann for P. Rebut, a cactus dealer.

KEY TO SPECIES.

Axils of scales on ovary and fruit naked.. 1. *R. minuscula*
Axils of scales on ovary and fruit hairy.
    Flowers from side of plant, near the middle.................................... 2. *R. fiebrigii*
    Flowers from the lower part of the plant.
        Central spines 1 to 4.. 3. *R. pseudominuscula*
        Central spines none.
            Areoles elliptic; spines spreading, swollen at base...................... 4. *R. pygmaea*
            Areoles circular; spines not widely spreading, not swollen at base......... 5. *R. steinmannii*

1. Rebutia minuscula Schumann, Monatsschr. Kakteenk. **5**: 102. 1895.

    Echinopsis minuscula Weber, Dict. Hort. Bois 471. 1896.
    Echinocactus minusculus* Weber in Schumann, Gesamtb. Kakteen 395. 1898.

Plants simple or tufted, globular, 2 to 5 cm. in diameter, covered with low tubercles arranged in 16 to 20 spirals, bright green; spines in clusters of 25 to 30, 2 to 3 mm. long, whitish; flowers often numerous, arising from the spine-areoles near the base of the plant, slightly bent just above the ovary, funnelform, 2.5 to 3 cm. long, bright crimson; scales on the small ovary ovate, acuminate, with naked axils; perianth-segments about 12, about 1 cm. long, linear-oblong, acute; stamens 15 to 30, whitish; stigma-lobes 4 or 5, whitish; fruit 3 mm. in diameter, scarlet.

---

*Weber (Dict. Hort. Bois 471. 1896) mentions it as a synonym, but does not describe it. It was introduced into cultivation in 1887 and has since been a favorite.

Type locality: Tucuman, Argentina.

Distribution: Northwestern Argentina.

This little plant has flowered frequently both in Washington and New York, often as early as March; the flowers open in the morning and close at night, opening for four days consecutively and then followed in a few days by the small scarlet fruits.

FIG. 57.—Rebutia minuscula.   FIG. 58.—Rebutia pseudominuscula.

This plant is so small that when grown alone it is quite inconspicuous, but de Laet has grown it very successfully as a graft on one of the cylindric cacti. When grown this way it gives off many new plants, forming a cespitose mass and flowering freely. De Laet also lists in his Catalogue the variety *cristatus* under *Echinocactus minusculus.*

Illustrations: Blühende Kakteen 1: pl. 31; Schumann, Gesamtb. Kakteen f. 67; Curtis's Bot. Mag. 140: pl. 8583; Monatsschr. Kakteenk. 26: 152, 153; 29: 141; Tribune Hort. 4: pl. 140; Kirtcht, Kakteen Zimmergarten 9; De Laet, Cat. Gen. 3. f. 2, as *Echinocactus minusculus;* Monatsschr. Kakteenk. 5: 103.

Plate IV, figure 5, shows a flowering plant sent to the New York Botanical Garden from the Missouri Botanical Garden in 1912. Figure 57 is from a photograph contributed by Dr. Spegazzini.

2. **Rebutia fiebrigii** (Gürke) Britton and Rose, Stand. Cycl. Hort. Bailey 5: 2915. 1916.

> *Echinocactus fiebrigii* Gürke, Notizbl. Bot. Gart. Berlin 4: 183. 1905.

Globose, depressed at apex, 5 cm. high, tuberculate; areoles elliptic; spines 30 to 40, setaceous, 1 cm. long, white, or the longest ones brownish at apex and 2 cm. long or more, porrect, acicular; flowers from the side of the plant, 2 cm. long, slender, funnelform, red, bent upwards; scales on the ovary small, woolly, and bristly; fruit small, purple; inner perianth-segments oblong, acute.

Type locality: Bolivia, at Escayacje, altitude 3,600 meters.

Distribution: Known only from the type locality.

This is a very attractive little plant which Dr. Rose saw in the Berlin Botanical Garden in 1912. A specimen was sent from the Berlin Botanical Garden to the New York Botanical Garden which we have also studied. The plant is named for Dr. C. Fiebrig, director of the Museum and Garden at Asuncion, Paraguay.

FIG. 59.—Rebutia fiebrigii.

Illustrations: Blühende Kakteen 2: pl. 109; Monatsschr. Kakteenk. 28: 139, as *Echinocactus fiebrigii*.

Figure 59 is copied from the first illustration above cited.

3. Rebutia pseudominuscula (Spegazzini).

Echinopsis pseudominuscula Spegazzini, Anal. Mus. Nac. Buenos Aires III. 4: 488. 1905.

Simple or somewhat cespitose, globular to short-cylindric, 5 cm. high, 3.5 cm. in diameter, strongly tubercled; the tubercles about 5 mm. high; areoles elliptic, gray-felted; radial spines 7 to 14, setaceous, 3 to 5 mm. long, at first yellowish or rose-colored; central spines 1 to 4; flowers from the lower half of the plant, red, 2.5 cm. long, with a slender funnelform tube; scales on ovary and flower-tube few, the axils hairy; perianth-segments oblanceolate to spatulate, 15 to 18 mm. long, obtuse, sometimes mucronate; stigma-lobes 6, white.

Type locality: Mountains, 3,500 meters altitude, province of Salta, Argentina.

Distribution: Known only from the type locality.

Echinocactus pseudominusculus Spegazzini (Anal. Mus. Nac. Buenos Aires III. 4: 488. 1905) was given as a synonym of this species when first described.

Figure 58 is from a photograph contributed by Dr. Spegazzini.

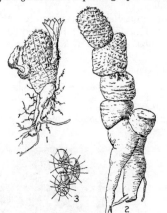
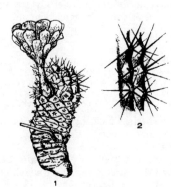

FIG. 60.—Rebutia pygmaea.

FIG. 60 a.—Plant and section of stem of Rebutia steinmannii.

4. Rebutia pygmaea (R. E. Fries).

Echinopsis pygmaea R. E. Fries, Nov. Act. Soc. Sci. Upsal. IV. 1¹: 120. 1905.

Simple, ovoid to short-cylindric, 1 to 3 cm. long, 1.2 to 2 cm. in diameter, or sometimes branched with 2 to many short joints from a much thickened root; tubercles small, more or less arranged into 8 to 12 spiraled rows; areoles narrow, somewhat lanate; spines all radial, 9 to 11, short, appressed, 2 to 3 mm. long, acicular, somewhat swollen at base; flowers from the lower part of the plant, somewhat curved at base, becoming nearly erect, 18 to 25 mm. long, rose-purple; scales on ovary and flower-tube small, hairy in their axils; fruit globular, 6 mm. in diameter.

Type locality: Yavi, province of Jujuy, Argentina.

Distribution: Bolivia and northwestern Argentina.

We have studied living plants collected by Dr. Rose at Oruro, Bolivia.

Illustrations: Nov. Act. Soc. Sci. Upsal. IV. 1¹: pl. 8, f. 1 to 3, as *Echinopsis pygmaea*.

Figure 60 is copied from the illustration above cited.

5. Rebutia steinmannii (Solms-Laubach).

Echinocactus steinmannii Solms-Laubach, Bot. Zeit. 65¹:133. 1907.

Small oblong plants (about 2 cm. high) 1 to 1.5 cm. in diameter; ribs low, often spiraled, tubercled; spines acicular, about 8; flowers from the side of the plant, erect, campanulate; outer perianth-segments oblong, apiculate; inner perianth-segments rounded.

Type locality: High mountains of Bolivia.

Distribution: Known only from the type locality.

This very strange plant is unknown to us except from the description and illustrations, but it seems to be of this relationship.

Illustrations: Bot. Zeit. 65[1]: pl. 2, f. 4, 10, as *Echinocactus steinmannii.*

Figure 60*a* is from a photograph of the illustrations cited above.

DESCRIBED SPECIES, PERHAPS OF THIS GENUS.

ECHINOPSIS DEMINUTA Weber, Bull. Mus. Hist. Nat. Paris 10: 386. 1904.

> *Echinocactus deminutus* Gürke, Monatsschr. Kakteenk. **16:** 103. 1906.

Plants globular to short-cylindric, 5 to 6 cm. high; ribs 11 to 13, somewhat spiraled, more or less tuberculate; spines numerous, somewhat rigid; flowers 3 cm. long, with a limb 3 cm. broad; outer perianth-segments lanceolate, purple, 4 to 5 mm. long; inner perianth-segments 15, red to orange, 5 to 6 mm. long; ovary bristly, 6 mm. in diameter; stigma-lobes white.

Type locality: Trancas, Argentina.

Distribution: Known only from the type locality.

We append the description of this plant to our genus *Rebutia*, to which it may belong, but we have not been able to study any specimens of it. Weber's reference of the species to *Echinopsis* certainly is erroneous, nor is it an *Echinocactus* in our understanding of that genus.

4. CHAMAECEREUS gen. nov.

Plants small, usually creeping and forming little clumps, sometimes some of the joints pendent, usually arising from the base, cylindric, with a few low ribs; spines acicular; flowers diurnal, solitary at the areoles, comparatively small, erect; tube cylindric, bearing acute scales with hairy axils; inner perianth-segments spreading, scarlet; stamens included; fruit small, globular, dry or nearly so, bearing long woolly hairs; seeds black, opaque, punctate.

Type species: *Cereus silvestrii* Spegazzini.

Only one species is known, inhabiting Argentina.

We are indebted to Mr. Alwin Berger for notes upon this interesting plant which have been freely used in our description.

The first part of the generic name is from χαμαί on the ground, referring to the creeping or depressed habit of the plant.

FIG. 61.—Chamaecereus silvestrii.

1. Chamaecereus silvestrii (Spegazzini).

> *Cereus silvestrii* Spegazzini, Anal. Mus. Nac. Buenos Aires III. **4:** 483. 1905.

Joints usually prostrate, sometimes 4 to 6 cm. high, pale green; ribs about 6 to 9, but usually 8, low; spines soft, white; flowers orange-scarlet, about 7 cm. long, the axils of the scales bearing long black and white hairs and a few bristles; flower-tube narrow, 2.5 to 3 cm. long; perianth-segments in 3 or 4 series, spreading, 1 to 2 cm. long, 4 mm. broad, lanceolate, the outer ones acute, the inner shorter and obtuse; filaments red to purple, short; style pale yellow to greenish white, longer than the stamens; stigma-lobes 8 or 9, connivent.

Type locality: Mountains between Tucuman and Salta, Argentina.

Distribution: Tucuman, Argentina.

This plant resembles in habit some of the creeping species of *Echinocereus*, but has very different flowers; it has no close relatives in Argentina; it is largely grown in European collections, where it is highly prized. Dr. Rose obtained specimens in Argentina in 1915, but so far none has flowered.

This species was named for Dr. Philip Silvestri, a friend of Dr. Spegazzini.

Illustrations: Curtis's Bot. Mag. **138**: pl. 8426; Gartenwelt **15**: 484; Haage and Schmidt, Haupt-Verz. 1919: 167, as *Cereus silvestrii.*

Figure 61 is from a photograph contributed by Dr. Spegazzini in 1915.

5. LOBIVIA gen. nov.

Plant globular to short-cylindric, either simple or in clusters, always ribbed, usually very spiny; flowers so far as known diurnal, short-funnelform to campanulate, lateral from old areoles, in some from near the top, in others well down on the side of the plant, with a short, broad tube, red in typical species, but in others yellow or white; scales on the ovary mostly bearing long hairs in their axils; fruit small, globular.

Type species: *Echinocactus pentlandii* Hooker.

The genus as here treated is composed of 20 species, mostly hitherto referred to *Echinopsis* and *Echinocactus.* It is made to include various anomalous species which can not properly be referred to any described genus, and it is questionable whether they are all congeneric. Some, however, we know only from descriptions or photographs and further knowledge of them may lead to a different arrangement. In form their flowers are much alike. The two species transferred from *Echinocactus* (*E. thionanthus* and *E. chionanthus*) are described as having a dense ring of hairs on the inside of the flower-tube below the stamens; this with other differences in the shape of the flower may be of generic value. The species all inhabit the highlands of Peru, Bolivia, and Argentina.

The generic name is an anagram of Bolivia.

KEY TO SPECIES.

```
A. Base of flower-tube naked within.
  B. Ribs 20 to 50 or more.
      Plant somewhat depressed; spines not long................................  1. L. bruchii
      Plant globose to cylindric; spines very long.
        Tubercles very large, 2 to 3 cm. long, 1 to 1.5 cm. high; central spines upwardly
            curved..........................................................  2. L. ferox
        Tubercles 1 to 2 cm. long or less, not over 1 cm. high; spines slender, nearly straight.
        Tubercles narrow, acute; spines subulate............................  3. L. longispina
        Tubercles broad, blunt; spines acicular.............................  4. L. boliviensis
  BB. Ribs up to 19, but usually fewer.
      Flowers yellow....................................................................  5. L. shaferi
      Flowers reddish.
        Scales on flower-tube few.
          Spines curved or somewhat hooked.
            Central spines 4, some hooked; radial spines weak, 4 to 5 mm. long, yellowish.  6. L. cachensis
            Central spine solitary, never hooked; radial spines 1 to 2 cm. long, brownish.  7. L. caespitosa
          Spines straight...................................................  8. L. saltensis
        Scales on flower-tube numerous.
          Ribs strongly undulate or broken into narrow tubercles.
            Inner perianth-segments broad.
              Flower-areoles short-hairy; tubercles long, acute.
                Spines grayish, short......................................  9. L. cinnabarina
                Spines yellowish brown, elongated..........................  10. L. pentlandii
              Flower-areoles long-hairy; tubercles short, blunt...........  11. L. lateritia
            Inner perianth-segments narrow.
              Flowers 5.5 to 6 cm. long, white-hairy.......................  12. L. pampana
              Flowers small, 3 cm. long, black-hairy.......................  13. L. corbula
          Ribs not strongly undulate, at least never tubercled.
            Flowers 6 cm. long or less.
              Central spine 1 to 2.5 cm. long..............................  14. L. andalgalensis
              Central spines up to 5 cm. long..............................  15. L. haematantha
            Flowers 10 cm. long...........................................  16. L. grandiflora
AA. Base of flower-tube with ring of hairs on inside; scales on outside of flower reflexed at
        apex.
      Flowers yellow....................................................  17. L. thionanthus
      Flowers white.....................................................  18. L. chionanthus
AAA. Species not grouped................................................ {19. L. grandis
                                                                         {20. L. cumingii
```

1. Lobivia bruchii sp. nov.

Plants simple, globular, more or less depressed in the center; ribs 50 or more, distinct but low, more or less tuberculate; areoles filled with short white wool; spines several, spreading, usually dark colored, those of the upper areoles connivent; flowers small, at areoles below the apex of plant, deep red; tube of flower short, its axils filled with wool; inner perianth-segments lanceolate, slightly spreading; filaments exserted beyond the throat, but shorter than the perianth-segments; fruit and seeds unknown.

Described from a photograph taken in March 1907, sent by Dr. C. Bruch, of a plant in Tafi del Valle, province of Tucuman, Argentina.

Figure 62 is from the photograph mentioned above.

Fig. 62.—Lobivia bruchii.

2. Lobivia ferox sp. nov.

Roots fibrous; plants globular, 3 dm. in diameter or more, almost hidden by the long upwardly curved spines; ribs numerous, often as many as 29, deeply undulate and broken into thin, acute tubercles 2 to 3 cm. long; spines light brown, sometimes mottled; radial spines 10 to 12, slender, 4 to 6 cm. long, somewhat curved; central spines 3 or 4, somewhat flattened in one vertical row, rather weak, curved upward, 10 to 15 cm. long; flower-buds woolly; flowers and fruit not seen.

Collected on dry hills east of Oruro, Bolivia, August 18, 1914, by J. N. Rose (No. 18918).

In cultivation, as is shown by our illustration, the long upturned spines are very poorly developed at the top of the plant which is nearly naked. This plant was observed in only one locality in Bolivia, although it is doubtless to be found elsewhere; it grows on very dry gravelly hills among low thorn bushes. It is easily detached from the soil having only fibrous roots and in this respect is very unlike another species of this genus (see No. 18919) which has fleshy deep-seated roots. In its long stout upturned spines it is unlike any other plant we know and has a very striking habit. Several living plants were sent to the New York Botanical Garden in 1914 by Dr. Rose, one of which still persists, but it has never flowered. We are disposed to refer here an illustration (Illustr. Hort. **6:** pl. 214) which was called *Echinopsis pentlandii.*

Figure 63 is from a photograph of a plant brought by Dr. Rose from the type locality in 1914.

FIG. 63.—Lobivia ferox.

3. Lobivia longispina sp. nov.

Globose to cylindric, 10 cm. in diameter, up to 25 cm. high, bluish green; ribs 25 to 50, rather low, deeply undulate, broken into acute tubercles 1 to 2 cm. long; spines 10 to 15, slender, elongated, nearly straight, the longest 7 to 8 cm. long, yellowish to brown; flowers funnelform, about 4 cm. long, slender, very hairy, the hairs long and white; limb short; fruit broadly obovoid, about 2 cm. thick, its scales distant, ovate, acuminate, 3 to 4 mm. long.

FIG. 64.—Lobivia longispina.

Collected by J. A. Shafer in crevices of rocks at La Quiaca, Jujuy, Argentina, altitude 3,450 meters, February 3, 1917 (No. 83).

We have studied living plants brought by Dr. Shafer to the New York Botanical Garden. Figure 64 is from a photograph of the type specimen; figure 65 shows the fruit.

4. Lobivia boliviensis sp. nov.

Cespitose, in clusters of about 6; plants globular, 8 to 10 cm. in diameter, almost hidden by the long, nearly straight spines; ribs about 20, undulate, broken into short, blunt tubercles; areoles 1 cm. apart; spines 6 to 8, brown, acicular, flexible, often 9 cm. long.

Collected by Dr. Rose at Oruro, Bolivia, in 1914 (No. 18919).

This plant was quite common on the low dry hills east of Oruro, associated with *Lobivia ferox,* but readily distingushed from it in its very slender spines and in its root system. This species has thick fleshy roots while the other has fibrous roots. It does not do well in cultivation.

Figure 67 is from a photograph of a plant brought by Dr. Rose from the type locality in 1914.

Fig. 65.—Fruit of   Fig. 66.—Flower of Lobivia
Lobivia longispina.   shaferi.   ×0.6.
  ×0.6.

5. Lobivia shaferi sp. nov.

Cespitose, at first globose, becoming cylindric, 7 to 15 cm. high, 2.5 to 4 cm. in diameter, densely covered with spines; ribs about 10, very low; areoles approximate; radial spines 10 to 15, acicular, white or brown, 1 cm. long or less; central spines several, one often much stouter than the others, 3 cm. long; buds very hairy; flowers 4 to 6 cm. long, funnelform, bright yellow, the tube stout, the limb 3 to 4 cm. broad; scales on ovary and flower-tube linear to ovate-linear, acute, their axils bearing long white hairs; style greenish white; stigma-lobes cream-colored.

Collected by J. A. Shafer in hillside thickets, Andalgala, province of Catamarca, Argentina, December 19, 1916 (No. 16).

Dr. Shafer says that this plant grows in firm leaf-mold underneath and entangled in shrubbery.

Figure 69 is from a photograph taken by Dr. Shafer at Andalgala; figure 66 shows a flower.

FIG. 67.—Lobivia boliviensis.   FIG. 68.—Lobivia cachensis.

6. Lobivia cachensis (Spegazzini).

Echinopsis cachensis Spegazzini, Anal. Mus. Nac. Buenos Aires III. **4**: 493. 1905.

Stems simple or tufted, 9 cm. high, 6.5 cm. in diameter; ribs about 19, about 5 mm. high; spines soft, hardly pungent, grayish, with yellowish tips; radial spines 7 to 20, straight, 4 to 5 mm. long;

central spines 4, 1 or 2 of them longer and hooked; flowers inodorous, 6 to 7 cm. long; inner perianth-segments linear-lanceolate, red; filaments dark purple; style yellowish; stigma-lobes 10, yellowish white, linear.

Type locality: Near Cachi, Argentina.
Distribution: In the high mountains of Salta, Argentina, altitude 2,500 meters.
We know the plant only from a photograph and the original description.
Spegazzini also gives the name *Echinocactus cachensis* (Anal. Mus. Nac. Buenos Aires III. 4: 493. 1905), but does not formally publish it.
Figure 68 is from a photograph contributed by Dr. Spegazzini.

7. Lobivia caespitosa (J. A. Purpus).

Echinopsis caespitosa J. A. Purpus, Monatsschr. Kakteenk. **27:** 120. 1917.

Cespitose, the joints erect or spreading, short, cylindric; ribs 10 to 12, somewhat undulate, acutish; areoles 1 to 1.5 cm. apart, white-woolly; radial spines acicular, 12, brownish, 1 to 2 cm. long; central spine solitary, brown, somewhat curved, 5 cm. long or less; flowers from the side of the plant near the middle, short-funnelform, 6.5 to 8 cm. long, reddish within, yellowish red without; perianth-segments oblong, obtuse.

Type locality: Bolivia.
Distribution: Bolivia.
We know this plant only from description and illustration. It is clearly not an *Echinopsis.*
Illustration: Monatsschr. Kakteenk. **27:** 121, as *Echinopsis caespitosa.*

FIG. 69.—Lobivia shaferi. FIG. 70.—Lobivia saltensis.

8. Lobivia saltensis (Spegazzini).

Echinopsis saltensis Spegazzini, Anal. Mus. Nac. Buenos Aires III. 4: 487. 1905.

Plants at first simple but becoming densely cespitose, light green, shining; ribs 17 or 18, low, obtuse, crenate; spines all short and straight; radial spines 12 to 14, 4 to 6 mm. long; central spines 1 to 4, stouter than the radials, 10 to 12 mm. long; flowers on the side of the plant near the middle, inodorous, 4 cm. long; perianth-segments red, short, obovate, 10 to 12 mm. long, obtuse; scales on ovary naked in their axils (according to Spegazzini).

Type locality: Near Amblaio, Argentina.
Distribution: Between Tucuman and Salta, Argentina.
Spegazzini also gives the name *Echinocactus saltensis* (Anal. Mus. Nac. Buenos Aires III. 4: 487. 1905), but does not formally publish it.
We know the plant only from a photograph and from the description; the character of the ovary-scales being without hairs in their axils is unusual in the genus. It is described as being only 6.5 cm. high.
Figure 70 is from a photograph contributed by Dr. Spegazzini.

9. Lobivia cinnabarina (Hooker).

Echinocactus cinnabarinus Hooker in Curtis's Bot. Mag. **73**: pl. 4326. 1847.
Echinocactus cinnabarinus spinosior Salm-Dyck, Cact. Hort. Dyck. 1849. 35, 176. 1850.
Echinopsis cinnabarina Labouret, Monogr. Cact. 288. 1853.
Echinopsis chereauniana Schlumberger, Rev. Hort. IV. **5**: 402. 1856.
Echinopsis cinnabarina spinosior Rümpler in Förster, Handb. Cact. ed. 2. 618. 1885.
Echinocereus cinnabarinus Schumann in Engler and Prantl, Pflanzenfam. 3⁶ᵃ: 185. 1894.

Stems simple, broader than high, usually depressed and unarmed at apex, bright green; ribs about 20, irregular and oblique, divided into acute tubercles; radial spines 8 to 10, all more or less curved backward, slender, grayish; central spines 2 or 3, somewhat curved; flowers from near top of the plant, rotate-campanulate, scarlet, about 4 cm. broad; inner perianth-segments broad, obtuse, spreading, the outer ones greenish; stamens and style much shorter than the perianth-segments; scales on ovary and flower-tube lanceolate, acute, hairy in their axils.

Type locality: Bolivia.
Distribution: In the higher Andes of Bolivia.

This species was collected first by Bridges in 1846 and sent to the Royal Gardens, Kew, where it flowered and was illustrated in Curtis's Botanical Magazine; the flowers are described as opening in the morning and closing the second day after.

Echinopsis cinnabarina cheroniana (Monatsschr. Kakteenk. **14**: 168. 1904) and var. *cristata* (Monatsschr. Kakteenk. **17**: 74. 1907) are mere garden names. Here Weber refers *Echinocactus chereaunianus* Cels (Dict. Hort. Bois 471. 1896) as does also Schumann.

Illustrations: Blühende Kakteen **1**: pl. 2; Schumann, Gesamtb. Kakteen f. 44; Möllers Deutsche Gärt. Zeit. 25: 475. f. 7, No. 13, as *Echinopsis cinnabarina;* Curtis's Bot. Mag. 73: pl. 4326; Loudon, Encycl. Pl. ed. 3. 1377. f. 19373, as *Echinocactus cinnabarinus.*

Plate IV, figure 2, shows a plant obtained in 1901 for the New York Botanical Garden from M. Simon, St. Ouen, Paris, which flowered in June 1912.

10. Lobivia pentlandii (Hooker).

Echinocactus pentlandii Hooker in Curtis's Bot. Mag. **70**: pl. 4124. 1844.
Echinopsis pentlandii Salm-Dyck in Dietrich, Allg. Gartenz. **14**: 250. 1846.
Echinopsis maximiliana Heyder in Dietrich, Allg. Gartenz. **14**: 250. 1846.
Echinopsis tricolor Dietrich, Allg. Gartenz. **16**: 210. 1848.
Echinopsis pentlandii coccinea Salm-Dyck, Cact. Hort. Dyck. 1849. 38. 1850.
Echinopsis scheeri Salm-Dyck, Cact. Hort. Dyck. 1849. 179. 1850.
Echinopsis pentlandii laevior Monville in Labouret, Monogr. Cact. 290. 1853.
Echinopsis pentlandii scheeri Lemaire, Illustr. Hort. **6**: with pl. 214. 1859.
Echinopsis pentlandii gracilispina Lemaire, Illustr. Hort. **6**: with pl. 214. 1859.
Echinopsis pentlandii pyracantha Lemaire, Illustr. Hort. **6**: with pl. 214. 1859.
Echinopsis pentlandii radians Lemaire, Illustr. Hort. **6**: with pl. 214. 1859.
Echinopsis colmarii Neubert, Gartenmag. 1878.*
Echinopsis pentlandii tricolor Rümpler in Förster, Handb. Cact. ed. 2. 612. 1885.
Echinopsis pentlandii longispina Rümpler in Förster, Handb. Cact. ed. 2. 612. 1885.
Echinopsis pentlandii neuberti Rümpler in Förster, Handb. Cact. ed. 2. 613. 1885.
Echinopsis pentlandii pfersdorffii Rümpler in Förster, Handb. Cact. ed. 2. 613. 1885.
Echinopsis pentlandii cavendishii Rümpler in Förster, Handb. Cact. ed. 2. 614. 1885.
Echinocereus pentlandii Schumann in Engler and Prantl, Pflanzenfam. 3⁶ᵃ: 185. 1894.
Echinopsis pentlandii maximiliana Schumann, Gesamtb. Kakteen 229. 1897.
Echinopsis pentlandii elegans Hildmann in Schumann, Gesamtb. Kakteen 230. 1897.
Echinopsis pentlandii ochroleuca R. Meyer in Schumann, Gesamtb. Kakteen 230. 1897.
Echinopsis pentlandii vitellina Hildmann in Schumann, Gesamtb. Kakteen 230. 1897.
Echinopsis pentlandii forbesii R. Meyer, Monatsschr. Kakteenk. **7**: 155. 1897.
Echinopsis cinnabarina scheeriana R. Meyer, Monatsschr. Kakteenk. **7**: 164. 1897.
Echinopsis cavendishii Hortus, Monatsschr. Kakteenk. **20**: 143. 1910.

Stems simple, higher than broad, somewhat umbilicate at apex, bright green or somewhat glaucous; ribs about 12, deeply crenate, rather high, broken into long acute tubercles, separated by acute intervals; spines 5 to 8, all radial, acicular, somewhat curved backward, yellowish brown, 3 cm. long or less; flowers short-funnelform, about 4 cm. long, the tube greenish; inner perianth-segments rose-

---

*This name and reference is taken from Schumann's monograph (Gesamtb. Kakteen 229). Schumann doubtless had in mind Deutsches Magazin für Garten und Blumenkunde 1878, but he gives no page nor do we find the name in this book. On pages 114 to 118 there is a short article on "Ein unbekannter (?) Cactus," with a colored illustration, but without legend. This illustration is reproduced by Rümpler (Förster, Hamb. Gartenz. 879. 1861) under the name of *Echinopsis colmarii.*

colored, narrowly obovate, abruptly acute, spreading; stamens and style much shorter than the inner perianth-segments; scales on ovary and flower-tube lanceolate, short-hairy in their axils; fruit subglobose, 10 to 12 mm. in diameter.

Type locality: Not cited.

Distribution: Bolivia.

We have followed Schumann who refers *Echinopsis colmarii* here. Rümpler (Förster, Handb. Cact. ed. 2. 615. 1885), however, says it comes from Mexico, which would exclude it from this alliance. Rümpler's illustrations (f. 79, 80) do not suggest this relationship, the latter resembling very much the flower of an *Echinocactus*. *Echinopsis colmariensis* (Schumann, Gesamtb. Kakteen 230. 1897) is a catalogue name for *E. colmarii*.

We have also followed Schumann in referring here *E. scheeri*, although it was originally described as having 13 to 19 ribs. Rümpler's illustration (f. 78) has long, linear perianth-segments and looks very unlike *Lobivia pentlandii*, though the plant body is very similar. *E. scheeriana* (Monatsschr. Kakteenk. 3: 127. 1893) is only a name and belongs here.

FIG. 71.—Lobivia pentlandii.   FIG. 72.—Lobivia andalgalensis.

E. pentlandii integra (Monatsschr. Kakteenk. 7: 139. 1897), *E. pentlandii achatina* (Monatsschr. Kakteenk. 14: 168. 1904), *E. columnaris, E. elegans vittata, E. ochroleuca, E. pfersdorffii,* and *E. achatina* (all in Monatsschr. Kakteenk. 20: 143. 1910) are unpublished names which probably belong here. *E. maximiliana longispina* (Monatsschr. Kakteenk. 20: 143. 1910) is of this relationship. *E. pentlandii cristata* (Monatsschr. Kakteenk. Index 58) is doubtless a crested form. *E. pentlandii pyrantha* Monville (Labouret, Monogr. Cact. 289. 1853) doubtless should be referred here as we have done.

Echinopsis pentlandii albiflora Weidlich (Gartenflora 69: 143. f. 17. 1920) has recently been briefly described as having white flowers, 5 cm. long and 4 cm. broad. We do not know it nor do we know its origin.

Illustrations: Curtis's Bot. Mag. 70: pl. 4124; Loudon, Encycl. Pl. ed. 3. 1376. f. 19367, as *Echinocactus pentlandii;* Blühende Kakteen 1: pl. 26; Dict. Hort. Bois f. 324; Monatsschr. Kakteenk. 13: 92. f. B; Palmer, Cult. Cact. 3; Rümpler, Sukkulenten 167. f. 91; Deutsche Gärt. Zeit. 5: 369; Möllers Deutsche Gärt. Zeit. 25: 475. f. 7, No. 1; Watson, Cact. Cult. 134. f. 52, as *Echinopsis pentlandii;* Förster, Handb. Cact. ed. 2. f. 78; Rümpler, Sukkulenten 166. f. 90, as *Echinopsis scheeri;* Förster, Handb. Cact. ed. 2. f. 79, 80, as *Echinopsis colmarii;* Lemaire, Cactées 70. f. 7; Rev. Hort. 1860: f. 109; Dict. Gard. Nicholson 1: 503. f. 698; Watson, Cact. Cult. 135. f. 53, as *Echinopsis pentlandii longispina;* (?) Lemaire, Cactées 70. f. 8; (?) Dict. Gard. Nicholson 1: 503. f. 699; Rev. Hort. 1860: f. 111, as *Echinopsis pentlandii scheeri;* Möllers Deutsche Gärt. Zeit. 25: 475. f. 7, No. 13, as *Echinopsis cinnabarina;* Rev. Hort. 1860: f. 110, as *Echinopsis pentlandii maximiliana;* Rev. Hort. 1860: f. 108, as *Echinopsis pentlandii levior scheeri;* Möllers Deutsche Gärt. Zeit. 25: 475. f. 7, No. 24, as *Echinopsis pentlandi colmari.*

Plate v, figure 3, shows a plant sent to the New York Botanical Garden from the Berlin Botanical Garden in 1902, which flowered in May 1913. Figure 71 is copied from the Botanical Magazine illustration above cited.

11. Lobivia lateritia (Gürke).

Echinopsis lateritia Gürke, Monatsschr. Kakteenk. 17: 151. 1907.

Simple, nearly globular or a little longer than broad, glaucous-green, 7.5 cm. high; ribs 18, broad at base, acute, more or less undulate, about 1 cm. high; areoles 1 to 2 cm. apart; radial spines 9 or 10, more or less curved, 10 mm. long, brownish; central spines 1 or 2, more or less curved upward, much longer than the radials, somewhat thickened at base; flowers 3 to 5 cm. long, short-funnelform, scarlet to brick-red; inner perianth-segments oblong, acute; scales on the ovary and flower-tube lanceolate, acute, bearing blackish hairs in their axils; filaments red; stigma-lobes 7 or 8.

Type locality: Bolivia.
Distribution: Bolivia.
Illustration: Blühende Kakteen 2: pl. 120, as *Echinopsis lateritia.*

Plate v, figure 4, shows a plant collected by Dr. Rose at La Paz, Bolivia, in 1914, which flowered the next year at the New York Botanical Garden; this plant differs a little from the one shown in the illustration above cited.

12. Lobivia pampana sp. nov.

More or less cespitose; plants globular, 5 to 7 cm. in diameter; ribs 17 to 21, more or less undulate; areoles distant, white-felted when young, very spiny except in cultivated plants and then often spineless; spines 5 to 20, often more or less curved, acicular, often 5 cm. long, puberulent; flowers 5.5 to 6 cm. long, red; outer perianth-segments linear-oblong, acuminate; inner perianth-segments oblong, acute to acuminate; scales on the ovary and fruit ovate, 4 to 6 mm. long, acuminate, with long white hairs in their axils.

Collected by J. N. Rose on the Pampa de Arrieros, southern Peru, August 23, 1914 (No. 18966).

13. Lobivia corbula (Herrera).

Mammillaria corbula Herrera, Rev. Univ. Cuzco 8: 61. 1919.

Nearly globular, growing in clumps of 5 to 8 plants; ribs 12 or more, strongly crenate; areoles filled with white wool; in cultivated plants few or no spines developing, but in wild plants appearing in clusters of 6 to 9, these yellowish, 3 to 5 cm. long; flowers opening in the evening, about 3 cm. long; tube short, bearing small scales, these hairy in their axils; outer perianth-segments lanceolate, acute, purplish; inner ones lanceolate, somewhat shorter and broader, acute, salmon-red; stamens and style greenish yellow, short, included; style 2.5 cm. long.

Type locality: Near Cuzco, Peru.
Distribution: On hills in the high Andean Valley of Peru.

Collected by J. N. Rose on hills at Juliaca, Peru, September 4, 1914 (No. 19090); also near Cuzco, September 2, 1914 (No. 19080), by O. F. Cook on the highlands of Peru in 1915 and by Fortunato L. Herrera near Cuzco in 1922.

Plate v, figure 2, shows a plant collected in 1914 by Dr. Rose at Juliaca, Peru, which flowered at the New York Botanical Garden in May 1916.

14. Lobivia andalgalensis (Weber).

Cereus huascha rubriflorus Weber in Schumann, Monatsschr. Kakteenk. 3: 151. 1893.
Cereus andalgalensis Weber in Schumann, Gesamtb. Kakteen 168. 1897.

Plants single or clustered, globular or a little flattened, 3 to 10 cm. in diameter, green; ribs about 13, stout, hardly crenate; areoles 5 to 10 mm. apart, circular; spines white, subulate, straight; radial spines 8 to 10, about equal, 5 to 7 mm. long; central spine solitary, stouter than the radials, 10 to 25 mm. long, porrect; flowers fugacious, short-funnelform, about 6 cm. long, green without; flower-tube and ovary bearing scales, with long, gray appressed hairs in their axils; inner perianth-segments red, oblanceolate or subspatulate, 1.8 to 2.5 cm. long, obtuse or retuse; filaments reddish purple; style pale red or yellow; stigma-lobes about 9.

M. E. Eaton del.

A. Hoen & Co.

1. Top of flowering plant of *Echinocereus engelmannii*.
2. Flowering plant of *Lobivia corbula*.
3. Flowering plant of *Lobivia pentlandii*.
4. Flowering plant of *Lobivia lateritia*.
(All natural size.)

Type locality: Andalgala, province of Catamarca, Argentina.

Distribution: Western Argentina.

This plant has passed as a variety of *Cereus huascha,* now *Trichocereus huascha,* but the two are so unlike that we have referred them to different genera. Both species are still little known and further study of them is much desired.

Figure 72 is from a photograph contributed by Dr. Spegazzini; figure 75 shows a flower collected by Dr. Shafer between Andalgala and Concepción, Argentina, in 1917 (No. 27).

15. Lobivia haematantha (Spegazzini).

Echinocactus haematanthus Spegazzini, Anal. Mus. Nac. Buenos Aires III. 4: 498. 1905.

Somewhat depressed, globose, 5 cm. high, 6 cm. in diameter, greenish, somewhat shining; ribs 11, somewhat tuberculate, obtuse; areoles nearly circular, 5 to 6 mm. in diameter, white-felted; radial spines 6 to 8, slender, 5 to 10 mm. long, more or less appressed; central spines 3, stouter than the radials, 5 cm. long, pale gray with yellowish tips; flowers 3 to 4 cm. broad; inner perianth-segments obovate to spatulate, obtuse, purplish; stigma-lobes 9 to 12, white; scales on ovary and flower-tube long-woolly.

FIG. 73.—Lobivia haematantha.   FIG. 74.—Lobivia thionanthus.

Type locality: Near Amblaio, province of Salta, Argentina.

Distribution: Known only from the province of Salta, Argentina.

We know the plant only from photographs and the description.

Figure 73 is from a photograph contributed by Dr. Spegazzini.

16. Lobivia grandiflora sp. nov.

Globose to short-cylindric, 7.5 to 10 cm. in diameter, 15 to 20 cm. long; ribs about 14; areoles about 1 cm. apart; spines about 15, slender, subulate, about 1 cm. long, yellowish; flowers funnel-form, 10 cm. long; perianth-segments narrow, acuminate, 4 to 5 cm. long, pink; scales on the ovary narrow, 10 to 12 mm. long, a little hairy in their axils.

Collected by J. A. Shafer between Andalgala and Concepción, Argentina, December 28, 1916, altitude 1,750 meters (No. 28).

The showy pink flowers of this plant are larger than those of any of the other species which we have included in this genus.

17. Lobivia thionanthus (Spegazzini).

Echinocactus thionanthus Spegazzini, Anal. Mus. Nac. Buenos Aires III. 4: 499. 1905.

Plants usually solitary, globular to short-cylindric, grayish green, 5 to 12 cm. high, 6 to 10 cm. in diameter; ribs usually 14, low, slightly undulate and divided into indistinct tubercles; areoles short-elliptic; spines subulate, grayish or with brownish tip, about equal, 10 to 15 mm. long; flowers obconic, 4.5 cm. long, not odorous; inner perianth-segments yellowish, 15 mm. long, elliptic, the apex subretuse or mucronate; filaments yellow; style greenish white; stigma-lobes 12, flesh-colored, a ring of dense brownish hairs within the flower-tube, below the stamens; ovary and flower-tube covered with scales, each with a reflexed cartilaginous somewhat pungent apex, very woolly in their axils.

Type locality: Near Cachi, province of Salta, Argentina.

Distribution: Known only from the type locality.

Figure 74 is from a photograph contributed by Dr. Spegazzini.

18. Lobivia chionanthus (Spegazzini).

> *Echinocactus chionanthus* Spegazzini, Anal. Mus. Nac.
> Buenos Aires III. 4: 499. 1905.

Plants elliptic to short-cylindric, grayish green, 6 to 7.5 cm. high, 5 to 6 cm. in diameter; ribs 13 to 15, low, rounded; areoles elliptic; spines 7 to 9, subulate, straight, grayish, all radial, somewhat appressed, 1.5 to 2 cm. long; flowers from the upper part of the plant, but not central, 4.5 cm. long; inner perianth-segments white; style green; stigma-lobes 13, white; scales on ovary and flower-tube cartilaginous, subspinescent, reflexed at apex; flower-tube with a ring of dense brown hairs near the base within.

Type locality: Near Cachi, province of Salta, Argentina.

Distribution: Known only from type locality.

FIG. 75.—Flower of Lobivia andalgalensis. ×0.6.　　FIG. 76.—Flower of Lobivia grandis. ×0.6.

Of this relationship are flowers collected by A. Dominguez from the Cerro de Macha, Córdoba, Argentina; in these specimens the scales are long, linear, chartaceous, and erect; the flowers are 5 to 6 cm. long; the outer perianth-segments and often the inner have acuminate or mucronate chartaceous tips. (For further discussion see *Echinocactus spiniflorus*, page 178.)

Figure 77 is from a photograph contributed by Dr. Spegazzini.

FIG. 77.—Lobivia chionanthus.　　　　　FIG. 78.—Lobivia grandis.

19. Lobivia grandis sp. nov.

Depressed-globose to short-cylindric, 2.5 dm. high, bright green; ribs 14 to 16, prominent, 2 cm. high, broad at base, rounded on margin; areoles somewhat depressed, circular, 2 to 3 cm. apart, white-felted when young; spines 10 to 15, yellow with brown tips, acicular to slender-subulate, elongated, 6 to 8 cm. long; flowers lateral, straight, 6 cm. long, with a stout funnelform tube 2 cm. thick and a rather small limb; inner perianth-segments ovate, acuminate, subulate-tipped, 1.5 cm.

long, probably white; scales on ovary and flower-tube ovate, 10 to 12 mm. long, narrowly ovate, acute, their axils filled with long black silky hairs.

Collected by J. A. Shafer on a cliff, at an altitude of 2,400 meters, between Andalgala and Concepción, Argentina, December 28, 1916 (No. 25, type). Dr. Shafer's No. 23 collected at the same locality is similar, but the flowers are much smaller, being only about 3 cm. long, and the plant is much larger, up to 12 dm. high. This plant is referred to this genus with hesitancy; it is much larger than any of the other species.

Figure 78 is from a photograph of the plant collected by Dr. Shafer; figure 76 shows its flower.

20. Lobivia cumingii (Hopffer).

Echinocactus cumingii Hopffer, Allg. Gartenz. **11**: 225. 1843.

Plants small, 5 to 6 cm. in diameter, simple, globular, bluish green, tubercled; tubercles arranged in about 18 spiraled rows; radial spines about 20, straight, 10 mm. long; central spines 2 to 8, 11 mm. long; flowers from the upper part of the plant but not central, orange-colored (sometimes shown as lemon-yellow), narrow, 2.5 cm. long; inner perianth-segments oblong, acute; scales on the ovary small, described as naked in their axils.

FIG. 79.—Lobivia cumingii.

Type locality: Mountains of Peru.
Distribution: Bolivia and Peru.

The first two illustrations cited below are so different in the shape of the tubercles and in the color and form of the flowers that we suspect that they may belong to different species. The one from the Botanical Magazine has lemon-yellow flowers, while the other has deep-orange or brick-red flowers.

We have not studied living plants of this species.

Schumann refers *Echinocactus rostratus* Jacobi (Allg. Gartenz. 24: 108. 1856) here; but it was based on specimens from Valparaiso, Chile, and is probably to be referred to *E. subgibbosus*, now taken up in another genus (see page 97).

Although this species was described by Hopffer in 1843, Salm-Dyck* much later (Cact. Hort. Dyck. 1849. 174. 1850) published it as a new species of his own. In his description he makes the significant remark that it is similar to *Echinocactus cinnabarinus* which confirms our conclusion that it is probably a *Lobivia*. In 1860 Regel and Klein (Ind. Sem. Hort. Petrop. 48. 1860) described also as a new species, *Echinocactus cumingii*. They say it was brought by Cuming from Chile and, if so, is doubtless different from our plant. They state that it was referred to *Echinocactus cinerascens*, a plant occurring in Chile, but it is certainly not that species. It may be a species of *Neoporteria*, but in any case the name is a homonym and its exact identification is not of much importance.

The type was collected by Thomas Bridges, but the plant was named for Hugh Cuming (1791-1865).

Echinocactus cumingii flavispinus (Monatsschr. Kakteenk. **14**: 77. 1904) is a form.

Illustrations: (?) Curtis's Bot. Mag. **100**: pl. 6097; (?) Blühende Kakteen **1**: pl. 19, as *Echinocactus cumingii*.

Figure 79 is a reproduction of the second illustration cited above.

*His original was *cummingii*.

LOBIVIA sp.

Dr. Shafer collected many specimens of a plant at Villazon, Bolivia, in February 1917 (No. 86), which may represent another species of this genus, but they were at that time without flowers or fruit, and none has flowered since brought by him to the New York Botanical Garden.

This cactus is tufted, forming clumps 1 to 2 dm. broad; its joints are short-cylindric to turbinate, 8 to 15 cm. high and 5 to 7.5 cm. thick, 14 to 18-ribbed; areoles few in each rib, white-felted when young, elliptic; spines 2 to 5, somewhat flattened and appressed, about 1 cm. long, white with black tips. Dr. Shafer was told that its flowers are white.

6. ECHINOPSIS Zuccarini, Abh. Bayer. Akad. Wiss. München **2**: 675. 1837.

Echinonyctanthus Lemaire, Cact. Gen. Nov. Sp. 10. 1839.

Stems usually low, rarely over 3 dm. high, usually much shorter, generally globular or short-cylindric, but some species large, columnar, either solitary or clustered, with ribs either continuous or more or less undulate; areoles usually circular, borne on the ribs, felted and spiny; flowers arising from old areoles just above the spine-clusters, with a very long narrowly funnelform tube; perianth-segments comparatively short and broad, more or less spreading, usually white, rarely yellow or rose-colored*; filaments and style projecting beyond the throat but not beyond the perianth-segments; stamens in 2 series, weak; stigma-lobes of various colors, narrow; fruit globose to ovoid or sometimes narrowly oblong, splitting open on one side; seeds minute, oblique, obovate, truncate at base.

Echinocactus eyriesii Turpin is the type of the genus.

The generic name is from ἐχῖνος hedgehog, and ὄψις appearance, referring to the armament of the plant.

Some of the species have been taken up in *Echinocactus* or *Echinonyctanthus*, many in *Cereus*, while one species, though excluded from *Echinopsis* in our treatment, has also been referred to *Cleistocactus* and *Pilocereus*.

In its flowers *Echinopsis* is like *Trichocereus* and somewhat like *Harrisia*, but in habit it is abundantly distinct from these genera. In habit, although not in flowers, it seems to be the South American counterpart of the North American genus *Echinocereus*. Gardeners and botanists generally have recognized it as a well-defined genus, but Bentham and Hooker in their Genera Plantarum reduced it to *Cereus*, and their course has been followed by some other English authors. While the genus as treated by Schumann contains mostly species of low stature there are some striking diversities in flowers and we have consequently segregated these under the generic names *Lobivia* and *Rebutia*. Schumann recognized 18 species of this genus. Von Rother states that he had 55 forms growing in his collection; some of these must have been hybrids of which there are many. We here recognize 28 species, but further field observations may prove that this number should be reduced. There are, however, more than 200 names published under *Echinopsis* to be accounted for. The known species inhabit southern South America, east of the Andes.

KEY TO SPECIES.

A. Tube of perianth distinctly longer than limb.
  B. Flowers white to red or pinkish.
    C. Spines all straight, subulate.
      Inner perianth-segments thread-like..................................... 1. *E. meyeri*
      Inner perianth-segments broad.
        Stems slender, cylindric, much longer than thick.
          Fruit very slender... 2. *E. mirabilis*
        Fruit (so far as known) globular.
          Flowers 10 cm. long.. 3. *E. forbesii*
        Flowers 17 to 20 cm. long.
          Central spines 1 to 4, 4 cm. long................................ 4. *E. huottii*
          Central spine solitary, 5 to 6 cm. long.......................... 5. *E. minuana*

*In *Echinopsis aurea* and *E. formosa* the flowers are yellow, in *E. multiplex* and *E. oxygona* red to rose.

KEY TO SPECIES—continued.

Stems globular or thicker than long or sometimes clavate, never slender.
  Flowers red.
      Flower-tube distinctly enlarged above, its scales distant, large......... 6. *E. multiplex*
      Flower-tube slender, nearly cylindric, its scales numerous, small......... 7. *E. oxygona*
  Flowers white.
      Inner perianth-segments acuminate.
        Spines very short or none.
            Areoles nearly spineless.. 8. *E. eyriesii*
            Areoles with several spines, 4 to 7 mm. long.................... 9. *E. turbinata*
        Spines subulate, 10 to 12 mm. long.............................10. *E. tubiflora*
      Inner perianth-segments not acuminate.
        Spines becoming white...11. *E. albispinosa*
        Spines yellow to gray or brown.
            Inner perianth-segments obtuse.............................12. *E. silvestrii*
            Inner perianth-segments acute.
              Plant small, 9 cm. in diameter or less; flower 16 cm. long........13. *E. calochlora*
              Plant 4 to 5 dm. high, 3 to 3.5 dm. thick; flower 20 to 22 cm. long. .14. *E. cordobensis*
  CC. Spines more or less curved.
      Spines very delicate, central one hooked..............................15. *E. ancistrophora*
      Spines stout.
        Central spine solitary.
            Radial spines straight.
              Plant about 9 cm. thick, 3 dm. high or less.........................16. *E. spegazziniana*
              Plant up to 1.5 meters high, 16 to 18 cm. in diameter...............17. *E. shaferi*
            Radial spines curved.
              Ribs strongly crenate...18. *E. fiebrigii*
              Ribs not strongly crenate.
                Flowers 15 cm. long or less..................................19. *E. rhodotricha*
                Flowers 20 cm. long or more.
                    Central spine up to 10 cm. long...........................20. *E. leucantha*
                    Central spine about 2.5 cm. long..........................21. *E. obrepanda*
        Central spines several.
            Ribs 16; spines at first rose......................................22. *E. intricatissima*
            Ribs 13 or 14; spines gray to blackish.
              Flowers straight..23. *E. molesta*
              Flowers curved...24. *E. baldiana*
    BB. Flowers yellow..25. *E. aurea*
  AA. Tube of perianth not longer than limb.
      Ribs not undulate ....26. *E. bridgesii*
      Ribs undulate ....27. *E. mamillosa*
AAA. Species not grouped...28. *E. formosa*

1. Echinopsis meyeri Heese, Gartenflora 56: 1. 1907.

Stems globose or somewhat depressed at apex, 10 cm. in diameter, pale green; ribs 14 to 16, acute, usually straight; spines subulate, all straight, rosy below, brown or black above, but in age nearly white; radial spines 7 or 8; central spine solitary; flowers numerous, lateral, 15 to 20 cm. long; all perianth-segments long, threadlike, twisted, the outer ones brownish, the inner dull white; axils of scales on ovary and flower-tube bearing many long hairs; stigma-lobes cream-colored.

Type locality: Paraguay.
Distribution: Paraguay.

We have not seen specimens of this species, but the type was illustrated; so far as we know it is not in cultivation. This should not be confused with the *Echinopsis meyeri* which is grown in gardens and which, according to Berger, is a hybrid between *E. eyriesii* and *E. leucantha*.

This plant is remarkable among cacti for its very narrow perianth-segments.

Haage and Schmidt offer a plant under this name for sale. It suckers very freely, both on the side and near the top of the plant and these begin to send out roots while still attached to the parent plant. They are covered with short brown spines. We do not know the origin of Haage and Schmidt's consignment and we have seen only very small plants from it. As these all show several central spines, while the *E. meyeri* Heese is known to have a single central spine, there may be doubt as to their identification.

Illustration: Gartenflora 56: pl. 1558.

Figure 80 is from a photographic copy of the illustration above cited.

2. Echinopsis mirabilis Spegazzini, Anal. Mus. Nac. Buenos Aires III. **4**: 489. 1905.

Simple, cylindric, 12 to 15 cm. high, 2 cm. in diameter, dull yellowish green; ribs 11, slightly undulate; areoles minute; spines all straight; radial spines 9 to 14, slender; central spine solitary, erect, 10 to 15 mm. long; flowers borne near the top of the plant, inodorous, 11 to 12 cm. long; inner perianth-segments white, acuminate; scales on ovary and flower-tube very woolly in their axils, thin, scarious at base, almost filiform, 8 mm. long; outer perianth-segments similar to the scales but longer; fruit 3.5 to 4 cm. long, 5 to 6 mm. in diameter; seeds globular, 1.5 mm. in diameter, with a depressed hilum.

FIG. 80.—Echinopsis meyeri.   FIG. 81.—Echinopsis mirabilis.

Type locality: Near Colonia Ceres, province of Santiago del Estero.

Distribution: Known only from province of Santiago del Estero, Argentina.

This plant is called flor de la oración.

Besides photographs of the type Dr. Spegazzini has presented us with a fruit from the type plant.

Figure 81 is from a photograph contributed by Dr. Spegazzini.

3. Echinopsis forbesii (Lehmann) A. Dietrich, Allg. Gartenz. **17**: 193. 1849.

> *Echinocactus forbesii* Lehmann in Walpers,* Repert. Bot. **2**: 319. 1843.
> *Echinopsis valida* Monville in Salm-Dyck, Cact. Hort. Dyck. 1849. 181. 1850.
> *Echinopsis valida forbesii* R. Meyer in Schumann, Gesamtb. Kakteen 239. 1897.

---

*Walpers says: "F. A. Lehm. in Terscheck, Suppl. Cact. 2." Förster (Handb. Cact. 520. 1846) used *Echinopsis forbesii* Hort. Angl. as a synonym of *Echinocactus forbesii*, but hardly publishes it.

Usually simple, columnar, claviform, sometimes 1 meter high, 20 cm. in diameter, glaucous-green; ribs 10 to 15, acute, separated by acute intervals; areoles circular, filled with spines and short white wool; spines 8 to 15, the longest 2 cm. long, acicular, straight, pale, nearly white, except the tips, these brown; central spines 1 to several, the longest 3 to 4 cm. long, stouter than the radials. horizontal; young joints borne near the top of the plant, densely covered with yellow and brown spines intermixed with soft white hairs; flowers borne near the top of the plant, about 10 cm. long; inner perianth-segments spreading, lanceolate, acute, white.

Type locality: Not cited.
Distribution: Paraguay (*fide* Weber).
This species is known to us only from descriptions and illustrations. Schumann follows Meyer in making *E. forbesii* a variety of *E. valida.* We have united the two and taken the older specific name.
The species was named for James Forbes (1773-1861), an enthusiastic student of cacti and gardener for the Duke of Bedford at Woburn Abbey.
Cereus validissimus Weber (Dict. Hort. Bois 473. 1896) is given as a synonym of *E. valida.*
Illustrations: Monatsschr. Kakteenk. **5**: 117; Palmer, Cult. Cact. 151; Möllers Deutsche Gärt. Zeit. **25**: 475. f. 7, No. 17, as *Echinopsis valida.*

4. **Echinopsis huottii** (Cels) Labouret, Monogr. Cact. 301. 1853.

> *Echinocactus huotti* Cels, Portef. Hort. 216. 1847.
> *Echinopsis apiculata* Linke, Wochenschr. Gärtn. Pflanz. 1: 85. 1858.

Plants simple, slender, up to 3.5 dm. high, short-columnar, dull green; ribs 9 to 11, crenate; radial spines 9 to 11, acicular, 2 cm. long or more; central spines normally 4, brown, 4 cm. long, subulate, porrect; flowers lateral, large, 17 to 20 cm. long, white; stamens included, greenish below, white above; style green; stigma-lobes 14, green.

Type locality: Cited as Chile (*fide* Labouret), but doubtless wrongly.
Distribution: Bolivia (*fide* Linke and Schumann).
It does not seem close to any of the other species. It is quite different from the Bolivian species collected by Dr. Rose at La Paz, Bolivia, which we have referred to *E. bridgesii.* (See page 74.)
We have studied a plant sent to the New York Botanical Garden from the Berlin Botanical Garden in 1902; in this there is only one central spine at each areole.
Schlumberger (Rev. Hort. IV. **3**: 348. 1854) calls this *Echinopsis kuottii,* doubtless a typographical error.
Cereus huottii Cels and *Echinopsis verschaffeltii* (Dict. Hort. Bois 471. 1896) are given as synonyms of this species by Weber.
Illustrations: Möllers Deutsche Gärt. Zeit. **25**: 475. f. 7, No. 11, as *Echinopsis apiculata;* Schumann, Gesamtb. Kakteen f. 45; Gartenwelt **17**: 145.

5. **Echinopsis minuana** Spegazzini, Anal. Mus. Nac. Buenos Aires III. **4**: 488. 1905.

Simple or rarely proliferous at base, columnar, 5 to 8 dm. high, 14 to 15 cm. in diameter; ribs 12, straight, a little undulate; spines all straight, dark brown to chestnut-colored; radial spines 4 to 7, short, 2 to 3 cm. long; central spine solitary, stouter than the radials, bulbose at base, 5 to 6 cm. long; flowers large, inodorous, 20 cm. long; inner perianth-segments oblanceolate, 4.5 cm. long; filaments and style greenish white; stigma-lobes 17 or 18, greenish white; fruit subglobose, 4.5 cm. long, greenish red.

Type locality: Bank of Paraná River, province of Entre Rios, Argentina.
Distribution: Province of Entre Rios, eastern Argentina.
We know this species only from description and a photograph taken by Dr. Spegazzini.

6. Echinopsis multiplex (Pfeiffer) Zuccarini in Pfeiffer and Otto, Abbild. Beschr. Cact. 1: pl. 4. 1839.

Cereus multiplex Pfeiffer, Enum. Cact. 70. 1837.
Echinonyctanthus multiplex Lemaire, Cact. Gen. Nov. Sp. 85. 1839.

Plants simple or very proliferous, globular to somewhat clavate, rounded at apex, 1.5 dm. high; ribs 13 to 15, broad at base, acute, slightly undulate; areoles large, filled with short white wool; spines brown, subulate; radial spines 5 to 15, ascending, 2 cm. long; central spines 2 to 5, 4 cm. long; flower 15 to 20 cm. long, its tube distinctly enlarged above, its scales large, distinct; inner perianth-segments broad, rose-colored, acuminate; stamens and style much shorter than perianth-segments, but exserted beyond the throat; stigma-lobes white, slender, 6 or 7.

Type locality: Southern Brazil.
Distribution: Southern Brazil.

This species may not be distinct from the following one. In collections of cacti, plants apparently intermediate in character are frequently found, as well as many hybrids.

Pfeiffer (Enum. Cact. 70. 1837) gives *Echinocactus multiplex* as a synonym of *Cereus multiplex*, while the name was in use in the Botanical Garden in Berlin in 1829 (Verh. Ver. Beförd. Gartenb. 6: 431. 1830). On the following page he describes the variety *monstrosus*. Other forms have been described as var. *cossa*, *picta*, and *cristata* under *Echinopsis multiplex*.

This plant is common in cultivation. Some of the illustrations cited for this and the two following species may represent hybrid plants with one of these species as one of the parents.

Illustrations: Watson, Cact. Cult. 80. f. 25; Dict. Gard. Nicholson 4: 512. f. 11, as *Cereus multiplex cristatus;* Dict. Gard. Nicholson Suppl. f. 365; Förster, Handb. Cact. ed. 2. f. 9; Grässner, Haupt-Verz Kakteen 1912: 16; Gard. Chron. III. 29: f. 80; Schelle, Handb. Kakteenk. f. 48; Möllers Deutsche Gärt. Zeit. 25: 475. f. 7, No. 12, as *Echinopsis multiplex cristata;* Dict. Gard. Nicholson 4: 512. f. 10; Curtis's Bot. Mag. 66: pl. 3789; Watson, Cact. Cult. 79. f. 24, as *Cereus multiplex;* Monatsschr. Kakteenk. 16: 89, as *Echinopsis multiplex monstrosa;* Monatsschr. Kakteenk. 6: 103; Pfeiffer and Otto, Abbild. Beschr. Cact. 1: pl. 4; Förster, Handb. Cact. ed. 2. 139. f. 8; Dict. Gard. Nicholson Suppl. f. 364; Rümpler, Sukkulenten 168. f. 92; Garden 84: 133; Möllers Deutsche Gärt. Zeit. 25: 475. f. 7, No. 6.

Plate VI, figure 2, shows a flowering plant in the collection of the New York Botanical Garden, received from the Missouri Botanical Garden.

7. Echinopsis oxygona (Link) Zuccarini in Pfeiffer and Otto, Abbild. Beschr. Cact. 1: under pl. 4. 1839.

Echinocactus oxygonus Link in Link and Otto, Verh. Ver. Beförd. Gartenb. 6: 419. 1830.
Cereus oxygonus Pfeiffer, Enum. Cact. 70. 1837.
Echinocactus octogonus G. Don in Sweet, Hort. Brit. ed. 3. 283. 1839.
Echinonyctanthus oxygonus Lemaire, Cact. Gen. Nov. Sp. 85. 1839.

Plants subglobose, simple or somewhat clustered, about 25 cm. in diameter, somewhat glaucous; ribs 14, broad at base, rounded on back; spines about 14, short and stout, 2 to 4 cm. long; flowers usually from areoles halfway up the side of the plant, sometimes nearly 3 dm. long, the tube slender, nearly cylindric, its scales numerous and small; inner perianth-segments pale red, acute or acuminate.

Type locality: Southern Brazil.
Distribution: Southern Brazil, Uruguay, and northeastern Argentina.

Pfeiffer (Enum. Cact. 70. 1837) and also Steudel referred *Echinocactus sulcatus* as a synonym of this species. *E. sulcatus* Link and Otto (Steudel, Nom. ed. 2. 1: 537. 1840) is supposed to be different from the last, but in any case it is only a name.

Echinopsis sulcata occurs as a name in a paper by Wercklé (Monatsschr. Kakteenk. 15: 180. 1905).

Echinopsis wilkensii (*E. eyriesii wilkensii* Linke), *E. rohlandii*, and *E. lagemannii* Dietrich (*E. eyriesii lagemannii* Dietrich) are all mentioned by Schumann (Gesamtb. Kakteen 235. 1897) as hybrids of which *E. oxygona* is one of the parents. Schelle (Handb.

PLATE VI

M. E. Eaton del.

1. Top of flowering plant of *Echinopsis turbinata*.
2. Flowering plant of *Echinopsis multiplex*.
(All natural size.)

A. Hoen & Co.

Kakteenk. 112. 1907) lists the following as hybrids with this species and *E. eyriesii: E. triumphans, E. nigerrima,* and *E. undulata. Echinopsis roehlandii* is figured in the Revue Horticole (**85**: pl. opp. 304).

Two varieties, *inermis* and *subinermis,* are sometimes given under this species. *E. oxygona turbinata* Mittler (Labouret, Monogr. Cact. 306. 1853) is considered a hybrid.

Illustrations: Möllers Deutsche Gärt. Zeit. **16**: 80; **25**: 475. f. 7, No. 3, as *Echinopsis lagemannii;* Curtis's Bot. Mag. **71**: pl. 4162; Edwards's Bot. Reg. **20**: pl. 1717; Verh. Ver. Beförd. Gartenb. 6: pl. 1, as *Echinocactus oxygonus;* Schelle, Handb. Kakteenk. 114. f. 49; Cact. Journ. 1: pl. 6; Pfeiffer, Abbild. Beschr. Cact. 2: pl. 4; Möllers Deutsche Gärt. Zeit. 25: 475. f. 7, No. 2; Gartenwelt 1: 283.

Figure 82 is from a photograph, contributed by Dr. Spegazzini.

FIG. 82.—Echinopsis oxygona.   FIG. 83.—Echinopsis tubiflora.

8. **Echinopsis eyriesii** (Turpin) Zuccarini in Pfeiffer and Otto, Abbild. Beschr. Cact. 1: under pl. 4. 1839.

> Echinocactus eyriesii Turpin,* Ann. Inst. Roy. Hort. Fromont 2: 158. 1830.
> Cereus eyriesii Pfeiffer, Enum. Cact. 72. 1837.
> Echinonyctanthus eyriesii Lemaire, Cact. Gen. Nov. Sp. 84. 1839.
> Echinopsis pudantii Pfersdorf, Monatsschr. Kakteenk. 10: 167. 1900.

Simple or clustered, globular to short-columnar; ribs 11 to 18, not tuberculate, rather thin above; areoles circular, filled with white or tawny wool; spines several, 14 to 18, very short, 5 mm. long or less; flower from the side of plant but above the middle, large, 17 to 25 cm. long; inner perianth-segments white, acuminate; stamens and style shorter than the perianth-segments; scales on the flower-tube small, ovate, brownish, hairy in their axils.

Type locality: Buenos Aires, according to Pfeiffer.

Distribution: Southern Brazil, Uruguay, and province of Entre Rios, Argentina.

The following varieties have been published, some well-known hybrids, others mere forms; var. *cristata* (Monatsschr. Kakteenk. **2**: 27. 1902); vars. *glauca* and *glaucescens* (Förster, Handb. Cact. 360. 1846); vars. *tettavii* and *triumphans* Jacobi (Förster, Handb. Cact. ed. 2. 626, 630. 1885), sometimes given as *Echinopsis triumphans* (Monatsschr. Kakteenk. **15**: 33. 1905) and var. *grandiflora* R. Meyer (Monatsschr. Kakteenk. **21**: 186. 1911). Schelle (Handb. Kakteenk. 111, 112. 1907) gives the following varieties besides two quadrinomials: *major, rosea* Link, *cristata, phyligera,* and *duvallii.*

---

* The reference for this species is usually cited as "Obs. Cact. 58," referring to a paper by Turpin entitled "Observations sur la Famille des Cactées, etc." in three parts which appeared in the above cited volume.

Echinopsis eyriesii inermis is in the trade (Grässner).

In cultivation this plant buds freely, sometimes producing at the same time a dozen or more small spiny buds which, dropping to the ground, start new plants. These appear only at the areoles, sometimes at the top and sometimes near the bottom of the old plant. This species has long been a favorite with gardeners.

This species was named for Alexander Eyries of Havre, France.

Illustrations: Edwards's Bot. Reg. **20**: pl. 1707; Curtis's Bot. Mag. **62**: pl. 3411; Loudon, Encycl. Pl. ed. 2 and 3. 1201. f. 17353; Ann. Inst. Roy. Hort. Fromont **2**: pl. 2, as *Echinocactus eyriesii;* Edwards's Bot. Reg. **24**: pl. 31, as *Echinocactus eyriesii glaucus;* Monatsschr. Kakteenk. **10**: 166; Schumann, Gesamtb. Kakteen Nachtr. f. 10, as *Echinopsis pudantii;* Deutsches Mag. Gart. Blumen. **1855**: opp. 112, as *Echinopsis tettavii;* Förster, Handb. Cact. ed. 2. 627. f. 83, as *Echinopsis eyriesii tettavii;* Förster, Handb. Cact. ed. 2. f. 84; Rümpler, Sukkulenten 171. f. 94, as *Echinopsis eyriesii triumphans;* Dict. Gard. Nicholson **4**: 541. f. 26; Suppl. 337. f. 363, as *Echinopsis eyriesii flore-pleno;* Schelle, Handb. Kakteenk. f. 46; Möllers Deutsche Gärt. Zeit. **25**: 475. f. 7, No. 4; Garten-Zeitung **4**: 182. f. 42, No. 12; Anal. Mus. Nac. Montevideo **5**: pl. 27; Blühende Kakteen **2**: pl. 72; Engler and Prantl, Pflanzenfam. 3^{6a}: f. 59, C; Förster, Handb. Cact. ed. 2. f. 82; Gartenflora **28**: 373; Schumann, Gesamtb. Kakteen f. 10; Martius, Fl. Bras. 4^2: pl. 47; Rümpler, Sukkulenten f. 93; Cact. Journ. **1**: pl. 6; **2**: 7.

9. Echinopsis turbinata Zuccarini in Pfeiffer and Otto, Abbild. Beschr. Cact. **1**: under pl. 4. 1839.

<small>*Cereus turbinatus* Pfeiffer, Allg. Gartenz. **3**: 314. 1835.
Echinonyctanthus turbinatus Lemaire, Cact. Gen. Nov. Sp. 84. 1839.
Echinonyctanthus turbinatus pictus Monville in Lemaire, Cact. Gen. Nov. Sp. 84. 1839.
Echinopsis gemmata Schumann in Martius, Fl. Bras. 4^2: 231. 1890.</small>

Simple or somewhat clustered, globose; ribs 13 or 14, broad at base, hardly undulate; spines several, 7 mm. long or less; flowers appearing from upper areoles, about 15 cm. long, with a strong odor of jasmine and citron; inner perianth-segments white, acuminate; stamens and style shorter than the perianth-segments, but projecting beyond the throat; scales on tube and ovary small, woolly in their axils.

Type locality: Not cited.

Distribution: Province of Entre Rios, Argentina.

Cereus jasmineus and *Echinocactus turbinatus* (Pfeiffer, Enum. Cact. 72. 1837), as synonyms for *Cereus turbinatus,* doubtless belong here.

Echinocactus gemmatus Link and Otto (Verh. Ver. Beförd. Gartenb. **6**: 431. 1830), only a name, is doubtless to be referred here, while *Cereus gemmatus* Otto (Allg. Gartenz. **3**: 314. 1835, not Verh. Ver. Beförd. Gartenb. **6**: 431. 1830, as cited by Schumann) was published as a synonym of *Cereus turbinatus.* For this reason we have substituted *Echinopsis turbinata* for *E. gemmata,* the name generally used for this plant.

Walpers refers to the following as an undescribed species: *Echinopsis picta* Walpers (Repert. Bot. **2**: 324. 1843; *Echinonyctanthus pictus* Lemaire, Cact. Gen. Nov. Sp. 84. 1839, *fide* Walpers, but in error), as synonym. This probably belongs here also. *Echinopsis turbinata picta* (Walpers, Repert. Bot. **2**: 275. 1843) is only a listed name.

Of this relationship are the following: *Echinopsis schelhasii* Pfeiffer and Otto (Abbild. Beschr. Cact. **1**: under pl. 4. 1839; *Echinonyctanthus schelhasii* Lemaire, Cact. Gen. Nov. Sp. 84. 1839*), *Cereus schelhasii* Pfeiffer (Allg. Gartenz. **3**: 314. 1835), *Echinopsis schelhasei rosea* Rümpler (Förster, Handb. Cact. ed. 2. 623. 1885), *Echinopsis gemmata schelhasei* (Schelle, Handb. Kakteenk. 113. 1907), *Echinopsis decaisneana* Walpers (Repert. Bot. **2**: 324. 1843; *Echinonyctanthus decaisnianus* Lemaire, Cact. Gen. Nov. Sp. 55. 1839; *Echinocactus decaisnei* Steudel, Nom. ed. 2. **1**: 536. 1840; *Echinopsis gemmata decaisneana* Schelle, Handb. Kakteenk. 113, 1907), *Echinopsis jamessiana* (Salm-Dyck, Cact. Hort. Dyck. 1849. 38. 1850) and *Echinopsis falcata* Rümpler (Förster, Handb. Cact. ed. 2. 622. 1885).

---

*According to Walpers, but in error.

Echinopsis decaisneana is a delicately fragrant, beautiful pink form with large flowers; the inner perianth-segments are oblong, acute or acuminate. It is a hybrid between this and some other species. The flowers open during the day and last usually for more than one day.

Illustrations: Cact. Journ. **1**: 59; **2**: 169, as *Cereus gemmatus;* Möllers Deutsche Gärt. Zeit. **25**: 475. f. 7, No. 14, as *Echinopsis gemmata cristata;* Cycl. Amer. Hort. Bailey **2**: f. 749; Stand. Cycl. Hort. Bailey **2**: f. 1377; Möllers Deutsche Gärt. Zeit. **25**: 475. f. 7, No. 23; Tribune Hort. **4**: pl. 139; Gartenwelt **7**: 289; U. S. Dept. Agr. Bur. Pl. Ind. Bull. **262**: pl. 10, as *Echinopsis gemmata;* Dict. Gard. Nicholson **1**: 502. f. 697; Förster, Handb. Cact. ed. 2. 621. f. 81, as *Echinopsis decaisneana;* Pfeiffer, Abbild. Beschr. Cact. **2**: pl. 7.

Plate VI, figure 1, shows a plant in the collection of the New York Botanical Garden.

10. Echinopsis tubiflora (Pfeiffer) Zuccarini in A. Dietrich, Allg. Gartenz. **14**: 306. 1846.

> *Cereus tubiflorus* Pfeiffer, Enum. Cact. 71. 1837.
> *Echinopsis zuccarinii* Pfeiffer in Pfeiffer and Otto, Abbild. Beschr. Cact. **1**: under pl. 4. 1839.
> *Echinocactus tubiflorus* Hooker in Curtis's Bot. Mag. **65**: pl. 3627. 1839.
> *Echinonyctanthus tubiflorus* Lemaire, Cact. Gen. Nov. Sp. 85. 1839.
> *Echinonyctanthus tubiflorus nigrispinus* Lemaire, Cact. Gen. Nov. Sp. 85. 1839.
> *Echinopsis nigrispina* Walpers, Repert. Bot. **2**: 324. 1843.
> *Echinopsis melanacantha* Dietrich, Allg. Gartenz. **14**: 306. 1846.
> *Echinopsis grandiflora* Linke, Allg. Gartenz. **25**: 239. 1857.
> *Echinopsis tubiflora paraguayensis* R. Meyer, Monatsschr. Kakteenk. **23**: 153. 1913.

Simple or clustered, subglobose, about 12 cm. in diameter; ribs about 12, prominent, slightly undulate; areoles circular, filled with white wool; spines subulate, black, 10 to 12 mm. long; flowers from the side of the plant, 15 to 20 cm. long; inner perianth-segments spreading, white, acuminate; filaments and style projecting a little beyond the throat; axils of scales on flower-tube bearing long wool.

Type locality: Not cited.

Distribution: Provinces of Tucuman, Catamarca, and Salta, Argentina; recorded from Brazil.

Pfeiffer (Enum. Cact. 71. 1837) gives *Echinocactus tubiflorus* as a synonym of *Cereus tubiflorus.*

Salm-Dyck (Cact. Hort. Dyck. 1849. 39. 1850) gives *Echinopsis zuccariniana* Pfeiffer instead of *E. zuccarinii* and Rümpler uses this spelling. Under *E. zuccariniana* several floral and abnormal forms have been described as varieties, among which are *rosea, cristata, monstrosa, picta, rohlandii,* and *nigrispina* and under *E. zuccarinii, monstruosa, nigrispina,* and *picta;* some of the same varieties appear under *E. tubiflora* including *nigrispina, rosea,* and *rohlandii.* Walpers (Repert. Bot. **2**: 324. 1843) credits the name *Echinonyctanthus nigrispinus* to Lemaire, but Lemaire used the name *nigrispinus* only as a variety of *E. tubiflorus.*

Echinopsis droegeana Berger (Monatsschr. Kakteenk. **1**: 24. 1891) is probably a hybrid with this species as one of the parents.

Echinopsis zuccarinii robusta is in the trade (Grässner).

Illustrations: Hartinger, Parad. **1**: 8, as *Cereus tubiflorus;* Curtis's Bot. Mag. **65**: pl. 3627, as *Echinocactus tubiflorus;* Möllers Deutsche Gärt. Zeit. **16**: 80, as *Echinopsis tubiflora* hybrid; Monatsschr. Kakteenk. **4**: 27, as *E. zuccariniana rohlandii;* Belg. Hort. **16**: pl. opp. 130, as *Echinopsis zuccariniana;* Schelle, Handb. Kakteenk. f. 50; Floralia **42**: 372.

Figure 83 is from a photograph contributed by Dr. Spegazzini.

11. Echinopsis albispinosa Schumann, Monatsschr. Kakteenk. **13**: 154. 1903.

Low, simple or somewhat cespitose, almost globular; ribs 10 or 11, slightly undulating; spines 11 to 14, at first reddish brown, becoming white, somewhat ascending; flowers white, 19.5 cm. long, as long or longer than the plant itself; scales on flower-tube and ovary bearing cobwebby hairs in their axils.

Type locality: Not cited.

Distribution: Supposed to have come from Bolivia or Paraguay, probably from the latter.

We have seen no specimens of this species, but the first illustration cited below is of the type specimen.

Illustrations: Monatsschr. Kakteenk. **13**: 155; Möllers Deutsche Gärt. Zeit. **25**: 475. f. 7, No. 22.

12. Echinopsis silvestrii Spegazzini, Anal. Mus. Nac. Buenos Aires III. **4**: 486. 1905.

Stems simple or somewhat clustered, 5 to 10 cm. high, 4 to 8 cm. in diameter; ribs 12 to 14; spines rather stout and short, grayish; radial spines 5 to 9, appressed; central spine solitary, erect; flowers inodorous, 20 cm. long; inner perianth-segments obtuse, white; style white; stigma-lobes 9, white.

Type locality: Mountains between the provinces of Tucuman and Salta, Argentina.

Distribution: Northwestern Argentina.

This species was named for Dr. Philip Silvestri, a friend of Dr. Spegazzini.

Plate VII, figure 1, shows a plant brought from Salta, Argentina, by Dr. Shafer in 1917 (No. 41) which flowered in the New York Botanical Garden in June 1918. Figure 84 is from a photograph contributed by Dr. Spegazznii.

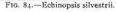
FIG. 84.—Echinopsis silvestrii.

FIG. 85.—Echinopsis calochlora.

13. Echinopsis calochlora Schumann, Monatsschr. Kakteenk. **13**: 108. 1903.

Plants small, nearly globular, 6 to 9 cm. in diameter, deep green; ribs 13, broad, strongly crenate, separated by narrow intervals; areoles 10 to 15 cm. long, sunken in the ribs; radial spines acicular, yellow, 10 to 14, ascending; central spines 3 or 4, similar to the radials; flowers lateral, appearing above the middle of the plant, 16 cm. long; the tube only a little broader at top than at base, greenish yellow; inner perianth-segments broad, acute, white; stamens exserted beyond the tube; stigma-lobes green.

M. E. Eaton del.   A. Hoen & Co.

1. Top of flowering plant of *Echinopsis silvestrii*.   2. Top of flowering plant of *Echinopsis leucantha*.
(All natural size.)

Type locality: Corumba, Brazil.

Distribution: Province of Goyaz, Brazil.

Illustrations: Blühende Kakteen 2: pl. 61; U. S. Dept. Agr. Bur. Pl. Ind. Bull. 262: pl. 3; Schelle, Handb. Kakteenk. f. 52.

Figure 85 is copied from the first illustration above cited.

14. Echinopsis cordobensis Spegazzini, Anal. Mus. Nac. Buenos Aires III. 4: 489. 1905.

Plants simple, large, 4 to 5 dm. high, 3 to 3.5 dm. thick, ellipsoid. dull green, somewhat glaucous; ribs 13, straight, acute, stout, not crenate; spines all straight, at first dark, then gray; radial spines 8 to 10, 10 to 20 mm. long; central spines 1 to 3, the lower one largest, 3 to 5 cm. long, bulbose at base; flowers erect, with little or no odor, 20 to 22 cm. long; axils of scales on ovary and flower-tube villous; inner perianth-segments white, acute; fruit globose, 2.5 cm. long, yellowish red.

Type locality: Near Villa Mercedes, province of Córdoba, Argentina.

Distribution: Rare in province of Córdoba, Argentina.

15. Echinopsis ancistrophora Spegazzini, Anal. Mus. Nac. Bueno; Aires III. 4: 492. 1905.

Stem simple, subglobose, 5 to 8 cm. in diameter, shining greens ribs 15 or 16, stout, 1 cm. high, broad at base, somewhat crenate; radial spines 3 to 7, slender, spreading backward, 5 to 15 mm. long; central spine solitary, more or less curved or hooked, 1 to 2 cm. long; flowers inodorous, 12 to 16 cm. long; outer perianth-segments green, linear, acuminate; inner perianth-segments white, oblong, acute; fruit ellipsoid, 1.6 cm. long, 8 mm. in diameter; scales on ovary and flower-tube small, their axils lanate.

Type locality: Between Tucuman and Salta, Argentina.

Distribution: The high mountains between the provinces of Tucuman and Salta, Argentina.

We have not seen specimens of this species, which Spegazzini says is rare.

Figure 86 is from a photograph contributed by Dr. Spegazzini.

Fig. 86.—Echinopsis ancistrophora.

16. Echinopsis spegazziniana sp. nov.

Stem simple, slender, dull green, about 3 dm. high and 9 cm. thick; ribs 12 to 14, low, slightly crenate; radial spines 7 or 8, straight, subulate, brown; central spine one, a little curved, much longer than the radials, 2 cm. long; flowers lateral, from near the middle of the plant, 15 to 17 cm. long; perianth-segments short, broad, acute, white; scales on ovary and tube small, very hairy in their axils.

This species which was found near Mendoza, Argentina, was first identified as *Echinopsis salpingophora* by Von Preinreich (Monatsschr. Kakteenk. 3: 163. 1893) and later as *E. campylacantha* by R. Meyer (Monatsschr. Kakteenk. 5: 27. 1895), both erroneously.

Illustrations: Monatsschr. Kakteenk. 3: 161, as *E. salpingophora;* Schumann, Gesamtb. Kakteen f. 46, as *E. campylacantha.*

Figure 88 is from a photograph contributed by Dr. Spegazzini.

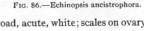

Fig. 87.—Fruit of Echinopsis shaferi. Natural size.

17. Echinopsis shaferi sp. nov.

Simple, erect, cylindric, up to 1.5 meters high, 16 to 18 cm. in diameter, dark green; ribs 10 to 12, 2 cm. high, separated by acute intervals; areoles approximate, 1 cm. apart or less; radial spines straight, at first brownish, but gray in age, slender, subulate, 6 to 9, 1.5 to 3.5 cm. long, somewhat spreading;

central spine solitary, 10 cm. long or less, ascending, somewhat curved, the upper ones more or less connivent over the top of the plant; flower slender, funnelform, 2 dm. long, white; filaments and style pale green; fruit ovoid, 3 cm. long, brick-red.

Collected by J. A. Shafer in sandy thickets, Trancas, Tucuman, Argentina, February 11, 1917 (No. 101).

This is the largest species of the genus known to us. It flowered at the New York Botanical Garden in June 1920. In the new growth the top is very woolly. The top

FIG. 88.—Echinopsis spegazziniana.

of the growing plant is covered with a mass of brown wool arising from the closely set young areoles.

John Adolph Shafer (1863–1918), an enthusiastic botanical collector, was commissioned by Dr. Britton to visit Argentina in the winter of 1916–1917 and he obtained plants and specimens of great importance in our studies of the cacti.

Figure 89 is from a photograph taken by Dr. Shafer at Trancas, Argentina, in 1917; figure 87 shows the fruit of the plant photographed.

18. Echinopsis fiebrigii Gürke, Notizbl. Bot. Gärt. Berlin **4:** 184. 1905.

Stems simple, depressed-globose, 9 cm. high, 15 cm. broad; ribs 18 to 24, strongly crenate, broken into long tubercles, 1.5 cm. high; radial spines 8 to 10, 10 to 25 mm. long, recurved; central spine one, curved, ascending; flowers 17 to 19 cm. long, the tube nearly cylindric; outer perianth-segments green, spreading; inner perianth-segments white, short, broad, obtuse or truncate; filaments white; style green; stigma-lobes 11, green, 15 to 17 mm. long.

Type locality: Bolivia.
Distribution: Bolivia.

The plant is known to us only from description and illustrations.

FIG. 89.—Echinopsis shaferi.

Illustrations: Blühende Kakteen **2**: pl. 100; Monatsschr. Kakteenk. **16**: 27; Möllers Deutsche Gärt. Zeit. **25**: 475. f. 7, No. 10.

Figure 90 is copied from the first illustration above cited.

19. Echinopsis rhodotricha Schumann, Monatsschr. Kakteenk. **10**: 147. 1900.

> *Echinopsis rhodotricha robusta* R. Meyer, Monatsschr. Kakteenk. **24**: 113. 1914.

Cespitose, dull grayish green, with 8 to 10 erect or ascending cylindric stems, 3 to 8 dm. high, 9 cm. in diameter, or sometimes simple in cultivation; ribs 8 to 13, rather low, a little sinuate; areoles 15 to 25 mm. apart; radial spines 4 to 7, widely spreading, a little curved, yellowish with brown tips, 2 cm. long; central spine one, 2.5 cm. long, shorter than the radials, or wanting, somewhat bent upward; flowers 15 cm. long; inner perianth-segments white, oblong, acute; stigma-lobes linear, 11, green.

FIG. 90—Echinopsis fiebrigii.   FIG. 91.—Echinopsis rhodotricha.

Type locality: Arroyo La Cruz, near San Salvador, Rio Tagatiya-mi, Paraguay.

Distribution: Paraguay and northeastern Argentina.

Spegazzini states (Anal. Mus. Nac. Buenos Aires III. **4**: 488. 1905) that Schumann first named this species *Echinopsis spegazzinii*, but as such it has not been formally published.

The variety *Echinopsis rhodotricha argentiniensis* R. Meyer (Monatsschr. Kakteenk. **21**: 188. 1911) seems to differ from the type in its shorter, darker stems with radial spines. It was introduced from Argentina and is now offered in the trade.

The variety *Echinopsis rhodotricha roseiflora* Schumann (Bull. Herb. Boiss. II. **3**: 251. 1903) comes from near Concepción, Paraguay, and is described as having pale rose-colored inner perianth-segments.

The variety *robusta* is offered for sale by R. Grässner.

The plant is known to us only from description and illustrations.

Illustrations: Blühende Kakteen **2**: pl. 76; Schumann, Gesamtb. Kakteen Nachtr.
f. 11; Monatsschr. Kakteenk. **11**: 139; Möllers Deutsche Gärt. Zeit. **25**: 475. f. 7, No. 8.
Figure 91 is copied from the first illustration above cited.

20. **Echinopsis leucantha** (Gillies) Walpers, Repert. Bot. **2**: 324. 1843.

> *Echinocactus leucanthus* Gillies in Salm-Dyck, Hort. Dyck. 341. 1834.
> *Cereus incurvispinus* Otto and Dietrich, Allg. Gartenz. 3; 244. 1835.
> *Cereus leucanthus* Pfeiffer, Enum. Cact. 71. 1837.
> *Echinonyctanthus leucanthus* Lemaire, Cact. Gen. Nov. Sp. 85. 1839.
> *Echinopsis campylacantha* Pfeiffer in Pfeiffer and Otto, Abbild. Beschr. 1: under pl. 4. 1839.
> *Echinopsis salpigophora** Lemaire in Salm-Dyck, Cact. Hort. Dyck. 1849. 181. 1850.
> *Echinopsis polyacantha* Monville in Labouret, Monogr. Cact. 302. 1853.
> *Echinopsis campylacantha leucantha* Labouret, Monogr. Cact. 305. 1853.
> *Echinopsis campylacantha stylodes* Monville in Labouret, Monogr. Cact. 305. 1853.
> *Echinopsis simplex* Niedt, Allg. Gartenz. **25**: 237. 1857.
> *Echinopsis melanopotamica* Spegazzini, Anal. Mus. Nac. Buenos Aires III. 4: 492. 1905.

Stems globose to oblong, about 3.5 dm. high; ribs 12 to 14, somewhat compressed; areoles close
together, oblong; radial spines 8, more or less curved, brownish; central spine 1, curved, elongated,
often 10 cm. long; flowers about 16 cm. long, described as up to 20 cm. long; the tube about 3 cm.
broad at the mouth, dark brown, with scattered areoles bearing small tufts of brown hairs; outer
perianth-segments brownish, spreading, 2 cm. long, with an acute scarious tip; inner perianth-segments in about 3 series, spreading, the outer ones purplish, the innermost ones nearly white, oblong,
acute, about 3 cm. long; filaments in many series of many lengths, the series at the mouth of the
flower-tube erect, 1.5 cm. long; style included; stigma-lobes numerous, green.

Type locality: Mendoza (*fide* Pfeiffer).
Distribution: Western Argentina.

Weber (Dict. Hort. Bois 471. 1896) gives *Echinopsis yacutulana* Weber as a synonym
of *E. leucantha*. *Echinocactus salpingophora* (Labouret, Monogr. Cact. 302. 1853) was
given as a synonym of *Echinopsis salpingophora*.

Although *Echinocactus leucanthus*, with *Melocactus ambiguus* Pfeiffer as a synonym,
appeared in 1833 (Allg. Gartenz. 1: 364), it was not actually published until the following
year.

Echinopsis polyacantha Monville (Labouret, Monogr. Cact. 302. 1853) and *E. stylosa*
Monville (Schumann, Gesamtb. Kakteen 241. 1897) are given by Schumann as synonyms
of this species, but neither was published and the latter was not cited at the place mentioned
by Schumann.   It has been briefly described as a variety and will be found in the synonymy
above as *stylodes*.   *E. campylacantha* with its two forms *longispina* and *brevispina* are
assigned to R. Meyer (Monatsschr. Kakteenk. **5**: 36. 1895), who as a matter of fact published only the names *Echinopsis poselgeri* var. *brevispina* and var. *longispina*.

Echinopsis leucantha aurea (Monatsschr. Kakteenk. **17**: 76. 1907), *E. salpingophora
aurea* (Monatsschr. Kakteenk. **12**: 63. 1902) and *E. leucantha salpingophora* Schumann
(Monatsschr. Kakteenk. **13**: 62. 1903) are not described at the places cited above. *Melocactus elegans* Pfeiffer (Allg. Gartenz. 3: 244. 1835) is usually referred here.

Echinopsis melanopotamica which comes from southern Argentina we have referred
here; if it belong here it represents the southern form of the species.   We have not seen the
type but we have seen fruits collected by Fischer and spines by Alex Wetmore (1920),
both from the Rio Negro region, presumably referable here.   They suggest the desirability
of further field study.

Illustrations: Edwards's Bot. Reg. **26**: pl. 13, as *Cereus leucanthus;* Curtis's Bot. Mag.
77: pl. 4567; Fl. Serr. **6**: pl. 635; Jard. Fleur. **1**: pl. 98; Loudon, Encycl. Pl. ed. 3. 1378.
f. 19385; Rümpler, Sukkulenten f. 95; Möllers Deutsche Gärt. Zeit. **25**: 475. f. 7, No. 15,
as *Echinopsis campylacantha;* Monatsschr. Kakteenk. **5**: 35; Kirtcht, Kakteen Zimmergarten 23, as *Echinopsis salpingophora;* Möllers Deutsche Gärt. Zeit. **25**: 475. f. 7, No. 7,

---

*The original spelling of this name was *salpigophora*, but Schumann says that it should be *salpingophora*.

as *Echinopsis salpingophora aurea;* Möllers Deutsche Gärt. Zeit. **25:** 475. f. 7, No. 21, as *Echinopsis leucantha aurea;* Addisonia **4:** pl. 147.

Plate VII, figure 2, shows a flowering plant brought from Mendoza to the New York Botanical Garden by Dr. Rose in 1915.

21. Echinopsis obrepanda (Salm-Dyck) Schumann in Engler and Prantl, Pflanzenfam. 3^{6a}: 184. 1894

> *Echinocactus obrepandus* Salm-Dyck, Allg. Gartenz. **13:** 386. 1845.
> *Echinocactus misleyi* Cels, Portef. Hort. 216. 1847.
> *Echinopsis cristata* Salm-Dyck, Cact. Hort. Dyck. 1849. 178. 1850.
> *Echinopsis cristata purpurea* Labouret in Curtis's Bot. Mag. **76:** pl. 4521. 1850.
> *Echinopsis misleyi* Labouret, Monogr. Cact. 291. 1853.

Plant globose or somewhat depressed, 15 to 20 cm. in diameter; ribs 17 or 18, rather prominent, thin, strongly undulate, pale bluish green; areoles somewhat immersed in the rib; spines rigid, brownish; radial spines 10, spreading, or somewhat recurved, 12 to 16 mm. long; central spine solitary, 25 mm. long, ascending, curved; flowers lateral, white or purplish, the tube 20 cm. long, green; scales on ovary and flower-tube acuminate, bearing an abundance of black hairs in their axils; inner perianth-segments large, serrate, mucronate.

Type locality: Bolivia.

Distribution: Bolivia.

This plant was collected by Mr. Thomas Bridges in Bolivia in 1844 and first described by Salm-Dyck in 1845 as *Echinocactus obrepandus,* but when in 1850 he transferred it to *Echinopsis* he changed the specific name to *cristata.* A part of Bridges's material went to Kew; one of the specimens produced purple flowers, and another nearly white flowers; there is a possibility that more than one species was collected by Bridges at this time. The figures given in Gartenflora (**38:** f. 47) and Monatsschrift für Kakteenkunde (**12:** 169) are not quite typical. Here Weber refers *Echinopsis obliqua* Cels (Dict. Hort. Bois 472. 1896).

The plant is known to us only from descriptions and illustrations.

Illustrations: Curtis's Bot. Mag. **78:** pl. 4687; Gartenflora **38:** f. 47; Jard. Fleur.

FIG. 92.—Echinopsis obrepanda.

1: pl. 73, 74; Loudon, Encycl. Pl. ed. 3. 1378. f. 19386; Cassell's Dict. Gard. **1:** 315, as *Echinopsis cristata;* Curtis's Bot. Mag. **76:** pl. 4521, as *Echinopsis cristata purpurea;* Möllers Deutsche Gärt. Zeit. **25:** 475. f. 7, No. 5; Monatsschr. Kakteenk. **12:** 169; Gartenwelt **16:** pl. opp. 106; 107.

Figure 92 is copied from the first illustration above cited.

22. Echinopsis intricatissima Spegazzini, Anal. Mus. Nac. Buenos Aires III. **4:** 491. 1905.

Simple, somewhat ovoid, 20 cm. high, not depressed at apex; ribs 16; spines at first rose-colored, in age gray, elongated, 3 to 6 cm. long, the lowest ones 8 to 10 cm. long; radial spines 8 to 13; central spines 4 to 6, curved upward; flowers 20 to 22 cm. long; inner perianth-segments lanceolate, white; fruit 3 cm. long.

Type locality: Near Mendoza, Argentina.

Distribution: Known only from the type locality.

23. Echinopsis molesta Spegazzini, Anal. Mus. Nac. Buenos Aires III. **4**: 490. 1905.

Plants simple, subglobose, 20 cm. in diameter, pale green, not shining; ribs 13, prominent, acute on edge and somewhat undulate; areoles large; spines all grayish, rather stout; radial spines 6 to 8, straight, 10 to 15 mm. long; central spines 4, bulbose at base, slightly incurved, the lower one the longest, 3 cm. long; flowers slightly odorous, large, 22 to 24 cm. long; inner perianth-segments lanceolate, white; stamens, style and stigma-lobes white.

Type locality: Province of Córdoba, Argentina.
Distribution: Córdoba, Argentina.
This species is known to us only from description.

24. Echinopsis baldiana Spegazzini, Anal. Mus. Nac. Buenos Aires III. **4**: 490. 1905.

Stems simple, cylindric, 2 to 3 dm. high, 12 to 15 cm. in diameter; ribs 13 or 14, not all crenate; areoles large; spines slender, blackish brown; radial spines 9 to 11, 15 mm. long; central spines 3 or 4, 3 to 5 cm. long; flowers odorous, very large; inner perianth-segments lanceolate, acute, white; fruit large, 4 to 5 cm. long.

Type locality: Near Ancasti, province of Catamarca, Argentina.
Distribution: The dry mountain regions of the province of Catamarca, Argentina.
This species is known to us only from description.

25. Echinopsis aurea sp. nov.

Plants solitary, small, globular to short-cylindric, 5 to 10 cm. high; ribs 14 or 15, acute on the edge, separated by deep intervals; areoles when young filled with short brown wool; radial spines about 10, 1 cm. long; central spines usually 4, much stouter than the radials, often flattened, 2 to 3 cm. long: flowers from the side of the plant; flower-tube slightly curved, funnelform, greenish white, its scales ovate-linear, 4 to 6 mm. long, pale green but reddish at base, their axils filled with black and white hairs; flower-bud 9 cm. long, when young covered with long silky hairs; expanded flower 8 cm. broad, the perianth-segments in about 3 series, lemon-yellow, the inner ones deeper colored, about 20, oblong, mucronately tipped; filaments in 2 series, the upper series attached at the top of the throat, stiff and erect, exserted; the lower series inserted near the base of the throat, included; style green, very short and included, only 3 cm. long; stigma-lobes linear, cream-colored; scales on ovary small, their axils filled with long hairs; fruit not known.

Collected by Dr. Rose near Cassafousth, Córdoba, Argentina, in 1915 (No. 21046) and flowered in the New York Botanical Garden, May 6, 1916.
Plate x, figure 1, shows the plant in flower.

26. Echinopsis bridgesii Salm-Dyck, Cact. Hort. Dyck. 1849. 181. 1850.

> *Echinocactus salmianus* Cels, Protef. Hort. 180. 1847. Not Link and Otto, 1827.
> *Echinopsis salmiana* Weber, Dict. Hort. Bois 472. 1896.
> *Cereus salmianus* Cels in Weber, Dict. Hort. Bois 472. 1896, as synonym.
> *Echinopsis salmiana bridgesii* Schumann, Gesamtb. Kakteen 237. 1897.

Plants usually in clumps of 3 to 6, low, 1 dm. in diameter or more; ribs 10 to 12, high, not undulate; areoles large, filled with spines and short brown wool; spines about 10, brown when young, unequal; flowers 15 to 18 cm. long, probably white; tube slender, about the length of the limb; inner perianth-segments 3 to 4 cm. long, acute; scales on the ovary and flower-tube filled with long gray and black hairs.

Type locality: Bolivia.
Distribution: Bolivia.
The species was originally described from barren specimens. We believe that we have its flowers in the specimens collected by Mr. Bang. Plants were collected by Miguel Bang near La Paz, Bolivia, in 1890 (No. 176) and at the same locality by J. N. Rose, August 11, 1914 (No. 18844), also by Mr. Bang near Cochibamba, Bolivia, in 1901 (No. 2051). Dr. Rose's plant was without flower and its reference here is only tentative, but the habit of the plant was clearly that of the cespitose species of *Echinopsis*. Schumann, who studied Bang's plant, compared it with *Cereus pasacana* (see Cactaceae

2: 133. 1920) from which it is very distinct although the flowers resemble very much those of a *Trichocereus*.

27. Echinopsis mamillosa Gürke, Monatsschr. Kakteenk. **17**: 128. 1907.

Stem simple, depressed-globose, 6 cm. high, 8 cm. in diameter, shining dark green, tubercled and unarmed at the apex; ribs 17, divided by deep furrows into acute tubercles; areoles 8 to 12 mm. apart, irregularly orbicular; radial spines 8 to 10, subulate, 5 to 10 mm. long; central spines 1 to 4, somewhat stronger and longer than the radials, all yellowish, brown at the apex; flowers 15 to 18 cm. long, white, rose-colored towards the apex of the segments; flower-tube funnelform, somewhat curved, green, bearing small ovate scales, these hairy in their axils; outer perianth-segments linear, brownish, spreading; inner perianth-segments oblong, apiculate; stigma-lobes yellow, linear, about 10.

Type locality: Bolivia.
Distribution: Bolivia.

We have not seen this plant, but have a colored sketch of the type made by Mrs. Gürke, July 16, 1907. Through some error, the Kew Bulletin (Kew Bull. Misc. Inf. App. 87. 1908) describes the flower as only three-fourths of an inch in length.

Although this species is formally described on page 135 of the Monatsschrift für Kakteenkunde, it is technically described a month earlier (p. 128). In fact, the flowers are much better and more fully characterized here than in the formal description.

Illustration: Monatsschr. Kakteenk. **31**: 153.

28. Echinopsis formosa (Pfeiffer) Jacobi in Salm-Dyck, Cact. Hort. Dyck. 1849. 39. 1850.

> *Echinocactus formosus* Pfeiffer, Enum. Cact. 50. 1837.
> *Echinopsis formosa spinosior* Salm-Dyck in Labouret, Monogr. Cact. 303. 1853.
> *Echinopsis formosa laevior* Monville in Labouret, Monogr. Cact. 303. 1853.
> *Echinopsis formosa rubrispina* Monville in Labouret, Monogr. Cact. 303. 1853.

Simple, oblong, 3 dm. high, pale green; ribs 15 to 35, vertical; areoles 8 to 10 mm. apart; spines acicular, reddish, 2 to 4 cm. long; radial spines 8 to 16, yellowish; central spines 2 to 4, brown; flowers golden-yellow, 8 cm. long, 8 cm. broad.

Type locality: Mendoza, Argentina.
Distribution: Western Argentina.

We know the species only from descriptions and from some very poor illustrations. H. J. Elwes (Gard. Chron. III. **70**: 199. 1921) states that there is a specimen in the Darrah Collection at Manchester that is 2 feet high. It has added but one inch to its height in the last 10 years. The specimen has been in England for 60 years.

Cereus gilliesii Weber (Dict. Hort. Bois 471. 1896) was given as a synonym of *Echinopsis formosa*.

Melocactus gilliesii (Otto, Allg. Gartenz. **1**: 364. 1833) and *Echinocactus gilliesii* and *Echinopsis formosa gilliesii* (Salm-Dyck, Cact. Hort. Dyck. 1844. 22. 1845) are usually referred to *Echinopsis formosa*.

Echinocactus formosus crassispinus Monville (Labouret, Monogr. Cact. 303. 1853) was published as a synonym of *Echinopsis formosa spinosior* and therefore doubtless belongs here.

Echinopsis formosa albispina Weber is mentioned by Schelle (Handb. Kakteenk. 118. 1907).

Illustrations: Schelle, Handb. Kakteenk. f. 51; Monatsschr. Kakteenk. **4**: 187. f. 1; Knippel, Kakteen pl. 16.

UNCERTAIN OR UNDESCRIBED SPECIES.

ECHINOPSIS MIECKLEYI R. Meyer, Monatsschr. Kakteenk. **28**: 122. 1918.

Simple, ellipsoid to short-columnar, pale grayish green, 16 cm. high, 10 cm. in diameter; ribs 14, high, somewhat sinuous; radial spines usually 10, but sometimes 9 or 11, straight, 2.5 cm. long; central spine solitary, stouter than the radials, pale brown, sometimes whitish at tips, 5 cm. long; flowers and fruit unknown.

ECHINOPSIS GIGANTEA R. Meyer, Monatsschr. Kakteenk. **29**: 58. 1919.

Simple, ellipsoid to columnar, pale grayish green; ribs 8 to 11, high, broad at base, somewhat sinuous; radial spines 5 to 10; central spines sometimes 2, but usually solitary; flowers unknown.

ECHINOPSIS SALUCIANA Schlumberger, Rev. Hort. IV. **5**: 402. 1856.

"Tube 15 to 16 cm. high, green, and covered with scales bearing tufts of brown hairs; sepals very numerous, lanceolate, 9 cm. long and 8 mm. wide at the base, dirty white with a central green stripe; petals 2 cm. wide and 6 cm. long, pure white; stamens yellowish; style short; stigma not projecting from the tube and having 12 yellowish-white stigma-lobes. The flower lasts but one day.

"With its large narrow sepals (?) and wide petals, the flower resembles very much more the flower of a *Cereus* than that of an *Echinopsis*."

A free translation of the original description is given above.

ECHINOPSIS DUCIS PAULII Förster, Handb. Cact. ed. 2. 641. 1885.

Simple, columnar, 6 to 7 cm. in diameter; ribs 18 to 21; radial spines 6 to 8, 2 cm. long; central spines 2 to 4; flowers and fruit unknown.

It is known only as a cultivated plant.

ECHINOPSIS TACUAREMBENSE Arechavaleta, Anal. Mus. Nac. Montevideo **5**: 254. 1905.

Dull green, 10 cm. high, about 15 cm. in diameter; ribs 13, vertical; areoles 1 cm. apart; spines 9 or 10, 1 to 1.5 cm. long; central spine solitary; flowers white.

Type locality: Not cited.
Distribution: Uruguay.

ECHINOPSIS ALBISPINA (Monatsschr. Kakteenk. **13**: 144. 1903) is described as a white-spined, very interesting form.

ECHINOPSIS BECKMANNII (Monatsschr. Kakteenk. **3**: 103. 1893) and E. BOECKMANNII (Monatsschr. Kakteenk. **3**: 165. 1893) are only names and have never been referred to any described species.

ECHINOPSIS BOUTILLIERI Parmentier (Förster, Handb. Cact. ed. 2. 622. 1885) is only a name.

ECHINOPSIS DUVALLII (Monatsschr. Kakteenk **1**: 54. 1891) is from a seedling of unknown origin with pale rose flowers.

ECHINOPSIS FOBEANA (Monatsschr. Kakteenk. **20**: 190. 1910) is without description. A poor illustration is published by Möllers (Deutche Gärt. Zeit. **25**: 475. f. 7, No. 20).

ECHINOPSIS FORMOSISSIMA Labouret (Rev. Hort. IV. **4**: 26. 1855) probably does not belong to this genus. It originally came from Chuquisaca, Bolivia, although it is credited to Mexico by the Index Kewensis. Schumann refers it to *Cereus pasacana* Weber. Two illustrations of barren juvenile plants have been published (Möllers Deutsche Gärt. Zeit. **25**: 475. f. 7, No. 9 and Monatsschr. Kakteenk. **4**: 187. f. 2).

ECHINOPSIS LONGISPINA (Monatsschr. Kakteenk. **3**: 157. 1893) is only a name.

ECHINOPSIS MUELLERI (Monatsschr. Kakteenk. **6**: 144. 1896) is a well-known garden form, presumably a hybrid. It is described in some detail in the Cactus Journal and illustrated (**2**: 7).

ECHINOPSIS NIGRICANS Linke (Allg. Gartenz. **25**: 239. 1857) is said to come from Chile. If so, it probably does not belong to this genus.

ECHINOPSIS PARAGUAYENSIS Mundt (Monatsschr. Kakteenk. **13**: 109. 1903) is briefly mentioned.

ECHINOPSIS POLYPHYLLA (Monatsschr. Kakteenk. **19**: 144. 1909) is perhaps a hybrid.

ECHINOPSIS PYRANTHA (Monatsschr. Kakteenk. **4**: 97. 1894) has not been described but has been said to be the most beautiful species of the genus.

ECHINOPSIS QUEHLII (Monatsschr. Kakteenk. **1**: 55. 1891) is said to have been grown from Mexican seed. It is said to have beautiful pale rose flowers. If native to Mexico it must be referable to some other genus. Schelle lists it with the hybrids of E. eyriesii and E. oxygona.

ECHINOPSIS SALM-DYCKIANA (Monatsschr. Kakteenk. **20**: 142. 1910) is only a name.

ECHINOPSIS TOUGARDII L. Herincq (Hort. Franc. **3**: 193. pl. 17. 1853) is, according to Schuman, only a hybrid. It has very beautiful flowers.

ECHINOPSIS TUBERCULATA Niedt (Allg. Gartenz. **25**: 237. 1857), said to come from Bolivia, we do not know.

ECHINOPSIS UNDULATA (Monatsschr. Kakteenk. **11**: 61. 1901) is briefly described as yellow-flowered and is probably a hybrid.

The three following names, although sometimes referred to *Echinopsis*, were evidently intended as species of *Echinocactus*. In all of them the generic name is abbreviated to "*Ech.*" and, as they follow a species of *Echinopsis*. this abbreviation has naturally been taken to refer to that name. In each case, however, the species are referred to definite sections of *Echinocactus* as outlined by Salm-Dyck. Then, too, the gender of the specific name agrees with *Echinocactus* and not with *Echinopsis*. The three names are referred to *Echinopsis* by the Index Kewensis:

> *Ech. nodosus* Linke, Wochenschr. Gärtn. Pflanz. 1:85. 1858.
> *Ech. setosus* Linke, Wochenschr. Gärtn. Pflanz. 1:86. 1858.
> *Ech. haageanus* Linke, Wochenschr. Gärtn. Pflanz. 1:86. 1858.

The first two have been taken up formally in *Echinocactus* by Hemsley (Biol. Centr. Amer. Bot. 1: 535, 537. 1880) and the third by Rümpler (Förster, Handb. Cact. ed. 2. 469. 1885).

Echinopsis fischeri tephracantha and *E. nigerrima* are in the trade.

Subtribe 4. ECHINOCACTANAE.

Plants usually low and small, but sometimes several meters tall and then of considerable size, simple or cespitose, terrestrial; stems normally one-jointed but sometimes budding or cespitose or making other joints when injured; ribs few to many, straight or spiral, usually spine-bearing; flowers always solitary at areoles near the apex of the plant, usually from the nascent areoles; fruit more or less scaly or naked, usually dry, in some cases a little fleshy and then somewhat edible, usually dehiscing by a basal pore, but sometimes irregularly breaking apart, or by a circumscissile opening; seeds black or sometimes brown, smooth or papillose.

We recognize 28 genera, most of which are taken from *Echinocactus* as circumscribed by previous authors.

The subtribe passes into the *Echinocereanae* on the one hand and into the *Coryphanthanae* on the other.

In most of the genera of this subtribe, as well as in a few other genera, such as *Oreocereus*, the seeds escape through a pore at the base of the fruit. If the fruit be gathered before it fully ripens, this pore will not be shown, but, as the fruit ripens, the basal part which is attached to the plant becomes absorbed and disappears and, when the fruit finally falls off, the large opening, sometimes 5 to 7 cm. in diameter, can be seen. In most of the genera the fruit becomes hollow and the seeds are attached on the inner surface until fully ripe, when they fall to the bottom and make their escape. In *Homalocephala texensis* the fruit bursts irregularly, while in *Mila* and a few other genera the fruit is a small juicy berry.

KEY TO GENERA.

```
A. Flower-tube bent; stamens long-exserted.................................  1. Denmoza (p. 78)
AA. Flower-tube straight, usually with a broad throat; stamens included.
    B. Ovary and fruit naked.  (See Copiapoa.)
       Plants spineless, except seedlings.
          Tubercles prominent, cartilaginous, flattened, more or less imbricated .  2. Ariocarpus (p. 80)
          Tubercles low, rounded above.................................  3. Lophophora (p. 83)
       Plants very spiny.
          Fruit crowned by sepal-like scales.............................  4. Copiapoa (85)
          Fruit naked at top.
             Fruit dry.
                Spines acicular .......................................  5. Pediocactus (p. 90)
                Spines flat, papery....................................  6. Toumeya (p. 91)
             Fruit fleshy, indehiscent..................................  7. Epithelantha (p. 92)
    BB. Ovary and fruit scaly.
       C. Flowers funnelform, often with a slender tube.
          Axils of flower-scales hairy or bristly.
             Axils of flower-scales with bristles and hairs ....................  8. Neoporteria (p. 94)
             Axils of flower-scales with hairs only.
                Flowers long-funnelform.................................  9. Arequipa (p. 100)
                Flowers short-funnelform ...............................  10. Oroya (p. 102)
```

KEY TO GENERA—continued.

1. DENMOZA gen. nov.

Plant cylindric, often elongated, the numerous, parallel straight ribs slightly undulate; spines in clusters at the areoles; flowers arising from the top of the plant, zygomorphic, scarlet, with a slender throat and very narrow limb; tube proper very short, its mouth closed with a mass of white wool; inner surface of the elongated throat covered with stamens; filaments and style long-exserted; ovary and tube bearing numerous scales, their axils filled with silky hairs; fruit globular, dry, splitting down from the top; seeds black, dull, pitted.

Type species: *Echinocactus rhodacanthus* Salm-Dyck.

The generic name is an anagram of Mendoza, the province in Argentina, where the plant is native. Only one species is known.

The peculiar mass of white wool near the base of the flower-tube on the inside is not known, as far as our observation goes, in any other cactus, except in two species of *Lobivia* of doubtful relationship, described by Dr. Spegazzini as species of *Echinocactus*, and in *E. spiniflorus*, all of which are otherwise quite different from *Denmoza*.

The genus here segregated was considered by Schumann as a species of *Echinopsis* but it has also been referred to *Cereus*, *Echinocactus*, *Cleistocactus*, and *Pilocereus*. In its

M. E. Eaton del.

A. Hoen & Co.

1. Top of flowering plant of *Pediocactus simpsonii*.
2. Flower of *Denmoza rhodacantha*.
3. Flowering plant of *Ariocarpus kotschoubeyanus*.
4. Top of flowering plant of *Neoporteria subgibbosa*.
(All natural size.)

rather narrow flowers and exserted stamens there are suggestions of *Cleistocactus*, but the plant body is very different. It is more like some species of *Echinopsis*, to which, however, its flowers show little resemblance. It has no close relationship to *Cereus* or *Cephalocereus*.

Denmoza differs from all other genera in this subtribe in producing long bristle-like spines from the flowering areoles of very old plants.

1. Denmoza rhodacantha (Salm-Dyck).

Echinocactus rhodacanthus Salm-Dyck, Hort. Dyck. 341. 1834.
Echinopsis rhodacantha Salm-Dyck, Cact. Hort. Dyck. 1849. 39. 1850.
Cleistocactus rhodacanthus Lemaire, Illustr. Hort. **8**: Misc. 35. 1861.
Pilocereus erythrocephalus Schumann, Gesamth. Kakteen 195. 1897.
Cereus erythrocephalus Berger, Rep. Mo. Bot. Gard. **16**: 69. 1905.
Pilocereus rhodacanthus Spegazzini, Anal. Mus. Nac. Buenos Aires III. **4**: 485. 1905.

Simple, at first globular, often 3 to 6 dm. long, but becoming elongated, and when of great age 1.5 meters high and 3 dm. in diameter; ribs 15 to 20 or even 30, broad at base, separated by narrow intervals, about 1 cm. high; young areoles felted, circular, when old 8 to 10 mm. in diameter, usually 1 to 2 cm. apart, but on very old plants approximate, perhaps confluent; spines very different on young and very old plants; spines on small plants 6 to 12 at
each areole, white or reddish, subulate, more or less curved, 4 cm. long or less; central spine, if present, solitary; spines on the top of old plants slender and longer, 7 cm. long, often accompanied by a row of 10 or more long brown bristles; flowers slender, 4 to 5 cm. long; ovary and flower-tube bearing small, triangular to lanceolate, appressed, acute scales with long white hairs in their axils; perianth-segments small, apparently connivent; filaments red, exserted for at least 1 cm. beyond the tube; style red, exserted; wool at base of throat matted, 6 to 8 mm. long; fruit 2 cm. in diameter, nearly smooth in age; seeds oblique, 1.5 mm. in diameter.

Fig. 93.—Denmoza rhodacantha.

Type locality: Not cited at place of publication, but doubtless Mendoza, Argentina.

Distribution: Western mountains of Argentina.

Vaupel (Monatsschr. Kakteenk. **31**: 13. 1921) gives an interesting description of the species which flowered in Berlin in 1920 where it is grown as *Cereus erythrocephalus*.

Pfeiffer gives *Echinocactus coccineus* (Enum. Cact. 50. 1837) as a synonym, while Weber gives *Cereus rhodacanthus* Weber (Dict. Hort. Bois 472. 1896) as a synonym, but neither is described. It is not at all unlikely that *Mammillaria coccinea* G. Don (Loudon, Hort. Brit. 194. 1830; *Cactus coccineus* Gillies), said to have come from Chile, is also to be referred here.

Schumann refers here *Echinopsis aurata* Salm-Dyck and *Echinocactus dumesnilianus* Cels but these references are very doubtful; they probably belong to *Eriosyce ceratistes*.

The following varieties have been referred to this species: *Echinocactus rhodacanthus coccineus* Monville (Labouret, Monogr. Cact. 304. 1853), *Echinopsis rhodacantha aurea* (Monatsschr. Kakteenk. **17**: 76. 1907) and *Echinopsis rhodacantha gracilior* Labouret (Monogr. Cact. 304. 1853).

Illustrations: Blühende Kakteen **1**: pl. 16; Monatsschr. Kakteenk. **4**: 187. f. 3; Möllers Deutsche Gärt. Zeit. **25**: 481. f. 12, as *Echinopsis rhodacantha*.

Plate VIII, figure 2, shows a flower of a plant brought by Dr. Rose to the New York Botanical Garden in 1915, which bloomed in 1917. Figure 93 is from a photograph taken by Paul G. Russell at Mendoza, Argentina, in 1915.

2. ARIOCARPUS Scheidweiler, Bull. Acad. Sci. Brux. 5: 491. 1838.

Anhalonium Lemaire, Cact. Gen. Nov. Sp. 1. 1839.

Plants spineless,* usually simple, low, with a flat or round top; tubercles tough, horny, or cartilaginous, triangular, imbricated, spirally arranged, the lower part tapering into a claw, the upper or blade-like part expanded; areoles terminal or at the bottom of a triangular groove near the middle of tubercle, filled with hair when young; flowers appearing from near the center on young tubercles, diurnal, rotate-campanulate, white to purple; fruit oblong, smooth; seeds black, tuberculately roughened, with a large basal hilum; embryo described as obovate, straight.

Type species: *Ariocarpus retusus* Scheidweiler.

This genus long passed under the name of *Anhalonium*, but it was found that *Ariocarpus* had priority and hence was taken up. Karwinsky proposed the name *Stromatocactus* for one of the species, but no description of it was ever published. The genus is usually considered as most closely related to *Mammillaria*, under which genus two of the species have been placed. Engelmann, who was greatly puzzled over the group, first considered it the same as *Mammillaria*, then as a subgenus of *Mammillaria*, and later as a distinct genus.

In its small, oblong, naked fruit and straight embryo, it suggests a *Mammillaria*, but in its tubercles, areoles, seeds, and absence of spines, it is very unlike any of the species of that genus.

The generic name is from the genus *Aria* and καρπός fruit, referring to the Aria-like fruit. We recognize three species, natives of southern Texas and northern Mexico.

KEY TO SPECIES.

Tubercles not grooved on upper side...1. *A. retusus*
Tubercles grooved on upper side.
    Plants small, 3 to 5 cm. broad...2. *A. kotschoubeyanus*
    Plants large, 10 to 15 cm. broad..3. *A. fissuratus*

1. Ariocarpus retusus Scheidweiler, Bull. Acad. Sci. Brux. 5: 492. 1838.

Anhalonium prismaticum Lemaire, Cact. Gen. Nov. Sp. 1. 1839.
Anhalonium retusum Salm-Dyck, Cact. Hort. Dyck. 1844. 15. 1845.
Anhalonium elongatum Salm-Dyck, Cact. Hort. Dyck. 1849. 77. 1850.
Anhalonium areolosum Lemaire, Illustr. Hort. 6: Misc. 35. 1859.
Anhalonium pulvilligerum Lemaire, Illustr. Hort. 16: Misc. 72. 1869.†
Mammillaria areolosa Hemsley, Biol. Centr. Amer. Bot. 1: 503. 1880.
Mammillaria elongata Hemsley, Biol. Centr. Amer. Bot. 1: 509. 1880. Not De Candolle, 1828.
Mammillaria prismatica Hemsley, Biol. Centr. Amer. Bot. 1: 519. 1880.
Mammillaria furfuracea‡ S. Watson, Proc. Amer. Acad. 25: 150. 1890.
Cactus prismaticus Kuntze, Rev. Gen. Pl. 1: 261. 1891.
Anhalonium trigonum Weber, Dict. Hort. Bois 90. 1893.
Anhalonium furfuraceum Coulter, Contr. U. S. Nat. Herb. 3: 130. 1894.
Ariocarpus pulvilligerus Schumann, Bot. Jahrb. Engler 24: 550. 1898.
Ariocarpus furfuraceus Thompson, Rep. Mo. Bot. Gard. 9: 130. 1898.
Ariocarpus trigonus Schumann, Gesamtb. Kakteen 666. 1898.
Ariocarpus prismaticus Cobbold, Journ. Hort. Home Farm. III. 46: 332. 1903.

Plants globular or more or less depressed, usually 10 to 12 cm. broad, grayish green to purplish, very woolly at the center; tubercles horny, imbricated, 5 cm. long or less, ovate, more or less 3-angled, acute to acuminate, often with a woolly areole on the upper side near the tip and this sometimes spinescent; flowers borne at the axils of young tubercles near the center, white or nearly so, up to 6 cm. long; outer perianth-segments pinkish, narrow, acute to acuminate; inner perianth-segments at first white, afterwards pinkish, narrowly oblanceolate, with a mucronate tip; stamens numerous, erect; style white; stigma-lobes 9, linear, white; fruit oblong, white, naked; seeds globular, 1.5 mm. in diameter, black, tuberculate-roughened.

---

*Sometimes in *Ariocarpus retusus* small spines are produced in the areoles near the tip of the tubercles.
†Lemaire gives for this species a reference (Herb. Génér. Amat. Nouvel Sér. Misc. 45) which we have not been able to locate. Coulter (Contr. U. S. Nat. Herb. 3: 130. 1894) refers this name to Lemaire "Cact. 1839." the Index Kewensis to "Hort. Monv. 1: 275," and Labouret to "Hort. Univ. 1: 275, figure," but we have not been able to confirm them. If this name were published in 1839, it would transfer the publication of *Anhalonium elongatum* Salm-Dyck back to 1845 (Cact. Hort. Dyck. 1844. 15).
‡Reported in the Index Kewensis (Suppl. 1. 263. 1906) as *Mammillaria purpuracea.*

Type locality: San Luis Potosí, Mexico.

Distribution: States of Coahuila, Zacatecas, and San Luis Potosí, Mexico.

This species, as here described, is extremely variable in the shape, size, color, and markings of the tubercles, and in the presence or absence of woolly areoles near the tips of the tubercles. Several species have been described from these various forms, but there seems to be no good ground for such a course. The plant is called chaute by the Mexicans.

The plant usually grows in the open in rocky places where it is nearly covered with broken stones and only its long tubercles are visible.

Mammillaria retusa Mittler (Handb. Liebh. 11) is referred here by Schumann, but we have not seen this reference.

Mammillaria aloides Monville (Cat. 1846) is referred by Labouret as a synonym of *Anhalonium prismaticum,* by Schumann as a synonym of *Ariocarpus retusus,* and by the Index Kewensis as a synonym of *Mammillaria prismatica. Anhalonium aloides pulvilligerum* Monville we know only from Lemaire (Illustr. Hort. 16: Misc. 72. 1869) who gives it

FIG. 94.—Ariocarpus retusus.   FIG. 95.—Ariocarpus fissuratus.

as a synonym of *A. pulvilligerum. Mammillaria pulvilligera* Monville (Förster, Handb. Cact. ed. 2. 231. 1885) is given by Rümpler as a synonym of *Anhalonium elongatum. Mammillaria aloidaea pulviligera* which appeared in Monville's Catalogue of 1846 is referred by Labouret to *Anhalonium elongatum.* To *Mammillaria trigona* is referred *Ariocarpus trigonus* by the Index Kewensis (Suppl. 2. 16. 1904).

Illustrations: Möllers Deutsche Gärt. Zeit. **29:** 76. f. 5, 6; 77. f. 7, 8, as *Ariocarpus trigonus;* Gartenwelt **15:** 538, as *Anhalonium trigonum;* Gartenwelt **15:** 538; Cact. Journ. **2:** 109; Hort. Univ. **1:** pl. 30; Balt. Cact. Journ. **2:** 266. f. 1; Herb. Génér. Amat. II. **2:** pl. 16; Arch. Exper. Path. **34:** pl. 1, f. 2; Journ. Amer. Chem. Soc. **18:** f. 5; Palmer, Cult. Cact. 123; Garten-Zeitung **4:** 541. f. 126; 182. f. 42, No. 16, as *Anhalonium prismaticum;* Curtis's Bot. Mag. **119:** pl. 7279, as *Mammillaria prismatica;* Cact. Journ. **1:** pl. for November; Rep. Mo. Bot. Gard. **9:** pl. 34; Ann. Rep. Smiths. Inst. **1908:** pl. 15, f. 2, as *Ariocarpus furfuraceus;* Möllers Deutsche Gärt. Zeit. **25:** 477. f. 11, No. 7; Blühende Kakteen **1:** pl. 48; Ann. Rep. Smiths. Inst. **1908:** pl. 15, f. 1; Bull. Acad. Sci. Brux. **5:** pl. 1; Rep. Mo. Bot. Gard. **9:** pl. 35; Cact. Journ. **1:** pl. for September and November; Schelle, Handb. Kakteenk. f. 199; Hort. Belge **5:** pl. 21, 22.

Plate IX, figure 2, is from a photograph of a plant sent Dr. Edward Palmer from San Luis Potosí, Mexico, in 1905, which afterwards flowered in Washington. Figure 94 is from a photograph of a plant sent by Professor Lloyd from Zacatecas in 1908.

2. Ariocarpus kotschoubeyanus (Lemaire) Schumann in Engler and Prantl, Pflanzenfam. Nachtr. 259. 1897.

Anhalonium kotschoubeyanum Lemaire, Bull. Cercle Confér. Hort. Dép. Seine. 1842.
Anhalonium sulcatum Salm-Dyck, Cact. Hort. Dyck. 1849. 5. 1850.
Cactus kotschubeyi* Kuntze, Rev. Gen. Pl. 1: 260. 1891.
Ariocarpus sulcatus Schumann, Monatsschr. Kakteenk. 7: 9. 1897.

Plants grayish green, 3 to 5 cm. broad, only the flat crown appearing above the surface of the ground, with a thickened, fleshy rootstock, and with several spindle-shaped roots from the base; upper part of tubercle flattened, triangular, 6 to 8 mm. long, grooved along its middle, almost to the tip, the groove very woolly; flowers 2.5 to 3 cm. long, originating in the center of the plant from the axils of the young tubercles, surrounded by a cluster of hairs; outer perianth-segments few, brownish, obtuse; inner perianth-segments up to 2 cm. long, oblanceolate, obtuse or apiculate, sometimes retuse, rose-colored to light purple, widely spreading; filaments, style, and stigma-lobes white; ovary naked; seeds oblong, 1 mm. long.

Type locality: Mexico.

Distribution: Central Mexico.

This species was collected in Mexico and sent to Europe by Karwinsky about 1840. Only three specimens were sent in the first shipment, one of which sold for $200. As a medium-sized plant weighs less than half an ounce, this price was somewhat in excess of its weight in gold! This plant was named for Prince Kotschoubey who was a prominent patron of horticulture. He paid a thousand francs for one of these plants.

We have not seen Lemaire's original reference to *Anhalonium kotschoubeyanum*, but in all his subsequent references the name is spelled as given here. Schumann, however, spells the name as follows: *Ariocarpus kotschubeyanus*.

Stromatocactus kotschoubeyi Karwinsky and *Anhalonium fissipedum* Monville were given by Lemaire (Illustr. Hort. 16: Misc. 72. 1869) as synonyms of *A. kotschoubeyanum* and by Rümpler (Förster, Handb. Cact. ed. 2. 232. 1885) as synonyms of *A. sulcatum*. *Ariocarpus mcdowellii* (Haage and Schmidt, Cat. 225. 1908), unpublished, belongs here. Dr. Rose obtained living specimens from McDowell in 1906.

Cactus kotschoubeyi Karwinsky (Hort. Univ. 6: 63. 1845) was recorded by Lemaire while the Index Kewensis refers the name to Otto Kuntze (Rev. Gen. Pl. 1: 206. 1891), where the transfer is technically made.

FIG. 96.—Ariocarpus kotschoubeyanus.

The plant, as *Mammillaria sulcata*, is described in the Gardeners' Chronicle (III. 30: 255. 1901) but no author is given and the article is unsigned. The name also occurs in the Index Kewensis (3: 160. 1894), credited to Salm-Dyck (Cact. Hort. Dyck. 1849. 78. 1850), but he never used this combination. The reference of Salm-Dyck which is cited is to *Anhalonium sulcatum*.

Illustrations: Gartenwelt 15: 538, as *Anhalonium kotschubeyanum;* Gard. Chron. III. 30: 255. f. 74, as *Mammillaria sulcata;* Monatsschr. Kakteenk. 7: 10; Cact. Journ. 1: 44, as *Ariocarpus sulcatus;* Bot. Jahrb. Engler 24: 544; Cact. Journ. 1: pl. for January and September; Schumann, Gesamtb. Kakteen f. 96; Ann. Rep. Smiths. Inst. 1908: pl. 3, f. 4; Journ. Hered. Washington 6[7]: f. 5; Monatsschr. Kakteenk. 10: 184; Rep. Mo. Bot. Gard. 9: pl. 33; Blühende Kakteen 1: pl. 52 a; Möllers Deutsche Gärt. Zeit. 25: 477. f. 11, No. 8; 29: 75. f. 4; Gartenwelt 15: 217.

---

*Sometimes spelled *kotschubei.*

Plate VIII, figure 3, shows a plant sent by Professor Lloyd from Zacatecas in 1908, which flowered at the New York Botanical Garden in 1911. Figure 96 is from a photograph of a plant sent by Dr. Elswood Chaffey from Zacatecas, Mexico, in 1910.

3. **Ariocarpus fissuratus** (Engelmann) Schumann in Engler and Prantl, Pflanzenfam. 3⁶ᵃ: 195. 1894.

> Mammillaria fissurata Engelmann, Proc. Amer. Acad. 3: 270. 1856.
> Anhalonium fissuratum Engelmann, Cact. Mex. Bound. 75. 1859.
> Anhalonium engelmannii Lemaire, Cactées 42. 1868.
> Ariocarpus lloydii Rose, Contr. U. S. Nat. Herb. 13: 308. 1911.

Plant body scarcely appearing above the ground, flat or somewhat rounded, sometimes 15 cm. broad; tubercles imbricated, ovate, the upper part 2 to 3 cm. broad at base, acute or obtuse, the whole surface more or less fissured and irregularly warty; areoles filled with a dense mass of hairs; flowers 3 to 4 cm. broad, white to purple; inner perianth-segments oblong-oblanceolate; style and stigma-lobes white; fruit oval, pale green, 10 mm. long; seeds black, tuberculate-roughened.

Type locality: Near the junction of the Pecos with the Rio Grande.

Distribution: Western Texas and northern Coahuila and Zacatecas, Mexico.

Engelmann refers here (Cact. Mex. Bound. 74) *Mammillaria heteromorpha* Scheer (*Anhalonium heteromorphum* Trelease in Engelmann's Botanical Works 537. 1887), basing his conclusions on a skeleton specimen so labeled in Salm-Dyck's collection. The species described under that name by Salm-Dyck (Cact. Hort. Dyck. 1849. 128. 1850) is certainly not of this genus.

This plant is generally known as living rock. It is dull gray to brown in color and grows on dry stony ground and, when not in flower, is easily mistaken for the rocks which surround it.

Illustrations: Cact. Mex. Bound. pl. 16; Dict. Gard. Nicholson 4: 563. f. 34, as *Mammillaria fissurata;* Illustr. Hort. 16: pl. [605a]; Förster, Handb. Cact. ed. 2. f. 20, as *Anhalonium engelmannii;* Cact. Journ. 2: 109; Gartenwelt 15: 538; Amer. Gard. 11: 465; Dict. Gard. Nicholson Suppl. 51. f. 48; Journ. Amer. Chem. Soc. 18: f. 4; Arch. Exper. Path. 34: 70. f. 1; 376; Goebel, Pflanz. Schild. 1: f. 14, 44; pl. 2, f. 7, as *Anhalonium fissuratum;* Contr. U. S. Nat. Herb. 13: pl. 63; Stand. Cycl. Hort. Bailey 1: f. 373; Möllers Deutsche Gärt. Zeit. 29: 73. f. 1, as *Ariocarpus lloydii;* Contr. U. S. Nat. Herb. 13: pl. 62; Rep. Mo. Bot. Gard. 9: pl. 32; Blühende Kakteen 1: pl. 52, b; Bull. Univ. Texas 82: pl. 4, f. 1; Engler and Prantl, Pflanzenfam. 3⁶ᵃ: f. 68; Cact. Journ. 1: pl. for January and September; Ann. Rep. Smiths. Inst. 1908: pl. 5, f. 1; Schelle, Handb. Kakteenk. f. 200; Möllers Deutsche Gärt. Zeit. 25: 477. f. 11, No. 6; 29: 74. f. 2, 3; Gartenwelt 15: 343; Alianza Cientifica Universal 3: opp. 150 (2 plates); Arch. Exper. Path. 34: 376; West Amer. Sci. 13: 2; Floralia 42: 369.

Figure 95 is from a photograph of a plant collected by Dr. Rose at Langtry, Texas, in 1908.

3. **LOPHOPHORA** Coulter, Contr. U. S. Nat. Herb. 3: 131. 1894.

Plant small, simple or proliferous, spineless (seedlings having a few weak pubescent bristle-like spines), very succulent; ribs very broad and rounded, bearing few low tubercles; areoles round, bearing flowers only when young, always filled with a tuft of erect matted hairs; flowers borne at the center of the plant, small, rotate-campanulate, white to rose-tinted; fruit club-shaped, naked, red to pinkish, maturing rapidly; seeds black, tuberculate-roughened.

Type species: *Echinocactus williamsii* Lemaire.

One species is here recognized but some writers have accepted two.

The generic name is from λόφος crest, and φορέω I bear, referring to the pencil of hairs borne at the areole.

This very curious little plant, although referred in turn to *Echinocactus*, *Mammillaria*, and *Anhalonium*, has very little in common with any of those genera. In the origin of the flower it is like *Echinocactus*, but otherwise it is very different. In its globular habit and the shape and size of the flowers it resembles many of the plants heretofore passing as *Mammillaria*, but it has very different seeds, flowers, areoles, and structure. In its fruits, seeds, and flowers it approaches *Ariocarpus*, but in other respects it is very different.

1. Lophophora williamsii (Lemaire) Coulter, Contr. U. S. Nat. Herb. **3**: 131. 1894.

> *Echinocactus williamsii* Lemaire in Salm-Dyck, Allg. Gartenz. **13**: 385. 1845.
> *Anhalonium williamsii* Lemaire in Förster, Handb. Cact. ed. 2. 233. 1885.
> *Anhalonium lewinii* Hennings, Gartenflora **37**: 410. 1888.
> *Mammillaria williamsii* Coulter, Contr. U. S. Nat. Herb. **2**: 129. 1891.
> *Lophophora williamsii lewinii* Coulter, Contr. Nat. Herb. **3**: 131. 1894.
> *Echinocactus lewinii* Hennings, Monatsschr. Kakteenk. **5**: 94. 1895.
> *Mammillaria lewinii* Karsten, Deutsche Fl. ed. 2. **2**: 457. 1895.
> *Lophophora lewinii* Thompson, Rep. Mo. Bot. Gard. **9**: 133. 1898.

Plants dull bluish green, globular to top-shaped or somewhat flattened at top, 5 to 8 cm. broad, with a thickened tap-root sometimes 10 cm. long or more; ribs 7 to 13, nearly vertical or irregular and indistinct, tubercled; flowers central, each surrounded by a mass of long hair, pale pink to white, 2.5 cm. broad when fully open, with a broad funnelform tube; outer perianth-segments and scales green on the back, callous-tipped; filaments much shorter than the perianth-segments, nearly white; style white below, pinkish above, shorter than the perianth-segments; stigma-lobes 5, linear, pinkish; ovary naked; fruit 2 cm. long or less; seeds 1 cm. in diameter, with a broad basal hilum.

Type locality: Not cited.

Distribution: Central Mexico to southern Texas.

This plant contains a narcotic and has been the subject of much study regarding its chemical, medicinal, and therapeutic properties. Dr. L. Lewin isolated an alkaloid which he named anhalonin. Since then one or more other alkaloids have been discovered. The active drug contained in this plant, however, it is claimed, does not lie in the alkaloids but in certain resinous bodies discovered by Dr. Erwin E. Ewell. The dried plants have been used since pre-Columbian times by certain North American Indians in some of their religious ceremonies and dances. The physiological effects which follow the eating of the dried plants are remarkable visions, and these

Fig. 97.—Lophophora williamsii.

have been described in considerable detail by writers who have visited the Indians and who have recorded laboratory experiences. There is considerable commerce carried on in this plant by some of the Indian tribes, although it is forbidden by law. The globular plants are sliced into 3 or 4 sections and then dried in the sun and these dried pieces form the mescal buttons of the trade.

According to Safford (Journ. Hered. Washington **8**: f. 5, 6, 7. 1916), Bernardo Sahagun in the sixteenth century spoke of its use by the Indians of Mexico; Sahagun, however, supposed the plant was a fungus, and called it teonanactl or "sacred mushroom."

This species is known variously as pellote, peyote, mescal button, devil's root, or sacred mushroom; it is sometimes also called the dumpling cactus and, according to Mr. Robert Runyon, challote in Starr County, Texas.

The name *Ariocarpus williamsii* Voss (Vilm. Blumengärtn. 368), according to the Monatsschrift für Kakteenkunde (**7**: 32. 1907), has been used, but whether it was formally published we do not know.

Anhalonium rungei Hildmann and *A. subnodusum* Hildmann (Monatsschr. Kakteenk. **3**: 68. 1893) are only names, but doubtless belong here; *A. visnagra* (Monatsschr. Kakteenk. **6**: 174. 1896) should perhaps also be referred here.

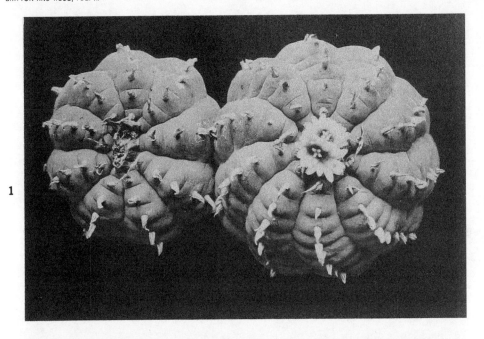

1. *Lophophora williamsii.*
2. *Ariocarpus retusus.*

Here probably belong *Anhalonium jourdanianum* Lewin (Ber. Deutsch. Bot. Gess. 12: 289. 1894), *Anhalonium jourdanianum* (Monatsschr. Kakteenk. 6: 180. 1896), and *Echinocactus jourdanianus* Rebut (Monatsschr. Kakteenk. 15: 122. 1905).

Illustrations: Journ. Hered. Washington 6[7]: f. 10, as *Lophophora;* De Laet, Cat. Gen. f. 13, as *Echinocactus williamsii lewinii;* Cact. Journ. 1: pl. for September, December; Journ. Hered. Washington 6[7]: f. 9; Rep. Mo. Bot. Gard. 9: pl. 37, as *Lophophora lewinii;* Journ. Amer. Chem. Soc. 18: f. 2, 7; Arch. Exper. Path. 34: pl. 1, f. 4; also 376. f. 1; Gartenflora 37: f. 92; Gartenwelt 15: 538; Journ. Hered. Washington 6[7]: f. 8; Monatsschr. Kakteenk. 1: facing 93, as *Anhalonium lewinii;* Arch. Exper. Path. 34: pl. 1, f. 3; Rümpler, Sukkulenten 190. f. 107; Gartenwelt 15: 538; Gartenflora 37: f. 93; Möllers Deutsche Gärt. Zeit. 29: 78. f. 9; Journ. Amer. Chem. Soc. 18: f. 1, 3; Cact. Journ. 2: 109, as *Anhalonium williamsii;* Curtis's Bot. Mag. 73: pl. 4296; Monatsschr. Kakteenk. 4:37;13:52; Pfeiffer, Abbild. Beschr. Cact. 2: pl. 21; Schumann, Gesamtb. Kakteen f. 55; Blühende Kakteen 3: pl. 149; Loudon, Encycl. Pl. ed. 3. 1377. f. 1937[2]; Schelle, Handb. Kakteenk. 150. f. 77; Möllers Deutsche Gärt. Zeit. 25: 477. f. 11, No. 23; De Laet, Cat. Gén. f. 12, as *Echinocactus williamsii;* Gartenwelt 15: 538, as *Anhalonium jourdanianum;* Smiths. Misc. Coll. 70: f. 111; Rep. Mo. Bot. Gard. 9: pl. 36; Journ. Hered. Washington 6[7]: f. 1 to 3, 6, 7, 9; Saunders, Useful Wild Plants U. S. Canada 253; Cact. Journ. 1: pl. for September, December; Ann. Rep. Smiths. Inst. 1908: pl. 3, f. 5; 1916: 424. pl. 5, 6, 7.

Plate IX, figure 1, is from a photograph of a plant sent from Zacatecas, Mexico, by Dr. Elswood Chaffey, in 1910; plate X, figure 3, shows a flowering plant from Zacatecas in the collection of the New York Botanical Garden; figure 4 shows another plant received from France in 1901. Figure 97 is from a photograph taken by Robert Runyon near Brownsville, Texas, in 1921.

4. COPIAPOA gen. nov.

Simple, globular to elongate-cylindric, or in one species forming large clumps or mounds containing hundreds of simple globular stems; areoles borne on definite ribs; top of plant covered with dense soft wool; flowers from the top of the plant, nearly hidden in the wool, campanulate to funnelform, yellow or sometimes tinged with red, with very short but broad tube; ovary short, turbinate, naked; fruit small, smooth, crowned with green, persistent, sepal-like scales; seed large, glossy, black, with large depressed hilum.

Type species: *Echinocactus marginatus* Salm-Dyck.

To this genus we are able to refer some 14 described species heretofore included in *Echinocactus* by authors. Most of the species are to be found in Salm-Dyck's section, *Cephaloidei,* while Schumann scatters them through his subgenus *Cephalocactus* which is a very unnatural group, containing 11 very diverse species. All the species of *Copiapoa* are from the coastal region of northern Chile. This region, although large and varied, does not possess this number of species. We recognize 6. Dr. Rose who collected here in 1914 obtained 3 of these of which he brought back living and herbarium specimens.

The generic name is derived from Copiapo, one of the provinces of Chile.

KEY TO SPECIES.

1. Copiapoa cinerea (Philippi).

Echinocactus cinereus Philippi, Fl. Atac. 23. 1860.

Simple, cylindric, 20 cm. high, 10 cm. in diameter, covered with wool at the apex; ribs 18, broad, obtuse; spine solitary or sometimes 5 or 6, terete, black; upper radials 4 mm. long; lower radials 12 to 16 mm. long; central spine 18 to 20 mm. long; flowers funnelform, 18 to 30 mm. long, 2.5 cm. broad, yellow; ovary naked; fruit 1.5 to 2 cm. long; seeds black and shining.

Type locality: Along the coast of Chile from Taltal to Cobre.

Distribution: Western Chile.

This species is similar to *Copiapoa marginata* but has more ribs and very different armament.

FIG. 98.—Copiapoa cinerea.   FIG. 99.—Copiapoa marginata.

We have seen no living specimens of this species, but Dr. Rose obtained a small piece of the type from the Philippi Herbarium. This agrees very well with Schumann's illustration cited below.

Illustrations: Schumann, Gesamtb. Kakteen Nachtr. f. 15; Monatsschr. Kakteenk. 11: 7, as *Echinocactus cinereus.*

Figure 98 is copied from the first illustration above cited.

2. Copiapoa marginata (Salm-Dyck).

Echinocactus marginatus Salm-Dyck, Allg. Gartenz. 13: 386. 1845.
Echinocactus columnaris Pfeiffer, Abbild. Beschr. Cact. 2: under pl. 14. 1847.
Echinocactus streptocaulon Hooker in Curtis's Bot. Mag. 77: pl. 4562. 1851.
Echinocactus melanochnus Cels in Labouret, Monogr. Cact. 174. 1853.

M. E. Eaton del.

1. Flowering plant of *Echinopsis aurea*.
2. Flowering plant of *Copiapoa coquimbana*.
3. Flowering plant of *Lophophora williamsii*.
4. Flowering plant of same.
   (All natural size.)

Plants subcylindric, growing in clusters of 2 to 9, usually erect, but when old often 6 dm. long and spreading with ascending tips, about 12 cm. in diameter; ribs 8 to 12, low, separated by broad intervals; young areoles and tops of flowering plants filled with masses of soft brown hair; areoles large, approximate, the adjoining ones usually touching; spines 5 to 10, unequal, subulate, stout, the longer one 3 cm. long; flowers small, 2.5 cm. long; outer perianth-segments broad, obtuse, with red tips; inner perianth-segments yellow; stamens included; fruit naked, small, 8 mm. long; seeds black, shining.

Type locality: Chile.

Distribution: Coastal hills of Antofagasta, Chile.

The four species, referred above as synonyms of this one, were described between 1845 and 1853 and may have come from the same source. Two of them are said to have been from Bolivia, but at the time they were described, Antofagasta, now a part of Chile, belonged to Bolivia. Dr. Rose, when collecting in Chile in 1914 (No. 19410), found these plants very common on the dry hills above Antofagasta, and a number of fine specimens were sent to the New York Botanical Garden.

We are following Pfeiffer in referring *E. columnaris* to this species. According to Pfeiffer, both species came from Valparaiso, Chile, but Dr. Rose could find no plant of this relationship about Valparaiso. Mr. Söhrens, whom he consulted, believes that Pfeiffer's station was wrongly recorded.

Illustrations: Pfeiffer, Abbild. Beschr. Cact. **2:** pl. 30, as *Echinocactus marginatus;* Curtis's Bot. Mag. **77:** pl. 4562; Loudon. Encycl. Pl. ed. 3. 1378. f. 19376, as *Echinocactus streptocaulon.*

Figure 99 is copied from the second illustration above cited.

3. Copiapoa coquimbana (Karwinsky).

Echinocactus coquimbanus Karwinsky in Förster, Handb. Cact. ed. 2. 601. 1885.

Plants clustered, forming mounds up to 1 meter broad and 6 dm. high, composed of several hundred heads; individual heads 12 cm. in diameter or less, pale green, at flowering time crowned by a dense mass of long white wool; ribs 10 to 17, obtuse, somewhat tubercled; radial spines 8 to 10, slender, straight or somewhat recurved; central spines 1 or 2, stouter, straight, 1.5 to 2.5 cm. long, black to gray; flowers campanulate, 3 cm. long; outer perianth-segments distinct, linear, acute, green; inner perianth-segments oblanceolate, yellow, obtuse; tube nearly or quite wanting; filaments, style, and stigma-lobes yellow; ovary small, turbinate, naked.

Type locality: Near the town of Coquimbo, Chile.

Distribution: Province of Coquimbo, Chile.

The Philippi Herbarium at Santiago de Chile has a specimen from Coquimbo, near La Serena, labeled *"Echinocactus cinerascens* Lemaire," which is doubtless to be referred here. *E. cinerascens* originally came from Copiapo, an interior town, much farther north than Coquimbo. Dr. Rose found this species very abundant on the hills near La Serena not far from Coquimbo (No. 19261).

Related to this species, and perhaps not distinct from it, is *Echinocactus fiedlerianus* Schumann (Gesamtb. Kakteen Nachtr. 121. 1903), but it grows farther north, not along the coast but in an interior valley. The type was collected by Mr. Söhrens near Vallenar, Huasco, Chile. Dr. Rose did not obtain specimens but he is now confident that this is the plant which he saw in great abundance just south of Vallenar. Schumann misunderstood the relationship for he places it between *Echinocactus megalothelos* and *E. schickendantzii,* two species of *Gymnocalycium.* It may be briefly characterized as follows:

Cespitose, with a turnip-like root, depressed-globose, grayish, covered with copious wool at the apex; ribs 13, tuberculate; areoles depressed; radial spines 4 to 7, 3 cm. long, subulate; flowers yellow, greenish without.

Illustration: Blühende Kakteen **3:** pl. 121, as *Echinocactus coquimbanus.*

Plate x, figure 2, shows one of the plants collected by Dr. Rose in flower.

4. Copiapoa cinerascens (Salm-Dyck).

Echinocactus cinerascens Salm-Dyck, Allg. Gartenz. **13**: 387.   1845.
Echinocactus copiapensis Pfeiffer, Abbild. Beschr. Cact. **2**: under pl. 14.   1847.
Echinocactus conglomeratus Philippi, Fl. Atac. 23.   1860.
Echinocactus ambiguus Hildmann in Schumann, Gesamtb. Kakteen 311.   1898.

Globose, about 8 cm. in diameter, green, the apex covered with gray wool; ribs 20 or 21, somewhat compressed; areoles 6 to 20 mm. apart; radial spines 8, usually 10 to 12 mm. long; central spines 1 or 2, 18 to 25 mm. long, stouter than the radials, all rigid, yellowish or grayish; flowers yellow; outer perianth-segments acute, often recurved; inner perianth-segments lanceolate, erose, or dentate.

Type locality: Copiapo, Chile.
Distribution: West coast of northern Chile.

In the original description of *Echinocactus ambiguus* it is stated that the ovary is probably scaly and woolly, but this is doubtless wrong.   In all the species of *Copiapoa*, the ovary is buried in a mass of wool but this arises from the areoles about the base of the flower.   This plant is known to us only from descriptions and figures.

Echinocactus intricatus longispinus Monville (Labouret, Monogr. Cact. 178. 1853) was referred here as a synonym.

Illustrations: Grässner, Haupt-Verz. Kakteen **1912**: 5; Möllers Deutsche Gärt. Zeit. **25**: 474. f. 6, No. 7; Monatsschr. Kakteenk. **14**: 89. f. a, as *Echinocactus cinerascens.*

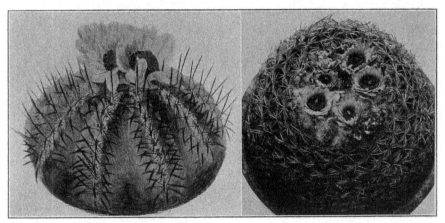

FIG. 100.—Copiapoa echinoides.   FIG. 101.—Pediocactus simpsonii.

5. Copiapoa echinoides (Lemaire).

Echinocactus echinoides Lemaire in Salm-Dyck, Allg. Gartenz. **13**: 386.   1845.
Echinocactus bridgesii Pfeiffer, Abbild. Beschr. Cact. **2**: pl. 14.   1847.
Echinocactus bolivianus Pfeiffer, Abbild. Beschr. Cact. **2**: under pl. 14.   1847.
? *Echinocactus salm-dyckianus* Pfeiffer, Abbild. Beschr. Cact. **2**: under pl. 14.   1847.

Simple, globose, very woolly at apex; ribs 8 to 13, straight, rounded, green; radial spines 5 to 7, stout, straight or somewhat curved; central spine solitary, porrect, 3 cm. long; flowers pale yellow; outer perianth-segments narrowly ovate, acute, reddish; inner perianth-segments broadly oblong, obtuse; scales of ovary and flower-tube described by Schumann as woolly in their axils, but undoubtedly he is wrong.

Type locality: Not cited.
Distribution: Reported from Bolivia, but perhaps from that part of Bolivia now belonging to Chile.

This name occurs first in Cels's Catalogue of 1845, but without description. We know the plant only from descriptions and illustrations; it may not belong to this genus.

Echinocactus macracanthus Salm-Dyck (Cact. Hort. Dyck. 1849. 143. 1850) may belong here. The varieties *Echinocactus macracanthus cinerascens* Salm-Dyck and *E. pepinianus affinis* Monville were both referred by Labouret (Monogr. Cact. 177. 1853) as synonyms of *Echinocactus macracanthus* Salm-Dyck.

Illustrations: Pfeiffer, Abbild. Beschr. Cact. 2: pl. 29, as *Echinocactus echinoides;* Pfeiffer, Abbild. Beschr. Cact. 2: pl. 14, as *Echinocactus bridgesii.*

Figure 100 is copied from the first illustration above cited.

6. Copiapoa megarhiza sp. nov.

Plants with large fleshy roots, sometimes 25 cm. long and 7 to 8 cm. in diameter, usually single, rarely in 2's and 3's, globular to elongate-cylindric, 8 to 26 cm. long, 4 to 9 cm. in diameter, dull green to almost white; ribs usually 13, very low; crown of plant covered with long white wool at flowering time; spines about 12, 1.5 cm. long, rather stout, at first yellow but soon gray; flowers yellow, 2.5 cm. long; fruit green, 6 to 8 mm. long, naked, crowned by 5 green scales; seeds black, 2 mm. long.

Collected by J. N. Rose on the very dry granitic hills near Copiapo, Chile, October 12, 1914 (No. 19323).

Two other species, *Echinocactus cinerascens* Lemaire and *E. copiapensis* Pfeiffer, were described as coming from Copiapo, but whether from the town or the province we do not know. Both have more ribs than the plant here described.

COPIAPOA sp.

A living specimen was collected by Dr. Rose, October 14, 1914, at Tres Cruces, north of Coquimbo, which was sent to the New York Botanical Garden under No. 19339. It may be described as follows:

Single, globular, about 1 cm. in diameter; ribs 11 or 12, obtuse; spines usually 10, brown at first, afterwards gray, subulate.

PUBLISHED SPECIES, PERHAPS OF THIS GENUS.

The four following species are probably of this relationship, but too little is known of them to place them definitely:

ECHINOCACTUS HUMILIS Philippi, Fl. Atac. 23. 1860. Not Pfeiffer, 1837.

Very small, depressed, subglobose, 2.5 cm. broad by 2 cm. high; ribs 10 to 12, tuberculate; radial spines 10 to 12, setaceous, spreading; central spine 1, 2.2 cm. long; flowers yellow, 2 cm. long.

Type locality: Paposo, Antofagasta, Chile.
Distribution: Antofagasta, Chile.

Paposo, the type locality of this species is on the coast north of Taltal. This species seems never to have been re-collected. It was not found in the Philippi Herbarium at Santiago and was unknown to Mr. Söhrens. The name being a homonym must be rejected.

ECHINOCACTUS FOBEANUS Mieckley, Monatsschr. Kakteenk. 17: 187. 1907.

Globose, 8 to 10 cm. in diameter, dark green, somewhat depressed and white-woolly at apex; ribs 14, spiraled; radial spines 8 or 9, black when young, 12 mm. long; central spines when present 1 or 2; flowers pale yellow.

This species is known only from the description of specimens which flowered in the Berlin Botanic Garden. It is supposed to have come from Chile and seems to be of this relationship although we can not definitely refer to it any species which we here recognize. A photograph of a small grafted plant is the only illustration we know (Möllers Deutsche Gärt. Zeit. 25: 474. f. 6, No. 15).

ECHINOCACTUS LINDLEYI Förster, Hamb. Gartenz. **17**: 162. 1861.

Nearly globular or a little broader than high, 7.5 cm. high; ribs 12, broad, rounded; areoles 1.5 cm. apart; spines yellowish brown at first, but in age only the tips brown; radial spines 9 to 11, spreading, 1.5 to 2 cm. long; central spines 2, the longest 3 to 3.5 cm. long.

Type locality: Probably Peru.

Its flowers were unknown when described and it has disappeared from collections although it is said to be very ornamental. It was referred to Salm-Dyck's group, *Cephaloidei*, to which most of the species of *Copiapoa* were referred.

ECHINOCACTUS PYRAMIDATUS Förster, Hamb. Gartenz. **17**: 162. 1861.

Short-pyramidal, about 18 cm. high, 13 cm. in diameter; ribs 15; areoles 4 to 6 mm. apart; spines stiff, reddish brown; radial spines 8; central spines 3, 3 to 3.5 cm. long, stout; flowers yellow.

Type locality: Probably Peru.

5. **PEDIOCACTUS** Britton and Rose in Britton and Brown, Illustr. Fl. ed. 2. **2**: 569. 1913.

Globular, single or cespitose, small, strongly tubercled cacti; tubercles borne on spiraled ribs; young areoles very woolly, but in age nearly naked; flowers small, with a rather indefinite funnel-shaped tube, pinkish, broadly campanulate; outer perianth-segments smaller than the inner and duller in color; inner perianth-segments oblong, numerous; scales on flower-tube few, naked in their axils; stamens numerous; ovary green, nearly globular, with a few scales towards the top, and a depressed scar at apex; fruit dry, greenish, splitting on one side; seeds dull black, tuberculate, keeled on the back with a large sub-basal hilum.

Type species: *Echinocactus simpsonii* Engelmann.

The plant has been described both as a *Mammillaria* and as an *Echinocactus*. In its globular shape and strongly tubercled surface it resembles very much many of the so-called *Mammillaria*, but the tubercles are really borne on ribs, while the flowers are borne near the center of the plant and originate just above the spines; therefore, this genus belongs to the *Echinocactanae* rather than to the *Coryphanthanae;* the seeds are unlike those of *Coryphanthanae*.

We recognize one species.

The generic name is from πεδίον a plain, and κάκτος cactus, referring to the general habitat of the plant.

1. **Pediocactus simpsonii** (Engelmann) Britton and Rose in Britton and Brown, Illustr. Fl. ed. 2. **2**: 570. 1913.

> *Echinocactus simpsonii* Engelmann, Trans. St. Louis Acad. **2**: 197. 1863.
> *Echinocactus simpsonii minor* Engelmann, Trans. St. Louis Acad. **2**: 197. 1863.
> *Mammillaria simpsonii* Jones, Zoe **3**: 302. 1893.
> *Mammillaria purpusii* Schumann, Monatsschr. Kakteenk. **4**: 165. 1894.
> *Echinocactus simpsonii robustior* Coulter, Contr. U. S. Nat. Herb. **3**: 377. 1896.

Plants depressed, globular, up to 15 cm. broad by 12 cm. high, strongly tuberculate; the tubercles contiguous; radial spines 15 to 20, spreading, white, acicular; central spines usually 5 to 7, more or less spreading, stouter and longer than the radials, 1 to 3 cm. long, the base white but the upper part reddish brown or brown throughout; flower-buds obtuse; flowers massed in the center and surrounded by brown or whitish wool; outer perianth-segments oblong, obtuse, their margins scarious and serrulate; inner perianth-segments linear-oblong, acutish; filaments golden yellow; style and stigma-lobes yellowish.

Type locality: Butte Valley in the Utah desert and Kobe Valley, farther west.

Distribution: Kansas to New Mexico, north to Nevada, Washington, Idaho, and Montana.

These plants often take on very weird shapes, very unlike the normal form, and then are called the snake cactus or brain cactus. We have photographs of some of these abnormal plants taken by Mr. M. E. Jones in Utah and Nevada.

The beautiful flowers close partially at night.

Mammillaria spaethiana Schumann, listed by Späth (Cat. 1894-1895), seems never to have been described. Schumann afterwards withdrew the name and Mrs. Brandegee (Zoe 5: 31. 1900) states that it has the seeds of *Echinocactus simpsonii* and she believes it to be one of the forms of this species.

The species as here treated covers a wide range and is represented by several striking forms. The one from the state of Washington has very dark, nearly black spines, the radials ascending and subulate. We have not seen this plant in flower but the flower-scar is at the spine-areole, as it always is in this genus. Mr. Charles V. Piper in his Flora of Washington says "quite certainly new." It is possibly a good species. Here we would also refer a plant collected by J. E. Edwards near Haycreek, Oregon. It is possible, as Coulter believed, that these are the same as the Nevada form which represents Engelmann's variety *robustior* and this view has been held by others. (See Cact. Journ. 2: 157.)

This Washington plant seems to have been collected more than 70 years ago, but the specimen has apparently been lost and the record overlooked. Our attention was called to this old record of Geyer, by Mr. C. V. Piper, here reproduced:

"A third species of *Mammillaria* I found on the Oregon plains while searching for a *Melocactus*. Of this I brought dry specimens to London and Mr. Scheer, at Kew, has already raised several from seeds. The above-mentioned *Melocactus* was gathered by Chief Factor MacDonald at Fort Colville, but the exact habitat was forgotten; the one specimen found was afterwards in possession of Dr. Tolmie on the lower Columbia. From the information I could gather at Fort Walla Walla, the true habitat of this cactus is at the Priests' Rapid,' on a rocky island in the Columbia River, about 60 miles above Fort Walla Walla. I received this intelligence too late, but hope that by publishing it other botanists may have the opportunity of getting the plant without loss of time."— Charles A. Geyer. (The London Journal of Botany 5: 25. 1846.)

Illustrations: Simpson's Rep. pl. 1, 2; Britton and Brown, Illustr. Fl. 2: f. 2524; Knippel, Kakteen pl. 11; Förster, Handb. Cact. ed. 2. 593. f. 76; Schelle, Handb. Kakteenk. 135; Gard. Chron. II. 6: 293. f. 60; III. 8: 166. f. 26, as *Echinocactus simpsonii;* Thomas, Zimmerkultur Kakteen 53; Gartenwelt 1: 85; Monatsschr. Kakteenk. 4: 167, as *Mammillaria purpusii;* Britton and Brown, Illustr. Fl. ed. 2. 2: f. 2983.

Plate VIII, figure 1, was painted from a plant collected by A. Nelson in Wyoming in 1914 and sent by Dr. Rose to the New York Botanical Garden. Figure 101 is from a photograph of the same plant.

6. TOUMEYA gen. nov.

A small, ovoid or short-cylindric cactus, the areoles borne on low spirally arranged tubercles; spines thin, flat, white, shining, papery, flexible, the central ones much longer than the radial; flowers central, about as wide as long, white, borne at the spine-areoles on nascent tubercles; ovary bearing a few minute scales, their axils naked; outer perianth-segments ovate, acute, the inner lanceolate, acuminate; perianth-tube short, bearing several papery lanceolate scales; fruit dry, globose, smooth; seeds compressed, oblique, black.

Type species: *Mammillaria papyracantha* Engelmann. A monotypic genus of New Mexico.

The generic name is in honor of Dean James W. Toumey, whose studies and collections of cacti have greatly aided our investigations.

1. Toumeya papyracantha (Engelmann).

Mammillaria papyracantha Engelmann, Pl. Fendl. 49. 1849.
Echinocactus papyracanthus Engelmann, Trans. St. Louis Acad. 2: 198. 1863.

Simple with fibrous roots, 5 to 10 cm. long; "ribs 8, oblique" but probably very indefinite even in living plants, bearing low distinct tubercles; areoles small, circular, pubescent when young, naked in age, the lower ones described as proliferous; spines chartaceous, the radials 8 to 10, unequal, 3 to 20 mm. long, spreading; central spines 1 to 4, 3 to 4 cm. long, the upper ones connivent over the top of the plant; flower 2.4 to 2.6 cm. long, a little broader when fully expanded, white; fruit nearly naked, globular, 4 to 6 mm. in diameter, thin-walled; seeds large, 2 to 2.5 mm. broad, somewhat pointed at base, angled on the back; hilum large, sub-basal.

Type locality: Between the lower hills near Santa Fé, New Mexico.

Distribution: Rare in isolated localities in northern New Mexico; reported from California by Watson (Cact. Journ. 1: 43), probably erroneously.

This is a remarkable plant whose generic position has been uncertain. Engelmann, who first described it as a *Mammillaria* and afterwards as an *Echinocactus*, associates it with *Echinocactus simpsonii*, that is, *Pediocactus simpsonii*, as representing a small group of *Echinocactus* "with the appearance of *Mammillaria*."

It has been reported only a few times and the fruit has not heretofore been described. Fendler reported it growing in loose red sandy fertile soil.

In 1893 (Zoe 3: 301) Mr. M. E. Jones published a note on this species and, on the basis of it, the plant has been admitted into the flora of Utah. He writes as follows:

"The flowers are an inch long, opening but little; stigma cleft a line deep into 6 anther-like divisions, papillose on the sides and upper surface; filaments 6 lines long; style almost as long as the petals, ½ a line thick, linear; the flowers open in the morning, and close in the afternoon, but apparently are not affected by cloudy weather. This grows in alkaline soil, and blooms in May. It is scarce everywhere."

He wrote in a letter (March 18, 1918) from Salt Lake City:

"The material that I thought was this species came from the desert west of here, towards Mount Ibapah. I remember very distinctly the appearances of the specimen but I did not collect it and I now have some doubts about its identity. The spines were papery. I have never seen it since, though I have hunted for it."

We have found a similar plant in the herbarium of the Philadelphia Academy of Sciences, collected by Siler in Utah, but it certainly is not the true *E. papyracanthus*. This, however, is only a slice from the plant and is without flowers or fruit; it may be described as follows:

Covered by a mass of spines; ribs numerous, low, tubercled; areoles close together, circular, white-felted when young; radial spines 10 to 12, white, about 10 mm. long, weak; central spines 3 to 5, weak and flexible, more or less twisted, 2 to 3 cm. long; some of them more or less flattened, pale or dark brown, one more or less hooked. A. L. Siler's note is as follows: "Only a few specimens have ever been found. Flowers of this were pink, not white, as described by Engelmann. Southern Utah, 1888."

Illustration: Cact. Journ. 1: pl. v, as *Echinocactus papyracanthus*.

7. EPITHELANTHA* Weber.

Plant globular, very small, the surface divided into numerous tubercles arranged in spiraled rows, mostly hidden by the numerous small spines; flowers very small, from near the center of the plant, arising from upper part of the spine-areole on the young tubercles; outer perianth-segments 3 to 5; inner perianth-segments few, often only 5; stamens few, usually 10, included; fruit small, clavate, red, few-seeded; seeds black, shining, rather large, with a large depressed hilum.

Type species: *Mammillaria micromeris* Engelmann.

We recognize one species, from western Texas and northern Mexico.

The generic name is from ἐπί on, θηλῆ nipple, and ἄνθος flower, indicating that the flower is borne on the tubercle.

This genus has heretofore been associated with the so-called *Mammillaria*, some of the species of which it resembles in its globose shape and small clavate red fruits. On account

---

*The name *Epithelantha* was given by Weber (Dict. Hort. Bois 804. 1898) as a synonym of *Mammillaria micromeris* and therefore was not formally published by him.

of the position of the flower, however, which is at the spine-areole, the genus is better referred to the sub-tribe, *Echinocactanae*, where in the genera *Lophophora* and *Ariocarpus* we have a similar fruit.

Mr. Charles Wright, in his field notes, first called attention to the central position of the flower of *Mammillaria micromeris*, and Engelmann, who discussed it (Cact. Mex. Bound. 4) in some detail, was in doubt as to its position. Dr. Weber seems to have been the first to determine the exact position of the flower and, recognizing its significance, proposed a new generic name for it, but he also referred it to *Echinocactus* and in still another place left it as a *Mammillaria*.

1. Epithelantha micromeris (Engelmann) Weber.

Mammillaria micromeris Engelmann, Proc. Amer. Acad. **3**: 260. 1856.
Mammillaria micromeris greggii Engelmann, Proc. Amer. Acad. **3**: 261. 1856.
Cactus micromeris Kuntze, Rev. Gen. Pl. **1**: 260. 1891.
Cactus micromeris greggii Coulter, Contr. U. S. Nat. Herb. **3**: 101. 1894.
Mammillaria greggii Safford, Ann. Rep. Smiths. Inst. **1908**: 531. pl. 4, f. 1. 1909.

Plants small, simple or cespitose, nearly globular, but depressed at apex, 6 cm. in diameter or less; tubercles very low, small, arranged in many spirals, 1 mm. long; spines numerous, white, the lower radials about 2 mm. long, the upper radials on the young tubercles 6 to 8 mm. long and connivent over the apex, narrowly clavate, the upper half finally falling off; flowers from near the center of the plant in a tuft of wool and spines; flower very small, whitish to light pink, 6 mm. broad; perianth-segments 8 to 10; stamens 10 to 15; stigma-lobes 3; fruit 8 to 12 mm. long; seed 1.5 mm. broad.

Type locality: Western Texas.

Distribution: Western Texas and northern Mexico.

Writers generally, as well as dealers of these plants, are disposed to treat the large forms of this species as a variety, var. *greggii*, but we have observed no reason except size for this conclusion. The large form seems to extend throughout the range of the species proper. In June 1921, Mrs. S. L. Pattison sent us from western Texas an unusually large plant which was nearly 8 cm. high and 6 cm. in diameter.

The plant is known as button cactus. Its fruits, called chilotos, are slightly acid and are edible.

FIG. 102.—Epithelantha micromeris.

The names *Epithelantha micromeris* Weber and *Echinocactus micromeris* Weber, although both mentioned by him (Dict. Hort. Bois 804. 1898), were not formally published.

Mammillaria micromeris fungifera (Monatsschr. Kakteenk. **19**: 140. 1909) is only a catalogue name.

Pelecyphora micromeris Poselger and Hildmann appears as a synonym of *Mammillaria micromeris* in Garten-Zeitung **4**: 322. 1885.

Illustrations: Cact. Mex. Bound. pl. 1; 2, f. 1 to 4; Cact. Journ. **1**: 43; pl. [2] for February in part; Rümpler, Sukkulenten 200. f. 115; Dict. Gard. Nicholson Suppl. 514. f. 545; Förster, Handb. Cact. ed. 2. 267. f. 26, 27; Cycl. Amer. Hort. Bailey **1**: 203. f. 302; Stand. Cycl. Hort. Bailey **5**: f. 3016; Amer. Garden **11**: 460; Monatsschr. Kakteenk. **20**: 126; **29**: 81; Schelle, Handb. Kakteenk. 248. f. 166, 167; Garten-Zeitung **4**: 323. f. 76; Watson, Cact. Cult. 167. f. 65; ed. 3. f. 42; Blanc, Cacti 71. f. 1394, as *Mammillaria micromeris;* Cact. Mex. Bound. pl. 2, f. 5 to 8; Blanc, Cacti 71. f. 1395, as *Mammillaria micromeris greggii;* Ann. Rep. Smiths. Inst. **1908**: pl. 4, f. 1, as *Mammillaria greggii.*

Figure 102 shows a plant in fruit, collected by Dr. Rose at Langtry, Texas, in 1908 (No. 11612).

8. NEOPORTERIA gen. nov.

Plants globose to cylindric, sometimes much elongated and then sprawling or pendent over cliffs; more or less hairy at the crown; ribs usually straight, more or less tubercled; flowers from the center of the plant, short-funnelform, usually pinkish or reddish; stigma-lobes cream-colored to reddish; scales on the flower-tube bearing wool and long bristles in their axils; fruit as far as known small, more or less globular, dehiscing by a basal pore; seeds brown, somewhat wrinkled, tuberculate with a somewhat depressed hilum.

There are several species along the coast and mountains of Chile of which *Echinocactus subgibbosus* Haworth is selected as the type. In Schumann's keys one would look for these species in his subgenus *Notocactus* of *Echinocactus* but, as a matter of fact, most of them have been assigned to the subgenus *Hybocactus*. The group as treated here is a natural one and deserves separation as a genus. Doubtless, still other species will be assigned here when better known. From *Malacocarpus*, to which we have referred most of Schumann's species grouped by him in *Notocactus*, it differs in its fruit, seeds, and in the shape and color of the flowers.

The genus is named for Carlos Porter of Chile, a well-known entomologist. Seven species are described.

KEY TO SPECIES.

```
Spines all weak, often thread-like....................................................1. N. nidus
Spines stouter, the central, at least, subulate.
    Spines black.
        Spines puberulent; outer perianth-segments reddish...........................2. N. occulta
        Spines glabrous; outer perianth-segments greenish............................3. N. nigricans
    Spines not black (or black when young in No. 7).
        Flowers pink.
            Plants bluish green; style and stigma-lobes reddish.......................4. N. jussieui
            Plants bright green; style and stigma-lobes white to greenish.
                Perianth-segments entire..........................................5. N. subgibbosa
                Perianth-segments toothed.........................................6. N. chilensis
        Flowers not pink..................................................................7. N. fusca.
```

1. Neoporteria nidus (Söhrens).

Echinocactus senilis Philippi, Gartenflora **35:** 485. 1886. Not Beaton, 1839.
Echinocactus nidus Söhrens in Schumann, Monatsschr. Kakteenk. **10:** 122. 1900.

Simple, somewhat glaucous, short-cylindric, 8 cm. long, 5 to 6 cm. in diameter, the top covered by the slender ascending connivent spines; ribs 16 to 18, obtuse, strongly tubercled; areoles large, circular; spines weak, numerous, sometimes 30 in a cluster, unequal, white, the longest 3 cm. long, bulbose at base; flowers rather slender, 4 cm. long, perhaps pinkish, but sometimes described as yellow; inner perianth-segments narrow, acute; scales on the flower-tube woolly and setose; fruit not known.

Type locality: East of Ovalle, Chile.
Distribution: Northern Chile.

Dr. Rose examined the type of *E. senilis* in the Philippi Herbarium in 1914 and it agrees with the first illustration cited below. He also saw a part of Mr. Söhrens's type of *Echinocactus nidus* at Santiago, Chile, in 1914. Its flowers were still unknown.

We have united *Echinocactus senilis* and *E. nidus*, using the later specific name since *E. senilis*, the older name, is a homonym. We have not seen living specimens, but we have two photographs sent us by Harvey Frank in 1905, labeled *E. senilis* and *E. nidus*, respectively, which are very much alike and led us to unite the two species. If two species are here involved, our description would apply to *E. senilis*.

Illustrations: Gartenflora **35:** pl. 1230, f. A, as *Echinocactus senilis;* Monatsschr. Kakteenk. **10:** 123, as *E. nidus.*

Figure 103 is copied from the first illustration above cited.

2. Neoporteria occulta (Philippi).

Echinocactus occultus Philippi, Fl. Atac. 23. 1860.

Plants globular to short-cylindric, 5 to 8 cm. high, somewhat depressed at apex; ribs about 14, obtuse, separated by acute intervals, strongly tubercled; tubercles somewhat rhombic in shape, with a chin at the base; areoles 1.5 cm. apart, rather narrow and depressed; spines wanting or when present 1 to 10, unequal, 1 to 4 cm. long, puberulent, blackish; flowers central, 2.5 cm. long but 5 cm. broad; outer perianth-segments spatulate, finely toothed near the apex; axils of the scales on the flower-tube bristly; scales on the ovary woolly in their axils.

Type locality: Seacoast from Copiapo to Cobre.
Distribution: Provinces of Copiapo and Antofagasta, Chile.

Schumann states that Philippi is not quite accurate in asserting that this species is found along the beach, but that it comes from the foot-hills.

Illustrations: Blühende Kakteen **1**: pl. 24; Monatsschr. Kakteenk. **11**: 92, 93; Schumann, Gesamtb. Kakteen Nachtr. f. 24, as *Echinocactus occultus*.

Figure 104 is copied from the first illustration above cited.

FIG. 103.—Neoporteria nidus.   FIG. 104.—Neoporteria occulta.

3. Neoporteria nigricans (Linke).

Echinopsis nigricans Linke, Allg. Gartenz. **25**: 239. 1857.
Echinocactus nigricans Dietrich in Schumann, Gesamtb. Kakteen 420. 1898.

Simple, short-cylindric, somewhat narrow at base; ribs 15, strongly tuberculate, glaucous-green, compressed; radial spines 8 or 9, somewhat curved, 7 mm. long; central spines 1 or 2, 1.2 cm. long; flowers 4.5 to 5 cm. long, white or yellowish green; inner perianth-segments spreading, somewhat toothed above, acute; stigma-lobes reddish or purplish; scales on ovary and flower-tube acute, bearing a few hairs or bristles in their axils.

Type locality: Chile or Bolivia.
Distribution: West coast of South America, doubtless Chile.

According to Mr. Juan Söhrens, this plant is found in the mountains of northern Chile. We know it only from description and the single illustration cited below.

According to Schumann (Gesamtb. Kakteen 420. 1898), *Echinocactus cupreatus* Poselger (Förster, Handb. Cact. ed. 2. 602. 1885) is related to this species. He states that it is distinguished by the darker brown color and fewer spines of a brown-black color, lighter at the base.

Illustration: Blühende Kakteen 1: pl. 45, as *Echinocactus nigricans*.

Figure 105 is copied from the illustration above cited.

FIG. 105.—Neoporteria nigricans.   FIG. 106.—Neoporteria fusca.

4. Neoporteria jussieui (Monville).

Echinocactus jussieui Monville in Salm-Dyck, Cact. Hort. Dyck. 1849. 170. 1850.

Simple, globose or short-cylindric, dark or bluish green; ribs 13 to 16, rather stout, divided into prominent tubercles; radial spines 7 or perhaps more, dark brown, somewhat spreading; central spines 1 or 2, 2.5 cm. long; flowers from near the center of the plant, 3 to 3.5 cm. long; perianth-segments linear-oblong, acute, pinkish, but sometimes described as yellow; style and stigma-lobes reddish; ovary bearing scales and these woolly in their axils; fruit not known.

Type locality: Not cited.

Distribution: Chile.

Echinocactus niger (Salm-Dyck, Cact. Hort. Dyck. 1849. 34. 1850) and *E. jussianus* Lemaire (Labouret, Monogr. Cact. 247. 1853), undescribed, belong here. A variety, *cristatus* (Rümpler in Förster, Handb. Cact. ed. 2. 581. 1885), is described as having 18 to 20 ribs.

The original description of this species is very meager while recent descriptions are scarcely more satisfactory. Both Schumann and Weber describe the flowers as yellow, while Gürke in Blühende Kakteen illustrates them as pinkish. The illustration cited below suggests relationship with *E. subgibbosus*. The flowers of both are pinkish. So far as we know, *N. jussieui* never has bristles in the axils of the scales on the ovary, and this seems to be true of all the species of the genus.

The plant illustrated in the Monatsschrift für Kakteenkunde (27: 53. 1917) seems to be some species of *Malacocarpus*.

FIG. 107—Neoporteria jussieui.   FIG. 108.—Neoporteria chilensis.

Illustrations: Blühende Kakteen **2**: pl. 67; (?) Monatsschr. Kakteenk. **27**: 53, as *Echinocactus jussieui.*

Figure 107 is copied from the first illustration above cited.

5. Neoporteria subgibbosa (Haworth).

Echinocactus subgibbosus Haworth, Phil. Mag. **10**: 419. 1831.
Cactus berteri Colla, Mem. Accad. Sci. Torino **37**: 77. 1833.
Echinocactus acutissimus Otto and Dietrich, Allg. Gartenz. **3**: 353. 1835.
Echinocactus exsculptus Otto in Pfeiffer, Enum. Cact. 65. 1837.
Cereus dichroacanthus Martius in Pfeiffer, Enum. Cact. 76. 1837.
Mammillaria atrata Hooker in Curtis's Bot. Mag. **65**: pl. 3642. 1839.
Mammillaria floribunda Hooker in Curtis's Bot. Mag. **65**: pl. 3647. 1839.
Echinocactus thrincogonus Lemaire, Cact. Gen. Nov. Sp. 22. 1839.
Echinocactus thrincogonus elatior Lemaire, Cact. Gen. Nov. Sp. 23. 1839.
Echinocactus berteri Remy in Gay, Fl. Chilena **3**: 15. 1847.
Echinocactus rostratus Jacobi, Allg. Gartenz. **24**: 108. 1856.
Cactus atratus Kuntze, Rev. Gen. Pl. **1**: 259. 1891.
Cactus floribundus Kuntze, Rev. Gen. Pl. **1**: 259. 1891.

Globose when young, soon cylindric, usually 3 dm. high and erect, but sometimes much elongated, a meter long or more, either prostrate or hanging over cliffs, very spiny; ribs numerous, often 20, 1 cm. high; areoles approximate, often large, sometimes 1 cm. in diameter; spines numerous, acicular, brownish in age, often paler at base, straight, the longest ones 3 cm. long; flowers usually abundant, 4 cm. long, the buds dark red, pointed; perianth-segments usually light pink, but sometimes darker, very numerous, the outer ones spreading, the central ones erect, concealing the stamens, acute; filaments attached below middle of flower-tube, erect, white, included; style slender, pale, slightly exserted; scales on ovary minute, acute, horny, those on tube hairy and bristly; fruit 1.5 to 2 cm. long, reddish; seeds brown, 1 mm. in diameter.

Type locality: Near Valparaiso, Chile.

Distribution: Along the seacoast of Chile, both north and south of Valparaiso.

Echinocactus exsculptus Otto, when first described, was a complex. Pfeiffer (Enum. Cact. 65. 1837) says it comes from Chile, Mexico, and Montevideo; he referred here several synonyms; one of them is *Echinocactus subgibbosus* which Haworth states in his original description comes from Valparaiso. This is doubtless Pfeiffer's Chilean element of the species. Another synonym is *Cereus montevidensis* Pfeiffer (Enum. Cact. 65. 1837) which is the Montevideo element. Two other names, but not described until later, *E. acanthion* and *E. interruptus*, seem to represent the Mexican element. In the Addenda (181. 1837) he adds two synonyms, *E. crenatus* and *E. guyannensis*. In addition to this synonymy Pfeiffer described plants in the Berlin Botanical Garden, the origin of which was not stated.

Förster's treatment (Handb. Cact. 291. 1846) is still more complex. He gives the distribution: Mexico, Buenos Aires, Chile, and Brazil. With Brazil he includes Montevideo, Para, and Guiana. Here he refers as synonyms those given by Pfeiffer and adds: *Echinocactus valparaiso*, *Cereus hoffmannseggii*, *Mammillaria hoffmannseggii*, and *M. gibbosa* (Salm-Dyck, Hort. Dyck. 343. 1834). He also mentions or describes the following varieties: *fulvispinus*, *dichroacanthus* (Salm-Dyck, Cact. Hort. Dyck. 1844. 18. 1845; *Cereus dichroacanthus* Martius in Pfeiffer, Enum. Cact. 76. 1837), *foveolatus* (Salm-Dyck, Cact. Hort. Dyck. 1844. 18. 1845; *Cereus foveolatus* Haage jr. in Pfeiffer, Enum. Cact. 77. 1837), *tenuispinus* and *thrincogonus* (Förster, Handb. Cact. 293. 1846; *Echinocactus thrincogonus* Lemaire, Cact. Gen. Nov. Sp. 22. 1839).

Echinocactus pseudo-cereus Meinshausen (Wochenschr. Gärtn. Pflanz. 1: 29. 1858) is described from a barren plant supposed to have been grown from Mexican seed obtained by Karwinsky. If related to *Echinocactus exsculptus*, as stated by Meinshausen, it is more likely to have come from South America.

Echinocactus acutissimus cristatus (Förster, Handb. Cact. ed. 2. 567. 1885) probably belongs here.

Echinocactus exsculptus gayanus Monville (Labouret, Monogr. Cact. 241. 1853) and *Echinocactus gayanus* (Lemaire, Cact. Gen. Nov. Sp. 22. 1839) were never described.

Schumann has also referred here *Echinocactus hoffmannseggii* (Gesamtb. Kakteen 426. 1898). He would also refer here *Cactus berteri* Colla and *Echinocactus rostratus*, both of which were also based on Valparaiso plants.

Echinocactus exsculptus fulvispinus (Förster, Handb. Cact. 292. 1846) was supposed to be a form of the species proper, while the variety *elatior* (Förster, Handb. Cact. 293. 1846) was referred as a synonym of one of its varieties; *Echinocactus exsculptus cristatus* (Förster, Handb. Cact. ed. 2. 566. 1885) is only an abnormal form; *Echinocactus foveolatus* Haage (Salm-Dyck, Cact. Hort. Dyck. 1849. 33. 1850) was never described but doubtless belongs here.

Echinocactus gayanus intermedius Monville (Labouret, Monogr. Cact. 240. 1853) appeared as a synonym of *E. thrincogonus*.

Mammillaria ambigua G. Don (Loudon, Hort. Brit. 194. 1830), to which *Cactus ambiguus* Gillies was referred, seems never to have been described. Schumann did not know it, but thought that it was some *Echinocactus*. If it actually came from Chile, as reported, it may possibly be referable here. It may be the same as *Melocactus ambiguus* Pfeiffer, which, however, is usually referred to *Echinopsis leucantha*.

Of this relationship is the plant described and illustrated by Walton (Cact. Journ. 1: 105. 1898) as *Echinocactus rubidus superbissimus* which he states is native of Chile.

Illustrations: Mem. Accad. Sci. Torino 37: pl. 17, f. 2, as *Cactus berteri;* Loudon, Encycl. Pl. ed. 3. 1201. f. 17360; Curtis's Bot. Mag. 65: pl. 3642, as *Mammillaria atrata;* Pfeiffer and Otto, Abbild. Beschr. Cact. 1: pl. 20; Blühende Kakteen 3: pl. 133; Monatsschr. Kakteenk. 30: 139, as *Echinocactus acutissimus;* Curtis's Bot. Mag. 65: pl. 3647, as *Mammillaria floribunda;* Martius, Fl. Bras. 4²: pl. 51, f. 1; Knippel, Kakteen pl. 7; Schelle, Handb. Kakteenk. 195. f. 128, as *Echinocactus exsculptus.*

Plate VIII, figure 4, shows the flowering top of a plant brought by Dr. Rose to the New York Botanical Garden from east of Las Vilas, Chile, in 1914.

6. Neoporteria chilensis (Hildmann).

Echinocactus chilensis Hildmann in Schumann, Gesamtb. Kakteen 423. 1898.

Simple or proliferous at base, globose to short-columnar, woolly at apex; ribs 20 or 21, crenate, pale green; radial spines about 20, somewhat acicular, 1 cm. long; central spines 6 to 8, 2 cm. long; flowers pink, 5 cm. broad when fully expanded; perianth-segments narrow, acute, the inner ones toothed toward the apex; filaments white, short; style white, with yellow stigma-lobes; the scales of the ovary and flower-tube bearing in their axils short wool and some of them a long white hair much longer than the scale.

Type locality: Chile.

Distribution: Western part of the Andes of Chile.

This species was originally described as having yellow flowers and naked scales but this was evidently an error.

E. chilensis confinis Hildmann (Schumann, Gesamtb. Kakteen 424. 1898) is said to differ from the species in its shorter yellow central spines.

Illustration: Blühende Kakteen 3: pl. 138, as *Echinocactus chilensis*.

Figure 108 is copied from the illustration above cited.

7. Neoporteria fusca (Mühlenpfordt).

Echinocactus fuscus Mühlenpfordt, Allg. Gartenz. 16: 10. 1848.
Echinocactus ebenacanthus Monville in Labouret, Monogr. Cact. 253. 1853.
Echinocactus humilis Rümpler in Förster, Handb. Cact. ed. 2. 471. 1885.

Globular to short-cylindric, about 10 cm. in diameter; ribs 12 or 13, dark green, somewhat tubercled; radial spines 5 to 7, more or less ascending, brownish; central spines 4, black when young, 3 cm. long; flowers 3 cm. long, described as yellow, certainly very pale and nearly white; scales on the flower-tube woolly and setose in their axils.

Type locality: Not cited.

Distribution: Andes of Chile.

Echinocactus hankeanus Förster (Handb. Cact. ed. 2. 471. 1885), referred here as a synonym by Schumann, was never described but first appeared as a synonym of *Echinocactus humilis*. The two varieties of *Echinocactus ebenacanthus, minor* and *intermedius*, were proposed by Labouret (Monogr. Cact. 254. 1853). To the former he referred as a synonym *Echinocactus ebenacanthus affinis* Cels.

Illustrations: Blühende Kakteen 1: pl. 51; Monatsschr. Kakteenk. 27: 135; Schelle, Handb. Kakteenk. 194. f. 127, as *Echinocactus ebenacanthus*.

Figure 106 is copied from the first illustration above cited.

DESCRIBED SPECIES, PERHAPS REFERABLE TO THIS GENUS.

ECHINOCACTUS CASTANEOIDES Cels in Salm-Dyck, Cact. Hort. Dyck. 1849. 165. 1850.

Simple, globose to short-columnar; ribs 15 to 20, tuberculate, light green; radial spines 18 to 20, acicular; central spines 6, larger than the radials; flowers very narrow, tubular.

Type locality: Not cited.

Distribution: Chile or Bolivia.

Schumann first placed this species near *Echinocactus acutissimus* and later referred it to a different section, placing it next to *Echinocactus clavatus*. He also says that it comes from Copiapo, Chile.

Mr. Söhrens sent a plant to Schumann, who called it *Echinocactus castaneoides*, but it was probably *E. acutissimus*.

ECHINOCACTUS KUNZEI* Förster, Handb. Cact. 293. 1846.

Spherical, sunken at top; ribs 16 to 21; spines bent upward, when young yellow, in age gray; radial spines 9 to 12; central spines 2 to 4, a little longer than the radials, 2.5 cm. long; flowers described as lateral, 6 to 8, 5.5 to 6 cm. long; scales of the ovary and flower-tube woolly and setose in their axils.

---

*Originally spelled *kunzii*.

Type locality: Chile.

Distribution: Mountain ridges in Chile.

In both the illustrations cited below the scales on the ovary and flower-tube are ovate and overlapping and are not shown as woolly or setose in their angles. Schumann, however, describes them as such and, if so, the species must be of this relationship.

Here we would refer the two varieties, *brevispinosus* Förster (Allg. Gartenz. **15**: 51. 1847) and *rigidior* Salm-Dyck (Cact. Hort. Dyck. 1849. 33. 1850).

Echinocactus neumannianus Labouret (Monogr. Cact. 245. 1853) is referred by Schumann as a synonym of this species. It comes from Copiapo, Chile, and may be a different species.

Echinocactus neumannianus rigidior (Salm-Dyck, Cact. Hort. Dyck. 1844. 18. 1845) is only a name.

Schumann also refers here *Echinocactus supertextus* Pfeiffer (Abbild. Beschr. Cact. **2**: under pl. 14. 1847), but the description reads more like that of *E. curvispinus*. This is the conclusion reached by Mr. Söhrens of Santiago. Specimens so named in the Philippi Herbarium we would certainly refer to *Neoporteria*.

The species was named for Dr. Gustave Kunze, at one time director of the Botanical Garden at Leipzig.

Illustrations: Förster, Handb. Cact. ed. 2. 571. f. 75; Gartenflora **31**: pl. 1082, a to c.

ECHINOCACTUS MALLETIANUS Lemaire, Allg. Gartenz. **13**: 387. 1845.

Stems simple, depressed-globose or somewhat cylindric, very woolly at the top, 1 dm. high; ribs 15 to 17, more or less; spines straight, acicular, black; radial spines 5 or 6, suberect; central spine solitary; flowers and fruit unknown.

Type locality: Not cited.

Dr. Rose obtained from L. Quehl a photograph of this species as it is now represented in collections. Its relationship is doubtful, but it should certainly not be placed just after *Echinocactus horizonthalonius* as it was by Schumann.

Illustration: Möllers Deutsche Gärt. Zeit. **25**: 474. f. 6, No. 16.

ECHINOCACTUS PEPINIANUS Schumann, Gesamtb. Kakteen 420. 1898.

This species is very different from the species so named by Lemaire. Its flowers and fruit are unknown and its relationship is not known to us. If it is from Chile or Peru, as Schumann suggests, it may be referable to one of the species of *Copiapoa*. For note on *Echinocactus pepinianus* Lemaire, see Britton and Rose (Cactaceae **2**: 137. 1920).

ECHINOCACTUS SUBNIGER Poselger in Förster, Handb. Cact. ed. 2. 588. 1885.

Simple, globose to short-columnar; ribs 16, grayish green; radial spines 8, 1.5 cm. long; central spines 1 to 3, 2 cm. long.

Type locality: Mexico.

This species is recognized by Schumann, but its flowers and fruit are unknown. It is impossible, without seeing a specimen, to make out its generic relationship. If it came from Mexico, as Rümpler thought, it does not belong to *Neoporteria*, but if it is from Chile, as Schumann believed, it should probably be placed there.

9. AREQUIPA gen. nov.

Either simple or cespitose, globular to short-cylindric, small cacti; ribs numerous, low, somewhat tubercled, very spiny; flowers central, funnelform, scarlet; ovary and flower-tube scaly; axils of scales long-hairy but not spiny nor bristly; fruit so far as known dry, dehiscing by a basal pore; seed black, pitted, with a broad basal hilum.

Type species: *Echinocactus leucotrichus* Philippi.

The genus as here treated consists of two species, one of which has heretofore appeared under several specific names, sometimes in *Echinocactus* and sometimes in *Echinopsis*, although it has little in common with either genus and, especially, as these genera are now understood. *Arequipa* is characterized by its slender, elongated, scarlet flower. The species are confined to the mountains of Peru and northern Chile. The generic name is that of the city in Peru near which the type species is found in great abundance.

KEY TO SPECIES.

Hairs of flower-tube white; spines acicular.. 1. *A. leucotricha*
Hairs of flower-tube brown; spines bristly.. 2. *A. myriacantha*

1. Arequipa leucotricha (Philippi).

Echinocactus leucotrichus Philippi, Anal. Mus. Nac. Chile 1891²: 27. 1891.
Echinocactus clavatus Söhrens, Monatsschr. Kakteenk. 10: 27. 1900.
Echinopsis hempeliana Gürke, Monatsschr. Kakteenk. 16: 94. 1906.
Echinocactus rettigii Quehl, Monatsschr. Kakteenk. 29: 129. 1919.

Plants simple or cespitose, sometimes branching, globular or sometimes elongated (4 to 6 dm. long) and then often prostrate, usually covered with spines but naked below when very old; ribs 10 to 20, closely set, low; areoles close together; spines* 6 to 20, slender; central spines much longer than the radials, 2 to 3 cm. long; flowers 5 to 6 cm. long, with a long slender tube, scarlet; scales on ovary and flower-tube small, with long white hairs in their axils; fruit globular, 2 cm. in diameter.

Type locality: Naquira, Chile.

Distribution: Vicinity of Arequipa, southern Peru to northern Chile.

Dr. Rose found this species very common about Arequipa and for a long time he was unable to identify it. Its habit and dry fruit suggested some of the so-called species of *Echinocactus* but its slender red flower was not typical of that genus. It did not suggest the genus *Echinopsis* in any particular. A careful examination of the description of *Echinopsis hempeliana* Gürke and the cited illustration, however, points definitely to this Arequipa plant. The home of *E. hempeliana* was unknown to Gürke, but it is not unlikely that it was collected in southern Peru, perhaps at the time *Opuntia hempeliana* was found.

Here belongs doubtless the Arequipa plant mentioned by Schumann under *Echinopsis rhodacantha.*

After leaving Arequipa Dr. Rose went to Santiago, Chile, where he found in the herbarium of Dr. Philippi the type specimen of *Echinocactus leucotrichus* which has flowers almost identical with those of the Arequipa plant and we therefore adopt this specific name for the group. We have not seen specimens of *Echinocactus clavatus* but the illustration and description of the flowers point strongly to this species.

Echinocactus hempelianus Schumann (Monatsschr. Kakteenk. 15: 178. 1905) is only a mentioned name.

Illustrations: Blühende Kakteen 2: pl. 85, as *Echinopsis hempeliana;* Schumann, Gesamtb. Kakteen Nachtr. f. 17; Monatsschr. Kakteenk. 10: 25, as *Echinocactus clavatus.*

2. Arequipa myriacantha (Vaupel)

Echinocactus myriacanthus Vaupel, Bot. Jahrb. Engler 50: Beibl. 111: 25. 1913.

Simple, depressed-globose, 10 cm. in diameter, 8 cm. high; ribs 26, strongly tubercled, separated by an acute sinus; areoles closely set, broadly elliptic; spines slender, bristle-like, when young brown, in age dark gray, 25 or more, the longer ones 3 cm. long; flowers slender, tubular, 5 to 6 cm. long; axils of the scales on the flower-tube and ovary bearing long silky brown hairs.

Type locality: Above Balsas in the provinces of Chachapoyas, Department of Amazonas, Peru, altitude 2,200 meters.

Distribution: Northeastern Peru.

This species is related to *Arequipa leucotricha*, but it has very different armament.

Through the kindness of F. Vaupel we have been able to examine a fragment of the type of this species which was collected by A. Weberbauer (No. 4272) and is now preserved in the herbarium of the Botanical Garden at Berlin.

---

*In seedlings and even in small 5 to 6 year-old plants the spines are pilose.

10. OROYA gen. nov.

Plants solitary, depressed-globose, low-ribbed; areoles elongated, narrow; spines widely spreading; flowers central, borne at the upper edge of the spine-areole, red to pink, short-funnelform; tube-proper very short, naked; stamens inserted on the throat of the flower, included; scales on the ovary small, bearing small tufts of hair in their axils; fruit short, clavate, glabrous.

Type species: *Echinocactus peruvianus* Schumann.

The generic name is that of the village in Peru in which the type species is found.

1. Oroya peruviana (Schumann).

Echinocactus peruvianus Schumann, Gesamtb. Kakteen Nachtr. 113. 1903.

Depressed and deep-seated in the ground, 10 to 14 cm. broad; ribs usually 21, low, obtuse, divided into low tubercles; areoles 8 to 12 mm. long; radial spines 18 to 20, yellowish with darker bases and reddish tips, unequal, the longer ones about 2 cm. long; central spines sometimes as many as 4, but often wanting, a little longer and stouter than the laterals and usually red in color; flowers 1.5 to 2 cm. long; outer perianth-segments acute, reddish, the inner pink, yellow at base, linear, obtuse or apiculate; style pink above; stigma-lobes pale yellow; seeds 2 mm. long, black.

Type locality: High mountains above Lima, Peru.

Distribution: High Andes of central Peru.

In 1914 Dr. Rose found this plant growing abundantly in the vicinity of Oroya, Peru, on a gravelly flat. The individual plants show only their flat tops above the surface of the soil. *Opuntia floccosa* was also common in the same locality.

Illustrations: Blühende Kakteen 2: pl. 88; Monatsschr. Kakteenk. 15: 191, as *Echinocactus peruvianus.*

DESCRIBED SPECIES, PERHAPS OF THIS GENUS.

ECHINOCACTUS AURANTIACUS Vaupel, Bot. Jahrb. Engler 50: Beibl. 111: 23. 1913.

Simple or cespitose, subglobose; ribs about 16; areoles elliptic; spines about 25, reddish brown, unequal, about 16, spreading, about 1 cm. long, somewhat more radial than the others; the subcentral one of them 5 cm. long; flowers narrowly funnelform, 7 cm. long; fruit 1 cm. in diameter, covered with small lanceolate scales; seeds black.

Type locality: San Pablo, department of Catamarca, Peru.

Distribution: Peru.

11. MATUCANA gen. nov.

Usually simple, small and globular, rarely elongated; ribs numerous, broad, low, somewhat tubercled; areoles approximate, very woolly when young, with numerous acicular or bristly spines; flower slender, tubular, scarlet, with a narrow limb; scales on the ovary and flower-tube scattered, naked in their axils; fruit not known.

Type species: *Echinocactus haynei* Otto.

The genus as known to us consists of one species, confined to the high mountains of central Peru. The generic name is that of the small village in central Peru near which the type species grows.

1. Matucana haynei (Otto).

Echinocactus haynii Otto in Salm-Dyck, Cact. Hort. Dyck. 1849. 165. 1850.
Cereus hayni Croucher, Garden 13: 290. 1878.

Stems generally single, usually globular but sometimes short-cylindric, normally 8 to 10 cm. in diameter but in cultivation sometimes 30 cm. high, densely covered and nearly concealed under the numerous spines; ribs 25 or more, tuberculate; areoles set closely together, with an abundance of wool when young, but without any when old; spines numerous, long and weak, the stouter ones pungent, up to 3.5 cm. long, usually gray with dark or blackish tips; flower with a long slender tube, 6 to 7 cm. long; stigma-lobes green; scales on ovary and tube few, small, ovate.

Type locality: Obrajillo, Peru, but not cited in the original place of publication.

Distribution: Central Peru.

Dr. Rose found this species fairly common among rocks just below Matucana, Peru. Its long, slender, scarlet flowers make it a very desirable plant for cultivation. This species was collected in abundance by Dr. Rose in 1914 but it has not yet flowered in cultivation, although it was flowering when collected. The original spelling of the specific name was *haynii*, named, according to Rümpler, for Friedrich Gottlieb Hayne, a professor of botany in Berlin, who was born in 1832. Schumann in his Monograph wrote the name *Echinocactus haynei* and we have adopted his spelling.

In color, shape, and size, the flowers resemble those of species of *Borzicactus* (Cactaceae **2**: 159); these, however, have the axils of scales on the corolla-tube hairy or woolly.

Echinocactus heynei (Monatsschr. Kakteenk. **20**: 190. 1910) was never described. It may be simply a misspelling for *E. haynei.*

We have referred here the illustration from the Garden because it is made from the same cut as the four other illustrations cited below. It is there described, however, as a foot high with white flowers!

Illustrations: Dict. Gard. Nicholson **1**: f. 689; Cact. Journ. **1**: 181; Watson, Cact. Cult. 103. f. 35; ed. 3. 54. f. 24, as *Echinocactus haynei;* Garden **13**: 291, as *Cereus hayni.*

Figure 109 is from a photograph of a plant collected by Dr. and Mrs. Rose at Matucana, Peru, in 1914 (No. 18651).

FIG. 109—Matucana haynei.

DESCRIBED SPECIES, PERHAPS OF THIS GENUS.

ECHINOCACTUS VILLOSUS (Monville) Labouret, Monogr. Cact. 239. 1853.

> *Cactus villosus* Monville, Hort. Univ. **1**: 223. 1839.
> *Echinocactus polyrhaphis* Pfeiffer in Förster, Handb. Cact. 297. 1846.
> *Echinocactus villosus crenatior* Monville in Labouret, Monogr. Cact. 240. 1853.

Simple, subglobose or short-columnar, somewhat depressed; ribs 13 to 15, somewhat tuberculate, subcompressed, glaucous-green or somewhat violet or even blackish; radial spines 12 to 16, setaceous to subulate; central spines usually 4, 3 cm. long; flowers rose-colored without, white within; scales of the ovary and flower-tube naked in their axils.

Type locality: Not cited.

Distribution: Chile (*fide* Schumann); Lima, Peru (*fide* Labouret).

This species is said to resemble *Echinocactus acutissimus* but is described by Schumann as having naked scales and the axils of the scales also naked. If it came from near Lima, Peru, as is usually stated, it is probably *Echinocactus haynei.* Söhrens claims it is from Huasco, Chile.

Gymnocalycium villosum Pfeiffer is given by Förster (Handb. Cact. 297. 1846) as a synonym of *E. polyrhaphis. E. polyrhaphis* is written *polygrhaphis* by Labouret and *polyraphis* by Salm-Dyck.

Illustration: Möllers Deutsche Gärt. Zeit. **25**: 474. f. 6, No. 26.

ECHINOCACTUS WEBERBAUERI Vaupel, Bot. Jahrb. Engler **50**: Beibl. **111**: 26. 1913.

Depressed, 10 cm. broad, 7 cm. high; ribs 21, divided into terete tubercles; areoles rather close together, broadly elliptic; spines about 30, the longer ones 3.5 cm. long, straight; flowers tubular,

5.5. cm. long; ovary and flower-tube bearing lanceolate acute scales, these without hairs in their axils.

Type locality: Above Balsas in the department of Amazonas, Peru.
Distribution: Northeastern Peru.

Through the kindness of F. Vaupel we have been able to study a fragment of this very interesting species. It much resembles *Matucana haynei*.

12. HAMATOCACTUS gen. nov.

Globose to short-cylindric, of flabby texture like an *Echinocereus*, distinctly ribbed, the ribs more or less spiraled; areoles circular; spines radial and central, one of them usually hooked; flower-bud pointed, covered with imbricating scales; flower-tube narrow, funnelform; limb broad; scales on the ovary few, fugacious, small, naked in their axils; fruit small, globular, red, dehiscing by a basal pore; seeds black, tuberculate; hilum large, basal, circular; embryo straight; cotyledons short and thick.

Echinocactus setispinus Engelmann, the only species here recognized, is the type.

Although this plant heretofore always passed as an *Echinocactus* its anomalous characters have been recognized, such as the texture of the fleshy stem, the fruit, and the seeds; Engelmann in his Synopsis of the Cactaceae thus spoke of it: "The compressed ribs, setaceous spines, small red berry, and tuberculated seeds easily distinguish it from all its allies."

The generic name is from *hamatus* hooked, and *cactus*, referring to the hooked central spine.

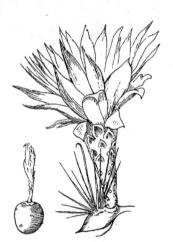

FIGS. 110 and 111.—Fruit and flower of
Hamatocactus setispinus. ×0.8.

FIG. 112.—Hamatocactus setispinus.

1. Hamatocactus setispinus (Engelmann).

Echinocactus setispinus Engelmann, Bost. Journ. Nat. Hist. **5**: 246. 1845.
Echinocactus muehlenpfordtii Fennel, Allg. Gartenz. **15**: 65. 1847.
Echinocactus hamatus Mühlenpfordt, Allg. Gartenz. **16**: 18. 1848. Not Forbes, 1837.
Echinocactus setispinus hamatus Engelmann, Bost. Journ. Nat. Hist. **6**: 201. 1850.
Echinocactus setispinus setaceus Engelmann, Bost. Journ. Nat. Hist. **6**: 201. 1850.
Echinocactus setispinus cachetianus Labouret, Monogr. Cact. 203. 1853.
Echinocactus hamulosus Regel, Ind. Sem. Hort. Petrop. 34. 1856.

Echinopsis nodosa Linke, Wochenschr. Gärtn. Pflanz. **1**: 85. 1858.
Echinocactus nodosus Hemsley, Biol. Centr. Amer. Bot. **1**: 535. 1880.
Echinocactus setispinus muhlenpfordtii Coulter, Contr. U. S. Nat. Herb. **3**: 370. 1896.
Echinocactus setispinus mierensis Schumann, Gesamtb. Kakteen 340. 1898.
Echinocactus setispinus orcuttii Schumann, Gesamtb. Kakteen 340. 1898.

Plants up to 15 cm. high, with long fibrous roots; ribs usually 13, more or less oblique, thin, high, undulate on the margin; radial spines 12 to 16, slender, often 4 cm. long, some white, others brownish; central spines 1 to 3, longer than radials; flower 4 to 7 cm. long, yellow, with a red center; inner perianth-segments oblong, acute, widely spreading; fruit 8 mm. in diameter, nearly naked; seeds 1.2 to 1.6 mm. in diameter.

Type locality: Thickets along the Colorado River, Texas.

Distribution: Southern Texas and northern Mexico.

Two species or very distinct forms pass under *E. setispinus;* both are common about Brownsville, Texas, and some very fine plants and photographs have recently been sent us by Mr. Robert Runyon. He believes that they are distinct species and the extreme forms are certainly very different. The two forms, however, grow at the same locality

Figs. 113 and 114.—Hamatocactus setispinus.

under the same conditions and so far as we know have the same flowers and fruits. Engelmann, too, had noted the differences and gave them the varietal names, *hamatus* and *setaceus*. It will require still further field work before we can reach a definite conclusion regarding the limits of this species.

According to Engelmann (Cact. Mex. Bound. 21), this species was sent to him by Berlandier as *Cactus bicolor* and Schumann refers to it as *Echinocactus bicolor* Berlandier (Gesamtb. Kakteen 339. 1898). It is, however, very different from *Cactus bicolor* Berlandier (Mem. Comm. Limites 1. 1832).

Echinocactus cachetianus (Labouret, Monogr. Cact. 202. 1853), although not described in the place cited, is probably to be referred here. It is described, however, in Garten-Zeitung (**4**: 173. 1885). *E. cachetianus orcuttii* is given as a synonym of *E. setispinus orcuttii* by Schelle (Handb. Kakteenk. 159. 1907). At the same place Schelle lists the variety *E. setispinus martelii* Garde frèr. *Echinocactus marisianus* Galeotti, a manuscript name, is referred here by Schumann (Gesamtb. Kakteen 339. 1898).

Illustrations: Cact. Journ. **1**: pl. for March; 181; Wiener Ill. Gart. Zeit. **29**: f. 22, No. 4; Monatsschr. Kakteenk. **8**: 131; **15**: 73; Förster, Handb. Cact. ed. 2. 522. f. 65; 523. f. 66; Cact. Mex. Bound. pl. 20; Meehan's Monthly **9**: pl. 6; Möllers Deutsche Gärt. Zeit. **25**: 474. f. 6, No. 23; Schumann, Gesamtb. Kakteen f. 59; Schelle, Handb. Kakteenk. 158. f. 87, as *Echinocactus setispinus;* Monatsschr. Kakteenk. **1**: pl. 9; Gartenwelt **9**: 266, as *Echinocactus cachetianus;* Möllers Deutsche Gärt. Zeit. **25**: 474. f. 6, No. 29, as *Echinocactus setispinus longispinus.*

Figures 110 and 111 show a flower and fruit copied from plate 20 of Cactaceae of the Mexican Boundary; figures 112 and 113 are from photographs made by Robert Runyon at Brownsville, Texas; figure 114 is from a photograph of a plant sent from Brownsville by Robert Runyon.

13. STROMBOCACTUS gen. nov.

A low, depressed, nearly spineless cactus, with imbricated chartaceous scale-like tubercles; flowers small, central, subcampanulate, nearly white; scales on the flower-tube with thin papery margins; scales on the ovary only near the top, small; fruit nearly naked; seeds small.

Type species: *Mammillaria disciformis* De Candolle.
One species is known, native of Mexico.
The generic name is from στρόμβος top, and κάκτος cactus, referring to the shape of the plant.

Figs. 115 and 116.—Strombocactus disciformis.

1. Strombocactus disciformis (De Candolle).

Mammillaria disciformis De Candolle, Mém. Mus. Hist. Nat. Paris **17**: 114. 1828.
Echinocactus turbiniformis Pfeiffer, Allg. Gartenz. **6**: 275. 1838.
Echinofossulocactus turbiniformis Lawrence in Loudon, Gard. Mag. **17**: 318. 1841.
Mammillaria turbinata Hooker in Curtis's Bot. Mag. **69**: pl. 3984. 1843.
Cactus disciformis Kuntze, Rev. Gen. Pl. **1**: 260. 1891.
Cactus turbinatus Kuntze, Rev. Gen. Pl. **1**: 261. 1891.
Anhalonium turbiniforme Weber, Dict. Hort. Bois 90. 1893.
Echinocactus disciformis Schumann in Engler and Prantl, Pflanzenfam. 3⁶ᵃ: 189. 1894.

Plants small, depressed, turbinate or semi-globose, 5 to 6 cm. broad; tubercles somewhat chartaceous, imbricate, more or less winged, bearing 1 to 4, white, acicular spines when young, naked when old; young areoles with white wool, naked in age; flowers from center of plant, 2 cm. long or less; scales and outer perianth-segments dark red, with whitish margins; inner perianth-segments white, lanceolate, acute, spreading; filaments much shorter than the inner perianth-segments, purple; stigma-lobes about 7, long, twisted; ovary naked except at the top, small; fruit 7 mm. long; seeds 3 mm. in diameter.

Type locality: Mineral del Monte, Mexico.
Distribution: Central Mexico.

This plant was collected by Dr. Rose in the state of Querétaro, Mexico, in 1905, and has repeatedly flowered each spring since 1906.   It is called pellote or peyote in Mexico.

Echinocactus helianthodiscus Lemaire (Salm-Dyck, Cact. Hort. Dyck. 1844. 17. 1845), given as a synonym of *Echinocactus turbiniformis*, was never described.

Illustrations: Blühende Kakteen 1: pl. 39, *a*; Schumann, Gesamtb. Kakteen f. 77; Monatsschr. Kakteenk. 5: 119; 12: 91; Pfeiffer, Abbild. Beschr. Cact. 2: pl. 3; Schelle, Handb. Kakteenk. 203. f. 136, as *Echinocactus turbiniformis;* Curtis's Bot. Mag. 69: pl. 3984, as *Mammillaria turbinata.*

Figure 115 is from a photograph of a plant collected by Dr. Rose at Higuerillas, Querétaro, Mexico, in 1905; figure 116 is copied from the plate in the Blühende Kakteen cited above.

FIG. 117.—Leuchtenbergia principis.

14. LEUCHTENBERGIA Hooker in Curtis's Bot. Mag. 74: pl. 4393. 1848.

A low, simple or cespitose cactus, with a thickened woody base; tubercles finger-like, slender, much elongated, arranged in indefinite spirals; areoles on the ends of the tubercles; spines several, weak, often papery; flowers from near the center of the plant, large, yellow, funnelform-campanulate; scales on the ovary few, broad, naked in their axils; fruit probably dehiscing by a basal pore.

Type species: *Leuchtenbergia principis* Hooker.

The genus contains but one species; both genus and species were named for Eugene de Bauharnais, Duke of Leuchtenberg and Prince of Eichstädt, a French soldier and statesman (1781–1824).   The generic name is usually credited to Hooker and Fischer, but a careful examination of the early literature indicates that the plant was first described by Sir William Hooker in 1848 who says, "I willingly adopt a name by which the plant is known on the continent." In 1850 it is described by Fischer as his own genus.

This genus is closely related to *Echinocactus* and its segregates, having very similar flowers and fruits, but in its elongated angled tubercles it looks very unlike any of them. Engelmann suggested, although he never saw the fruit, that it might be a subgenus of *Mammillaria*.

1. Leuchtenbergia principis Hooker in Curtis's Bot. Mag. **74**: pl. 4393. 1848.

Plants up to 5 dm. high, 5 to 7 cm. in diameter, with a large simple or branched tap-root, often 12 cm. long; tubercles erect, ascending or widely spreading, very woolly in their axils, bluish green, 10 to 12.5 cm. long, more or less 3-angled, nearly truncate at apex, gradually dying off below and leaving broad scars on the trunk; spines papery, thin; radial spines 8 to 14, about 5 cm. long; central spines 1 or 2, sometimes 10 cm. long; flowers lasting for several days, fragrant, solitary, from just below the tips of the young tubercles, more or less funnelform, the limb when widely expanded 10 cm. broad; outer perianth-segments reddish with a brown stripe down the middle; inner perianth-segments oblong, acute, serrate at apex; stamens and style somewhat exserted; stigma-lobes 9 to 12, linear; fruit probably dry; seeds dark brown, minutely tuberculate.

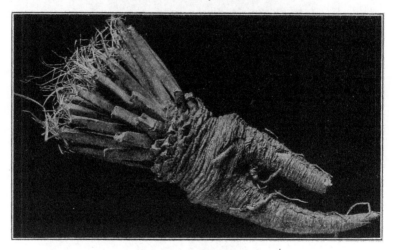

FIG. 117 a.—Leuchtenbergia principis.

Type locality: Real del Monte (not Rio del Monte), Hidalgo, Mexico.
Distribution: Central to northern Mexico.

Hooker's plant came from Real del Monte, Hidalgo, where it was obtained by John Taylor. This is the only locality cited by Hemsley in the Biologia. It has been reported from the states of San Luis Potosí, Guanajuato, Zacatecas, and Coahuila. This is a wide distribution for the species. We have never seen plants from near the type locality.

In appearance this plant is very unlike any of the other cacti. Hooker speaks of its resemblance to some aloid plant with stems like those of some cycads. It is said to be used by the Mexicans as a medicine.

The plant was called by Hooker noble leuchtenbergia and also agave cactus.

Dr. C. A. Purpus writes that he found this plant in slate and lime formation in the Sierra de la Parras near Parras, Coahuila, and still more abundant in the Sierra de la Paila, also in Coahuila. This last station is a very inaccessible desert mountain range, almost without water; vegetation is here very scanty. This species is associated with other cacti and with *Agave lophantha,* which it resembles in its habit more than it does that of its own relatives.

Illustrations: Palmer, Cult. Cact. 125; Curtis's Bot. Mag. **74**: pl. 4393; Cact. Journ. **1**: 149; Dict. Gard. Nicholson Suppl. f. 515; Ann. Rep. Smiths. Inst. 1908: f. 23; Krook, Handb. Cact. 30; Monatsschr. Kakteenk. **4**: 9; Schumann, Gesamtb. Kakteen f. 5, 78; Förster, Handb. Cact. ed. 2. f. 77; Engler and Prantl, Pflanzenfam. 3^{6a}: f. 66; Cycl. Amer. Hort. Bailey **2**: f. 1269; Amer. Gard. **11**: 464; Schelle, Handb. Kakteenk. f. 137; Möllers, Deutsche Gärt. Zeit. **25**: 477. f. 11, No. 5; **29**: 90. f. 12; 91. f. 13; Garten-Zeitung **4**: 182. f. 42, No. 9; 286. f. 66; Gard. Chron. **1873**: 1116. f. 240; III. **29**: f. 63; Belg. Hort. **5**: pl. 40; Stand. Cycl. Hort. Bailey **2**: 610. f. 720; **4**: f. 2139; Blühende Kakteen **3**: pl. 158; Rümpler, Sukkulenten 192. f. 108; Goebel, Pflanz. Schild. **1**: pl. 2, f. 1; Gartenwelt **5**: 110; Watson, Cact. Cult. 186. f. 74; ed. 3. f. 51; Thomas, Zimmerkultur Kakteen 44; Remark, Kakteenfreund 21.

Figure 117 is copied from plate 4393 of the Botanical Magazine; figure 117*a* is from a photograph of a plant sent by Dr. Elswood Chaffey from Zacatecas, Mexico, in 1910.

15. ECHINOFOSSULOCACTUS Lawrence in Loudon, Gard. Mag. **17**: 317. 1841.

Mostly rather small plants, rarely over 10 cm. in diameter, but generally much smaller, usually solitary, rarely clustered, deep-seated in the ground, globular or depressed, or very old plants becoming short-cylindric; ribs usually numerous, in one species as few as 10, in others 50 to 100, usually very thin, more or less wavy; areoles on each rib sometimes only 1 or 2, always felted when young; spines in numerous clusters often covering the plant, some of them strongly flattened and ribbon-like; flowers small, campanulate to subrotate with a very short tube; stamens numerous, shorter than the perianth-segments; scales on the perianth and ovary few to numerous, scarious, naked in their axils; fruit globular to short-oblong, bearing a few papery scales, these perhaps deciduous in age; seeds black with a broad basal truncated hilum.

About 22 species, all native of Mexico, are here recognized, although more than three times as many species of this relationship have been described in *Echinocactus.* From our field observations the number of species must be larger than here recognized, but the herbarium material is so scanty and the species already described are so many that for the present we have contented ourselves chiefly in describing those which have been illustrated or are represented by preserved material.

Although this genus appears to be very distinct, the species are so little known that we can give only a few of the characters. In the case of one plant which recently fruited in the New York Botanical Garden the fruit splits down one side as in *Pediocactus.* This may be a common character in the genus and should be looked for whenever possible.

This genus was established by George Lawrence, gardener to the Rev. Theodore Williams at Hendon Vicarage, Middlesex, England, but so far has been overlooked by catalogues. We came upon it while looking through Loudon's Magazine of Gardening for new cactus names. Its publication, however, had been observed and noted by that keen bibliophile, James Britten (Journ. Bot. **54**: 338. 1916).

Lawrence numbers 35 species and varieties, most of which are named and briefly described. The genus as he defines it is not a very natural one. He arranges the species in three sections and each section is divided into two subsections.

His first section, *Gladiatores,* corresponds to Schumann's subgenus *Stenocactus* of the genus *Echinocactus* and represents *Echinofossulocactus* in our treatment, with *Echinocactus coptonogonus* Lemaire as its type.

The species belonging to his second section (*Latispineae*) and to his third section are referred elsewhere as synonyms, except the following which we are not able to associate with any of the names of *Echinocactus: E. harrisii* and *E. ignotus-venosus.*

Schumann described the group briefly, as follows:

"Ribs mostly moderately high, laterally compressed, almost like cardboard, very many (*E. coptonogonus* with only 13 to 15); flowers small, like a *Mammillaria* flower; ovary with scales and glabrous."

KEY TO SPECIES.

Ribs thick at base, triangular in cross-section.
  Ribs 10 to 14... 1. *E. coptonogonus*
  Ribs about 35.. 2. *E. hastatus*
Ribs always numerous, very thin even at base.
  Ribs 100 or more... 3. *E. multicostatus*
  Ribs 25 to 55.
    Some or all of radial spines acicular or setaceous.
      Radial spines all acicular, white, straight.
        Flowers greenish yellow.
          Central spines terete.. 4. *E. wippermannii*
          Central spines narrow, but flattened..................................... 5. *E. heteracanthus*
        Flowers not greenish yellow.
          Central spines 4.. 6. *E. albatus*
          Central spines 3.
            Central spines annulate; apex of plant not depressed..................... 7. *E. lloydii*
            Central spines not annulate; apex of plant umbilicate.................... 8. *E. zacatecasensis*
      Upper radial spines subulate, some flattened.
        Spines yellow or white.
          Spines only 5 or 6... 9. *E. lamellosus*
          Spines 8 to 11... 10. *E. grandicornis*
        Central and upper spines brownish... 11. *E. arrigens*
    None of the spines acicular.
      Perianth-segments rather short.
        All spines appressed against plant.. 12. *E. violaciflorus*
      Some spines erect or porrect.
        Ribs about 25.
          Four upper spines much elongated..................................... 13. *E. obvallatus*
          Spines all somewhat similar.
            Spines only 5.. 14. *E. pentacanthus*
            Spines 10 or more... 15. *E. crispatus*
        Ribs 30 or more.
          Radial spines white... 16. *E. dichroacanthus*
          Radial spines brown.
            Flowers purplish.. 17. *E. anfractuosus*
            Flowers yellow.
              Upper and flattened spines 3, rather short, red...................... 18. *E. tricuspidatus*
              Upper and flattened spines usually 1, rarely 2...................... 19. *E. phyllacanthus*
Perianth-segments much elongated and widely spreading or recurved .... 20. *E. lancifer*
Species not grouped .... {21. *E. gladiatus*
                                                                                 {22. *E. confusus*

1. Echinofossulocactus coptonogonus (Lemaire) Lawrence in Loudon, Gard. Mag. **17**: 317. 1841.

    Echinocactus coptonogonus Lemaire, Cact. Aliq. Nov. 23. 1838.
    Echinocactus coptonogonus major Lemaire, Cact. Gen. Nov. Sp. 87. 1839.
    Echinofossulocactus coptonogonus major Lawrence in Loudon, Gard. Mag. **17**: 317. 1841.

Simple or perhaps cespitose, globular or a little depressed, 7 to 10 cm. high, glaucous-green; ribs stout, 1.5 cm. high, 10 to 14, acute; areoles about 2 cm. apart, when young abundantly floccose, but in age naked; spines 3 to 5, stout, a little incurved, the longest 3 cm. long, flattened; flowers 3 cm. long, 4 cm. broad; inner perianth-segments numerous, linear-oblong, acute, purple with white margins; ovary brownish violet, bearing thin scales.

Type locality: Mexico.

Distribution: Mexico, near San Luis Potosí and Pachuca, according to Schumann; the plant found at the latter locality is probably to be referred elsewhere.

This species is very abundant about San Luis Potosí from which place we have received considerable material from Orcutt and Palmer. It does not do well in cultivation. Only one plant is now alive in our collection and this has never flowered.

Echinocactus interruptus Scheidweiler (Cact. Hort. Dyck. 1849. 29. 1850) was referred here but never published. It was also used by Pfeiffer (Enum. Cact. 65. 1837) as a synonym of *E. exsculptus*.

Illustrations: Pfeiffer, Abbild. Beschr. Cact. **2**: pl. 19; Blühende Kakteen **1**: pl. 28; Dict. Gard. Nicholson **4**: 538. f. 18; Suppl. f. 353; Förster, Handb. Cact. ed. 2. 527. f. 67; Lemaire, Icon. Cact. pl. 7; Schelle, Handb. Kakteenk. 168. f. 99; Watson, Cact. Cult. 95. f. 30, as *Echinocactus coptonogonus.*

2. Echinofossulocactus hastatus (Hopffer).

Echinocactus hastatus Hopffer in Schumann, Gesamtb. Kakteen 376. 1898.

Simple, depressed-globose, 10 cm. high, 12 cm. in diameter; ribs 35, triangular in section, light green, somewhat crenate; radial spines 5 or 6, very short, straight, yellow, the upper ones flattened, often 3 cm. long; central spine solitary, 4 cm. long, porrect; flowers white (the largest in this genus); fruit becoming dry; seeds obovate, 1.5 mm. long, brownish gray, shining, finely punctate.

Type locality: Mexico.

Distribution: Mexico, Hidalgo, north of Pachuca.

This species and its variety *fulvispinus* Allardt were only mentioned by Förster (Handb. Cact. 315. 1846). We know this species only from description.

3. Echinofossulocactus multicostatus (Hildmann).

Echinocactus multicostatus Hildmann in Mathsson, Gartenflora **39**: 465. 1890.

Simple, usually globose, but sometimes depressed, 6 to 10 cm. in diameter; ribs 100 or more, very thin, wavy, each bearing only a few areoles; areoles pubescent when young; spines usually 6 to 9, divided into two classes, the 3 upper spines elongated, 4 to 8 cm. long, erect or ascending, flexible, rather thin but not very broad, yellowish to brownish; lower spines spreading, weak-subulate, 5 to 15 mm. long; flowers 2.5 cm. long; outer perianth-segments oblong, acuminate; inner perianth-segments oblong, acute or obtuse; scales on the flower-tube oblong, acuminate; scales on the ovary broadly ovate, acute to acuminate, very thin, more or less papery, early deciduous.

Type locality: Mexico.

Distribution: Eastern part of Mexico.

According to Schumann, this species was collected by Mathsson at Saltillo, Mexico, and this place is doubtless the type locality for the species. Here, fine specimens were collected by Dr. Edward Palmer in 1905 and by W. E. Safford in 1907, both in flower, so we are now able to describe the flowers.

Illustrations: Schumann, Gesamtb. Kakteen f. 64; Ann. Rep. Smiths. Inst. **1908**: pl. 4, f. 3; Schelle, Handb. Kakteenk. 175. f. 104, 105, as *Echinocactus multicostatus.*

4. Echinofossulocactus wippermannii (Mühlenpfordt).

Echinocactus wippermannii Mühlenpfordt, Allg. Gartenz. **14**: 370. 1846.

Simple, obovoid, 15 cm. high, 5 to 6 cm. in diameter, dull green; ribs 35 to 40, compressed, slightly undulate; areoles 12 mm. apart, hairy when young, glabrate in age; radial spines 18 to 22, setaceous, white, 15 mm. long; central spines 3 or 4, erect, elongated, 2 to 5 cm. long, subulate, terete, blackish; flowers 1.5 mm. long, dull yellow.

Type locality: Mexico.

Distribution: Hidalgo, Mexico.

We know this species only from description. Schumann states that both Ehrenberg and Mathsson collected it in the state of Hidalgo, but the species collected there by Dr. Rose, while of this relationship, is certainly distinct (Rose, No. 8717, in part).

Schumann refers here *Echinocactus acifer spinosus* Wegener (Linnaea **19**: 355. 1847), an unpublished variety, which is the same as *Echinocactus spinosus* Wegener (Allg. Gartenz. **12**: 66. 1844). This name is older, and if the same as *E. wippermannii* it would replace it; *E. acifer* Hopffer (Förster, Handb. Cact. 520. 1846) is said to belong here according to Labouret.

Illustrations: Schelle, Handb. Kakteenk. 169. f. 100; Möllers Deutsche Gärt. Zeit. **25**: 474. f. 6, No. 12, as *Echinocactus wippermannii.*

5. Echinofossulocactus heteracanthus (Mühlenpfordt).

Echinocactus heteracanthus Mühlenpfordt, Allg. Gartenz. **13:** 345. 1845.
Echinocactus tetraxiphus Otto in Schumann, Gesamtb. Kakteen 363. 1898.

Globose to short-cylindric, light green, nearly hidden by the closely set spines; ribs 30 to 34, much compressed, somewhat undulate; areoles white, hairy when young; radial spines 11 to 13 (16 to 18, according to Schumann), acicular, white, spreading; central spines 4, brownish to flesh-colored, more or less annulate, compressed; flowers and fruit unknown to us, but greenish yellow, according to Schumann.

Type locality: Real del Monte, Mexico.
Distribution: Real del Monte, Hidalgo, Mexico.

We have had this species in cultivation, but it never flowered and the plant has died. Schumann has given a good illustration of it. Rose, No. 8717, in part, from near Real del Monte, may belong here.

Schumann took up an unpublished name of Otto (Salm-Dyck, Cact. Hort. Dyck. 1844. 20. 1845) for this species.

Illustrations: Schumann, Gesamtb. Kakteen f. 63; Monatsschr. Kakteenk. **15:** 160, as *Echinocactus tetraxiphus.*

Fig. 118.—Echinofossulocactus lloydii.

6. Echinofossulocactus albatus (Dietrich).

Echinocactus albatus Dietrich, Allg. Gartenz. **14:** 170. 1846.

Simple, depressed-globose, 10 to 12 cm. in diameter, glaucous, the apex covered with spines; ribs about 35, flat, undulate; spines yellowish white; radial spines 10, setaceous, 1 cm. long; central spines 4, the uppermost flat and annulate (according to Schumann), the central terete, porrect; flowers white, 2 cm. long.

Type locality: Mexico.
Distribution: Mexico.

We do not know this species. The plant described by Schumann may be different from the original of Dietrich.

7. Echinofossulocactus lloydii sp. nov.

Nearly globular, 12 cm. in diameter or more, crowned by the long overtopping connivent spines; ribs very numerous, thin, more or less folded; areoles brown, woolly when young; radial spines acicular, 10 to 15, white, 2 to 8 mm. long, spreading; central spines 3, light brown, much elongated, somewhat incurved and connivent, the two lateral ones similar and not so papery, the middle one very thin, annulate, 4 to 9 cm. long; flowers small, nearly white; outer perianth-segments with a green stripe on the mid-vein; inner perianth-segments thin, narrowly oblong, acute; scales on the ovary ovate, acute, very thin.

Collected by F. E. Lloyd in Zacatecas, Mexico, August 14, 1908 (No. 7), and flowered in Washington the same year.

Figure 118 is from a photograph of the type plant.

8. Echinofossulocactus zacatecasensis sp. nov.

Plants solitary, globular, 8 to 10 cm. in diameter; ribs pale green; very thin, about 55; radial spines 10 to 12, spreading, acicular, white, 8 to 10 mm. long; central spines 3, brownish, 2 of them terete, but the middle one flattened, erect or connivent, longer than the other 2, sometimes 3 to 4 cm. long, never annulate; flowers 3 to 4 cm. broad, nearly white; inner perianth-segments linear-oblong, with an ovate apiculate tip, slightly tinged with lavender, 15 mm. long; style slender; stigma-lobes bifid(!); scales on the ovary broadly ovate, apiculate, scarious.

Collected by F. E. Lloyd in northern Zacatecas, Mexico, in 1908 (No. 58). Living specimens were sent to Washington and these flowered in March 1909.

FIG. 119.—Echinofossulocactus zacatecasensis.   FIG. 120.—Echinofossulocactus arrigens.

This species is perhaps nearest *E. multicostatus*, but has fewer ribs and fewer radial spines.

Figure 119 is from a photograph of the type plant.

9. Echinofossulocactus lamellosus (Dietrich).

Echinocactus lamellosus Dietrich, Allg. Gartenz. **15:** 177. 1847.

Subglobose to short-cylindric, more or less depressed at the apex; ribs about 30, strongly flattened, more or less undulate; areoles remote, tomentose when young; spines 5 or 6, white (Schumann says yellow) with brown tips; flowers tubular, 3.5 to 4 cm. long; inner perianth-segments linear to linear-lanceolate, acute; stigma-lobes 5 to 8, linear, yellow.

Type locality: Mexico.
Distribution: Hidalgo, Mexico.

This species is known to us only from descriptions.

Echinocactus lamellosus fulvescens Salm-Dyck (Cact. Hort. Dyck. 1849. 30, 159. 1850) seems never to have been described.

10. Echinofossulocactus grandicornis (Lemaire).

Echinocactus grandicornis Lemaire, Cact. Gen. Nov. Sp. 30. 1839.
Echinocactus grandicornis fulvispinus Salm-Dyck in Labouret, Monogr. Cact. 210. 1853.
Echinocactus grandicornis nigrispinus Labouret, Monogr. Cact. 210. 1853.

Plants simple, globose to slender-cylindric, 10 cm. high, 5 to 6 cm. in diameter, glaucous-green, the apex hidden by the spines; ribs 34 or 35, much compressed, acute, undulate; areoles only a few to each rib, tomentose when young, naked in age; spines 8 to 11, at first yellowish; upper spine erect, stout, flat, 5 cm. long, the 2 lateral ones not so stout, a little shorter and nearly terete, the other spines slender; flowers whitish purple.

Type locality: Mexico.
Distribution: Mexico.
This species is known to us only from descriptions.

11. Echinofossulocactus arrigens (Link).

Echinocactus arrigens Link in Dietrich, Allg. Gartenz. 8: 161. 1840.
Echinocactus sphaerocephalus Mühlenpfordt, Allg. Gartenz. 14: 370. 1846.
Echinocactus allardtianus Dietrich, Allg. Gartenz. 15: 178. 1847.
Echinocactus arrigens atropurpureus Salm-Dyck, Cact. Hort. Dyck. 1849. 31, 162. 1850.

Plants simple, deep-seated in the soil, globular, 5 to 7 cm. in diameter, glaucescent, more or less depressed at the apex; ribs 24, thin and wavy; spines 8 to 11, yellow (according to Schumann); uppermost spine elongated, 2 to 4 cm. long, flattened, brownish; central spines 2 or 3, more slender and not quite so long as the uppermost one; radial spines 6 to 8, acicular, usually pale, spreading; flowers small, 2 to 2.5 cm. long; inner perianth-segments oblong, apiculate, with a deep purple stripe running down the center and with pale, nearly white margins.

Type locality: Mexico.
Distribution: Mexico, but definite locality is unknown.

Echinocactus xiphacanthus Miquel (Linnaea 12: 1. pl. 1, f. 1. 1838) is referred here by Schumann; if correctly, the name would replace *E. arrigens.* It is described as having 34 ribs; radial spines 4 or 5, short, pale; central spines 1 to 3, the upper one flat and long.

Echinocactus ensiferus Lemaire (Cact. Aliq. Nov. 26. 1838; *E. anfractuosus ensiferus* Salm-Dyck, Cact. Hort. Dyck. 1849. 31. 1850) is also referred here by Schumann. It, too, has priority over *E. arrigens.* It is described as globose; ribs 30 to 40.

Echinocactus arrectus Otto (Förster, Handb. Cact. 346. 1846), without description, is referred here as a synonym by Schumann.

Echinocactus ensiferus pallidus (Förster, Handb. Cact. 306. 1846) may also belong here.

Echinofossulocactus ensiformis (Lawrence in Loudon, Gard. Mag. 17: 317. 1841) may or may not belong here.

Illustrations: Pfeiffer, Abbild. Beschr. Cact. 2: pl. 27; Förster, Handb. Cact. ed. 2. f. 69; Schelle, Handb. Kakteenk. 173. f. 103, as *Echinocactus arrigens;* (?) Linnaea 12: pl. 1, f. 1, as *E. xiphacanthus.*

Figure 120 is a reproduction of the first illustration cited above.

12. Echinofossulocactus violaciflorus (Quehl).

Echinocactus violaciflorus Quehl, Monatsschr. Kakteenk. 22: 102. 1912.

Simple, at first globose, but becoming columnar, 8 to 10 cm. in diameter; ribs about 35, thin, deeply crenate; spines about 7, the 4 or 5 lower ones 7 to 12 mm. long, appressed or incurved, white, subulate, the 3 upper spines flattened, 3 to 6 cm. long, ascending and the uppermost ones connivent over the top of the plant; flowers 2 to 2.5 cm. long; perianth-segments narrow, acuminate, white with a violet or purplish stripe down the middle; scales on the ovary more or less imbricated, in 3 or 4 rows, broadly ovate, apiculate with scarious margins.

Type locality: Zacatecas, Mexico.

Distribution: Zacatecas and Aguas Calientes.

Our description is drawn in part from a plant sent to the New York Botanical Garden by Mr. H. Donnerstein in 1908, which flowered in April 1921, and in part from specimens collected by W. E. Safford at Aguas Calientes, Mexico, in 1907 (No. 1359).

Illustrations: Monatsschr. Kakteenk. **22:** 103, as *Echinocactus violaciflorus;* Ann. Rep. Smiths. Inst. **1908:** pl. 4, f. 5, as *Echinocactus crispatus.*

Plate XXIII, figure 5, shows the plant collected by William E. Safford, February 21, 1907 (No. 1359), which flowered in Washington and was painted May 9, 1907. Figure 121 is from a photograph of the same plant.

13. **Echinofossulocactus obvallatus** (De Candolle) Lawrence in Loudon, Gard. Mag. **17:** 317. 1841.

    Echinocactus obvallatus De Candolle, Prodr. **3:** 462. 1828.

Obovoid to globose, depressed at apex; ribs about 25, rather thin and undulate; spines about 8, 4 spines subulate, ascending or spreading, 4 spines short, perhaps not one-fourth the length of the longer ones; flowers central, very large; perianth-segments linear-oblong.

FIG. 121.—Echinofossulocactus violaciflorus.

Type locality: Mexico.

Distribution: Hidalgo, Mexico.

This species is based on Mociño and Sessé's illustration of *Cactus obvallatus.* We have been unable to refer here, with any degree of approximation, any Mexican material we have seen. Pfeiffer's plate 22 (Abbild. Beschr. Cact. **2:**), originally referred here, must be quite distinct, for it has very differently shaped flowers, spines, and ribs. The Index Kewensis refers this illustration to *Echinocactus lancifer* of which it seems to be the type.

Echinocactus obvallatus spinosior of Lemaire (Salm-Dyck, Cact. Hort. Dyck. 1849. 30. 1850) and also of Monville (Salm-Dyck, Cact. Hort. Dyck. 1844. 20. 1845) as well as variety *pluricostatus* Monville (Salm-Dyck, Cact. Hort. Dyck. 1844. 20. 1845) are all names without description.

Echinocactus coptonogonus obvallatus Salm-Dyck (Cact. Hort. Dyck. 1844. 20. 1845), unpublished, doubtless belongs here.

Illustrations: Cact. Journ. **2:** 102; Dict. Gard. Nicholson **1:** 500. f. 692; Förster, Handb. Cact. ed. 2. 535. f. 68; Schelle, Handb. Kakteenk. 172. f. 101; Watson, Cact. Cult. 113. f. 41; Karsten and Schenck, Vegetationsbilder **2:** pl. 19, *a;* Mém. Mus. Hist. Nat. Paris **17:** pl. 9, as *Echinocactus obvallatus.*

Figure 122 is copied from the last illustration cited above.

14. **Echinofossulocactus pentacanthus** (Lemaire).

    Echinocactus pentacanthus Lemaire, Cact. Aliq. Nov. 27. 1838.
    Echinocactus biceras Jacobi, Allg. Gartenz. **16:** 370. 1848.
    Echinocactus anfractuosus pentacanthus Salm-Dyck, Cact. Hort. Dyck. 1849. 31. 1850.

Simple, depressed-globose to short-cylindric, more or less glaucous; ribs about 25, perhaps even 40 to 50; areoles only a few to the rib; spines 5, unequal, grayish red, hardly angled, flattened; 3 upper spines erect or spreading; 2 lower spines much slenderer and shorter than the upper; flowers large for this group, deep violet; perianth-segments with white margins.

Type locality: Mexico.

Distribution: Mexico, in the states of Hidalgo and San Luis Potosí, according to Schumann.

We know this species definitely only from Lemaire's plate which is doubtless typical. According to Schumann, it has the same range as *E. obvallatus,* a species which, judging from the illustrations, must be very near it, if not actually the same.

A specimen collected by Rose in San Juan del Rio in 1905 has 5 spines in a cluster, but the whole plant is more heavily armed than Lemaire's illustration would indicate.

Echinocactus anfractuosus laevior Monville (Labouret, Monogr. Cact. 220. 1853) was given as a synonym of *E. anfractuosus pentacanthus.*

Illustration: Lemaire, Icon. Cact. pl. 11, as *Echinocactus pentacanthus.*

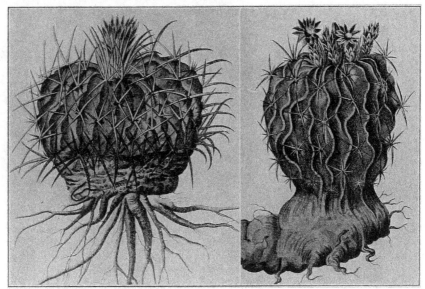

FIG. 122.—Echinofossulocactus obvallatus.　　　FIG. 123.—Echinofossulocactus crispatus.

15. Echinofossulocactus crispatus (De Candolle) Lawrence in Loudon, Gard. Mag. **17**: 317. 1841.

　　Echinocactus crispatus De Candolle, Prodr. **3**: 461. 1828.

Plant obovoid, somewhat depressed at apex; ribs about 25, more or less folded, somewhat undulate; spines 10 or 11, rigid, unequal; flowers central, rather small; perianth-segments in 2 series, purplish, oblong-linear, acute; flower-tube covered with imbricating scales.

Type locality: Mexico.

Distribution: Hidalgo, Mexico, according to Schumann.

This species was based on Mociño and Sessé's illustration which De Candolle reproduced and *Cactus crispatus* Mociño and Sessé (De Candolle, Prodr. **3**: 462. 1828) was the first name given to it, but it was never formally published. De Candolle, himself, does not compare it with *E. obvallatus* which is of this series, but with *E. cornigerus,* belonging to a very different series. After De Candolle had described the species he states (Mém. Mus. Hist. Nat. Paris **17**: 115. 1828) that the ribs vary from 30 to 60; this was doubtless drawn from new material, perhaps sent by Thomas Coulter who was then collecting in eastern Mexico. At the same place he describes var. *horridus,* based on Coulter's plant, and says that the spines are stouter, erect, long, and grayish brown.

Echinocactus stenogonus, first mentioned by Schumann (Monatsschr. Kakteenk. **5**: 107. 1895), who credits the name to Weber, seems to have been only a garden name for *Echinocactus crispatus*.

Watson (Cact. Cult. 99) states that "it is apparently closely allied to *E. longihamatus*," but this is hardly warranted.

The specimen figured in the Dictionary of Gardening, referred to below, may belong elsewhere. It is described as having about 20 ribs; radial spines 8 or 9, spreading, setaceous, white with brown tips; central spines 4, reddish and much larger than the radials. The plant came from the collections of F. A. Haage jr., of Erfurt. It may belong to *E. heteracanthus*. *E. crispatus cristatus* Gürke (Monatsschr. Kakteenk. **16**: 188. 1906) is also different.

E. flexispinus Salm-Dyck (Cact. Hort. Dyck. 1849. 159. 1850. Not Engelmann, 1848) is referred here by Schumann. It is described, however, as follows:

Globose or obovate, light green; ribs numerous, 30 or 31, strongly compressed, undulate, interrupted; areoles remote, when young bearing yellow wool; 3 upper spines recurved, ascending, somewhat flattened, a central one porrect and subulate, the 4 lower spines elongated, subulate, but flexible, white when young, pale brown when old; flowers unknown.

E. undulatus Dietrich (Allg. Gartenz. **12**: 187. 1844) is referred here also by Schumann, but Salm-Dyck thought it was unlike his *E. flexispinus* for the 3 upper spines are large and flat and the 4 lower ones are rigid.

Illustrations: Mém. Mus. Hist. Nat. Paris **17**: pl. 8; (?)Dict. Gard. Nicholson **1**: f. 688; Schelle, Handb. Kakteenk. 172. f. 102; Watson, Cact. Cult. 99. f. 33, as *Echinocactus crispatus;* Monatsschr. Kakteenk. **16**: 189, as *Echinocactus crispatus cristatus*.

Figure 123 is copied from the first illustration above cited.

16. Echinofossulocactus dichroacanthus (Martius).

Echinocactus dichroacanthus Martius in Pfeiffer, Enum. Cact. 62. 1837.
Echinocactus dichroacanthus spinosior Monville in Labouret, Monogr. Cact. 213. 1853.

Plant obovoid, dull green, 15 cm. high, 10 cm. in diameter, somewhat umbilicate at apex; ribs 32, thin, acute, undulate, somewhat wavy; areoles only a few on each rib, white-tomentose; upper spines 3, erect, flattened, purplish; radial spines 4 to 6, white; flowers and fruit unknown.

Type locality: Mexico.

Distribution: Hidalgo, Mexico.

Schumann's description differs somewhat from the original, but the only plant he refers to is that of Karwinsky, which is probably the type.

17. Echinofossulocactus anfractuosus (Martius) Lawrence in Loudon, Gard. Mag. **17**: 317. 1841.

Echinocactus anfractuosus Martius in Pfeiffer, Enum. Cact. 63. 1837.
Echinocactus anfractuosus spinosior Lemaire, Cact. Gen. Nov. Sp. 89. 1839.
Echinocactus anfractuosus orthogonus Monville in Labouret, Monogr. Cact. 220. 1853.

Plant simple, somewhat longer than broad, 12.5 cm. long, 6 cm. in diameter, dull green; ribs many (about 30, according to Schumann), compressed, wavy, each bearing only a few areoles; spines somewhat curved, straw-colored with brown tips; radial spines 7, stout, the 3 upper radials much larger, about 3 cm. long, the 4 lower radials slender; central spine solitary, 2.5 cm. long, brownish; perianth-segments purple with white margins.

Type locality: Mexico.

Distribution: Mexico, in Hidalgo at Pachuca and Ixmiquilpan.

18. Echinofossulocactus tricuspidatus (Scheidweiler).

Echinocactus tricuspidatus Scheidweiler, Allg. Gartenz. **9**: 51. 1841.
Echinocactus melmsianus Wegener, Allg. Gartenz. **12**: 65. 1844.
Echinocactus phyllacanthus tricuspidatus Förster, Handb. Cact. 311. 1846.

Globose to short-cylindric, 5 to 8 cm. broad; ribs numerous, 30 to 55, thin, wavy; areoles at first lanate, afterwards naked; spines 5 (Schumann says 9 to 11), the upper one thin, compressed, sometimes 3-toothed at apex, 8 to 33 mm. long, reddish with a black tip; the other 4 spines spreading, more or less appressed, straight or recurved, gray or reddish with black tips, much shorter than the upper one; flowers greenish yellow, 1.5 cm. long; inner perianth-segments short-oblong, obtuse, the outer ones more or less acute or apiculate; scales on the ovary broadly ovate with a scarious margin and a more or less prominent cusp.

Type locality: Not cited.

Distribution: San Luis Potosí, Mexico.

Our description is based on a large series of specimens all from near San Luis Potosí where they were collected by Dr. Edward Palmer in 1902 and 1905 and by C. R. Orcutt about 1915. This species is unlike any of its relatives, being characterized by the very short foliaceous upper spine.

Schumann's description of this species does not read much like the original and must represent a different species.

19. Echinofossulocactus phyllacanthus (Martius) Lawrence in Loudon, Gard. Mag. 17: 317. 1841.

> *Echinocactus phyllacanthus* Martius, Allg. Gartenz. **4**: 201. 1836.
> *Echinocactus phyllacanthoides* Lemaire, Cact. Gen. Nov. Sp. 28. 1839.
> *Echinofossulocactus phyllacanthus macracanthus* Lawrence in Loudon, Gard. Mag. **17**: 317. 1841.
> *Echinofossulocactus phyllacanthus micracanthus* Lawrence in Loudon, Gard. Mag. **17**: 317. 1841.

Simple, depressed-globose to short-cylindric, 3 to 15 cm. high, 4 to 10 cm. in diameter, dull green; ribs 30 to 35, thin, undulate; areoles only a few to a rib, white-tomentose when young; spines 5 to 9; upper spine, or rarely 2 spines, much elongated, erect or connivent over the top of the plant, flattened, thin, somewhat annulate, 4 cm. long; other spines weak-subulate, usually pale and spreading; flowers 15 to 20 mm. long, yellowish; inner perianth-segments acute.

Type locality: Mexico.

Distribution: Central Mexico; also reported from Mazatlan on the Pacific Coast of Mexico.

A good illustration of this species is published by Pfeiffer and Otto which is doubtless typical. Karwinsky's plant, from which the illustration was made, came from near Pachuca, Mexico. At this locality Dr. Rose collected flowers in 1905 (No. 8717) and these correspond to Pfeiffer's illustrations. At the same time and under the same number was collected a second species of this genus which is very distinct, showing how easily the species of this group can be confused.

Several varieties have been described, but these may not all belong here. These are as follows: *laevior* Monville, *laevis* Lemaire, *macracanthus*, *micracanthus*, *pentacanthus*, *tenuiflorus* (*E. tenuiflorus* Link in Salm-Dyck, Cact. Hort. Dyck. 1844. 20. 1845, name only) and *tricuspidatus* Förster.

Echinocactus stenogoni occurs occasionally as a legend for illustrations. This of course refers to *Echinocactus* series *stenogoni*. The first one which we have examined (Krook, Handb. Cact. 71. f. a) is figure 2 taken from Pfeiffer and Otto's plate 9, cited below. It occurs also in de Laet (Cat. Gen. f. 50, No. 6) for this or some closely related species. A third reference also occurs (Wiener Ill. Gart. Zeit. **29**: f. 22, No. 6).

Illustrations: Pfeiffer and Otto, Abbild. Beschr. Cact. **1**: pl. 9; Abh. Bayer. Akad. Wiss. München **2**: (see 738) pl. 2, f. 3, as *Echinocactus phyllacanthus.*

Figure 124 shows the first illustration cited above.

20. Echinofossulocactus lancifer (Dietrich).

> *Echinocactus lancifer* Dietrich, Allg. Gartenz. **7**: 154. 1839.
> *Echinocactus dietrichii* Heynhold, Nom. **2**: 92. 1846.

Nearly ovoid, somewhat depressed at apex; ribs numerous, strongly compressed, undulate; areoles few to each rib, when young tomentose; spines 8, white or brownish at apex; some of them broad and flat; flowers rather large, rose-colored; flower-tube described as long; perianth-segments linear-oblong, widely spreading.

Type locality: Mexico.

Distribution: Mexico.

Pfeiffer in 1837 attempted to identify *Echinocactus obvallatus* with certain plants then in the Schelhase collection, but later when he figured his plant he questioned this identification, although he did not rename it. Dietrich, however, in 1839, named the *Echinocactus obvallatus* Pfeiffer in part, as above.

Echinocactus lancifer was used by Reichenbach (in Terscheck, Suppl. Cact. 2), but whether properly described or not we do not know. The name was, however, formally published by Walpers in 1843 (Repert. Bot. **2**: 320), but this date was after Dietrich had published his name. Heynhold evidently considered Reichenbach's name properly published for he changed Dietrich's name to *Echinocactus dietrichii.*

Illustration: Pfeiffer, Abbild. Beschr. Cact. **2**: pl. 22 (*fide* Index Kewensis), as *Echinocactus obvallatus.*

Figure 125 shows the illustration cited above.

FIG. 124.—Echinofossulocactus phyllacanthus.   FIG. 125.—Echinofossulocactus lancifer.

21. Echinofossulocactus gladiatus (Link and Otto) Lawrence in Loudon, Gard. Mag. **17**: 317. 1841.

> *Echinocactus gladiatus* Link and Otto, Verh. Ver. Beförd. Gartenb. **3**: 426. 1827.
> *Melocactus gladiatus* Link and Otto, Verh. Ver. Beförd. Gartenb. **3**: pl. 17. 1827.
> *Echinocactus gladiatus ruficeps* Lemaire in Labouret, Monogr. Cact. 215. 1853.
> *Echinocactus gladiatus intermedius* Lemaire in Labouret, Monogr. Cact. 215. 1853.

Plant glaucescent, ovoid to oblong, 12.5 cm. high, 10 cm. in diameter with a depressed apex covered with connivent spines; ribs prominent, rather broad, obtuse, 14 to 22; spines 10, gray, 4 upper spines subulate, of these 3 usually ascending, the central spreading or porrect, the largest 5 cm. long, 4 lower spines acicular; flowers and fruit unknown.

Type locality: Mexico.

Distribution: Probably eastern Mexico.

Schumann in his monograph does not take for this name the plant as originally described and figured by Link and Otto, but a later description used by Pfeiffer and adopted by Salm-Dyck. The original *Echinocactus gladiatus,* based on Deppe's plant, is very different from Schumann's plant.

The plant is known to us only from description and illustration. We are following Lawrence in including this little-known species in *Echinofossulocactus*, but we are very doubtful about its true relationship.

Illustration: Verh. Ver. Beförd. Gartenb. **3**: pl. 17, as *Melocactus gladiatus*.

Figure 126 is a reproduction of the illustration cited above.

22. Echinofossulocactus confusus sp. nov.

Simple, pale green, stout, columnar to short-clavate, 6 to 15 cm. high, 6 to 8 cm. in diameter; ribs 26 to 30, thin, low, wavy; areoles 4 or 5 on each rib, 2 to 3 cm. apart; spines all yellow, subulate; radial spines 4 or 5, only slightly flattened, 7 to 10 mm. long; central spine solitary, up to 4 cm. long, usually porrect; flowers purplish, 4 cm. broad; perianth-segments oblong, acute.

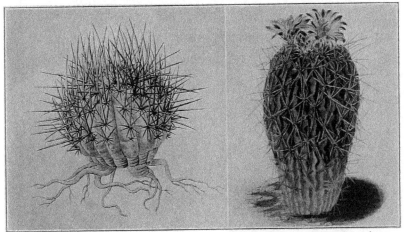

FIG. 126.—Echinofossulocactus gladiatus.   FIG. 127.—Echinofossulocactus confusus.

Our description is based on plate 159 of Blühende Kakteen, there called *Echinocactus gladiatus* Salm-Dyck.

This is also *Echinocactus gladiatus* Schumann (Gesamtb. Kakteen 374) in most part. Schumann states, however, that the flowers are yellow. Salm-Dyck never described *E. gladiatus* as a new species although he seems to have described a different plant than Link and Otto under their name. This plant is therefore without any true synonyms. Living plants, doubtless from Mexico, were in the Botanical Garden at Berlin.

Figure 127 is a reproduction of the illustration mentioned above.

PUBLISHED SPECIES, PERHAPS OF THIS GENUS.

The following species, judging from the brief unsatisfactory descriptions, belong here. They are all Mexican. None has been illustrated and the flowers of only a few of them are known.

ECHINOCACTUS ACANTHION Salm-Dyck, Cact. Hort. Dyck. 1849. 161. 1850.

Stems globose, 10 cm. in diameter or more, light green; ribs 35 to 40; upper spines 3, flattened, the central one stouter, the longest 35 mm. long; lower radial spines 8, slender, spreading, white; central spines 2, subulate; flowers unknown.

ECHINOCACTUS ACROACANTHUS Stieber, Bot. Zeit. **5**: 491. 1847.

Almost globose; ribs 27; areoles when young white-woolly, in age naked; spines 7, the 3 upper ones flattened, 2.5 to 3 cm. long, yellowish brown, with black tips; the 4 lower spines smaller than the upper, 8 to 10 mm. long, whitish; flowers and fruit unknown.

Distribution: Mexico.

The original spelling of this name was *E. acrocanthus,* doubtless a typographical error.

ECHINOCACTUS ADVERSISPINUS Mühlenpfordt, Allg. Gartenz. **16:** 10. 1848.

Obovoid; ribs 34, acute; areoles 3 cm. apart, white, lanate when young, naked when old; radial spines 7; central spine solitary, the 3 upper radials and central spine elongated and very different from the lower radials, 3 cm. long.

ECHINOCACTUS BRACHYCENTRUS Salm-Dyck, Cact. Hort. Dyck. 1849. 160. 1850.

    Echinocactus brachycentrus olygacanthus Salm-Dyck, Cact. Hort. Dyck. 1849. 160. 1850.
    Echinocactus oligacanthus Salm-Dyck in Schumann, Gesamtb. Kakteen 374. 1898. Not Pfeiffer, 1837.

Plant simple, short-cylindric, 20 cm. high, 15 cm. in diameter, very stout, yellowish green; ribs 30 to 35, strongly compressed, more or less undulate; spines all radial or sometimes one short central, the 3 upper spines erect, brownish (yellow, according to Schumann), the 2 lower ones white, smaller, and slenderer than the others.

Dr. Rose collected in San Juan del Rio in August 1905 a plant which answers very well this description, at least as far as it goes. It has 3 erect, stout, brown spines, the middle one much flattened. The two lower spines are small and reflexed, but brown. Other specimens collected by Dr. Rose in this region are somewhat similar, but the lower spines are in four's instead of two's.

Echinocactus brachiatus (Labouret, Monogr. Cact. 636. 1853) was written by mistake for this species.

ECHINOCACTUS CEREIFORMIS De Candolle, Mém. Mus. Hist. Nat. Paris **17:** 115. 1828.

Somewhat cylindric, 10 cm. high, green; ribs 13, flattened, separated by acute intervals, somewhat obtuse; areoles somewhat velvety, 3 to each rib; spines grayish, slender, but rigid; radial spines 7; central spine solitary, straight.

This species was based on a defective plant brought by Coulter from Mexico and the type has not been preserved. De Candolle put a question mark after the genus and also asked whether it might not be a young plant of some *Cereus,* but we do not know any species of *Cereus* with so few areoles on the ribs; in this respect it is like *Echinofossulocactus,* perhaps *E. coptonogonus.* The plant was unknown to Labouret and to Schumann and is listed by both among the unknown species.

ECHINOCACTUS DEBILISPINUS Berg, Allg. Gartenz. **8:** 131. 1840.

Subglobose to clavate, 16 cm. high, somewhat umbilicate at apex; ribs 34, flattened, wavy, acute; areoles few to each rib, white-tomentose when young, naked in age; spines 7 to 9, the 3 upper spines flat, yellowish white, brown at tip, the uppermost one longer, thinner, annulate, 4 to 6 lower spines subterete, subulate, yellowish white.

ECHINOCACTUS ELLEMEETII Miquel, Nederl. Kruidk. Arch. **4:** 337. 1858.

This species is supposed to be of this relationship, although it has only 13 ribs.

ECHINOCACTUS FLEXUOSUS Dietrich, Allg. Gartenz. **19:** 347. 1851.

Subglobose, umbilicate at apex; ribs strongly compressed, undulate, wavy; areoles few on each rib, when young tomentose; spines white, spotted; upper spine ensiform; central spines 3-angled, a little curved.

ECHINOCACTUS FLUCTUOSUS Dietrich, Allg. Gartenz. **19:** 154. 1851.

Subglobose, 5 cm. high; ribs numerous, strongly compressed, undulate; areoles few to each rib, tomentose; spines 7, grayish, subulate, unequal, some of them flattened; central spine solitary, erect, terete.

ECHINOCACTUS FOERSTERI Stieber, Bot. Zeit. **5:** 491. 1847.

Nearly globular, dark green; ribs 21, sharp on the edge; spines up to 9, the 3 upper longer and stronger than the others, 12 to 18 mm. long, dark red, the middle one very thin; the lower spines 4 to 6, very small, bristly, 2 to 8 mm. long.

Known only from the type locality.

ECHINOCACTUS GRISEISPINUS Jacobi, Allg. Gartenz. **24**: 99.   1856.

Stems clavate, 15 cm. high, 10 cm. in diameter, glaucescent, somewhat umbilicate at apex; ribs 34 to 38; radial spines 7; 3 upper spines erect, when young purple with black tips, about 2.5 cm long; lower spines white; central spine solitary, 2.5 cm. long or more.

ECHINOCACTUS HEXACANTHUS Mühlenpfordt, Allg. Gartenz. **14**: 369.   1846.

Obovate, umbilicate at apex; ribs 34, flattened; radial spines 5; central spine solitary.

ECHINOCACTUS HEYDERI Dietrich, Allg. Gartenz. **14**: 170.   1846.

Obovate, pale green, rounded at apex; ribs numerous, strongly compressed; spines 8, grayish white; central spine solitary, terete, porrect.

ECHINOCACTUS HOOKERI Mühlenpfordt, Allg. Gartenz. **13**: 345.   1845.

Obovate; ribs numerous; upper spines flattened, incurved, 5 cm. long; radial spines whitish, 16 mm. long.

ECHINOCACTUS HYSTRICHOCENTRUS Berg, Allg. Gartenz. **8**: 131.   1840.

Echinocactus hystrichodes Monville in Labouret, Monogr. Cact. 215.   1853.

Clavate, 10 cm. high, somewhat umbilicate at apex; ribs 39, compressed, wavy; areoles only a few to each rib; spines 8, all flat and thin; the 3 upper larger, whitish gray with black tips; central spine solitary, erect, incurved, annulate.

ECHINOCACTUS LINKEANUS Dietrich, Allg. Gartenz. **16**: 298.   1848.

Spines 6, white; radial spines 5; central spine solitary, flat, variously curved.

ECHINOCACTUS MACROCEPHALUS Mühlenpfordt, Allg. Gartenz. **14**: 370.   1846.

Subglobose, light green, depressed at apex; ribs 34, somewhat acute, undulate; areoles 5 cm. apart; radial spines 7 or 8; the 3 upper ones elongated; the 4 lower ones short; central spine solitary, erect.

ECHINOCACTUS MAMMILLIFER Miquel, Linnaea **12**: 8.   1838.

This seems to be a seedling and may belong to some species of this genus. Its strongly compressed ribs certainly suggest this genus. Miquel does not state where his plant came from, but suggests a relationship with Pfeiffer's second group to which *Echinocactus scopa* and other South American species belong; it does not seem to us to belong with it. Schumann did not know the species. *Echinocactus theiacanthus* Lemaire (Cact. Gen. Nov. Sp. 86.   1839) which was originally spelled *E. theionacanthus* Lemaire (Cact. Aliq. Nov. 22. 1838) was taken up by Lemaire as *E. mammillifer.* This is further discussed on page 137.

ECHINOCACTUS OCHROLEUCUS Jacobi, Allg. Gartenz. **24**: 101.   1856.

Stems cylindric to clavate, 10 cm. high, 7.5 cm. in diameter, light green, somewhat umbilicate at apex; ribs 33 to 36, acute; areoles only a few to each rib; spines 7, yellowish; 3 upper spines 3 cm. long, erect; 4 lower spines very short.

ECHINOCACTUS OCTACANTHUS Mühlenpfordt, Allg. Gartenz. **16**: 10.   1848.

Obovate; ribs 40 to 44, compressed; areoles with white hairs when young; radial spines 7; central spine solitary.

ECHINOCACTUS QUADRINATUS Wegener, Allg. Gartenz. **12**: 66.   1844.

Echinocactus wegeneri Salm-Dyck, Cact. Hort. Dyck. 1849. 31.   1850.

Plant subcylindric, 7.5 cm. high, 5 cm. in diameter; ribs 34 to 36, strongly compressed; areoles remote, somewhat velvety when young; spines usually 7; 3 upper spines subulate, angled, the middle one shorter; central spine solitary, grayish brown, sometimes wanting; lower spines 4, recurved, reddish white, the 2 lowermost ones a little longer.

ECHINOCACTUS RAPHIDACANTHUS Salm-Dyck, Cact. Hort. Dyck. 1849. 160.   1850.

Stems globose, light green; ribs 35 to 40, strongly compressed; young areoles white, velvety. upper and central spines 4, stout, yellowish red, the uppermost one broad and flat; lower spines 6; slender, white.

ECHINOCACTUS RAPHIDOCENTRUS Jacobi, Allg. Gartenz. **24**: 101.   1856.

Plant depressed-globose, 5 cm. high, 6 cm. in diameter, light green, somewhat umbilicate; ribs 24 to 28, acute; radial spines 7, the 3 upper ones reddish brown.

PLATE XI

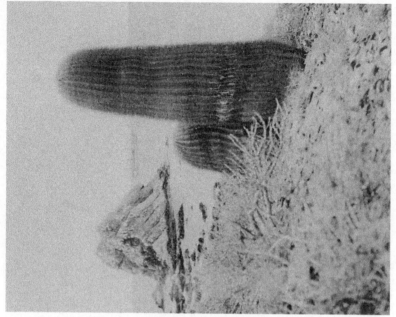

2. *Ferocactus diguetii.*

1. *Ferocactus pringlei.*

ECHINOCACTUS SULPHUREUS Dietrich, Allg. Gartenz. **13**: 170. 1845.

Globose, 7 to 10 cm. in diameter, green, depressed at the apex, very spiny; ribs numerous, much compressed, undulate, each bearing a few areoles; spines 8 or 9, white, when young with brown tips, compressed at base, subulate; central spine solitary, porrect, long; flowers probably yellow.

The plant described by Schumann as *E. gladiatus* has yellow spines, a character not referred to either under that species, as originally described, or under *E. sulphureus*. According to Salm-Dyck, this plant is near *Echinocactus anfractuosus*.

ECHINOCACTUS TELLII Hortus, Monatsschr. Kakteenk. **11**: 161. 1901.

Said to be of this relationship.

ECHINOCACTUS TERETISPINUS Lemaire, Hort. Univ. **6**: 60. 1845.

"We have observed, in the rich collection of M. Odier, a distinguished amateur of Bellevue, a species belonging to our section *Stenogoni* of *Echinocactus*, with cylindrical spines, a character peculiar to this section, which is possessed by only one other species, the *E*. [name not given], but in the latter the upper spine is flat, and anyway these two plants are quite distinct. Our specific name signifies this character. We intend to refer again to this curious species and at the same time to point out many other novelties of M. Odier, which he received directly from Mexico."

ECHINOCACTUS TRIBOLACANTHUS Monville in Labouret, Monogr. Cact. 221. 1853.

Cylindric; ribs numerous, flat; spines 8; flowers red.

ECHINOCACTUS TRIFURCATUS Jacobi, Allg. Gartenz. **24**: 100. 1856.

Plant pyriform, 6 cm. in diameter near the apex, umbilicate at apex, glaucous-green; ribs 32, membranaceous, compressed; spines 5.

16. FEROCACTUS gen. nov.

Globular to cylindric, often large cacti; ribs thick and prominent; spines well developed, either straight or hooked; areoles usually large, bearing flowers only when young and then only just above the spine-clusters, more or less felted when young; flowers usually large, broadly funnel-shaped to campanulate, usually with a very short tube; stamens numerous, borne on the throat, short; ovary and flower-tube very scaly; scales naked in their axils; fruit oblong, usually thick-walled and dry, dehiscing by a large basal pore; seeds black, pitted, never tuberculate; embryo curved.

Type species: *Echinocactus wislizeni* Engelmann.

The oldest species in this genus is *Ferocactus nobilis* which was collected by William Houston in Mexico before 1733. It was described by Miller in the Gardeners' Magazine 7th ed. 1759. Upon this description Linnaeus in 1767 (Mantissa 243) based his *Cactus nobilis* and Miller in 1768 (Gard. Mag.) his *Cactus recurvus*.

The generic name is from *ferus* wild, fierce, and *cactus*, referring to the very spiny character of the plants.

We recognize 30 species, heretofore treated under *Echinocactus*, all from North America. The genus differs from *Echinocactus* proper in its fruits and flowers.

KEY TO SPECIES.

A. Giant species, often 1 meter high or more (except apparently 3, 4, and 5).
  B. Areoles with a marginal row of bristles or hairs.
    Areoles with marginal weak hairs.
      Central spines yellowish; flowers yellow............................... 1. *F. stainesii*
      Central spines bright red; flowers red................................. 2. *F. pringlei*
    Areoles with marginal bristles.
      Central spine hooked.
        Central spines 8 cm. long or less, 4 to 6 mm. wide.
          Inner perianth-segments pink.
            Inner perianth-segments linear .... 3. *F. fordii*
            Inner perianth-segments oblong................................. 4. *F. townsendianus*
          Inner perianth-segments yellow to red, the outer pinkish.
            Inner perianth-segments about 2 cm. long; spines yellow to red...... 5. *F. chrysacanthus*
            Inner perianth-segments 4 to 5 cm. long; spines white to reddish..... 6. *F. wislizeni*
        Central spines up to 12 cm. long and 8 mm. wide...................... 7. *F. horridus*

1. Ferocactus stainesii (Hooker).

> *Echinocactus stainesii* Hooker in Audot, Rev. Hort. **6**: 248. 1845.
> *Echinocactus pilosus* Galeotti in Salm-Dyck, Cact. Hort. Dyck. 1849. 148.
>   1850.
> *Echinocactus pilosus steinesii** Salm-Dyck, Cact. Hort. Dyck. 1849. 149.
>   1850.

Simple or proliferous, globular to columnar, up to 1.5 meters high; ribs 13 to 20, compressed, more or less undulate; areoles distant, circular; radial spines reduced to long white hairs; central spines several, subulate, at first purplish, becoming pale yellow in age; flowers yellow; fruit unknown.

Type locality: Not cited.

Distribution: San Luis Potosí, Mexico.

This species differs from the following one in having more distant ribs, the areoles more widely separated, the spines duller colored, more numerous, somewhat curved, two of them decidedly flattened and the hairs white. We know the plant only from description and illustrations.

FIG. 128.—Section of rib of Ferocactus stainesii.

*This plant was named for Fred. Staines whose name is sometimes wrongly spelled Staine and Steins and hence the specific name *steinsii* used by Salm-Dyck is incorrect.

M. E. Eaton del.

1. Top of flowering plant of *Ferocactus townsendianus*.
2. Top of flowering plant of *Ferocactus wislizeni*.
3. Flower of *Ferocactus diguetii*.
        (All natural size.)

A. Hoen & Co.

Echinocactus piliferus Lemaire (Labouret, Monogr. Cact. 186. 1853) is usually referred here, but was not described in the place cited.

Illustrations: Schumann, Gesamtb. Kakteen f. 52; Monatsschr. Kakteenk. **22:** 39, Möllers Deutsche Gärt. Zeit. **25:** 484. f. 16; Schelle, Handb. Kakteenk. 147. f. 72; Gartenwelt **7:** 277, as *Echinocactus pilosus;* Rev. Hort. II. **4:** 1; Belg. Hort. **4:** pl. 2, as *Echinocactus stainesii.*

Figure 128 is copied from the first illustration above cited.

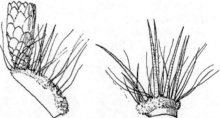

FIGS. 129 and 130.—Flower and cluster of spines of Ferocactus pringlei. ×0.5.

FIG. 131.—Ferocactus pringlei.

FIG. 131 *a.*—Ferocactus wislizeni.   Natural size.

2. Ferocactus pringlei (Coulter).

Echinocactus pilosus pringlei Coulter, Contr. U. S. Nat. Herb. **3:** 365.   1896.
Echinocactus pringlei Rose, Contr. U. S. Nat. Herb. **10:** 127.   1906.

Growing in clumps, becoming cylindric, sometimes 3 meters high and 3 to 4 dm. in diameter; ribs usually 16 to 18, more or less compressed; areoles numerous, closely set or contiguous, the outer margin with a row of white or straw-colored hairs, 2 to 4 cm. long; spines red, various, the three lower ones slender, almost acicular, the innermost much stouter, somewhat flattened, angular, curved or nearly straight; flowers red without, yellow within, 2.5 cm. long; scales on the ovary numerous, orbicular, imbricated; inner perianth-segments oblanceolate, obtuse or apiculate; fruit yellow, somewhat succulent, dehiscing by a basal pore, 3 to 4 cm. long, crowned by the persisting perianth; seeds 1.5 mm. long, brownish, pitted, with a small basal hilum.

Type locality: Jimulco, Coahuila, Mexico.

Distribution: Mountains of Coahuila and Zacatecas, Mexico.

Illustrations: Monatsschr. Kakteenk. **22**: 87, as *Echinocactus pilosus;* Ann. Rep. Smiths. Inst. **1908**: pl. 13, f. 5, as *Echinocactus pringlei.*

Plate XI, figure 1, is from a photograph of the plant, taken by F. E. Lloyd in Zacatecas, Mexico, in 1907. Figure 129 shows the flower of a plant collected by F. E. Lloyd at Zacatecas in 1908; figure 130 shows its spines and hairs; figure 131 is from a photograph taken by Robert Runyon near Saltillo, Mexico, in 1921.

FIG. 132.—Ferocactus fordii.   FIG. 133 —Ferocactus townsendianus.

3. Ferocactus fordii (Orcutt).

Echinocactus fordii Orcutt, Rev. Cact. **1**: 56. 1899.

Globose to short-cylindric, grayish green, 12 cm. in diameter; ribs usually 21, about 1 cm. high; areoles about 2 cm. apart; radial spines whitish, acicular, widely spreading, about 15; central spines usually 4; one of the centrals flattened, porrect, longer than the others, with a curved or hooked tip, about 4 cm. long; the other centrals subulate, somewhat angled; flowers rose-colored, 3.5 to 4 cm. long; outer perianth-segments ovate to ovate-oblong, acute; inner perianth-segments linear, acuminate; scales on the ovary broadly ovate; filaments pink; style and stigma-lobes greenish yellow to whitish.

Type locality: Not cited, but Mr. C. R. Orcutt's type specimen is labeled Lagoon Head, Lower California.

Distribution: Lower California.

Herbarium and living specimens of this species were obtained by Dr. Rose at San Bartolomé Bay, Lower California, in 1911 (No. 16188), some of which afterwards flowered in cultivation and were used in preparing the above description. The largest plants seen by him were up to about 4 dm. high, but the species may reach greater development.

Dr. Rose also collected this species at Abreojos Point, Lower California, March 16, 1911 (No. 16249). It is apparently common along the west coast of Lower California, usually growing at low elevations and as a rule forming the dominant feature of the landscape.

1. *Ferocactus flavovirens.*
2. *Ferocactus latispinus.*

The original reference to this species was very brief, as follows: "*E. fordii* is a name proposed for an allied form with ashy gray spines." The plant was later described by Mr. Orcutt (Rev. Cact. **2**: 81. 1890) in detail and this fuller description was republished still later (West Amer. Sci. **13**: 31. 1902). It was named for Lyman M. Ford of San Diego, California.

Illustration: Blühende Kakteen **1**: pl. 11, as *Echinocactus fordii*.

Figure 132 is copied from the illustration above cited.

4. Ferocactus townsendianus sp. nov.

Short-cylindric, 4 dm. high or more; ribs about 16, often spiraled, somewhat undulate; areoles large, distant; radial spines widely spreading, 14 to 16, 3 to 4 cm. long, most of them thread-like, but often 2 or more above and below, subulate; central spines subulate, grayish, usually one curved or hooked at apex, the others straight, all annulate; flowers large, 5 to 6 cm. long; outer perianth-segments ovate, reddish, with narrow yellow margins; inner perianth-segments oblong-lanceolate with a narrow pink stripe down the center with greenish-yellow margins; filaments and style dark pink; stigma-lobes pale greenish brown.

Collected by J. N. Rose on San Josef Island, Gulf of California, March 15, 1911 (No. 16570).

This species is common near the coast, growing chiefly on the dry mesa along with desert shrubs.

This species is named for Dr. Charles H. Townsend, Director of the New York Aquarium, who was in charge of the scientific work of the *Albatross* during the cruise into Lower California waters in 1911, when this plant was discovered.

Plate XII, figure 1, shows the flowering type-plant sent by Dr. Rose to the New York Botanical Garden in 1911, where it immediately bloomed. Figure 133 is from a photograph of the same plant.

5. Ferocactus chrysacanthus (Orcutt).

Echinocactus chrysacanthus Orcutt, Rev. Cact. **1**: 56. 1899.

Globose to cylindric; ribs about 18, tubercled; radial spines 4 to many, slender, white; central spines sometimes as many as 10, 5 cm. long, either red or yellow, curved; flowers from near the center of the plant, 5 cm. broad when fully open; scales naked in the axils, closely set and overlapping, the lower one orbicular and green, the upper ones more oval, brownish or with brown tips, the margin thin, sometimes ciliate or ragged; outer perianth-segments rather stiff, pinkish brown; inner perianth-segments 2 cm. long, satiny yellow with a jagged or toothed margin; fruit yellow, 3 cm. long; seeds large, black.

Type locality: Cedros Island, Lower California.

Distribution: Only on Cedros Island, and the adjacent coast of Lower California.

This plant is common on Cedros Island in the broad dry valleys which run back from the coast. It was re-collected by Dr. Rose at the type locality, in 1911. Dr. Rose obtained plants on Cedros Island about 2 dm. high, but we presume the plant becomes much larger.

According to C. R. Orcutt, "*E. rubrispinus* (Rev. Cact. **1**: 56. 1899) is a name proposed by L. M. Ford for the red-spined form and so distributed"; otherwise we know nothing of it.

Echinocactus emoryi chrysacanthus (Schumann, Gesamtb. Kakteen Nachtr. 99. 1903) is a garden name and so far as we are aware is not published.

6. Ferocactus wislizeni (Engelmann).

Echinocactus wislizeni Engelmann in Wislizenus, Mem. Tour North. Mex. 96. 1848.
Echinocactus emoryi Engelmann in Emory, Mil. Reconn. 157. 1848.
Echinocactus wislizeni decipiens Engelmann in Rothrock, Rep. U. S. Geogr. Surv. **6**: 128. 1878.
Echinocereus emoryi Rümpler in Förster, Handb. Cact. ed. 2. 804. 1885.
Echinocactus wislizeni albispinus Toumey, Gard. and For. **8**: 154. 1895.
Echinocactus falconeri Orcutt, West Amer. Sci. **12**: 162. 1902.
Echinocactus arizonicus Kunze, Monatsschr. Kakteenk. **19**: 149. 1909.

At first globular but becoming cylindric, when very old much elongated, 2 meters long or more, usually simple, but when injured often giving off several heads or branches; ribs numerous, often 25, 3 cm. high; areoles elliptic, large, sometimes 2.5 cm. long, brown-felted, 2 to 3 cm. apart, or the flowering ones often approximate; spines variable; radials, absent in young plants, thread-like to acicular, the longest 5 cm. long; central spines several, white to red, annular, all subulate, one of them much stouter, usually strongly flattened, strongly hooked; flowers yellow, some red, 5 to 6 cm. long; fruit yellow, oblong, scaly, 4 to 5 cm. long; seeds dull black, the surfaces covered with shallow indistinct pits.

Type locality: Doñana, New Mexico.

Distribution: El Paso, Texas, west through southern New Mexico and Chihuahua to Arizona and Sonora and perhaps south along the Gulf of California into Sinaloa. Reported also from Utah, perhaps erroneously, and from Lower California.

A peculiar form was collected by J. W. Toumey at Dudleyville, Arizona, September 25, 1896. The spine-clusters lack the marginal bristles, the spines are shorter and the flowers smaller. Mr. Toumey says that it is quite different from *Echinocactus wislizeni.*

Echinocactus wislizeni latispinus (Schelle, Handb. Kakteenk. 168) is without description.

Echinocactus wislizeni purpureus is in the trade (Grässner).

Echinocactus sclerothrix Lehmann (Del. Sem. Hort. Hamb. 1838) is a very doubtful species; Schumann thought that it might be referred to *E. wislizeni;* if so, the name has priority.

Illustrations: Schelle, Handb. Kakteenk. 167. f. 97; Emory, Mil. Reconn. 157. No. 4; Gard. Chron. III. 8: 159. f. 25; 35: 181. f. 75; Rümpler, Sukkulenten 112. f. 62; Cact. Journ. 1: pl. for March; July; Anh. Rep. Smiths. Inst. 1911: pl. 6, A; Journ. Hort. Home Farm. III. 60: 144; Monatsschr. Kakteenk. 14: 185; 20: 57; 30: 19; Cact. Mex. Bound. pl. 25, 26; Pac. R. Rep. 4: pl. 3, f. 1, 2; Hornaday, Campfires on Desert and Lava, facing 216; Plant World 9: f. 47, 48; 11⁶: f. 2; Förster, Handb. Cact. ed. 2. 510. f. 61; Gard. and For. 8: 154. f. 24; Dict. Gard. Nicholson 4: 541. f. 25; Suppl. 337. f. 362; Watson, Cact. Cult. 126. f. 49, as *Echinocactus wislizeni;* Monatsschr. Kakteenk. 19: 151, as *Echinocactus arizonicus;* Saunders, Useful Wild Pl. opp. 158; Gard. Chron. II. 7: 749. f. 119; Kunze, Cactaceae 1909, 1910, as *Echinocactus;* Emory Mil. Reconn. 157. No. 5, as *Echinocactus emoryi;* Contr. U. S. Nat. Herb. 16: pl. 123, A, as *Echinocactus falconeri.*

Plate 1 is from a photograph taken by Dr. MacDougal near Pima Canyon, Arizona, in 1910; plate XII, figure 2, shows the flowering top of a plant sent by Dr. MacDougal to the New York Botanical Garden from Torres, Mexico, in 1902. Figure 131 *a* is from a painting of *Ferocactus wislizeni* made for Dr. MacDougal, February 28, 1911.

7. Ferocactus horridus sp. nov.

Globular, 3 dm. in diameter or more; ribs 13, broad, 2 cm. high, obtuse, not tubercled; areoles 1.5 to 2.5 cm. apart, large; radial spines 8 to 12, acicular, spreading, white, 3 to 4 cm. long; central spines 6 to 8, very diverse, all reddish, either spreading or porrect, all straight except 1, this much elongated, often 12 cm. long, much flattened, very strongly hooked; flowers and fruit unknown.

Collected by J. N. Rose on San Francisquito Bay, Lower California, April 9, 1911 (No. 16746).

Only very small plants of this species were obtained by Dr. Rose during his hurried visit at this locality in 1911. Larger plants will doubtless be found in this same region. Indeed, Ivan M. Johnston has reported seeing a plant there of this relationship which was a meter high. This species, while most closely related to *F. wislizeni,* is much more strongly armed. It has, perhaps, the most formidable spine-armament of any species of this genus; the central spine is not so long as in No. 14, but is stouter and more strongly hooked.

M. E. Eaton del.

A Hoen & Co.

1. Flowering plant of *Ferocactus viridescens*.
2. Top of flowering plant of *Ferocactus rectispinus*.
(All natural size.)

Type locality: California.

Distribution: Deserts of southeastern California, northern Lower California, and southern Nevada.

Living plants from southern California, apparently referable to this species, do not have any bristle-like radial spines, and the stout spines vary greatly in color, from red to nearly white.

The range of this species seems not to be very extensive; Dr. Coulter records it from New Mexico and Texas but this must refer to *F. wislizeni.* It is nearest *F. wislizeni* but the spines are never hooked and the seeds are more shining, with stronger reticulations.

FIGS. 134 and 135.—Ferocactus acanthodes.

In America this species has long passed under the name of *Echinocactus cylindraceus.* The above name, however, is much older and we are following Weber in using it. Weber states that it has long been known as such in French collections and we believe that we are justified in taking up the older name.

It is also found in collections under the name of *E. californicus* and *E. copoldi* (Schumann, Gesamtb. Kakteen 357. 1898).

Echinocactus cylindraceus albispinus is in the trade (Grässner).

Illustrations: Garten-Zeitung 4: 241. f. 54; 242. f. 55; Gard. Chron. II. 7: 241. f. 39; III. 8: 167. f. 27; Deutsche Gärt. Zeit. 5: 209; Schelle, Handb. Kakteenk. 165. f. 95; West Amer. Sci. 7: 68; Gartenwelt 9: 249; Förster, Handb. Cact. ed. 2. 474. f. 55; Schumann, Gesamtb. Kakteen Nachtr. f. 18; Monatsschr. Kakteenk. 12: 123; Cact. Journ. 2: 115; Gartenflora 26: pl. 905 b; 30: 414; Cact. Mex. Bound. pl. 30; Ill. Gärt. Zeit. 21: 65, as *Echinocactus cylindraceus;* Cact. Journ. 2: pl. for February, as *Echinocactus cylindraceus longispinus.*

Plate xv shows two views of the Coachella Desert, three miles northeast of White Water, California, from a photograph taken by Dr. Wm. S. Cooper. Figure 134 is from a

PLATE XV

1, 2. *Ferocactus acanthodes* as seen in the Coachella Desert, California.

photograph taken by Dr. Wm. S. Cooper, between Indio and Palm Springs, California; figure 135 is from a photograph taken also by Dr. Cooper from the east base of the Laguna, May 12, 1919; figure 136 is from a photograph taken by Dr. MacDougal near Palm Springs, California, in 1913; figure 137 is from a photograph taken by S. C. Mason near Andreas Canyon, California, in 1918.

10. Ferocactus santa-maria sp. nov.

Cylindric, 6 dm. high or more; ribs about 14; outer spines several, thread-like; central spines in 2 series, all straight, grayish, all annulate, subulate, the central one stouter, flatter, ascending; somewhat curved at tip; old flowers persisting, 6 to 7 cm. long; fruit 3 to 4 cm. long, bearing orbicular scales; seeds 2 mm. long, finely reticulated.

Collected by J. N. Rose on the shores of Santa Maria Bay, Lower California, May 18, 1913 (No. 16279).

FIGS. 136 and 137.—Ferocactus acanthodes.

This plant was seen at only one locality and only small specimens were observed, but much larger ones may be expected. This is a densely armed plant, peculiar in having all the straight dagger-like strong central spines ascending.

According to letters from F. Vaupel, this plant has been in cultivation in Germany for several years, grown from seed, perhaps from the type collection. It is briefly mentioned in the Monatsschrift für Kakteenkunde (**29:** 13. 1919) as *Echinocactus santa-maria* Rose.

11. Ferocactus diguetii (Weber).

Echinocactus diguetii Weber, Bull. Mus. Hist. Nat. Paris **4:** 100. 1898.

Plants very stout, usually 1 to 2 meters, but sometimes 3 and 4 meters, high, 6 to 8 dm. in diameter or more; ribs numerous, sometimes as many as 39, rather thin; areoles large, 1 to 1.5 cm. long, somewhat elliptic, approximate or on old plants coalescent; spines 6 to 8, yellow, subulate, 3 to 4 cm. long, slightly curved and a little spreading; flowers numerous, 3 to 3.5 cm. long; scales on ovary and flower-tube ovate, closely imbricate, thin on the margin and somewhat lacerate; inner perianth-segments red with yellow margins, oblong, 2 cm. long; filaments pink, numerous; tube of flower below stamens very short; style yellow; fruit scaly.

Type locality: Santa Catalina Island, off Lower California.
Distribution: Islands of the Gulf of California.

This species which is common on several of the islands in the Gulf of California is perhaps the largest of all the visnagas or barrel cacti. On Santa Catalina Island, especially, enormous individuals are to be found and here it is the most conspicuous plant. It seems to have no very definite habitat, growing both on the mountain sides among the large igneous rocks as well as along the old shell beaches. These plants have an enormous display of surface roots with only a few weak supporting ones and consequently large plants can easily be toppled over. Its spines are all very much alike.

Mr. Ivan M. Johnston, botanist of the California Academy of Sciences' Expedition to the Gulf of California, who explored all the islands in the Gulf in 1921, writes of the distribution of this species, as follows: "I found the distribution of *Echinocactus diguetii* to be peculiar; I saw it at the following disconnected points: Angel de la Guardia, Carmen, Coronado, Dansante, and San Diego Islands. It seems to skip hither and thither over the Gulf Islands without rhyme or reason."

Illustrations: Bull. Mus. Hist. Nat. Paris **4**: 99. f. 1; Contr. U. S. Nat. Herb. **16**: pl. 123 B; Journ. N. Y. Bot. Gard. **12**: f. 47; Bull. Soc. Acclim. **52**: 53. f. *12*, as *Echinocactus diguetii*.

Plate XI, figure 2, is from a photograph taken by Dr. Rose on Carmen Island, Gulf of California; plate XII, figure 3, shows the flower of a plant collected by Dr. Rose on Carmen Island in 1911.

12. Ferocactus covillei sp. nov.

Plant simple, globular to short-cylindric, often 1.5 meters high; ribs 22 to 32, 2 to 4 cm. high, rather thin, when young more or less tubercled, but when old hardly undulate; areoles on small plants distant, often 3 to 4 cm. apart, but on old and flowering plants approximate or contiguous, densely brown-felted when young, naked in age, the spine-bearing areoles large and circular; the flowering areoles more elongated and complex, divided into three parts, the lower part bearing spines, the central part spinescent glands, and the upper part the flower; spines variable as to color, sometimes red to white; radial spines 5 to 8, somewhat spreading, subulate, straight or more or less curved backward, 3 to 6 cm. long, annulate; central spine always solitary, very variable, straight or with the tip bent or even strongly hooked, annulate, terete to strongly flattened or 3-angled, 3 to 8 cm. long; upper areoles of old plants bearing 5 to 7 glands, becoming spinescent, 5 to 6 mm. long; flowers described as red, tipped with yellow, sometimes reported as yellow throughout, 6 to 7 cm. long; inner perianth-segments linear-oblong, acuminate, often serrate; throat broad, covered with stamens; tube-proper short, 2 to 3 mm. long; fruit oblong, 5 cm. long, bearing a few broad scales; seeds black, dull or shining, nearly smooth or slightly pitted, 2 mm. long.

Collected on hills and mesas near Altar, Sonora, Mexico, by C. G. Pringle, August 11, 1884 (type), by Rose, Standley, and Russell on plain near Empalme, Sonora, March 11, 1910 (No. 12642), and by F. V. Coville, 10 miles west of Torres, Sonora, February 10, 1903 (No. 1657).

This species ranges from southern Arizona to Guaymas, Sonora. It has heretofore passed as *Echinocactus emoryi;* the type of that species, however, came from southwestern New Mexico and has been referred by us as a synonym of *Ferocactus wislizeni.* Dr. Engelmann in his synopsis of the Cactaceae and in his later references transferred the name *emoryi* to the plant here described. This species needs further study; the color of the flowers is not definitely known and there is considerable variation in the markings of the seeds. The species as here considered has a wide range altitudinally and may include more than one species. We have reluctantly referred here two specimens (Nos. 4154 and 4155), collected by J. C. Blumer from the Comobabi Mountains, Arizona.

From this and related species water is often obtained by travelers in the great deserts of western Mexico and the southwestern United States. This has been described and illustrated by Dr. F. V. Coville in an article "Desert Plants as a Source of Drinking Water." He tells how by slicing off the top of a large plant and mashing the pulp three quarts of drinkable water were obtained (Ann. Rep. Smiths. Inst. **1903**: 499 to 505. 1904).

In Mexico a candy is made from the flesh of this and other large species. The spines and epidermis are all cut off; the flesh is cut into slices of various shapes and sizes and then cooked in sugar. This candied product is sold in all the towns and markets of Mexico.

Illustrations: Journ. N. Y. Bot. Gard. **3**: 95. f. 14, as *Cereus* sp.; Carnegie Inst. Wash. **6**: pl. 18; MacDougal, Bot. N. Amer. Des. pl. 8, 62; Monatsschr. Kakteenk. **25**: 93; Nat. Geogr. Mag. **21**: 712; Amer. Gard. **11**: 459; Möllers Deutsche Gärt. Zeit. **25**: 474. f. 6, No. 25; Schelle, Handb. Kakteenk. 161. f. 90; Dict. Gard. Nicholson **4**: 539. f. 20; Suppl.

354. f. 335; Gard. Chron. III. **35**: 181. f. 76; Engler and Prantl, Pflanzenfam. 3^{6a}: f. 56, D; Strand Mag. 626, 627; Goebel, Pflanz. Schild. **1**: f. 47; Förster, Handb. Cact. ed. 2. 208. f. 16; Ann. Rep. Smiths. Inst. **1903**: 500. f. 1; pl. 1, 2; Cact. Mex. Bound. pl. 28; Pac. R. Rep. **4**: pl. 3, f. 3; Watson, Cact. Cult. 101. f. 34; ed. 3. 52. f. 22, as *Echinocactus emoryi;* Bull. Geol. Surv. 613: pl. 38 A, without name.

Figure 138 is from a photograph of the plant, taken by F. E. Lloyd in the Quijotoa Mountains, Arizona, in 1906; figure 139 is from a photograph taken by Dr. MacDougal near Torres, Sonora, in 1903.

FIGS. 138 and 139.—Ferocactus covillei.

13. Ferocactus peninsulae (Weber).

Echinocactus peninsulae Weber, Bull. Mus. Hist. Nat. Paris **1**: 320. 1895.

Simple, erect, 2.5 meters high, clavate to cylindric; ribs 12 to 20, prominent; areoles 4 cm. apart or even less in old plants; spines red with yellow tips; radial spines 11, spreading, straight, terete, more or less annulate, the lower ones stouter and more colored; central spines 4.

Type locality: Lower California, but no definite locality cited.

Distribution: Southern Lower California.

Engelmann and Weber seemed to have been in agreement regarding this species being new, but Engelmann's name (Contr. U. S. Nat. Herb. **3**: 361. 1896) was based on Gabb's specimen (No. 11), now preserved in the Missouri Botanical Garden, while Weber's name is based on Diguet's plant. The plants of these two collections may or may not be conspecific. We have seen only Engelmann's specimen which we have used in making our illustration.

In December 1920, Dr. William S. W. Kew sent us fruit and a small living plant from near Boca de Guadalupe on the west coast of Lower California which we believe belongs here. His plant is less than 10 cm. high with 8 broad ribs; young areoles brown-felted, circular; radial spines spreading, brownish or white; central spines 4, grayish brown, the lower one flattened, strongly hooked, annulate; flowers yellow; fruit yellowish, 2.5 cm. long, bearing broad rounded scales; seeds 2 mm. long, reticulate. The plant is known as bisnaga or visnaga, as are also other species of this relationship. Dr. Kew states that the Mexicans on the peninsula of Lower California often cut off the spines of this plant and use it as feed for cattle.

Illustrations: Bull. Mus. Hist. Nat. Paris 4: 101; Bull. Soc. Acclim. 52: f. 11; Möllers Deutsche Gärt. Zeit. 25: 474. f. 10, as *Echinocactus peninsulae.*

Figure 140 shows the spines of a specimen in the Engelmann herbarium collected by William M. Gabb in Lower California.

FIG. 140.—Ferocactus peninsulae.   FIG. 141.—Ferocactus robustus.

14. Ferocactus rectispinus (Engelmann).

Echinocactus emoryi rectispinus Engelmann in Coulter, Contr. U. S. Nat. Herb. 3: 362. 1896.
Echinocactus rectispinus Britton and Rose, Journ. N. Y. Bot. Gard. 12: 269. 1911.

Globose to cylindric, 1 to 2 meters high; radial spines 8 to 12, the three upper spines stouter and sometimes curved; central spine one, 9 to 13 cm. long (not 30 to 32 cm. long), rather slender, nearly straight, never hooked; flowers 6 cm. long, yellowish; scales on ovary rounded, thin-margined, sometimes ciliate, naked in the axils; inner perianth-segments lemon-yellow, lanceolate, 5 cm. long, acuminate.

Type locality: Vicinity of Muleje, Lower California.
Distribution: Central Lower California.

This species was described as a variety of *Echinocactus emoryi* (i. e. *Ferocactus covillei*) by Dr. Coulter, but it seems distinct, especially in its spines.

Mr. C. H. Thompson in Bailey's Cyclopedia of American Horticulture assigned it to the southern United States, but this is erroneous.

The type is Gabb's No. 12, from Lower California. It consists of two clusters of spines and is now deposited in the herbarium of the Missouri Botanical Garden.

Palmer's specimen from Sonora, referred here by Coulter, may be *F. wislizeni.*

Illustrations: Cycl. Amer. Hort. Bailey 2: 513. f. 745, as *Echinocactus emoryi rectispinus;* Stand. Cycl. Hort. Bailey 2: f. 1372, as *Echinocactus rectispinus.*

Plate XIV, figure 2, shows the flowering top of a plant sent by Dr. Rose from the head of Concepción Bay, Lower California, to the New York Botanical Garden in 1911. Figure 142 is from a photograph of the type specimen obtained by William M. Gabb in 1867 (No. 12).

15. Ferocactus orcuttii (Engelmann).

Echinocactus orcuttii Engelmann, West Amer. Sci. 2: 46. 1886.

Single, or cespitose in clusters of 15 to 20 stems, 6 to 13 dm. high, 2.5 to 4.5 dm. in diameter; ribs 13 to 30, somewhat spiraled, obtuse, somewhat tuberculate; areoles approximate; spines reddish, straight or simply curved, all annulate, angled or flat; radial spines 9 to 13, spreading; central spines 4, stouter than the radials; flower 3 to 5 cm. long, dull crimson; perianth-segments short-oblong, rounded at apex with a more or less erose-margin; scales on the ovary orbicular, small; stigma-lobes 16 to 20, green; fruit described as pulpy, crimson, scaly; seeds numerous, small.

Type locality: Palm Valley, Lower California.

Distribution: Definitely known only from type locality. Recorded from San Diego.

Palm Valley is not shown on ordinary maps. We wrote to Mr. E. C. Rost who recently returned from northern Lower California to locate the place and he writes as follows: "Palm Valley is about 40 miles southeast of Tia Juana, and is probably not noted on any of the maps as it is not a pueblo, but merely the watershed of the Rio Tia Juana." He subsequently visited Palm Valley and sent us two plants, neither of them quite agreeing with the descriptions of the species.

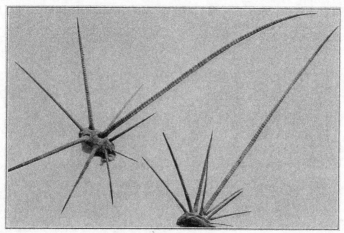

Fig. 142.—Ferocactus rectispinus.

We know this plant only from descriptions and illustrations. After our manuscript was in type Mr. C. R. Orcutt wrote us as follows: "You know, I suppose that I consider *Echinocactus orcuttii* only a luxuriant development of *E. viridescens?* At Palm Valley, Lower California, I have seen it very large and one cluster contained 25 heads, forming quite a large mass. I found one large cristate." The published illustrations do not indicate close relationship with *F. viridescens.*

Illustrations: West Amer. Sci. 2: 47; 7: 69; Schelle, Handb. Kakteenk. 166. f. 96; Blanc, Cacti 49. f. 575, as *Echinocactus orcuttii.*

16. Ferocactus robustus (Link and Otto).

Echinocactus robustus Link and Otto, Allg. Gartenz. 1: 364. 1833.
Echinofossulocactus robustus Lawrence in Loudon, Gard. Mag. 17: 318. 1841.

In large clumps, often 3 meters, rarely 5 meters, in diameter, 1 to 1.3 meters high, with hundreds of branches; ribs 8, prominent in young growth, but becoming indistinct in age, somewhat undulate; areoles brown-felted when young; radial spines ascending, about 10, often thread-like; central spines subulate, about 6, brown at first, somewhat flattened, annulate, often 6 cm. long; flowers 3.5 to 4 cm. long; inner perianth-segments oblong, acute, yellowish; scales on the ovary broad, rounded at tip; fruit 2 to 2.5 cm. long; seeds black, oblong, 1.5 mm. long.

Type locality: Mexico.

Distribution: Tehuacán, Puebla, Mexico.

The large mounds formed by this plant are striking features of the landscape; the individual heads are globose or short-oblong, 1 to 2 dm. in diameter.

Echinocactus robustus prolifer (Pfeiffer, Enum. Cact. 61. 1837) to which *Echinocactus agglomeratus* Karwinsky was assigned as a synonym seems to be the normal form of the species, while var. *monstrosus* (Pfeiffer, Enum. Cact. 61. 1837) is an abnormal form. To it were referred also as synonyms *Echinocactus spectabilis* and *E. subuliferus* (Pfeiffer, Enum. Cact. 61. 1837). This latter name, according to Pfeiffer, is different from *E. subuliferus* Link and Otto (Verh. Ver. Beförd. Gartenb. 3: 427. pl. 27. 1827) collected in Mexico by Deppe. The flowers and fruit were unknown.

Illustrations: Nov. Act. Nat. Cur. 19: pl. 16, f. 3, 6; Contr. U. S. Nat. Herb. 10: pl. 16, f. A; Möllers Deutsche Gärt. Zeit. 25: 474. f. 6, No. 9; 29: 441. f. 17; Karsten and Schenck, Vegetationsbilder 1: pl. 44, as *Echinocactus robustus;* Verh. Ver. Beförd. Gartenb. 3: pl. 27, as [*Echinocactus*] *subuliferus.*

Figure 141 shows the flower, copied from the first illustration above cited. Figure 143 is from a photograph of the plant taken by Dr. Rose at El Reago, Puebla, Mexico, in 1905.

FIG. 143.—Ferocactus robustus.

FIG. 144.—Ferocactus echidne.

17. Ferocactus echidne* (P. De Candolle).

Echinocactus echidne P. De Candolle, Mém. Cact. 19. 1834.
Echinocactus vanderaeyi Lemaire, Cact. Aliq. Nov. 20. 1838.
Echinocactus dolichacanthus Lemaire, Cact. Aliq. Nov. 25. 1838.
Echinofossulocactus vanderaeyi Lawrence in Loudon, Gard. Mag. 17: 318. 1841.
Echinofossulocactus vanderaeyi ignotus-longispinus Lawrence in Loudon, Gard. Mag. 17: 318. 1841.
Echinofossulocactus echidne Lawrence in Loudon, Gard. Mag. 17: 318. 1841.
Echinocactus gilvus Dietrich, Allg. Gartenz. 13: 170. 1845.
Echinocactus echidne gilvus Salm-Dyck, Cact. Hort. Dyck. 1849. 27. 1850.
Echinocactus victoriensis Rose, Contr. U. S. Nat. Herb. 12: 291. 1909.

Depressed-globose, 12.5 cm. high, 18 cm. in diameter, green; ribs 13, acute, broad at base; areoles remote, velvety when young, oval in shape; radial spines rigid, about 7, about 2 cm. long, yellow; central spine solitary, porrect, 3 cm. long or more; flowers lemon-yellow; perianth-segments linear-oblong, acute, sometimes toothed near the apex; stigma-lobes about 10, elongated, spreading or reflexed; scales on the ovary ovate, acute.

Type locality: Mexico.
Distribution: Hidalgo, Mexico.

Echinocactus dolichocentrus Salm-Dyck (Cact. Hort. Dyck. 1844. 22. 1845) is usually referred here as a synonym but it was never described.

Illustrations: Blühende Kakteen 3: pl. 146; Schelle, Handb. Kakteenk. 156. f. 83; De Candolle, Mém. Cact. pl. 11, as *Echinocactus echidne.*

Figure 144 is copied from the last illustration above cited.

---

*Schumann has changed the spelling of this name to *E. echidna.*

18. Ferocactus alamosanus Britton and Rose.

Echinocactus alamosanus Britton and Rose, Contr. U. S. Nat. Herb. **16:** 239. 1913.

Plants usually single, sometimes in clusters, somewhat flattened above, green, 30 cm. in diameter or more; ribs about 20, narrow; spines all yellow; radials usually 8, 3 to 4 cm. long, more or less spreading; central single, porrect or erect, somewhat flattened laterally, 6 cm. long and a little longer than the radials; flower-buds covered with ovate, ciliate scales, these brownish except in the margin; fruit unknown.

Type locality: Alamos Mountain, Mexico.
Distribution: Southern Sonora, Mexico.

This species is quite unlike anything we have yet seen from the west coast of Mexico. A small living plant of the type collection was brought to Washington by Dr. Rose in 1910, but is still quite small, being only 10 cm. in diameter. It has now been sent to Mr. Wm. Hertrich, superintendent of the Huntington Estate near Los Angeles, where it will be planted in the open and given a chance to develop. The illustration cited below was made from this plant.

Mr. Ivan M. Johnston has collected this species or a closely related one on the hill-sides about San Carlos Bay, Sonora (No. 4348). He says that it is 6 dm. high, 5 dm. in diameter and has 23 ribs. He describes the flower as clear lemon-yellow with the outer segments greenish red at tip; the scales on the ovary are broadly ovate, apicular, ciliate.

Illustration: Contr. U. S. Nat. Herb. **16:** pl. 66, as *Echinocactus alamosanus.*

Figure 145 is from a photograph of the living plant collected by Dr. Rose.

FIG. 145 —Ferocactus alamosanus.

19. Ferocactus glaucescens (De Candolle).

Echinocactus glaucescens De Candolle, Mém. Mus. Hist. Nat. Paris **17:** 115. 1828.
Echinocactus pfeifferi Zuccarini in Pfeiffer, Enum. Cact. 58. 1837.
Echinofossulocactus pfeifferi Lawrence in Loudon, Gard. Mag. **17:** 318. 1841.

Globular, 2 to 4 dm. in diameter, or a little higher than broad, glaucous; ribs 11 to 15, somewhat flattened, acute, 2 to 3 cm. high; areoles 8 to 12 mm. apart, oblong, 12 to 20 cm. long, yellowish, tomentose when young; radial spines 6, nearly equal, rigid, only slightly spreading, straight, 2.5 to 3 cm. long, pale yellow at first, when old blackish, more or less banded; central spine solitary, similar to the radials; flowers yellow, 2 cm. long, perhaps broader when fully expanded; outer perianth-segments ovate, acuminate, sometimes brownish on the back, ciliate on the margins; inner perianth-segments oblong, usually only acute, somewhat toothed or lacerate; stigma-lobes slender, cream-colored; scales on the ovary brownish, ovate, acute, ciliate on the margins, imbricate.

Type locality: Toliman, Mexico.
Distribution: Eastern central Mexico.

Our knowledge of this species is drawn not only from illustrations and published descriptions but also from a plant obtained by Dr. Rose from Hidalgo, Mexico, in 1915, which has since been grown in the cactus house of the U. S. Department of Agriculture, but has never flowered. Dr. Rose, however, found it in flower at La Mortola, Italy, in 1912, and a few flowers were obtained.

Schumann thought that *Echinocactus dietrichianus* Förster (Hamb. Gartenz. **17:** 160. 1861) was probably referable to *E. pfeifferi.*

Schumann refers here *E. theionacanthus* Lemaire (Cact. Aliq. Nov. 22. 1838) and *E. theiacanthus* Lemaire (Cact. Gen. Nov. Sp. 86. 1839) but he has the names interchanged. The latter seems to have been based on *E. mammilifer* Miquel (Linnaea **12:** 8. 1838), a name which Schumann listed among his unknown species. Hemsley (Biol. Centr. Amer. Bot. 1: 536. 1880) refers here *E. mammillarioides* Hooker, a different plant, native of Chile. Although there are some slight differences in the descriptions it is not at all unlikely that these last two species had a common origin, the names being similar and published about the same time. It is almost certain that all four names should be excluded from this species. For a description of *E. mammillarioides* see *Malacocarpus* page 203.

Illustrations: Pfeiffer, Abbild. Beschr. Cact. **2**: pl. 2; Abh. Bayer. Akad. Wiss. München **2**: (see p. 739) pl. 3, f. 6; (see p. 740) pl. 5, sec. 1, f. 1 to 5, as *Echinocactus pfeifferi*.

20. Ferocactus flavovirens (Scheidweiler).

Echinocactus flavovirens Scheidweiler, Allg. Gartenz. **9**: 50. 1841.

Plant cespitose, forming great masses, pale green, 3 to 4 dm. high; stems 1 to 2 dm. in diameter; ribs 13, rarely 11 or 12, 1 to 2 cm. high, acute, somewhat sinuate; areoles 2 cm. apart, large, grayish, woolly; spines pale brown, becoming gray in age, long and stout; centrals 4, much longer than the radials, somewhat unequal, the longer ones 5 to 8 cm. long; flowers and fruit not seen; flower-buds globular, covered with long linear imbricating scales, their margins with long ciliate hairs.

Type locality: Tehuacán, Mexico.

Distribution: Known only from about Tehuacán, Puebla, Mexico.

This species was introduced into cultivation from the type locality by Dr. Rose in 1906 but it has never flowered. It grows with *F. robustus* and may easily be mistaken for it, but the color of the stems, number of ribs, and color of spines are quite different.

Here is referred *Echinocactus polyocentrus* Lemaire (Salm-Dyck, Cact. Hort. Dyck. 1844. 22. 1845), but unpublished.

The Index Kewensis refers *E. flavovirens* to *E. orthacanthus* Link and Otto (Verh. Ver. Beförd. Gartenb. **3**: 427. 1827; *Melocactus orthacanthus* Link and Otto, Verh. Ver. Beförd. Gartenb. **3**: pl. 18. 1827), a much earlier name, but the description suggests a very different plant, with 17 ribs and one stout central spine. The original description states definitely that it comes from Montevideo, but the Index Kewensis refers it to Mexico.

Illustrations: MacDougal, Bot. N. Amer. Des. pl. 18; Nat. Geogr. Mag. **21**: 700, as *Echinocactus flavescens.*

Plate XIII, figure 1, is from a photograph taken by Dr. MacDougal near Esperanza, Puebla, in 1900.

21. Ferocactus melocactiformis (De Candolle).

Echinocactus melocactiformis De Candolle, Prodr. **3**: 462. 1828.
Echinocactus histrix De Candolle, Mém. Mus. Hist. Nat. Paris **17**: 115. 1828.
Echinocactus coulteri G. Don, Gen. Syst. **3**: 162. 1834.
Echinocactus oxypterus Zuccarini in Pfeiffer, Enum. Cact. 57. 1837.
Echinocactus electracanthus Lemaire, Cact. Aliq. Nov. 24. 1838.
Echinocactus lancifer Reichenbach in Terscheck, Cact. Suppl. 2. .
Echinofossulocactus oxypterus Lawrence in Loudon, Gard. Mag. **17**: 318. 1841.
Echinocactus electracanthus capuliger Monville in Labouret, Monogr. Cact. 184. 1853.

Simple, cylindric, 5 to 6 dm. in diameter, bluish green; ribs about 24; areoles 2 to 3 cm. apart; spines usually 10 to 12, a little curved, yellow, becoming brown, of these 6 to 8 slender-subulate, 2 to 3 cm. long, more or less spreading; 3 or 4 spines more central than the others, but usually only one definitely so, much stouter and longer, 4 to 6 cm. long, porrect or ascending, annulate; flowers 2.5 to 3.5 cm. long, bright yellow, sometimes reddish without; inner perianth-segments linear-oblong, acute, somewhat spreading; stigma-lobes 6, linear, green; scales on the ovary ovate, acute, small, 2 to 4 mm. long, somewhat ciliate; fruit short-oblong, about 2 cm. long, somewhat edible; seeds minute, 1 mm. long, brown.

Type locality: Mexico.

Distribution: Eastern Mexico.

The numerous thin ribs of this plant, as shown in the original illustration, resemble those of some species of *Echinofossulocactus*, but its flowers appear to be like those of *Ferocactus*.

Echinocactus pfersdorffii Hortus may be referable here. It is probable that *E. pfersdorffii* Hildmann Catalogue (Monatsschr. Kakteenk. **5**: 92. 1905) is the same, but neither was accompanied by a description.

In cultivation this plant is simple, depressed-globose, 4.5 dm. in diameter, but in the wild state sometimes cylindric and up to 6 dm. high, and described as proliferous; ribs 20, perhaps even more, acute; areoles rather large, distant; radial spines usually 8, subulate, somewhat curved, 4.5 cm. long; central spines usually solitary, but as many as 4 reported, all yellow.

We have followed Schumann in referring here various synonyms, but the indications are that we have more than one
species. Dr. Rose obtained flowers of this
species at La Mortola in 1912 and his notes
were used in drawing up the description.

Here belongs *Cactus multangularis*
Mociño and Sessé (Mém. Mus. Hist. Nat.
Paris **17**: 38. 1828), but never described
by them. *Echinocactus electracanthus
rufispinus* (Monatsschr. Kakteenk. **3**: 70.
1893) we would also refer here.

Echinocactus hystrichacanthus Lemaire
(Cact. Gen. Nov. Sp. 17. 1839) may be of
this relationship. This species as well as
E. pycnoxyphus Lemaire (Cact. Gen. Nov.
Sp. 16. 1839) Weber considered as only
varieties of *E. hystrix*.

Illustrations: De Candolle, Mém. Mus.
Hist. Nat. Paris **17**: pl. 10, as *Echinocactus
melocactiformis;* Monatsschr. Kakteenk. **3**:
158. f. 2; **21**: 171; Schumann, Gesamtb.

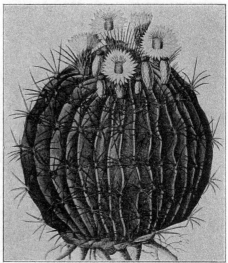

FIG. 146.—Ferocactus melocactiformis.

Kakteen f. 58, as *Echinocactus electracanthus;* Blühende Kakteen **1**: pl. 22, as *Echinocactus ingens.*

Figure 146 is copied from the first illustration above cited.

22. Ferocactus macrodiscus (Martius).

Echinocactus macrodiscus Martius, Nov. Act. Nat. Cur. 16; 341. 1832.
Echinocactus macrodiscus laevior Monville in Labouret, Monogr. Cact. 197. 1853.
Echinocactus macrodiscus decolor Monville in Labouret, Monogr. Cact. 197. 1853.
Echinocactus macrodiscus multiflorus R. Meyer, Monatsschr. Kakteenk. 24: 150. 1914.

Simple, depressed-globose or sometimes short-cylindric, sometimes 4.5 dm. in diameter; ribs 16, perhaps more in some specimens, somewhat flattened, sometimes acute on the margin, somewhat depressed at the distant areoles; spines all yellow, more or less curved backward; radial spines 6 to 8, mostly 2 to 3 cm. long; central spines 4, stouter and flatter than the radials, 3.5 cm. long; flowers 5 cm. long, dark red to purple, obconic; inner perianth-segments linear-oblong, acute; stamens and style included.

Type locality: Not definitely cited but probably on the Cumbre at about 10,000 feet, in a place called El Renosco, Mexico.

Distribution: San Luis Potosí and southward.

We do not know this species definitely although it is supposed to have a rather wide distribution in Mexico. The only specimen which we can refer with any confidence is one obtained through Professor Conzatti in 1910 from Oaxaca, Mexico.

The plant illustrated in Blühende Kakteen as cited below has flowers of different color and shape, and hence is referred here with some doubt.

Schumann (Gesamtb. Kakteen 349. 1898), following Labouret, refers as a synonym of this species *E. campylacanthus* Scheidweiler (Allg. Gartenz. **8:** 337. 1840), which is described as having 21 ribs and only one central spine. It should probably be referred elsewhere. The specimens distributed by de Laet under this name seem to be *Echinopsis leucantha*.

Illustrations: Nov. Act. Nat. Cur. **16:** pl. 26; Blühende Kakteen **3:** pl. 134; Schelle, Handb. Kakteenk. 162. f. 92; Gard. Chron. III. **50:** 135. f. 64, E, as *Echinocactus macrodiscus;* Monatsschr. Kakteenk. **24:** 151, as *Echinocactus macrodiscus multiflorus.*

Figure 147 is copied from the first illustration above cited.

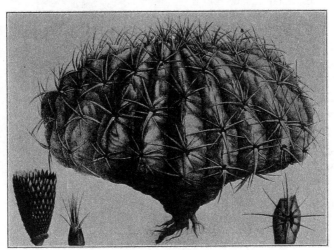

Fig. 147.—Ferocactus macrodiscus.

23. Ferocactus viridescens (Torrey and Gray).

Echinocactus viridescens Torrey and Gray, Fl. N. Amer. **1:** 554. 1840.
Melocactus viridescens Nuttall in Teschemacher, Bost. Journ. Nat. Hist. **5:** 293. 1845.
Echinocactus limitus Engelmann in Coulter, Contr. U. S. Nat. Herb. **3:** 374. 1896.

At first nearly globose or somewhat depressed, in age becoming cylindric, 3 to 4.5 dm. high, 2.5 to 3.5 dm. in diameter, simple or cespitose, deep green, somewhat glossy; ribs 13 to 21, somewhat rounded, 1 to 2 cm. high, obtuse, undulate; areoles narrow, elliptic, 1 to 2 cm. long, spine-bearing in the lower part, felted in upper part, flower-bearing and also with several reddish glands, these becoming elongated and spinescent in age; spines at first bright red, becoming duller by age or turning yellow or horn-colored; radial spines 9 to 20, more or less spreading, 1 to 2 cm. long; central spines 4, the lower one stouter and more flattened, up to 3.5 cm. long; flowers yellowish green, 4 cm. long; perianth-segments oblong, obtuse, sometimes apiculate, more or less serrulate on the margins; flower-tube bearing stamens almost to the top of the ovary; scales on the ovary orbicular, imbricate; fruit 1.6 to 2 cm. long, reddish with a pleasant acid taste; seeds 1.6 mm. long, pitted.

Type locality: Near San Diego, California.

Distribution: California and Lower California near the International Boundary Line, not far from the sea coast and in the foothills.

Echinocactus viridescens is usually credited to Nuttall, but he referred it in manuscript to *Melocactus*, and Torrey and Gray, who revised and published his manuscript, referred it doubtfully to *Echinocactus*.

Echinocactus californicus Monville (Labouret, Monogr. Cact. 199. 1853), first grown from seed supposed to have come from California, but without definite locality, may belong here although it has been referred to other species such as *F. orcuttii*. *E. californicus* Hortus is referred here by Rümpler (Förster, Handb. Cact. ed. 2. 472. 1885).

Illustrations: Gard. Chron. II. 7: 172. f. 26; Cact. Mex. Bound. pl. 29, as *Echinocactus viridescens*.

Plate XIV, figure 1, shows a flowering plant sent to the New York Botanical Garden from southern California by W. T. Schaller in 1909. Figure 148 is from a photograph of plants collected by C. R. Orcutt in southern California in 1917.

FIG. 148.—Ferocactus viridescens.

24. Ferocactus johnsonii (Parry).

Echinocactus johnsonii Parry in Engelmann, Bot. Kings's Surv. 117. 1871.
Echinocactus johnsonii octocentrus Coulter, Contr. U. S. Nat. Herb. 3: 374. 1896.

Simple, oblong, 10 to 20 cm. high, up to 11.5 cm. in diameter, often hidden under its mass of spines; ribs 17 to 21, low, somewhat tuberculate; spines reddish gray; radial spines 10 to 14, spreading, 10 to 20 mm. long; central spines 4 to 8, longer and stouter than the radials, somewhat curved, the upper ones connivent, 3.5 to 4 cm. long; flowers deep red to pink, 5 to 6.5 cm. long, sometimes 10 cm. broad when fully expanded; inner perianth-segments oblong to spatulate, obtuse; ovary bearing a few broad, scarious, fimbriate, margined scales; fruit oblong, 10 to 15 cm. long, nearly naked; seeds finely reticulated.

Type locality: Near St. George, Utah.

Distribution: Northwestern Arizona, eastern California, western Utah, and southern Nevada.

This species was named for Joseph Ellis Johnson (1817–1882), an amateur botanist of St. George, Utah, who, according to Professor Vasco M. Tanner, was once awarded a gold medal for having the best garden in the state of Utah.

Illustrations: Förster, Handb. Cact. ed. 2. 558. f. 71; Schelle, Handb. Kakteenk. 202. f. 134; Cact. Journ. 1: pl. 5; Deutsche Gärt. Zeit. 7: 53; Gartenflora 32: 58, as *Echinocactus johnsonii*.

Figure 149 (single plants in foreground) is from a photograph taken by M. E. Jones at Searchlight, Nevada, April 1907.

25. Ferocactus nobilis (Linnaeus).

Cactus nobilis Linnaeus, Mantissa 243. 1767.
Cactus recurvus Miller, Dict. Gard. ed. 8. No. 3. 1768.
Echinocactus recurvus Link and Otto, Verh. Ver. Beförd. Gartenb. 3: 426. 1827.
Melocactus recurvus Link and Otto, Verh. Ver. Beförd. Gartenb. 3: pl. 20. 1827.
Echinocactus spiralis Karwinsky in Pfeiffer, Enum. Cact. 60. 1837.

Echinocactus curvicornis Miquel, Linnaea **12**: 5. 1838.
Echinocactus stellatus Scheidweiler, Allg. Gartenz. **8**: 338. 1840.
Cereus recurvus Steudel, Nom. ed. 2. **1**: 335. 1840.
Echinocactus solenacanthus Scheidweiler, Allg. Gartenz. **9**: 50. 1841.
Echinofossulocactus recurvus campylacanthus Lawrence in Loudon, Gard. Mag. **17**: 318. 1841.
Echinocactus recurvus spiralis Schumann, Gesamtb. Kakteen 348. 1898.

Globular; ribs 15; radial spines straight, widely spreading; central spine solitary, erect, 7 cm. long, broad and flat, recurved at the tip, brownish red; flowers 2.5 to 4 cm. long; perianth-segments narrow, acute, red with white margins; ovary covered with ovate imbricated scales; fruit short, oblong, 2 cm. long, 12 mm. in diameter.

Type locality: Mexico.
Distribution: Eastern Mexico.
This species is referred both to Mexico and Peru. It is undoubtedly from eastern Mexico for it is based on *Cactus recurvus* of Miller. In the original description Miller thus speaks of it, "The third sort was brought into England by the late Dr. William Houston who procured the plant from Mexico." We do not know this species definitely, but plants collected by Dr. MacDougal and Dr. Rose in Tomellín Canyon, Oaxaca, answer the description, but have flowers up to 4 cm. long.

FIG. 149.—Ferocactus johnsonii.   FIG. 150.—Ferocactus nobilis.

We have referred here the synonymy given by Schumann, but suspect some of it should be referred elsewhere. Our description is based on Miller's original of *Cactus recurvus* for the stem and spines and on Pfeiffer's original description of *Echinocactus spiralis* for the flower and fruit. Schumann's description is somewhat different.

Echinocactus spiralis stellaris Salm-Dyck (Cact. Hort. Dyck. 1844. **21**. 1845), *Echinocactus stellaris* Karwinsky, also mentioned here by Salm-Dyck as a synonym and by Hemsley (Biol. Centr. Amer. Bot. **1**: 538. 1880) as a synonym of *Echinocactus spiralis*, and *Melocactus besleri affinis* (Förster, Handb. Cact. 320. 1846) doubtless are to be referred here.

Echinocactus multangularis Voigt we do not know. It is cited by Schumann (Gesamtb. Kakteen 348) as a synonym of *Echinocactus recurvus*, but no place of publication is given. In the only list of Voigt which we have consulted (Hort. Suburb. Calcutt. 1845) he lists three species of this genus, viz. *ottonis*, *eyriesii*, and *cornigerus*. These are followed by *Cereus multangularis* which suggests that a mistake has been made. Dr. John Hendley Barnhart suggests a different origin for the name of *Echinocactus multangularis*. It is to be noted that Förster's Handbuch appeared the next year after the appearance of Voigt's Calcutta List. Dr. Barnhart's note is as follows:

"I do not think that you have the correct explanation of the name *Echinocactus multangularis* Voigt. Schumann's citation of this name as a synonym of *E. recurvus* appears to me to have been copied from the first edition of Förster's Handbuch (1846), page 316, where under *E. recurvus* you will find the synonym '*C*.' (i. e., '*Cactus*') '*multangularis* Voigt.' In other words Schumann has simply made the slip of writing '*Echinocactus*' instead of '*Cactus*' for the Voigt name. The name '*Cactus multangularis* Voigt' seems to go back in literature as one of the synonyms of this species (and as a synonym only) to Steud. Nom. Bot. Phan. 132. 1821, where it appears as a synonym of *Cactus nobilis*. I doubt if the name was ever *published* anywhere, but do not think that at any time in its history it had anything to do with *Cereus multangularis*."

Echinocactus glaucus Karwinsky (Pfeiffer, Enum. Cact. 57. 1837), although never described, was referred to this species as a synonym. Dr. Rose examined the type of *Echinocactus spiralis* in Munich in 1912 and believes that it belongs here; *Echinocactus agglomeratus* (Pfeiffer, Enum. Cact. 60. 1837) was referred as a synonym of *Echinocactus spiralis*.

The following varieties seem to be only color or spine forms: *Echinocactus recurvus latispinus*, *E. recurvus solenacanthus*, and *E. recurvus tricuspidatus* (Förster, Handb. Cact. 317. 1846) and *E. recurvus bicolor* (Monatsschr. Kakteenk. 20: 144. 1910).

Illustrations: Schelle, Handb. Kakteenk. 162. f. 91; Monatsschr. Kakteenk. 16: 73; 21: 149; Möllers Deutsche Gärt. Zeit. 29: 440. f. 16; Abh. Bayer. Akad. Wiss. München 2: (see 738) pl. 1, sec. 7, f. 4; Knippel, Kakteen pl. 10; R. Grässner, Haupt-Verz. Kakteen 1912: 13, as *Echinocactus recurvus;* Verh. Ver. Beförd. Gartenb. 3: pl. 20, as *Melocactus recurvus;* Rev. Hort. 61: f. 140; Nov. Act. Nat. Cur. 19: pl. 16, f. 4, 7, as *Echinocactus spiralis*.

Figure 150 is copied from the last illustration above cited.

26. Ferocactus latispinus (Haworth).

Cactus latispinus Haworth, Phil. Mag. **63**: 41. 1824.
Echinocactus cornigerus P. De Candolle, Mém. Mus. Hist. Nat. Paris **17**: 36. 1828.
Mammillaria latispina Tate in Loudon, Gard. Mag. **16**: 26. 1840.
Echinofossulocactus cornigerus Lawrence in Loudon, Gard. Mag. **17**: 318. 1841.
Echinofossulocactus cornigerus elatior Lawrence in Loudon, Gard. Mag. **17**: 318. 1841.
Echinofossulocactus cornigerus rubrospinus Lawrence in Loudon, Gard. Mag. **17**: 318. 1841.
Echinofossulocactus cornigerus angustispinus Lawrence in Loudon, Gard. Mag. **17**: 318. 1841.
Echinocactus latispinus Hemsley, Biol. Centr. Amer. Bot. **1**: 533. 1880.
Echinocactus latispinus flavispinus Weber, Dict. Hort. Bois 467. 1896.

Plant simple, globular or somewhat depressed, 2.5 to 4 dm. high, 4 dm. in diameter; ribs 15 to 23, but usually 21, prominent; areoles large; radial spines 6 to 10, slender, annulate, white to rose, 2 to 2.5 cm. long; central spines 4 or more, stouter and more highly colored than the radials, all straight except one, this much flattened and hooked; flowers campanulate, 2.5 to 3.5 cm. long, rose to purple; perianth-segments narrowly oblong, acute; scales on the ovary closely imbricated, thin and papery, ovate, with thin ciliate margins; scales on flower-tube similar to those on ovary but more elongated; fruit elongated, 4 cm. long (dehiscence not known); seeds described as reniform, slightly pitted, 1.5 mm. long.

Type locality: Mexico.

Distribution: Widely distributed in Mexico; reported from Guatemala by De Candolle.

A plant sent to the New York Botanical Garden by A. de Lautreppe from Mexico in 1905 flowered in November 1913, the same flower opening successively for four days.

Echinocactus cornigerus var. *flavispinus* and var. *latispinus* (Förster, Handb. Cact. 318. 1846), published as synonyms, belong here.

Melocactus latispinus (Pfeiffer, Enum. Cact. 56. 1837) is also to be referred here. *Echinocactus cornigerus* Mociño and Sessé (De Candolle, Prodr. 3: 461. 1828) occurs as a synonym of this plant.

Echinocactus corniger rubrispinosus (Monatsschr. Kakteenk. 12: 59. 1902) is probably a form of this species.

Illustrations: Abh. Bayer. Akad. Wiss. München 2: pl. 3, f. 2; Schumann, Gesamtb. Kakteen f. 4, 62; Mém. Mus. Hist. Nat. Paris 17: pl. 7; De Candolle, Mém. Cact. pl. 10; Cact. Journ. 1: 54; 2: 173; Schelle, Handb. Kakteenk. 164. f. 94; Ann. Rep. Smiths. Inst. 1908: pl. 13, f. 6; Förster, Handb. Cact. ed. 2. 507. f. 60; Dict. Gard. Nicholson 4: 538. f. 19; Suppl. 334. f. 354; Rümpler, Sukkulenten 184. f. 102; Watson, Cact. Cult. 96. f. 31, as *Echinocactus cornigerus;* Cact. Journ. 1: pl. for March; Monatsschr. Kakteenk. 21: 11, as *E. corniger flavispinus*.

Plate XIII, figure 2, is from a photograph taken by Dr. MacDougal at El Riego, Tehuacán, Mexico, in 1906; plate XVI, figure 3, shows the flowering top of the plant sent by M. de Lautreppe, above alluded to.

27. Ferocactus crassihamatus (Weber).

> *Echinocactus crassihamatus* Weber, Dict. Hort. Bois 468. 1896.
> *Echinocactus mathssonii* Berge, Monatsschr. Kakteenk. **7**: 76. 1897.

Simple, globose to short-cylindric, pale green, somewhat glaucous; ribs 13, rather prominent, obtuse, strongly undulate; areoles large, only a few on each rib; radial spines 8, spreading, the upper ones straight, 2 or 3 of the lower ones hooked; central spines 5, longer and stouter than the radials, usually red, the stoutest one porrect and hooked; flowers small, about 2 cm. long, purple; inner perianth-segments linear-oblong, acute.

FIG. 151.—Ferocactus crassihamatus.   FIG. 152.—Ferocactus hamatacanthus.

Type locality: Querétaro.

Distribution: Querétaro, Mexico.

We know this plant only from descriptions and illustrations; its size is not recorded.

The original place of publication of *Echinocactus mathssonii*, as given by Schumann (Monatsschr. Kakteenk. **3**: 45. 1893), is without description and must be rejected. In the place we cite above there is only a phrase of description, but he later described it in detail.

Illustrations: Blühende Kakteen **1**: pl. 8; Schumann, Gesamtb. Kakteen f. 61, as *Echinocactus mathssonii*.

Figure 151 is copied from the first illustration above cited.

28. Ferocactus hamatacanthus (Mühlenpfordt).

> *Echinocactus hamatocanthus** Mühlenpfordt, Allg. Gartenz. **14**: 371. 1846.
> *Echinocactus flexispinus* Engelmann in Wislizenus, Mem. Tour North. Mex. 111. 1848.
> *Echinocactus longihamatus* Galeotti in Pfeiffer, Abbild. Beschr. Cact. **2**: pl. 16. 1848.
> *Echinocactus sinuatus* Dietrich, Allg. Gartenz. **19**: 345. 1851.
> *Echinocactus setispinus sinuatus* Poselger, Allg. Gartenz. **21**: 119. 1853.
> *Echinocactus setispinus robustus* Poselger, Allg. Gartenz. **21**: 119. 1853.
> *Echinocactus setispinus longihamatus* Poselger, Allg. Gartenz. **21**: 119. 1853.
> *Echinocactus longihamatus hamatacanthus* Labouret, Monogr. Cact. 201. 1853.
> *Echinocactus treculianus* Labouret, Monogr. Cact. 202. 1853.
> *Echinocactus longihamatus gracilispinus* Engelmann, Proc. Amer. Acad. **3**: 273. 1856.

**Originally spelled thus by Mühlenpfordt.

Echinocactus longihamatus crassispinus Engelmann, Proc. Amer. Acad. 3: 273. 1856.
Echinocactus longihamatus brevispinus Engelmann, Proc. Amer. Acad. 3: 274. 1856.
Echinocactus flavispinus Meinshausen, Wochenshr. Gärtn. Pflanz. 1: 28. 1858.
Echinocactus haematochroanthus Hemsley, Biol. Centr. Amer. Bot. 1: 532. 1880.
Echinocactus hamatacanthus longihamatus Coulter, Contr. U. S. Nat. Herb. 3: 365. 1896.
Echinocactus hamatacanthus brevispinus Coulter, Contr. U. S. Nat. Herb. 3: 366. 1896.
Echinocactus longihamatus sinuatus Weber in Schumann, Gesamtb. Kakteen 342. 1898.

Solitary, globular to oblong, up to 60 cm. high; ribs usually 13, sometimes 17, strongly tubercled, 2 to 3 cm. high; areoles large, 1 to 3 cm. apart; radial spines about 12, acicular, terete, 5 to 7 cm. long; central spines 4, elongated, angled, sometimes 15 cm. long, one of them hooked at apex; flowers large, 7 to 8 cm. long, yellow, in some forms said to be scarlet within; fruit oblong, 2 to 5 cm. long, fleshy, edible, dark brown to drab-colored (not red); seeds pitted.

Type locality: Mexico.
Distribution: Southern Texas, New Mexico, and northern Mexico.

This species develops elongated glands, 2 to 4 mm. long, in the areoles between the flower and the spines, as do some of the others; these at first are soft, but in age become hard and spine-like. The fruit of this species is unlike that of most other species of the genus; the skin is thin and the flesh juicy and edible.

Echinocactus insignis Haage jr. (Monatsschr. Kakteenk. 5: 76. 1905), a name only, was referred by Schumann as a synonym of *E. longihamatus.*

The following names (not described) are usually referred to this species or one of

FIGS. 153 and 153a.—Ferocactus uncinatus.

its synonyms: *Echinocactus longihamatus sinuatus* Weber (Monatsschr. Kakteenk. 12: 69. 1902), *Echinocactus longihamatus bicolor* (Monatsschr. Kakteenk. 3: 140. 1893), *E. longihamatus deflexispinus* (Monatsschr. Kakteenk. 12: 69. 1902), *E. longihamatus insignis* (Monatsschr. Kakteenk. 12: 69. 1902), and *E. texensis treculianus* (Förster, Handb. Cact. ed. 2. 504. 1885).

Echinocactus deflexispinus Gruson (Schumann, Gesamtb. Kakteen 343. 1898) was never described; it was considered by Schumann to be only a form of this species.

Illustrations: Blanc, Cacti 47. No. 556; Pfeiffer, Abbild. Beschr. Cact. 2: pl. 16; Schelle, Handb. Kakteenk. 159. f. 88; Ann. Rep. Smiths. Inst. 1908: pl. 9, f. 4; Förster,

Handb. Cact. ed. 2. 513. f. 63; Curtis's Bot. Mag. **78**: pl. 4632; Blühende Kakteen 1: pl. 9; Schumann, Gesamtb. Kakteen f. 3, 60; Cact. Mex. Bound. pl. 21 to 24; Watson, Cact. Cult. 109. f. 39, as *Echinocactus longihamatus;* Monatsschr. Kakteenk. **16**: 57, as *Echinocactus longihamatus sinuatus;* Cact. Mex. Bound. pl. 74, f. 11 to 14, as *Echinocactus sinuatus.*

Plate XVI, figure 1, shows the flowering top of a plant sent by Dr. Rose from near Devil's River, Texas, in 1913, which flowered at the New York Botanical Garden in 1916. Figure 152 is from a photograph of a plant sent by Dr. Edward Palmer from Victoria, Tamaulipas, Mexico, in 1907.

29. Ferocactus uncinatus (Galeotti).

Echinocactus uncinatus Galeotti in Pfeiffer, Abbild. Beschr. Cact. **2**: pl. 18.   1848.
Echinocactus ancylacanthus Monville in Labouret, Monogr. Cact. 201.   1853.
Echinocactus uncinatus wrightii Engelmann, Proc. Amer. Acad. **3**: 272.   1856.
Echinocactus wrightii Coulter, Cycl. Amer. Hort. Bailey **2**: 513.   1900.

Plant short-cylindric, 10 to 20 cm. high, bluish, slightly glaucous, with spindle-shaped roots; ribs usually 13, straight, strongly tubercled, undulate; flowering areoles narrow, extending from the spine-clusters to the base of the tubercles with the flower at the opposite end, felted; areoles also bearing one or more large flat yellow glands, these surrounded by a ring of short yellow hairs; central spine usually solitary, 12 cm. long or less, erect, yellow below, reddish above, hooked at tip; 3 lower radial spines spreading or reflexed, hooked; upper radials straight; flowers brownish, 2 to 2.5 cm. long, widely spreading; perianth-segments numerous, linear-oblong; filaments numerous, short; scales on ovary and flower-tube triangular, scarious-margined, in age broadly auriculate at base; fruit small, oblong, 2 cm. long, at first green, turning brown to crimson and finally scarlet, naked except the appressed scales, somewhat fleshy, edible; seeds black, small, oblong, 1 to 1.5 mm. long, with basal hilum; cotyledons foliaceous.

Type locality: Mexico.

Distribution: Rocky ridges and foothill-slopes in western Texas to central Mexico.

This species is doubtfully included in *Ferocactus*, for it is not closely related to any of those described above.   Technically it is different from all the other species in having the tubercles grooved on the upper side and the flower borne at the opposite end of the groove from the spine-cluster.   It might be better to segregate it as a generic type.

The glands in the areole described above secrete small drops of a honey-like substance much sought after by bees.   While usually found in the groove above the spines and below the flower they are also found on the outer side of the spine-areoles proper.   While these glands are usually sessile, they are sometimes elongated and suggest stunted spines.   One which we have preserved is 8 mm. long.   This species in its short groove above the spine-areole with its sessile gland suggests a relationship with some of the *Coryphanthanae.*

Illustrations: Dict. Gard. Nicholson Suppl. 336. f. 361 (with flowers of an *Echinocereus!*); Pfeiffer, Abbild. Beschr. Cact. **2**: pl. 18; Cact. Mex. Bound. pl. 74, f. 9; Watson, Cact. Cult. 123. f. 47; ed. 3. f. 29, as *Echinocactus uncinatus;* Cact. Mex. Bound. pl. 74, f. 10; Monatsschr. Kakteenk. **20**: 105, as *Echinocactus uncinatus wrightii.*

Figure 153 is from a photograph of a plant collected by F. E. Lloyd on Escondido Creek near Tuna Springs, Texas, in 1910, which flowered in 1911; figure 153a shows the same plant photographed in December 1920.

30. Ferocactus rostii sp. nov.

Sometimes growing in clumps of 8 to 10 heads but usually slender-cylindric, up to 3 meters high; ribs 16 to 22, rather low (hardly 1 cm. high), obtuse, somewhat tubercled; areoles large, white-felted, approximate; spine-clusters closely set, the spines interlocking and almost hiding the body of the plant; radial bristles sometimes wanting but when present 2 to 8, white or yellowish; spines about 12, sometimes fewer, 3 or 4 central, those on the lower part of the plant more or less spreading, those at or near the top erect, somewhat flexible, flattened, annulate, pungent, either straight or curved at apex, perhaps never hooked, usually yellow but sometimes reddish on young plants but also turning yellow in age; flowers dark yellow; fruit red.

PLATE XVI

Eaton del.

1. Top of flowering plant of *Ferocactus hamatacanthus*.
2. Flowering plant of *Sclerocactus whipplei*.
3. Top of flowering plant of *Ferocactus latispinus*.
(All natural size.)

According to Mr. E. C. Rost, for whom the plant is named, this species extends from the western fringe of the Imperial Valley, California, almost to Jacumba and down Lower California for about 40 miles.

Mr. Rost's note on a plant sent to the New York Botanical Garden is as follows:

"This cluster of yellow-spined plants shows in the main plant the appearance of being wrapped in straw. All of the mature plants of this variety have the same peculiarity. Note that the young offshoots of this specimen show a number of bright red spines, which disappear in the mature plants. One specimen I found to be 8 feet in height as shown in photograph. Some of the plants are single, but many are clustered."

This is a very striking plant, perhaps nearest *F. acanthodes*, but with a much more slender stem, and more appressed spines and these straw-colored.

The type is based on a plant collected in Lower California, 40 miles south of the International Boundary Line (Rost, No. 327).

Figure 153*b* is from a photograph taken by E. C. Rost at the type locality in 1921.

FIG. 153*b*.—Ferocactus rostii.

DESCRIBED SPECIES, PERHAPS OF THIS GENUS.

ECHINOCACTUS HAEMATACANTHUS Monville in Weber, Dict. Hort. Bois 466. 1896.

> *Echinocactus electracanthus haematacanthus* Salm-Dyck, Cact. Hort. Dyck. 1849. 150. 1850.

Simple, sometimes perhaps proliferous, short-cylindric, 5 dm. high, 3 dm. in diameter; ribs 12 to 20, stout, light green; spines all straight, reddish with yellowish tips, the radials 6, the centrals 4, 3 to 6 cm. long; flowers funnelform, 6 cm. long, purple; scales of the ovary round, white-margined; fruit ovoid, 3 cm. long.

Type locality: Not cited.

Distribution: Between Puebla and Tehuacán, according to Weber.

We do not know this species and our description is based on Weber's. His differs from the original where the central spine is described as solitary and reflexed. Schumann does not seem to have understood this species, as he first placed it after *E. pilosus*. *Echinocactus gerardii* Weber (Dict. Hort. Bois 466. 1896) is referred here by Schumann but it seems never to have been described.

ECHINOCACTUS RAFAELENSIS Purpus, Monatsschr. Kakteenk. 22: 163. 1912.

In clusters of 8 to 10, globose to short-cylindric, light green, at the apex slightly depressed and woolly; ribs 13 to 20, prominent; areoles elliptic; radial spines 7 to 9, 3 cm. long, the upper ones somewhat connivent; central spine solitary, 4 to 6 cm. long; flowers and fruit unknown.

Type locality: Minas de San Rafaél, San Luis Potosí, Mexico.

Distribution: Known only from the type locality.

We do not know this species and place it here on the statement of Quehl, who writes that it is similar to *E. robustus* and *E. flavovirens*.

Illustration: Monatsschr. Kakteenk. 23: 35, as *Echinocactus rafaelensis*.

17. ECHINOMASTUS gen. nov.

Plants small, globular or short-cylindric, ribbed, the ribs low, more or less spiraled, divided into definite tubercles; areoles bearing several acicular spines with or without stouter central ones; flowers central, medium-sized, borne at the spine-areoles, usually purple; fruit small, short-oblong, scaly, becoming dry, dehiscing by a basal opening; scales few, their axils naked; seed large, muricate, black, with a depressed ventral hilum.

Type species: *Echinocactus erectocentrus* Coulter.

The species which we have referred to this genus resemble in size, form, and habit the species of *Coryphantha* much more than they do the species of *Echinocactus* or *Ferocactus*. This resemblance is strengthened by definite tubercles on the ribs. Schumann referred them all to *Thelocactus*, his very complex subgenus of *Echinocactus*.

The generic name is from ἐχῖνος hedgehog, and μασπός breast, referring to the spiny tubercles of the plant. We recognize 6 species, all from northern Mexico and the adjacent parts of the United States. They are closely interrelated.

KEY TO SPECIES.

Areoles elongated, with more or less pectinate spines.
  One or two of the central spines different from the others.
    One central spine elongated, erect... 1. *E. erectocentrus*
    One central spine short, conic... 2. *E. intertextus*
  Central spines several, nearly alike.. 3. *E. dasyacanthus*
Areoles circular.
  Central spines subulate, some strongly recurved.................................. 4. *E. unguispinus*
  Central spines acicular.
    Globular; ribs 20 to 25; radial spines white.................................... 5. *E. macdowellii*
    Ovoid; ribs 18 to 21; radial spines with black tips............................ 6. *E. durangensis*

1. Echinomastus erectocentrus (Coulter).

    Echinocactus erectocentrus Coulter, Contr. U. S. Nat. Herb. **3**: 376. 1896.

Plants broadly ovoid to short-cylindric, 8 to 14 cm. high, sometimes 10 cm. broad, pale bluish green; ribs 21, somewhat oblique, very low, made up of closely set tubercles; radial spines 14, straight, terete, pale below, red above (in old dead plants dense and interwoven above but pectinate-appressed on lower part of plant); central spines 1 or 2, elongated, erect, slightly swollen at base, more conspicuous in dead than in living plants, usually ascending, one sometimes very short and porrect; flowers pinkish, 3 to 5 cm. long; stamens short, greenish yellow; style longer than the stamens, pale green; stigma-lobes 8, pinkish to deep red; ovary bearing a few ovate scarious scales. (This description is from living plants and differs somewhat from Dr. Coulter's which was made from dead ones.)

FIG. 154.—Echinomastus erectocentrus.   FIG. 155.—Echinomastus unguispinus.

Type locality: Near Benson, Arizona.
Distribution: Southeastern Arizona.

Unfortunately, Dr. Coulter associated with this species a plant from Saltillo, Mexico, collected by Weber; this plant has been published as *Echinocactus beguinii* Weber, to which

Dr. Schumann erroneously referred *Echinocactus erectocentrus*. *E. beguinii* is described as having a naked ovary and is a quite different plant of the *Coryphanthanae*.

The flowers on various plants differ somewhat in color, the deeper colored flowers being associated with higher colored spines. This difference in color extends also to the stigma-lobes. The flowers give off a delicate odor; they open in the morning and close at night, lasting for four days.

Echinocactus horripilus erectocentrus is credited by Schumann (Gesamtb. Kakteen 443. 1898) to Weber although he never formally published the variety.

Figure 154 is from a photograph of a plant collected by Kirk Bryan in southeastern Arizona in March 1921.

2. Echinomastus intertextus (Engelmann).

Echinocactus intertextus Engelmann, Proc. Amer. Acad. 3: 277. 1856.
Cereus pectinatus centralis Coulter, Contr. U. S. Nat. Herb. 3: 386. 1896.
Echinocereus pectinatus centralis Schumann, Gesamtb. Kakteen 271. 1898.
Echinocereus centralis Rose, Contr. U. S. Nat. Herb. 12: 293. 1909.

Simple, globular or nearly so, 2.5 to 10 cm. in diameter; ribs 13, somewhat acute, more or less divided into tubercles; areoles 5 to 6 mm. apart, somewhat elliptic; spines rigid, red with darker tips; radial spines 16 to 25, appressed, 8 to 15 mm. long, 3 or 4 of the upper radial spines white or nearly so, more slender than the others, almost bristle-like; central spines 4, subulate, 3 of them turned upward and similar to the radials, 10 to 18 mm. long, the other one very short, porrect; flowers 2.5 cm. long, nearly as broad as long, purplish; outer perianth-segments about 20, broadly ovate, white-margined; inner perianth-segments 20 to 25, oblong, mucronate; fruit nearly globular, 8 to 10 mm. in diameter, with a few scarious scales; seeds black, shining, 2 mm. in diameter.

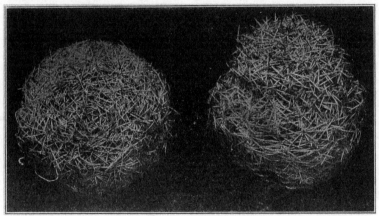

Fig. 156.—Echinomastus intertextus.

Type locality: Not definitely cited.

Distribution: Southwestern Texas, to southeastern Arizona and northern Mexico.

Engelmann states that the scales on the fruit are with or without some wool in their axils. The fruit is always in a mass of wool, but so far as we have seen the scales are always naked in their axils.

When Engelmann described this species he also briefly characterized a variety *dasya-canthus* which we have treated here as a distinct species. He says that *Echinocactus intertextus* in this broad sense ranges from El Paso to the Limpio and southward to Chihuahua and adds that the variety is more common about El Paso. We have seen only

Echinomastus dasyacanthus about El Paso while we have the true *Echinomastus intertextus* from Chihuahua. This latter station may be the type locality for this species.

In making this study we have at last been able to place definitely *Cereus pectinatus centralis* from near Fort Huachuca, Arizona. This *Echinocereus*-like plant was described from two sterile specimens whose flowers and fruit were not known. In 1921 J. W. Gidley sent a single specimen from southeastern Arizona. This flowered a few months afterwards, showing clearly that it was not an *Echinocereus*, but that it belonged to *Echinomastus*. Further study shows that it is referable to *Echinomastus intertextus*, although coming from west of the hitherto known range of the species.

Echinocactus krausei Hildmann (Schumann, Gesamtb. Kakteen 446. 1898) which came from Dragoon Summit, eastern Arizona, may belong here, but Schumann states that the ovary bears spines; it is known to us only from his description.

Illustrations: Schelle, Handb. Kakteenk. 201. f. 133, as *Echinocactus krausei;* Förster, Handb. Cact. ed. 2. 561. f. 72; Cact. Mex. Bound. pl. 34; Blanc, Cacti 46. f. 524, as *Echinocactus intertextus.*

Figure 156 is from a photograph of the type specimen of *Cereus pectinatus centralis.*

3. Echinomastus dasyacanthus (Engelmann).

Echinocactus intertextus dasyacanthus Engelmann, Proc. Amer. Acad. 3: 277. 1856.

Plants cylindric, 10 to 15 cm. high; ribs somewhat spiraled, made up of numerous compressed tubercles; spines slender, more or less purplish; radials 19 to 25, 12 to 22 mm. long; centrals about 4, nearly equal; top of flowering plant and young areoles very woolly; scales and outer perianth-segments red with white margins; inner perianth-segments white or purplish, about 2.5 cm. long, acute or acuminate; ovary bearing a few ovate scales, these naked in their axils; stigma-lobes 9, erect, truncate at apex, deep purple.

Type locality: Near El Paso, Texas.
Distribution: Southwestern Texas.

Most writers, including Engelmann, have treated this species as a variety of *Echinocactus intertextus* but in the light of a fuller series of specimens we believe it deserves specific rank. In the past many plants which we now know are true *Echinomastus dasyacanthus* have been passing as *Echinocactus intertextus.*

Besides the difference brought out by Engelmann this species has much larger flowers than *Echinomastus intertextus* and the inner perianth-segments are acute or acuminate. This species has also a more northern and eastern range.

Coulter (Contr. U. S. Nat. Herb. 3: 375. 1896) refers to *Echinocactus intertextus dasyacanthus*, a plant from San Luis Potosí, which we have not seen but which we suspect belongs elsewhere.

Illustrations: Cact. Mex. Bound. pl. 35, f. 1 to 5, as *Echinocactus intertextus dasyacanthus.*

Figure 157 is from a photograph of a plant sent by F. C. Platt from El Paso, Texas, in 1908.

4. Echinomastus unguispinus (Engelmann).

Echinocactus unguispinus Engelmann in Wislizenus, Mem. Tour North. Mex. 111. 1848.
*Echinocactus trollietii** Rebut, Balt. Cact. Journ. 2: 147. 1895.

Plants simple, usually globular, sometimes short-cylindric, 10 to 12 cm. high when mature, pale bluish green; ribs low; areoles woolly when young, circular; armament very peculiar, at times almost hiding the plant itself, most of the spines being erect or connivent; radial spines widely spreading, often as many as 25, usually white, except the tips, these darker, the upper ones 2 cm. long, a little longer than the lower; central spines 4 to 8, stouter than the radials, at first reddish or black, but becoming grayish blue in age, the lowermost turned outward and downward and all more or less curved; flowers 2.5 cm. long, reddish.

*The usual reference to the first publication of this name is the Monatsschrift für Kakteenkunde (5: 184. 1895). This appeared, however, in December while the Baltimore Cactus Journal reference appeared in July of the same year.

Type locality: About Pelayo, Chihuahua, between Chihuahua City and Parras.
Distribution: States of Chihuahua and Zacatecas, Mexico.

This species was described by Dr. Engelmann in 1848 from a single specimen collected by Dr. Wislizenus in Pelayo, Chihuahua. No other material was known to Dr. Coulter in 1896 when he wrote his monograph and the species was not in cultivation in this country in 1900. In 1908 Professor F. E. Lloyd sent material from the state of Zacatecas, and since then both Dr. Elswood Chaffey and Dr. C. A. Purpus have sent living plants from central Mexico.

Illustrations: Cact. Mex. Bound. pl. 35, f. 6 to 8; Schumann, Gesamtb. Kakteen f. 61, C; Monatsschr. Kakteenk. 5: 185; Knippel, Kakteen pl. 12, f. 1, as *Echinocactus unguispinus;* Balt. Cact. Journ. 2: 147; Orcutt, Rev. Cact. 54, as *Echinocactus trollietii;* West Amer. Sci. 8: 119, as *Echinocactus* No. 79.

Figure 155 is from a photograph of a specimen sent in by Dr. C. A. Purpus from Cerro de Movano, Mexico, which has more slender and less curved central spines than the type.

FIG. 157.—Echinomastus dasyacanthus.   FIG. 158.—Echinomastus macdowellii.

5. Echinomastus macdowellii* (Rebut).

Echinocactus macdowellii Rebut in Quehl, Monatsschr. Kakteenk. 4: 133. 1894.

Simple, globular or a little depressed, about 7 cm. high, 12 cm. in diameter, covered with a mass of interlocking spines; ribs 20 to 25, pale green, 5 to 7 mm. high, divided into tubercles; radial spines 15 to 20, white, spreading, up to 1.8 cm. long; central spines 3 or 4, dark colored, the longest up to 5 cm. in length; flowers rose-colored, up to 4 cm. long; ovary globose, said to be scaly.

Type locality: Not cited.
Distribution: Northern Mexico.

We have had this species in cultivation, but it has never flowered in this country. According to Mr. McDowell, it comes from Nuevo Leon near the border of Coahuila, Mexico.

In addition to the synonyn cited above the Index Kewensis cites a homonym, credited to C. R. Orcutt (West Amer. Sci. 8: 118. 1894). Perhaps both names refer to the same plant since they appeared in the same year, the first in September and the second in November.

Illustrations: Knippel, Kakteen pl. 9; Cact. Journ. 1: pl. for March; Monatsschr. Kakteenk. 4: 134; West Amer. Sci. 8: 118; Schelle, Handb. Kakteenk. 199. f. 131; Orcutt, Rev. Cact. 54, as *Echinocactus macdowellii.*

Figure 158 is from a photograph obtained by Dr. Rose from L. Quehl in 1912.

*The original spelling of the specific name was *E. mcdowellii,* but Schumann has corrected it to *E. macdowellii.*

6. Echinomastus durangensis (Rünge).

Echinocactus durangensis Rünge, Hamb. Gartenz. 46: 231. 1890.

Simple, ovoid, about 8 cm. long, 7 cm. in diameter; ribs 18 to 21, low; areoles white-woolly when young, but without wool when old; radial spines 15 to 30, the lower ones shorter than the upper, more or less incurved, white except the black tips, 1.5 cm. long; central spines 3 or 4, a little longer than the radials, acicular, about 2 cm. long; flowers and fruit not known.

Type locality: Not cited, but Schumann reports it only from Rio Nazas, west of Villa Lerdo, Durango, Mexico.

Distribution: Zacatecas and Durango, Mexico.

This species is similar to *Echinomastus unguispinus*, but not so large, with more slender and lighter-colored spines, none of them strongly recurved. We know it only from a specimen collected in Zacatecas by Dr. Elswood Chaffey in 1910.

Illustration: Schumann, Gesamtb. Kakteen f. 61, B, as *Echinocactus durangensis.*

18. GYMNOCALYCIUM Pfeiffer, Abbild. Beschr. Cact. **2**: under pl. 1 and pl. 12.   1845.*

Plants globular, simple or cespitose, strongly ribbed; ribs divided into tubercles often protruding at the base; flowers campanulate to short-funnelform, from upper and normally from the nascent areoles, usually large for size of plant, white, pink, or rarely yellow; flower-tube bearing broad scales, these with naked axils; fruit oblong, red so far as known, scaly; seeds cap-shaped or dome-shaped, brownish, tuberculate.

The species of this genus which were treated by Schumann are found in his subgenus *Hybocactus* of *Echinocactus*. We recognize about 23 species, all from South America, east of the Andes and chiefly from Argentina, with a few species from Bolivia, Paraguay, and Uruguay.   The generic name is from γυμνός naked, and κάλυξ bud, referring to the glabrous flower-bud.

The genus was originally based on three species of which *G. denudatum* was the first, and this is taken by us as the generic type.   Heynhold (Nom. **2**: 103. 1846) uses the three names of Pfeiffer.

The tubercles on the ribs have an enlargement more or less conspicuous just below the spine-areole which Schumann calls a "chin."   So far as our observation goes this is present in all the species, although it is very small in *G. saglione*, and it may be of considerable diagnostic importance.   By this character plants belonging to species of *Gymnocalycium* can be referred generically when not in flower.

The flower in this genus, as in the other genera of this tribe, normally comes from the center of the plant, borne on nascent areoles; but sometimes, especially in greenhouse plants, the flowers of some are lateral and borne on old areoles as in *E. gibbosus* (see Blühende Kakteen **1**: pl. 55), *E. stellatus*, and *E. schickendantzii* (see Schumann, Gesamtb. Kakteen Nachtr. f. 29).

KEY TO SPECIES.

```
A. Inner perianth-segments yellow to yellowish green.
    Ribs acute.....................................................................   1. G. mihanovichii
    Ribs rounded.
       Ribs 11 to 14; inner perianth-segments broadly oblong.
          Ribs very definite; tubercles broader than high.........................   2. G. netrelianum
          Ribs low, rather indefinite; tubercles subglobose.......................   3. G. leeanum
       Ribs 9; inner perianth-segments narrowly oblong.............................   4. G. guerkeanum
AA. Inner perianth-segments red, pink, or white.
    B. Ribs hardly tubercled.
       Ovary and tube bearing few scales.
          Scales on ovary rounded.............................................   5. G. spegazzinii
          Scales on ovary acute...............................................   6. G. denudatum
       Ovary and tube very scaly.............................................   7. G. hyptiacanthum
```

* Pfeiffer says: "Pfeiff. in Catal. Hort. Schelh. 1843" but this we do not credit as place of publication.

KEY TO SPECIES—continued.

BB. Ribs with prominent tubercles.
C. Flowers with a very short tube.
    Spines brown to black.
        Central spines one or more.
            Tubercles without a distinct chin.................................. 8. *G. saglione*
            Tubercles with a large chin....................................... 9. *G. mostii*
        Central spines wanting... 10. *G. gibbosum*
    Spines yellow.
        Ribs 10 to 15... 11. *G. multiflorum*
        Ribs 22.. 12. *G. brachyanthum*
CC. Flowers with more or less elongated tube.
    Spines yellow, at least when young, or white.
        Spines 5 to 7, tortuous, up to 6 cm. long........................... 13. *G. anisitsii*
        Spines a little curved, 1 to 4 cm. long.
            Spines 10 to 13.
                Spines slender, not appressed..................................... 14. *G. monvillei*
                Spines stout, appressed... 15. *G. melanocarpum*
            Spines 3, rarely more... 16. *G. uruguayense*
    Spines brown.
        Central spines present.
            Ribs acute... 17. *G. megalothelos*
            Ribs obtuse.. 18. *G. kurtzianum*
        Central spines none.
            Spines slender, acicular.
                Plant dark green... 19. *G. damsii*
                Plant reddish or bronze... 20. *G. platense*
            Spines stout, subulate.
                Flower-tube as long as limb; flowers sublateral...................... 21. *G. schickendantzii*
                Flower-tube much shorter than limb; flowers central................. 22. *G. stuckertii*
AAA. Species not grouped... 23. *G. joossensianum*

FIG. 159.—Gymnocalycium mihanovichii.   FIG. 160.—Gymnocalycium netrelianum.

1. Gymnocalycium mihanovichii (Frič and Gürke).

Echinocactus mihanovichii Frič and Gürke, Monatsschr. Kakteenk. **15**: 142. 1905.

Plant somewhat depressed, 5 cm. in diameter or less, grayish green; ribs 8, prominent, acute; areoles small, 12 mm. apart; spines 5 or 6, spreading, yellowish; flowers about 3 cm. long; outer perianth-segments brownish green; inner perianth-segments green to yellowish green, sometimes tinged with red; scales on the slender flower-tube and ovary broad.

Type locality: Paraguay.

Distribution: Paraguay.

We have not grown this plant, but Dr. Rose studied it in Berlin, in 1912.

This plant is successfully grown as a graft on the top of some of the *Cereus* allies.

Illustrations: Blühende Kakteen 2: pl. 101; Monatsschr. Kakteenk. 29: 67, as *Echino-cactus mihanovichii.*

Figure 159 is copied from the first illustration above cited.

2. Gymnocalycium netrelianum (Monville).

Echinocactus netrelianus Monville in Labouret, Monogr. Cact. 248. 1853.

Simple or sometimes proliferous, globular or somewhat depressed, 3 cm. in diameter, naked at apex; ribs 14, broad, rounded, tuberculate, somewhat glaucous; spines 5 to 8, all radial, brownish, setaceous, flexible, less than 1 cm. long; flowers pale citron-yellow, 5 cm. long; inner perianth-segments broadly oblong, acute.

Type locality: Not cited.

Distribution: Probably Uruguay or Argentina, according to Schumann, but not reported from the former by Arechavaleta or from the latter by Spegazzini.

According to Dr. Weber, this species is very similar to *E. hyptiacanthus,* but it is much smaller and the flowers are yellow, not white.

Illustration: Blühende Kakteen 1: pl. 39, b, as *Echinocactus netrelianus.*

Figure 160 is copied from the illustration above cited.

3. Gymnocalycium leeanum (Hooker).

Echinocactus leeanus Hooker in Curtis's Bot. Mag. 71: pl. 4184. 1845.

Globose or somewhat depressed, glaucous-green; tubercles hemispherical but usually 6-angled at base, not definitely arranged; areoles oval; spines about 11, slender; radial spines somewhat curved, appressed, 12 mm. long; central spine 1, straight, porrect; flowers large; outer perianth-segments green, tinged with purple; inner perianth-segments pale yellow.

Type locality: Argentina.

Distribution: Argentina and Uruguay.

This plant was originally obtained from Messrs. Lee, of the Hammersmith Nursery, who grew it from seed sent by Mr. John Tweedie from Argentina. Schumann referred it to *Echinocactus hyptiacanthus,* a white-flowered species from which we believe that it must be distinct. We have found no records of the rediscovery of this species, but we are inclined to refer here J. A. Shafer's No. 123, collected at Salto, Uruguay, March 7, 1917. This plant flowered in the New York Botanical Garden in 1918 and has beautiful yellow flowers.

Illustrations: Curtis's Bot. Mag. 71: pl. 4184; Loudon, Encycl. Pl. ed. 3. 1377. f. 19370, as *Echinocactus leeanus.*

Figure 164 is copied from the first illustration cited above.

4. Gymnocalycium guerkeanum (Heese).

Echinocactus guerkeanus Heese, Monatsschr. Kakteenk. 21: 132. 1911.

Usually simple but sometimes cespitose, about 5 cm. in diameter; ribs 9, broad and obtuse, somewhat tuberculate; spines all radial, usually 5, unequal, the longest 12 mm. long, yellowish, with brownish bases, rough, usually spreading or appressed; flowers near center of plant, 5 cm. long, yellow, nearly as broad as long when expanded; inner perianth-segments narrowly oblong, acute, sometimes toothed; scales on the ovary acute; fruit and seeds not known.

Type locality: Bolivia.

Distribution: Bolivia.

This species is said to be near *E. netrelianus* but apparently quite distinct. We know it only from description and illustrations.

Illustrations: Monatsschr. Kakteenk. **21**: 133; Blühende Kakteen **3**: pl. 144, as *Echinocactus guerkeanus.*

Figure 161 is copied from the last illustration above cited.

5. Gymnocalycium spegazzinii nom. nov.

Echinocactus loricatus Spegazzini, Anal. Mus. Nac. Buenos Aires III. **4**: 502. 1905. Not Poselger, 1853.

Depressed, globular, 6 cm. high, 14 cm. in diameter, grayish green; ribs 13, broad and low, rounded on the margin; areoles elliptic; spines usually 7, subulate, rigid, appressed to the ribs, sometimes recurved, grayish brown, 2 to 2.5 cm. long; flowers 7 cm. long; inner perianth-segments more or less rose-tinted; filaments and style violaceous; stigma-lobes 16, white to rose-colored; scales on the ovary few, broad.

Type locality: La Viña, province of Salta, Argentina.

Distribution: Known only from the type locality.

As this plant requires a new name it gives us great pleasure to dedicate it to such an enthusiastic cactus student as Dr. Carlos Spegazzini of La Plata, Argentina.

Figure 162 is from a photograph of the type specimen, contributed by Dr. Spegazzini.

FIG. 161.—G. guerkeanum.　　　　FIG. 162.—G. spegazzinii.

6. Gymnocalycium denudatum (Link and Otto) Pfeiffer, Abbild. Beschr. Cact. **2**: under pl. 1. 1845.

Echinocactus denudatus Link and Otto, Icon. Pl. Rar. 17. 1828.
Cereus denudatus Pfeiffer, Enum. Cact. 73. 1837.
Echinocactus denudatus typicus Schumann in Martius, Fl. Bras. 4²: 248. 1890.

Simple, subglobose or somewhat depressed, 5 to 15 cm. in diameter; ribs 5 to 8, very broad and low, obtuse, hardly tubercled; spines usually only 5, sometimes 8, all radial, appressed, slender, sometimes curved; flowers white or pale rose-colored; perianth-segments oblong, acute; ovary and flower-tube bearing only an occasional acute scale.

Type locality: Southern Brazil.

Distribution: Southern part of Brazil and reported from Argentina and Uruguay.

Echinocactus intermedius (Monatsschr. Kakteenk. **8**: 36. 1898) is, according to Dr. Schumann, a hybrid between this species and *Echinocactus multiflorus*, while Hildmann (Garten-Zeitung **4**: 479. f. 111. 1885) states that it is a cross between this species and *E. monvillei.* Numerous varieties have been described, some belonging here, while others have been referred to other species, and some are doubtless mere forms not deserving of a name. The following we do not know and they are therefore left under this species: var. *octogonus* Schumann (Martius, Fl. Bras. 4²: 248. 1890), var. *golzianus* Mundt (Monatsschr. Kakteenk. **7**: 187. 1897), vars. *wieditzianus, andersohnianus, heuschkelianus, meiklejohnianus, delaetianus, wagnerianus, scheidelianus* (F. Haage jr., Monatsschr. Kakteenk.

8: 36, 37. 1898), var. *roseiflorus* Hildmann (Schumann, Gesamtb. Kakteen 414. 1898), and *bruennowii* and *flavispinus* (Schelle, Handb. Kakteenk. 189. 1907).

Echinocactus denudatus paraguayensis, sometimes credited to Mundt and sometimes to Haage jr., has not been formally published, but has been illustrated as follows: Schelle, Handb. Kakteenk. 190. f. 123; Tribune Hort. 4: pl. 140; Gartenwelt 9: 266; De Laet, Cat. Gen. f. 3. This plant has been referred to by Schumann (Monatsschr. Kakteenk. 13: 50, 51, 109) as *Echinocactus paraguayensis*.

FIG. 163.—Gymnocalycium denudatum.   FIG. 164.—Gymnocalycium leeanum.

Illustrations: Martius, Fl. Bras. 4²: pl. 50, f. 1; Möllers Deutsche Gärt. Zeit. 25: 474. f. 6, No. 3; Blühende Kakteen 1: pl. 59; Schumann, Gesamtb. Kakteen f. 72; Monatsschr Kakteenk. 3: 158. f. I; 14: 41; 29: 141; Link and Otto, Icon. Pl. Rar. pl. 9; Schelle, Handb Kakteenk. 188. f. 119, as *Echinocactus denudatus;* De Laet, Cat. Gen. f. 10; Tribune Hort. 4: pl. 139, as *E. denudatus* var.; Schelle, Handb. Kakteenk. 189. f. 120, as *E. denudatus bruennowii;* Schelle, Handb. Kakteenk. 189. f. 121, as *E. denudatus delaetii;* Schelle, Handb. Kakteenk. 189. f. 122, as *E. denudatus heuschkehlii;* De Laet, Cat. Gen. f. 18, as *Echinocactus denudatus paraguayensis;* Garten-Zeitung 4: 479. f. 111, as *E. denudatus intermedius.*

Figure 163 is copied from the third illustration above cited.

7. Gymnocalycium hyptiacanthum (Lemaire).

Echinocactus hyptiacanthus Lemaire, Cact. Gen. Nov. Sp. 21. 1839.
Cactus hyptiacanthus Lemaire in Steudel, Nom. ed. 2. 1: 246. 1840.
Echinocactus hyptiacanthus eleutheracanthus Monville in Labouret, Monogr. Cact. 249. 1853.
Echinocactus hyptiacanthus megalotelus Monville in Labouret, Monogr. Cact. 249. 1853.
Echinocactus hyptiacanthus nitidus Monville in Labouret, Monogr. Cact. 249. 1853.

Simple, globose or sometimes depressed, 5 to 7 cm. in diameter, dull green, the apex somewhat umbilicate; ribs 9 to 12, broad, obtuse, somewhat tuberculate; radial spines 5 to 9, spreading or appressed, 10 to 12 mm. long, flexible, sometimes pubescent; central spine solitary or wanting; flowers white, 4.5 to 5 cm. long.

Type locality: Not cited.
Distribution: Uruguay.

This plant is sometimes distributed as *Echinocactus multiflorus*, but it is very different from that species. It was first described from barren plants, but afterwards Labouret described the flowers as white as did also Arechavaleta and Gürke. We believe, therefore, that the yellow-flowered species (*G. leeanum*), referred here by Schumann, should be excluded.

Illustrations: Schumann, Gesamtb. Kakteen f. 70; Blühende Kakteen **3**: pl. 164, as *Echinocactus hyptiacanthus.*

8. Gymnocalycium saglione (Cels).

Echinocactus saglionis Cels, Portef. Hort. 180. 1847.
Echinocactus hybogonus Salm-Dyck, Cact. Hort. Dyck. 1849. 167. 1850.
Echinocactus hybogonus saglionis Labouret, Monogr. Cact. 257. 1853.

Plants simple, globular, often very large, sometimes 3 dm. in diameter, dull green; ribs 13 to 32 according to the size of the plant, low, very broad, sometimes 4 cm. long, separated by wavy intervals, divided into large, low, rounded tubercles; areoles 2 to 4 cm. apart, large, felted when young; spines dark brown to black, at first ascending, afterwards more or less curved outward, 8 to 10 on small plants but on old plants often 15 or more, 3 to 4 cm. long; central spines 1 to several; flowers white or slightly tinged with pink, 3.5 cm. long, the tube short and broadly funnelform; inner perianth-segments spatulate, acute; scales of the ovary nearly orbicular, rounded, with a scarious margin.

Fig. 165.—Gymnocalycium saglione.   Fig. 166.—Gymnocalycium gibbosum.

Type locality: Catamarca, Tucuman, Argentina.
Distribution: Northern Argentina and perhaps southern Bolivia.

Our Bolivian reference is based on a living specimen and flowers collected by P. L. Porte at Lagunillas, southeastern Bolivia, July 1920, and delivered to us in good condition March 10, 1921; this may or may not belong here; it flowered May 7 and again on June 21, 1921. It may be described as follows:

Ribs 8, obtuse; flower 3 to 3.5 cm. long; flower-tube proper very short, only 1 to 2 mm. long; throat of flower broad, funnelform, 15 mm. long, bearing many stamens; inner surface of throat and tube deep reddish purple; filaments short, purple; style and stigma-lobes purple; inner perianth-segments short-oblong, obtuse, ochre-yellow, but drying pinkish.

According to Labouret, *Echinocactus hybogonus* which we refer here as a synonym is a native of Chile, but probably came from Argentina.

Illustrations: De Laet, Cat. Gén. f. 14, 17; Gartenwelt **7**: 279; Blühende Kakteen **1**: pl. 58; Monatsschr. Kakteenk. **12**: 27; Schumann, Gesamtb. Kakteen Nachtr. f. 30; Schelle, Handb. Kakteenk. f. 125, as *Echinocactus saglionis.*

Plate XVII, figure 1, shows a plant brought by Dr. Shafer from near Tapía, Argentina, in 1917 (No. 94), which flowered in the New York Botanical Garden in May 1919. Figure 165 is from a photograph of an Argentine specimen contributed by Dr. Spegazzini.

9. Gymnocalycium mostii (Gürke) Britton and Rose, Addisonia **3:** 5.   1918.

Echinocactus mostii Gürke, Monatsschr. Kakteenk. **16:** 11.   1906.

Plant depressed-globose, 6 to 7 cm. high, 13 cm. in diameter, dark green; ribs 11 to 14, broad and obtuse, more or less tubercled, often strongly; spines subulate, brownish; radial spines 7 to 9, unequal, 6 to 22 mm. long; central spine solitary, 18 to 20 mm. long; flower campanulate, pale red, 6 to 8 cm. broad; perianth-segments spreading, oblong; scales on the ovary 8 to 10, broad.

Type locality: Córdoba, Argentina.

Distribution: Province of Córdoba.

Illustrations: Blühende Kakteen **2:** pl. 93, as *Echinocactus mostii;* Addisonia **3:** pl. 83, B.

Plate XVII, figure 2, shows the top of a plant in bloom, brought by Dr. Rose from Cassafousth, Córdoba, Argentina, to the New York Botanical Garden, in 1915; figure 3 shows a flower.

10. Gymnocalycium gibbosum (Haworth) Pfeiffer, Abbild. Beschr. Cact. **2:** under pl. 1.   1845.

Cactus gibbosus Haworth, Syn. Pl. Succ. 173.   1812.
Cactus nobilis Haworth, Syn. Pl. Succ. 174.   1812.   Not Linnaeus, 1767.
Cactus reductus Link, Enum. **2:** 21.   1822.
Echinocactus gibbosus De Candolle, Prodr. **3:** 461.   1828.
Cereus reductus De Candolle, Prodr. **3:** 463.   1828.
Echinocactus nobilis Haworth, Phil. Mag. **7:** 115.   1830.
Cereus gibbosus Pfeiffer, Enum. Cact. 74.   1837.
Echinocactus mackieanus Hooker in Curtis's Bot. Mag. **64:** pl. 3561.   1837.
Echinocactus gibbosus nobilis Monville in Lemaire, Cact. Gen. Nov. Sp. 91.   1839.
Gymnocalycium reductum Pfeiffer, Abbild. Beschr. Cact. **2:** pl. 12.   1847.
Echinocactus reductus Schumann, Monatsschr. Kakteenk. **5:** 107.   1895.
Echinocactus gibbosus chubutensis Spegazzini, Anal. Mus. Nac. Buenos Aires II. **4:** 285.   1902.
Echinocactus gibbosus ventanicola Spegazzini, Anal. Mus. Nac. Buenos Aires III. **2:** 7.   1903.
Echinocactus spegazzinii Weber in Spegazzini, Anal. Mus. Nac. Buenos Aires III. **2:** 7.   1903.
Echinocactus gibbosus typicus Spegazzini, Anal. Mus. Nac. Buenos Aires III. **4:** 503.   1905.

Plants simple, sometimes depressed, but usually taller than thick, sometimes 20 cm. high; ribs 12 to 14, broad, strongly tubercled; spines 7 to 12, all radial, straight or somewhat curved, usually light brown; flowers white to pinkish, 6 to 6.5 cm. long; inner perianth-segments oblong; ovary-scales ovate, acutish.

Type locality: Not cited.

Distribution: Argentina.

Numerous varieties have been described under this species, among which are the following: *celsianus* Labouret (Förster, Handb. Cact. ed. 2. 583. 1885), *cerebriformis* Spegazzini (Anal. Soc. Cient. Argentina **48:** 50. 1899), *fennellii* F. A. Haage jr. (Monatsschr. Kakteenk. **9:** 115. 1899; *Echinocactus fennellii* F. A. Haage jr. in Schumann, Gesamtb. Kakteen 409. 1898), *ferox* Labouret (Förster, Handb. Cact. ed. 2. 583. 1885; *Echinocactus ferox,* Monatsschr. Kakteenk. **4:** 193. 1894), *leonensis* Hildmann (and *Echinocactus leonensis* Cels in Schumann, Gesamtb. Kakteen 409. 1898), *leucanthus* Rümpler (Förster, Handb. Cact. ed. 2. 583. 1885), *leucodictyus* Salm-Dyck (Cact. Hort. Dyck. 1849. 34. 1850; *Echinocactus leucodictyus* Salm-Dyck, Cact. Hort. Dyck. 1849. 34. 1850), *pluricostatus* (Förster, Handb. Cact. ed. 2. 584. 1885), *polygonus* Schumann (Gesamtb. Kakteen 409. 1898), and *schlumbergeri* Rümpler (Förster, Handb. Cact. ed. 2. 584. 1885; *Echinocactus schlumbergeri* Cels in Schumann, Gesamtb. Kakteen 409. 1898).

Echinocactus towensis Cels (Schumann, Gesamtb. Kakteen 409) which comes from Towa, an island off the coast of Argentina, we do not know. It may be of this relationship.

Schumann (Gesamtb. Kakteen 409. 1898) uses the binomial *Echinocactus celsianus* Labouret, but says it is a variety of this species. *Echinocactus globosus cristatus* is only a gardener's name for a monstrosity. *Echinopsis gibbosa* Pfeiffer (Förster, Handb. Cact. 291. 1846) was given as a synonym of this species.

Illustrations: Blühende Kakteen **1:** pl. 55; Schumann, Gesamtb. Kakteen f. 71; Lemaire, Icon. Cact. pl. 13 (?); Gartenwelt **15:** 536; Möllers Deutsche Gärt. Zeit. **25:** 474. f. 6, No. 28; Monatsschr. Kakteenk. **26:** 21, as *Echinocactus gibbosus;* Edwards's Bot. Reg. **2:**

M. E. Eaton del.

A.Hoen & Co.

1. Flowering plant of *Gymnocalycium saglione*.
2. Top of flowering plant of *Gymnocalycium mostii*.
3. Flower of same.
　　　　　　　(All natural size.)

pl. 137; Loddiges, Bot. Cab. 16: pl. 1524; Reichenbach, Fl. Exot. pl. 326, as *Cactus gibbosus;* Pfeiffer, Abbild. Beschr. Cact. 2: pl. 12, as *Gymnocalycium reductum;* Curtis's Bot. Mag. 64: pl. 3561, as *Echinocactus mackieanus;* Curtis's Bot. Mag. 75: pl. 4443, as *Cereus reductus;* Monatsschr. Kakteenk. 30: 181, as *Echinocactus gibbosus nobilis.*

Figure 166 is from a photograph obtained by Dr. Rose from Dr. Spegazzini in 1915.

11. Gymnocalycium multiflorum (Hooker) Britton and Rose, Addisonia 3: 5. 1918.

Echinocactus multiflorus Hooker in Curtis's Bot. Mag. 71: pl. 4181. 1845.

Simple or cespitose, globular or somewhat depressed or sometimes short-columnar, 9 cm. high or more, sometimes 12 cm. in diameter; ribs 10 to 15, broad at base, somewhat tubercled, especially above, acutish; areoles elliptic, 10 mm. long; spines 7 to 10, all radial, spreading, somewhat flattened, stout, yellowish, the longest one 3 cm. long; flower-bud ovoid, covered with imbricate scales; flowers 3.5 to 4 cm. long, pinkish to nearly white, short-campanulate; inner perianth-segments oblong, 3 cm. long, obtuse or acute; scales on the ovary broad and rounded, their margins scarious.

Type locality: Not cited.

Distribution: Reported from Brazil, Uruguay, Paraguay, and Argentina. We know it definitely from Argentina, where it was collected by Dr. Rose in 1915, in Córdoba.

Schumann (Gesamtb. Kakteen 405. 1898) describes briefly the three following varieties: *albispinus, parisiensis,* and *hybopleurus.*

Echinocactus ourselianus Monville (Salm-Dyck, Cact. Hort. Dyck. 1849. 34. 1850) is cited by Schumann as a synonym of this species, but it was never published; the name was attributed to Cels by Salm-Dyck. Its variety *albispinus* (Monatsschr. Kakteenk. 5: 111. 1895) is sometimes met with.

Illustrations: Möllers Deutsche Gärt. Zeit. 25: 474. f. 6, No. 22; Loudon, Encycl. Pl. ed. 3. 1376. f. 19369; Monatsschr. Kakteenk. 26: 67; Curtis's Bot. Mag. 71: pl. 4181; Blühende Kakteen 1: pl. 30, as *Echinocactus multiflorus;* Addisonia 3: pl. 83, A.

Plate XVIII, figure 3, shows a flowering plant brought by Dr. Rose from Cosquin, Argentina, to the New York Botanical Garden in 1915, where it promptly bloomed. Figure 167 is from a photograph of a plant from Catamarca, Argentina, contributed by Dr. Spegazzini.

FIG. 167.—Gymnocalycium multiflorum.

12. Gymnocalycium brachyanthum (Gürke).

Echinocactus brachyanthus Gürke, Monatsschr. Kakteenk. 17: 123. 1907.

Stem simple, depressed-globose, 7 cm. high, 18 cm. in diameter; ribs 22, strongly tubercled; tubercles 5 or 6-sided; areoles elliptic; spines 5 to 7, all radial, subulate, yellowish, 10 to 25 mm. long; flowers, including the ovary, 3 to 5 cm. long, campanulate; inner perianth-segments white to rose-colored; scales of the ovary few, broader than long, rounded, the margin scarious.

Type locality: Argentina.

Distribution: Northern Argentina.

We have studied a plant sent to the New York Botanical Garden from Berlin in 1914 which has not yet flowered.

13. Gymnocalycium anisitsii (Schumann).

Echinocactus anisitsii Schumann, Blühende Kakteen 1: pl. 4. 1900.

Simple, short-cylindric, about 1 dm. long, pale green; ribs 11, strongly tubercled, acute; spines 5 to 7, yellowish, slender, usually all radial, subulate, somewhat angled, tortuous, sometimes 6 cm. long; flower 4 cm. long, somewhat funnelshaped; scales and outer perianth-segments broad, greenish white; inner perianth-segments white, broadly oblong, acute.

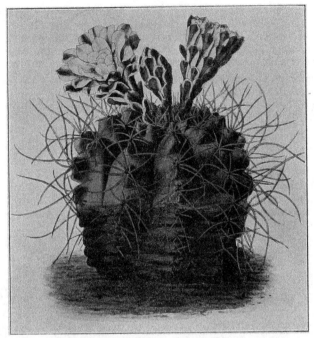

FIG. 168.—Gymnocalycium anisitsii.

FIG. 169.—Gymnocalycium monvillei.

M. E. Eaton del.   A.Hoen &Co.

1. Flowering plant of *Gymnocalycium megalothelos*.
2. Top of flowering plant of *Gymnocalycium platense*.
3. Flowering plant of *Gymnocalycium multiflorum*.
(All natural size.)

Type locality: Rio Tigatigami, Paraguay.

Distribution: Paraguay.

Illustrations: Blühende Kakteen 1: pl. 4; Monatsschr. Kakteenk. 29: 81; Schumann, Gesamtb. Kakteen Nachtr. f. 26, as *Echinocactus anisitsii.*

Figure 168 is copied from the first illustration above cited.

14. Gymnocalycium monvillei Pfeiffer, ined.*

    Echinocactus monvillei Lemaire, Cact. Aliq. Nov. 14. 1838.

Globose, large, 20 cm. in diameter or more; ribs 13 to 17, broad and obtuse, strongly tubercled; areoles elliptic; spines 12 or 13, all radial, yellowish except the purplish bases, subulate, spreading, 3 to 4 cm. long, slightly curved; flowers large, nearly white, 6 to 8 cm. long; inner perianth-segments oblong, acute; scales on the ovary orbicular.

FIG. 170.—Gymnocalycium monvillei.

Type locality: Paraguay.

Distribution: Mountains of Paraguay.

Echinocactus contractus Hildmann (Monatsschr. Kakteenk. 1: 15, with plate. 1891) is a hybrid between this species and *G. gibbosum.*

Illustrations: Blühende Kakteen 1: pl. 10; Lemaire, Icon. Cact. pl. 1; Lemaire, Cact. Aliq. Nov. pl. 1, f. 1, 2; Möllers Deutsche Gärt. Zeit. 25: 474. f. 6, No. 4; Garten-Zeitung 4: 182. f. 42, No. 17; Monatsschr. Kakteenk. 27: 171; 29: 81, as *Echinocactus monvillei.*

Figure 169 is copied from the first illustration above cited; figure 170 is from a photograph contributed by Dr. Spegazzini, showing his understanding of this species. This may be different from the plant shown by figure 169, but we know the species only from descriptions and the illustrations cited above. We are unable to determine which of the figures is the more nearly correct.

FIG. 171.—Gymnocalycium melanocarpum.   FIG. 172.—Gymnocalycium uruguayense.

15. Gymnocalycium melanocarpum (Arechavaleta).

    Echinocactus melanocarpus Arechavaleta, Anal. Mus. Nac. Montevideo 5: 220. 1905.

Simple, globose, 7 to 9 cm. in diameter; ribs 15, broad and rounded, strongly tubercled; spines all radial, 10 to 12, yellow when young, in age grayish, 2 to 2.5 cm. long; flowers nearly central; ovary nearly globular, bearing a few broad scales.

---

*The only reference to this binomial which we have seen is that of Schumann (Gesamtb. Kakteen 411) where it is used as a synonym.

Type locality: Near Paysandú, Uruguay.
Distribution: Northwestern Uruguay.
Illustration: Anal. Mus. Nac. Montevideo 5: pl. 15, as *Echinocactus melanocarpus.*
Figure 171 is copied from the illustration of the type plant above cited.

16. Gymnocalycium uruguayense (Arechavaleta).

Echinocactus uruguayensis Arechavaleta, Anal. Mus. Nac. Montevideo 5: 218. 1905.

Usually much depressed; ribs 12 to 14, strongly tubercled; areoles orbicular, grayish tomentose when young; spines 3, 1.5 to 2 cm. long, usually white; flowers 4 cm. long; inner perianth-segments linear-lanceolate.

Type locality: Paso de los Toros, Uruguay.
Distribution: Known only from the type locality.
Illustration: Anal. Mus. Nac. Montevideo 5: pl. 14, as *Echinocactus uruguayensis.*
Figure 172 is copied from the illustration of the type plant above cited.

FIG. 173.—Gymnocalycium megalothelos.　　FIG. 174.—Gymnocalycium kurtzianum.

17. Gymnocalycium megalothelos (Sencke).

Echinocactus megalothelos Sencke in Schumann, Gesamtb. Kakteen 415. 1898.

Plant simple, somewhat depressed-globose, 10 cm. in diameter, but sometimes said to be short-columnar, in cultivation becoming bronzed; ribs 10 to 12, prominent, 10 to 12 mm. high, acute, deeply divided into tubercles; spines acicular, brownish; radial spines 7 or 8, spreading or sometimes ascending, 1 to 1.5 cm. long; central spine solitary, ascending or porrect, more or less curved, 2 to 3 cm. long; flower campanulate-funnelform, erect, 3 to 4 cm. long; outer perianth-segments broad, greenish purple with a broad ovate acute tip; inner perianth-segments pinkish white; scales on the ovary very broad with a pinkish tip.

Type locality: Paraguay.
Distribution: Paraguay.
This species was collected in Paraguay by Professor R. Chodat in 1915 and has flowered repeatedly in the New York Botanical Garden.
Plate XVIII, figure 1, shows the plant collected by Professor Chodat. Figure 173 is from a photograph of the same plant.

18. Gymnocalycium kurtzianum (Gürke).

Echinocactus kurtzianus Gürke, Monatsschr. Kakteenk. **16:** 55. 1906.

Simple, globose but depressed, 10 to 15 cm. in diameter, naked at apex; ribs 10 to 18, divided into tubercles; radial spines 8, spreading, brownish, with recurved tips, 2.5 to 4 cm. long; central spine solitary, 3 cm. long; flowers large, white, reddish at base; scales on ovary large; inner perianth-segments obtuse.

Type locality: Probably Córdoba, Argentina.

Distribution: Argentina.

The plant is known to us only from descriptions and illustrations.

This species was named for Dr. Fritz Kurtz (1854–1920) who lived for many years in Argentina.

Illustrations: Monatsschr. Kakteenk. **17:** 126; Blühende Kakteen **2:** pl. 97, as *Echinocactus kurtzianus.*

Figure 174 is copied from the second illustration above cited.

19. Gymnocalycium damsii (Schumann).

Echinocactus denudatus bruennowianus Haage jr., Monatsschr. Kakteenk. **8:** 37. 1898.
Echinocactus damsii Schumann, Gesamtb. Kakteen Nachtr. 119. 1903.

Simple, globular or somewhat depressed; ribs 10, green, tuberculate; spines all radial, straight, short, the longest 12 mm. long; flowers narrow, funnelform, 6 cm. long; inner perianth-segments oblong, white to light pinkish, spreading; scales on the ovary and flower-tube small, scattered; fruit oblong, 2.5 cm. long, 6 mm. in diameter, red.

Type locality: Northern part of Paraguay.

Distribution: Paraguay.

The plant is known to us only from descriptions and illustrations.

Illustrations: Blühende Kakteen **2:** pl. 83; Schumann, Gesamtb. Kakteen Nachtr. f. 27; Monatsschr. Kakteenk. **14:** 76, as *Echinocactus damsii;* Cact. Journ. **1:** 53, as *Echinocactus denudatus brunnowianus.*

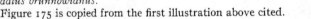

FIG. 175.—Gymnocalycium damsii.

Figure 175 is copied from the first illustration above cited.

20. Gymnocalycium platense (Spegazzini).

Echinocactus platensis Spegazzini, Contr. Fl. Vent. 28. 1896.
Echinocactus quehlianus F. Haage jr. in Quehl, Monatsschr. Kakteenk. **9:** 43. 1899.
Echinocactus stenocarpus Schumann, Monatsschr. Kakteenk. **10:** 181. 1900.
Echinocactus gibbosus platensis Spegazzini, Anal. Mus. Nac. Buenos Aires III. **2:** 8. 1902.
Echinocactus platensis typicus Spegazzini, Anal. Mus. Nac. Buenos Aires III. **4:** 504. 1905.
Echinocactus platensis leptanthus Spegazzini, Anal. Mus. Nac. Buenos Aires III. **4:** 504. 1905.
Echinocactus platensis quehlianus Spegazzini, Anal. Mus. Nac. Buenos Aires III. **4:** 504. 1905.
Echinocactus platensis parvulus Spegazzini, Anal. Mus. Nac. Buenos Aires III. **4:** 505. 1905.
Echinocactus stellatus Spegazzini, Anal. Mus. Nac. Buenos Aires III. **4:** 505. 1905. Not Scheidweiler, 1840.
Echinocactus baldianus Spegazzini, Anal. Mus. Nac. Buenos Aires III. **4:** 505. 1905.

Plants small, depressed, half-hidden in the earth, 4 to 9 cm. broad, dull bluish green or purple or bronzed; ribs 8 to 12, broad and low, divided by cross lines into tubercles; tubercles with a horizontal or ascending chin-like projection; areoles when young white-felted; spines 3 to 6, 1 cm. long or less, brown with white tips, acicular, more or less appressed; flower inodorous, 6 cm. long, dull bluish green; tube and ovary bearing a few broad, short, rounded scales, these more or less purplish on the edge; outer perianth-segments white with a broad green stripe down the center; inner perianth-segments pure white; throat broad, purple within; filaments numerous, scattered over the throat; style short and thick, 2 cm. long; stigma-lobes cream-colored; ovary oblong.

Type locality: Argentina.

Distribution: Argentina.

This species has a wide range in southern Argentina and consists evidently of several races differing in armament and relative length of the perianth-tube and limb. A plant from the Berlin Botanical Garden sent as *Echinocactus quehlianus* produced flowers identical with those of a plant of *E. platensis* brought by Dr. Rose from Córdoba.

Illustrations: Blühende Kakteen **2**: pl. 105; Möllers Deutsche Gärt. Zeit. **25**: 474. f. 6, No. 21; Monatsschr. Kakteenk. **10**: 152; Schumann, Gesamtb. Kakteen Nachtr. f. 28, as *Echinocactus quehlianus;* Monatsschr. Kakteenk. **17**: 9; **29**: 141, as *Echinocactus platensis.*

FIGS. 176 and 177.—Gymnocalycium platense.

Plate XVIII, figure 2, shows a plant in flower, brought by Dr. Rose in 1915 to the New York Botanical Garden from near Córdoba, Argentina; plate XIX, figure 1, shows the flowering top of a plant received at the New York Botanical Garden from the Berlin Botanical Garden as *Echinocactus quehlianus.* Figure 176 is from a photograph of a plant from Argentina determined by Dr. Spegazzini as *Echinocactus platensis leptanthus.* Figure 177 is from a photograph contributed by Dr. Spegazzini; figure 178 is from a photograph also from Dr. Spegazzini, illustrating *Echinocactus baldianus.*

21. Gymnocalycium schickendantzii (Weber).

Echinocactus schickendantzii Weber, Dict. Hort. Bois 470. 1896.
Echinocactus delaetii Schumann, Monatsschr. Kakteenk. **11**: 186. 1901.

Usually simple, sometimes depressed, up to 10 cm. in diameter; ribs usually 7, broad, more or less tuberculate; spines 6 or 7, all radial, more or less spreading, the larger ones flattened; flowers often from the side of the plant as well as central, white or pinkish in age, 5 cm. long; inner perianth-segments oblong to spatulate, obtuse; scales on the ovary and flower-tube purplish, broad and rounded.

Type locality: Catamarca, Tucuman, Argentina.

Distribution: Northern Argentina.

M. E. Eaton del.   A.Hoen.&Co.

1. Top of flowering plant of *Gymnocalycium platense*.
2. Top of flowering plant of *Gymnocalycium schickendantzii*.
3. Top of flowering plant of *Homalocephala texensis*.
4. Fruit of same.
5. Seed of same.

(All natural size, except seed.)

Shafer's No. 103 from Trancas flowered in Washington in June 1920. The flowers were erect and the perianth-segments waxy, becoming pinkish; the ribs were strongly tubercled.

FIG. 178.—Gymnocalycium
platense.

FIG. 179.—Gymnocalycium
schickendantzii.

Illustrations: De Laet, Cat. Gén. f. 19; Schumann, Gesamtb. Kakteen Nachtr. f. 29; Schelle, Handb. Kakteenk. 191. f. 124, as *Echinocactus schickendantzii;* Monatsschr. Kakteenk. 11: 187; Gartenwelt 7: 279; Gard. Chron. III. 33: suppl. plate, as *Echinocactus delaetii.*

Plate XIX, figure 2, shows a plant collected by Dr. Shafer at Andalgala, Argentina, in 1917 (No. 15), which flowered in the New York Botanical Garden in July 1918. Figure 179 is from a photograph contributed by Dr. Spegazzini, showing a plant from Córdoba, Argentina; figure 181, also from one of Dr. Spegazzini's photographs, shows another plant from Córdoba.

FIG. 180.—Gymnocalycium stuckertii.

FIG. 181.—Gymnocalycium schickendantzii.

22. Gymnocalycium stuckertii (Spegazzini).

Echinocactus stuckertii Spegazzini, Anal. Mus. Nac. Buenos Aires III. 4: 502. 1905.

Plant globose, sometimes depressed, dull green, 6 to 6.5 cm. in diameter, 3.5 to 4 cm. high; ribs 9 to 11, obtuse; spines all radial, pinkish to brown, flattened, puberulent, 1 to 2.5 cm. long, somewhat spreading; flowers 4 cm. long, the tube rather short; inner perianth-segments nearly white; scales on the ovary and flower-tube scattered, broadly ovate, scarious-margined.

Type locality: Province of San Luis Potosí, Argentina.
Distribution: Northern Argentina.

This species was named for T. Stuckert who aided Dr. Spegazzini in his studies of the cacti of Argentina.

Figure 180 is from a photograph of an Argentine plant contributed by Dr. Spegazzini.

23. Gymnocalycium joossensianum (Bödeker).

Echinocactus joossensianus Bödeker, Monatsschr. Kakteenk. 28: 40. 1918.

Simple, depressed-globose, somewhat umbilicate at apex; ribs 6 to 9, obtuse, straight, somewhat tubercled; spines 6 to 9, the lower ones a little longer than the upper; flowers wine-red, nearly central, campanulate with a short tube; inner perianth-segments longer than the outer, oblong-obtuse; stigma-lobes 6; fruit fusiform; scales on the fruit few, red-tipped; seeds brownish yellow.

Type locality: Not definitely cited.

Distribution: Paraguay or northern Argentina.

We have not seen specimens of this plant, but from the illustration it is clearly a species of *Gymnocalycium*.

Illustration: Monatsschr. Kakteenk. 28: 41, as *Echinocactus joossensianus*.

19. ECHINOCACTUS Link and Otto, Verh. Ver. Beförd. Gartenb. 3: 420. 1827.

Plants very large, thick, cylindric and many-ribbed, or low and several-ribbed, the top clothed with a dense mass of wool or nearly naked; areoles very spiny, large, those on the upper part of old plants sometimes united; flowers from the crown of the plant, often partly hidden by the dense wool at the top, these usually yellow, rarely pink, of medium size; outer perianth-segments narrow, sometimes terminating in pungent tips; inner perianth-segments oblong, thinner than the outer, the inner ones obtuse; scales on the flower-tube numerous, imbricate, persistent, pungent; scales on the ovary small, often linear, their axils filled with matted wool; fruit densely covered with white wool, thin-walled, oblong; seeds blackish, smooth, shining, or rarely papillose, with a small subbasal hilum.

The generic name is from ἐχῖνος hedgehog, and κάκτος cactus, referring to the spiny armament.

The genus *Echinocactus*, as treated by Karl Schumann in his monograph (Gesamtb. Kakteen 290 to 452. 1898), contains 138 species, while more than 1,025 names have been used in the genus. Our review of *Echinocactus* convinces us that there is a number of distinct genera, several of which have already been proposed and others entirely new. Before making these segregates it was necessary to establish the type of the genus which was proposed in 1827. Before that time the species of *Echinocactus* were usually considered as belonging to the genus *Melocactus*. In that year Link and Otto[*] established the genus *Echinocactus*, describing and illustrating 14 species. The illustrations, however, must have been made and engraved before it was decided to establish the genus for they all bear a *Melocactus* legend. Since these 14 species do not belong to the same genus, it is important to establish the type.

In their introduction Link and Otto state that *Echinocactus tenuispinus* and *E. platya-canthus* have the flowers of a *Cereus* and, for this reason, as well as the absence of a cephalium, were separated as *Echinocactus*. The other 12 species referred there, whose flowers they did not know, were evidently thus referred from the supposed lack of a cephalium. It seems, therefore, the type of the genus *Echinocactus* should be either *E. tenuispinus* or *E. platyacanthus*. In the last paragraph of their paper they state that *E. tenuispinus* should probably be referred to *Cereus* and that *E. platyacanthus* and 7 other species belong to *Echinocactus*. We therefore designate *Echinocactus platyacanthus* Link and Otto as the type of the genus. We recognize 9 species of *Echinocactus*.

This group to which *E. platyacanthus* belongs is characterized by a densely woolly crown to the plant, very woolly, thin-skinned fruit, and smooth seed, with a lateral hilum. As thus characterized, the genus contains at least 6 Mexican species, although there are

---

[*] Verh. Ver. Beförd. Gartenb. 3: 420. 1827.

about a dozen species of this relationship which have been described; to these we append 3 species of the southwestern United States and border states of northern Mexico, one with smooth, the others with papillose seeds.

Echinocactus texensis has a similar woolly ovary, but the fruit is fleshy, with different seeds and purple flowers; this we regard as a new generic type.

Astrophytum with its 4 species, usually classed as *Echinocactus*, also has pubescent fruit, but is very different in other respects.

There are 2 other cacti from North America which bear wool on the ovary, *E. whipplei* and *E. polyancistrus*. These have only small scales on the ovary, bearing minute tufts of hairs in their axils and have very different seeds. We refer them to a new genus (see p. 212).

In South America there are 2 old genera with woolly fruit which have been associated with *Echinocactus*, namely, *Malacocarpus* and *Eriosyce*, both of which, in our opinion, are generically distinct.

KEY TO SPECIES.

```
A. Plants very large, often becoming cylindric (see No. 5).
    Spines all bright yellow...................................................  1. E. grusonii
    Spines brown to gray, rarely some of them yellowish.
        Inner perianth-segments linear-oblong, entire...........................  2. E. ingens
        Inner perianth-segments oblong, more or less toothed or lacerate.
            Spines all of one kind............................................  3. E. visnaga
            Spines both radial and central.
                Central spine solitary.
                    Flowers 4 to 5 cm. long; central spine 4 to 5 cm. long, nearly black.  4. E. grandis
                    Flowers 3 cm. long; central spine 3 cm. long, grayish in age......  5. E. platyacanthus
                Central spines several.........................................  6. E. palmeri
AA. Plants relatively small, subglobose.
    Seeds smooth and shining..................................................  7. E. xeranthemoides
    Seeds papillose.
        Flowers yellow.......................................................  8. E. polycephalus
        Flowers pink.........................................................  9. E. horizonthalonius
```

1. Echinocactus grusonii Hildmann, Monatsschr. Kakteenk. 1:4. 1891.

Plants single, depressed-globose, large, 2 to 13 dm. high or more, often 4 to 8 dm. in diameter, light green; ribs 21 to 37, rather thin and high; spines when young golden yellow, becoming pale and nearly white, but in age dirty brown; radial spines 8 to 10, subulate, 3 cm. long; central spines usually 4, up to 5 cm. long; flowers 4 to 6 cm. long, opening in bright sunlight, 5 cm. broad at top, the segments never widely spreading; flower-tube 3 cm. broad, covered with lanceolate, long-acuminate scales; outer perianth-segments long-acuminate, brownish on the outside, yellowish within; inner perianth-segments cadmium-yellow, with a silky luster, erect, narrowly lanceolate, acuminate, much shorter than the outer segments; stamens numerous, yellow, connivent, forming a thick cylinder in the center of the perianth; style yellow; stigma-lobes 12; ovary spherical, bearing acuminate scales with an abundance of wool in their axils; fruit oblong to spherical, 12 to 20 mm. long, thin-walled, covered with white wool or becoming naked below; seeds smooth, dark chestnut-brown, shining, 1.5 mm. long.

Type locality: Central Mexico.
Distribution: San Luis Potosí to Hidalgo, Mexico.

This is a very attractive species and is much grown in collections, but usually only small plants are seen.

We are greatly indebted to Mr. E. C. Rost, a private grower of cacti in southern California and a very keen observer, not only for procuring for us flowers, fruits, and good photographs, but also for valuable observations. He writes that the flowers are deeply imbedded in the dense felt cushion and must actually be dug out. The depth to which the flowers are sunk is shown by a definite band near the top of the ovary. The flowers open in sun-light and the perianth-segments are nearly erect or slightly spreading. The stamens and style are erect. Under date of October 9, 1919, Mr. Rost sent us the following statement regarding this plant:

"In my garden these plants bloom at irregular intervals for a period of about six months each year.   The first flower of the current season opened on May 15 and one is in blossom today, while a number of well-developed buds will open unless killed by unseasonable frosts.   The hour of the day that the flower opens varies according to the time that the warm rays of the sun reach the plant. Just as soon, however, as the sun-light leaves the flower, it closes whether it be in the forenoon or after-noon.   Clouds obscuring the sun for more than a few minutes or any artificial shade will cause the flowers to close.   If conditions are suitable, the flowers will open for three consecutive days, closing each night.   The perianth-segments of the flower separate very little.

"New plants can easily be obtained either by means of seeds or from cuttings.   I have been very successful in obtaining cuttings by slicing off the top of a large plant which causes it to bud freely, and these buds can be cut off and will develop into good plants."

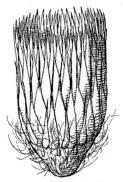

FIGS. 182 and 183.—Echinocactus grusonii.

FIG. 184.—Flower of Echinocactus grusonii. ✕0.8.

Echinocereus grusonii azureus is a form incidentally mentioned by Von Zeisold (Monats-schr. Kakteenk. **3**: 141. 1893), while Nicholson refers here, as a synonym, *Echinocactus aureus* (Dict. Gard. Suppl. 334. 1900).

Echinocactus corynacanthus Scheidweiler and *Echinocactus galeottii* Scheidweiler (Allg. Gartenz. **9**: 50. 1841), while doubtless referable to this genus, are more likely to belong to *Echinocactus grusonii* than to *E. ingens* where they are referred by Schumann.

Illustrations: Monatsschr. Kakteenk. **1**: 4, 7; Gartenwelt **1**: 429; Dict. Gard. Nicholson Suppl. 335. f. 356; Cact. Journ. **1**: pl. for March; 165; **2**: 42; Wiener Ill. Gart. Zeit. **29**: f. 22, No. 1; Journ. Hort. Home Farm. III. **60**: 144; Journ. Intern. Gard. Club **3**: 10; Schelle, Handb. Kakteenk. f. 74; West Amer. Sci. **13**: 6; Gartenwelt **7**: 277; Möllers Deutsche Gärt. Zeit. **25**: 474. f. 6, No. 8; De Laet, Cat. Gén. f. 6; f. 50, No. 1; Watson, Cact. Cult. ed. 2. 250. f. 94.

Figure 182 is from a photograph of a large plant grown by Mr. E. C. Rost at Alhambra, California, photographed by Miles E. Rost; figure 183 shows the flowering top of the plant; figure 184 is from a drawing of a flower from the same collection.

2. Echinocactus ingens Zuccarini in Pfeiffer, Enum. Cact. 54.   1837.

Globular to short-oblong, 15 dm. high, 12.5 dm. in diameter (but reported by Karwinsky to be 5 to 6 feet in diameter), glaucescent, somewhat purplish, very woolly at the top; ribs 8, obtuse, tuberculate; areoles large, distant, 2.5 to 3 cm. apart, bearing copious yellow wool; spines brown, straight, rigid, 2 to 3 cm. long; radial spines 8; central spine 1; perianth 2 cm. long, 3 cm. broad; inner perianth-segments linear-oblong, yellow, entire, obtuse; fruit ovoid, 3 cm. long, copiously cov-ered by wool coming from the axils of small scales; stigma-lobes brick-red, about 8; seeds large, black, shining, reniform.

Type locality: Mexico.

Distribution: Mexico.

We refer here the plant collected at Ixmiquilpan by Dr. Rose in 1905 but we have seen no authentic material. The original description is based upon small juvenile plants but, according to Karwinsky, it is a very large plant fully 2 meters high. Pfeiffer's illustration of the flower, doubtless of the type, indicates that it is a true *Echinocactus*, but the narrow, entire, obtuse perianth-segments are very unlike those of any species we know. Schumann has referred here numerous names as synonyms, some of which may belong here while others do not.

Echinocactus karwinskii Zuccarini (Pfeiffer, Enum. Cact. 50. 1837) is referred here by Schumann. It is doubtless of this relationship. It is described as only 20 cm. high, with 13 to 20 ribs. Its very woolly apex would suggest this relationship. The species came from Pachuca. If it were identical with *Echinocactus ingens*, it would replace it as it has page priority. *E. karwinskianus* (Monatsschr. Kakteenk. 1: 126. 1891) is undoubtedly the same. *Melocactus ingens* Karwinsky (Pfeiffer, Enum. Cact. 54. 1837) is given as a synonym, but never published, *Echinocactus macracanthus* De Vriese (Tijdschr. Natuurl. Geschild. 6: 49. pl. 2. 1839) is referred here also by Schumann. It, too, has been described from a juvenile plant. From the illustration we would judge that it was of this relationship,

Figs. 185 and 186.—Echinocactus ingens.

but certainly a different species. *Echinocactus minax* Lemaire (Cact. Aliq. Nov. 18. 1838) is referred by Schumann to *E. ingens*. Its spotted stem suggests a young plant of *E. grandis*. It is indeed a small plant, being only 5 inches high and is doubtless only a juvenile. It is described as globose, depressed, subumbilicate, green, with 13 ribs. The flowers were unknown and it is impossible to identify it definitely. *Echinocactus platyceras* Lemaire (Cact. Aliq. Nov. 19. 1838; *Echinofossulocactus platyceras* Lawrence in Loudon, Gard. Mag. 17: 318. 1841) is also described from a juvenile plant, but Lemaire states that it and *E. minax* are sometimes 6 and even 10 feet high. From his illustrations (f. 3 and 4) it is evidently related to *E. grandis* and *E. ingens*. *E. minax laevior* Lemaire (Labouret, Monogr. Cact. 192. 1853), *E. platyceras laevior* (Förster, Handb. Cact. 325. 1846), and *E. platyceras minax* Salm-Dyck (Förster, Handb. Cact. 324. 1846) must be different names for this plant. *Echinocactus helophorus* Lemaire (Cact. Gen. Nov. Sp. 12. 1839; *E. ingens helophorus* Schumann, Gesamtb. Kakteen 317. 1898) and its two varieties *laevior* and *longifossulatus* Lemaire (Cact. Gen. Nov. Sp. 13. 1839) are possibly the same as *E. minax* but all are without flowers and without definite habitat. *E. aulacogonus* Lemaire (Cact. Gen. Nov. Sp. 14. 1839) and the variety *diacopaulax* Lemaire (Cact. Gen. Nov. Sp.

15. 1839) are doubtless of this relationship. The origin was unknown and the flowers have not been described. *E. haageanus* Linke (Wochenschr. Gärtn. Pflanz. 1: 86. 1858) is referred here by Schumann. It is described, however, as only 8-ribbed and as coming from Peru. Its flowers are not known and it doubtless can never be identified. *Echinocactus tuberculatus* Link and Otto (Verh. Ver. Beförd. Gartenb. 3: 425. 1827; *Melocactus tuberculatus* Link and Otto, Verh. Ver. Beförd. Gartenb. 3: pl. 26. 1827), referred by Schumann to *E. ingens*, undoubtedly belongs elsewhere; it is described as having only 8 ribs and these obtuse; the spines are only 8. It was collected by Deppe in Mexico. Here Schumann refers *E. hystrix* Monville (Labouret, Monogr. Cact. 183. 1853), an unpublished homonym. *Echinocactus ingens edulis* Labouret (Monogr. Cact. 193. 1853; *Echinocactus edulis* Haage in Förster, Handb. Cact. 346. 1846) is of this relationship. It is well known that these species are used in central Mexico in making candy and hence the name, *edulis*. The variety *subinermis* (Schumann, Gesamtb. Kakteen 317. 1898) we do not know. *Echinocactus ingens grandis* (Monatsschr. Kakteenk. 17: 116. 1907) is only a name. *Echinocactus irroratus* Scheidweiler (Bull. Acad. Sci. Brux. 6¹: 90. 1839; *Echinocactus ingens irroratus* Monville in Labouret, Monogr. Cact. 191. 1853) and *Echinocactus oligacanthus* Martius (Pfeiffer, Enum. Cact. 53. 1837) are usually referred here.

Echinocactus tuberculatus spiralis De Candolle (Mém. Mus. Hist. Nat. Paris 17: 115. 1828), a Mexican plant, is of uncertain relationship.

Echinofossulocactus macracanthus, *E. helophora*, *E. helophora longifossulatus*, and *E. karwinskianus*, all briefly described by Lawrence (Loudon, Gard. Mag. 17: 318. 1841), may belong here.

Illustrations: ? Goebel, Pflanz. Schild. 1: f. 46, as *Echinocactus aulacogonus* (seedling); Nov. Act. Cur. Nat. 19¹: pl. 16, f 2,5; Engler and Drude, Veg. Erde 13: f. 30; Contr. U. S. Nat. Herb. 10: pl. 17, f. B; Abh. Bayer. Akad. Wiss. München 2: pl. 1, sec. 3, f. 6; Rev. Hort. 61: 568. f. 139; Karsten and Schenck, Vegetationsbilder 1: pl. 45, 47.

Figure 185 is copied from the second illustration cited above; figure 186 is from a photograph taken by Dr. Rose at Ixmiquilpan, Mexico, in 1905.

3. Echinocactus visnaga Hooker in Curtis's Bot. Mag. 77: pl. 4559. 1851.

Echinocactus ingens visnaga Schumann, Gesamtb. Kakteen 317. 1898.

Very large, 2 to 3 meters high, 7 to 10 dm. in diameter, glaucous-green, the summit covered with a mass of tawny wool; ribs 15 to 40, somewhat undulate but hardly tubercled, acute; areoles large, approximate and sometimes almost touching one another; spines 4, stout, subulate, all radial, the upper one erect, 5 cm. long, the 3 lower spreading, pale brown; flowers large, yellow, 7 to 8 cm. broad when fully expanded; inner perianth-segments numerous, oblong, spatulate, acute, serrate, 3.5 cm. long; stigma-lobes about 12, filiform; ovary elongated, 8 to 10 cm. long, crowned by the persistent perianth, densely lanate; scales on the upper part of ovary, at least, narrow, subpungent.

Type locality: Near San Luis Potosí.

Distribution: Highlands of San Luis Potosí, Mexico.

This species, one of the giant echinocacti of Mexico, was sent to Kew before 1846; several specimens were sent, one weighing a ton and estimated as being a thousand years old; Schumann refers this species to *Echinocactus ingens* and uses Hooker's plate to illustrate that species, but his description is different.

The first place of publication of this species is in some doubt. The Index Kewensis, as well as Hooker himself, the author of this species, cites Illustrated London News 1846, but although the species is illustrated here, accompanied by a popular account under the title of "Monster Cactus at Kew," in the text it is referred to as "*Cactus (Echinocactus)* of [or] *visnager*." Hooker (Curtis's Bot. Mag. 77: pl. 4559) cites also the Kew Garden Guide ed. 7. 53. 184(9?), but Mr. S. A. Skan writes us that it also appeared in the first edition 1847, p. 43.

A few years ago the governor of Tamaulipas sent a large plant to the City of Mexico, of which we have a photograph. This plant was 3 meters high, 1.3 meters in diameter, and weighed 2,000 kilograms.

Illustrations: Schumann, Gesamtb. Kakteen 54, as *Echinocactus ingens;* Illustr. London News 9: 245. 1846, as monster cactus; (?) Schelle, Handb. Kakteenk. 149. f. 76; Gartenwelt 7: 277; De Laet, Cat. Gén. f. 9, as *Echinocactus ingens visnaga;* Curtis's Bot. Mag. 77: pl. 4559; Fl. Serr. 6: pl. 616; Amer. Garden 11: 461; Dict. Gard. Nicholson 1: 501. f. 694; Jard. Fleur. 2: pl. 123; Watson, Cact. Cult. 125. f. 48; ed. 3. 47. f. 20.

Figure 187 is a reproduction of the plate in Curtis's Botanical Magazine above cited.

4. Echinocactus grandis Rose, Contr. U. S. Nat. Herb. 10: 126. 1906.

FIG. 187.—Echinocactus visnaga.

Simple, large, cylindric, 1 to 2 meters high, 6 to 10 dm. in diameter, dull green and, when young, with broad horizontal bands, very woolly at the crown; ribs on young plants as few as 8, broad, high, and more or less undulate, but in old plants very numerous and rather thin; areoles remote on young plants, confluent in old flowering plants; spines stout, subulate, distinctly banded, especially the stouter ones, at first yellowish but soon reddish brown; radial spines usually 5 or 6, 3 to 4 cm. long, central spine solitary, 4 to 5 cm. long, straight; flowers numerous, yellow, 4 to 5 cm. long; scales on the ovary linear, their axils bearing an abundance of wool covering the ovary with a dense felty mass; upper scales narrow, rigid, more or less spiny-tipped; outer perianth-segments ovate, long-apiculate, with ciliate margins; inner segments oblong, obtuse, retuse or apiculate, serrulate; fruit hidden in a mass of soft white wool, oblong, 4 to 5 cm. long; seeds black, shining, 2.5 cm. long.

Type locality: Hills near Tehuacán, Puebla, Mexico.

Distribution: Limestone hills of Puebla, Mexico.

This is one of the very large species of *Echinocactus* and is very characteristic of the deserts of Puebla where it is often the most conspicuous plant of the landscape. The juvenile plants appear very different from the mature ones.

Illustrations: Monatsschr. Kakteenk. 12: 73; U. S. Dept. Agr. Bur. Pl. Ind. Bull. 262: pl. 18; Bull. Soc. Acclim. 52: 54. f. 13; Schelle, Handb. Kakteenk. f. 75, as *Echinocactus ingens;* Reiche, Elem. Bot. f. 163, as *Cereus ingens;* Nat. Geogr. Mag. 21: 701; Plant World 11[6]: f. 3; Journ. N. Y. Bot. Gard. 8: f. 3; MacDougal, Bot. N. Amer. Des. pl. 17, in part; Monatsschr. Kakteenk. 27: 87; Möllers Deutsche Gärt. Zeit. 29: 439. f. 15.

5. Echinocactus platyacanthus Link and Otto, Verh. Ver. Beförd. Gartenb. 3: 423. 1827.

Stems nearly globular, 5 dm. high, 6 dm. broad, light green, very woolly at apex; ribs 21 to 30, acute; spines brownish at first, grayish in age; radial spines 4, spreading, 12 to 16 mm. long; central spines 3 or 4, spreading, 3 cm. long; flowers 3 cm. long, long-woolly; outer perianth-segments lanceolate, mucronate; inner perianth-segments obtuse, yellow; stigma-lobes 10.

Type locality: Mexico.

Distribution: Eastern Mexico.

Unfortunately, the type of the genus *Echinocactus* is now known only from the early descriptions and a single illustration. It seems to be quite distinct from the other species of the genus. The large giant cacti are very common in eastern Mexico, but it will require some very careful field work to disentangle the species.

Illustration: Link and Otto, Verh. Ver. Beförd. Gartenb. **3**: pl. 14, as *Melocactus platyacanthus.*

Fig. 188.—Echinocactus palmeri.

6. **Echinocactus palmeri** Rose, Contr. U. S. Nat. Herb. **12**: 290.   1909.

> *Echinocactus saltillensis* Hortus, Cact. Journ. **1**: 100.   1898.   Not Poselger, 1853.
> *Echinocactus ingens saltillensis* Schumann, Gesamtb. Kakteen 317.   1898.
> *Echinocactus ingens subinermis* Schumann, Gesamtb. Kakteen 317.   1898.

Stems 1 to 2 meters high, 4 to 5 dm. in diameter; ribs 12 to 26, or perhaps more in large plants; central spines 4, annular, the upper one erect, 6 to 8 cm. long, stout, straight, yellow above, brownish and somewhat swollen at base, the 3 lower ones shorter, spreading, similar in color and markings but

Fig. 189.—Echinocactus palmeri.

flattened; radial spines 5 to 8, much smaller, lighter colored and weaker; flowers yellow, rather small; perianth-segments about 2 cm. long, more or less lacerated along the margin; fruit about 3 cm. long, hidden in a dense covering of soft white wool; scales weak and bristle-tipped.

Type locality: Concepción del Rio, Zacatecas.

Distribution: Eastern northern Mexico.

This is the well-known *Echinocactus saltillensis* of horticultural collections, but is not the species first described under that name.

Not uncommon from southern Coahuila to Zacatecas. Professor F. E. Lloyd states that it is the most striking cactus in northern Zacatecas where it is found on the higher foothill slopes and on the hills on the slopes facing the south, with only very few exceptions.

Illustrations: Cact. Journ. **1**: pl. for August, as *Echinocactus saltillensis;* Möllers Deutsche Gärt. Zeit. **25**: 485. f. 18, as *Echinocactus ingens subinermis;* Contr. U. S. Nat. Herb. **12**: pl. 23; Ann. Rep. Smiths. Inst. **1908**: pl. 8; pl. 13, f. 1; Stand. Cycl. Hort. Bailey **2**:f. 1373.

Figure 188 is from a photograph of a potted plant from San Luis Potosí, Mexico, collected by Dr. E. Palmer and photographed by Coney Doyle; figure 189 is from a photograph of a piece of a rib from near the top of an old plant collected by Dr. E. Palmer in 1904 (No. 314).

FIG. 190.—Echinocactus xeranthemoides.

7. **Echinocactus xeranthemoides** (Coulter) Engelmann in Rydberg, Fl. Rocky Mountains 579. 1917.

Echinocactus polycephalus xeranthemoides Coulter, Contr. U. S. Nat. Herb. **3**: 358. 1896.

Cespitose, the stems globose, 2.5 to 18 cm. high, light green; ribs 13, interrupted or somewhat tubercled, sharp on the margin; areoles circular, about 1 cm. in diameter, often less than 2 cm. apart; spines 10 to 15, when young whitish pink, but in age a dirty gray, slender and rather stiff, more or less annulate; radial spines 3 to 4 cm. long, more or less curved backwards; central spines 4, 3 to 6 cm. long; one of them longer than the others, somewhat curved, rather stiff; flowers bright yellow, 5 cm. long; scales on the ovary and flower-tube linear, pink, papery, stiff, but not pungent, the longer ones 2 to 3 cm. long, persistent; perianth-segments narrowly oblong, more or less serrate, apiculate or cuspidate; stamens included; style yellow (?), included; fruit shortly oblong, 3 cm. long, densely and permanently white-woolly, dehiscing by a basal pore; seeds brownish black, shining, delicately reticulate, 2.5 mm. long.

9. **Echinocactus horizonthalonius** Lemaire, Cact. Gen. Nov. Sp. 19. 1839.

Echinocactus equitans Scheidweiler, Bull. Acad. Sci. Brux. 6¹: 88. 1839.
Echinocactus horizonthalonius curvispinus Salm-Dyck, Cact. Hort. Dyck. 1849. 146. 1850.
Echinocactus horizonthalonius centrispinus Engelmann, Proc. Amer. Acad. 3: 276. 1856.
Echinocactus laticostatus Engelmann and Bigelow, Pac. R. Rep. 4: 32. 1856.
?Echinocactus parryi Engelmann, Proc. Amer. Acad. 3: 276. 1856.
Echinocactus horizonthalonius obscurispinus R. Meyer, Monatsschr. Kakteenk. 21: 181. 1911.

Simple, globular or sometimes depressed or short-cylindric, 4 to 25 cm. high, glaucous; ribs 7 to 13,*obtuse, often spirally arranged; spines 6 to 9, somewhat curved or straight, 2 to 4 cm. long, often very stout, more or less flattened, often annulate, reddish or sometimes blackish at base; central spine solitary, stouter than the radials; flowers pale rose to pink, 5 to 7 cm. long before expanding, broader than long when fully open; outer perianth-segments linear with more or less pungent tips; inner perianth-segments narrowly oblong, about 3 cm. long; throat of flower short and broad, covered with numerous stamens; tube of flower wanting or nearly so; filaments white; style pink; stigma-lobes pinkish to olive; ovary and fruit bearing linear scales, their axils very woolly; fruit dehiscing by a basal pore, oblong, red, 3 cm. long, clothed with long white wool; seeds 2 mm. long, more or less angled, brownish black, papillose; hilum large, lateral but below the middle.

Type locality: Not cited.
Distribution: Western Texas, southern New Mexico to Arizona,† and northern Mexico.
Echinocactus horizontalis (Förster, Handb. Cact. 327. 1846) is given as a synonym of this species but it was never described. According to F. E. Lloyd, it is known in Mexico as manca caballo.

This cactus is said to be used in making a Mexican candy.

Illustrations: Cact. Mex. Bound. pl. 32, f. 6, 7, as *Echinocactus parryi;* Schelle, Handb. Kakteenk. f. 71, as *E. horizonthalonius curvispinus;* Blühende Kakteen 2: pl. 117; Schumann, Gesamtb. Kakteen f. 51; Monatsschr. Kakteenk. 21: 179; Ann. Rep. Smiths. Inst. 1908: pl. 2, f. 5; Cact. Journ. 1: pl. for March; Dict. Gard. Nicholson 4: 539. f. 21; Suppl. 335. f. 357; Cact. Mex. Bound. pl. 31; 32, f. 1 to 5; Förster, Handb. Cact. ed. 2. f. 56, 57; Goebel, Pflanz, Schild. 1: f. 48; Lemaire, Icon. Cact. pl. 3; De Laet, Cat. Gén. f. 15; Orcutt, Rev. Cact. opp. 41; Watson, Cact. Cult. 106. f. 37; Remark, Kakteenfreund 14; Balt. Cact. Journ. 1: 68.

Plate xx, figure 3, shows a flowering plant collected by E. O. Wooton in the Tortiegas Mountains, Las Cruces, New Mexico, in 1905, which bloomed at the New York Botanical Garden in July 1917; figure 4 shows the fruit and figure 5 the seed of a plant collected by Mrs. S. L. Pattison in southern Texas in 1920.

UNIDENTIFIED SPECIES DESCRIBED AS ECHINOCACTUS.

The following species are recorded here because their generic relationship can not be determined. They are mostly described without flower or fruit.

ECHINOCACTUS AMAZONICUS Witt, Monatsschr. Kakteenk. 12: 29. 1902.

Cespitose, each plant 8 to 10 cm. broad, 4 to 5 cm. high, dark green, shining; ribs 11 to 13, separated by short intervals; areoles 2 cm. apart, when young woolly, becoming glabrate; spines 8, when young chestnut-brown, the lower one longest, sometimes 3 cm. long, the upper ones often only 6 mm. long; flowers and fruit unknown.

Type locality: In the Serra de Tucunaré on the Rio Tacutú in northern Brazil.
This plant is known only from the single collection of Alfred Wauer, and as flowers and fruit are both unknown its generic position is in doubt. It is certainly not an *Echinocactus* as we now treat the genus. It may be a young form of some species of *Cactus (Melocactus)*.

---

*The number of ribs is almost always 8; in small plants we have seen as few as 7; Coulter has reported 10. *Echinocactus parryi* which we have referred here doubtfully was described as having 13.

†The Arizona record is based on a plant collected by Dr. Forrest Shreve in Pine County in 1918. Professor F. E. Lloyd reports finding the plant at Silver Bell Mountain but we have never seen his specimen.

ECHINOCACTUS ARACHNOIDEUS Scheidweiler, Bull. Acad. Sci. Brux. 6[1]: 90. 1839.*

Ovoid, 7.5 to 10 cm. long, 7.5 cm. in diameter; ribs 9 or 10, rounded, somewhat gibbose between the areoles, separated by acute intervals but these disappearing below; radial spines 10 to 12, spreading, equal, 1 cm. long; central spines 4, stouter than the radials, purplish at base; flowers and fruit unknown.

ECHINOCACTUS ARANEOLARIUS Reichenbach in Terscheck, Suppl. Cact. Verz. 2.

Oblong, obtuse; ribs 12; areoles prominent, white-lanate; radial spines 15 to 17, spreading, slender, straight, yellowish; central spines 5 to 7, porrect, very short, purplish; flowers and fruit not known.

According to Walpers (Repert. Bot. 2: 317. 1843) the species came from Montevideo.

ECHINOCACTUS ARMATISSIMUS Förster, Hamb. Gartenz. 17: 162. 1861.

"Normal specimens, which are before me, are an original stalk and a cutting raised therefrom. The first is 10 inches high, by 3½ inches in diameter, brownish green, lighter colored toward the crown, and is 14-ribbed. Ribs are rounded, furrowed rather deep, almost sharp, the areoles not far apart, sparsely covered with short gray wool, and not very deep. Spines: pearl gray, stiff, straight; radial spines: 9 to 11, spreading, extended, very dissimilar, 4 to 10 lines long; central spines: only 1, stiff, upright, 1½ to 1¾ inches long.

"The cutting on the other hand is 1 inch high, by 1½ inches in diameter, light green, and only 11 ribbed. The ribs, furrows, and areoles are like the original stalk. The spines upon the crown are brown, the rest pearl-gray. Radial spines: only 7 to 8, radiating, extended, 3 or 4 of the lower are longer (up to 9 lines long). Central spines: likewise only 1, upright, stiff, and up to 10 lines long.

"This belongs to the *Cephaloidei* and because of its strong spreading bundles of spines presents a peculiarly interesting aspect. Country: Peru and Colombia."

The above is a free translation of Förster's original description. We are not able to identify the plant. Schumann (Gesamtb. Kakteen 313. 1898) associates it with *Echinocactus ceratistes* but the descriptions of these two species do not read much alike.

ECHINOCACTUS CHRYSACANTHION Schumann, Gesamtb. Kakteen 396. 1898.

Usually simple, globose to short-cylindric, 5 to 6 cm. high, 5 cm. in diameter; ribs spiraled, tubercled; spines 30 to 40, setaceous, golden yellow to brown, 1.4 cm. long, straight; flower yellow, 17 to 18 mm. long; ovary naked.

Type locality: Province of Jujuy, Argentina.
Distribution: Known only from the type locality.
Schumann placed this species between *Echinocactus minusculus* and *E. microspermus.* From the description we judge that it is near the former species. We know it only from description. It may belong to *Rebutia.*

ECHINOCACTUS CUPULATUS Förster, Hamb. Gartenz. 17: 161. 1861.

We do not know this species. It is supposed to have come from Chile.

ECHINOCACTUS CUPREATUS Poselger in Förster, Handb. Cact. ed. 2: 602. 1885.

This species was described without flower and fruit and we are not able to suggest its relationship. Schumann discusses it briefly under *E. nigricans;* it is also Chilean. The name was listed in Seitz's Catalogue.

ECHINOCACTUS DEPRESSUS De Candolle, Prodr. 3: 463. 1828.

Subglobose, depressed at apex; ribs 20, somewhat tuberculate; radial spines 10 to 20; central spines 3 or 4; flowers and fruit unknown.

This plant from tropical America has never been identified. De Candolle cites *Melocactus* (?) *depressus* Salm-Dyck, as a synonym, and also refers here with a question

---

* Although we have never seen the reprint, this paper by Scheidweiler was repaged and issued separately, judging by references to it by Walpers (Repert. Bot. 2: 323. 1843) and Hemsley. According to Scheidweiler this plant came from Buenos Aires but Walpers and Hemsley refer it to Mexico. Schumann did not know it.

Cactus depressus Haworth (Syn. Pl. Succ. 173. 1812). Haworth's plant, however, was described as having 10 ribs and was believed to be related to *Cactus gibbosus*, that is to *Malacocarpus.*

ECHINOCACTUS ECHINATUS Forbes, Journ. Hort. Tour Germ. 152. 1837.

Like a hedgehog cactus; ribs 19; spines light brown, elongated.

This species is so briefly described that it can not be definitely identified. It is perhaps a *Ferocactus*. It is said to have come from Mexico.

ECHINOCACTUS GEISSEI Poselger in Schumann, Gesamtb. Kakteen 406. 1898.

Echinocactus geissei albicans Hildmann in Schumann, Gesamtb. Kakteen 406. 1898.

Neither has this species nor its variety been known to flower and as we have not seen the plants we are not able to suggest their relationship. The only illustration which we have seen (Möllers Deutsche Gärt. Zeit. 25: 474. f. 6, No. 14) indicates that it is not a true *Echinocactus*. It is recorded as from Chile or Bolivia.

ECHINOCACTUS HAMATUS Forbes, Journ. Hort. Tour Germ. 152. 1837.

Undoubtedly different from Mühlenpfordt's species of the same name. It presumably had hooked spines judging from the name; and was briefly described as having a depressed stem, 21 ribs, and 7 gray spines. Forbes reports it was introduced from Buenos Aires in 1833. It does not answer to any Argentine plant we know.

ECHINOCACTUS MALLETIANUS Lemaire, Allg. Gartenz. 13: 387. 1845.

Stems simple, depressed-globose or somewhat cylindric, very woolly at the top, 1 dm. high; ribs 15 to 17, more or less repand; spines straight, acicular, black; radial spines 5 or 6, suberect; central spine solitary; flowers and fruit unknown.

Type locality: Not cited.
Distribution: According to Schumann, Chile or Bolivia.

Dr. Rose obtained from L. Quehl a photograph of this species as it is now represented in collections. Its relationship is doubtful but it should certainly not be placed just after *E. horizonthalonius* as it was by Schumann.

ECHINOCACTUS MUTABILIS Förster, Hamb. Gartenz. 17: 161. 1861.

Simple, globose; ribs 10, sinuate and tuberculate, yellowish green to violet; radial spines 7 or 8, spreading, straight or somewhat curved, dull yellow; central spine solitary, straight, porrect.

Type locality: Peru.
Distribution: Reported from Peru, but since this species was discovered the southern provinces have been annexed by Chile. It is not of this genus.

ECHINOCACTUS ODIERI Lemaire in Salm-Dyck, Cact. Hort. Dyck. 1849. 174. 1850.

Echinocactus araneifer Labouret, Monogr. Cact. 248. 1853.
Echinocactus odierianus Monville in Weber, Dict. Hort. Bois 470. 1896.

Small, nearly globular, 5 cm. in diameter, purplish; ribs indefinite, broken up into tubercles, more or less spiraled; spines all radial, brownish to gray, 6 to 9, appressed, small, about 2 mm. long; flowers white, rose-colored without, about 5 cm. broad; outer perinth-segments narrowly lanceolate, acuminate, reddish green with dark purple tips; inner perianth-segments broadly lanceolate, acute, the margins serrate, white within, pale rose without; filaments white; style red, longer than the stamens; stigma-lobes 14, erect, flesh-colored.

Type locality: Not cited.
Distribution: Copiapo, Chile.

We do not know this species. It seems to have been well known in Europe at one time. Mr. Söhrens of Chile tells us he has seen it near Huasco, Chile. We are very uncertain as to its generic relationship. It may be a near relative of *Lobivia cumingii*.

Variety *mebbesii* Hildmann (Schumann, Gesamtb. Kakteen 413. 1898) is said to have more spreading, stouter, and lighter-colored spines and is more inclined to sucker than the typical forms. Other names associated with this species are the following: *E. odieri spinis nigris* Labouret (Monogr. Cact. 248. 1853) and *E. odieri magnificus* Hildmann (Monatsschr. Kakteenk. **5**: 184. 1895). It is said to have a very large flower as compared with the size of the plant.

ECHINOCACTUS PACHYCORNIS Mühlenpfordt, Allg. Gartenz. **14**: 371. 1846.

Depressed; ribs 7, thick, obtuse; areoles 18 mm. apart; radial spines 5; central spine 1; spines all reddish except the uppermost one and this horn-colored; flowers and fruit unknown.

Type locality: Mexico.
This seems to be only a juvenile plant of which all record is lost.

ECHINOCACTUS PULVERULENTUS Mühlenpfordt, Allg. Gartenz. **16**: 9. 1848.

Green, ovate or oblong-ovate; ribs 13, obtuse; areoles 4 mm. apart, grayish woolly; radial spines 6 or 7, stiff, 4 to 5 mm. long; central spine 1, 1¼ cm. long.

Type locality: Bolivia.
Both Mühlenpfordt and Schumann considered this species related to *Echinocactus ceratistes*, but it surely must be very different, judging from the brief description.

ECHINOCACTUS SPINA-CHRISTI Zuccarini in Pfeiffer, Enum. Cact. 59. 1837.

Globose, 15 cm. in diameter, dull green; ribs 13 or 14, acute, crenate; areoles large, oval, 15 or 20 mm. apart, when young white, velvety; spines stout, rigid, curved, black when young, paler at base, becoming yellow; radial spines 6 to 8, the lower spines stouter than the upper, 3.5 cm. long; central spine solitary, erect; flowers and fruit unknown.

Type locality: Southern Brazil.
To this species Pfeiffer referred *E. fischeri* as a synonym. Schumann lists it among the species unknown to him and afterwards (Gesamtb. Kakteen 468. 1898) under *Melocactus ferox* Pfeiffer, following Förster (Handb. Cact. 519. 1846).

Here also both Förster and Schumann refer *Echinocactus armatus* Salm-Dyck (Hort. Dyck. 341. 1834), a reference we very much question. Labouret (Monogr. Cact. 16. 1853) who takes the same view of this species says it is native of Mexico coming from Santa Rosa de Toliman.

Melocactus spina-christi Cels (Förster, Handb. Cact. 279. 1846) was never described but doubtless belongs here.

ECHINOCACTUS SPINIFLORUS Schumann, Gesamtb. Kakteen Nachtr. 88. 1903.

Usually simple, globose to cylindric, up to 6 dm. high, 1.5 dm. in diameter; ribs 20 or more, 1 to 1.5 cm. high; areoles circular, 4 to 5 mm. in diameter or near the crown of the plant 8 mm. in diameter, at first white-tomentose but in age naked; spines 14 to 20, spreading, straight, stiff, subulate, reddish yellow, unequal, the longest 2.5 cm. long; flowers 4 cm. long, 3.5 to 4 cm. broad, campanulate, rose-red; outer perianth-segments 8 mm. long, thin, spine-tipped; inner perianth-segments somewhat diverse, the outermost ones spinescent, the innermost ones not armed; stamens about two-thirds the length of the perianth; style about one-half the length of the longest stamens, surrounded at base with yellow wool; stigma-lobes 19; ovary turban-shaped, covered with awl-shaped prickly-tipped amber scales, these woolly in their axils.

Type locality: On Cerro Morro or Cerro Bianco.
Distribution: Argentina.
We know this species only from description and we are in doubt as to the relationship. It must, however, be excluded from *Echinocactus* and *Malacocarpus* and probably does not belong to the sub-tribe *Echinocactaneae*. According to Dr. Vaupel, the type can not be found in the herbarium of the Botanical Garden at Berlin, and the collector is unknown.

In the spinescent scales of the ovary and flower-tube and in the mass of wool at the base of the style it is similar to the two anomalous species which we have referred to *Lobivia*, viz., *L. thionanthus* and *L. chionanthus*. We have seen flowers in the herbarium of the Instituto de Botanica y Farmacologia collected by Dr. A. Dominguez on Cerro de Macha which probably belong to *Echinocactus spiniflorus* or to a closely related species. Unfortunately, we know nothing about the plant body from which these flowers come. A very similar plant was collected by Dr. C. Spegazzini at Jujuy, Argentina. This we know only from a photograph which is labeled *Echinocactus hylainacanthus*.

ECHINOCACTUS SPINOSISSIMUS Forbes, Journ. Hort. Tour Germ. 152. 1837.

Ribs 14 or 15; spines numerous; radial spines white; central spines 7 or 8, reddish brown, elongated.

The above description is compiled from Forbes's abbreviations and while it can not be definitely identified we suspect it refers to the so-called *Mammillaria spinosissima*.

Forbes did not have much knowledge of the cacti. He was the gardener of the Duke of Bedford who sent him to the Continent of Europe in 1835 where he obtained many cacti and on his return to England published a list of them, sometimes with brief descriptions.

The names had been given to him by Pfeiffer and others who were studying this family. As he published his list very promptly, many names appear there first or in the same year as in Pfeiffer's lists. *Mammillaria spinosissima* must have been in cultivation at the time of Forbes's visit, for it was published in 1838.

The following are mostly names which have been printed, but were unaccompanied by descriptions or, when described, so poorly or briefly characterized that no one has been able to identify them:

Echinocactus acutispinus Hildmann is only a catalogue name which Schumann (Monatsschr. Kakteenk. **5**: 44. 1895) lists without description of any kind.

Echinocactus castaniensis (Monatsschr. Kakteenk. **5**: 75. 1895) is only a name said to have come from Rünge's Catalogue.

Echinocactus cerebriformis Macloskie (Rep. Princeton Univ. Exped. Patagonia **8**: 593. 1905) we do not know. Specimens could not be found in the Herbarium of Princeton University. It was described briefly as follows: A monstrosity, the ribs greatly contorted, and the spines short. It comes from the Rio Negro, northern Patagonia.

Echinocactus confertus, a garden name, appeared (Förster, Handb. Cact. 346. 1846) without description.

Echinocactus corrugatus Steudel (Nom. ed. 2. **1**: 536. 1840) was based on *Cactus corrugatus* Loudon (Hort. Brit. 194. 1830) and is said to have come from Chile. Schumann did not know it. It was described originally as simply "corrugated." It may be referable to *Opuntia corrugata* (Cactaceae **1**: 95).

Echinocactus dadakii Frič, we do not know, but it is said to come from South America. We find it offered for sale by Johnsens of Odesse, Denmark, in "Succulenta" for November 1920. It is stated (Monatsschr. Kakteenk. **31**: 15. 1921) that the plant is small with few spines. The flowers are unknown.

Echinocactus flavicoma (Monatsschr. Kakteenk. **19**: 139. 1909) is only mentioned in the place here cited. It is also advertised for sale by Frantz de Laet, but we have seen no description.

Echinocactus foliosus Steudel (Nom. ed. 2. **1**: 536. 1840) is based on *Cactus foliosus* Gillies (G. Don in Loudon, Hort. Brit. 194. 1830. Not Willdenow, 1813). Schumann did not know it. It is said to come from Chile. If leafy, as described, it may be *Opuntia subulata*.

Echinocactus gigas Pfeiffer is only a name listed by Förster (Handb. Cact. 347. 1846).

Echinocactus glabrescens Weber (Monatsschr. Kakteenk. 8: 98. 1898) is only a name.
Echinocactus hemifossus Lemaire and its variety *gracilispinus* (Illustr. Hort. 5: Misc. 10. 1858) which came from Peru or Bolivia have never been identified.

Echinocactus intricatus Salm-Dyck (Allg. Gartenz. 13: 387. 1845) is a homonym which no student has been able to identify. The flowers and fruit were unknown and it is of doubtful origin. To it the name *Echinocactus criocerus* Lemaire (Labouret, Monogr. Cact. 178. 1853) has been referred.

Echinocactus junori (Monatsschr. Kakteenk. 8: 107. 1898), sometimes spelled *E. juori*, seems never to have been described.

Echinocactus latispinosus (Link and Otto, Verh. Ver. Beförd. Gartenb. 6: 431. 1830) is only a name.

Echinocactus longispinus Scheidweiler (Förster, Handb. Cact. 347. 1846) is only a name.
Echinocactus mamillosus Lemaire (Salm-Dyck, Cact. Hort. Dyck. 1844. 19. 1845) is only a name.

Echinocactus merckeri Hildmann (Monatsschr. Kakteenk. 5: 92. 1905) is only a catalogue name.

Echinocactus micracanthus Fennell (Förster, Handb. Cact. 347. 1846) is only a name.
Echinocactus montevidensis G. Don (Sweet, Hort. Brit. ed. 3. 283. 1839) is without description and probably does not belong to this genus.

Echinocactus olacogonus Audot (Rev. Hort. 6: 248. 1845) is briefly described as 58 cm. in diameter, flattened at the top. Galeotti is said to have had specimens in Brussells and Cels in Paris. The description follows a discussion of *Echinocactus stainesii*. Schumann makes no reference to the name and it is omitted from the Index Kewensis.

Echinocactus oreptilis Haage (Förster, Handb. Cact. 347. 1846) is only a name.
Echinocactus oxyacanthus (Forbes, Journ. Hort. Tour Germ. 152. 1837) is too briefly described to be definitely identified.

Echinocactus pelachicus (Monatsschr. Kakteenk. 20: 39. 1910) has never been described.

Echinocactus platycarpus (Förster, Handb. Cact. 347. 1846) is only a name.
Echinocactus plicatilis (Monatsschr. Kakteenk. 1: 22. 1891) is only a name.
Echinocactus pluricostatus (Monatsschr. Kakteenk. 4: 193. 1894) is only a name.
Echinocactus punctulatus (Monatsschr. Kakteenk. 5: 106. 1895) is only a garden name of Weber.

Echinocactus rebutii Weber (Monatsschr. Kakteenk. 5: 107. 1895) was listed by Schumann as growing in the Botanical Garden at Paris. We find no description or other information regarding it. It is possible that this was Weber's first name for *Rebutia minuscula*.

Echinocactus retusus Scheidweiler (Förster, Handb. Cact. 347. 1846) is only a name.
Echinocactus salmii Jacobi (Allg. Gartenz. 19: 9. 1851) is of unknown origin and we find no other mention of it.

Echinocactus sickmannii is unknown to us except from the description in Linnaea (12: Litt. 83. 1838). The only other reference which we have seen is that in the Index Kewensis (p. 1281) where it is credited to Lehmann. Steudel (Nom. ed. 2. 1: 537. 1840) credits it to South America. *E. sickmannii* of Förster (Handb. Cact. 347. 1846) and Labouret (Monogr. Cact. 266. 1853), without descriptions, probably refers to the same plant. Labouret credits the plant to the Berlin Gardens. Its relationship is doubtful while its definite origin is unknown. It is described as follows:

Depressed-globose, dull green, umbilicate at apex; ribs 20 or 21, acute, divided into oblong oblique tubercles; areoles white-tomentose; spines white, rigid; radial spines about 7, the upper ones smaller and more slender than the lower ones; central spine solitary, straight.

Echinocactus sparathacanthus Martius (Förster, Handb. Cact. 344. 1846), supposed to be from Mexico, is without description.

Echinocactus subgrandicornis Haage (Förster, Handb. Cact. 347. 1846) is only a name.

Echinocactus thelephorus (Hortus in Forbes, Journ. Hort. Tour Germ. 152. 1837) is very briefly described and we can not identify the plant.

Echinocactus verutum (Förster, Handb. Cact. 344. 1846), from Mendoza, was once grown in English gardens. It is only a name.

Echinocactus villiferus Scheidweiler (Förster, Handb. Cact. 347. 1846) is only a name.

Echinocactus wilhelmii is listed by Schumann (Monatsschr. Kakteenk. 5: 108. 1895) as from Hildmann's Catalogue.

20. HOMALOCEPHALA gen. nov.

Low, depressed or subglobose plants, strongly ribbed; spines stout; flowers central, rather large, day-blooming; outer perianth-segments very narrow, pungent; inner perianth-segments narrow, widely spreading; ovary covered with numerous linear pungent scales bearing in their axils masses of white wool; fruit globular, scarlet, becoming naked, at first juicy, bursting irregularly; seeds large, black, smooth, reniform.

The generic name is from ὁμαλός level, and κεφαλή head, referring to the depressed plant body.

We recognize only one species, first published as *Echinocactus texensis* Hopffer.

1. **Homalocephala texensis** (Hopffer).

> *Echinocactus texensis* Hopffer, Allg. Gartenz. 10: 297. 1842.
> *Echinocactus lindheimeri* Engelmann, Bost. Journ. Nat. Hist. 5: 246. 1845.
> *Echinocactus platycephalus* Mühlenpfordt, Allg. Gartenz. 16: 9. 1848.
> *Echinocactus texensis gourgensii* Cels in Labouret, Monogr. Cact. 196. 1853.
> *Echinocactus texensis longispinus* Schelle, Handb. Kakteenk. 161. 1907.

Fig. 192.—Homalocephala texensis.

Usually simple, sometimes globose, but generally much depressed, in large plants 30 cm. broad, 10 to 15 cm. high; ribs 13 to 27, very prominent, acute; areoles only 2 to 6 to a rib, densely white-felted when young, large; radial spines usually 6, rarely 7, spreading or recurved, more or less flattened, unequal, 1.2 to 4 cm. long, or rarely 5 cm. long, reddish, more or less annulated; central spine solitary, longer than the radials, 3 to 6.5 cm. long, 3 to 8 mm. broad, much flattened, strongly annulate; flowers broadly campanulate, 5 to 6 cm. long and fully as broad, scarlet and orange below, pink to nearly white above; outer perianth-segments linear with more or less lacerate margins and terminated by long spinose tips; inner perianth-segments with less pungent tip or without any, but with strongly lacerate margins; filaments red; stigma-lobes 10, linear, pale pink; scales on the ovary and flower-tube linear, pungent; fruit scarlet, globular, 16 to 40 mm. in diameter, nearly smooth when mature, at first pulpy but becoming dry and apparently splitting open unequally; seeds large, uniform, black, smooth, shining, somewhat flattened, angled on the back, 3 mm. broad; hilum lateral, large, depressed; "embryo curved or hooked with the foliaceous cotyledons buried in the large albumen" (Engelmann).

Type locality: Texas; type grown in a botanical garden from seed.

Distribution: Southeastern New Mexico, Texas, and northern Mexico.

The flowers of this species open for four days in bright sunlight, closing at night; they are delicately fragrant.

This plant shows great variation in the size of the fruit and in the way it ripens and dehisces the seeds. In 1921 Mr. Robert Runyon sent us a box of very large fruits, almost twice as large as any previously studied; none of these fruits split open as it ripened.

Dr. C. R. Ball writes of this plant as follows: "This plant is extremely abundant on the high plains of western and northern Texas. In establishing farms in this section large numbers of this cactus are plowed out in the breaking of the sod land. Occasionally, the farmers gather them and haul them to the margins of the field and there build fences much like the stone walls so familiar in New England. The plants are easily corded and the strong sharp spines make the fences quite formidable.''

Echinocactus courantianus Lemaire is given as a synonym of this species by Labouret (Monogr. Cact. 196. 1853) and seems never to have been described. *Melocactus laciniatus* Berlandier is only mentioned by Engelmann (Cact. Mex. Bound. 27. 1859).

The plant is called devil's pincushion and devil's head cactus.

Illustrations: Alianza Cientifica Universal **3**: opp. 222; Cact. Mex. Bound. pl. 33, f. 1 to 6; Blühende Kakteen **1**: pl. 50; Gartenflora **32**: 20; **37**: pl. 1286; Monatsschr. Kakteenk. **12**: 57; Dict. Gard. Nicholson **1**: 501. f. 693; Rümpler, Sukkulenten 185. f. 103; Ann. Rep. Smiths. Inst. 1908: pl. 2, f. 1; Förster, Handb. Cact. ed. 2. f. 58, 59; Orcutt, Rev. Cact. 56; Schulz, 500 Wild Fl. San Antonio pl. 13; Thomas, Zimmerkultur Kakteen 39; Schelle, Handb. Kakteenk. f. 89; Watson, Cact. Cult. 121. f. 46, as *Echinocactus texensis*.

Plate XIX, figure 3, shows a plant sent by Dr. MacDougal to the New York Botanical Garden from Austin, Texas, in 1902, which flowered in 1904 and 1905; figure 4 shows a fruit, painted by D. G. Passmore in Washington, D. C., of a plant collected by F. E. Upton near Fort Worth, Texas, in 1907; figure 5 shows a seed from a plant collected by Robert Runyon at Brownsville, Texas, in 1920. Figure 192 is from a photograph of a plant which flowered and fruited in Washington, D. C. This was sent from Fort Worth, Texas, by F. E. Upton in 1907.

21. ASTROPHYTUM Lemaire, Cact. Gen. Nov. Sp. 3. 1839.

Plants globular or more or less flattened to short-cylindric; ribs few, very prominent, more or less covered with white, radiating, hairy scales; spines usually wanting, weak or subulate in two species; flowers borne at the top of the plant, large, yellowish with a reddish center, soon fading, persistent, campanulate to short-funnelform; fruit globular, covered with brown, scarious, imbricating scales, these woolly in their axils, and more or less pungent; seeds dark brown, smooth and shining, with a large depressed hilum having inturned margins.

Four species, all Mexican, are here recognized; the type species is *Astrophytum myriostigma* Lemaire.

The generic name is from ἀστήρ star, and φυτόν plant, referring to the star-like shape of the plant.

KEY TO SPECIES.

Spines wanting.
    Plants globular to columnar; flowers 4 to 6 cm. long............................... 1. *A. myriostigma*
    Plants much depressed; flowers 3 cm. long..................................... 2. *A. asterias*
Spines present.
    Spines flat, ribbon-like, hardly pungent... 3. *A. capricorne*
    Spines subulate.. 4. *A. ornatum*

1. Astrophytum myriostigma Lemaire, Cact. Gen. Nov. Sp. 4. 1839.

Cereus callicoche Galeotti in Scheidweiler, Bull. Acad. Sci. Brux. 6¹: 88. 1839.
Echinocactus myriostigma Salm-Dyck, Cact. Hort. Dyck. 1844. 22. 1845.
Astrophytum prismaticum Lemaire, Cactées 50. 1868.
Echinocactus myriostigma columnaris Schumann, Gesamtb. Kakteen 321. 1898.
Echinocactus myriostigma nudus R. Meyer, Monatsschr. Kakteenk. **22**: 136. 1912.

Plants solitary or cespitose, globular to cylindric, up to 6 dm. high; ribs usually 5, sometimes 6, 8, or rarely even 10, very broad, acute, usually covered with white woolly scales but sometimes naked; spines wanting, at least on old plants; flowers 4 to 6 cm. long; outer perianth-segments narrow, with brown scarious tips; inner perianth-segments oblong; scales on ovary and flower-tube scarious, imbricated, narrow, often bristly tipped, with long wool in their axils.

Type locality: Not cited.
Distribution: Northern central Mexico.

Dr. C. A. Purpus who knows this species very well writes that it has two very different forms. The gray or grayish-white form grows near Torreon, in Cerro de la Bola, and in the mountains near Viesca, all in Coahuila. The more greenish lower form is abundant in the Sierra la Tabla, near Guascama or Minas de San Rafaél, San Luis Potosí. It usually grows in the open mesa among broken stones, but is sometimes associated with other plants, such as *Opuntia leptocaulis.*

Cereus inermis Scheidweiler (Bull. Acad. Sci. Brux. 6[1]: 88. 1839), usually referred here as a synonym, was never published.

Echinocactus myriostigma hybridus is advertised by Haage and Schmidt, but we do not know its origin; the varieties, *columnaris* and *nudus*, are in the trade.

Many hybrids are produced with this species as one of the parents. In 1896 (Monatsschr. Kakteenk. 6: 20) Fl. Radl* named and described twelve hybrids, while in 1907 Schellet (Handb. Kakteenk. 151, 152) listed 59 hybrid names under *Echinocactus myriostigma.*

Illustrations: Monatsschr. Kakteenk. 29: 81; Gartenwelt 15: 537, as *Echinocactus myriostigma columnaris;* Lemaire, Icon. Cact. pl. 16; Schelle, Handb. Kakteenk. f. 78; Möllers Deutsche Gärt. Zeit. 25: 474. f. 6, No. 11; 486. f. 19; 29: 89. f. 11; Cycl. Amer. Hort. Bailey 2: 515. f. 746; Stand. Cycl. Hort. Bailey 2: f. 1374; Gardening 9: 617; Gard. Chron. III. 52: f. 103; Loudon, Encycl. Pl. ed. 3. 1376. f. 19368; Blühende Kakteen 2: pl. 110; Engler and Prantl, Pflanzenfam. 3[6a]: f. 56, E; f. 62; Monatsschr. Kakteenk. 6: 22; 12: 4; 18: 9; 29: 141; Schumann, Gesamtb. Kakteen f. 1; Curtis's Bot. Mag. 71: pl. 4177; Gartenwelt 15: 537; 17: pl. opp. 412; Journ. Hort. Home Farm. III. 59: 631; De Laet, Cat. Gén. f. 16, 22; Watson, Cact. Cult. 112. f. 40; ed. 3. f. 27; Gard. Chron. III. 12: 789. f. 129, as *Echinocactus myriostigma;* Förster, Handb. Cact. ed. 2. 461. f. 54; Cact. Journ. 1: pl. for September; 164; Illustr. Hort. 8:

FIG. 193.—Astrophytum myriostigma.

pl. 292; Rümpler, Sukkulenten 188. f. 106; Orcutt, West Amer. Sci. 13: 3; Gartenflora 34: 56. f. 1885; Monatsschr. Kakteenk. 3: 159. f. 111; 7: 170; Lemaire, Cactées 50. f. 4; Deutsche Gärt. Zeit. 5: 369; Orcutt, Rev. Cact. opp. 41.

Plate XXII, figure 3, shows a flowering plant in the collection of the New York Botanical Garden, received in 1901 from M. Simon, St. Ouen, Paris, France, which has since bloomed several times. Figure 193 is from a photograph of a plant collected by C. A. Purpus in northern Mexico in 1905.

2. Astrophytum asterias (Zuccarini) Lemaire, Cactées 50. 1868.

Echinocactus asterias Zuccarini, Abh. Bayer. Akad. Wiss. München 4[2]: 13. 1845.

Plant much depressed, only 2 to 3 cm. high, about 8 cm. broad; ribs 8, very low, almost flat on top, the surface bearing numerous depressions, containing tufts of wool; areoles prominent, circular, felted, 4 to 5 mm. apart, spineless; flowers 3 cm. long, yellow.

---

*The names given by Radl are as follows: *amabile, bedinghausii, beguinii, conspicuum, imperiale, lapaixii, lesaunieri, mirabile, octogonum, princeps, regale,* and *schilinzkyi.*

†Under this species Schelle lists the following hybrids: *amabilis, amoenus, bedinghausi, beguinii, bellus, candidus, cereiformis, cinerascens, cinerascens brevispinus, cinerascens crassisipinus, cinerascens longispinus, cinerascens parvimaculatus, conspicuus, cornutus, cornutus candidus, crenatus, darrahii, delaeti, diadematus, elegantissimus, erectus, formosus, gardei, glabrescens, hanburyi, imperialis, incanus, incomparabilis, inermis, insignis, jusberti, lapaixi, laurani, lesaunieri, lophothele, lophothele cereiformis, martini, mirabilis, nobilis, octagonus, pentagonus, pictus, princeps, quadratus, rebuti, regalis, regulare, regulare spinosum, robustum, schilinskyi, schumannii, speciosus, spectabilis, spiralis, splendidus, variegatus, weberi,* and *zonatus.*

Type locality: Mexico.

Distribution: Northern Mexico.

This species has, until now, been known only from the type collection of Karwinsky which should be at Munich. In 1912 Dr. Rose obtained a specimen from Dr. Radlkofer, but without label, which we now suspect is a part of the original material of Karwinsky.

Señor Octavio Solis wrote us that in 1919 he obtained specimens of this plant at Barretillas, Nuevo León, and also at Ciudad Guerrero, Tamaulipas. The four specimens which he took back to the City of Mexico soon died. In May 1921 Señor Solis sent one of the specimens from Ciudad Guerrero which had been collected by Professor Francisco Contreras and we have been able to confirm his identification. Señor Solis says that the plant is known as peyote.

Illustration: Abh. Bayer. Akad. Wiss. München 4²: pl. 3, as *Echinocactus asterias.*

Figs. 194 and 195.—Astrophytum asterias.

Figure 194 is from a photograph of the plant from Munich referred to above; figure 195 is copied from the illustration above cited.

3. Astrophytum capricorne (Dietrich).

Echinocactus capricornis Dietrich, Allg. Gartenz. 19: 274. 1851.
Echinocactus capricornis minor Rünge, Monatsschr. Kakteenk. 2: 82. 1892.

Subglobose or short-cylindric, up to 25 cm. high; ribs 7 or 8, high, acute; areoles distant, 2 to 3 cm. apart; spines several, more or less flattened, weak, hardly pungent, brown, 3 to 5 cm. long; flowers 6 to 7 cm. long, widely spreading when in full bloom; outer perianth-segments reddish, gradually passing into the lemon-yellow inner perianth-segments with papery tips, orange at base, spatulate, acute or cuspidate at the apex, entire or more or less toothed; stamens numerous, attached over all the inner surface of the flower-tube; style slender, cream-colored; stigma-lobes linear, somewhat spreading, 5 to 9, cream-colored; seeds 2.5 mm. broad, shining.

Type locality: La Rinconada, Mexico.

Distribution: Northern Mexico.

Dr. C. A. Purpus writes that this plant is found on the hills of lime and slate formation south of Parras. It is very scarce and grows associated with *Lophophora williamsii* and *Ariocarpus furfuraceus.* He believes that the variety *minor* is specifically distinct; this he found at Peña and Villareal, Coahuila, and also on Cerro de la Bola and in the Sierra de la Paila.

Various hybrids have been produced by crossing this species with *Astrophytum ornatum* and *A. myriostigma.*

Echinocactus capricornis major (Monatsschr. Kakteenk. 19: 139. 1909) has never been described.

Illustrations: Ann. Rep. Smiths. Inst. 1908: pl. 5, f. 2; Monatsschr. Kakteenk. 14: 183; 26: 135; Gartenwelt 15: 537; Schelle, Handb. Kakteenk. f. 80, 81; Alianza Cientifica Universal 3: pl. opp. 190; Möllers Deutsche Gärt. Zeit. 25: 474. f. 6, No. 1; Blanc, Cacti 41. No. 420; Karsten and Schenck, Vegetationsbilder 2: pl. 20, as *Echinocactus capricornis;* Monatsschr. Kakteenk. 2: 82; Floralia 42: 372, as *Echinocactus capricornis minor;* De Laet, Cat. Gén. f. 5, as *Echinocactus capricornis major.*

Plate XXI, figure 1, is from a painting by E. I. Schutt of a plant collected by C. A. Purpus at Parras, Mexico.

4. Astrophytum ornatum (De Candolle) Weber.*

Echinocactus ornatus De Candolle, Mém. Mus. Hist. Nat. Paris 17: 114. 1828.
Echinocactus mirbelii Lemaire, Cact. Aliq. Nov. 22. 1838.
Echinocactus holopterus Miquel, Linnaea 12: 2. 1838.
Echinocactus tortus Scheidweiler, Bull. Acad. Sci. Brux. 5: 493. 1838.
Echinofossulocactus mirbelii Lawrence in Loudon, Gard. Mag. 17: 318. 1841.
Echinocactus ghiesbrechtii Salm-Dyck, Allg. Gartenz. 18: 395. 1850.
Echinopsis haageana Linke, Wochenschr. Gärtn. Pflanz. 1: 86. 1858.
Echinocactus ornatus mirbelii Croucher, Gard. Chron. 1873: 983. 1873.
Echinocactus haageanus Rümpler in Förster, Handb. Cact. ed. 2. 469. 1885.
Echinocactus ornatus glabrescens Schumann, Gesamtb. Kakteen 324. 1898.

FIG. 196.—Astrophytum ornatum.

Subglobose to cylindric, 3 dm. high or more, the surface more or less white-floccose; ribs 8, rather prominent, 2 cm. high or more, acute; areoles 1 to 5 cm. apart, felted; spines 5 to 11, subulate, yellow at first, becoming brown, often 3 cm. long; flowers large, lemon-yellow, 7 to 9 cm. broad; inner perianth-segments broadly oblong, with a broad, more or less serrated apex; scales on ovary very narrow.

Type locality: Mexico.
Distribution: Hidalgo and Querétaro, Mexico.
Dr. Rose collected this species in the deserts of eastern Querétaro, Mexico, in 1905 (No. 10286).

Astrophytum glabrescens Weber (Dict. Hort. Bois 467. 1896) is given as a synonym of this species, although it has never been described.

Echinopsis haageana Linke (Wochenschr. Gärtn. Pflanz. 1: 86. 1858), although originally described as probably from Peru, doubtless belongs to this species.

Illustrations: Blühende Kakteen 2: pl. 113; Cact. Journ. 1: pl. for September; 54; Ann. Rep. Smiths. Inst. 1908: pl. 13, f. 4; Schumann, Gesamtb. Kakteen f. 56; Gard. Mag. 4: 279; Möllers Deutsche Gärt. Zeit. 25: 474. f. 6, No. 19, as *Echinocactus ornatus;* Blanc, Cacti 50. f. 581; Schelle, Handb. Kakteenk. f. 79, as *Echinocactus ornatus mirbelii;* Möllers Deutsche Gärt. Zeit. 25: 485. f. 17, as *Echinocactus ornatus glabrescens;* Gartenwelt 15: 537, as *Echinocactus mirbelii ornatus;* Gard. Chron. 1873: 983. f. 196, as *Echinocactus mirbelii;* Cact. Journ. 2: 173.

Figure 196 is from a photograph furnished by Dr. W. E. Safford of the plant collected by Dr. Rose near Higuerillas, Mexico, in 1905.

*This binomial has several times been credited to Weber, but has never been formally published.

22. ERIOSYCE Philippi, Anal. Univ. Chile 41: 721. 1872.

A very large, globular to thick-cylindric cactus; ribs numerous, very spiny; flowers from the apex of the plant, campanulate, the tube longer than the perianth-segments; outer perianth-segments linear, more or less pungent; inner perianth-segments narrow, acutish; stamens borne near the base of the flower-tube, included; ovary densely clothed with matted wool; fruit oblong, becoming dry, dehiscing by a basal pore, very spiny above; seeds rather large, dull, black-pitted with a sub-basal sunken hilum.

This very interesting plant, well known to the Chileans under the name of sandillon, is not very well understood botanically. It has no near relatives in South America but resembles in habit and fruit some of our giant species of *Echinocactus* in Mexico. It has good technical differences and we have no hesitancy in following the late Dr. Rudolph Philippi in regarding it as constituting a distinct genus.

Only one species is here recognized, a native of Chile, although Mr. Söhrens states that there are two very definite forms, one of which is more slender, with narrow fruit, the other nearly globular and with globular fruit. The genus was based on *Echinocactus sandillon* Remy.

The generic name is from ἔριον wool, and σῦκον fig, referring to the woolly fruit.

1. Eriosyce ceratistes (Otto).

> *Echinocactus ceratistes* Otto in Pfeiffer, Enum. Cact. 51. 1837.
> *Echinocactus sandillon* Remy in Gay, Fl. Chilena 3: 14. 1847.
> *Echinocactus auratus* Pfeiffer, Abbild. Beschr. Cact. 2: under pl. 14. 1847.
> *Echinopsis aurata* Salm-Dyck, Cact. Hort. Dyck. 1849. 39. 1850.
> *Eriosyce sandillon* Philippi, Anal. Univ. Chile 41: 721. 1872.

Simple, 3 to 10 dm. high, usually 2 to 3 dm. in diameter or even more, very woolly at apex; ribs numerous, 21 to 35, but fewer in young plants, while in old ones sometimes more; areoles large, usually 3 cm. apart; spines 11 to 20, nearly equal, straight or somewhat curved, 2.5 to 3.5 cm. long, subulate, yellowish when young; flowers 3 to 3.5 cm. long, yellowish red, opening for 3 or 4 hours and then whitening; inner perianth-segments 1.5 cm. long; fruit 4 cm. long; seeds 3 mm. long.

FIG. 197.—Eriosyce ceratistes.

Type locality: Bellavista, Chile.

Distribution: Provinces of Santiago, Aconcagua, and Coquimbo, Chile.

Dr. Rose did not see wild plants of this species but he obtained fruit through Mr. Söhrens and also obtained a photograph of a fine plant growing in the Botanical Garden at Santiago.

The plant is found only in the mountains, growing at an altitude of 2,000 meters or more, and flowers abundantly. Dr. Philippi states that he counted 74 flowers and fruits on one plant.

Although this plant was first described as an *Echinocactus*, Pfeiffer questions whether it might not be a *Melocactus*. The original spelling of the specific name, *ceratistes*, was changed by Salm-Dyck to *ceratitis*.

The two varieties *Echinocactus ceratistes melanacanthus* (Labouret, Monogr. Cact. 246. 1853; *E. melanacanthus* Monville) and *Echinocactus ceratistes celsii* (Labouret, Monogr. Cact. 246. 1853) may possibly belong here. To the latter variety Labouret doubtfully refers *Echinocactus copiapensis* Monville, not Pfeiffer. To the latter binomial seems to have been applied the name *Ceratistes copiapensis* which we have seen mentioned only by Labouret.

Illustrations: Schelle, Handb. Kakteenk. 147. f. 73; Engler and Drude, Veg. Erde 8: pl. 15, f. 30; Schumann, Gesamtb. Kakteen f. 53, as *Echinocactus ceratites;* Cact. Mex. Bound. pl. 33, f. 7, as *Echinocactus sandillon.*

Figure 197 is from a photograph of a plant in the Botanical Garden at Santiago, Chile, taken by Mrs. J. N. Rose, in 1914.

23. MALACOCARPUS Salm-Dyck, Cact. Hort. Dyck. 1849. 24. 1850.

Plants globose to short-cylindric, either simple or clustered; ribs definite, usually straight, either entire or broken up into more or less definite tubercles; areoles felted, especially when young, spine-bearing; flowers from the center of the plant, broad and short, mostly yellow; perianth funnelform to subrotate; stigma-lobes in typical species red; ovary densely covered with scales bearing an abundance of wool and usually bristles in their axils; fruit soft, rose-red or crimson; seeds brown or black, tuberculate with a broad truncate base; hilum white.

Prince Salm-Dyck, who established the genus, assigned 6 species of *Echinocactus* to it, of which *E. corynodes* Pfeiffer was the first and is therefore taken by us as the generic type.

Schumann treats the group as a subgenus of *Echinocactus;* he assigns 3 species to it, all from the State of Rio Grande do Sul, Brazil; Arechavaleta, who follows Schumann's treatment, describes 6 species from Uruguay. Besides those heretofore treated in the subgenus *Malacocarpus,* we refer here most of the species assigned by Schumann to the subgenus *Notocactus.*

We recognize 29 species, all from South America and all found south of the Equator. The generic name is from μαλακός soft, and καρπός fruit, referring to the fleshy fruit.

KEY TO SPECIES.

```
A. Plants globular to stout-cylindric.
  B. Areoles of the ovary and flower-tube long-hairy or long-woolly.
    C. Spines 4 cm. long or less, straight.
      D. Flowers yellow.
        Ribs acute.
          Spines subulate.....................................................  1. M. tephracanthus
        Spines acicular.
          Spines yellow.......................................................  2. M. schumannianus
        Spines white or becoming silvery.
          Spines 3 to 7.......................................................  3. M. grossei
          Spines 9 or 10......................................................  4. M. nigrispinus
      Ribs obtuse or rounded.
        Ribs spirally arranged, broken into tubercles.
          Spirals many; plant gray............................................  5. M. reichei
          Spirals few; plant brown............................................  6. M. napinus
        Ribs straight or nearly so, undulate or continuous.
          Perianth short-funnelform.
            Perianth-tube very stout..........................................  7. M. apricus
          Perianth-tube relatively slender.
            Plant deeply umbilicate; spines slender...........................  8. M. concinnus
            Plant slightly umbilicate; spines short...........................  9. M. tabularis
        Perianth campanulate to subrotate.
          Spines setaceous or acicular.
            Ribs 30 to 40; radial spines up to 40 or more..................... 10. M. scopa
            Ribs 21 or fewer; radial spines much fewer than 40.
              Ribs very low and rounded....................................... 11. M. pulcherrimus
              Ribs prominent.
                Areoles only 4 to 7 mm. apart................................. 12. M. muricatus
                Areoles more separated.
                  Inner perianth-segments obtuse or merely apiculate.......... 13. M. linkii
                  Inner perianth-segments acute or acuminate................. 14. M. ottonis
          Spines stouter, subulate.
            Inner perianth-segments 2 to 3 cm. long.
              Spines terete.
                Spines slender, slightly curved.............................. 15. M. catamarcensis
                Spines stout, rigid.......................................... 16. M. patagonicus
              Spines flattened.
                Central spines not much longer than the radials.............. 17. M. erinaceus
```

KEY TO SPECIES—continued.

Central spines definitely longer than the radials.
Spines strongly curved.. 18. *M. langsdorffii*
Spines straight... 19. *M. mammulosus*
Inner perianth-segments about 1 cm. long......................... 20. *M. islayensis*
DD. Flowers salmon or red.
Ribs about 13, obtuse; flowers salmon......................... 21. *M. strausianus*
Ribs 30 or more, acutish; flowers red......................... 22. *M. haselbergii*
CC. Spines elongated, the central ones 1.5 to 7 cm. long, curved; flowers orange-red.. 23. *M. maassii*
BB. Areoles of ovary and flower-tube with tufts of short hairs.
Spines stout, subulate....................................... 24. *M. tuberisulcatus*
Spines slender, acicular.
Spines long, much curved................................. 25. *M. curvispinus*
Spines short, nearly straight............................. 26. *M. mammillarioides*
AA. Plants becoming slender-cylindric and much elongated.................... 27. *M. leninghausii*
AAA. Species not grouped .... {28. *M. graessneri*
{29. *M. escayachensis*

1. Malacocarpus tephracanthus (Link and Otto) Schumann, Fl. Bras. 4²: 243. 1890.

Echinocactus tephracanthus Link and Otto, Verh. Ver. Beförd. Gartenb. 3: 422. 1827.
Echinocactus acuatus* Link and Otto, Verh. Ver. Beförd. Gartenb. 3: 424. 1827.
Echinocactus sellowii Link and Otto, Verh. Ver. Beförd. Gartenb. 3: 425. 1827.
Melocactus tephracanthus Link and Otto, Verh. Ver. Beförd. Gartenb. 3: pl. 16, f. 2. 1827.
Melocactus sellowii Link and Otto, Verh. Ver. Beförd. Gartenb. 3: pl. 22. 1827.
Melocactus acuatus Link and Otto, Verh. Ver. Beförd. Gartenb. 3: pl. 23. 1827.
Echinocactus sellowianus Pfeiffer, Enum. Cact. 55. 1837.
Echinocactus sessiliflorus Mackie in Curtis's Bot. Mag. 64: pl. 3569. 1837.
Echinocactus tetracanthus Lemaire, Cact. Aliq. Nov. 15. 1838.
Echinocactus courantii Lemaire, Cact. Aliq. Nov. 20. 1838.
Echinocactus sessiliflorus pallidus Monville in Lemaire, Cact. Gen. Nov. Sp. 88. 1839.
Echinocactus sessiliflorus tetracanthus Monville in Lemaire, Cact. Gen. Nov. Sp. 88. 1839.
Cereus tephracanthus Steudel, Nom. ed. 2. 1: 336. 1840.
Malacocarpus sellowianus Salm-Dyck, Cact. Hort. Dyck. 1849. 25. 1850.
Malacocarpus sellowianus tetracanthus Salm-Dyck, Cact. Hort. Dyck. 1849. 25. 1850.
Malacocarpus courantii Salm-Dyck, Cact. Hort. Dyck. 1849. 25. 1850.
Echinocactus tephracanthus spinosior Labouret, Monogr. Cact. 171. 1853.
Echinocactus courantii spinosior Monville in Labouret, Monogr. Cact. 171. 1853.
Echinocactus sellowianus tetracanthus Labouret, Monogr. Cact. 172. 1853.
Malacocarpus martinii Rümpler in Förster, Handb. Cact. ed. 2. 454. 1885.
Malacocarpus sellowii Schumann, Fl. Bras. 4²: 238. 1890.
Malacocarpus sellowii tetracanthus Schumann, Fl. Bras. 4²: 239. 1890.
Malacocarpus tetracanthus Meyer, Monatsschr. Kakteenk. 4: 143. 1894.
Echinocactus sellowii martinii Schumann, Gesamtb. Kakteen 297. 1898.
Echinocactus acuatus sellowii Spegazzini, Anal. Mus. Nac. Buenos Aires III. 4: 494. 1905.
Echinocactus acuatus tetracanthus Spegazzini, Anal. Mus. Nac. Buenos Aires III. 4: 494. 1905.
Echinocactus sellowii macrocanthus Arechavaleta, Anal. Mus. Nac. Montevideo 5: 230. 1905.
Echinocactus sellowii macrogonus Arechavaleta, Anal. Mus. Nac. Montevideo 5: 232. 1905.
Echinocactus sellowii acutatus Arechavaleta, Anal. Mus. Nac. Montevideo 5: 234. 1905.
Echinocactus sellowii turbinatus Arechavaleta, Anal. Mus. Nac. Montevideo 5: 235. 1905.
Echinocactus fricii Arechavaleta, Anal. Mus. Nac. Montevideo 5: 244. 1905.
Echinocactus pauciareolatus Arechavaleta, Anal. Mus. Nac. Montevideo 5: 246. 1905.
Echinocactus sellowii courantii Gürke, Monatsschr. Kakteenk. 18: 149. 1908.
Echinocactus sellowii typicus Gürke, Monatsschr. Kakteenk. 18: 149. 1908.

Simple, globular or somewhat depressed, up to 15 cm. in diameter, woolly at apex; ribs 18 to 22, acute, rather high, hardly undulate on the margin, light green; areoles 1.5 to 2 cm. apart; spines 4 to 6, straight or curved backward, the longest 2 cm. long; flowers from the woolly apex of the plant, 4 to 4.5 cm. long, broader than long when fully expanded; perianth-segments yellow, narrowly oblong, mucronate-tipped; stamens and style slightly exserted; stigma-lobes red; perianth deciduous; fruit small, 1 cm. long, purple, fleshy, scaly; scales ovate, bearing hairs and bristles in their axils; seeds black, 1 mm. long.

Type locality: "Rio Grande," perhaps better, Rio Grande do Sul, Brazil.

Distribution: Brazil, Argentina, and Uruguay.

Walpers (Repert. Bot. 2: 275. 1843) referred *Echinocactus acuatus spinosior* Lemaire to *E. courantii.* This variety was not described, however, until 1839 when Lemaire (Cact. Gen. Nov. Sp. 87. 1839) assigned it to Monville and referred to it, as a synonym, *Echinocactus suberinaceus* Lemaire.

---

*The specific name was originally given by Link and Otto as above, but Schumann changed it, writing both *Echinocactus acutatus* and *Melocactus acutatus.* Don (Gen. Syst. 3: 163. 1834) writes the name *E. arcuatus.*

Schumann refers *Echinocactus martinii* Cels (Gesamtb. Kakteen 297. 1898) as a synonym of *E. sellowii*, but we do not know that the binomial has been formally made.

Illustrations: Anal. Mus. Nac. Montevideo 5: pl. 18; Pfeiffer and Otto, Abbild. Beschr. Cact. 1: pl. 1, as *Echinocactus sellowii;* Schelle, Handb. Kakteenk. 143. f. 68, as *E. sellowii martinii;* Anal. Mus. Nac. Montevideo 5: pl. 21, as *E. sellowii turbinatus;* Anal. Mus. Nac. Montevideo 5: pl. 19, as *E. sellowii macrocanthus;* Anal. Mus. Nac. Montevideo 5: pl. 20, as *E. sellowii macrogonus;* Link and Otto, Verh. Ver. Beförd. Gartenb. 3: pl. 16, f. 2, as *Melocactus tephracanthus;* Pfeiffer, Abbild. Beschr. Cact. 2: pl. 6, as *Echinocactus tetracanthus;* Curtis's Bot. Mag. 64: pl. 3569, as *E. sessiliflorus;* Anal. Mus. Nac. Montevideo 5: pl. 25, as *E. fricii;* Anal. Mus. Nac. Montevideo 5: pl. 26, as *E. pauciareolatus;* Link and Otto, Verh. Ver. Beförd. Gartenb. 3: pl. 23, as *Melocactus acuatus;* Lemaire, Icon. Cact. pl. 12; Martius, Fl. Bras. 4²: pl. 49; Monatsschr. Kakteenk. 4: 141, as *Malacocarpus sellowii;* Link and Otto, Verh. Ver. Beförd. Gartenb. 3: pl. 22, as *Melocactus sellowii.*

FIG. 198.—Malacocarpus tephracanthus.   FIG. 199.—Malacocarpus schumannianus.

Plate xx, figure 1, shows the top of a plant collected by Dr. Shafer at Concordia, Argentina, in 1917 (No. 119), which flowered in the New York Botanical Garden in 1918; plate xxi, figure 2, shows a plant obtained by Dr. Rose from Dr. Spegazzini in 1915, labeled *Echinocactus sellowianus*, which flowered in the New York Botanical Garden in 1917. Figure 198 is copied from Pfeiffer and Otto's illustration of *Echinocactus sellowii;* figure 203 is copied from the illustration given by Link and Otto of *Melocactus acuatus.*

2. **Malacocarpus schumannianus** (Nicolai).

Echinocactus schumannianus Nicolai, Monatsschr. Kakteenk. **3**: 175. 1893.
Echinocactus schumannianus longispinus Haage jr. in Quehl, Monatsschr. Kakteenk. **9**: 43. 1899.

Simple, globose or elongated, becoming bent or procumbent, sometimes over a meter long and 1 to 4 dm. in diameter; ribs about 30, low, acute, dull green; spines 4 to 7, setaceous, brownish to yellow; flowers central, large, citron-yellow, 4.5 cm. long; perianth-segments oblong, obtuse, spreading; scales on ovary with wool and bristles in their axils.

Type locality: Said to be in the territory of Misiones, Paraguay.

Distribution: Paraguay and northeastern Argentina.

Illustrations: Schumann, Gesamtb. Kakteen f. 65; Monatsschr. Kakteenk. **7**: 55; Schelle, Handb. Kakteenk. 178. f. 111; Chodat, Veg. Paraguay 1: f. 87, 88, 89, as *Echinocactus schumannianus.*

Figure 199 is copied from the first illustration cited above.

3. Malacocarpus grossei (Schumann).

Echinocactus grossei Schumann, Monatsschr. Kakteenk. **9**: 44. 1899.

Globose to depressed or sometimes cylindric, sometimes up to 1.7 meters high; ribs usually 16, acute, somewhat crenate; areoles small, circular; spines 3 to 7, spreading, acicular, curved, white, the longer ones 4 cm. long; flower large, funnelform, citron-yellow, 4 cm. long, when fully expanded broader than long; perianth-segments oblanceolate to spatulate, obtuse, serrate above; stamens numerous, short; style slender, longer than the stamens; stigma-lobes 12 to 17, linear, white, recurved; scales on the ovary numerous, linear, purplish, with wool and bristles in their axils; fruit short-oblong, 2.5 cm. long, 2 cm. in diameter; seeds black, 2 mm. long.

FIG. 200.—Malacocarpus grossei.   FIG. 201.—M. napinus.

Type locality: Paraguay.
Distribution: Paraguay, between Carepegua and Acaay.

The species is known to us only from illustrations and description; it and the preceding one are much the largest of the genus, as known to us. The Blühende Kakteen shows the spines as yellow, but they were originally described as white.

Illustrations: Blühende Kakteen **2**: pl. 89; Möllers Deutsche Gärt. Zeit. **25**: 474. f. 6, No. 18; Monatsschr. Kakteenk. **9**: 44; Schumann, Gesamtb. Kakteen Nachtr. f. 19, as *Echinocactus grossei*.

Figure 200 is copied from the first illustration cited above.

4. Malacocarpus nigrispinus (Schumann).

Echinocactus nigrispinus Schumann, Monatsschr. Kakteenk. **9**: 45. 1899.

Cespitose, globose to short-columnar, green; ribs 20 or more, acute; spines 9 or 10, somewhat curved, slender, reddish when young, afterward silvery; flowers yellow, funnelform; scales of the ovary filled with hairs and bristles.

Type locality: Between Carepegua and Acaay, Paraguay.
Distribution: Paraguay.

We have had small plants of this species growing which do not differ very much, if any, from *Malacocarpus schumannianus*.

I. Schutt del.

M. E. Eaton del.

A Hoen & Co.

1. Flowering plant of *Astrophytum capricorne*.
2. Top of flowering plant of *Malacocarpus tephracanthus*.
(All natural size.)

Echinocactus schumannianus nigrispinus Haage jr. (Monatsschr. Kakteenk. 9: 45. 1899) was given as a synonym of *E. nigrispinus*, but has never been published otherwise.

Illustrations: Weinberg, Cacti 11; Knippel, Kakteen pl. 9; Schelle, Handb. Kakteenk. 179. f. 112; Chodat, Veg. Paraguay 1: f. 90, as *Echinocactus nigrispinus.*

5. Malacocarpus reichei (Schumann).

Echinocactus reichei Schumann, Gesamtb. Kakteen Nachtr. 110. 1903.

Simple, globular, 6 to 7 cm. in diameter; ribs spiraled, broken into very regular tubercles; spines minute, appressed, 7 to 9, about equal; flowers small, light yellow, 2.5 cm. long or more; inner perianth-segments linear-oblong, acute; style slender, longer than the filaments, red; stigma-lobes red; ovary and tube with small scales, pilose and setose in the axils.

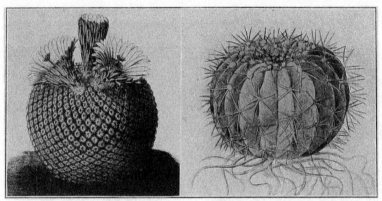

FIG. 202.—Malacocarpus reichei.   FIG. 203.—Malacocarpus tephracanthus.

Type locality: Not cited.
Distribution: Chile.

This species was sent from Santiago to Dr. Schumann by Dr. Karl Reiche in 1900 and does not seem to have been very much distributed. It is a very remarkable plant, judging from the illustration below cited, and may not be of this alliance. We know it only from description and illustration.

Illustration: Blühende Kakteen 1: pl. 42, as *Echinocactus reichei.*

Figure 202 is copied from the illustration cited above.

6. Malacocarpus napinus (Philippi).

Echinocactus napinus Philippi, Anal. Univ. Chile 41: 720. 1872.
Echinocactus mitis Philippi, Anal. Univ. Chile 85: 493. 1894.

Plant 2 to 9 cm. high with a very large root, larger than the globose stem itself; ribs broken into rounded tubercles; spines about 9, minute, 3 mm. long, appressed; flower small, about 3 cm. long, pale yellow to nearly white; flower-tube covered with minute scales, the axils long-woolly and bristly; stigma-lobes reddish.

Type locality: Huasco, Chile.
Distribution: Northern Chile.

Echinocactus napinus and *E. mitis* both came from Huasco, and Schumann is probably right in uniting them under the older name.

Illustrations: Monatsschr. Kakteenk. 11: 93, in part; Blühende Kakteen 2: pl. 77; Gartenflora 21: pl. 721, f. 1; Schumann, Gesamtb. Kakteen f. 69, A, as *Echinocactus napinus;* Schumann, Gesamtb. Kakteen f. 69, B, as *Echinocactus mitis.*

Figure 201 is copied from the third illustration cited above.

7. Malacocarpus apricus (Arechavaleta).

Echinocactus apricus Arechavaleta, Anal. Mus. Nac. Montevideo **5**: 205. 1905.

Cespitose, in clusters of 2 to 10, subglobose, 3 to 5 cm. in diameter, umbilicate at apex, densely covered with interlocking spines; ribs 15 to 20, somewhat curved, more or less tuberculate; areoles orbicular, 3 to 4 mm. apart, tomentose when young, becoming naked in age; radial spines 18 to 20, grayish yellow, flexible; central spines several, 4 of the larger ones reddish at base; flowers yellow, 8 cm. long; flower-tube densely woolly and setose on the outside, very stout.

Type locality: Punta de la Ballena, Uruguay.

Distribution: Uruguay.

We know this plant only from description and illustration, from which the above description has been drawn.

Illustration: Anal. Mus. Nac. Montevideo **5**: pl. 10, as *Echinocactus apricus*.

Figure 204 is copied from the illustration above cited.

FIG. 204.—Malacocarpus apricus.   FIG. 205.—Malacocarpus tabularis.

8. Malacocarpus concinnus (Monville).

Echinocactus concinnus Monville, Hort. Univ. **1**: 222. 1839.
Echinocactus joadii Hooker in Curtis's Bot. Mag. **112**: pl. 6867. 1886.
Echinocactus concinnus joadii Arechavaleta, Anal. Mus. Nac. Montevideo **5**: 204. 1905.

Simple, globular or somewhat depressed, 5 to 7.5 cm. in diameter; ribs about 16 to 20, somewhat tuberculate, light green; young areoles white-felted; spines 10 to 12, spreading, setaceous; radial spines 5 to 7 mm. long; central spines 1 to 4, one much longer, spreading or turned downward; flowers large, 7 cm. long; outer perianth-segments narrow, acute, reddish; inner perianth-segments oblong, yellow, except the reddish tips, acute; stigma-lobes bright red; scales on the ovary hairy in their axils; perianth-tube slender.

Type locality: Not definitely cited.

Distribution: Southern Brazil and Uruguay.

We know this species only from description and illustrations.

Illustrations: Lemaire, Icon. Cact. pl. 6; Loudon, Encycl. Pl. ed. 3. 1376. f. 19366; Förster, Handb. Cact. ed. 2. 551. f. 70; Curtis's Bot. Mag. 70: pl. 4115; Pfeiffer, Abbild. Beschr. Cact. 2: pl. 11; Blühende Kakteen 2: pl. 94; Anal. Mus. Nac. Montevideo 5: pl. 9; Monatsschr. Kakteenk. 29: 141; Schelle, Handb. Kakteenk. 179. f. 113; 180. f. 114; Wiener Ill. Gart. Zeit. 29: f. 104; Rümpler, Sukkulenten 178. f. 97; Palmer, Cult. Cact. 129; De Laet, Cat. Gén. f. 8; Engler and Prantl, Pflanzenfam. 3^{6a}: f. 63; Watson, Cact. Cult. 94. f. 29; ed. 3. 50. f. 21, as *Echinocactus concinnus;* Curtis's Bot. Mag. 112: pl. 6867, as *Echinocactus joadii.*

9. Malacocarpus tabularis (Cels).

Echinocactus concinnus tabularis Cels in Förster, Handb. Cact. ed. 2. 552.   1885.
Echinocactus tabularis Cels in Schumann, Gesamtb. Kakteen 389.   1898.

Simple, globose or short-columnar; ribs 16 to 18, somewhat crenate, obtuse, glaucous; radial spines 16 to 18, acicular; central spines 4; flowers yellow, 6 cm. long; perianth-segments narrow, acute; scales of ovary bearing dense wool and long brown bristles in their axils; seeds hemispheric or dome-shaped with a broad truncate base, brownish, papillose-roughened, about 1 mm. broad.

Type locality: Not cited definitely.
Distribution: Brazil or Uruguay.
In the first two illustrations cited, the flowers are not shown as coming from the apex of the plant as one would expect.
The illustration given by Schumann (Gesamtb. Kakteen f. 66) suggests *Malacocarpus concinnus.*
Echinocactus tabularis cristatus Rebut seems to be only a garden form.
Illustrations: Blühende Kakteen 1: pl. 23; (?) Schumann, Gesamtb. Kakteen f. 66; Monatsschr. Kakteenk. 26: 57; 29: 141; Anal. Mus. Nac. Montevideo 5: pl. 6, as *Echinocactus tabularis.*
Figure 205 is copied from the first illustration cited above.

10. Malacocarpus scopa (Sprengel).

Cactus scopa Sprengel, Syst. 2: 494. 1825.*
Cereus scopa Salm-Dyck in De Candolle, Prodr. 3: 464. 1828.
Echinocactus scopa Link and Otto, Icon. Pl. Rar. 81. 1830.
Echinocactus scopa candidus Pfeiffer, Enum. Cact. 64. 1837.
Echinopsis scopa Carrière, Rev. Hort. 47: 374. 1875.
Echinocactus scopa albicans Arechavaleta, Anal. Mus. Nac. Montevideo 5: 199. 1905.

At first globular but becoming cylindric to clavate, 1 to 4.5 dm. high; ribs 30 to 40, low, obtuse, almost hidden by the spines; radial spines 40 or more, white, setaceous, spreading; central spines about 4, brown or purple, much stouter than the radials; flowers lemon-yellow, widely spreading and then 6 cm. broad; inner perianth-segments in 2 series, spatulate, somewhat toothed above; stigma-lobes about 10, bright red; scales on the ovary bearing wool and conspicuous brown bristles.

Type locality: Not cited.
Distribution: Southern Brazil and Uruguay.
Echinocactus scopa candidus cristatus, E. scopa cristatus Hortus, *E. scopa ruberrimus,* and *E. scopa rubrinus* Link and Otto may or may not be published varietal names.
Illustrations: Cact. Journ. 1: 57; Gartenwelt 15: 539; Watson, Cact. Cult. 119. f. 45, as *Echinocactus scopa cristatus;* Cact. Journ. 1: 67; Gartenwelt 9: 267; Schelle, Handb. Kakteenk. 176. f. 107; 177. f. 109, as *E. scopa candidus cristatus;* Anal. Mus. Nac. Montevideo 5: pl. 8, as *E. scopa albicans;* Möllers Deutsche Gärt. Zeit. 25: 474. f. 6, No. 13; Schelle, Handb. Kakteenk. 175. f. 106, as *E. scopa candidus;* Rev. Hort. 47: 374. f. 60, as *Echinopsis*

---

*We have credited the name, *Cactus scopa,* to Sprengel, as above.   He marks it with an asterisk (*) as he does all his new names.   The usual citation is to Link (Enum. Hort. Berol. 2: 21. 1822) who in the place cited does list a number of species of *Cactus* but not *C. scopa.*   It is remarkable how general this error has become for we find it in De Candolle (Prodr. 3: 464. 1828), Pfeiffer (Enum. Cact. 64. 1837), Förster (Handb. Cact. 304. 1846), Labouret (Monogr. Cact. 238. 1853), Hooker (Curtis's Bot. Mag. 90: pl. 5445), Schumann (Gesamtb. Kakteen 381. 1898), the Index Kewensis, and elsewhere.

scopa; Rev. Hort. **47:** 375. f. 61; Rümpler, Sukkulenten 181. f. 100; Förster, Handb. Cact. ed. 2. 137. f. 7; Dict. Gard. Nicholson **4:** 540. f. 24; Suppl. 336. f. 360 (these last five illustrations are the same, and are sometimes called *Echinopsis scopa candida cristata, Echinocactus scopa candidus, E. scopa candidus cristatus,* and *E. scopa cristatus); Loudon, Encycl. Pl. ed. 3. 1378. f. 19383; Curtis's Bot. Mag. **90:** pl. 5445; Edwards's Bot. Reg. **25:** pl. 24; Förster, Handb. Cact. ed. 2. 136. f. 6; Anal. Mus. Nac. Montevideo **5:** pl. 7; Blühende Kakteen **3:** pl. 155; Abh. Bayer, Akad. Wiss. München **2:** pl. 1, sec. 3, f. 5; Link and Otto, Icon. Pl. Rar. pl. 41; Rümpler, Sukkulenten f. 99; Gartenflora 56: 20. f. 5; Watson, Cact. Cult. 118. f. 44; ed. 3. 59. f. 28, as *Echinocactus scopa.*

11. Malacocarpus pulcherrimus (Arechavaleta).

Echinocactus pulcherrimus Arechavaleta, Anal. Mus. Nac. Montevideo **5:** 222. 1905.

Small, 3 to 5 cm. high, 1.5 to 2 cm. in diameter; ribs 19 to 21, low and broad, tuberculate; radial spines 10 to 12, acicular, white, 1 to 2 mm. long; flowers 1.5 to 2 cm. long, 2.5 to 3 cm. broad, yellow; perianth-segments oblong, acute, sometimes mucronate; ovary and flower-tube densely white-woolly and setose; fruit turbinate, 1 cm. long, fleshy.

FIG. 206.—Malacocarpus pulcherrimus.   FIG. 207.—Malacocarpus muricatus.

Type locality: Paso de los Toros.
Distribution: Uruguay, but known only from the type collection.
We know this little plant from the original description and illustration only.
Illustration: Anal. Mus. Nac. Montevideo **5:** pl. 16, as *Echinocactus pulcherrimus.*
Figure 206 is copied from the illustration above cited.

12. Malacocarpus muricatus (Otto).

Echinocactus muricatus Otto in Pfeiffer, Enum. Cact. 49. 1837.

Simple or sometimes proliferous, either globular or columnar, said to be depressed at apex, 2 dm. in diameter; ribs 16 to 20, obtuse, crenate, dull, glaucous; radial spines 15 to 20, white, setaceous, 8 mm. long; central spines 3 or 4, brown at tips, 13 mm. long; areoles approximate; flowers 3 cm. long, yellow; inner perianth-segments acute; style longer than the stamens; stigma-lobes 7 to 9, purple; scales of the ovary with their axils filled with wool and bristles.

Type locality: Brazil.
Distribution: Southern Brazil.

We know this species from description and illustration only.

Illustration: Martius, Fl. Bras. 4²: pl. 50, f. 2, as *Echinocactus muricatus.*
Figure 207 is copied from the illustration above cited.

13. Malacocarpus linkii (Lehmann).

Cactus linkii Lehmann, Ind. Sem. Hamburg 16. 1827.
Echinocactus linkii Pfeiffer, Enum. Cact. 48. 1837.

Oval to short-cylindric, 7 to 15 cm. high; ribs 13, obtuse; areoles somewhat sunken into the ribs, 8 mm. apart; spines weak, spreading; radial spines 10 to 12, white with brownish tips; central spines 3 or 4, brownish; flowers yellow, 2.5 cm. long, 5 cm. broad when fully expanded; inner perianth-segments broad, obtuse; scales of the ovary woolly and setose in their axils; stigma-lobes red.

FIG. 208.—Malacocarpus linkii.   FIG. 209.—Malacocarpus ottonis.

Type locality: Cited as Mexico, but in error.

Distribution: Southern Brazil.

This species must be close to *Malacocarpus ottonis* and the two are often confused. The original illustrations are so different, however, that we believe they must be distinct.

Echinocactus linkii spinosior (Förster, Handb. Cact. 301. 1846) is only a name.

The name *Cereus linkii* Lehmann appears in Pfeiffer's Enumeratio (48. 1837) as a synonym of *Echinocactus linkii*, but it does not occur thus where he cites it (Nov. Act. Nat. Cur. 16: 316. 1828) but as *Cactus (Cereus) linkii.*

Illustration: Nov. Act. Nat. Cur. 16: pl. 14, as *Cactus linkii.*
Figure 208 is copied from the illustration cited above.

14. Malacocarpus ottonis (Lehmann).

Cactus ottonis Lehmann, Ind. Sem. Hamburg 16. 1827.
Echinocactus tenuispinus Link and Otto, Verh. Ver. Beförd. Gartenb. 3: 421. 1827.
Echinocactus tenuispinus minor Link and Otto, Verh. Ver. Beförd. Gartenb. 3: 422. 1827.
Echinocactus tortuosus Link and Otto, Icon. Pl. Rar. 29. 1829.
Echinocactus ottonis Link and Otto, Icon. Pl. Rar. 31. 1830.
Opuntia ottonis G. Don, Hist. Dichl. Pl. 3: 172. 1834.
Echinocactus ottonis tenuispinus Pfeiffer, Enum. Cact. 48. 1837.
Echinocactus ottonis pallidior Monville in Lemaire, Cact. Gen. Nov. Sp. 88. 1839.
Echinocactus ottonis spinosior Monville in Lemaire, Cact. Gen. Nov. Sp. 88. 1839.
Echinocactus ottonis tortuosus Schumann, Gesamtb. Kakteen 392. 1898.

Echinocactus ottonis paraguayensis Heese, Gartenwelt 9: 266.　1905.
Echinocactus ottonis uruguayus Arechavaleta, Anal. Mus. Nac. Montevideo 5: 213.　1905.
Echinocactus arechavaletai Spegazzini, Anal. Mus. Nac. Buenos Aires III. 4: 496.　1905.
Echinocactus spegazzinii Gürke, Monatsschr. Kakteenk. 15: 110.　1905.
Echinocactus ottonis brasiliensis Haage jr., Monatsschr. Kakteenk. 24: 41.　1914.

Simple or cespitose, globular or somewhat depressed, more or less glossy green, 5 to 6 cm. in diameter; ribs 10, broad and rounded below; areoles few, usually distant, 1 cm. apart or more, small, circular; spines acicular, brown, 1 cm. long or less; flowers from the uppermost areoles, one or more appearing at a time, 5 to 6 cm. long, bright yellow; perianth-segments linear-oblong, acute; stamens about half the length of the perianth-segments; style yellow; stigma-lobes red; axils of scales filled with long brown wool and brown bristles.

Type locality: Supposed to be Mexico, but the species was described from a garden plant.
Distribution: Southern Brazil, Uruguay, and adjacent parts of Argentina.

The varietal name, *Echinocactus ottonis paraguayensis,* is usually credited to Schumann who used it in 1900 (Monatsschr. Kakteenk. 10: 179).　The name *Cereus ottonis* appears in Pfeiffer's Enumeratio (47. 1837) as a synonym of *Echinocactus ottonis,* but it does not occur thus where Pfeiffer cites it (Nov. Act. Nat. Cur. 16: 316. 1828), but as *Cactus (Cereus) ottonis.*

The following varieties are sometimes met with: *E. ottonis brasiliensis* (Monatsschr. Kakteenk. 18: 48. 1908), *pfeifferi* Monville (Salm-

FIGS. 210 and 211.—Malacocarpus ottonis.

Dyck, Cact. Hort. Dyck. 1844.　19.　1845), and *minor* (Förster, Handb. Cact. 302. 1846), and *Echinocactus muricatus hortatani* (Labouret, Monogr. Cact. 232. 1853).

A hybrid has been produced with this species and a plant called *Echinopsis zuccarinii.*
Illustrations: Cact. Journ. 1: 43, 54; Monatsschr. Kakteenk. 12: 158; 29: 125; Martius, Fl. Bras. 4²: pl. 51, f. 3; Edwards's Bot. Reg. 24: pl. 42; Rev. Hort. 1861: 270. f. 62; Curtis's Bot. Mag. 58: pl. 3107; Link and Otto, Icon. Pl. Rar. pl. 16; De Laet, Cat. Gén. f. 11; Möllers Deutsche Gärt. Zeit. 25: 474. f. 6, No. 17; Rümpler, Sukkulenten 179. f. 98, as *Echinocactus ottonis;* Anal. Mus. Nac. Montevideo 5: pl. 12; De Laet, Cat. Gén. f. 4; Tribune Hort. 4: pl. 140, as *Echinocactus ottonis tenuispinus;* Link and Otto, Icon. Pl. Rar. pl. 15, as *E. tortuosus;* Nov. Act. Nat. Cur. 16: pl. 15, as *Cactus ottonis;* Verh. Ver. Beförd. 3: pl. 19, f. 1, 2, as *Melocactus tenuispinus;* Anal. Mus. Nac. Montevideo 9: 267; Schelle, Handb. Kakteenk. 182. f. 116, as *Echinocactus ottonis paraguayensis;* Anal. Mus. Nac. Montevideo 5: pl. 11, as *Echinocactus arechavaletai;* Curtis's Bot. Mag. 68: pl. 3963, as *Echinocactus tenuispinus;* Dict. Hort. Bois 465. f. 323, as *Echinocactus tenuispinus ottonis;* Karsten, Deutsche Fl. 887. f. 501, No. 12; ed. 2. 2: 456. f. 605, No. 12, as *Echinocactus tenuissimus.*

Plate xx, figure 2, shows the plant obtained by Dr. Rose from W. Mundt, in 1912, which has since flowered repeatedly in the New York Botanical Garden; plate xxiii, figure 2, shows a plant obtained by Dr. Shafer at Concordia, Argentina, in 1917 (No. 118) which afterwards flowered in the New York Botanical Garden. Figure 209 is copied from the illustration of Link and Otto, cited above as *Cactus ottonis;* figure 210 is from a photograph furnished by Dr. Spegazzini of a plant cultivated by him as *Echinocactus arechavaletai;* figure 211 shows a plant collected by Dr. Shafer at Concordia, Argentina, in 1917 (No. 118).

15. Malacocarpus catamarcensis (Spegazzini).

Echinocactus catamarcensis Spegazzini, Anal. Mus. Nac. Buenos Aires III. 4: 500. 1905.
Echinocactus catamarcensis pallidus Spegazzini, Anal. Mus. Nac. Buenos Aires III. 4: 500. 1905.
Echinocactus catamarcensis obscurus Spegazzini, Anal. Mus. Nac. Buenos Aires III. 4: 501. 1905.

Simple, elliptic to short-cylindric, 10 to 50 cm. high, 8 to 12 cm. in diameter, grayish green; ribs 11 to 13, obtuse, tuberculate; spines terete, more or less erect, grayish with brown tips, subulate, slightly curved; radial spines 14 to 21, 10 to 20 mm. long; central spines 4 to 7, 25 to 30 mm. long; flowers 4.5 cm. long, citron to golden; stigma-lobes yellowish; scales of the ovary filled with wool and bristles.

Type locality: Argentina.
Distribution: Western Argentina.

FIG. 212.—Malacocarpus catamarcensis.

FIG. 213.—Malacocarpus patagonicus.

We know this species chiefly from the original description and photograph obtained by Dr. Rose in 1915 from Dr. Spegazzini. To it we have referred a living plant collected by Dr. Aleš Hrdlička in Argentina in 1910, which has flowered with us on one or two occasions. Figure 212 is from a photograph of the plant collected by Aleš Hrdlička.

16. Malacocarpus patagonicus (Weber).

Echinocactus intertextus Philippi, Linnaea **33**: 81. 1864. Not Engelmann, 1856.
Cereus patagonicus Weber in Spegazzini, Rev. Agron. La Plata **3**: 604. 1897.
Echinocactus coxii Schumann, Gesamtb. Kakteen 422. 1898.
Cereus duseni Weber, Anal. Soc. Cient. Argentina **48**: 49. 1899.

Usually simple and erect, slender, cylindric, 6 dm. long or less, 3 to 5 dm. in diameter, very spiny, green or somewhat glaucous-green; ribs 6 to 10, straight or spiraled, somewhat undulate; areoles approximate; radial spines 6 to 10, spreading; central spines 1 to 3, much stouter, subulate, some of them sometimes more or less hooked; flowers from near the top of plant, 3.5 cm. long, fully as broad when expanded, inodorous; inner perianth-segments pale rose-colored, spatulate, 18 mm. long, 8 mm. broad, mucronate; fruit about 2 cm. long, greenish; style thick, 15 mm. long; stigma-lobes black-purple; ovary turbinate, 8 mm. in diameter, the axils of its scales woolly and bristly; seeds 2.5 mm. broad.

Type locality: Chubut, Argentina.

Distribution: Southern Argentina.

This species has been very confusing not only as to its identification, but as to its generic relationship.   It is possible that more than one species has been treated here.   The plant grows in barren regions, often among boulders where there is no other vegetation.

Figure 213 is from a photograph taken by Mr. Walter Fischer at General Roca, Rio Negro, showing how the plant grows in its natural surroundings; figure 214, showing a potted plant, and figure 215, the top of a flowering plant, are from photographs contributed by Mr. C. Bruch.

FIGS. 214 and 215.—Malacocarpus patagonicus.

17. Malacocarpus erinaceus (Haworth) Rümpler in Förster, Handb. Cact. ed. 2. 455.   1885.

Cactus erinaceus Haworth, Pl. Succ. Suppl. 74.   1819.
Echinocactus poliacanthus Link and Otto, Verh. Ver. Beförd. Gartenb. 3: 422.   1827.
Melocactus poliacanthus Link and Otto, Verh. Ver. Beförd. Gartenb. 3: pl. 16, f. 1.   1827.
Echinocactus corynodes Otto in Pfeiffer, Enum. Cact. 55.   1837.
Echinocactus erinaceus Lemaire, Cact. Aliq. Nov. 16.   1838.
Malacocarpus corynodes Salm-Dyck, Cact. Hort. Dyck. 1849. 25.   1850.
Malacocarpus corynodes erinaceus Salm-Dyck, Cact. Hort. Dyck. 1849. 25.   1850.
Malacocarpus polyacanthus Salm-Dyck, Cact. Hort. Dyck. 1849. 25.   1850.
Malacocarpus acuatus Salm-Dyck, Cact. Hort. Dyck. 1849. 25.   1850.
Echinocactus corynodes erinaceus Labouret, Monogr. Cact. 170.   1853.
Echinocactus acuatus corynodes Spegazzini, Anal. Mus. Nac. Buenos Aires III. 4: 494.   1905.
Echinocactus acuatus depressus Spegazzini, Anal. Mus. Nac. Buenos Aires III. 4: 494.   1905.
Echinocactus acuatus erinaceus Spegazzini, Anal. Mus. Nac. Buenos Aires III. 4: 495.   1905.
Echinocactus leucocarpus Arechavaleta, Anal. Mus. Nac. Montevideo 5: 239.   1905.

Simple, globular to short-cylindric, 15 cm. high, very woolly at top, up to 15 cm. high; ribs 15 to 20, obtuse, strongly undulate; areoles borne in the depressions on ribs, felted when young; radial spines 6 to 8, subulate, yellowish, 1 to 2 cm. long; central spine solitary; flowers yellow, 4 to 5 cm. long, 7 cm. broad when fully open; inner perianth-segments spreading, oblong to spatulate, acute, often serrate above; stigma-lobes bright red.

Type locality: Not cited.

Distribution: Southern Brazil and adjacent parts of Argentina and Uruguay.

Schumann was inclined to refer here *Echinocactus aciculatus* Salm-Dyck (Hort. Dyck. 341. 1834; *Malacocarpus aciculatus* Salm-Dyck, Cact. Hort. Dyck. 1849. 25. 1850) and

Echinocactus terscheckii Reichenbach (Terscheck, Suppl. 3; also Walpers, Repert. Bot. 2: 315. 1843).

Echinocactus rosaceus (Otto, Allg. Gartenz. 1: 364. 1833), *E. acutangulus* Zuccarini (Pfeiffer, Enum. Cact. 55. 1837), and *E. conquades* (Förster, Handb. Cact. 338. 1846) have usually been referred to *Echinocactus corynodes* but were never described.

Echinocactus erinaceus elatior Monville (Salm-Dyck, Cact. Hort. Dyck. 1844. 22. 1845), without description, must be referred here.

Illustrations: Schumann, Gesamtb. Kakteen f. 50; Schelle, Handb. Kakteenk. f. 69; De Laet, Cat. Gén. f. 21, as *Echinocactus erinaceus;* Verh. Ver. Beförd. Gartenb. 3: pl. 16, f. 1, as *Melocactus poliacanthus;* Abh. Bayer. Akad. Wiss. München 2: pl. 1, sec. 3, f. 1 to 4; Curtis's Bot. Mag. 68: pl. 3906; Anal. Mus. Nac. Montevideo 5: pl. 22, as *Echinocactus corynodes;* Monatsschr. Kakteenk. 4: 141; Förster, Handb. Cact. ed. 2. f. 52; Rümpler, Sukkulenten 174. f. 96; Garten-Zeitung 4: 182. f. 42, No. 18; Krook, Handb. Cact. 67, as *Malacocarpus corynodes;* Gartenflora 31: 216, as *Malacocarpus corynodes erinaceus;* Anal. Mus. Nac. Montevideo 5: pl. 23, as *Echinocactus leucocarpus;* Deutsche Gärt. Zeit. 7: 312; Dict. Gard. Nicholson 2: 317. f. 504; Monatsschr. Kakteenk. 4: 141; Förster, Handb. Cact. ed. 2. 455. f. 53.

Figure 216 is copied from plate 3906 of Curtis's Botanical Magazine, cited above.

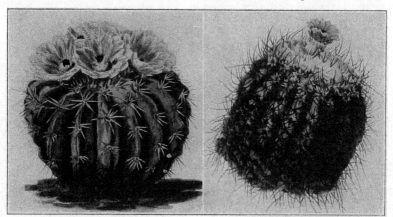

FIG. 216.—Malacocarpus erinaceus.   FIG. 217.—Malacocarpus langsdorfii.

18. Malacocarpus langsdorfii (Lehmann).

Cactus langsdorfii Lehmann, Ind. Sem. Hamburg 17. 1826.
Melocactus langsdorfii De Candolle, Prodr. 3: 461. 1828.
Echinocactus langsdorfii Link and Otto, Icon. Pl. Rar. 79. 1830.

Oblong, 10 cm. high or more, very woolly at apex; ribs 17, obtuse, strongly tubercled; radial spines about 6, more or less unequal, somewhat spreading; central spine usually solitary, 2.5 cm. long; flower yellow, 1.5 cm. broad, campanulate; inner perianth-segments oblong, obtuse, about 20; filaments yellow; stigma-lobes numerous, purple.

Type locality: Central Brazil.
Distribution: Central and southern Brazil.

This plant was first described in the Seed Catalogue of the Botanical Garden of Hamburg. The next year Lehmann published two descriptions of it under the name of *Cactus (Echinocactus) langsdorfii*, one of which was accompanied by a colored illustration (see second illustration cited above). We have seen no living plants or other illustrations which we are disposed to refer here and we have therefore kept the species distinct, although we are aware that Schumann refers it first to *Malacocarpus polyacanthus* and afterwards to *Echinocactus erinaceus*.

Illustrations: Link and Otto, Icon. Pl. Rar. pl. 40, as *Echinocactus langsdorfii;* Nov. Act. Nat. Cur. **16:** pl. 13, as *Cactus langsdorfii.*

Figure 217 is copied from the second illustration cited above.

19. Malacocarpus mammulosus (Lemaire).

Echinocactus mammulosus Lemaire, Cact. Aliq. Nov. 40. 1838.
Echinocactus hypocrateriformis Otto and Dietrich, Allg. Gartenz. **6:** 169. 1838.
Echinocactus submammulosus Lemaire, Cact. Gen. Nov. Sp. 20. 1839.
Echinocactus pampeanus Spegazzini, Contr. Fl. Vent. 27. 1896.
Echinocactus acuatus arechavaletai Spegazzini, Anal. Mus. Nac. Buenos Aires III. **4:** 494. 1905.
Echinocactus mammulosus submammulosus Spegazzini, Anal. Mus. Nac. Buenos Aires III. **4:** 496. 1905.
Echinocactus mammulosus pampeanus Spegazzini, Anal. Mus. Nac. Buenos Aires III. **4:** 496. 1905.
Echinocactus mammulosus hircinus Spegazzini, Anal. Mus. Nac. Buenos Aires III. **4:** 496. 1905.
Echinocactus mammulosus typicus Spegazzini, Anal. Mus. Nac. Buenos Aires III. **4:** 496. 1905.
Echinocactus floricomus Arechavaleta, Anal. Mus. Nac. Montevideo **5:** 183. 1905.
Echinocactus pampeanus charruanus Arechavaleta, Anal. Mus. Nac. Montevideo **5:** 193. 1905.
Echinocactus pampeanus rubellianus Arechavaleta, Anal. Mus. Nac. Montevideo **5:** 194. 1905.
Echinocactus pampeanus subplanus Arechavaleta, Anal. Mus. Nac. Montevideo **5:** 194. 1905.
Echinocactus arechavaletai Schumann in Arechavaleta, Anal. Mus. Nac. Montevideo **5:** 208. 1905. Not Spegazzini, Jan. 1905.

Simple, nearly globose, about 8 cm. high, light shining green; ribs 18 to 25, strongly tuberculate, almost covered by the numerous interlocking spines; radial spines 20 to 30, 5 cm. long; central spines 2 to 4, about 2 cm. long; flowers yellow, 3.5 to 4 cm. long; scales of the ovary woolly and setose in their axils.

FIGS. 218 and 219.—Malacocarpus mammulosus.

Type locality: Not cited.

Distribution: Brazil, Uruguay, and Argentina.

Labouret (Monogr. Cact. 228, 229. 1853) mentioned three varieties of this species as follows: *spinosior* Haage, *cristatus* Monville, and *minor* Monville.

Echinocactus hypocrateriformis spinosior Haage probably should be referred here.

Spegazzini (Anal. Mus. Nac. Buenos Aires III. **4:** 496. 1905) referred the species *pampeanus* and *submammulosus* as varieties of this species.

Illustrations: Schelle, Handb. Kakteenk. 181. f. 115; Knippel, Kakteen pl. 9; Cact. Journ. **2:** 102; Anal. Mus. Nac. Montevideo **5:** pl. 3, as *Echinocactus mammulosus;* Martius, Fl. Bras. **4²:** pl. 51, f. 2, as *Echinocactus hypocrateriformis;* Anal. Mus. Nac. Montevideo **5:**

pl. 2, as *Echinocactus floricomus;* Anal. Mus. Nac. Montevideo **5:** pl. 5, as *Echinocactus pampeanus;* Anal. Mus. Nac. Montevideo **5:** pl. 24; Monatsschr. Kakteenk. **15:** 107, as *Echinocactus arechavaletai;* Monatsschr. Kakteenk. **27:** 18; **29:** 141; Anal. Mus. Nac. Montevideo **5:** pl. 4, as *Echinocactus submammulosus.*

Plate XXII, figure 1, shows a plant collected by Dr. Shafer near Salto, Uruguay, in 1917 (No. 124) which has flowered repeatedly in the New York Botanical Garden. Figure 218 is copied from Arechavaleta's illustration of *Echinocactus floricomus* cited above; figure 219 is copied from Arechavaleta's illustration of *Echinocactus arechavaletai* cited above.

20. Malacocarpus islayensis (Förster).

> *Echinocactus islayensis* Förster, Hamb. Gartenz. **17:** 160. 1861.
> *Echinocactus molendensis** Vaupel, Bot. Jahrb. Engler **50:** Beibl. **111:** 24. 1913.

Simple, 5 to 7 cm. in diameter, almost entirely hidden under a mass of spines, woolly at the apex; ribs numerous, 19 to 25, low and obtuse; areoles approximate, 2 to 4 mm. apart, brown-felted when young; radial spines 8 to 22, spreading, 1 to 10 mm. long; central spines 4 to 7, stouter than the radials, straight, 12 to 16 mm. long, grayish to horn-colored; flowers small, 1.5 to 2 cm. long, yellow; outer perianth-segments reddish; scales on ovary and flower-tube bearing in their axils long hairs and long reddish bristles.

Type locality: Province of Islay, southern Peru.
Distribution: Southern Peru.

In 1914, while traveling in Peru, Dr. Rose made a special trip to Mollendo to re-collect *Echinocactus molendensis* which he found quite common on the hills above the town (Rose, No. 18999). A careful study of this material, some of which was sent home alive, leads us to believe that it is the same as *Echinocactus islayensis.*

Plate XXII, figure 4, is from a plant collected by Dr. Rose near Mollendo, Peru, in 1914.

21. Malacocarpus strausianus (Schumann).

> *Echinocactus strausianus* Schumann, Monatsschr. Kakteenk. **11:** 112. 1901.

Globular to short-cylindric, dull grayish green, up to 16 cm. high, very spiny; ribs about 13, more or less tuberculate, obtuse; spines 9 to 20, subulate, the longest 3 cm. long, reddish brown; central spines 1 to several; flower 1.5 cm. long, opening for at least 2 days, closing at night; outer perianth-segments brownish, 2.5 cm. long; inner perianth-segments about 20, linear-oblong, acute, deep salmon; filaments erect; style white; stigma-lobes linear, cream-colored; scales on ovary and flower-tube white-woolly and bristly in their axils.

Type locality: Argentina.
Distribution: Western Argentina.

This species is common about Mendoza, Argentina; specimens sent from there by Dr. Rose in 1915 (No. 21019) first flowered in the New York Botanical Garden in May 1917. It is named for Kaufmann Straus.

Illustrations: Monatsschr. Kakteenk. **11:** 107; Schumann, Gesamtb. Kakteen Nachtr. f. 16, as *Echinocactus strausianus.*

22. Malacocarpus haselbergii (F. Haage).

> *Echinocactus haselbergii* F. Haage in Förster, Handb. Cact. ed. 2. 563. 1885.

Simple, bright green, globose or somewhat depressed, 7 to 8 cm. in diameter; ribs 30 or more, somewhat tuberculate, more or less spiraled; radial spines about 20, acicular, white, almost bristle-like, 1 cm. long; central spines 3 to 5, pale yellow; flowers small, 2.5 to 3 cm. broad, red without, variegated within; flower-tube very short or none; stamens yellow, included; stigma-lobes 6, erect; scales of ovary small, woolly, and setose in their axils.

Type locality: Not cited in original description, but afterwards said to be the state of Rio Grande do Sul, Brazil.

---

*This species was named for the town, Mollendo, and, therefore, should have been spelled with two l's.

Distribution: Southern Brazil.

There is much uncertainty regarding the limitations of this species and also regarding its generic disposal. It was first described incidentally by Rümpler who considered it a bare form of *Echinocactus scopa*. Hooker, a few years after Rümpler, described and figured it and expressed his belief that it was a distinct species, well separated from *E. scopa*. Later on, it was described and illustrated by Gürke; the flower is shown with a slender elongated tube which is very unlike the flower illustrated by Hooker. Whether we have a plant with a very variable flower or two distinct species we can not determine without further field study.

Dr. Rose saw a plant in the Berlin Botanical Garden in 1912, just after it flowered, which he believed then was a generic type. He noted that the ovary was covered with clusters of spines as in the species of *Echinocereus*. It was first supposed to be a form of *Echinocactus scopa*, under which species it was incidentally first described. Hooker, when he described and illustrated it, stated that while it belonged to the same section of the genus it differed from it in the form of the plant and in the perianth. He placed it in Salm-Dyck's section, *Microgoni*.

This plant was named for Dr. von Haselberg of Stralsind, a cultivator of cacti. The variety *cristatus* is in the trade.

Illustrations: Curtis's Bot. Mag. **114**: pl. 7009; Blühende Kakteen **2**: pl. 98; Schelle, Handb. Kakteenk. **176**. f. **108**; Monatsschr. Kakteenk. **26**: 171; Tribune Hort. **4**: pl. 139; De Laet, Cat. Gén. f. 1; Möllers Deutsche Gärt. Zeit. **25**: 474. f. 6, No. 6, as *Echinocactus haselbergii*.

23. Malacocarpus maassii (Heese).

Echinocactus maassii Heese, Gartenflora **56**: 410. 1907.

Globular to short-cylindric, 10 to 15 cm. in diameter, yellowish green; ribs 13, spiraled, prominent near the apex, almost wanting at base, somewhat undulate or tubercled; radial spines 8 to 10, white, long, and weak, or sometimes 1 or 2 stouter; central spine 4 to 7 cm. long, much stouter than the radials, much curved and often hooked; flowers 14 cm. broad, orange-red; segments numerous, linear-oblong, 10 mm. long; filaments yellow; style stout, white; stigma-lobes yellow; ovary long and densely soft-woolly; fruit 5 to 6 cm. in diameter, dry, dehiscing by abscission above the base; umbilicus broad, circular; scales on the ovary minute, their axils filled with long white hairs; seeds black, globular, 2 mm. in diameter, tuberculate-roughened, with a prominent white aril at base.

Type locality: Bolivia.

Distribution: Southern Bolivia and northern Argentina.

The original description and illustration are poor. We believe, however, that this is the plant collected by J. A. Shafer in crevices of rocks, altitude 3,450 meters, at La Quiaca, Jujuy, Argentina, February 13, 1917 (No. 81). Our description is drawn entirely from Dr. Shafer's plant.

Illustrations: Gartenflora **56**: 410. f. 50; Monatsschr. Kakteenk. **25**: 45, as *Echinocactus maassii*.

24. Malacocarpus tuberisulcatus (Jacobi).

Cactus horridus Colla, Mem. Accad. Sci. Torino **37**: 76. 1833. Not Humboldt, Bonpland, and Kunth, 1823.
Echinocactus horridus Remy in Gay, Fl. Chilena **3**: 15. 1847.
Echinocactus tuberisulcatus Jacobi, Allg. Gartenz. **24**: 108. 1856.
Echinocactus soehrensii Schumann, Monatsschr. Kakteenk. **11**: 75. 1901.

Simple or sometimes in clusters of 9 or fewer, globular, often 2 dm. in diameter, dull green, depressed at apex; ribs 14 to 20, prominent, obtuse, strongly tubercled, separated by narrow intervals; tubercles with a flattened acute chin; areoles at first small, spineless, with an abundance of white wool but when old large, sometimes 1.5 cm. in diameter; radial spines not all developing the first year, brown when young, dark gray in age; radial spines 10 to 12; central spines 4 or 5, similar to the radials but a little stouter and longer, at most 2.5 cm. long; flowers 4.5 cm. long, yellowish, their areoles described as sparingly woolly.

M. E. Eaton del.

A.Hoen & Co.

1. Top of flowering plant of *Malacocarpus mammulosus*.
2. Fruiting plant of *Mila caespitosa*.
3. Flowering plant of *Astrophytum myriostigma*.
4. Fruiting plant of *Malacocarpus islayensis*.
   (All natural size.)

Type locality: Stony hillslopes near Valparaiso, Chile.

Distribution: Along the coastal hills of central Chile.

Dr. Rose found this plant very common in two localities in central Chile. One was in pasture on the hills above Valparaiso, altitude about 1,000 feet; the other was on the edge of cliffs about Valparaiso Harbor and only about 20 feet above the water. In the latter locality it was associated with *Neoporteria subgibbosa.*

Under *Echinocactus soehrensii,* Haage and Schmidt (1920) offer for sale the varieties *albispinus, brevispinus,* and *niger.*

Cactus horridus Colla, *Echinocactus tuberisulcatus* Jacobi, and *E. soehrensii* were all based upon plants from Valparaiso and we believe we are justified in combining them under the oldest specific name available.

Illustrations: Mem. Accad. Sci. Torino 37[1]: pl. 17, f. 1, as *Cactus horridus;* Schumann, Gesamtb. Kakteen Nachtr. f. 25; Monatsschr. Kakteenk. 11: 73, as *Echinocactus soehrensii.*

25. Malacocarpus curvispinus (Bertero).

Cactus curvispinus Bertero, Merc. Chil. 598. No. 13. 1829; Colla, Mem. Accad. Sci. Torino 37: 76. 1833.
Echinocactus curvispinus Remy in Gay, Fl. Chilena 3: 16. 1847.
Echinocactus froehlichianus Schumann, Gesamtb. Kakteen Nachtr. 124. 1903.

Simple or clustered, subglobose or short-columnar, 15 cm. high, pale green; ribs 16, broad and obtuse, divided into large tubercles; spines 15, all radials, or at least no very definite central ones, straight or somewhat curved, flexuous; flowers yellow or reddish brown, large, 3 to 6.5 cm. long; perianth-segments lanceolate, acute; stigma-lobes green; scales on the flower-tube and ovary small, scattered, bearing short wool and setae in their axils.

Type locality: Chile.

Distribution: Chile.

Mr. Söhrens tells us that he obtained the specimens, which were named *Echinocactus froehlichianus,* from the mountains south of Santiago and that he now considers the species the same as *Echinocactus curvispinus.*

Bertero's type of *Cactus curvispinus* is preserved in the museum at Santiago. It consists of one small fragment bearing two clusters of spines and one flower; the spine-cluster contains 10 or 11 spines, the longest of which is 2 cm. long; the flower is 3 cm. long with the ovary bearing small scales with woolly axils and the uppermost scales bearing bristles in their axils.

Illustrations: Blühende Kakteen 2: pl. 63; Schumann, Gesamtb. Kakteen Nachtr. f. 31, as *Echinocactus froehlichianus;* Mem. Accad. Sci. Torino 37[1]: pl. 16, f. 2, as *Cactus curvispinus;* Schelle, Handb. Kakteenk. 193. f. 126, as *Echinocactus curvispinus.*

26. Malacocarpus mammillarioides (Hooker).

Echinocactus mammillarioides Hooker in Curtis's Bot. Mag. 64: pl. 3558. 1837.
Echinocactus hybocentrus Lehmann in Pfeiffer, Enum. Cact. 65. 1837.
Echinocactus centeterius Lehmann in Pfeiffer, Enum. Cact. 65. 1837.
Echinocactus pachycentrus Lehmann in Pfeiffer, Enum. Cact. 66. 1837.
Echinocactus centeterius major Lemaire and Monville in Lemaire, Cact. Gen. Nov. Sp. 91. 1839.
Echinocactus nummularioides Steudel, Nom. ed. 2. 1: 536. 1840.
Echinocactus centeterius pachycentrus Salm-Dyck, Cact. Hort. Dyck. 1849. 33. 1850.
Echinocactus centeterius grandiflorus Labouret, Monogr. Cact. 244. 1853.

Subglobose to short-cylindric, bright green; ribs about 14 to 16, broad, obtuse, strongly tubercled; areoles rather large, felted; spines about 7, short, spreading, slender; flowers large, yellowish red; perianth-segments oblong, obtuse; ovary bearing small scales with a little wool in their axils.

Type locality: Chile.

Distribution: Chile.

This species was introduced by a Mr. Hitchen from Chile and flowered in 1836.

The status of this species is very confusing. It was described very briefly by Hooker who had never seen the living plant; its exact habitat was not given and it has never with certainty been re-collected.

It was referred by Lemaire in 1840 as a synonym of *Echinocactus centeterius*, but as a matter of fact it must have had a prior publication, although both names first appeared the same year, 1837.*

In spite of the difference in size and shape of the flowers, Lemaire (Hort. Univ. 2: 161. 1841) is positive that it is the same as the plant which he illustrates and yet he designates his illustration as var. *major!*

Echinocactus centeterius, to which it is generally referred, is scarcely less confusing. It seems to have been named and distributed by Lehmann, but was first described by Pfeiffer in 1837 and figured by him about 1843. His plant is small with rather small flowers; the perianth-segments are broad and abruptly acute. He states definitely that the plant comes from Minas Geraes, Brazil. As figured by Lemaire, where mentioned above, the flowers are very large with narrow elongated perianth-segments; in 1843 it was figured and described by Hooker. He describes a larger plant than the type and the flowers are somewhat different, the perianth-segments being spatulate and toothed above.

Förster (Handb. Cact. 296. 1846) redescribes the species, referring here *Echinocactus mammillarioides* Hooker (Curtis's Bot. Mag. 64: pl. 3558) as a large form under the name of *E. centeterius major* of Cels.

In 1853 Labouret redescribes the species and makes *Echinocactus pachycentrus* a variety and synonymous with *Echinocactus mammillarioides* and *Echinocactus centeterius major* Monville and also the variety *grandiflorus* based on the description of Lemaire (Hort. Univ. 2: 161). He also states that the species comes from Mexico and Minas Geraes.

In 1882 Regel in the Gartenflora (30: 258. pl. 1094) describes and figures this species and the variety *major* from plants sent by Dr. Philippi from Chile.

Rümpler (Förster, Handb. Cact. ed. 2. 568. f. 73, 74. 1885) redescribes the species and also reproduces the illustrations from Gartenflora, but adds nothing new except the statement that the species is also found in Peru and Chile.

Finally, Schumann (Gesamtb. Kakteen 418, 419. 1898) states that it surely comes from the Andes of Argentina, but he would exclude it from Mexico and Brazil. After a careful study of the works mentioned above and a comparison of all the illustrations cited below we are convinced that several species may be involved, but we have not been able to disentangle them.

Illustrations: Pfeiffer and Otto, Abbild. Beschr. Cact. 1: pl. 2; Förster, Handb. Cact. ed. 2. f. 73, 74; Curtis's Bot. Mag. 69: pl. 3974; Gartenflora 31: f. 1094, a; Schumann, Gesamtb. Kakteen f. 73; Monatsschr. Kakteenk. 27: 60; Loudon, Encycl. Pl. ed. 3. 1378. f. 19378; Herb. Génér. Amat. II. 2: pl. 56, as *Echinocactus centeterius;* Pfeiffer and Otto, Abbild. Beschr. Cact. 1: pl. 21, as *Echinocactus hybocentrus;* Curtis's Bot. Mag. 64: pl. 3558; Loudon, Encycl. Pl. ed. 2 and 3. 1201. f. 17354, as *Echinocactus mammillarioides;* Hort. Univ. 2: pl. 16; Gartenflora 31: f. 1094, b, as *Echinocactus centeterius major.*

27. Malacocarpus leninghausii (Haage jr.).

Pilocereus leninghausii Haage jr., Monatsschr. Kakteenk. 5: 147. 1895.
Echinocactus leninghausii Schumann, Monatsschr. Kakteenk. 5: 189. 1895.

Stem slender, cylindric, sometimes 1 meter long, 10 cm. in diameter; ribs about 30, low, obtuse; radial spines about 15, setaceous; central spines 3 or 4, longer than the radials, 4 cm. long, yellow; flowers 5 cm. broad, citron-yellow; scales on the ovary bristly in their axils.

Type locality: Not cited.
Distribution: Southern Brazil.

This is a very curious plant which has long been cultivated in Europe, but has only recently flowered in cultivation. Dr. Rose found it being widely propagated, but saw

---

*Curtis's Botanical Magazine, volume 64, plate 3558, appeared in March 1837 while Pfeiffer's Enumeratio Diagnostica Cactearum appeared probably after August and before November of the same year.

neither flowers nor fruit. We have taken it up in *Malacocarpus* since its relationship appears nearer to species of this genus than to those of other described genera.

The name *Echinocactus leninghausii cristatus* is given by Schelle (Handb. Kakteenk. 178. 1907).

Illustrations: Cact.·Journ. **2:** 4, as *Pilocereus leninghausii;* Schelle, Handb. Kakteenk. 178. f. 110; Möllers Deutsche Gärt. Zeit. **25:** 474. f. 6, No. 5; 27; Haage and Schmidt, Haupt-Verz. 1919: 169; 1920: 127. f. 10779; De Laet, Cat. Gén. f. 7, as *Echinocactus leninghausii.*

28. Malacocarpus graessneri (Schumann).

Echinocactus graessneri Schumann, Monatsschr. Kakteenk. **13:** 130. 1903.

Stems simple, depressed, 5 to 6 cm. high. 9 to 10 cm. in diameter, somewhat umbilicate at apex; ribs very numerous (more than 60), low, usually arranged in spirals, tuberculate; areoles bearing numerous bright yellow spines, 2 cm. long, the 3 to 6 central spines stouter and darker yellow; flowers small, narrow, from near the center of the plant.

Type locality: State of Rio Grande do Sul, Brazil.

Distribution: Southern Brazil.

This species must be a near relative of *M. haselbergii,* but with more ribs and slightly different spines and flowers. Both species are referred to this genus only tentatively. They very much resemble *Rebutia fiebrigii* in their form, ribs, and spines. *Echinocactus graessneri* is offered in the trade catalogues of Europe, but we know it only from descriptions and illustrations and from some small plants sent us by Haage and Schmidt.

Illustrations: Monatsschr. Kakteenk. **23:** 3; Gartenwelt **15:** 536; Möllers Deutsche Gärt. Zeit. **25:** 474. f. 6, No. 20; 489. f. 23, as *Echinocactus graessneri.*

29. Malacocarpus escayachensis (Vaupel).

Echinocactus escayachensis Vaupel, Monatsschr. Kakteenk. **26:** 125. 1916.

Globose, 12 cm. in diameter, very woolly at apex; ribs about 15; areoles approximate, short-tomentose; spines about 20, unequal, some curved, others straight; flower 2.5 cm. long.

Type locality: Escayache near Tarija, Bolivia.

Distribution: Southern Bolivia.

MALACOCARPUS sp.

Simple, or in clusters, short-cylindric, 1 dm. high; ribs about 15, low, broad, somewhat tubercled; radial spines 10 to 12, acicular, ascending, 2 to 3 cm. long, brownish, darker toward the tip; central spine usually solitary and hooked, a little stouter than the radials; flowers dull red, small, extending only a short distance beyond the spines.

Collected by Juan Söhrens in Tacna, Chile, altitude 3,000 meters, in 1911. Dr. Rose obtained a few clusters of spines from Mr. Söhrens's specimen in 1914. Mr. Söhrens has a photograph of it also.

The above species does not agree with any of the known species of South America. It is the only one of the *Echinocactus* relationship on the Pacific side of South America which has hooked spines. Mr. Söhrens believed it was a new species of *Echinocactus.*

PUBLISHED OR RECORDED SPECIES, PROBABLY REFERABLE TO THIS GENUS.

ECHINOCACTUS ELACHISANTHUS Weber, Bull. Mus. Hist. Nat. Paris **10:** 387. 1904.

Stem cylindric, 25 cm. high, 12 cm. in diameter; ribs very numerous, 45 or more, spiraled, divided into tubercles; radial spines 12 to 15, setaceous, white, 5 to 12 mm. long; flowers very small, 12 to 15 mm. long; flower-tube very short; perianth-segments yellowish green; ovary green, small, spiny; fruit greenish, 5 to 6 mm. in diameter; seeds small, dark brown.

Type locality: Northeast of Maldonado, Uruguay.

Distribution: Uruguay.

ECHINOCACTUS INTRICATUS Link and Otto, Verh. Ver. Beförd. Gartenb. **3:** 428. 1827.

Melocactus intricatus Link and Otto, Verh. Ver. Beförd. Gartenb. **3:** pl. 24. 1827.

Ovoid, 10 cm. high, green; ribs 20, obtuse; radial spines 14 to 16, spreading, 8 mm. long; central spines 4, stouter than the radials, 16 mm. long; flowers and fruit unknown.

The plant is said to have come from Montevideo and to have been collected by Sellow.

ECHINOCACTUS WEINGARTIANUS Haage jr. in Quehl, Monatsschr. Kakteenk. **9:** 73. 1899.

Short-columnar, rounded above; crown lightly tubercled, with sparse wool, exceeded by the upright spines, but can be seen, 8 cm. high by 5.5 cm. in diameter, bright green when young, later turning gray; ribs 13, straight, separated by sharp furrows, about 1 cm. high, divided into tubercles, broadening out with age; areoles 10 to 12 mm. apart, very large, covered with yellow-ish, curly wool turning gray and finally disappearing; radial spines 5 to 10, of various sizes, when young brownish black on the tips and base, horn-colored in the middle, more or less upright, later becoming white, more spreading; central spines up to 4, upright, when young a few ebony-black, the rest shading into the color of the radial spines, the uppermost and stoutest always black, up to 3 cm. long, bent above; flowers (according to traces) numerous in the region of the crown; seeds small, kidney-shaped, dark grayish brown, tubercled.

Distribution: Argentina.

ECHINOCACTUS SANJUANENSIS Spegazzini, Anal. Mus. Nac. Buenos Aires III. **4:** 501. 1905.

Nearly globose, 8 to 9 cm. in diameter; ribs 13, rather broad and low, strongly tubercled; areoles nearly circular, prominent, 4 mm. in diameter; spines 15 to 19, slender, rigid, 1 to 2.5 cm. long, at first rose-colored, becoming in age blackish.

Type locality: Province of San Juan.
Distribution: Western Argentina.

We know this species definitely only from the original description and a photograph obtained by Dr. Rose in 1915 from Dr. Spegazzini.

ECHINOCACTUS ROTHERIANUS Haage jr. in Quehl, Monatsschr. Kakteenk. **9:** 74. 1899.

Simple, slender, crooked, up to 26 cm. high and 5 cm. in diameter near the crown, constricted below; crown somewhat sunken and somewhat woolly, exceeded by the central spines of the new areoles; ribs 23, straight, separated by sharp furrows about 5 cm. deep, tubercled; tubercles slightly bent downward and bearing areoles, bright olive-green; areoles 5 mm. apart, naked; radial spines about 10, radiating, the lateral longer than the upper and lower, up to 5 mm. long, bright amber in color; central spines about 4, upright, stouter and longer (up to 10 mm.) than the radial spines, thickened at the base, reddish when young, otherwise colored like the radial spines; later all the spines bending downward and mixing together; these stiff, brittle, and finally disappearing, leaving the body naked; flowers unknown.

Type locality: Paraguay.

The following descriptions are translated from Fries who described three plants which he referred to *Echinocactus* but gave no specific names.   They are evidently better referred to *Malacocarpus* or to some other South American genus.

"ECHINOCACTUS sp.

"Prov. Jujuy; in rocky places in Yavi, 3,400 m. alt.

"Spherical, up to 1 dm. high; ribs 13, more or less divided into conical tubercles; radial spines about 8, directed obliquely outward, 3 to 3.5 cm. long; central spine 1, stouter, projecting straight outward, 3 to 8 cm. long, twisted-round, crooked above; flowers collected on the summit, 2.5 to 3 cm. long, with thick wool on the outside."   (Fries, Nov. Act. Reg. Soc. Upsal IV. **1:** 121. 1905.)

"ECHINOCACTUS sp.

"Prov. Jujuy; in rocky places at Moreno, 3,500 m. alt.

"Plant 2 to 3 dm. high, 1 to 2 dm. in diameter, short-cylindric; ribs about 30, plainly divided into tubercles; areoles covered with thick wool; spines up to 17, bent outward, up to 5 cm. long, stiff, twisted, straight; the areoles on the summit are spineless; the flowers are 5 to 10 cm. from the growing point, are 4 to 5 cm. long, are covered with thick wool; petals reddish brown."   (Fries, Nov. Act. Reg. Soc. Upsal IV. **1:** 122. 1905.)

"Echinocactus sp.

"Prov. Jujuy; among rocks at Moreno, 3,500 m. alt.
"Very similar in habit to *Echinocactus nidus* Söhrens, as shown in Monatsschr. Kakteenk. 10: 122; since, however, no description of that species is furnished, I can not tell whether or not these are identical.
"Spherical, 3 to 4 dm. high; ribs running spirally, divided into well-marked tubercles; areoles oval, covered with wool; radial spines and central spines little differentiated, all together 20 to 25, up to 45 cm. long, the outer ones more or less pressed on the plant-body, forming an extraordinarily thick network; the summit is naked, the spines closing in over it; the finer spines are white, the stouter ones golden-yellow at the base, lilac-colored in the middle, violet at the tips; the flowers appear about 2 cm. from the growing-point; they are yellowish green, 3.5 to 4 cm. long." (Fries, Nov. Act. Reg. Soc. Upsal. IV. 1: 121, 122. 1905.)

24. HICKENIA gen. nov.

Small, usually globular, very spiny cacti; ribs more or less definite, sometimes spiraled, divided into low, rounded tubercles; spines radial and central, one of the latter strongly hooked; flowers central, large for the size of the plant, borne at the top of the very young tubercles, subcampanulate, with a broad spreading limb; scales on ovary and flower-tube small, their axils filled with wool and bristles; fruit small, oblong, thin-walled, many-seeded; seeds minute, brown, shining, smooth, with a prominent white corky hilum.

Figs. 220 and 221.—Hickenia microsperma.

Type species: *Echinocactus microspermus* Weber.
The genus is named for Dr. C. Hicken, professor in the University of Buenos Aires.
Only one species is here recognized, a native of Argentina, but we find such great diversity in the spines, arrangement of the tubercles, and the color and size of the flowers that we suspect that more material would lead to some segregations.

1. Hickenia microsperma (Weber).

Echinocactus microspermus Weber, Dict. Hort. Bois 469. 1896.
Echinocactus microspermus macrancistrus Schumann, Monatsschr. Kakteenk. 12: 157. 1902.

Simple or in small clusters, usually globular, sometimes short-cylindric and 2 dm. high, 5 to 10 cm. in diameter, the surface divided into low tubercles; tubercles arranged in definite straight or spiraled ribs or very indefinitely arranged; radial spines 11 to 25, white, acicular, spreading, 4 to 6 mm. long; central spines 3 or 4, red to brown, subulate, glabrous or pubescent, the lowest one hooked at apex, ascending, spreading, or reflexed; flowers variable in color, bright yellow to red, 2 to 4 cm. broad; filaments red; style and stigma-lobes light reddish yellow; seeds 0.5 mm. long.

Type locality: Tucuman, Argentina.
Distribution: Northern Argentina.

Dr. Spegazzini has described two varieties, *erythranthus* and *thionanthus* (Anal. Mus. Nac. Buenos Aires III. **4**: 498. 1905) which have different-sized flowers and different-colored filaments and which suggest the probability of there being more than one species in this genus.

The varieties *brevispinus* Haage jr. and *elegans* Haage jr. are in the trade.

Two forms were also collected by Dr. Shafer in Argentina in 1917, both of which have flowered in the New York Botanical Garden. No. 9 has small yellow flowers, less than 2 cm. long, with yellow filaments and style; No. 18 has large red flowers, 4 cm. long with red filaments.

Illustrations: Gartenwelt **7**: 281; De Laet, Cat. Gén. f. 20; Monatsschr. Kakteenk. **7**: 105; **12**: 155; **16**: 48; Blühende Kakteen **1**: pl. 1; Schumann, Gesamtb. Kakteen f. 68; Nachtr. 109. f. 22; Curtis's Bot. Mag. **128**: pl. 7840, as *Echinocactus microspermus;* Monatsschr. Kakteenk. **12**: 155; **31**: 59; Schumann, Gesamtb. Kakteen Nachtr. 110. f. 23, as *Echinocactus microspermus macrancistrus.*

Plate XXIII, figure 1, shows a plant collected by Dr. Shafer at Andalgala, Argentina, in 1917 (No. 9), which flowered in the New York Botanical Garden, January 4, 1920. Figures 220, 221, and 222 are from photographs furnished by Dr. Spegazzini.

Fig. 222.—Hickenia microsperma.

25. FRAILEA gen. nov.

Plants small, globular or cylindric, with the apex rounded or depressed, usually cespitose; ribs numerous, low, divided into tubercles, these bearing small spines; flowers small, often cleistogamous, arising from the apex of the central tubercles; fruit small, spherical to ellipsoid, bearing narrow yellow scales with hair-like bristles in their axils, these forming a crest to the flower; seeds black or brown, smooth or pubescent, shining, with a triangular, deeply concave face; embryo straight (!), splitting the testa on the back of the seed in germinating; endosperm wanting; cotyledons minute, if at all developed.

Type species: *Echinocactus cataphractus* Dams.

These plants very much resemble some of the small species of *Mammillaria*, but the low tubercles are more definitely arranged on ribs and the flowers and fruit are very different; the seeds are much like those of *Epithelantha*, but the flowers are different and the fruit is scaly. The flowers of this genus are not very well known; several of the species have been described as having the flowers often cleistogamous and it may be that they all are. We have had two or three of the species under observation for several years. Flower-buds are often formed and occasionally ripe fruit with fertile seed is produced, but we have never seen open flowers or indications that they had opened. The flowers may open at night, but we doubt it; they certainly do not in day-light.

While the species of this group which we know from living plants clearly represent a very distinct generic type, we are not quite certain whether *Echinocactus caespitosus* belongs here or not. Its larger flowers suggest a possibility of its being related to *Malacocarpus* and indeed Spegazzini states that it is near *E. concinnus.*

The genus is named for Manuel Fraile who was born at Salamanca, Spain, in 1850, and who for years has diligently cared for the cactus collection in the U. S. Department of Agriculture, Washington, D. C. Eight species are recognized.

PLATE XXIII

M. E. Eaton del.

A. Hoen & Co

1. Flowering plant of *Hickenia microsperma*.
2. Flowering plant of *Malacocarpus ottonis*.
3. Fruit of *Sclerocactus polyancistrus*.
4. Seed of same.
5. Flowering plant of *Echinofossulocactus violaciflorus*.
(All natural size, except seed.)

KEY TO SPECIES.

Stems cylindric, usually simple.. 1. *F. gracillima*
Stems globular, more or less cespitose.
    Seeds puberulent.. 2. *F. grahliana*
    Seeds smooth.
        Ribs more distinct than in the other species............................. 3. *F. pumila*
        Ribs very indistinct.
            Radial spines 12 to 14.. 4. *F. schilinzkyana*
            Radial spines 5 to 9.
                Plants usually simple; fruit red.......................... 5. *F. cataphracta*
                Plants cespitose; fruit black........................... 6. *F. pygmaea*
Uncertain relationship.. { 7. *F. caespitosa*
                                                   { 8. *F. knippeliana*

1. Frailea gracillima (Lemaire).

> *Echinocactus gracillimus* Monville in Lemaire, Cact. Gen. Nov. Sp. 24. 1839.
> *Echinocactus pumilus gracillimus* Schumann, Gesamtb. Kakteen 394. 1898.

Cylindric, simple, 10 cm. high, 2.5 cm. in diameter, grayish green; ribs about 13, but indistinct, more or less spiraled, tuberculate; areoles small, with a purple blotch beneath each one; radial spines about 16, setaceous, white, 2 mm. long, more or less appressed; central spines 2 to 4, more or less unequal, 4 to 8 mm. long; flowers yellow, 3 cm. long; scales on the ovary and flower-tube woolly and bristly in their axils; fruit 6 mm. in diameter; seeds 1.5 mm. long.

Type locality: Not cited.
Distribution: Paraguay.
Echinocactus gracilis (Förster, Handb. Cact. 304. 1846) is given as a synonym of this species, but was not described. This is probably the *E. gracilis* Lemaire of collections (Monatsschr. Kakteenk. 10: 16. 1900).
Illustration: Monatsschr. Kakteenk. 9: 55, A, as *Echinocactus gracillimus*.

2. Frailea grahliana (Haage jr.).

> *Echinocactus grahlianus* Haage jr. in Schumann, Monatsschr. Kakteenk. 9: 54. 1899.
> *Cactus grahlianus* Kuntze, Deutsch. Bot. Monatsschr. 21: 193. 1903.

Cespitose; plants small, depressed-globose, 3 to 4 cm. in diameter; ribs about 13, low and indistinct; spines all radial, 9 or 10, subulate, appressed, somewhat curved backward, 3.5 mm. long; flowers 4 cm. long, yellowish; fruit 6 mm. in diameter; seeds brown, puberulent.

Type locality: Paraguari, Paraguay.
Distribution: Paraguay; also Argentina, according to Spegazzini.
Echinocactus grahilianus adustior (Monatsschr. Kakteenk. 14: 156. 1904) is only mentioned.
Illustrations: Schumann, Gesamtb. Kakteen Nachtr. f. 20; Monatsschr. Kakteenk. 9: 55, B; 12: 141, as *Echinocactus grahlianus*.

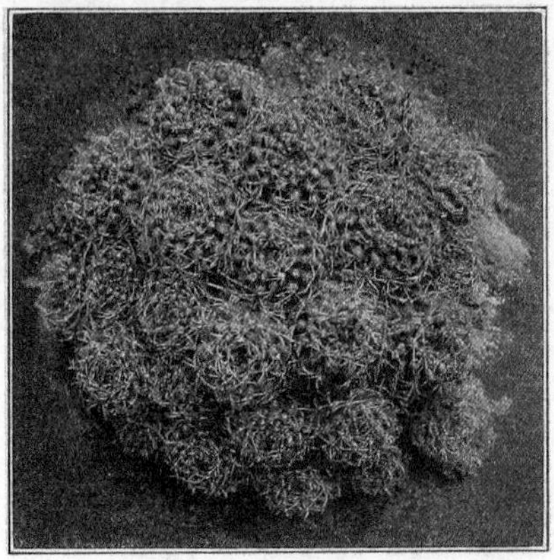

3. Frailea pumila (Lemaire).

> *Echinocactus pumilus* Lemaire, Cact. Aliq. Nov. 21. 1838.

Plant cespitose, small, globose, umbilicate at apex, deep green, sometimes becoming reddish; ribs 13 to 15, more distinct than in the related species, more or less tuberculate; areoles small, nearly circular; spines all pubescent, yellowish brown; radial spines 9 to 14, setaceous, more or less appressed; central spines 1 or 2, flower 2 cm. long, yellow; axils of

FIG. 223.—Frailea pumila.

scales on the ovary and flower-tube woolly and setose; seeds smooth, brown, obovate, 1.5 mm. long, angled on the back; depressed hilum much smaller than in *Frailea cataphracta*.

Type locality: Not cited.
Distribution: Paraguay and Argentina.

This species, the first of the group to be described, is not even now very well understood. The description of the seeds given above is drawn from Dr. Shafer's plant from Concordia, Argentina (No. 125); he notes that it grows in gravel and, being inconspicuous, is hard to find.

In 1920 Dr. C. Fiebrig, Director of the Botanical Garden at Asuncion, Paraguay, sent us a fine specimen; it is very cespitose, with 20 heads or more, forming a low mound nearly 10 cm. in diameter.

Illustrations: Knippel, Kakteen pl. 8; Monatsschr. Kakteenk. 9: 55, C, as *Echinocactus pumilus.*

Figure 223 shows a potted plant sent by Dr. Fiebrig in 1920 (No. 7).

4. Frailea schilinzkyana (Haage jr.).

> *Echinocactus schilinzkyanus* Haage jr. in Schumann, Monatsschr. Kakteenk. 7: 108. 1897.
> *Cactus schilinzkyanus* Kuntze, Deutsch. Bot. Monatsschr. 21: 193. 1903.

Simple or somewhat cespitose, usually globular, somewhat flattened above, about 3 cm. in diameter, but umbilicate at apex; ribs 10 to 13, but very indistinct, more or less spiraled, strongly tubercled; radial spines 12 to 14, 2 to 3 mm. long, more or less appressed and reflexed; central spine solitary, stouter than the radials; flowers small, often cleistogamous; fruit yellowish.

Type locality: Meadows near the River Paraguari, Paraguay.

Distribution: Paraguay; also Argentina, according to Spegazzini.

Nicholson (Dict. Gard. Suppl. 336. 1900) states that "this may be a *Mammillaria*."

Echinocactus schilinzkyanus grandiflorus Haage jr. (Monatsschr. Kakteenk. 8: 143. 1898) is only mentioned.

Illustrations: Cact. Journ. 1: 45; Schelle, Handb. Kakteenk. 183. f. 117; Schumann, Gesamtb. Kakteen Nachtr. f. 21; Monatsschr. Kakteenk. 9: 55, D, as *Echinocactus schilinzkyanus.*

5. Frailea cataphracta (Dams).

> *Echinocactus cataphractus* Dams, Monatsschr. Kakteenk. 14: 172. 1904.

Small, globose plants, 1 to 2 cm. in diameter, deeply umbilicate at apex, simple or sometimes proliferous, dull green; ribs low and broad, 10 to 15; tubercles flattened above, each with a purple lunate band near the margin; radial spines 5 to 9, straight, 1 to 2 mm. long, appressed, yellowish or white; central spines none; flowers evidently minute, but unknown; fruit small; seeds comparatively large, 2 mm. broad.

Type locality: Described from greenhouse plants.   Supposed to have come from Paraguay.

Distribution: Paraguay.

Seeds planted November 24, 1912, developed into plants which flowered in March 1915.   In the case of a plant which fruited in 1914 the seeds were carried away by ants and were found germinating in the sand in March 1915.

6. Frailea pygmaea (Spegazzini).

> *Echinocactus pygmaeus* Spegazzini, Anal. Mus. Nac. Buenos Aires III. 4: 497. 1905.

Simple or cespitose, half buried in the ground, globose, umbilicate at apex, with a turbinate base, 1 to 3 cm. in diameter, dull green; ribs 13 to 21, obtuse, low, divided by transverse depressions into tubercles; spines 6 to 9, white, setaceous, 1 to 4 mm. long, appressed; flowers from the apex of the plant, often cleistogamous, with dense, rose-colored pubescence without, 2 to 2.5 cm. long; inner perianth-segments lanceolate, acute, yellow; filaments and style white; stigma-lobes yellowish; seeds 2 mm. long, black, shining, with a large oblong hilum nearly as long as the body.

Type locality: Mountains about Montevideo, Uruguay.

Distribution: Uruguay and province of Entre Rios, Argentina.

Echinocactus pygmaeus phaeodiscus (Spegazzini, Anal. Mus. Nac. Buenos Aires III. 4: 498. 1905) is similar to the type, but has lower ribs, blackish areoles, and 6 to 12 spines from an areole; flowers said to be like those of the type.   It also suggests *Frailea cataphracta.*

We obtained a flower and seeds of this species from Dr. Spegazzini and have also examined a sketch made by him.

Illustration: Anal. Mus. Nac. Montevideo 5: pl. 17, as *Echinocactus pygmaeus.*

7. Frailea caespitosa (Spegazzini).

Echinocactus caespitosus Spegazzini, Anal. Mus. Nac. Buenos Aires III. 4: 495.   1905.

Simple or densely cespitose, half buried in the ground; separate plants small, turbinate to clavate or even oblong, 4 to 7 cm. long, 1.5 to 4.5 cm. in diameter, deeply umbilicate at apex; ribs 11 to 22, low, obtuse, 4 to 5 mm. broad, somewhat crenate; areoles orbicular to short-elliptic, 3 to 4 mm. apart; radial spines 9 to 11, setaceous, appressed, yellowish, very short, 3 to 6 mm. long; central spines 1 to 4, unequal, more or less curved, the longest one 10 to 15 mm. long; flowers small, 3.5 to 4 cm. long; inner perianth-segments yellow, lanceolate, acute; filaments yellow; style white; stigma-lobes purplish violet; axils of scales on ovary and flower-tube densely gray-tomentose and setose.

Type locality: Mountains near Montevideo.
Distribution: Uruguay.

8. Frailea knippeliana (Quehl).

Echinocactus knippelianus Quehl, Monatsschr. Kakteenk. 12: 9.   1902.

Simple, small, cylindric, 6 cm. high, 2 cm. in diameter; ribs 15, low, divided into tubercles; spines about 16, yellowish; flowers 2.5 cm. long, yellowish; wool in the axils of the flower-scales long, white; seed 1.5 cm. long.

Type locality: Paraguay.
Distribution: Paraguay.

We know this species from descriptions and illustrations and from a single flower sent by L. Quehl in 1912.

Quehl and Schumann place it near *Echinocactus gracillimus.*

Illustrations: Knippel, Kakteen pl. 8; Schelle,. Handb. Kakteenk. f. 118, as *Echinocactus knippelianus.*

26. MILA gen. nov.

Plants growing in small clumps, more or less cylindric, resembling in habit and texture some of the species of *Echinocereus;* ribs low, bearing closely set areoles with acicular spines; flowers small, campanulate, yellow, borne at the apex of the plant; scales on the ovary and flower-tube minute, bearing a few long white hairs in their axils; fruit small, globular, green, shining, nearly naked, at first juicy; seeds black, tuberculate, longer than broad, hilum large, subbasal, white.

Type species: *Mila caespitosa* Britton and Rose.

The generic name is an anagram of Lima, the city in Peru near which the plant is found. Only one species is known.

1. Mila caespitosa sp. nov.

Plants low, rarely as much as 15 cm. high, 2 to 3 cm. in diameter; ribs usually 10, 3 to 5 mm. high, the margins nearly straight; areoles at first densely brown-felted, 2 to 4 mm. apart; spines at first yellowish with brown tips, in age becoming brown throughout; radial spines 20 or more, usually about 10 mm. long; central spines several, the longer ones up to 3 cm. long; flowers about 1.5 cm. long, yellow but drying reddish; inner perianth-segments oblong; tube-proper very short; stamens shorter than the perianth-segments; style 8 mm. long; fruit 5 to 10 mm. in diameter; seed 1 mm. long.

Collected by J. N. Rose near Santa Clara, Peru, July 3, 1914 (No. 18555, type).   This plant is common near the mouth of the narrow valleys between the low hills bordering on the Remac Valley.   It does not extend into the main valley which is here devoid of all vegetation.   In early July Dr. Rose found it both in flower and fruit.   It is not closely related in flower, fruit, or appearance to any of the so-called species of *Echinocactus.*

Dr. and Mrs. Rose collected, July 9, 1914, at Matucana, much higher up the mountains than Santa Clara, two other specimens (Nos. 18652 and 18653); in the field these seemed different from each other as well as different from the Santa Clara plant, but as neither was in flower nor fruit and as they have not flowered in cultivation we are unable to identify them, but in any case they are doubtless associated with the above.

The plant much resembles in its habit some of the species of *Echinocereus*, but it has very different flowers and fruit. Rümpler describes a species of *Echinocereus* (*E. flavescens* Förster, Handb. Cact. ed. 2. 826. 1885) which from its name, description, and locality suggests that it might be this species. Rümpler credits the name to Otto who described a *Cereus flavescens* (Pfeiffer, Enum. Cact. 79. 1837), but without citing a locality for it. A little later the Wilkes's Expedition collected near Lima a plant which we believe is the same as ours and which Engelmann referred to *Mammillaria flavescens*, a quite different species. The use of the same specific name is simply a coincidence for the indications are that Engelmann did not know of Otto's name.

The following account by Dr. Asa Gray appeared in Wilkes, U. S. Exploring Expedition (13: 660. 1854).

"There are no specimens in the collection; but there is a good drawing, made from the living plant by the late Mr. Agate; from which Dr. Engelmann has drawn up the characters given above and the subjoined description and remarks.—'*Stems several* from the same very *thick root*, or proliferous at the base, 2½ to 3 inches high, an inch and a half or less in diameter, *ovoid-cylindrical; the setaceous straight prickles* half an inch in length, *brown*. *Flowers upright* from the summit of the stems, 14 lines long, 9 lines in diameter; the spreading sepals about 20 in number, *linear-oblong, obtuse*, yellowish; *petals* about the same number, *ovate-oblong, obtusish, yellow*. Style half an inch long; *stigmas 9, radiate*.'

"*M. flavescens* is one of the very few species coming from tropical South America. The descriptions which I find in different works agree tolerably well with our plant; though the stems are said to be proliferous towards the summit, the spines are generally lighter-colored and the yellow flowers appear in a ring around the top."

Inquiry regarding Agate's drawing, referred to above, at the Gray Herbarium and at the Missouri Botanical Garden, as well as an examination of Wilkes's manuscripts in the U. S. National Museum at Washington, D. C., have been without results.

Plate XXII, figure 2, shows the type plant collected by Dr. Rose in 1914 which fruited at the New York Botanical Garden.

27. SCLEROCACTUS gen. nov.

Usually simple but sometimes clustered, spiny cacti; ribs rather prominent, more or less undulate or tubercled; spine-clusters well developed, some of the central ones hooked, the others straight; flowers forming on the young areoles above and adjacent to the spine-cluster, subcampanulate, purplish; ovary oblong, bearing thin scattered scales, each with a tuft of short wool in its axils; fruit oblong to pyriform, nearly naked, dehiscing by a basal pore; seeds large, black, tuberculate; hilum lateral, large; embryo strongly curved; endosperm abundant.

Two species are known from the deserts of California, Utah, Colorado, Arizona, and southern Nevada, of which *Echinocactus polyancistrus* Engelmann and Bigelow is the type. The habit of the plants resembles somewhat that of *Ferocactus*, but the fruit is nearly naked and the scales bear small tufts of wool in their axils. The seeds, too, are not smooth or pitted as in *Ferocactus* but are tuberculate.

The generic name is from σκληρός hard, cruel, obstinate, and κάκτος cactus, referring to the formidable hooked spines which hold on in a most aggravating manner.

KEY TO SPECIES.

Style puberulent; flower 3 to 4 cm. long.. 1. *S. whipplei*
Style glabrous; flower 7 to 8 cm. long.. 2. *S. polyancistrus*

1. Sclerocactus whipplei (Engelmann and Bigelow).

Echinocactus whipplei Engelmann and Bigelow in Engelmann, Proc. Amer. Acad. **3**: 271.   1856.
Echinocactus whipplei spinosior Engelmann, Trans. St. Louis Acad. **2**: 199.   1863.

Usually single, but sometimes in small clusters, globose, 7.5 cm. in diameter or oblong and up to 15 cm. long; ribs 13 to 15, often spiraled, prominent, more or less tubercled; spines on seedlings all radials but on old plants both radials and centrals; radial spines 7 to 11, somewhat flattened, spreading or recurved, 12 to 18 mm. long, mostly white, but some black; central spines usually 4, the uppermost one flattened and straight, all or only one of the 3 lower ones hooked, usually brown or black, stouter than the radials; flowers from near the center of the plant, often abundant, short-campanulate, purplish to rose-colored, 3 to 4 cm. long; outer perianth-segments green with pale margins, broad, obtuse or acute; inner perianth-segments lavender, oblong, acuminate; tube-proper very short; filaments lavender; style reddish, puberulent throughout; fruit oblong, 1.5 cm. long, red, nearly naked; scales on the fruit small, hyaline, each bearing in its axil a small tuft of hair; seeds 3 to 3.4 mm. long, much larger at the upper end than at the lower; hilum large, lateral on the lower half of the seed; "embryo curved, about three-fourths around a rather copious albumen."

Type locality: On the Little Colorado in Arizona.

Distribution: Northern Arizona, southeastern Utah, and western Colorado.

According to Mr. M. E. Jones who knows this species very well it is found only on the high mesas growing in clayey soil.   It is an inconspicuous plant usually found singly under small bushes and is easily overlooked except when in flower.   The species has a wide range and shows considerable variation in number, color, and shape of spines, but we are not disposed to recognize the variety *spinosior* which seems to grade into the type.

This cactus is remarkable in having a puberulent style and is the only one we recall of the many species of cacti examined in which the style is puberulent throughout.

E. whipplei nanus (Monatsschr. Kakteenk. **10**: 119.   1900) we do not know; it is doubtless only a form.

Echinocactus glaucus Schumann (Gesamtb. Kakteen 438.   1898), of which *E. subglaucus* Rydberg (Fl. Rocky Mountains 580.   1917) is a change of name, based on Purpus's plant from Dry Creek, Mesa Grande, Colorado, probably belongs here, although we have not seen the type specimen.   Here we would refer specimens so named by Standley from northwestern Arizona.

Echinocactus pubispinus Engelmann (Trans. St. Louis Acad. **2**: 199.   1863), which came from Pleasant Valley near Salt Lake Desert, is known only from the type which was without flowers or fruit.   It is certainly related to this species if not identical with it.   It is described as having pubescent spines, a character also possessed by *E. whipplei* in the early stages of its growth.

Echinocactus spinosior Brandegee, referred to by Purpus (Monatsschr. Kakteenk. **10**: 97.   1900), although never published, doubtless refers to the variety of this species with the same name.

Illustrations: Pac. R. Rep. **4**: pl. 1; Cact. Journ. **1**: pl. V, in part; Stand. Cycl. Hort. Bailey **2**: f. 1371; Möllers Deutsche Gärt. Zeit. **25**: 474. f. 6, No. 2; Meehans' Monthly **9**: pl. 3, as *Echinocactus whipplei*; Gartenwelt **1**: 89, as *Echinocactus glaucus*.

Plate XVI, figure 2, shows the plant collected by Dr. P. A. Rydberg in 1911 at Moab, Utah, which flowered in the New York Botanical Garden in April 1912.

2. Sclerocactus polyancistrus (Engelmann and Bigelow).

Echinocactus polyancistrus Engelmann and Bigelow in Engelmann, Proc. Amer. Acad. **3**: 272.   1856.

Simple, globular to oblong, 1 to 4 dm. high; ribs 13 to 17, 1 to 1.5 cm. high, obtuse, strongly undulate; areoles 1 to 1.5 cm. apart; spines about 20; radial spines acicular, white, 1 to 2.5 cm. long; central spines several, rather unequal, up to 12.5 cm. long, the upper ones erect, white, flattened, the others brown, spreading, terete and often hooked; flowers magenta-colored, nearly 8 cm. long, and perhaps as broad; inner perianth-segments oblong, about 4 cm. long; throat of flower broad, covered with stamens; tube-proper very short, 1 to 2 mm. long; style longer than the stamens, stout, glabrous; stigma-lobes about 8, 4 mm. long; scales on the base of the flower-tube large, obtuse, on the ovary scattered, minute, 1 to 2 mm. long; fruit at first bright magenta, with fleshy walls, but becoming dry and thin-walled, oblong to pyriform, almost destitute of scales, 3.5 to 4 cm. long, dehiscing by a large basal pore; seeds large, 4 mm. long, black, tuberculate; hilum sublateral, large.

Type locality: At the head of the Mojave River, California.

Distribution: Deserts of California and Nevada; reported from western Arizona.

Engelmann in his original description states that the flowers are yellow, as Schumann also states, while Coulter describes them as yellow or red. In our description based on several collections we describe the flowers as magenta. In order to clear up the matter we wrote to Mr. E. C. Rost, a very keen observer, who had recently had this plant under observation; he replied under date of September 2, 1921, as follows:

"In regard to the flowers of *E. polyancistrus*, I will say that I have seen *many* of these plants in blossom. My own color notes record the flower as having petals of a pure transparent 'madder-lake'; stamens bright yellow; pistil crimson. The mature buds are the shade of burnt sienna, or

FIG. 224.—Sclerocactus polyancistrus.

brownish red. These flowers, however, could be described as both purple and red, for on their first day of bloom they are 'purple' (madder-lake), while on the second day—and even more so on the third day—they would popularly be termed 'red.' A yellow *polyancistrus* I have never seen, although the bright-yellow stamens sometimes protrude from the blossom in such a manner that a person—at a distance—might think he saw a yellow flower.

"I have never found the *polyancistrus* growing in a dense colony, but have always seen the plants widely scattered, then for miles not a plant, and again a few growing here and there, rather far apart. I have found them in various different localities."

This species, while found over a large area, is said never to be abundant and is found chiefly on the mesas.

Illustrations: Pac. R. Rep. 4: pl. 2, f. 1, 2; Cact. Journ. 1: pl. V, in part; Monatsschr. Kakteenk. 20: 131; 31: 21, as *Echinocactus polyancistrus*.

Plate XXIII, figure 3, shows a fruit and figure 4 a seed from a specimen collected by Ivar Tidestrom in Goldfield, Nevada, June 1919. Figure 224 shows the plant from which the fruit and seed were obtained.

28. UTAHIA gen. nov.

A small globose cactus, prominently ribbed, the ribs tubercled, the areoles felted and bearing several subulate spines; flowers small, nearly rotate, yellow, borne at the areoles of the upper part of the plant; ovary and perianth-tube densely covered with dry imbricated fimbriate-lacerate scales; perianth-segments short, narrow.

Type species: *Echinocactus sileri* Engelmann.
Named with reference to its type locality in the state of Utah. A monotypic genus.

1. Utahia sileri (Engelmann).

Echinocactus sileri Engelmann in Coulter, Contr. U. S. Nat. Herb. **3**: 376. 1896.

Globose, 10 cm. in diameter; ribs 13 to 16, prominent, densely crowded, with short rhombic-angled tubercles; radial spines 11 to 15, stiff, white; central spines 3 or 4, black with pale base, most of them curved upward, 18 mm. long, the upper one slightly longer, the lower ones sometimes stouter and porrect; flowers scarcely 2.5 cm. long; fruit unknown.

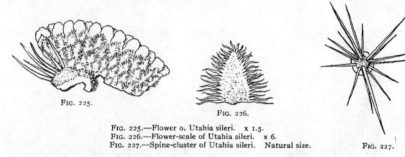

FIG. 225.
FIG. 226.
FIG. 225.—Flower o. Utahia sileri. x 1.5.
FIG. 226.—Flower-scale of Utahia sileri. x 6.
FIG. 227.—Spine-cluster of Utahia sileri. Natural size.
FIG. 227.

Type locality: Cottonwood Springs and Pipe Springs, southern Utah.
Distribution: Southern Utah.

Through the kindness of Dr. J. M. Greenman, we have studied the type specimen of this rare plant, preserved in the herbarium of the Missouri Botanical Garden. Flowers, of some other cactus have been erroneously identified as of this species in other collections, but so far as we are aware this plant is known only from the type specimen.

Illustration: Schumann, Gesamtb. Kakteen f. 61, A, as *Echinocactus sileri*.

Figure 225 shows the outside of a flower cut on one side and spread out; it also shows the origin of the flower at the young spine-areole; figure 226 shows a flower-scale, enlarged; figure 227 shows a spine-cluster.

Subtribe 5. CACTANAE.

One-jointed plants, usually stout, globose to oblong, either solitary or cespitose, terrestrial; ribs usually straight, their areoles nearly or always spine-bearing; the flower-bearing areoles forming a terminal cephalium composed of a central woody core surrounded by a dense mass of long wool, bristles or both, often elongated; flowers regular, salverform or funnelform, opening in the afternoon or at night; fruit a small naked berry; seeds small.

We recognize two genera, which are not very closely related.

KEY TO GENERA.

Flowers large, white or rose, night-blooming, the limb of many segments...................... 1. *Discocactus*
Flowers small, rose or pinkish, opening in the late afternoon, the limb of few or several segments.. 2. *Cactus*

1. DISCOCACTUS Pfeiffer, Allg. Gartenz. 5: 241. 1837.

Plants rather small, globose or flattened, ribbed; ribs rather low, tubercled; spines borne at the areoles in clusters, more or less curved; flowers from the center of the plant, appearing from a cephalium similar to, but usually not so prominent as, that of *Cactus* (*Melocactus*); flowers rather large for the plants, opening at night, with a definite tube, the limb broad, composed of many segments, usually white or pinkish; fruit small, naked; seeds black, roughened.

FIG. 228.—Discocactus subnudus.   FIG. 229.—Discocactus alteolens.

Discocactus was made a subgenus of *Echinocactus* by Schumann. It is a valid genus, however, confined to eastern South America, although one of the species was originally described, in error, as coming from the West Indies. It is characterized by its plant body, by its cephalium, by its naked fruits, all these suggesting *Cactus*, and by its flowers which open at night and become limp by the next morning.

Three species were described by Pfeiffer, of which the first, *Discocactus insignis*, is the type, but these were afterwards combined into *D. placentiformis*. Three other described species we believe belong here, only one heretofore referred to *Discocactus*, the other two having been described under *Echinocactus* and *Malacocarpus* respectively. Three more are here added.

The generic name is from δίσκος disk, and κάκτος, referring to the flattened, disk-like shape of the plant body.

KEY TO SPECIES.

Plant-body spineless or nearly so.. 1. *D. subnudus*
Plant-body very spiny.
  Cephalium composed of long wool and many erect bristles.
    Ribs 12 to 16; perianth-segments acute.. 2. *D. hartmannii*
    Ribs 9 to 11; perianth-segments so far as known obtuse.
      Ribs made up of large tubercles.. 3. *D. heptacanthus*
      Ribs made up of indistinct tubercles.. 4. *D. alteolens*
  Cephalium with few bristles or none.
    Spines acicular.. 5. *D. zehntneri*
    Spines stout.
      Spines flattened.. 6. *D. placentiformis*
      Spines terete... 7. *D. bahiensis*

1. Discocactus subnudus sp. nov.

Plant simple; ribs few (perhaps 8 or 9), somewhat flattened, divided into tubercles, spineless or nearly so; cephalium small, containing many erect bristles; flowers large.

This species, according to Dr. Albert Löfgren, is to be found in the sands along the coast at Bahia, Brazil, but, unfortunately, Dr. Rose did not find it while he was there in 1915, although he searched for it repeatedly. Dr. Löfgren has furnished us, however, a very characteristic photograph from which the above description is drawn.

Figure 228 is from a photograph taken by Dr. Löfgren.

Fig. 230.—Discocactus hartmannii.   Fig. 231.—Cactus broadwayi.

2. Discocactus hartmannii (Schumann).

Echinocactus hartmannii Schumann, Monatsschr. Kakteenk. **10**: 170. 1900.

Plants simple, broader than high, green, shining, crowned by a brown bristly cephalium; ribs 12 to 16, broad at base, obtuse, divided transversely into large tubercles; spines yellow, subulate; radial spines 6 to 12, curved backward and appressed; central spines solitary; flower-tube funnelform, bearing a few spreading scales; outer perianth-segments green; inner perianth-segments lanceolate to nearly oblong, white; fruit yellow, thin-skinned; seeds globular, 2 mm. in diameter.

Type Locality: Campos am Capivary, Paraguay.
Distribution: Paraguay.
The plant is known to us from description and illustrations only.
Illustrations: Blühende Kakteen **2**: pl. 69; Schumann, Gesamtb. Kakteen Nachtr. f. 12, 13; Monatsschr. Kakteenk. **10**: 171; **11**: 184, as *Echinocactus hartmannii*.

Figure 230 is a reproduction of the first illustration cited above.

3. Discocactus heptacanthus (Rodrigues).

Malacocarpus heptacanthus Rodrigues, Pl. Mattogr. 29. 1898.

Globose or slightly depressed, usually solitary, 8 to 10 cm. in diameter; ribs 10 or 11, very broad, broken up into a few large, rounded tubercles; areoles circular, at first tomentose; spines usually 7, all radial, stout, recurved; cephalium small but definite, white, containing many erect bristles; flowers and fruit unknown.

Type locality: Serra da Chapada, near Cuyabá, Matto-Grosso, Brazil.

Distribution: Known only from the type locality.

Schumann in his Nachträge referred this to *Echinocactus alteolens*, but it is certainly different from that species. We have not seen the type of this species, which has not been preserved; but we have seen Mr. F. C. Hoehne's specimen, which came from Rodrigues's locality, Cuyabá, Matto-Grosso, and we have a photograph of it.

Illustration: Rodrigues, Pl. Mattogr. pl. 11, as *Malacocarpus heptacanthus*.

Figure 232 is a copy of the illustration cited above.

4. Discocactus alteolens Lemaire in Dietrich, Allg. Gartenz. **14**: 202. 1846.

Discocactus tricornis Monville in Pfeiffer, Abbild. Beschr. Cact. **2**: pl. 28. 1846–1850.
Echinocactus alteolens Schumann in Martius, Fl. Bras. **4²**: 246. 1890.
Cactus alteolens Kuntze, Deutsch. Bot. Monatsschr. **21**: 173. 1903.

Plant solitary, broader than high, dull gray, crowned by a broad cephalium; ribs 9 or 10, broad at base, low; radial spines 5 or 6, the 3 upper ones very short, ascending; central spines none or rarely

FIG. 232.—Discocactus heptacanthus.   FIG. 233.—Discocactus placentiformis.

solitary and porrect; flowers salverform; tube slender, bearing slender scales with naked axils; inner perianth-segments white, numerous, obtuse; ovary naked, glabrous.

Type locality: Not cited.

Distribution: Doubtless eastern central Brazil.

This species is known to us from description and illustration only.

Echinocactus tricornis Monville is given by A. Dietrich (Allg. Gartenz. **14**: 203. 1846) as a synonym.

Illustration: Pfeiffer, Abbild. Beschr. Cact. **2**: pl. 28, as *Discocactus tricornis*.

Figure 229 is a reproduction of the illustration here cited.

5. Discocactus zehntneri sp. nov.

Small globose plants, 5 to 7 cm. in diameter, entirely covered by the numerous interlocking spines; spines 12 to 14 in a cluster, gray to nearly white at least when old, slender, acicular, curved backward, 1.5 to 2.5 cm. long; central spine usually solitary, similar to the radials; cephalium small, made up of long soft wool and few bristles if any; flowers about 3 cm. long, 4 cm. broad when fully expanded; inner perianth-segments numerous, white, acute; fruit a small naked red clavate berry, 2.5 cm. long; seeds globular, tuberculate.

The material upon which this species is based was obtained from Dr. Leo Zehntner, who procured it at Sentocé, Bahia, Brazil. A box of living material, fruit, seeds, and several photographs of flowering plants were contributed by him. It is a very distinct species.

6. Discocactus placentiformis (Lehmann) Schumann in Engler and Prantl, Pflanzenfam. 3^{6a}: 190. 1894.

> *Cactus placentiformis* Lehmann, Delect. Sem. Hamb. 1826.
> *Melocactus besleri* Link and Otto, Verh. Ver. Beförd. Gartenb. **3:** 420. 1827.
> *Melocactus placentiformis* De Candolle, Prodr. **3:** 460. 1828.
> *Discocactus insignis* Pfeiffer, Allg. Gartenz. **5:** 241. 1837.
> *Discocactus lehmannii* Pfeiffer, Nov. Act. Nat. Cur. **19:** 120. 1839.
> *Discocactus linkii* Pfeiffer, Nov. Act. Nat. Cur. **19:** 120. 1839.
> *Echinocactus placentiformis* Schumann, Fl. Bras. 4^2: 246. 1890.
> *Discocactus besleri* Weber, Dict. Hort. Bois 450. 1896.

Fig. 234.—Discocactus placentiformis.

Plant broad and low, solitary, blue-green; ribs 10 to 14, broad and low; areoles 6 or 7 on each rib; radial spines dark, flattened (?), 6 or 7, stout, more or less recurved; central spine usually wanting, sometimes solitary and porrect; flowers large; outer perianth-segments rose-colored; inner perianth-segments white, acute; fruit white, globular, juicy.

Type locality: Brazil, but no definite locality cited.

Distribution: Brazil.

This species was figured by Besler in 1613 and Besler's figure was afterwards made the type of *Melocactus besleri.*

Mammillaria besleri is credited by the Index Kewensis to Link and Otto in Rümpler (Förster, Handb. Cact. ed. 2. 1020. 1885) which is the index of that work where the name *Melocactus besleri* is intended.

We have referred to this species the plant collected by Dr. Rose at Joazeiro, Bahia, in 1915 (No. 19764), although we are not quite certain of this reference. Dr. Rose found the plant on the dry mesa east of the town. It is deep-seated in the hard ground and appears only a little above the surface. Old plants produce a large white-woolly cephalium from near the center of which the flowers appear. Birds are said to be very fond of the fruit of this plant.

Illustrations: Nov. Act. Nat. Cur. 16[1]: pl. 16, as *Cactus placentiformis;* Nov. Act. Nat. Cur. 18: Suppl. 1. pl. 4, f. 2; Verh. Kon. Akad. Wetensch. II. 5[3]: pl. 1, f. 2, 2a, 2b; Verh. Ver. Beförd. Gartenb. 3: pl. 21, as *Melocactus besleri;* Krook, Handb. Cact. 63; Nov. Act. Nat. Cur. 19: Suppl. 1. pl. 15; Palmer, Cult. Cact. 127; Förster, Handb. Cact. ed. 2. 449. f. 51; Pfeiffer, Abbild. Beschr. Cact. 2: pl. 1, as *Discocactus insignis;* Besler, Hort. Eystett. 4. Ord. f. 1, as *Melocactus;* Engler and Prantl, Pflanzenfam. 3[6a]: f. 64.

Figure 234 is a reproduction of the first illustration cited above; figure 233 is from a photograph, taken by Paul G. Russell, of Dr. Rose's plant at Joazeiro, Brazil.

7. Discocactus bahiensis sp. nov.

Small, about 6 cm. in diameter, somewhat flattened; ribs about 10, but nearly hidden under the mass of spines; cephalium prominent, made up of a mass of white wool, almost devoid of bristles; spines 7 to 9, slightly flattened, stout, somewhat curved backward, rose-colored, 1.5 to 3 cm. long; flowers 4 to 5 cm. long with a slender tube; perianth-segments oblong, white, tinged with yellow; fruit a small naked berry; seeds globular, tuberculate.

Obtained by Dr. J. N. Rose through Dr. Leo Zehntner near Joazeiro, Bahia, Brazil (Rose's No. 19783, type, and No. 19742).

Somewhat similar to *D. zehntneri* in size, but with different spines, a more prominent cephalium, and coarser tuberculations on the seeds.

Plate xxiv, figure 4, shows the type which flowered in the New York Botanical Garden in April 1916.

2. CACTUS Linnaeus, Sp. Pl. 466. 1753.

Melocactus Link and Otto, Verh. Ver. Beförd. Gartenb. 3: 417. 1827.

Plants solitary or clustered, globular to short-cylindric, sometimes depressed, at least when young, 1-jointed, bearing clusters of spines on the ribs; ribs 9 to 20, mostly straight, rarely spiraled in some individuals; inflorescence a compact mass of hairs and bristles forming a cephalium borne at the top of the plant, this often very large and elongated; flowers small, pinkish, appearing in the mid-afternoon from the top of the cephalium, tubular-salverform, the few perianth-segments spreading; stamens attached near the top of the slender flower-tube; style slender; stigma-lobes linear, few; fruit clavate, naked, red or rarely white; seeds black.

Type species: *Cactus melocactus* Linnaeus.

The terminal cephalium is a woody axis, the wood-elements arranged in interlocking spirals, showing a characteristic, complicated pattern, in cross-section, well illustrated in that of *Cactus intortus;* on this axis the flower-bearing areoles are densely aggregated, spineless, but woolly and often bristly.

The fruits in this genus are very much alike in shape, varying from white to scarlet, always smooth, naked, and of an agreeable acid taste. They are often eaten in the West Indies being slightly juicy when first ripe.

About 224 names occur under *Melocactus* and 282 under *Cactus;* many of the latter, however, are referable to other genera. We recognize 18 species in the genus *Cactus* as here circumscribed.

Plants of this genus were among the earliest cacti known to Europeans, by whom they were first called *Echinomelocactus*. Tournefort shortened the name to *Melocactus* and Linnaeus again shortened it to *Cactus*. Under this name these plants generally passed until 1827 when Link and Otto restored the name *Melocactus*.

At the time that Linnaeus wrote his Species Plantarum only one species of this genus, as we now understand it, was known, although Linnaeus included with it all the other members of the family; this he called *Cactus melocactus*. Under this name, however, were included several species. For instance, we know that the plants described and figured by Bradley came from Nevis and St. Christopher and are different from the Jamaican species. Philip Miller in his Gardeners' Dictionary, published in 1768, described five species in the genus *Cactus*, but only the first two (*C. melocactus* and *C. intortus*) belong to the genus as now delimited.

These plants have various common names, but the ones most generally known among English speaking people are Turk's cap or Turk's head. In Brazil they are called cabeça de frade; in Cuba and Porto Rico, melones.

The name *Cactus* is of Greek origin, meaning thistle, with reference to the spiny armament, but, as here used, comes directly from *Melocactus*, that name having been shortened by Linnaeus.

The species of *Cactus* have a wide distribution; none of them reaches the continental United States, but one inhabits Porto Rico and the Virgin Islands. Hemsley (Biol. Centr. Amer. Bot. 1: 502) lists four species of *Melocactus* (*M. curvispinus*, *M. delessertianus*, *M. ferox*, and *M. mammillariaeformis*) from Mexico, some of which may belong to this genus, but the last one, however, must be referred elsewhere. We have seen only a barren Mexican plant, but Dr. C. A. Purpus has reported a species from Tehuantepec. *M. salvador* Murillo, from near Jalapa, was published a few years ago; photographs of this plant are at Kew. One species (*Cactus maxonii*) is known from Guatemala; we have this in cultivation. One species (*Cactus ruestii*) is known from Honduras; this we have seen. No species have been reported from the other Central American countries, including Panama.

Three species have been reported from Colombia (*M. amoenus*, *M. rubens*, and *M. obtusipetalus*). *M. obtusipetalus* is reported from near Bogotá, which is a very unusual habitat. We have seen two species from Colombia, one obtained by Mr. Pittier in the Cauca Valley and the other brought from Santa Marta by Mr. Curran. No species have been found in Ecuador or Chile, but one is known from the foothills of central Peru. No species are known from Bolivia, Argentina, Uruguay, or Paraguay. In central and northern Brazil, especially along the coast and in the desert regions of the interior, several species occur. Our collections indicate four species, although some 14 names, based on Brazilian material, have been published. We have seen no specimens from any of the Guianas, although a species has been reported recently from Dutch Guiana and is shown in one of our illustrations (fig. 245). Four species have been reported from Venezuela (*M. caesius*, *M. griseus*, *M. lehmannii*, and *M. cephalenoplus*). We have seen one or two species from Venezuela.

These plants are very common in the West Indies. We have collected them ourselves on the four larger islands and many of the smaller ones. Two species have been reported from Cuba (*Cactus harlowii* and *Melocactus havannensis*), one from Jamaica (*M. communis* or *C. melocactus*), and several species from Santo Domingo, but we know only one from Porto Rico. We also know one from St. Thomas, St. Croix, St. Christopher, and Antigua respectively, while Boldingh reports one species on St. Eustatius, Saba, and St. Martin respectively. We have seen a plant from Tobago but this is not referable to any species known to us. The genus is found also on Curaçao, Bonaire, and Aruba, and many names have been given to what seems to represent a single species on those islands.

Because of the interest in the species they have been much planted in tropical America.

KEY TO SPECIES.

A. Radial spines acicular, appressed, very unlike the central ones; species of Curaçao,
    Aruba, and Bonaire.. 1. *C. macracanthus*
AA. Radial spines subulate, spreading, or diverging.
    B. Flowers 3 to 4 cm. long; fruit 5 to 6 cm. long; species of Jamaica.............. 2. *C. melocactus*
    BB. Flowers 2.5 cm. long or less; fruit 1 to 2 cm. long.
      C. Spines all very stout; species of Hispaniola............................. 3. *C. lemairei*
      CC. Spines relatively slender.
        D. Spines flexible, greatly elongated; species of Brazil..................... 4. *C. oreas*
        DD. Spines stiff, not greatly elongated.
          E. Central American and Mexican species.
            Spines up to 3 cm. long... 5. *C. ruestii*
            Spines shorter.
              Central spines about as long as the radial ones..................... 6. *C. maxonii*
              Central spines twice as long as the radials or longer................ 7. *C. salvador*
          EE. West Indian species.
            Radial spines very slender, widely spreading, slightly curved.......... 8. *C. broadwayi*
            Radial spines stouter, divergent, straight or nearly so.
              Ribs thick; plants up to 4 dm. in diameter....................... 9. *C. intortus*
              Ribs thinner; plants 1 to 1.5 dm. in diameter.................... 10. *C. harlowii*
          EEE. South American species.
            Spines white, at least when young.
              Spines curved; species of Colombia.
                Flowers 1 cm. wide or less................................ 11. *C. amoenus*
                Flowers 2.5 cm. wide....................................... 12. *C. obtusipetalus*
              Spines straight; species of Venezuela and Patos Island, Trinidad...... 13. *C. caesius*
            Spines gray, brown or reddish.
              Spines dark brown; species of Peru........................... 14. *C. townsendii*
              Spines gray to reddish; species of Brazil.
                Spines straight, up to 3 cm. long.......................... 15. *C. bahiensis*
              Spines curved, shorter.
                Berry white or pale pink................................. 16. *C. melocactoides*
              Berry red to carmine.
                  Spines very stiff, curved backward........................ 17. *C. zehntneri*
                  Spines slender, curved upward........................... 18. *C. neryi*

1. Cactus macracanthus* Salm-Dyck, Observ. Bot. 1: 3. 1820.

    Cactus pyramidalis Salm-Dyck, Observ. Bot. 1: 4. 1820.
    Melocactus macracanthus Link and Otto, Verh. Ver. Beförd. Gartenb. 3: 418. 1827.
    Melocactus pyramidalis Link and Otto, Verh. Ver. Beförd. Gartenb. 3: 419. 1827.
    Echinocactus salmianus Link and Otto, Verh. Ver. Beförd. Gartenb. 3: 423. 1827.
    Melocactus salmianus Link and Otto, Verh. Ver. Beförd. Gartenb. 3: pl. 13. 1827.
    Melocactus spatangus Pfeiffer, Enum. Cact. 45. 1837.
    Melocactus lehmanni Miquel, Linnaea 11: 642. 1837.
    Melocactus zuccarinii Miquel, Linnaea 11: 645. 1837.
    Melocactus microcephalus Miquel, Nov. Act. Nat. Cur. 18: Suppl. 1. 156. 1841.
    Melocactus pyramidalis carneus Miquel, Nov. Act. Nat. Cur. 18: Suppl. 1. 166. 1841.
    Melocactus parvispinus Suringar, Versl. Med. Akad. Wetensch. III. 2: 183. 1886. Not De Candolle, 1828.
    Melocactus koolwijckianus Suringar, Versl. Med. Akad. Wetensch. III. 2: 184. 1886.
    Melocactus rubellus Suringar, Versl. Med. Akad. Wetensch. III. 2: 184. 1886.
    Melocactus stramineus Suringar, Versl. Med. Akad. Wetensch. III. 2: 185, 186. 1886.
    Melocactus (rubellus) ferox Suringar, Versl. Med. Akad. Wetensch. III. 2: 185. 1886. Not Pfeiffer, 1837.
    Melocactus (rubellus) hexacanthus Suringar, Versl. Med. Akad. Wetensch. III. 2: 185. 1886.
    Melocactus (stramineus ?) trichacanthus Suringar, Versl. Med. Akad. Wetensch. III. 2: 186. 1886.
    Melocactus reversus Suringar, Versl. Med. Akad. Wetensch. III. 2: 187. 1886.
    Melocactus reliusculus Suringar, Versl. Med. Akad. Wetensch. III. 2: 187. 1886.
    Melocactus angusticostatus Suringar, Versl. Med. Akad. Wetensch. Amst. III. 2: 188. 1886.
    Melocactus reliusculus angusticostatus Suringar, Versl. Med. Akad. Wetensch. III. 2: 188. 1886.
    Melocactus approximatus Suringar, Versl. Med. Akad. Wetensch. III. 2: 189. 1886.
    Melocactus evertszianus Suringar, Versl. Med. Akad. Wetensch. III. 2: 190. 1886.
    Melocactus patens Suringar, Versl. Med. Akad. Wetensch. III. 2: 190. 1886.
    Melocactus cornutus Suringar, Versl. Med. Akad. Wetensch. III. 2: 191. 1886.
    Melocactus intermedius Suringar, Versl. Med. Akad. Wetensch. III. 2: 192. 1886.
    Melocactus pusillus Suringar, Versl. Med. Akad. Wetensch. III. 2: 192. 1886.
    Melocactus spatanginus Suringar, Versl. Med. Akad. Wetensch. III. 2: 193. 1886.
    Melocactus koolwijkianus adustus Suringar, Versl. Med. Akad. Wetensch. III. 6: 438. 1889.
    Melocactus argenteus Suringar, Versl. Med. Akad. Wetensch. III. 6: 439. 1889.
    Melocactus argenteus tenuispinus Suringar, Versl. Med. Akad. Wetensch. III. 6: 439. 1889.
    Melocactus roseus Suringar, Versl. Med. Akad. Wetensch. III. 6: 439. 1889.
    Melocactus limis Suringar, Versl. Med. Akad. Wetensch. III. 6: 440. 1889.
    Melocactus obliquus Suringar, Versl. Med. Akad. Wetensch. III. 6: 440. 1889.
    Melocactus flexus Suringar, Versl. Med. Akad. Wetensch. III. 6: 441. 1889.
    Melocactus incurvus Suringar, Versl. Med. Akad. Wetensch. III. 6: 441. 1889.
    Melocactus nanus Suringar, Versl. Med. Akad. Wetensch. III. 6: 441. 1889.
    Melocactus rudis Suringar, Versl. Med. Akad. Wetensch. III. 6: 442. 1889.

*This name was originally spelled *Cactus macrocanthos* by Salm-Dyck.

Melocactus capillaris Suringar, Versl. Med. Akad. Wetensch. III. 6: 442. 1889.
Melocactus extensus Suringar, Versl. Med. Akad. Wetensch. III. 6: 442. 1889.
Melocactus martialis Suringar, Versl. Med. Akad. Wetensch. III. 6: 443. 1889.
Melocactus compactus Suringar, Versl. Med. Akad. Wetensch. III. 6: 444. 1889.
Melocactus ferus Suringar, Versl. Med. Akad. Wetensch. III. 6: 444. 1889.
Melocactus (radiatus) contortus Suringar, Versl. Med. Akad. Wetensch. III. 6: 445. 1889.
Melocactus pentacanthus Suringar, Versl. Med. Akad. Wetensch. III. 6: 445. 1889.
Melocactus radiatus Suringar, Versl. Med. Akad. Wetensch. III. 6: 446. 1889.
Melocactus albispinus Suringar, Versl. Med. Akad. Wetensch. III. 6: 446. 1889.
Melocactus eburneus Suringar, Versl. Med. Akad. Wetensch. III. 6: 447. 1889.
Melocactus euryacanthus Suringar, Versl. Med. Akad. Wetensch. III. 6: 447. 1889.
Melocactus baarsianus Suringar, Versl. Med. Akad. Wetensch. III. 6: 448. 1889.
Melocactus uncinatus Suringar, Versl. Med. Akad. Wetensch. III. 6: 450. 1889.
Melocactus arcuatus Suringar, Versl. Med. Akad. Wetensch. III. 6: 450. 1889.
Melocactus elongatus Suringar, Versl. Med. Akad. Wetensch. III. 6: 451. 1889.
Melocactus (stellatus ?) sordidus Suringar, Versl. Med. Akad. Wetensch. III. 6: 451. 1889.
Melocactus stellatus Suringar, Versl. Med. Akad. Wetensch. III. 6: 452. 1889.
Melocactus obovatus Suringar, Versl. Med. Akad. Wetensch. III. 6: 453. 1889.
Melocactus (stellatus) flavispinus Suringar, Versl. Med. Akad. Wetensch. III. 6: 453. 1889.
Melocactus flexilis Suringar, Versl. Med. Akad. Wetensch. III. 6: 453. 1889.
Melocactus reticulatus Suringar, Versl. Med. Akad. Wetensch. III. 6: 453. 1889.
Melocactus (stellatus) inflatus Suringar, Versl. Med. Akad. Wetensch. III. 6: 454. 1889.
Melocactus (stellatus) dilatatus Suringar, Versl. Med. Akad. Wetensch. III. 6: 454. 1889.
Melocactus leucacanthus Suringar, Versl. Med. Akad. Wetensch. III. 6: 454. 1889.
Melocactus trachycephalus Suringar, Versl. Med. Akad. Wetensch. III. 6: 455. 1889.
Melocactus trigonus Suringar, Versl. Med. Akad. Wetensch. III. 6: 456. 1889.
Melocactus ovatus Suringar, Versl. Med. Akad. Wetensch. III. 6: 456. 1889.
Melocactus flammeus Suringar, Versl. Med. Akad. Wetensch. III. 6: 457. 1889.
Melocactus pulvinosus Suringar, Versl. Med. Akad. Wetensch. III. 6: 458. 1889.
Melocactus armatus Suringar, Versl. Med. Akad. Wetensch. III. 6: 458. 1889.
Melocactus salmianus adauctus Suringar, Versl. Med. Akad. Wetensch. III. 6: 460. 1889.
Melocactus salmianus contractus Suringar, Versl. Med. Akad. Wetensch. III. 6: 461. 1889.
Melocactus salmianus aciculosus Suringar, Versl. Kon. Akad. Wetensch. 6: 187. 1897.
Melocactus rotifer Suringar, Versl. Kon. Akad. Wetensch. Amst. 6: 188. 1897.
Melocactus exsertus Suringar, Versl. Kon. Akad. Wetensch. 6: 189. 1897.
Melocactus grollianus Suringar, Versl. Kon. Akad. Wetensch. 6: 190. 1897.
Melocactus pyramidalis pumilus Suringar, Versl. Kon. Akad. Wetensch. 6: 191. 1897.
Melocactus rotatus Suringar, Versl. Kon. Akad. Wetensch. 6: 191. 1897.
Melocactus bargei Suringar, Verh. Kon. Akad. Wetensch. Amst. II. 8: 9. 1901.
Melocactus pinguis Suringar, Verh. Kon. Akad. Wetensch. Amst. II. 8: 11. 1901.
Melocactus intermedius laticostatus Suringar, Verh. Kon. Akad. Wetensch. Amst. II. 8: 15. 1901.
Melocactus intermedius tenuispinus Suringar, Verh. Kon. Akad. Wetensch. Amst. II. 8: 17. 1901.
Melocactus firmus Suringar, Verh. Kon. Akad. Wetensch. Amst. II. 8: 17. 1901.
Melocactus inversus Suringar, Verh. Kon. Akad. Wetensch. Amst. II. 8: 19. 1901.
Melocactus appropinquatus Suringar, Verh. Kon. Akad. Wetensch. Amst. II. 8: 19. 1901.
Melocactus salmianus spectabilis Suringar, Verh. Kon. Akad. Wetensch. Amst. II. 8: 22. 1901.
Melocactus aciculos J. V. Suringar, Verh. Kon. Akad. Wetensch. Amst. II. 8: 23. 1901.
Melocactus gilvispinus Suringar, Verh. Kon. Akad. Wetensch. Amst. II. 8: 28. 1901.
Melocactus gilvispinus planispinus Suringar, Verh. Kon. Akad. Wetensch. Amst. II. 8: 30. 1901.
Melocactus aciculosus adauctus J. V. Suringar, Verh. Kon. Akad. Wetensch. Amst. II. 8: 33. 1901.
Melocactus pyramidalis costis-angustioribus J. V. Suringar, Verh. Kon. Akad. Wetensch. Amst. II. 8: 35. 1901.
Melocactus buysianus Suringar, Verh. Kon. Akad. Wetensch. Amst. II. 8: 38. 1901.
Melocactus microcarpus Suringar in J. V. Suringar, Verh. Kon. Akad. Wetensch. Amst. II. 16³: 3. 1910.
Melocactus trigonaster Suringar in J. V. Suringar, Verh. Kon. Akad. Wetensch. Amst. II. 16³: 4. 1910.
Melocactus pyramidalis compressus Suringar in J. V. Suringar, Verh. Kon. Akad. Wetensch. Amst. II. 16³: 4. 1910.
Melocactus cordatus J. V. Suringar, Verh. Kon. Akad. Wetensch. Amst. II. 16³: 5. 1910.
Melocactus tenuissimus J. V. Suringar, Verh. Kon. Akad. Wetensch. Amst. II. 16³: 6. 1910.
Melocactus pinguis areolosus J. V. Suringar, Verh. Kon. Akad. Wetensch. Amst. II. 16³: 7. 1910.
Melocactus rotula angusticostatus J. V. Suringar, Verh. Kon. Akad. Wetensch. Amst. II. 16³: 8. 1910.
Melocactus intermedius rotundatus J. V. Suringar, Verh. Kon. Akad. Wetensch. Amst. II. 16³: 9. 1910.
Melocactus rotula validispinus J. V. Suringar, Verh. Kon. Akad. Wetensch. Amst. II. 16³: 10. 1910.
Melocactus grandis J. V. Suringar, Verh. Kon. Akad. Wetensch. Amst. II. 16: 11. 1910.
Melocactus grandispinus J. V. Suringar, Verh. Kon. Akad. Wetensch. Amst. II. 16³: 12. 1910.
Melocactus lutescens J. V. Suringar, Verh. Kon. Akad. Wetensch. Amst. II. 16³: 13. 1910.
Melocactus rotifer angustior J. V. Suringar, Verh. Kon. Akad. Wetensch. Amst. II. 16³: 14. 1910.
Melocactus pinguis planispinus J. V. Suringar, Verh. Kon. Akad. Wetensch. Amst. II. 16³: 15. 1910.
Melocactus gracilis J. V. Suringar, Verh. Kon. Akad. Wetensch. Amst. II. 16³: 16. 1910.
Melocactus microcephalus olivascens J. V. Suringar, Verh. Kon. Akad. Wetensch. Amst. II. 16³: 17. 1910.
Melocactus cylindricus J. V. Suringar, Verh. Kon. Akad. Wetensch. Amst. II. 16³: 19. 1910.
Melocactus pinguis laticostatus J. V. Suringar, Verh. Kon. Akad. Wetensch. Amst. II. 16³: 19. 1910.
Melocactus pinguis tenuissimus J. V. Suringar, Verh. Kon. Akad. Wetensch. Amst. II. 16³: 20. 1910.

Plants pale green or dull green, globular or somewhat broader than high, often 3 dm. in diameter and in time crowned by a cephalium; cephalium elongated, becoming 2 dm. high or more, 1 dm. in diameter, composed of white felt and brown bristles longer than the felt; ribs 11 to 15, broad at base, rounded ; spines all brown to yellow at first; radial spines 12 to 15, acicular, widely spreading, 3 cm. long or more; central spines usually 4 but sometimes more, subulate, much stouter than the radials, unequal in length, sometimes 7 cm. long; flower, including the ovary, about 2 cm. long, swollen at the base, the segments linear-lanceolate, acutish, about 5 mm. long; fruit broadly clavate, shining, 1.5 to 2 cm. long, capped by the more or less persistent perianth; seeds short-oblong, 1 mm. long, dull black.

Type locality: Not cited.
Distribution: Curaçao and the adjacent Dutch Islands.

Dr. Britton, while studying the vegetation of Curaçao in 1913 with special relation to the cacti,* closely examined this species at many localities. Many slight variations in size and shape of the plant-body, length and color of the spines, and development of the cephalium were observed, but the conclusion was reached that all the individuals were, doubtless, referable to but one species, notwithstanding Professor Suringar's contrary opinion; Dr. Shafer who accompanied Dr. Britton agreed with this result. Dr. Rose, visiting Curaçao in 1916, arrived at the conclusions reached by Dr. Britton and independently by Dr. I. Boldingh, who studied the vegetation of Bonaire, Curaçao, and Aruba in 1909 and 1910.

The synonymy of the species is the largest of any of the cacti and perhaps not exceeded by that of any other plant.

Illustrations: Verh. Ver. Beförd. Gartenb. 3: pl. 12; Nov. Act. Nat. Cur. 18: Suppl. 1. pl. 4, f. 4; Mus. Bot. Leide 3: pl. 17; 18, C, as *Melocactus macracanthus;* Verh. Ver. Beförd. Gartenb. 3: pl. 13; Mus. Bot. Leide 3: pl. 10, F; Nov. Act. Nat. Cur. 18: Suppl. 1. pl. 4, f. 6, as *Melocactus salmianus;* Mus. Bot. Leide 3: pl. 9, 11, as *Melocactus salmianus* (forma); Mus. Bot. Leide 3: pl. 9, as *Melocactus salmianus trispinus;* Mus. Bot. Leide 3: pl. 11, as *Melocactus salmianus quadrispinus;* Verh. Ver. Beförd. Gartenb. 3: pl. 25; Nov. Act. Nat. Cur. 18: Suppl. 1. pl. 3; pl. 4, f. 5, as *Melocactus pyramidalis;* Nov. Act. Nat. Cur. 18: Suppl. 1. pl. 6, as *Melocactus dichroacanthus;* Nov. Act. Nat. Cur. 18: Suppl. 1. pl. 1. f. 1; pl. 2, f. 1, g, h; f. 2; pl. 4, f. 7; pl. 8; Förster, Handb. Cact. ed. 2. 440. f. 49; Mus. Bot. Leide 3: pl. 16; 18, D, as *Melocactus lehmanni;* Nov. Act. Nat. Cur. 18: Suppl. 1. pl. 1, f. 3; pl. 4, f. 1, a, b, e, f; pl. 9, as *Melocactus microcephalus;* Verh. Kon. Akad. Wetensch. II. 5³: pl. 1. f. 3, 3a, 3b, as *Melocactus uncinatus;* Verh. Kon. Akad. Wetensch. II. 5³: pl. 1, f. 4, 4a, 4b, 4c, as *Melocactus arcuatus;* Nov. Act. Nat. Cur. 18: Suppl. 1. pl. 10; Förster, Handb. Cact. ed. 2. 444. f. 50, as *Melocactus zuccarinii;* Mus. Bot. Leide 3: pl. 6, 10, C, as *Melocactus patens;* Mus. Bot. Leide 3: pl. 7, 10, D, as *Melocactus cornutus;* Mus. Bot. Leide 3: pl. 8, 10, E, as *Melocactus pusillus;* Mus. Bot. Leide 3: pl. 10, B, as *Melocactus parvispinus;* Mus. Bot. Leide 3: pl. 14, 18, A, as *Melocactus communiformis;* Mus. Bot. Leide 3: pl. 15, 18, B, as *Melocactus rotula;* Mus. Bot. Leide 3: pl. 18, E; 21, as *Melocactus grollianus;* Mus. Bot. Leide 3: pl. 18, F, 22, as *Melocactus exsertus;* Mus. Bot. Leide 3: pl. 18, G; 23, as *Melocactus spatanginus;* Mus. Bot. Leide 3: pl. 24, as *Melocactus bargei.*

2. Cactus melocactus Linnaeus, Sp. Pl. 466. 1753.

> *Cactus coronatus* Lamarck, Encycl. 1: 537. 1783.
> *Cactus melocactus communis* Aiton, Hort. Kew ed. 2. 3: 175. 1811.
> ? *Cactus lamarckii* Colla, Mem. Accad. Sci. Torino 33: 127. 1826.
> *Melocactus meonacanthus* Link and Otto, Verh. Ver. Beförd. Gartenb. 3: pl. 15. 1827.
> *Melocactus communis* Link and Otto, Verh. Ver. Beförd. Gartenb. 3: 417. 1827.
> *Echinocactus meonacanthus* Link and Otto, Verh. Ver. Beförd. Gartenb. 3: 428. 1827.
> ? *Melocactus lamarckii* G. Don, Hist. Dichl. Pl. 3: 160. 1834.
> ? *Melocactus rubens* Pfeiffer, Enum. Cact. 43. 1837.
> ? *Melocactus communis laniferus* Pfeiffer, Enum. Cact. 43. 1837.
> *Echinocactus melocactoides* Lemaire, Cact. Aliq. Nov. 28. 1838.
> ? *Melocactus communis acicularis* Monville in Lemaire. Cact. Gen. Nov. Sp. 103. 1839.
> ? *Melocactus communis magnisulcatus* Lemaire, Cact. Gen. Nov. Sp. 103. 1839.
> *Melocactus communis spinosior* Monville in Lemaire, Cact. Gen. Nov. Sp. 103. 1839.
> *Cactus communis* Steudel, Nom. ed. 2. 1: 245. 1840.
> *Cactus meonacanthus* Steudel, Nom. ed. 2. 1: 246. 1840.
> *Melocactus melocactus* Karsten, Deutsche Fl. 888. 1882.

---

*Journ. N. Y. Bot. Gard. 14: 105, 106. 1913.

Short-cylindric, 5 to 10 dm. high; cephalium up to 10 cm. broad, 3 to 5 cm. high, rounded, composed of white wool and long brown bristles or spines, much longer than the wool; ribs 10 or 11, up to 3 dm. in diameter, prominent, 2 to 3 cm. high; spines about 10 to 12, stout, 3 to 5 cm. long, terete, yellowish to brown; flowers 3 to 4 cm. long; the tube narrowly cylindric, the lobes oblong, obtuse or mucronulate; fruit clavate, much elongated, 5 to 6 cm. long; seeds numerous, black, shining.

Type locality: Jamaica.

Distribution: Arid southern parts of Jamaica.

Dr. Britton studied this plant in the Healthshire Hills, south of Spanish Town, Jamaica, in company with Mr. William Harris in 1908. Here it is abundant on limestone with other cacti and xerophytic shrubs and trees and travelers use it for drinking water by cutting off the top and scraping out the watery pulp.

Mammillaria communis (Steudel, Nom. ed. 2. 1: 245. 1840) appeared as a synonym of *Cactus communis.*

Melocactus brongniartii Lemaire (Cact. Aliq. Nov. 12. 1838), of unknown origin, has never been definitely identified. Lemaire states that it is related to *Melocactus communis.*

FIG. 235.—Cactus lemairei.   FIG. 236.—Cactus broadwayi.

Illustrations: Gerarde, Herball ed. 1. 1013; ed. 2 and 3. 1177, as *Melocarduus echinatus;* L. Obel. Kruydboeck 2: 27. (24?) 1581, as *Echinomelocactus;* Abh. Bayer. Akad. Wiss. München 2: pl. 2, f. 10 (?), as *Echinocactus leucacanthus;* Mus. Bot. Leide 3: pl. 14; pl. 18, A, as *M. communiformis;* Mus. Bot. Leide 3: pl. 8; pl. 10, E, as *M. pusillus;* Nov. Act. Nat. Cur. 19¹: pl. 16, f. 9: Abh. Bayer. Akad. Wiss. München 2: pl. 1, II. f. 3, as *M. rubens;* Nov.Comm. Acad. Scient. 3: pl. 26, as *M. rufispinus;* Thomas, Zimmerkultur Kakteen 43, as *M. brongniartii;* Clusues, Exot. pl. 92; Verh. Ver. Beförd. Gartenb. 3: pl. 15, as *M. meonacanthus;* (?) Mem. Accad. Sci. Torino 33: pl. 7, as *Cactus lamarckii;* Verh. Ver. Beförd. Gartenb. 3: pl. 11; Förster, Handb. Cact. ed. 2. 433. f. 48; Mém. Mus. Hist. Nat. Paris 17: pl. 6; Watson, Cact. Cult. 140. f. 54; ed. 3. f. 33; Balt. Cact. Journ. 1: pl. 30; Cycl. Amer. Hort. Bailey 2: f. 1389; Möllers Deutsche Gärt. Zeit. 25: 477. f. 11, No. 1, 10; Schelle, Handb. Kakteenk. 206. f. 138; Verh. Kon. Akad. Wetensch. II. 5³: pl. 2, f. 3; The Garden 64: 337; Ann. Inst. Roy. Hort. Fromont 2: pl. 1, f. C; Garten-Zeitung 4: 182. f. 42, No. 15; Gartenwelt 7: 277;

11: 498; Karsten, Deutsche Fl. 887. f. 501, No. 1; ed. 2. 2: 456. f. 605, No. 1; Dict. Hort. Bois 826. f. 578; Nov. Act. Nat. Cur. 18: Suppl. 1. pl. 1, f. 2; Lemaire, Cactées 29. f. 1; Palmer, Cult. Cact. 105; Rev. Hort. 1857: f. 124, as *Melocactus communis;* Hort. Ripul. App. 3: pl. 7, as *Cactus communis;* Besler, Hort. Eystett. 4. Ord. f. 1; De Candolle, Pl. Succ. 2: pl. 112; De Tussac, Fl. Antill. 2: pl. 27; Loudon, Encycl. Pl. 410. f. 6848; Stand. Cycl. Hort. Bailey 2: 613. f. 731.

3. Cactus lemairei (Monville).

Melocactus communis oblongus Link and Otto, Verh. Ver. Beförd. Gartenb. 3: 418. 1827.
Melocactus communis macrocephalus Link and Otto, Verh. Ver. Beförd. Gartenb. 3: 418. 1827.
Echinocactus intortus purpureus De Candolle, Prodr. 3: 462. 1828.
? *Melocactus communis conicus* Pfeiffer, Enum. Cact. 43. 1837.
*Echinocactus lemarii** Monville in Lemaire, Cact. Aliq. Nov. 17. 1838.
Melocactus crassispinus Salm-Dyck, Allg. Gartenz. 8: 10. 1840.
*Melocactus lemarii** Miquel in Lemaire, Hort. Univ. 1: 286. 1839.
Melocactus hispaniolicus Vaupel, Monatsschr. Kakteenk. 29: 121. 1919.

Plant usually rather slender, 2 to 3 dm. high and sometimes 2 dm. in diameter, but young plants sometimes broader than high; flowering plants crowned by a slender cephalium sometimes 1 dm. high, made up of white wool and brown bristles; areoles large, very woolly when young; ribs 9 or 10; spines 8 to 13 in a cluster, all stout, more or less flattened, 2 to 3 cm. long, horn-colored or somewhat brownish; central spines usually one on young plants, but 2 or 3 on old ones; flowers rose, about 2 cm. long, 15 mm. broad when fully expanded, exserted about 12 mm. above the cephalium; outer perianth-segments obtuse, the inner acute, serrate near the tip; fruit pinkish, slender, 2 cm. long, naked; seeds black, tuberculate.

Type locality: Santo Domingo.
Distribution: Deserts of Hispaniola.

Collected at Azua, Santo Domingo, in March 1913, by J. N. Rose (No. 3832). Living material was sent both to the New York Botanical Garden and to Washington.

The Santo Domingan plant, although it has long been known, has been confused with another species. The trouble began in 1827 when Link and Otto redescribed *Cactus macracanthus* under *Melocactus* and gave the habitat of the species as Santo Domingo. Salm-Dyck, who described *Cactus macracanthus* first in 1820, did not give a definite locality. Haworth, however, in 1821 told about seeing a specimen of this species from Holland, which had been sent from South America, and which might have been sent from the Dutch Colony at Curaçao; while Salm-Dyck, in redescribing his species in 1834, gives Curaçao definitely as the locality for it. In 1837 Pfeiffer referred material from both Curaçao and Santo Domingo to this species and he has been followed by most writers since. It seems almost certain that the Santo Domingan species can not be the *Cactus macracanthus* of Salm-Dyck.

In 1839 Lemaire figured and described a species of Miquel's, *Melocactus lemarii†* (Hort. Univ. 1: 286), which was redescribed by Miquel the next year in his monograph on the genus *Melocactus.* This species came from Santo Domingo and is undoubtedly the same as the one which has been passing as *Melocactus macracanthus.*

Melocactus communis oblongus and *M. communis macrocephalus* (Link and Otto, Verh. Ver. Beförd. Gartenb. 3: 418. 1827) were assigned to Santo Domingo by Pfeiffer and, if from that island, are presumably to be referred here.

This plant is very common in the arid plain about Azua in southern Santo Domingo. It grows with other cacti and desert shrubs.

Illustrations: Hort. Univ. 1: pl. 35; Herb. Génér. Amat. II. 2: pl. 36, as *Melocactus lemarii;* Descourtilz, Fl. Med. Antill. 7: pl. 515, as *Cactier rouge main (fide* Urban); Nov. Act. Nat. Cur. 18: Suppl. 1. pl. 4, f. 3, as *Melocactus communis macrocephalus.*

---

*This name was originally published as *lemarii*, although named for Charles Lemaire.
†This is evidently the same plant which he described in 1838 as *Echinocactus lemarii.*

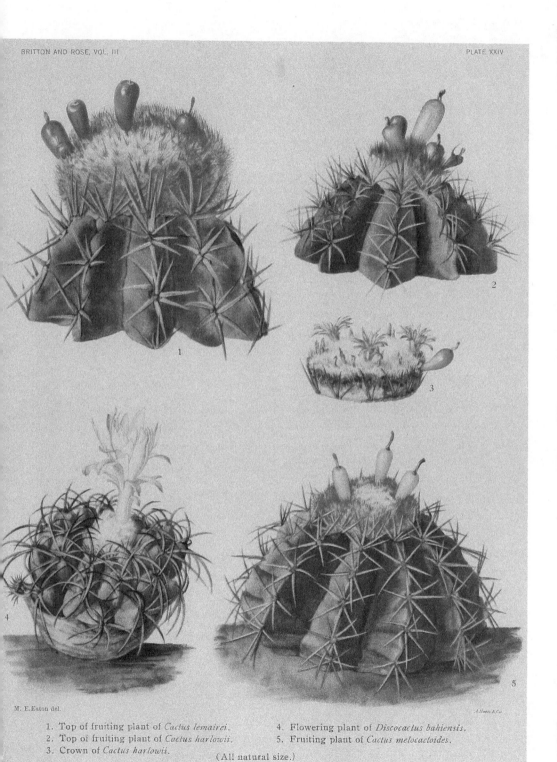

M. E.Eaton del.

A.Hoen & Co.

1. Top of fruiting plant of *Cactus lemairei*.
2. Top of fruiting plant of *Cactus harlowii*.
3. Crown of *Cactus harlowii*.
4. Flowering plant of *Discocactus bahiensis*.
5. Fruiting plant of *Cactus melocactoides*.

(All natural size.)

Plate XXIV, figure 1, shows the top of a fruiting plant collected by Rose, Russell, and Fitch at Azua, Santo Domingo, March 12, 1913 (No. 6324). Figure 235 is a reproduction of the plate published in L'Horticulteur Universel and cited above.

4. Cactus oreas (Miquel).

Melocactus oreas Miquel, Nov. Act. Nat. Cur. **18**: Suppl. 1. 192. 1841.
Melocactus ernesti Vaupel, Monatsschr. Kakteenk. **30**: 8. 1920.

Plant 2 to 4 cm. in diameter; cephalium 4 to 6 cm. high, composed of white wool and many soft brown hairs; ribs 10 to 12, somewhat acute, about 20 cm. high; spines 14 to 20, of which 4 to 6 are central, brown, but paler toward the tips, slender, nearly straight, subulate, much elongated, the longer ones 12 cm. long and flexible, terete; flowers pinkish; fruit clavate, 2 cm. long, bright red; seeds black, minute, 1 mm. broad, covered with low depressed tubercles.

Type locality: Bahia, Brazil.
Distribution: Dry parts of Bahia.
Common on the hills of the interior parts of Bahia, Brazil. Dr. Rose found it in Dr. Leo Zehntner's collection at Joazeiro and later collected it at Machado Portella, June 20, 1915 (No. 19729).

This plant is characterized by its greatly elongated spines, these being longer than those of any other species of the genus.

Illustrations: Karsten and Schenck, Vegetationsbilder **6**: pl. 17, as *Melocactus;* Monatsschr. Kakteenk. **30**: 8, 9, as *Melocactus ernesti.*

5. Cactus ruestii (Schumann).

Melocactus ruestii Schumann, Verzeichn. Kult. Kakt. 26. 1896.

Fig. 237.—Cactus ruestii.

Plants mostly globular, 5 to 15 cm. in diameter; ribs 11 to 19, rather high, separated by sharp intervals; young areoles brown-felted; spines dark brown at first, gray in age; radial spines 5 to 8, spreading or reflexed, subulate, 2.5 to 3 cm. long; central spine solitary, erect, 2.5 to 3 cm. long, subulate; cephalium a small crown of brown stiff bristles and white wool; flowers small, red.

Type locality: Honduras.
Distribution: Honduras.
It has been recently collected by F. J. Dyer in Comayagua Valley, Honduras, May 7, 1917 (No. 261) and also by George W. Ellis in rocky soil, altitude 1,500 meters, south of Comayagua, Honduras.

The species is characterized by its closely set areoles and by the very stiff bristles in the cephalium.

Illustration: Monatsschr. Kakteenk. **2**: 89, as *Melocactus brongnartii.*
Figure 237 is a reproduction of the illustration cited above.

6. Cactus maxonii Rose, Smiths. Misc. Coll. 50: 63. 1907.

Melocactus guatemalensis Gürke and Eichlam, Monatsschr. Kakteenk. **18**: 37. 1908.
Melocactus maxonii Gürke, Monatsschr. Kakteenk. **18**: 93. 1908.

Depressed-globose, 10 to 15 cm. high; cephalium small; ribs 11 to 15, broad at base; radial spines 7 to 11, spreading or recurved, pale red or rose-colored, 1 to 1.5 cm. long; central spine usually solitary, porrect or ascending; flowers rose-colored; fruit clavate; seeds black, shining.

Type locality: Near El Rancho, Guatemala.
Distribution: Guatemala.

228 THE CACTACEAE.

Soon after this species was described E. von Tuerckheim, who had spent many years in Guatemala, wrote that the plant was an old acquaintance and that some 20 years before he had sent several hundred from Salama to a European dealer.

The habitat of this plant is various; Mr. Maxon reported finding it on barren stony hillsides while Charles C. Deam found it in flat exposed rocky places and sometimes in open woods. The plants grow singly or in pairs at an altitude of about 300 meters.

Illustrations: Stand. Cycl. Hort. Bailey **2**: 612. f. 729, as melon-cactus; Monatsschr. Kakteenk. **23**: 179; Möllers Deutsche Gärt. Zeit. **25**: 477. f. 11, No. 2, 9, as *Melocactus maxonii;* Smiths. Misc. Coll. **50**: pl. 6; Ann. Rep. Smiths. Inst. **1908**: pl. 2, f. 2.

Figure 239 is from a photograph of the type plant.

7. Cactus salvador (Murillo).

Melocactus salvador Murillo, Circular [about 1897].

Simple, globose, 3 to 4 dm. in diameter; ribs 13; radial spines 8, somewhat recurved; central spines 1 to 3, longer and stouter than the radials, those near the center of the plant nearly erect, those on the side somewhat curved downward; cephalium 8 cm. in diameter; flowers rose-pink; seeds black.

FIG. 238.—Cactus melocactoides.   FIG. 239.—Cactus maxonii.

Type locality: Not cited.

Distribution: High mountains above Jalapa, Vera Cruz.

This plant was first described in a circular issued by Louis Murillo of Jalapa about 1897. It was offered for sale in March 1898 in the Cactus Journal (p. 28) and in several subsequent numbers. It was later described by Walton (Cact. Journ. **2**: 103. 1899) and is referred to incidentally in the Monatsschrift für Kakteenkunde (**9**: 178; **18**: 61, 62, 64, 93; **19**: 81), sometimes as *M. san-salvador*, and is also incidentally mentioned by Schumann (Gesamtb. Kakteen 454. 1898).

Dr. Rose in 1912 found four photographs of the plant at Kew which had been sent by Professor Murillo. Murillo's original description was accompanied by an illustration of four potted plants; as his description is accessible to only a few it is reproduced here:

"MELOCACTUS SALVADOR.

"A new and very scarce Cactus. Discovered by Louis Murillo.—Jalapa, Mexico.

"This is a new and beautiful plant, of spherical form, with 13 symmetrical furrows, which are deeply marked and covered with long and conic spines that are arranged in the shape of a crown, 8 radial, 1 to 3 central, the latter much longer and all slightly curved. It reaches a diameter of from

30 to 40 centimeters when full grown. The cephalium has a diameter of about 8 cm. and a height of 3 inches and is formed by clusters of short spines strongly set together and which are of a reddish hue. In the middle of the clusters beautiful flowers sprout out and are followed by purplish fruit that give the whole plant a fair and elegant aspect. This *Melocactus* is found in the fissures of lofty, perpendicular mountain passes, but in very limited numbers, and in a region not exceeding a square mile in extent. Therefore it is impossible for me to collect large numbers of the same.

"Cactus lovers have now an opportunity to enrich their collections with this new specimen which has already excited the cordial admiration of amateurs both in the United States and Germany."

We wrote to Mr. S. A. Skan, of the Kew Library, regarding this publication and name and he replied as follows:

"I have made an effort to ascertain the date of its publication but regret that I have not obtained any definite information. The name *Melocactus salvador* has not been taken up in any of the supplements to the Index Kewensis. In the Kew Hand-List of Tender Dicotyledons (1899) there is '*M* [*elocactus*] *salvatoris*, Hort. Mexico.' I supposed that to be a mistake for *M. salvador* and communicated with Mr. Watson about it. He tells me that a plant named *M. salvatoris* was purchased from Prof. L. Murillo in 1898. It is not now at Kew."

Fig. 240.—Cactus salvador.

It must be different from *Melocactus curvispinus* (Pfeiffer, Enum. Cact. 46. 1837) also from Mexico, a species which we know only from description. It is described as globose, 10 cm. high, 7.5 cm. in diameter, depressed; ribs 10 to 12; areoles large, round, white-velvety; radial spines 7, curved, brownish or white, 12 to 16 mm. long; central spines 2, erect, 2.5 cm. long, blackish.

Another species, *Melocactus delessertianus* Lemaire (Hort. Univ. 1:225. 1839), has been described from Mexico which may or may not be this plant. It was described as slightly depressed, about 10 cm. high; ribs 12 to 15; radial spines 8 or 9; central spines 2; flowers and fruit unknown.

Figure 240 is reproduced from a photograph, taken by Louis Murillo, now on deposit in the Library at Kew, a copy of which was sent us by the Director, Captain Arthur W. Hill.

8. Cactus broadwayi sp. nov.

Plant a little longer than thick, 1 to 2 dm. high, yellowish green; ribs 14 to 18, sometimes branching above, rather low, 1 to 1.5 cm. high, 1 to 2 cm. broad at base, rounded, separated from one another by acute intervals; areoles small, depressed, 1 cm. apart; spines horn-colored, but often with brownish tips or some, especially the central ones, brown throughout, at least when young; radial spines 8 to 10, 1 to 1.5 cm. long, more or less curved inwards; central spines usually one, sometimes 2 or 3, a little stouter than the radials; cephalium small, 6 to 7 cm. broad at base, 2 to 3 cm. high, made up of soft brown bristles and white wool; flowers small, purplish; fruit clavate, 2.5 cm. long, purple; seeds black.

Collected on Tobago Island, West Indies, by W. R. Broadway in 1914 and through W. G. Freeman from the same island in 1921, type; also obtained on Grenada, British West Indies, by R. O. Williams in 1921.

Figure 231 is from a photograph of the plant sent by W. G. Freeman from Tobago; figure 236 is from a photograph of the plant sent by R. O. Williams from Grenada.

FIGS. 241 and 242.—Cactus intortus.

9. Cactus intortus Miller, Gard. Dict. ed. 8. No. 2.   1768.

Echinocactus intortus De Candolle, Prodr. 3: 462.   1828.
Melocactus communis atrosanguineus Link and Otto, Verh. Ver. Beförd. Gartenb. 6: 430.   1830.
Melocactus communis ovatus Hooker in Curtis's Bot. Mag. 58: pl. 3090.   1831.
Melocactus communis viridis* Pfeiffer, Enum. Cact. 42.   1837.
Melocactus communis grengeli Pfeiffer, Enum. Cact. 43.   1837.
Melocactus communis havannensis Pfeiffer, Enum. Cact. 43.   1837.
Melocactus atrosanguineus Pfeiffer, Enum. Cact. 44.   1837.
Melocactus grengelii Forbes, Journ. Hort. Tour Germ. 151.   1837.
Echinocactus xanthacanthus Miquel, Linnaea 11: 155.   1837.
Melocactus macrocanthus Miquel, Linnaea 11: 157.   1837.   Not Link and Otto, 1827.
Melocactus miquelii Lehmann, Del. Sem. Hort. Hamb.   1838.
Melocactus havannensis Miquel, Nov. Act. Nat. Cur. 18: Suppl. 1. 144.   1841.
Melocactus wendlandii Miquel, Nov. Act. Nat. Cur. 18: Suppl. 1. 146.   1841.
Melocactus dichroacanthus Miquel, Nov. Act. Nat. Cur. 18: Suppl. 1. 147.   1841.
Melocactus xanthacanthus Miquel, Nov. Act. Nat. Cur. 18: Suppl. 1. 169.   1841.
Melocactus macracanthoides Miquel, Nov. Act. Nat. Cur. 18: Suppl. 1. 173.   1841.
Melocactus schlumbergerianus Lemaire, Illustr. Hort. 8: Misc. 32.   1861.
Melocactus portoricensis Suringar, Versl. Med. Akad. Wetensch. III. 9: 408.   1891.
Melocactus bradleyi Suringar, Verh. Kon. Akad. Wetensch. Amst. II. 5³: 23.   1896.
Melocactus hookeri Suringar, Verh. Kon. Akad. Wetensch. Amst. II. 5³: 31.   1896.
Melocactus eustachianus Suringar, Verh. Kon. Akad. Wetensch. II. 5³: 37.   1896.
Melocactus linkii Suringar, Verh. Kon. Akad. Wetensch. Amst. II. 5³: 39.   1896.
Melocactus croceus Suringar, Verh. Kon. Akad. Wetensch. Amst. II. 5³: 40.   1896.
Melocactus communis bradleyi Monatsschr. Kakteenk. 6: 142.   1896.
Melocactus communis croceus Monatsschr. Kakteenk. 6: 142.   1896.
Melocactus communis eustachianus Monatsschr. Kakteenk. 6: 142.   1896.
Melocactus communis hookeri Monatsschr. Kakteenk 6: 142.   1896.
Melocactus intortus Urban, Repert. Sp. Nov. Fedde 16: 35.   1919.

*This name first appeared in 1830 (Link and Otto, Verh. Ver. Beförd. Gartenb. 6: 430) and was assigned to a plant of Curaçao, but was without description.

Globose to cylindric, sometimes nearly a meter high; cephalium cylindric, sometimes nearly as long as the plant-body, made up of white wool and soft brown bristles; ribs 14 to 20, thick, large, 2 to 3 cm. high; spines 10 to 15, stout, yellow to brown, 2 to 7 cm. long; flowers pinkish, 1.5 to 2 cm. long; outer perianth-segments acutish or obtuse and mucronulate; inner perianth-segments acute; stigma-lobes 6 or 7, apiculate; fruit oblong to broadly clavate, 2 to 2.5 cm. long; seeds dull black, strongly tubercled, especially at the distal end.

Type locality: Antigua, West Indies.

Distribution: Southern Bahamas, Porto Rico, Virgin Islands, St. Christopher, Antigua, Montserrat, and Dominica.

Urban, who has followed us in restoring Miller's old specific name *intortus*, although using it under *Melocactus*, has applied the name to the Hispaniolan plant while, as a matter of fact, Miller's plant came from Antigua and represents a very different species.

The plant is abundant along and near the coast in southwestern Porto Rico and grows also on the Porto Rican Islands Culebra, Vieques, Mona, and Desecheo; a headland near Cabo Rojo, Punta Melones, has taken its name from this cactus. On the islands Mona and Desecheo in the Mona Passage a race with elongated slender spines exists; and through the Virgin Islands, east to Anegada, the species shows much variability in its armament. It grows on several islands in the southern part of the Bahamas, north to Acklin's Island and Long Island, called Turk's cap or Turk's head here as in the Lesser Antilles; the Turk's Islands have taken their name from this plant which appears on their postage stamps.

Fig. 243.—Cactus intortus.

Illustrations: Bradley, Hist. Succ. Pl. ed. 2. pl. 32, as *Echinomelocactus;* Journ. N. Y. Bot. Gard. 6: 7. f. 3; 9: 46. f. 11, as *Melocactus* sp.; Curtis's Bot. Mag. 58: pl. 3090, as *Melocactus communis ovatus;* Monatsschr. Kakteenk. 6: 87, 135; 26: 115; Schumann, Gesamtb. Kakteen f. 6; Zool. Soc. Bull. 22: 1466; Dict. Gard. Nicholson 2: 347. f. 539, as *Melocactus communis;* Dict. Gard. Nicholson 4: 568. f. 42; Suppl. 530. f. 568; Nov. Act. Nat. Cur 18: Suppl. 1. pl. 7; Förster, Handb. Cact. ed. 2. 432. f. 47; Watson Cact. Cult. 141. f. 55., as *Melocactus miquelii;* Nov. Act. Nat. Cur. 18: Suppl. 1. pl. 11, as *Melocactus macracanthoides;* Bradley, Pl. Succ. pl. 32, as *Echinomelocactus;* Linnaea 11: pl. 4, as *Echinocactus xanthacanthus;* Mus. Bot. Leide 3: pl. 1, 2, 4, A, as *Melocactus eustachianus;* Mus. Bot. Leide 3: pl. 3, 4 E, as *Melocactus portoricensis;* Wendland, Coll. Pl. Succ. 1: pl. 5, *fide* Miquel; Verh. Kon. Akad. Wetensch. Amst. II. 5³: pl. 2, f. 4, 4a, 4b, 6; Mus. Bot. Leide 3: pl. 4, B, 12, 13, 19, as *Melocactus linkii;* Mus. Bot. Leide 3: pl. 4, C, as *Melocactus linkii* (form) Mus. Bot. Leide 3: pl. 11, as *Melocactus linkii* (seedlings); Verh. Kon. Akad. Wetensch. Amst. II. 5³: pl. 2, f. 5 to 5d; Mus. Bot. Leide 3: pl. 4, D, 12, 20, as *Melocactus croceus;* Mus. Bot. Leide 3: pl. 11 (seedlings); Mus. Bot. Leide 3: pl. 4, C, as *Melocactus linkii trispinus.*

The following illustrations, more or less diagrammatic, while generally showing the characters of this genus remarkably well, do not bring out the specific differences and we have been unable to distribute them. It seems desirable to make a record of them here:

Krook, Handb. Cact. 57; Rev. Hort. Belge 40: after 186; Descourtilz, Fl. Med. Antilles ed. 2. 7: pl. 515, as Cactier rouge; Dict. Gard. Nicholson 2: 347. f. 539, as *Melocactus communis;* Remark, Kakteenfreund 18, as pertinato.

Figure 241 shows a barren plant and figure 242 flowering plants, both sent from Turk's Island, British West Indies, through the courtesy of the Director of the New York Aquarium, July 1916; figure 243 shows a single flowering plant from Mona Island, off Porto Rico, photographed by Frank E. Lutz in 1914; the other smaller plant is the snowy cactus.

10. **Cactus harlowii** Britton and Rose, Torreya 12: 16.   1912.

> *Melocactus harlowii* Vaupel, Monatsschr. Kakteenk. 22: 66.   1912.

Plants usually solitary, light green, rather slender, 2.5 dm. high; ribs 12, narrow; areoles closely set, usually less than 1 cm. apart; radial spines about 12, slender, slightly spreading, 1 to 2 cm. long, reddish, becoming straw-colored in age; central spines usually 4, similar to the radial, but usually a little stouter and longer; cephalium small; flowers small, 2 cm. long, deep rose-red; fruit deep red, obovoid, 2 cm. long; seeds black, shining.

Type locality: Coastal cliffs, Guantánamo Bay, Oriente, Cuba.

Distribution: Eastern Cuba.

This seems to be the only species of this genus in Cuba, although *Melocactus havannensis* was based upon a plant from Cuba, now supposed, however, to have been a garden plant.   In the Sauvalle Herbarium there is, however, a fragment of a plant so named but its exact locality we do not know.

The plant is locally abundant on coastal cliffs from Point Maysi to Guantánamo and grows also on river cliffs near Ensenada de Mora, farther west.   Dr. Felipe Garcia Cañizares, Director of the Havana Botanical Garden, has sent us fine photographs of the plant from Point Maysi.   It was recorded by Grisebach and by Sauvalle as *Melocactus communis;* its Cuban name is cardon.

The specific name was given in honor of Captain Charles Henry Harlow, U. S. N., Commandant of the Guantánamo Naval Station at the time Dr. Britton studied the flora of that reservation in 1909.

Illustration: Cañizares, Jardin Bot. Inst. Habana 61.

Of plate xxiv, figure 2 shows a fruiting plant and figure 3 shows the crown of the plant in flower and fruit, all from the type collection.

11. **Cactus amoenus** Hoffmannsegg, Preiss. Verz. ed. 7. 22.   1833.

> *Melocactus amoenus* Pfeiffer, Enum. Cact. 43.   1837.

Simple, 2 dm. high; ribs 10 to 15, 2 cm. high; radial spines 9, spreading, more or less curved, 2 cm. long or less; central spine solitary, 2.5 cm. long or less; cephalium 7 to 8 cm. broad, 2 to 3 cm. high; flowers small, red.

Type locality: Colombia.

Distribution: Coast of northern Colombia.

We are referring here plants collected on the low dry hills near Santa Marta, Colombia. We have seen the following specimens: H. H. Smith, 1898 (No. 2611), Sinclair, 1914, H. M. Curran, 1916 (No. 358) and Pennell, 1918.

Melocactus communis joerdensii Otto (Pfeiffer, Enum. Cact. 43. 1837) and *M. communis joerdensis* (Förster, Handb. Cact. ed. 2. 425. 1885) are only names.   If this variety comes from Venezuela as Rümpler suggested the name would be referred to *Cactus caesius.*

12. **Cactus obtusipetalus** (Lemaire).

> *Melocactus obtusipetalus* Lemaire, Cact. Aliq. Nov. 11.   1838.
> *Melocactus crassicostatus* Lemaire, Cact. Aliq. Nov. 13.   1838.
> *Melocactus obtusipetalus crassicostatus* Lemaire in Miquel, Nov. Act. Nat. Cur. 18: Suppl. 1. 136.   1841.

Ribs 10, vertical, stout, 5 cm. high, somewhat repand, acute, somewhat inflated at the areoles; areoles 5 cm. apart; spines up to 11, about 2.5 cm. long, rigid; radial spines light brown with transverse striations; lateral spines more or less recurved or reflexed; cephalium small; flowers rose-colored, twice the size of those of *Cactus meloctacus;* perianth-segments oblong, rounded or obtuse at apex; style white; stigma-lobes 6.

Type locality: Santa Fé de Bogotá, Colombia.
Distribution: Colombia.

The probabilities are that this species is not native at Bogotá but that it was in cultivation there, specimens having been sent from lower altitudes. Mr. Pittier found a species at Venticas del Dagua, Cauca, in the western cordillera of Colombia, altitude 1,000 meters, which we are disposed to refer here. Dr. F. W. Pennell and Mr. E. P. Killip sent us from this same locality in June 1922 (No. 5415) a single living plant. This specimen is over 2 dm. in diameter and has 14 ribs; the radial spines are usually 8, spreading or a little curved backward, 2 cm. long or more; in addition to these spines there are 2 or 3 short ones (3 to 7 mm. long) from the upper part of the spine-areoles which make the upper stout radial appear sub-central; in addition to these there is one stout central spine, porrect or ascending, 2 to 2.5 cm. long; the spines were probably brown at first, but in age are pale, almost gray. The cephalium is very small.

Dr. Rusby collected in 1917 specimens of this or a related species on an arid plain near Cabrello on the Cabrero River, Colombia. We have studied a fragmentary specimen collected by J. F. Holton at Opia, Colombia, in 1852 (No. 728), preserved in the Torrey Herbarium, which has only 6 spines at each areole and these all radial.

13. Cactus caesius (Wendland) Britton and Rose, Bull. Dept. Agr. Trinidad **19**: 86. 1921.

Melocactus caesius Wendland in Miquel, Nov. Act. Nat. Cur. 18: Suppl. 1. 184. 1841.
Melocactus griseus Wendland in Miquel, Nov. Act. Nat. Cur. 18: Suppl. 1. 185. 1841.
Melocactus cephalenoplus Lemaire, Hort. Univ. 2: 128. 1841.
Melocactus caesius griseus Förster, Handb. Cact. 263. 1846.
Melocactus humilis Suringar, Versl. Med. Akad. Wetensch. III. 6: 459. 1889.

Plants globose, depressed, or narrowed above, 1 to 2 dm. high; ribs 10 to 15, prominent, acute, 2 cm. high; areoles circular, white-woolly when young, 2 to 3 cm. apart; radial spines about 8, spreading, horn-colored; central spines similar to the radials, about 2 cm. long; cephalium broad and low, composed of white wool and brown bristles, broader than the apex of the plant body; fruit obovoid, wine-colored, up to 3 cm. long, 1 to 1.5 cm. thick.

Type locality: La Guayra, Venezuela.
Distribution: Coast of Venezuela and Colombia; Patos Island, Trinidad; and perhaps Dutch Guiana.

This species has recently been collected by Henry Pittier at the type locality. His specimens show a variation in the number of ribs of from 10 to 14 and probably represent the two forms which Wendland described as different species. Mr. Pittier's plants show considerable variation but hardly seem to warrant the recognition of two species.

In September 1920 Dr. Gerold Stahel of Paramaribo, Surinam, sent us a photograph of a group of *Cactus* plants which we tentatively refer here. The living specimens which he collected for us were lost a week after he had collected them through the indifference of his camp helpers. According to Dr. Stahel, the plants were found near the big Raleigh Falls on the Upper

Fig. 244.—Cactus caesius.

Coppename River. They grow on the entirely nude rocks of the Volkberg and are usually found solitary.

Illustrations: Blühende Kakteen **2**: pl. 92, as *Melocactus caesius;* Mus. Bot. Leide **3**: pl. 5, 10, A; Versl. Kon. Akad. Wetensch. Amst. 6: opp. p. 192. pl. [4]; Gartenflora **46**: pl. 1439; **52**: 61. f. 8, as *Melocactus humilis.*

Figure 245 shows the Surinam plant growing on a mass of rock on the nearly bare summit of a hill; figure 246 is from a photograph of a plant obtained by Mr. Pittier near La Guayra in 1913; figure 247 was also obtained by Mr. Pittier at Barquisimeto in 1913; figure 244 shows a plant of Patos Island, Trinidad, taken by Professor Tracy E. ·Hazen in 1921.

14. Cactus townsendii nom. nov.

> *Melocactus peruvianus* Vaupel, Bot. Jahrb. Engler **50**: Beibl. **111**: 28. 1913. Not *Cactus peruvianus* Linnaeus, 1753.

Usually solitary, but sometimes several plants together forming a clump, nearly globular, 1 to 1.5 dm. in diameter; ribs usually 12 or 13, prominent; areoles 1 to 1.5 cm. apart, somewhat elliptic; spines usually 8 or 9, brown or brownish, long and spreading or recurved; central spine, if present, porrect, sometimes 4 cm. long; cephalium usually 6 to 8 cm. high, composed of reddish brown bristles and white wool; flowers pinkish, 2.5 cm. long, persistent on the ovary; fruit red, narrowly clavate, 12 to 16 mm. long; seeds black, roughened.

FIG. 245.—Cactus caesius

Type locality: Chosica, on the Lima and Oroya Railroad, central Peru.

Distribution: Mountains of western central Peru, from above Lima to above Eten.

Some years ago Dr. C. H. Tyler Townsend sent us a specimen of *Cactus* from Peru which we studied and described, but before our description could be printed Dr. F. Vaupel published his *Melocactus peruvianus.* His specific name can not be used under *Cactus* and we have, therefore, substituted the one which we first gave the plant.

This is the most southern species of the genus *Cactus* on the west coast of America. Although described as a distinct species only in 1913, the presence of a so-called *Melocactus* has been known in Peru for a long time; various travelers, including Roezl (1874), mention such a plant.

15. Cactus bahiensis sp. nov.

Dull green, 1 dm. high, 1.5 dm. in diameter; ribs 10 to 12, broad at base, 2.5 cm. high, each bearing 6 or 7 areoles; spines all brown; radial spines about 10, the longest 2.5 cm. long; central spines usually 4, the longest 3.5 cm. long; cephalium low, with many dark brown bristles; flowers pinkish; fruit red, clavate, 1.5 cm. long; seed black, shining, 1 mm. in diameter.

Collected by Rose and Russell near Machado Portella, Bahia, Brazil, in 1915 (No. 19935). The plant was found only at a single locality in central Bahia, but it was there very common and will doubtless be obtained from other localities. It grows on the tops of nearly barren hills and is very different from *Cactus zehntneri*, from northern Bahia.

16. Cactus melocactoides Hoffmannsegg, Verz. Pfl. Nachtr. **3**: 24. 1826.

Melocactus melocactoides De Candolle, Prodr. **3**: 461. 1828.
Melocactus violaceus Pfeiffer, Allg. Gartenz. **3**: 313. 1835.
Melocactus goniodacanthus Lemaire, Cact. Aliq. Nov. 11. 1838.
Melocactus pentacentrus Lemaire, Cact. Gen. Nov. Sp. 108. 1839.
Melocactus depressus Hooker in Curtis's Bot. Mag. **65**: pl. 3691. 1839. Not Salm-Dyck, 1828.

Somewhat depressed, 8 cm. high by 15 cm. broad, light green; ribs usually 10* (rarely 9 or 11), broad, obtuse, a little "crenate"; areoles only 5 or 6 on a rib; radial spines 5 to 8, sometimes a little curved, angled, usually pale brown, in age grayish; central cephalium (so far as known) small; flowers pinkish; perianth-segments with toothed margins; fruit white to very pale rose-color, oblong or club-shaped, 1.5 to 2.5 cm. long; seeds black, reticulated.

FIGS. 246 and 247.—Cactus caesius.

Type locality: Brazil, but no definite locality cited.

Distribution: Coast of Brazil, especially Rio de Janeiro, Bahia, and Pernambuco.

Dr. Rose collected a plant along the coast of Bahia (No. 19691) which he would refer here. A somewhat similar plant, but smaller, was collected by him at Cabo Frio (No. 20698) which we have tentatively referred here. Schumann, however, kept the Bahia and Rio de Janeiro plants distinct, referring the plant from Bahia to *Melocactus depressus* and the one from Rio de Janeiro to *M. violaceus.*

All the Brazilian species of this genus are called cabeça de frade on account of the cephalium; this plant is sold for use in the preparation of some household remedy.

Melocactus gardenerianus Booth was given by Förster (Handb. Cact. 277. 1846) as a synonym of *M. depressus* Hooker. The name, *M. depressus* Salm-Dyck, was given by De Candolle (Prodr. **3**: 463. 1828) as a synonym of *Echinocactus depressus.* *M. parthoni* (Miquel, Nov. Act. Nat. Cur. **18**: Suppl. 1. 190. 1841) was given as a synonym of *M. violaceus.* Schumann also refers it here, giving the name to Cels (Cat. et. Hortul.).

Illustrations: Curtis's Bot. Mag. **65**: pl. 3691; Nov. Act. Nat. Cur. **18**: Suppl. 1. pl. 2, f. 1, c, d; pl. 4, f. 1; Monatsschr. Kakteenk. **32**: 39, as *Melocactus depressus;* Engler and Prantl, Pflanzenfam. **3**^(6a): f. 65, A; Martius, Fl. Bras. **4**^2: pl. 48, as *Melocactus violaceus.*

---

*Schumann describes *M. goniodacanthus* (he spells it *goniacanthus*) with 16 to 20 ribs.

Plate XXIV, figure 5, shows a plant collected by Rose and Russell at Cabo Frio, August 8, 1915 (No. 20698). Figure 238 is reproduced from the first illustration cited above.

17. Cactus zehntneri sp. nov.

Often cylindric, sometimes 2 to 3 dm. high; ribs 12 to 15, rather thin, acutish; radial spines terete, stout, dark brown, more or less incurved, 2.5 cm. long or less; central spine one, similar to the radials, erect or ascending; flowers pinkish; fruit red.

Very common on the flats near Joazeiro, Bahia, where it was collected by Dr. Rose and P. G. Russell in 1915 (No. 19728).

These plants grow in the open on the flats in the semiarid part of Bahia, often associated with *Cephalocereus gounellei* and other cacti.

Illustration: Vegetationsbilder **6:** pl. 15, as *Melocactus* sp.

Figure 248 is from a photograph of the type plant.

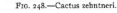
FIG. 248.—Cactus zehntneri.   FIG. 249.—Cactus sp.

18. Cactus neryi (Schumann).

     Melocactus neryi Schumann, Monatsschr. Kakteenk. **11:** 168. 1901.

More or less depressed, 10 to 11 cm. high, 13 to 14 cm. in diameter, crowned by a small cephalium a little broader than high; ribs 10, broad and low; radial spines 7 to 9, terete, spreading outward, 2.5 cm. long; flowers 2.2 cm. long; stigma-lobes greenish; fruit clavate, red.

Type locality: Aracá-Fluss, Brazil.

Distribution: State of Amazonas, Brazil.

The plant is known to us only from description and illustrations.

The specific name for this plant was originally spelled *negryi* instead of *neryi*. The plant was named for Herr Nery, at one time Governor of Amazonas, Brazil. Schumann describes the species and cites himself as the author of the name in the original place of publication while in his Keys of the Monograph he credits the name to Witt.

Illustrations: Monatsschr. Kakteenk. **11:** 169; Schumann, Gesamtb. Kakteen Nachtr. 130. f. 32, as *Melocactus neryi*.

Figure 250 is a reproduction of the original illustration cited above.

CACTUS sp.

Plant small, globose, 1 dm. in diameter; ribs about 11 to 13, rounded, low; spines usually 10 to 12, subulate, more or less recurved; central spine 1 or sometimes 2; crown 10 to 12 cm. in diameter; flowers small; fruit small.

We know this only from a barren plant but it is evidently of this genus.

A living plant was sent to Dr. Rose by Professor C. Conzatti in October 1913 (No. 151a), from Salina Cruz, Oaxaca, and it has been reported by Dr. C. A. Purpus from San Geronimo. Dr. Purpus has written to us as follows:

"The *Melocactus* from San Geronimo is indeed a most interesting and remarkable cactus. When I saw the cactus, I mean to say without a crown, very few specimens ever having one, I thought it was an *Echinocactus*, but of course it is undoubtedly a small *Melocactus*, the smallest which I ever saw. Later I found some specimens with a woolly crown and with it flower and fruit. The flower and fruit resemble flower and fruit of a *Mammillaria*."

Here may belong *Melocactus curvispinus* Pfeiffer (Enum. Cact. 46. 1837) and *M. delessertianus* Lemaire (Hort. Univ. 1: 225. 1839).

Figure 249 shows a barren plant by C. Conzatti, referred to above.

FIG. 250.—Cactus neryi.

DESCRIBED SPECIES, PERHAPS OF THIS GENUS.

CACTUS HEPTAGONUS Linnaeus, Sp. Pl. 466. 1753.

Cereus heptagonus Miller, Gard. Dict. ed. 8. No. 6. 1768.

The name has not been definitely associated with any known cactus by authors subsequent to Miller. Linnaeus indicates that the plant was of American origin and states (Hort. Cliff. 161) that "it is exactly ovate, with 7 angles deeply sculptured; some say they have seen the same thing 1 or 2 feet high, but our plant did not change its shape in growth." Miller's account of it is not more satisfactory; he indicates that he received it, among other kinds of *Cereus* from the British Islands of America and that it has 7 or 8 ribs and several very long white spines. He also says: "Upright, thickest torch thistle, having many angles, several very long white spines and yellow down." There may be doubt whether Miller's plant was the same as that of Linnaeus.

Cactus heterogonus (De Candolle, Prodr. 3: 470. 1828) is a misspelling for *C. heptagonus*.

CACTUS PARVISPINUS Haworth, Suppl. Pl. Succ. 73. 1819.

Echinocactus parvispinus De Candolle, Prodr. 3: 463. 1828.

Ribs about 12; spines 6 to 8 mm. long, white with brownish tips; flowers unknown.

This plant was in cultivation in England in 1815 and is said to have come from the West Indies. It was probably a young plant and is doubtless of this alliance. *Melocactus parvispinus* Haworth (De Candolle, Prodr. 3: 463. 1828) was given only as a synonym.

MELOCACTUS EXCAVATUS Forbes, Journ. Hort. Tour Germ. 151. 1837.

This plant is said to come from Mexico and is probably not of this relationship. Forbes briefly described it as hollow, crowned with 13 ribs; radial spines 7 or 8; central spine solitary, reddish yellow.

MELOCACTUS HYSTRIX Parmentier in Miquel, Nov. Act. Nat. Cur. 18: Suppl. 1. 138. 1841.

The origin of this plant is unknown and it has never been definitely referred to any of the described species.

MELOCACTUS MONVILLEANUS Miquel, Nov. Act. Nat. Cur. 18: Suppl. 1. 133. pl. 5. 1841.

This species can not be identified and its origin is unknown. The illustration shows a barren plant which does not suggest any of the species of *Cactus* known to us, but rather some stubby *Cereus* relative such as *Cereus lormata*.

MELOCACTUS LEOPOLDII Gard. Chron. II. **5**: 603. 1876.

The only information regarding this plant which we have is an account of an International Exhibition at Brussels in 1876 in which it was awarded a prize with this comment: The third prize was given to M. de Smet for *Melocactus leopoldii*, a globular plant with very numerous spines of reddish hue, paler at base, the largest ones 2.5 inches long.

MELOCACTUS LEUCASTER Hoffmannsegg, Preiss. Ver. ed. 7. 22. 1833.

This species is described as having short white spines. We do not know it nor have we seen the original description. Walpers thought it was a variety of *Melocactus communis*.

MELOCACTUS OCTOGONUS Forbes, Journ. Hort. Tour Germ. 151. 1837.

This plant came from Mexico in 1834 and can not now be definitely identified but it probably is not of this genus. The original description says it has 8 remote ribs and 8 to 10 brownish yellow spines. The name occurs also in Sweet's Hortus Britannicus (ed. 3. 282. 1839). It is probably different from *Cactus octogonus* which Steudel refers as a synonym of *Cereus hexagonus*.

Cactus aculeatissimus Zeph (Steudel, Nom. 131. 1821) is only a name.

Cactus aurantiiformis Thiery, catalogued in both editions of Steudel (Nom. 131. 1821; ed. 2. 1: 245. 1840) and also catalogued by the Index Kewensis, was without synonymy and so far as we know has never been described.

Cactus luteus Thiery (Steudel, Nom. 132. 1821; ed. 2. 1: 246. 1840) is only a name which Steudel himself questioned.

Cactus mensarum Thiery (Steudel, Nom. ed. 2. 1: 246. 1840) is only a name.

Cactus proteiformis Desfontaines (Tabl. Bot. ed. 3. 276. 1829; *Cereus proteiformis* Steudel, Nom. ed. 2. 1: 246. 1840) can not be identified.

Cactus pseudotuna (Steudel, Nom. ed. 2. 1: 246. 1840) was said to be *Opuntia pseudotuna*, a very doubtful plant.

Cactus sylvestris Thiery (Steudel, Nom. 132. 1821; ed. 2. 1: 246. 1840) is only a name.

Cactus trichotomus Tenore (Steudel, Nom. ed. 2. 1: 246. 1840) is only a name.

Cactus verticillatus Brotero (Heynhold, Nom. **2**: 103. 1846) is only a name.

Melocactus atrovirens Hortus (Förster, Handb. Cact. 279. 1846) is only a name.

Melocactus coronatus Cels (Förster, Handb. Cact. 279. 1846) is only a name.

Melocactus ferox Pfeiffer (Förster, Handb. Cact. 519. 1846) which was supposed to come from southern Brazil we do not know. With this is also referred *Echinocactus armatus* Salm-Dyck (Pfeiffer, Enum. Cact. 61. 1837) and *Echinocactus spina-christi* Zuccarini (Pfeiffer, Enum. Cact. 59. 1837) and to the latter Pfeiffer refers *Echinocactus fischeri* as a synonym and Förster gives *Melocactus spina-christi* Cels (Handb. Cact. 279. 1846) as simply a name.

Melocactus fluminensis Poselger (Monatsschr. Kakteenk. **2**: 50. 1892; **3**: 68. 1893) is only a name.

Melocactus hookerianus Forbes (Förster, Handb. Cact. 279. 1846) is only a name.

Melocactus nigro-tomentosus (Monatsschr. Kakteenk. **3**: 1. 1893), *Melocactus lobelii* (Monatsschr. Kakteenk. **6**: 142. 1896), illustrated (Verh. Kon. Akad. Wetensch. II. 5³: pl. 1. f. 1, 1, a, 1, b), *M. communiformis* (Monatsschr. Kakteenk. **8**: 31. 1898) and *Melocactus repens* (Monatsschr. Kakteenk. **18**: 167. 1908) seem never to have been described.

INDEX.

(Pages of principal entries in heavy-face type.)

239

1. *Coryphantha runyonii.*
2. *Dolichothele sphaerica.*

THE CACTACEAE

DESCRIPTIONS AND ILLUSTRATIONS OF PLANTS OF THE CACTUS FAMILY

BY

N. L. BRITTON AND J. N. ROSE

VOLUME IV

THE CARNEGIE INSTITUTION OF WASHINGTON
WASHINGTON, DECEMBER 24, 1923

CARNEGIE INSTITUTION OF WASHINGTON

PUBLICATION No. 248, VOLUME IV

Pages 1-80, text only, were distributed
under date of October 9, 1923

PRESS OF GIBSON BROTHERS, Inc
WASHINGTON

CONTENTS.

III

ILLUSTRATIONS.

PLATES.

IV

PLATES—*continued*.

TEXT-FIGURES.

TEXT-FIGURES—*continued.*

TEXT-FIGURES—*continued.*

THE CACTACEAE

Descriptions and Illustrations of Plants of the Cactus
Family

DESCRIPTIONS AND ILLUSTRATIONS OF PLANTS OF THE CACTUS FAMILY.

Tribe 3. CEREEAE.

Subtribe 6. CORYPHANTHANAE.

Terrestrial, spiny, low cacti, mostly globose, sometimes cylindric, rarely elongated, 1-jointed, solitary or cespitose, tuberculate, the tubercles numerous; tubercles usually arranged in spirals; juice watery or milky; flowers always solitary at areoles, either at top or side of plant, but never at spine-areoles, large or small, regular (except in the genus *Cochemiea*); ovary naked or bearing a few scales; fruit a green or red indehiscent berry (except in the genus *Bartschella*); seeds small, brown or black.

We recognize 14 genera.

KEY TO GENERA.

A. Ovary more or less scaly (not known in *Mamillopsis*).
  Flower campanulate with short tube.
    Some of spines hooked.. 1. *Ancistrocactus* (p. 3)
    None of spines hooked (see species No. 2 in *Neolloydia*).
      Tubercles not deeply grooved; fruit scaly .... 2. *Thelocactus* (p. 6)
      Tubercles deeply grooved; fruit nearly naked .... 3. *Neolloydia* (p. 14)
    Flower-tube elongated, scaly... 4. *Mamillopsis* (p. 19)
AA. Ovary naked or nearly so.
  B. Flowers irregular... 5. *Cochemiea* (p. 21)
  BB. Flowers regular.
    C. Flowers central, borne in axils of young, usually nascent, tubercles, large (except in genus No. 8); tubercles containing a watery juice; fruit dull green or red; seeds brown or black.
      D. Tubercles grooved on upper side; flowers borne at base of groove.
        Seeds mostly light brown; fruit greenish or yellowish even when mature, ripening slowly................................... 6. *Coryphantha* (. 23)
        Seeds black to dark brown; fruit red, maturing rapidly.
          Tubercles long, not numerous, not persisting as woody knobs; aril of seed large.................................... 7. *Neobesseya* (p. 51)
          Tubercles short, numerous, persisting after spines fall off as woody knobs; aril of seed small.............................. 8. *Escobaria* (p. 53)
      DD. Tubercles not grooved above.
        Fruit circumscissile; tubercles fleshy; spines acicular........... 9. *Bartschella* (p. 57)
        Fruit not circumscissile; tubercles woody; spines pectinate....... 10. *Pelecyphora* (p. 59)
    CC. Flowers lateral, borne in axils of old and mature tubercles; these never grooved above.
      Seeds with a large corky aril................................. 11. *Phellosperma* (p. 60)
      Seeds without a corky aril.
        Flowers large with an elongated tube; tubercles elongated, flabby.... 12. *Dolichothele* (p. 61)
        Flowers small, campanulate; tubercles not flabby.
          Hilum of seed large; tubercles lactiferous; spines pectinate...... 13. *Solisia* (p. 64)
          Hilum of seed minute; tubercles sometimes lactiferous, but not in species with black seeds; spines not pectinate............. 14. *Neomammillaria* (p. 65)

1. ANCISTROCACTUS gen. nov.

Small, globular or short-cylindric plants, indistinctly ribbed, strongly tubercled, very spiny, one of central spines always hooked; flowering tubercles more or less grooved on upper side; flowers rather small, short, funnelform, borne at top of plant; ovary small, bearing a few thin scales, these always naked in their axils; fruit oblong, greenish, juicy, thin-walled, usually naked below but with a few broad cordate, thin-margined scales above; seeds globular, rather large, brownish to black, the papillae low, flattened; hilum large, depressed, sub-basal, surrounded by a thick rim.

Type species: *Echinocactus megarhizus* Rose.

Engelmann in describing *Echinocactus scheeri*, one of the species of this genus, refers to its anomalous characters when he says:

"Seeds are large, about 1 line long, 0.8 line in diameter, with very minute and flattened tubercles, *brown* (the only *Echinocactus* with seeds of that color known to me); hilum large and circular, surrounded by a thick rim; albumen very small; embryo curved but cotyledons small, connate, more like those of a *Mammillaria*, separating on the curvature and not at the end of the hook, as in all other hooked embryos of *Cactaceae* known to me." (Cact. Mex. Bound. 19. 1859.)

3

The generic name is from ἄγκιστρον fish-hook, and κάκτος cactus, referring to the .long, hooked central spines.

Ancistrocactus was used by Schumann for a subgenus of *Echinocactus*. We recognize three species in the genus, occurring in southern Texas and northern Mexico.

Coulter (Contr. U. S. Nat. Herb. **3**: 368, 369) calls attention to grooved areoles of *Echinocactus brevihamatus* resembling those of *Coryphantha* and *Echinocactus scheeri*.

KEY TO SPECIES.

Radial spines 20 or more, strongly appressed, pectinate; flowering areoles naked. . . . .   1. *A. megarhizus*
Radial spines 18 or fewer, more or less spreading, hardly pectinate; flowering areoles woolly.
   Groove half length of tubercle; flower greenish; radial spines 15 to 18. . . . .   2. *A. scheeri*
   Groove extending full length of tubercle; flower rose-colored; radial spines usually 12.   3. *A. brevihamatus*

1. Ancistrocactus megarhizus (Rose).

    Echinocactus megarhizus Rose, Contr. U. S. Nat. Herb. **12**: 290. 1909.

Solitary or in clusters of 3 or 4; plant body nearly globular or a little elongated, 5 to 8 cm. high, usually solitary, from large and fleshy roots; ribs spiral, divided into dark-green tubercl˞s, 4 to 5

FIG. 1.—Ancistrocactus megarhizus.

cm. high; radial spines 20 or more, pectinate, at first pale yellow, in age white; in seedlings the spines pubescent; central spines usually 4, the 3 upper similar to the radials, although a little stouter and in young areoles not easily distinguished from them, the lower central spines stout and strongly hooked, 15 mm. long; flowers not seen; fruit green, suggesting that of a *Coryphantha*, clavate, bearing a few naked scales near top; seed black, smooth, shining.

    Type locality: Near Victoria, Mexico.

    Distribution: Known only from the type locality.

Text-figure 1 is from a photograph of the type specimen collected by Dr. Edward Palmer.

2. Ancistrocactus scheeri (Salm-Dyck).

    Echinocactus scheeri * Salm-Dyck, Cact. Hort. Dyck. 1849. 155. 1850.

Globular to clavate, 3.5 to 5 cm. long; ribs usually 13, indistinct, somewhat spiraled, strongly divided into stout, terete tubercles grooved only to middle; radial spines 15 to 18, spreading, 12

---

        * This name was originally spelled *Echinocactus scheerii*.

M. E. Eaton del.

1. Fruiting plant of *Coryphantha neo-mexicana*.
1a. Fruit of same.
2. Top of flowering plant of *Ancistrocactus scheeri*.
3. Flowering plant of *Cochemiea poselgeri*.
3a. Fruit of same.
3b. Seed of same.
4. Flowering plant of *Coryphantha cornifera*.

mm. long or less, white to straw-colored; central spines 3 or 4, the lowest one strongly hooked; flowers small, 2.5 cm. long, greenish yellow; ovary small, nearly naked; seeds large (about 2 mm, long), brown and minutely tuberculate (according to Coulter).

Type locality: Not cited.

Distribution: Southern Texas and northern Mexico.

It is probable that this species is based on Potts's specimen from Chihuahua and, if so, may be a different species from the one described by Engelmann, which he said was "a most elegant little species, one and a half to two inches high; larger spines black and white variegated." We have not seen Potts's plant, but it was referred here by Hemsley.

Illustrations: Cact. Mex. Bound. pl. 17; Rümpler, Sukkulenten f. 105; Cact. Journ. 1: pl. for March; Schelle, Handb. Kakteenk. 156. f. 84, as *Echinocactus scheeri.*

Plate 11, figure 2, shows a plant collected by Dr. Rose at Laredo, Texas, in 1913, which flowered in the New York Botanical Garden in 1914. Text-figure 2 is from a photograph taken by Robert Runyon of a plant collected in 1921 near Brownsville, Texas.

FIG. 2.—Ancistrocactus scheeri.　　　　FIG. 3.—Ancistrocactus brevihamatus.

3. Ancistrocactus brevihamatus (Engelmann).

Echinocactus brevihamatus Engelmann, Proc. Amer. Acad. 3: 271. 1856.
Echinocactus scheeri brevihamatus Weber in Schumann, Gesamtb. Kakteen 336. 1898.

Globular to obovoid, 5 to 10 cm. high, 5 to 7.5 cm. in diameter, dark green; ribs usually 13, compressed, strongly tubercled; tubercles grooved on upper side from spine-cluster to base, the groove woolly; radial spines 10 to 14, terete, white, 10 to 20 mm. long; central spines 4, the lower one porrect, hooked at apex; flowers rose-colored, 25 to 32 mm. long, not so broad as long; inner perianth-segments 15 mm. long, 4 mm. broad; mid-rib darker colored than margins; fruit about 1.5 cm. long, thin-walled, nearly naked; seeds brownish black, about 2 mm. long, smooth or with low flattened papillae, with a deep-set basal hilum.

Type locality: On the San Pedros, Texas.

Distribution: Southern Texas.

Illustrations: Haage and Schmidt, Haupt-Verz. Cact. **1908**: 226; Schelle, Handb. Kakteenk. 157. f. 85, as *Echinocactus scheeri brevihamatus;* Cact. Mex. Bound. pl. 18, 19; Ann. Rep. Smiths. Inst. **1908**: pl. 3, f. 3; Förster, Handb. Cact. ed. 2. 516. f. 64; Rümpler, Sukkulenten 186. f. 104; Blanc, Cacti 41. No. 414; Engler and Prantl, Pflanzenfam. 3[6a]: 162. f. 56, c, as *Echinocactus brevihamatus.*

Text-figure 3 is a reproduction of plate 18 of the Cactaceae of the Mexican Boundary Survey.

2. THELOCACTUS (Schumann) Britton and Rose, Bull. Torr. Club 49: 251. 1922.

Cacti of medium size, globular or somewhat depressed, spiny, often densely so; ribs few, low or even indefinite, divided into large, often spiraled, tubercles; flowering tubercles more or less grooved above; flowers from near center of plant, borne on very young tubercles, rather large for the subtribe, campanulate, diurnal; scales on ovary usually few, their axils naked; fruit so far as known dry, dehiscing by a basal pore; seeds black, finely tuberculate, with a large basal hilum.

Type species: *Echinocactus hexaedrophorus* Lemaire.

The generic name is from θηλή nipple, and cactus, referring to the tubercled ribs. *Thelocactus* was used for a subgenus of *Echinocactus* by Schumann; he described it with "ribs mostly divided into spirally disposed tubercles or mamillae, not protruding like a chin at base; spines straight or slightly curved." He referred to the group a number of diverse species representing several generic types, some of which we took up in Volume III.

We recognize 12 species, all native of Mexico.

To this genus we have referred the *Echinocacti* of previous authors which seem to intergrade with the *Coryphanthanae*. The group is perhaps complex and may contain two or more distinct genera, but most of the species are little known.

KEY TO SPECIES.

```
Ribs indefinite, strongly tubercled.
  Spines all straight.
    Tubercles not flattened laterally; radial spines 6 to 9........................   1. T. hexaedrophorus
    Tubercles flattened laterally; spines 1 to 5.
      Flowers white....................................................................   2. T. rinconensis
      Flowers not white.
        Flowers salmon to yellow.......................................................   3. T. lophothele
        Flowers rose-purple...........................................................   4. T. phymatothele
  Some of spines curved outward.......................................................   5. T. buekii
Ribs definite, but more or less divided into tubercles.
  Flowers yellowish.
    Ribs 8 to 13......................................................................   6. T. leucacanthus
    Ribs 20 to 25.....................................................................   7. T. nidulans
  Flowers red to purple.
    Spines all straight.
      Spines subulate.................................................................   8. T. fossulatus
      Spines acicular.................................................................   9. T. tulensis
    Spines more or less curved.
      Spines 8 or fewer..............................................................  10. T. lloydii
    Spines numerous.
      Central spines flexible, usually straight, porrect or ascending...........  11. T. bicolor
      Central spines subulate, rigid, some of them curved and reflexed........  12. T. pottsii
```

1. Thelocactus hexaedrophorus (Lemaire) Britton and Rose, Bull. Torr. Club 49: 251. 1922.

Echinocactus hexaedrophorus Lemaire, Cact. Gen. Nov. Sp. 27. 1839.
Echinocactus hexaedrophorus roseus Lemaire in Labouret, Monogr. Cact. 251. 1853.
Echinocactus hexaedrophorus labouretianus Schumann, Gesamtb. Kakteen 438. 1898.
Echinocactus hexaedrophorus major Quehl in Schumann, Gesamtb. Kakteen 438. 1898.

Globose or somewhat flattened above or umbilicate, glaucous, strongly tubercled, not ribbed, 13 to 14 cm. in diameter; tubercles prominent, somewhat 6-sided, 27 mm. broad at base, arranged in indefinite spirals; radial spines 6 to 9, spreading, unequal, 11 to 18 mm. long, rigid, straight, subulate, annulate; central spine much stouter than the radials, erect, 2.3 to 3 cm. long; flowers large, 5.5 cm. long and broader than long when expanded; perianth-segments oblong, purplish; stigma-lobes yellowish white.

Type locality: Tampico, Mexico.

Distribution: Central Mexico.

Schumann refers a plant from San Luis Potosí* to this species. The type, however, is said to have come from Tampico on the coast, while San Luis Potosí is on the table-land at an altitude of 7,000 feet or more, and such an altitudinal distribution is not to be expected. It is possible, but hardly probable, that the plant was actually collected at San Luis Potosí but shipped from Tampico, the port of San Luis Potosí, as such mistakes were common in the early shipments of cacti. Thus, species are attributed to Buenos Aires which came

---

* This plant of the table-land is *Echinocactus fossulatus* Scheidweiler.

from northwestern Argentina, and *Echinocactus insculptus*, referred to below, although reported from Buenos Aires, is really of Mexican origin.

Echinocactus insculptus Scheidweiler (Hort. Belge **4**: 120. pl. 7. 1837) is referred here by Schumann, but the illustration indicates a very different plant.

Echinocactus labouretianus, referred by Schumann (Gesamtb. Kakteen 438. 1898) to Cels's Catalogue, probably never described, is to be referred here.

Illustrations: Cact. Journ. **1**: 181; Lemaire, Icon. Cact. pl. 4; Dict. Gard. Nicholson **1**: f. 690; Balt. Cact. Journ. **2**: 196; Rümpler, Sukkulenten 182. f. 101; Knippel, Kakteen pl. 12; Amer. Gard. **11**: 461; Blanc, Cacti 45. No. 508; Schumann, Gesamtb. Kakteen 437. f. 76; Watson, Cact. Cult. 105. f. 36; ed. 3. f. 25, as *Echinocactus hexaedrophorus*.

FIG. 4.—Thelocactus rinconensis.　　　FIG. 5.—Thelocactus phymatothele.

2. Thelocactus rinconensis (Poselger).

Echinocactus rinconensis * Poselger, Allg. Gartenz. **23**: 18. 1855.

Simple, globose or somewhat depressed, 6 to 8 cm. high, 12 cm. in diameter; ribs somewhat spiraled, strongly tubercled; tubercles more or less flattened laterally, somewhat angled; spines usually only 3, acicular, 1.5 cm. long; flowers white, 4 cm. long; inner perianth-segments lanceolate, acute.

Type locality: Near Rinconada, Mexico.

Distribution: Nuevo León, Mexico.

We do not know this species definitely, but we suspect that the plant collected and illustrated by Safford as *Echinocactus lophothele* belongs here.

Illustrations: Schumann, Gesamtb. Kakteen 433. f. 75; Schelle, Handb. Kakteenk. 197. f. 130, as *Echinocactus rinconadensis*; (?) Ann. Rep. Smiths. Inst. **1908**: pl. 3, f. 1, as *Echinocactus lophothele*.

Text-figure 4 is reproduced from the first illustration cited above.

3. Thelocactus lophothele (Salm-Dyck) Britton and Rose, Bull. Torr. Club **49**: 251. 1922.

Echinocactus lophothele Salm-Dyck, Allg. Gartenz. **18**: 395. 1850.

Simple, or in its native state cespitose, globose, sometimes depressed or short-cylindric, up to 25 cm. high, glaucous; ribs indefinite, strongly tuberculate; tubercles flattened; areoles depressed,

---

* Because this species came from Rinconada, Schumann (Engler and Prantl, Pflanzenfam. 3⁶ᵃ: 189. 1894) has changed the name to *Echinocactus rinconadensis*.

grayish lanate when young; radial spines 3 to 5, stout, purplish brown, 1 to 3 cm. long; central spines wanting or solitary; flowers salmon to yellow, about 5 cm. broad; perianth-segments nearly linear, acute; scales of ovary glabrous, 6 mm. long.

Type locality: Near Chihuahua.
Distribution: Chihuahua, Mexico.
Our description is drawn mostly from the figure in Blühende Kakteen, plate 126.

We have seen flowering specimens of what is called this species at La Mortola, Italy. Although the type came from Chihuahua, we have seen no plant from that region which answers it.

There is a plant in collections, passing as *Echinocactus lophothele longispinus* (Monatsschr. Kakteenk. **15:** 138. 1905), which we do not know.

Illustrations: Schelle, Handb. Kakteenk. 196. f. 129; Blühende Kakteen **3:** pl. 126; Weinberg, Cacti 12; Blanc, Cacti 48. No. 560, as *Echinocactus lophothele.*

4. Thelocactus phymatothele (Poselger).

  Echinocactus phymatothelos * Poselger in Förster, Handb. Cact. ed. 2. 602. 1885.

Simple, depressed-globose, 5 cm. high, 9 to 10 cm. in diameter; ribs 13, glaucous-green, divided into low irregular tubercles, these somewhat flattened and pointed; spines usually 1 to 3, sometimes wanting, subulate, rigid, 2 cm. long, brown, spreading; flowers 6 cm. broad; inner perianth-segments rose-purple to pinkish, narrow, acute; scales on ovary and flower-tube acute.

Type locality: Not cited.
Distribution: Mexico.
This plant is evidently related to *Thelocactus lophothele.*

Illustrations: Möllers Deutsche Gärt. Zeit. **25:** 474. f. 6, No. 24; Blühende Kakteen **3:** pl. 130, as *Echinocactus phymatothelos.*

Text-figure 5 is reproduced from the second illustration above cited.

5. Thelocactus buekii (Klein).

  Echinocactus buekii † Klein, Gartenflora 8: 257. 1859.

Stems simple, deep green; tubercles distinct, somewhat pointed, angled; spines about 7, reddish, unequal, some of them outwardly curved, the longer ones much elongated; flowers dark red; inner perianth-segments narrow.

Type locality: Mexico.
Distribution: Known only from the type locality.
Schumann refers this species to *Echinocactus tulensis,* but it is clearly different from his illustration of that species. Its relationship must be rather with *Thelocactus rinconensis* (see Schumann's figure, No. 75).

This plant is probably named for Dr. Johannes Nicolaus Bück, a botanist and physician of Frankfurt, Germany, and author of the Index to De Candolle's Prodromus.

Illustration: Gartenflora **8:** pl. 266, as *Echinocactus buekii.*

Text-figure 6 is reproduced from the illustration cited above.

6. Thelocactus leucacanthus (Zuccarini).

  Echinocactus leucacanthus Zuccarini in Pfeiffer, Enum. Cact. 66. 1837.
  Cereus tuberosus Pfeiffer, Enum. Cact. 102. 1837.
  Cereus maelenii Pfeiffer, Allg. Gartenz. **5:** 378. 1837.
  Echinocactus porrectus Lemaire, Cact. Aliq. Nov. 17. 1838.
  Echinocactus subporrectus Lemaire, Cact. Aliq. Nov. 25. 1838.
  Echinocactus maelenii ‡ Salm-Dyck, Cact. Hort. Dyck. 1842. 18. 1843.
  Mammillaria maelenii Salm-Dyck, Cact. Hort. Dyck. 1844. 14. 1845.
  Echinocactus leucacanthus tuberosus Förster, Handb. Cact. 287. 1846.
  Echinocactus leucacanthus crassior Salm-Dyck, Cact. Hort. Dyck. 1849. 35. 1850.
  Echinocactus theloideus Salm-Dyck, Allg. Gartenz. 18: 396. 1850.

---

 * This is the original spelling of the name, but it is sometimes written *Echinocactus phymatothele,* the ending being the usual one for specific names of this kind.
 † The original spelling of this name was *buckii,* but on the accompanying plate it was *buekii.*
 ‡ This name is spelled *macleanii* by Hemsley (Biol. Centr. Amer. Bot. **1:** 534. 1880).

Densely cespitose, short-cylindric, 10 to 15 cm. long; ribs 8 to 13, sometimes spiraled, obtuse, tubercled; radial spines 7 to 20, at first light yellow, in age gray, spreading or recurved, unequal, the longer ones 4 cm. long, more or less annulate; central spine solitary, at first blackish, but in age gray, up to 5 cm. long; flowers yellow, 5 cm. long; inner perianth-segments numerous, lanceolate, acute; ovary and flower-tube bearing broad imbricated scales.

Type locality: Near Zimapán, Mexico.
Distribution: Zimapán and Ixmiquilpan, Mexico.

We are inclined to refer here *Echinocactus ehrenbergii* Pfeiffer (Allg. Gartenz. **6:** 275. 1838), which, according to Schumann, also came from Ixmiquilpan, Mexico. In his monograph Schumann describes the flowers as yellow like those of *E. leucacanthus,* but in his English Keys he says that the flowers are rose-red. Dr. Rose, who collected in this region in 1905, found only one species of this relationship.

FIG. 6.—Thelocactus buekii.   FIG. 7.—Thelocactus leucacanthus.

Echinocactus tuberosus Salm-Dyck (Förster, Handb. Cact. 287. 1846) is known only as a synonym.

Echinocactus tuberosus subporrectus (Förster, Handb. Cact. 523. 1846) belongs here.

Illustrations: Pfeiffer and Otto, Abbild. Beschr. Cact. **1:** pl. 14; Abh. Bayer. Akad. Wiss. München **2:** pl. 2, f. 10; pl. 3, f. 4, as *Echinocactus leucacanthus.*

Figure 7 is from a photograph of the plant collected by Dr. Rose at Ixmiquilpan in 1905.

7. Thelocactus nidulans (Quehl).

Echinocactus nidulans Quehl, Monatsschr. Kakteenk. **21:** 119. 1911.

Simple, depressed-globose, 10 cm. high, sometimes 20 cm. in diameter, gray, usually glaucous; ribs 20 to 25, rather indistinct, divided into tubercles; spines about 15, all similar, 2 to 6 cm. long; flowers 4 cm. long, yellowish white.

Type locality: Mexico.
Distribution: Mexico, but known only from cultivated plants.
Illustrations: Monatsschr. Kakteenk. **22:** 51; Alianza Científica Universal **3:** 114, as *Echinocactus nidulans.*

Figure 8 is from a photograph given to Dr. Rose by Frantz de Laet in 1912.

8. Thelocactus fossulatus (Scheidweiler).

Echinocactus fossulatus Scheidweiler, Allg. Gartenz. **9**: 49. 1841.
Echinocactus hexaedrophorus subcostatus Salm-Dyck, Cact. Hort. Dyck. 1849. 34. 1850.
Echinocactus hexaedrophorus fossulatus Salm-Dyck in Labouret, Monogr. Cact. 251. 1853.

Globose to much depressed, 10 to 15 cm. in diameter; ribs usually 13, slightly glaucous, bronzed; tubercles large, somewhat flabby, more or less compressed, dorsally somewhat angled; flowering areoles narrow, sometimes extending forward to next tubercle; radial spines 4 or 5, unequal, 1 to 3.5 cm. long, brown; central spine solitary, 3 to 4.5 cm. long, subulate, annulate; flowers nearly white or slightly tinged with pink; scales on flower-tube ovate, their scarious margins slightly ciliate.

Type locality: Near San Luis Potosí, Mexico.
Distribution: San Luis Potosí, Mexico.

Fig. 8.—Thelocactus nidulans.   Fig. 9.—Thelocactus fossulatus.

Somewhat similar to the foregoing species is C. A. Purpus's No. 15 from Minas de San Rafael, Mexico. This plant has more rounded tubercles, only 4 spines, these all radial and 2 cm. long or less, somewhat flattened.

Thelocactus fossulatus is certainly distinct from *Thelocactus hexaedrophorus*, differing in the arrangement of the tubercles and in the color of the flowers. The former is from an altitude of 7,000 feet, while the other is from near sea-level.

Echinocactus drageanus (Moerder, Rev. Hort. **67**: 186. 1895) and *E. droegeanus* Hildmann (Schumann, Gesamtb. Kakteen 438. 1898) probably belong here, although the latter is referred by Schumann to *Echinocactus hexaedrophorus*. This may be the plant, judging from the name and authorities mentioned, which Schelle (Handb. Kakteenk. 257. 1907) refers to as *Mammillaria rhodantha droegeana* Schumann (*M. droegeana* Hildmann). Schelle questions whether it may not be a distinct species, presumably a *Mammillaria*.

Illustrations: Scientific Amer. **124**: 492, as *Echinocactus*; Curtis's Bot. Mag. **73**: pl. 4311; Ann. Rep. Smiths. Inst. **1908**: pl. 13, f. 3, as *Echinocactus hexaedrophorus*; Pfeiffer, Abbild. Beschr. Cact. **2**: pl. 13, as *Echinocactus fossulatus*; Monatsschr. Kakteenk. **27**: 41, as *Echinocactus hexaedrophorus droegeanus*.

Figure 9 is from a photograph of a plant collected by Dr. Edward Palmer at San Luis Potosí, Mexico, in 1905.

9. Thelocactus tulensis (Poselger).

Echinocactus tulensis Poselger, Allg. Gartenz. **21**: 125. 1853.

Plant simple to abundantly cespitose, globular to short-cylindric, up to 25 cm. high; ribs 8 to 13, strongly tubercled; radial spines 6 to 8, more or less spreading, 10 to 15 mm. long, brownish; central spines solitary or sometimes 2, 3 cm. long; flowers 2.5 cm. long, rose-colored; inner perianth-segments linear-oblong, acute.

Type locality: Near Tula, Tamaulipas, Mexico.
Distribution: Tamaulipas, Mexico.
We have not seen this plant but we have seen two good illustrations. It is closely related to *Thelocactus hexaedrophorus*.
Illustrations: Blühende Kakteen **1**: pl. 18; Schumann, Gesamtb. Kakteen 431. f. 74, as *Echinocactus tulensis*.
Figure 10 is reproduced from the first illustration cited above.

FIG. 10.—Thelocactus tulensis.   FIG. 11.—Thelocactus bicolor.

10. Thelocactus lloydii sp. nov.

Plants simple, depressed-globose, 8 to 12 cm. broad, pale bluish green, strongly tubercled and strongly armed; tubercles conspicuous but low, often wider than long, sometimes 4 cm. wide; flowering groove rather conspicuous but narrow, extending from spines to about half-way to axil of tubercle; spines usually 8, sometimes with a smaller accessory one, all ascending from base and curved outward from middle, terete or somewhat angled at base, often highly colored below with sharp yellowish-crimson tips, the longer ones 6 cm. long; outer perianth-segments very pale purple, never deep purplish pink; filaments white; anthers deep yellow; style yellowish, pinkish at top; stigma-lobes pinkish yellow.

Collected by F. E. Lloyd in northern Zacatecas, Mexico, May 25, 1908 (No. 33).

11. Thelocactus bicolor (Galeotti) Britton and Rose, Bull. Torr. Club **49**: 251. 1922.

Echinocactus bicolor Galeotti in Pfeiffer, Abbild. Beschr. Cact. **2**: pl. 25. 1848.
Echinocactus rhodophthalmus Hooker in Curtis's Bot. Mag. **76**: pl. 4486. 1850.
Echinocactus rhodophthalmus ellipticus Hooker in Curtis's Bot. Mag. **78**: pl. 4634. 1852.
Echinocactus ellipticus Lemaire, Jard. Fleur. **3**: pl. 270. 1853.
Echinocactus bicolor schottii Engelmann, Proc. Amer. Acad. **3**: 277. 1856.
Echinocactus bolansis Rünge, Gartenflora **38**: 106. 1889.
Echinocactus bicolor bolansis Schumann, Gesamtb. Kakteen 303. 1898.
Echinocactus bicolor tricolor Schumann, Gesamtb. Kakteen 303. 1898.
Echinocactus schottii Small, Fl. Southeast. U. S. 814. 1903.

Plants simple, globose to conic, glaucous, small, up to 3 cm. high, very spiny; ribs usually 8, broad, somewhat tubercled; areoles approximate; spines highly colored, sometimes bright red or yellowish or red and yellow; radial spines 9 to 18, widely spreading or sometimes bent backward at tip, 3 cm. long or less; central spines usually 4, ascending or porrect, all straight, 3 to 5 cm. long, subulate; flowers large, 5 to 6 cm. long and fully as broad when expanded; outer perianth-segments pale purple; inner perianth-segments deep purplish pink, oblong, acute; scales on ovary and flower-tube imbricated, ovate, with scarious and ciliate margins; filaments white to purple; stigma-lobes pale to pinkish yellow; fruit small, about 1 cm. long, dehiscing by a large irregular basal opening; seeds 2 mm. long, black, broader at apex, tuberculate with a circular and depressed basal hilum.

Type locality: Mexico.

Distribution: Southern Texas to central Mexico.

Echinocactus tricolor, E. castaniensis, and *E. bicolor montemorelanus* Weber (all in Dict. Hort. Bois 465. 1896) are usually referred here but were never described.

Illustrations: Jard. Fleur 3: pl. 270, as *Echinocactus ellipticus;* Gartenflora 38: 106. f. 21, as *Echinocactus bolansis;* Curtis's Bot. Mag. 76: pl. 4486; Jard. Fleur 1: pl. 101; Loudon, Encycl. Pl. ed. 3. 1377. f. 19375; Gard. Mag. Bot. 1: 40, as *E. rhodophthalmus;* Curtis's Bot. Mag. 78: pl. 4634, as *E. rhodophthalmus ellipticus;* Karsten and Schenck, Vegetationsbilder 2: pl. 20, c; Pfeiffer, Abbild. Beschr. Cact. 2: pl. 25; Schumann, Gesamtb. Kakteen Nachtr. 87. f. 14; Ann. Rep. Smiths. Inst. 1908: pl. 13, f. 2; Blühende Kakteen 2: pl. 74; Monatsschr. Kakteenk. 12: 7; 29: 81; Schelle, Handb. Kakteenk. 157. f. 86; Blanc, Cacti 41. No. 412, as *E. bicolor.*

Figure 11 is from a photograph taken by Robert Runyon at Saltillo, Mexico, in 1921.

12. Thelocactus pottsii (Salm-Dyck).

> *Echinocactus pottsii* * Salm-Dyck, Allg. Gartenz. 18: 395. 1850.
> *Echinocactus bicolor pottsii* Salm-Dyck, Cact. Hort. Dyck. 1849. 173. 1850.
> *Echinocactus heterochromus* Weber, Dict. Hort. Bois 466. 1896.

Globular or somewhat depressed, 10 to 15 cm. in diameter, somewhat glaucous, yellowish; ribs 8 or 9, broad and obtuse, more or less distinctly tubercled; areoles large, closely set on old plants, densely felted when young, naked in age; spines variable as to number, shape, size, and color; radial spines 7 to 10, acicular, usually terete, straight or incurved, more or less banded with red and white or pale yellow, 1 to 3 cm. long; central spines several, stout-subulate, more or less flattened, 3 or 4 cm. long, often white, but sometimes banded with red; flowers 5 to 6 cm. long; scales on ovary and flower-tube ovate, greenish; margins thin and ciliate; inner perianth-segments light purple, darker at base, oblong; stigma-lobes yellow; fruit globose, small, 1.5 cm. in diameter; seed tuberculate, black, truncate at base, ridged on back; hilum basal, white, circular.

Type locality: Near Chihuahua City.

Distribution: Chihuahua to Coahuila, Mexico.

There are three illustrations passing as *Echinocactus pottsii,* none of which agrees with the original description of Salm-Dyck.   Two of these are in Nicholson's Dictionary (Dict. Gard. 4: 540. f. 23 and Suppl. f. 359) where the species is described as follows: flowers yellow, about 2 inches across, short-tubed, several expanding together at the top of the stem; stem globular, 1½ feet in diameter: ridges about a dozen, rounded and even, with acute sinuses; spines 1 inch long, bristle-like, arranged in clusters of 7 or 9, with a cushion of white wool at the base.

Nicholson indicates that his plant of *E. pottsii* was from California and introduced into cultivation in 1840.   There is no Californian species which answers this description or illustration.

The other illustration is Schumann's (Gesamtb. Kakteen 328. f. 57), which is somewhat similar to the above.   Schumann states that the radial spines are commonly 6, spreading and yellow; central spines solitary.   We are not able to identify this illustration; it suggests some *Echinocereus* as much as it does an *Echinocactus.*

---

* Salm-Dyck (Cact. Hort. Dyck. 1849. 35. 1850) credits this name to Scheer.

Illustrations: Knippel, Kakteen pl. 7, in part; Schelle, Handb. Kakteenk. 144. f. 70, as *Echinocactus heterochromus*; Dict. Gard. Nicholson **4:** 540. f. 23; Suppl. 336. f. 359; Schumann, Gesamtb. Kakteen 328. f. 57; Garden **2:** 521; Monatsschr. Kakteenk. **30:** 53; Schelle, Handb. Kakteenk. 155. f. 82; Watson, Cact. Cult. 117. f. 43, as *Echinocactus pottsii.*

PUBLISHED SPECIES, POSSIBLY OF THIS RELATIONSHIP.

ECHINOCACTUS CONOTHELOS Regel and Klein, Ind. Sem. Hort. Petrop. **1860:** 48. 1860.

Ovoid to subcylindric, 10 cm. high, 7.5 cm. in diameter, grayish green; ribs somewhat spiraled, somewhat tubercled at base, the lower tubercles 12 to 20 mm. long; upper areoles oblique, white-tomentose; radial spines 14 to 16, white, spreading to recurved, 8 to 10 mm. long; central spines 2 to 4, erect or a little spreading and recurved, stouter and longer than the radials, 13 to 34 mm. long; flowers and fruit unknown.

Type locality: Near Tanquicillos and Jaumave, Mexico.

This plant was collected by Karwinsky and is known only from his collection. The authors refer the species to Salm-Dyck's section of the *Theloidei*, which, however, is a very diverse group containing representatives of several genera. Schumann was unable to place the species; it may be related to some species of *Thelocactus.*

ECHINOCACTUS HEXAEDRUS Scheidweiler, Bull. Acad. Sci. Brux. **6:** 89. 1839.

Globose to oblong-ovate, glaucous; ribs 18, tuberculate; tubercles 6-angled, gibbous below areoles; areoles oblong, lanate; spines 13, white with purplish bases; lowermost spine longest; central spines 2, either straight or recurved; flowers and fruit unknown.

Type locality: San Luis Potosí.

ECHINOCACTUS SAUSSIERI Weber, Dict. Hort. Bois 468. 1896.

Depressed-globose, 15 to 20 cm. in diameter; ribs spiraled, strongly tubercled; radial spines 9, grayish white, 15 mm. long; central spines 4, acicular, 3 to 4 cm. long; flowers purplish, 4 cm. in diameter; inner perianth-segments lanceolate; stamens and style yellow.

Type locality: Matehuala, state of San Luis Potosí, Mexico.

We know this species from the brief description only and are unable to determine its relationship.

ECHINOCACTUS SMITHII Mühlenpfordt in Otto and Dietrich, Allg. Gartenz. **14:** 370. 1846.

Simple, globose to cylindric, 7 cm. in diameter; ribs 21, often spiraled, strongly tubercled, glaucous; radial spines 20 to 27, setaceous, white, 16 mm. long; central spines 4, the upper one flattened, white with brown or black tips; flowers reddish, 3.5 cm. long; fruit globular, 8 mm. in diameter; seed nearly globular, flattened at the hilum.

Type locality: San Luis Potosí.

We know this species from description only and are unable to assign it to any genus.

ECHINOCACTUS VARGASII Regel and Klein, Ind. Sem. Hort. Petrop. **1860:** 48. 1860.

Globose, 5 cm. high, 6 cm. in diameter; tubercles rather large, somewhat angled, arranged in spirals; radial spines 5 or 6, terete, subulate, brownish, 2 to 6 mm. long; central spine 1, erect, 12 mm. long; flowers and fruit unknown.

Type locality: Mexico, near Río Blanco.

Schumann did not know this plant nor do we, but to us it suggests a *Thelocactus.* The authors of the species compared it with *Echinocactus poselgerianus*, now referred to *Coryphantha*, and with *E. phymatothelos.*

3. NEOLLOYDIA Britton and Rose, Bull. Torr. Club **49**: 251.   1922.

Small, more or less cespitose cacti, fibrous-rooted, cylindric, densely spiny, tubercled; tubercles more or less arranged on spiraled ribs, grooved above; radial spines numerous, widely spreading; central spines one to several, much stouter and longer than radials; flowers large, pink or purple, subcentral from axils of nascent tubercles, their segments widely spreading; fruit compressed-globose, dull-colored, thin-walled, becoming papery, dry, with few scales or none; seeeds globose, black, dull, tuberculate-roughened, with a large white basal scar; embryo straight in typical species.

Type species: *Mammillaria conoidea* De Candolle.

We recognize 7 species from central and northern Mexico and Texas, which have been transferred from *Echinocactus* and *Mammillaria*.   The genus is dedicated to Professor Francis E. Lloyd, whose collections and observations have contributed highly important information to our investigations.

KEY TO SPECIES.

Plants 3 cm. in diameter or less; central spines sometimes wanting..................... 1. *N. pilispina*
Plants larger; central spines always present.
    Central spines curved or hooked... 2. *N. clavata*
    Central spines all straight.
        Central spine solitary.
            Central spine stiff, porrect... 3. *N. horripila*
            Central spine weak, ascending or connivent........................... 4. *N. beguinii*
        Central spines several.
            Spines white or sometimes dark above................................ 5. *N. ceratites*
            Central spines or some of them black.
                Radial spines 25 or more; Mexican species........................ 6. *N. conoidea*
                Radial spines 15 or less... 7. *N. texensis*

1. Neolloydia pilispina (J. A. Purpus).

    Mammillaria pilispina J. A. Purpus, Monatsschr. Kakteenk. **22**: 150.   1912.

Plants cespitose, about 3 cm. in diameter; ribs indistinct, made up of very definite, somewhat angled tubercles young spine-areoles clothed with abundant, long, white wool covering top of

FIG. 12.—Neolloydia pilispina.   FIG. 13.—Neolloydia horripila.

plant; radial spines 6 or 7, 5 to 6 mm. long, weak and spreading, the upper ones longer and connivent over top of plant, 2 cm. long or more, white with blackish tips; central spines often wanting, sometimes one; flowers small, 1.5 to 2 cm. long, purplish; outer perianth-segments brownish.

Type locality: Minas de San Rafael, San Luis Potosí, Mexico.
Distribution: Known only from the type locality.
Figure 12 is from a photograph of a plant collected by C. A. Purpus at the type locality.

2. Neolloydia clavata (Scheidweiler).

Mammillaria clavata Scheidweiler, Bull. Acad. Sci. Brux. **5**: 494. 1838.
Mammillaria stipitata Scheidweiler, Bull. Acad. Sci. Brux. **5**: 495. 1838.
Mammillaria rhaphidacantha Lemaire, Cact. Gen. Nov. Sp. 34. 1839.
Mammillaria ancistracantha Lemaire, Cact. Gen. Nov. Sp. 36. 1839.
Mammillaria rhaphidacantha humilior * Salm-Dyck in Förster, Handb. Cact. 244. 1846.
Mammillaria scolymoides raphidacantha Salm-Dyck, Cact. Hort. Dyck. 1849. 132. 1850.
Echinocactus corniferus rhaphidacanthus Poselger, Allg. Gartenz. **21**: 102. 1853.
? Mammillaria potosiana Jacobi, Allg. Gartenz. **24**: 92. 1856.
Mammillaria sulcoglandulifera Jacobi, Allg. Gartenz. **24**: 92. 1856.
Coryphantha raphidacantha Lemaire, Cactées 34. 1868.
Coryphantha ancistracantha Lemaire, Cactées 34. 1868.
Cactus ancistracanthus Kuntze, Rev. Gen. Pl. **1**: 261. 1891.
Cactus rhaphidacanthus Kuntze, Rev. Gen. Pl. **1**: 261. 1891.
Cactus brunneus Coulter, Contr. U. S. Nat. Herb. **3**: 117. 1894.
Cactus maculatus Coulter, Contr. U. S. Nat. Herb. **3**: 117. 1894.
Mammillaria raphidacantha † *ancistracantha* Schumann, Gesamtb. Kakteen 506. 1898.
Mammillaria radicantissima Quehl, Monatsschr. Kakteenk. **22**: 164. 1912.

Plants simple, elongated, cylindric, 10 to 15 cm. high, dark bluish green; tubercles in rows of 5, 8, and 13, conic, grooved above, the axils when young bearing short white wool; glands in the groove 1 to several, large, red; radial spines 6 to 12, with reddish or black tips; central spine 1, some-what longer than radials, curved or even hooked; flowers small for the genus, about 2 cm. long; outer perianth-segments linear, acute, entire, with broad brownish midrib; inner perianth-segments linear, entire, narrow, creamy white; stamens pinkish, much shorter than the perianth-segments; style pinkish; stigma-lobes 5 or 6, short, greenish.

FIG. 14.—Neolloydia clavata.   FIG. 15.—Neolloydia conoidea.

Type locality: Not cited.
Distribution: San Luis Potosí, Mexico.

The two species of Coulter, *Cactus brunneus* and *Cactus maculatus*, as well as *Mammillaria radicantissima*, came from San Luis Potosí, and all seem to be so much alike that we do not hesitate to reduce them as above.

Echinocactus raphidacanthus is credited by Schumann to Poselger, but he used the name *raphidacanthus* only as a variety of *E. corniferus*. This binomial was used in 1850 by Salm-Dyck for a very different plant.

Mammillaria humilior Förster we have seen only in Schumann's Index (Gesamtb. Kakteen 824. 1898). He refers it to *M. raphidacantha ancistracantha*.

---

* Schumann (Gesamtb. Kakteen 506, 824, Index, 1898), perhaps not intentionally, gives this name specific rank.
† Schumann has dropped the first "h" in *Mammillaria rhaphidacantha* and he is followed by the Monatsschrift für Kakteenkunde.

Illustrations: Blühende Kakteen **1**: pl. 7; Schumann, Gesamtb. Kakteen 505. f. 83, as *Mammillaria rhaphidacantha*; Blühende Kakteen **3**: pl. 163; Monatsschr. Kakteenk. **22**: 165, as *M. radicantissima.*

Figure 14 is from a photograph of a plant collected by Dr. Edward Palmer at San Luis Potosí, Mexico, in 1908 (No. 814).

3. Neolloydia horripila (Lemaire).

> *Mammillaria horripila* Lemaire, Cact. Aliq. Nov. 7.   1838.
> *Echinocactus horripilus* Lemaire, Cact. Gen. Nov. Sp. 91.   1839.
> *Echinocactus horripilus longispinus* Monville in Labouret, Monogr. Cact. 265.   1853.

Simple or somewhat cespitose, globular to short-cylindric, 10 to 12 cm. high; tubercles glaucous, prominent, rounded at apex; radial spines 8 to 10, acicular, spreading, 15 mm. long, grayish; central spine solitary, straight, a little longer than the radials; flowers deep purple, 3 cm. long; inner perianth-segments narrowly oblong, acute; stigma-lobes 5, white.

Type locality: Not cited.
Distribution: Hidalgo, Mexico.

Lemaire first referred this plant to *Mammillaria*, but finally described it as an *Echinocactus* on account of its grooved tubercles; he believed that it was an intergrade between these two genera.   As he states, its general appearance is that of a species of the so-called *Mammillaria.*

Echinocactus caespititius Pfeiffer is usually given as a synonym of this species, but it seems never to have been described.   Schumann cites the place of publication as Salm-Dyck's Cactaceae of 1850 (p. 35), but it is given only as a synonym.   It appeared also in Salm-Dyck's Cactaceae of 1845 (p. 17) and in Förster's Handbuch (p. 283), but also as a synonym.

Illustration: Blühende Kakteen **1**: pl. 6, as *Echinocactus horripilus.*
Figure 13 is reproduced from the illustration above cited.

4. Neolloydia beguinii (Weber) Britton and Rose, Bull. Torr. Club **49**: 252.   1922.

> *Echinocactus beguinii* Weber in Schumann, Gesamtb. Kakteen 442.   1898.

Plant-body cylindric, 10 to 15 cm. high; ribs spiraled and divided at regular intervals into low tubercles resembling geometric figures, pale bluish green in color but nearly hidden by the dense covering of spines; radial spines 20 or more, white, but with dark tips; centrals usually single, longer and ascending; flowers appearing from top of plant, large, 3 to 4 cm. long, bright pink; stigma-lobes 7, long, white; ovary without scales; seeds black, tubercled, with a broad triangular hilum.

Type locality: Probably at Saltillo, in Coahuila, Mexico.
Distribution: Zacatecas and Coahuila, Mexico.

This plant is very distinct from *Echinomastus erectocentrus*, with which it was confused both by Coulter and by Schumann.

Mammillaria beguinii and *Echinocactus beguinii* Weber are referred by Weber (Dict. Hort. Bois 466. 1896) as synonyms of *Echinocactus erectocentrus.*   The Index Kewensis (Suppl. 5) refers the former name to Schelle (Handb. Kakteenk. 200. 1907).   The name *E. beguinii* has been previously used in Rebut's Catalogue and by Schumann (Montasschr. Kakteenk. 5: 44. 1905), but not described.

5. Neolloydia ceratites (Quehl).

> *Mammillaria ceratites* Quehl, Monatsschr. Kakteenk. **19**: 155.   1909.

Simple or in small clusters, short-cylindric, 6 to 10 cm. high; tubercles somewhat 4-angled, more or less arranged in ribs; young areoles very woolly but becoming naked; radial spines 15 to 20, more or less spreading, white, 1.5 cm. long; central spines 5 or 6, longer and stouter than the radials, blackish above; flowers purple, 3 to 3.5 cm. long; perianth-segments oblong, acute.

Type locality: Mexico.
Distribution: Mexico.
Illustration: Monatsschr. Kakteenk. **19**: 155, as *Mammillaria ceratites.*
Figure 16 is from a photograph of the type plant sent us by Mr. Quehl.

6. Neolloydia conoidea (De Candolle) Britton and Rose, Bull. Torr. Club **49:** 252. 1922.

Mammillaria conoidea De Candolle, Mém. Mus. Hist. Nat. Paris **17:** 112. 1828.
Mammillaria grandiflora Otto in Pfeiffer, Enum. Cact. 33. 1837.
Mammillaria diaphanacantha Lemaire, Cact. Aliq. Nov. 39. 1838.
Mammillaria inconspicua Scheidweiler, Bull. Acad. Sci. Brux. **5:** 495. 1838.
Mammillaria echinocactoides Pfeiffer, Allg. Gartenz. **8:** 281. 1840.
Mammillaria scheeri Mühlenpfordt, Allg. Gartenz. **13:** 346. 1845.
Mammillaria strobiliformis Engelmann in Wislizenus, Mem. Tour North. Mex. 113. 1848.
Echinocactus conoideus Poselger, Allg. Gartenz. **21:** 107. 1853.
Cactus conoideus * Kuntze, Rev. Gen. Pl. **1:** 260. 1891.
Cactus echinocactoides * Kuntze, Rev. Gen. Pl. **1:** 260. 1891.
Cactus grandiflorus Kuntze, Rev. Gen. Pl. **1:** 260. 1891. Not Linnaeus, 1753.

Sometimes simple, but usually cespitose, sometimes forming large clusters, often branching or budding above, short-cylindric; tubercles in 5 or 8 spiral rows, obtuse, their axils very woolly; spines very numerous, often completely covering the plant; radial spines white, 25 or more, widely spread-

FIG. 16.—Neolloydia ceratites.   FIG. 17.—Neolloydia conoidea.

ing, 8 to 10 mm. long; central spines several, stouter and longer than the radials, 1 to 3 cm. long, blackish; flowers large; outer perianth-segments dull purple without, lighter toward the margins; inner perianth-segments rich purple; anthers orange; filaments pale yellow, purplish at base; style and stigma-lobes pale yellow, the latter 5 or 6; fruit compressed-globose, dull yellow, mottled with red, becoming dry and papery, then brown; seeds 1 mm. in diameter.

Type locality: Mexico.
Distribution: Northern Mexico.

Mammillaria canescens, listed by De Candolle (Prodr. **3:** 460. 1828) as hardly known and given by Pfeiffer (Enum. Cact. 33. 1837) as a synonym of *M. grandiflora*, doubtless belongs here. A plant of this name was in the Berlin Botanical Garden in 1829 (Verh. Ver. Beförd. Gartenb. **6:** 430. 1830).

The name *Coryphantha conoidea* occurs in C. R. Orcutt's Circular to Cactus Fanciers 1922.

Illustrations: De Candolle, Mém. Cact. pl. 2; Pfeiffer, Abbild. Beschr. Cact. **2:** pl. 26; Blühende Kakteen **2:** pl. 96; Schelle, Handb. Kakteenk. 238. f. 155; Ann. Rep. Smiths.

---

* Kuntze's spelling of these two names is as follows: *C. conodeus* and *C. echinocactodes.*

Inst. 1908: pl. 14, f. 1; Thomas, Zimmerkultur Kakteen 46, as *Mammillaria conoidea*; Monatsschr. Kakteenk. 6: 119, as *Mammillaria grandiflora*.

Figure 15 is from a photograph of a barren plant collected by Dr. Safford in Mexico in 1907 (No. 1334); figure 17 is from a photograph of a flowering plant collected by Dr. Chaffey in the state of Zacatecas, Mexico, July 4, 1910.

Related to the preceding is:

MAMMILLARIA CREBRISPINA De Candolle, Mém. Mus. Hist. Nat. Paris 17: 111.   1828.

    Cactus crebrispinus Kuntze, Rev. Gen. Pl. 1: 260.   1891.

This plant was collected by Thomas Coulter but its identification is very uncertain. Pfeiffer thought that it was related to *Mammillaria conoidea* and perhaps it should be referred there.

Mammillaria polychlora Scheidweiler (Förster, Handb. Cact. 205. 1846) was given as a synonym of *M. crebrispina*.

FIG. 18.—Neolloydia texensis.

7. Neolloydia texensis sp. nov

Globular to short-oblong, 4 to 6 cm. long; tubercles arranged in long spirals, somewhat imbricated, a little flattened dorsally; radial spines 10 to 15, white, widely spreading, about 1 cm. long; central spines 1 to 3, much stouter than the radials, elongated, 2 to 3 cm. long, black; flowers not seen; fruit small, globular, almost hidden by the spines, greenish, thin-walled, dry; seeds black, tuberculate, 1.5 mm. in diameter; hilum large, basal, white lunate.

Collected by MacDougal and Shreve at Sanderson, Texas, December 1920.

This seems to be the plant from Texas referred by Engelmann to *Mammillaria scolymoides* but it probably is not that species which came from central Mexico.   *M. scolymoides* probably should be considered a synonym of *Coryphantha cornifera*, the species of which Engelmann once thought that it might be only a form.   Coulter (Contr. U. S. Nat. Herb. 3: 115. 1894) treats the Texan plant under the name of *Cactus scolymoides* but the range which he gives is too wide, and doubtless more than one species is involved, both in his description and range.   The only specimen which we have seen of this species, except MacDougal and Shreve's plant, is one collected by Walter M. Evans in 1891, which is mixed with *Cactus echinus* and labeled as from near El Paso, Texas.

Figure 18 is from a photograph of plants collected by Dr. MacDougal and Dr. Shreve.

4. MAMILLOPSIS * (Morren) Weber.

Cespitose cacti, often forming large clusters, globular or short-cylindric, completely hidden under a mass of long, soft, white, hair-like spines; tubercles not arranged in ribs, more or less conic, not grooved above, spine-bearing at apex, their axils pubescent and bristly; radial spines numerous, weak, straight; central spines 4 to 6, with yellow, hooked tips; flowers from near top of plant but apparently from axils of old areoles, with a regular, straight, slender, scaly tube and a broad, spreading limb; perianth-segments oblong, obtuse; stamens and style erect, long-exserted beyond tube; scales on flower-tube orbicular, obtuse.

Schumann associated *Mammillaria senilis* Loddiges, the type of the genus, with species now referred to *Cochemiea*, treating them all as a subgenus of *Mammillaria*, but *Cochemiea* has an irregular flower and otherwise is different from this genus.

Morren first proposed the subgeneric name *Mamillopsis*, but Weber, we believe, was justified in recognizing the genus. He states, very properly, that the flowers are very unlike those of any of the species of *Mammillaria*. He also calls attention to the long-exserted stamens, and long and scaly flower-tube, and also to the fact that the filaments are borne in two series, one series being on the flower-tube. The ovary, too, seems to be scaly, and doubtless other differences will be recorded when the species are better known. Two species are here recognized, both from the high mountains of Mexico.

The generic name, *Mamillopsis*, means *Mammillaria*-like.

KEY TO SPECIES.

Flowers 6 to 7 cm. long, orange-yellow.. 1. *M. senilis*
Flowers 3 cm. long, deep red.. 2. *M. diguetii*

FIGS. 19 and 20.—Mamillopsis senilis.

1. Mamillopsis senilis (Loddiges) Weber.

Mammillaria senilis Loddiges in Salm-Dyck, Cact. Hort. Dyck. 1849. 82. 1850.
Cactus senilis Kuntze, Rev. Gen. Pl. 1: 261. 1891. Not Haworth, 1824.

Stems 6 to 15 cm. high, 3 to 6 cm. in diameter, the flesh juicy and drying red; tubercles 3 to 4 mm. long; spines 30 to 40, 2 cm. long; flowers 6 to 7 cm. long, 6 cm. broad, orange-yellow; perianth-segments oblong, acute, with serrated margin; stigma-lobes 6, spreading; fruit not known.

Type locality: Not cited.
Distribution: High mountains of Chihuahua and Durango.

---

* *Mamillopsis* has never been formally published as a genus, but it is mentioned by Weber as a synonym of *Mammillaria senilis* (Dict. Hort. Bois 803. 1898). It was proposed as a subgenus by Morren in 1874 (Belg. Hort. 24: 33).

This species was probably first collected by Seemann in the Sierra Madre of Mexico, where it was collected by Dr. Rose in 1897. It has frequently been introduced into cultivation but does not do well, soon dying out. It is able to stand considerable cold and in its home is usually covered with snow during the winter.

Salm-Dyck gave two varieties without descriptions, based on two unpublished names, when he first listed *Mammillaria senilis*, as follows: *M. senilis haseloffii* (Salm-Dyck, Cact. Hort. Dyck. 1849. 8. 1850; *M. haseloffii* Ehrenberg, Allg. Gartenz. **17**: 303. 1849) and *M. senilis linkei* (Salm-Dyck, Cact. Hort. Dyck. 1849. 8. 1850; *M. linkei* Ehrenberg). The former, however, was published the previous year as *M. haseloffii* and has priority.

Illustrations: Fl. Serr. **21**: pl. 2159; Rev. Hort. IV. **2**: pl. 334; Belg. Hort. 24: pl. 3; Cact. Journ. **1**: pl. for March; Contr. U. S. Nat. Herb. **5**: pl. 62; Schelle, Handb. Kakteenk. 245. f. 163; Tribune Hort. **4**: pl. 140; De Laet, Cat. Gén. 28. f. 41; Gartenwelt **14**: 331; Möllers Deutsche Gärt. Zeit. **25**: 475. f. 8, No. 31; Succulenta **4**: 80, as *Mammillaria senilis*.

Figure 19 is from a photograph of a flowering plant; figure 20 is from a photograph of two flowers of a plant obtained in the Sierra Madre, Mexico, by I. Ochoterena in 1911; figure 21 is reproduced from the third illustration cited above.

FIG. 21.—Mamillopsis senilis.

2. Mamillopsis diguetii (Weber).

Mammillaria senilis diguetii Weber, Bull. Mus. Hist. Nat. Paris **10**: 383. 1904.

Plants densely cespitose, forming a hemispheric clump of about 35 globular heads, each 25 cm. in diameter; radial spines numerous, dark straw-colored; flowers 3 cm. long, about 2 cm. broad, deep red; ovary bearing small scales.

Type locality: Sierra de Nayarit, Jalisco.

Distribution: Jalisco to Sinaloa, Mexico.

This species, until recently, was known only from the single collection of L. Diguet made in March 1900; he found it in the mountains of Jalisco at an altitude of 2,500 meters. It has again been collected by J. G. Ortega in the Sierra de Chabarra, Concordia, Sinaloa, in 1921.

The type is in the Museum of Natural History of Paris and was studied there by Dr. Rose in May 1912; he believes that it is distinct from *M. senilis*, the spines being of a different color and much more rigid than in that species.

5. COCHEMIEA (K. Brandegee) Walton, Cact. Journ. 2: 50. 1899.

Plant-body cylindric, often much elongated, the surface covered with spirally arranged tubercles, these not milky; tubercles not grooved above; spines both central and radial; flowers borne from axils of upper old tubercles, narrowly tubular, curved and bilabiate; perianth-segments in 2 series; stamens and style red, exserted; ovary naked; fruit indehiscent, globular, red, naked, bearing a large scar at top; seeds black, reticulated.

Type species: *Mammillaria halei* Brandegee.

The genus was named for an Indian tribe which once inhabited Lower California. Mrs. Brandegee, who first separated these species as a subgenus, describes the flowers as "scarlet, tubular, slender, somewhat curved, and oblique, with spreading unequal petaloid sepals, so making the flower apparently double as in *Cereus flagelliformis*."

Four species are known, all inhabiting Lower California.

FIG. 22.—Cochemiea halei. FIG. 23.—Cochemiea poselgeri.

The fact that *Cochemiea* had been raised to generic rank, to which four species had been transferred, has been overlooked by all our botanical indexes. Walton's remarks in this connection are interesting:

"The plants so classed have flowers very elongated, tubular, with sepals placed as a second ring, removed some distance below the petals; they are oblique like *Epiphyllum truncatum* and *Cereus flagelliformis* and in fact more resemble those flowers than they do those of any *Mammillaria*, so much so that I think it would be best to drop the generic name of *Mammillaria* and simply adopt Mrs. Brandegee's name of *Cochemiea* as a generic name."

Mrs. Brandegee suggested (Erythea 5: 117), "It is possible that some of the elongated species of Mexico proper will be found to belong to this section when the flowers are better known." But we have seen no plants from the mainland of Mexico which suggest this relationship.

KEY TO SPECIES.

Spines all straight... 1. *C. halei*
Some or all of central spines hooked.
   Central spine normally solitary.. 2. *C. poselgeri*
   Central spines normally 2 to 11 (sometimes only 1 in *C. setispina*).
      Central spines 1 to 4.. 3. *C. setispina*
      Central spines 8 to 11... 4. *C. pondii*

1. **Cochemiea halei** * (Brandegee) Walton, Cact. Journ. **2:** 50. 1899.

   Mammillaria halei Brandegee, Proc. Calif. Acad. II. **2:** 161. 1889.
   Cactus halei Coulter, Contr. U. S. Nat. Herb. **3:** 106. 1894.

Cespitose; stems nearly upright, often 30 to 50 cm. high, 5 to 7.5 cm. in diameter, almost entirely covered by the spines; tubercles short; axils of tubercles woolly but not setose; radial spines 10 to 20, 10 to 12 mm. long; central spines 3 or 4, 25 mm. long, all straight; flowers central or nearly so, 4 to 5 cm. long; filaments yellow; stigma-lobes scarlet; fruit scarlet, 12 mm. long; seeds reticulated.

Type locality: Magdalena Island, Lower California.

Distribution: Islands of southern Lower California.

This species was observed first by Mr. T. S. Brandegee in 1889, while making a botanical excursion through Lower California, and described by him the same year. It has been reported from only two islands off the coast of Lower California but it is there very abundant. It has been introduced into Europe and is sometimes offered in the trade. It is remarkable for its very large slender flowers. An abundance of material was collected by Dr. Rose in 1911. The plant does not do well in cultivation.

The species was named for Mr. J. P. Hale, who had extensive domains in Lower California and who assisted Mr. Brandegee while making explorations in 1889.

Illustrations: Proc. Calif. Acad. II. **2:** pl. 6; Monatsschr. Kakteenk. **5:** 89; Schumann, Gesamtb. Kakteen 510. f. 84; Thomas, Zimmerkultur Kakteen 47, as *Mammillaria halei*.

Figure 22 is from a photograph of a barren shoot of a specimen collected by C. R. Orcutt at Magdalena Bay, Lower California, 1917.

2. **Cochemiea poselgeri** (Hildmann).

   Mammillaria poselgeri Hildmann, Garten-Zeitung **4:** 559. 1885.
   Mammillaria roseana Brandegee, Zoe **2:** 19. 1891.
   Mammillaria radliana Quehl, Monatsschr. Kakteenk. **2:** 104. 1892.
   Cactus roseanus Coulter, Contr. U. S. Nat. Herb. **3:** 105. 1894.
   Cochemiea rosiana Walton, Cact. Journ. **2:** 50. 1899.

Stems numerous from a central root, spreading or sometimes pendent from rocks or creeping over the ground, often 2 meters long, 4 cm. thick; areoles and upper axils white-woolly, the latter rarely setose; tubercles remote, somewhat flattened; radial spines 7 to 9, 9 to 12 mm. long, straw-colored; central spine 1, hooked, 25 mm. long; flowers appearing in the upper axils, 3 cm. long, scarlet; stamens and style exserted; fruit globular, 6 to 8 mm. in diameter.

Type locality: Cape Region, Lower California.

Distribution: At lower elevations in southern Lower California.

This cactus, according to Mr. Brandegee, is one of the most showy of this region.

Mammillaria longihamata Engelmann was a manuscript name taken up by Coulter (Contr. U. S. Nat. Herb. **3:** 105. 1894) as a synonym of *Cactus roseanus*.

Illustrations: Thomas, Zimmerkultur Kakteen 49; Monatsschr. Kakteenk. **2:** 105, as *Mammillaria radliana;* Garten-Zeitung **4:** 559. f. 131; Schelle, Handb. Kakteenk. 246. f. 164, as *M. poselgeri*.

Plate 11, figure 3, shows a plant collected by Dr. Rose at Cape San Lucas, Lower California, which flowered in the New York Botanical Garden in 1915; figure 3a shows the fruit and figure 3b the seed from a plant collected by Dr. Wm. S. W. Kew near La Junta, Lower California, November 10, 1920. Figure 23 is from a photograph of a plant collected by C. R. Orcutt near Magdalena, Lower California, and sent to the Bureau of Chemistry, U. S. Department of Agriculture, in 1917.

?. **Cochemiea setispina** (Coulter) Walton, Cact. Journ. **2:** 51. 1899.

   Cactus setispinus Coulter, Contr. U. S. Nat. Herb. **3:** 106. 1894.
   Mammillaria setispina Engelmann in K. Brandegee, Erythea **5:** 117. 1897.

Stems ascending, 30 cm. high; tubercles short; axils of tubercles woolly but not setose; radial spines 10 to 12, white with black tips, widely spreading, unequal, 10 to 34 cm. long, slender; central spines 1 to 4, stouter than the radials; one of them strongly hooked; flowers not definitely known but probably large; fruit obovoid, 3 cm. long, scarlet; seeds black and pitted.

---

* Walton published this name as *Cochemiea hallei*.

Type locality: San Borgia, Lower California.

Distribution: Interior of southern Lower California.

We have not seen living specimens of the species. Dr. Rose obtained a small specimen from L. Quehl at Halle in 1912.

The type of this species, now in the herbarium of the Missouri Botanical Garden, was collected by William Gabb in 1867, while Brandegee obtained specimens in 1889. Dr. C. A. Purpus found it near Calmalli and wrote of it as follows (Cact. Journ. **2**: 54. 1899):

"My next trip was to a chain of granite mountains about 20 miles from Calmalli.

"I was very much surprised to find on the slope of the mountains *Mamillaria setispina* Engelmann, which until now I had not been able to collect as a living specimen. I came upon it afterwards also in gneiss, trachyt, porphur, and in a sandstone conglomerate. Ground composed of granite gravel appears to suit it best."

4. Cochemiea pondii (Greene) Walton, Cact. Journ. **2**: 51.   1899.

> *Mammillaria pondii* Greene, Pittonia **1**: 268.   1889.
> *Cactus pondii* Coulter, Contr. U. S. Nat. Herb. **3**: 102.   1894.

Stems at first upright, cylindric, simple or few-branched, 7 cm. to 3 dm. high, hidden under a dense covering of spines; axils of tubercles setose; young areoles white-tomentose; radial spines white, whitish or sometimes brownish, 15 to 25, spreading; central spines 8 to 11, much longer and stouter than the radials, the longest 3 cm. long, 1 or 2 hooked; flowers slender, 5 cm. long, bright scarlet; stamens exserted; fruit purplish red, 18 mm. long, ovoid to obovoid.

Type locality: Cedros Island.

Distribution: Islands off the western coast of northern Lower California.

This plant was found in great abundance on Cedros Island by Dr. Rose in 1911 (No. 16090) and a number of living specimens was brought to Washington and New York. These have been in cultivation for more than ten years but have never flowered. It is not often met with in cultivation.

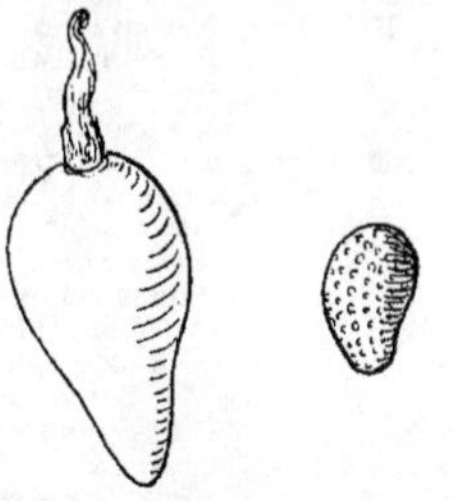

FIGS. 24 and 24a.—Fruit and seed of Cochemiea pondii.

The species was named for Charles Fremont Pond, U. S. N., who collected plants on Cedros and other islands off the coast of Lower California in 1889.

Figure 24 shows the fruit and figure 24a the seed from specimens obtained at the type ocality by Dr. Rose in 1911.

6. CORYPHANTHA (Engelmann) Lemaire, Cactées 32.   1868.

Plant body globular to cylindric, either solitary or cespitose; tubercles, except the very earliest ones, grooved on upper surface* from apex to base; flowers from near top of plant and from base of young and growing tubercles, large and showy, generally yellow, sometimes purple or red; ovary naked or, occasionally, bearing a few scales in some species; perianth long-persistent†; fruit large,‡ ripening slowly, ovoid to oblong, greenish or yellowish; seeds brown (black and angled in *Coryphantha cubensis*), lightly reticulated or nearly smooth, thin-shelled, with a central or subventral hilum; embryo curved, at least in some species.

Type species:§ *Mammillaria sulcolanata* Lemaire.

The generic name is from κορυφή top, and ἄνθος flower, referring to the insertion of the flowers at the top of the plant. We recognize 37 species in the genus. The genus *Coryphantha* was proposed by Lemaire in 1868, but he did not designate a type. The

---

* In *C. macromeris* the tubercle is grooved only for about half its length.

† We quote the following observation of Engelmann in this connection: "I have repeatedly observed, and in a considerable number of species, that the red berries of the *Mammillariae* are always destitute of the remnants of the perigone, but the green fruits always are topped with it (Mem. Tour North. Mex. 21).

‡ The only fruit which we have seen of *C. nickelsae* was globose and small, 5 to 7 mm. in diameter, but the species otherwise of this alliance.

§ See Britton and Millspaugh, Bahama Flora 295.   1920.

name, however, comes from Engelmann, who first used it as a subgenus of *Mammillaria* (Proc. Amer. Acad. **3**: 264. 1856).

The position of this group has always been puzzling to cactus students. Dr. Poselger believed that it was a section of *Echinocactus* and transferred certain of these species which had been described under *Mammillaria* to *Echinocactus*. In its vertical, nearly central flowers it does approach the *Echinocactanae*, but otherwise it is quite distinct.

In the origin of their large flowers, in the shape and structure of their fruit, and in the color and form of their seeds the species compose a rather natural group, but they are diverse in form and armament. The species are most common in central Mexico, a few extending into the southern United States, and one extending into southern Canada.

The groove on the upper side of the tubercle which is so characteristic of the genus does not occur on seedlings or on very young plants, but it is always found on old flowering plants and seems to be associated with the inflorescence, for the flowers appear only in the axils of grooved tubercles and originate at the bottom of this groove. Plants which grow in conservatories for a long time without flowering lose this groove;* we have had one plant of this kind under observation for fifteen years.

KEY TO SPECIES.

A. Seeds brown, not angled; flowers usually large.
  B. Tubercles grooved to middle or a little below; ovary bearing scales with woolly
    axils. Series *Macromeres.*
      Tubercles elongated, bright green.. 1. *C. macromeris*
      Tubercles short, grayish green.. 2. *C. runyonii*
  BB. Tubercles grooved from tip to base except in young plants; ovary naked.
    C. Grooves of tubercles bearing large yellow or red glands. Series *Recurvatae.*
      Flowers white.. 3. *C. ottonis*
      Flowers not white.
        Stems globular.
          Radial spines more or less recurved............................. 4. *C. recurvata*
          Radial spines spreading or ascending.
            Spines dark, sometimes black................................ 5. *C. poselgeriana*
            Spines yellow or sometimes tinged with red.
              Central spines slender and flexible........................... 6. *C. muehlenpfordtii*
              Central spines stout and rigid.
                Radial spines subulate..................................... 7. *C. guerkeana*
                Radial spines acicular..................................... 8. *C. echinoidea*
        Stems cylindric.
          Stems bluish green.. 9. *C. clava*
          Stems yellowish green.
            Central spine usually one.
              Glands in groove red..................................... 10. *C. octacantha*
              Glands in groove yellow.................................. 11. *C. exsudans*
            Central spines 2... 12. *C. erecta*
    CC. Grooves of tubercles without large glands. Series *Sulcolanatae.*
      D. Outer perianth-segments not ciliate.
        E. Flowers purplish or rose................................... 13. *C. elephantidens*
        EE. Flowers yellow or white.
          F. Tubercles very large, broader than high.................... 14. *C. bumamma*
          FF. Tubercles of medium size, if large, longer than broad.
            G. Plants large for this genus (often 8 cm. in diameter); seeds
              3 mm. in diameter............................... 15. *C. robustispina*
            GG. Plants much smaller than in *C. robustispina;* seeds 2 mm.
              in diameter or less.
              H. Central spines usually wanting.
                Secondary cluster of spines developed in upper part of
                  areoles and connivent at top................. 16. *C. connivens*
                Secondary cluster of spines not developed.
                  Spines pectinate............................. 17. *C. pectinata*
                  Spines not pectinate.
                    Spines 14 or more.
                      Spines slender with long black tips......... 18. *C. nickelsae*
                      Spines rather short with light tips.
                          Spines subulate....................... 19. *C. compacta*
                          Spines acicular........................ 20. *C. radians*
                    Spines fewer than 15.
                      Spines slender and weak.................. 21. *C. sulcolanata*
                      Spines not slender....................... 22. *C. retusa*

---

* *Mammillaria potosiana* and *M. polymorpha* seem to have been based on such plants.

HH. Central spines one to several.
   I. Central spines strongly hooked................ 23. *C. palmeri*
   II. Central spines straight or at most curved.
      J. Central spines more or less curved.
         Central spine one, sometimes more in No. 25.
            Radial spines nearly as long as central.... 24. *C. cornifera*
            Radial spines about half as long as central.. 25. *C. salm-dyckiana*
         Central spines several.
            Radial spines 20 or more............... 26. *C. pallida*
            Radial spines 12 or fewer............... 27. *C. pycnacantha*
      JJ. Central spines straight.
         Radial spines, two kinds (to be looked for here). 5. *C. poselgeriana*
         Radial spines of one kind.
            Plant almost hidden under mass of spines;
               fruit oblong........................ 28. *C. echinus*
            Plant not hidden under mass of spines; fruit
               globular........................... 29. *C. durangensis*
  DD. Outer perianth-segments ciliate.
    Flowers yellow.................................. 30. *C. chlorantha*
    Flowers purplish to pink.
      Inner perianth-segments linear or lanceolate.
        Stigma-lobes purple, apiculate............... 31. *C. vivipara*
        Stigma-lobes white, obtuse or notched.
          Flowers 4 to 7 cm. broad, rose to purple.
            Plants mostly solitary; inner perianth-segments broadly linear 32. *C. neo-mexicana*
            Plants mostly cespitose; inner segments linear-lanceolate.... 33. *C. arizonica*
         Flowers very short, 3 cm. broad, light pink............... 34. *C. deserti*
      Inner perianth-segments oblanceolate................... 35. *C. aggregata*
AAA. Seeds black, angled; flowers minute. Series *Cubenses*.......................... 36. *C. cubensis*
 AA. Ungrouped species.. 37. *C. sulcata*

1. Coryphantha macromeris (Engelmann) Lemaire, Cactées 35. 1868.

    Mammillaria macromeris Engelmann in Wislizenus, Mem. Tour North. Mex. 97. 1848.
    Mammillaria heteromorpha Scheer in Salm-Dyck. Cact. Hort. Dyck. 1849. 128. 1850.
    Echinocactus macromeris Poselger, Allg. Gartenz. 21: 102. 1853.
    Echinocactus heteromorphus Poselger, Allg. Gartenz. 21: 126. 1853.
    Mammillaria dactylithele Labouret, Monogr. Cact. 146. 1853.
    Cactus macromeris Kuntze, Rev. Gen. Pl. 1: 260. 1891.
    Cactus heteromorphus Kuntze, Rev. Gen. Pl. 1: 260. 1891.

Plant branching at base, often many-headed, up to 2 dm. long; tubercles large, soft, loosely arranged, elongated, 12 to 30 cm. long, grooved on upper side about two-thirds of their length; spines 10 to 17, slender, the radials white; central spines several, the longer ones 5 cm. long; flowers large, purple, 6 to 8 cm. broad; scales on flower-tube ciliate; ovary bearing a few scales with hairy axils; fruit 15 to 25 mm. long; seeds globose, brown but sometimes described as yellow, smooth.

    Type locality: Near Doñana, New Mexico.
    Distribution: Southern New Mexico, western Texas, and Chihuahua, south to Zacatecas, Mexico.

    This species and the following one are not closely related to the others of this genus. The tubercles are much more elongated and flattened, and the groove on the upper surface never extends to the base. Sometimes a branch or bulblet is produced instead of a flower.

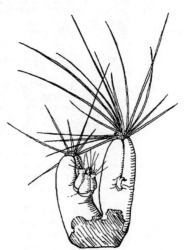

FIG. 25.—Tubercles of Coryphantha macromeris.

    Here may belong *Coryphantha heteromorpha* Lemaire (Cactées 34. 1868); this name is apparently erroneously referred to in the Index Kewensis (1: 624) as *Coryphantha heterophylla* (see *Ariocarpus fissuratus*, Cactaceae 3: 83).

    Mammillaria brownii Toumey was erroneously referred here by Schumann.

    Mammillaria macromeris var. *longispina* and var. *nigrispina* are mentioned by Schelle (Handb. Kakteenk. 237. 1907).

Illustrations: Cact. Journ. **1**: 43; Förster, Handb. Cact. ed. 2. 399. f. 41; Rümpler, Sukkulenten 205. f. 116; Dict. Gard. Nicholson **4**: 564. f. 36; Suppl. 517. f. 552; Goebel, Pflanz. Schild. **1**: pl. 1, f. 6; Amer. Gard. **11**: 460; West Amer. Sci. **13**: 39; Cact. Mex. Bound. pl. 14, 15; Cycl. Amer. Hort. Bailey **2**: f. 746a, 1355; Stand. Cycl. Hort. Bailey **4**: f. 2314; Gartenflora **42**: 543. f. 111; Schelle, Handb. Kakteenk. 237. f. 152; Balt. Cact. Journ. **1**: 21; Watson, Cact. Cult. 165. f. 64; ed. 3. f. 41, as *Mammillaria macromeris.*

Figure 25 is from a drawing of two tubercles, showing the grooves on the upper side, of a plant sent by Mrs. S. L. Pattison from western Texas. At the base of one is shown the flower-scar; in the other is a small bud.

2. Coryphantha runyonii sp. nov.

Forming low clumps, sometimes 5 dm. in diameter, grayish green, with a thick, elongated tap-root; tubercles rather short, 1 to 2 cm. long, terete or somewhat flattened, grooved on the upper half, rarely more, but never to the base; radial spines 6 or more, spreading, acicular, very variable in length, 3 cm. long or less, sometimes all yellow or sometimes one or more in a cluster brown, otherwise yellow; central spines on young plant solitary, dark brown to black but in old plants sometimes 2 or 3, somewhat angled, up to 6 cm. long; flowers large, purple, 5 cm. broad; outer perianth-segments ciliate; inner perianth-segments spatulate, oblong, acute; fruit green; seeds brown.

Found along the Rio Grande from Brownsville to Rio Grande City. This species has been repeatedly observed by Robert Runyon, from whom we received living plants in 1921 (No. 15, type) and 1922.

Mr. Runyon wrote us about the plant as follows:

"I also inclose you herewith two photographs of the plant you have called *Coryphantha runyonii.* I first became interested in this plant about two years ago when I saw it growing near Rio Grande, Texas. It was found at one place only, but in abundance. It grows on the gravel hillside and down in the lower land in a kind of white silt soil.

"The fruit is green and the flowers are a very pretty pink to a purple with a delicate fringed petal. The tubercles are very irregular. The largest plants are about 18 inches in diameter and would weigh not less than fifty pounds."

Plate 1, figure 1, is from a photograph sent us by Robert Runyon.

3. Coryphantha ottonis (Pfeiffer) Lemaire, Cactées 34. 1868.

> *Mammillaria ottonis* Pfeiffer, Allg. Gartenz. **6**: 274. 1838.
> *Echinocactus ottonianus* Poselger, Allg. Gartenz. **21**: 102. 1853.
> *Cactus ottonis* Kuntze, Rev. Gen. Pl. **1**: 261. 1891. Not Lehmann, 1827.
> *Mammillaria bussleri* Mundt in Schumann, Monatsschr. Kakteenk. **12**: 47. 1902.
> *Mammillaria golziana* Haage jr., Monatsschr. Kakteenk. **19**: 100. 1909.

Simple, globular to short-cylindric, 12 cm. high or less, 8 cm. in diameter, glaucous to grayish green; radial spines 8 to 12, nearly equal, 8 to 10 mm. long; central spines 3 or 4, longer and a little stouter than the radials; axils of flowering tubercles woolly; flowers white, 4 cm. long; outer perianth-segments oblong, obtuse; inner perianth-segments apiculate; stigma-lobes 10, green.

Type locality: Mineral del Monte, Mexico.
Distribution: Central Mexico.

The name here used was proposed by Lemaire (Cactées 34) in 1868 but not formally published. *Mammillaria ottonis tenuispina* Pfeiffer is sometimes used but we have seen no formal description.

Nicholson (see also Watson, Cact. Cult. 168. f. 66; and ed. 3. f. 40) describes and illustrates (Dict. Gard. Nicholson Suppl. 517. f. 553) under this name a very peculiar specimen in which the flowers are borne away from the top of the plant; it is doubtless not congeneric with this species. Nicholson's description is here quoted:

"Flowers white, large for the size of the plant. May and June. Stem small, compressed, 3 in. across, with numerous compressed tubercles, and short hair-like spines (Mexico. 1834. See fig. 553). There is another species called *M. ottonis*, having a large spiny stem."

Here we believe belong some of the plants which are passing as *Mammillaria golziana*. Very different, however, are the two published illustrations of Kunze (Cact. 1910 and Monatsschr. Kakteenk. **19**: 101. 1909), which also seem to differ from each other.

Illustrations: Monatsschr. Kakteenk. **12**: 47, as *Mammillaria bussleri*; Monatsschr. Kakteenk. **27**: 3. f. a, as *Mammillaria golziana*; Monatsschr. Kakteenk. **27**: 3. f. b, as *Mammillaria ottonis*.

4. Coryphantha recurvata (Engelmann).

Mammillaria recurvispina Engelmann, Proc. Amer. Acad. **3**: 266.   1856.   Not De Vriese, 1839.
Mammillaria recurvata Engelmann, Trans. St. Louis Acad. **2**: 202.   1863.
Cactus recurvatus Kuntze, Rev. Gen. Pl. **1**: 259.   1891.
Cactus engelmannii * Kuntze, Rev. Gen. Pl. **1**: 260.   1891.

FIGS. 26 and 27.—Coryphantha recurvata.

Plant-body depressed-globose, 10 to 20 cm. in diameter, often forming large masses 30 to 90 cm. in diameter and sometimes with over 50 heads; tubercles low; radial spines about 20, yellow to gray, with dark tips, pectinate, recurved; central spines 1, rarely 2, longer and darker than the radials, 12 to 20 mm. long, more or less reflexed, often appressed; flowers 25 to 35 mm. long, said to be brownish outside; inner perianth-segments lemon-yellow; fruit not known.

Type locality: Sonora.   Explained in the Cactaceae of the Mexican Boundary to be eastern parts of Pimeria Alta in Sonora, especially in the Sierra del Pajarito.

Distribution: Arizona and Mexico, especially along the United States-Mexican Boundary near Nogales.

Engelmann describes † a peculiar flowering habit for *Coryphantha* when he says that the flowers originate in the base of the grooves of full-grown tubercles, being scattered over the top of the plant. We have also noticed this character; not only are the flowers borne in the axils of mature tubercles, but they are produced in great abundance in a circle 5 to 6 cm. in diameter.

---

* It is possible that Lemaire also gave the name *Coryphantha engelmannii* for *Mammillaria recurvispina*, though this is not shown by the text.
† See Cact. Mex. Bound. **12**.   1859.

Otto Kuntze made the binomial *Cactus engelmannii* because, as he states, the name *Mammillaria recurvispina* De Vriese had priority over Engelmann's name. Engelmann, however, had long before renamed his plant.

Mammillaria nogalensis Rünge (Schumann, Gesamtb. Kakteen 494. 1898) has been referred here as a synonym, but this name had already been used by Walton.

Illustrations: Schelle, Handb. Kakteenk. 239. f. 156, as *Mammillaria recurvata*; Cact. Journ. 1: pl. for March; 2: 148; pl. for September, as *M. nogalensis*.

Figure 26 is from a photograph by Dr. MacDougal at Calabasas, showing a clump; figure 27 is from a photograph of a plant sent by F. J. Dyer from Nogales.

5. Coryphantha poselgeriana (Dietrich).

Echinocactus poselgerianus Dietrich, Allg. Gartenz. 19: 346. 1851.*
Echinocactus saltillensis Poselger, Allg. Gartenz. 21: 101. 1853.
Echinocactus salinensis Poselger, Allg. Gartenz. 21: 106. 1853.
Mammillaria difficilis Quehl, Monatsschr. Kakteenk. 18: 107. 1908.
Mammillaria valida J. A. Purpus, Monatsschr. Kakteenk. 21: 97. 1911. Not Weber, 1898.

Plant-body large for the genus, globular, bluish green; tubercles large, closely packed together and at base strongly angled; radial spines of two kinds, the 4 or 5 lower ones spreading, subulate, reddish to black, about as long as the single central one (2 to 4 cm. long); the upper radials, 5 to 8, ascending together, yellowish with black tips, weak, acicular; flower large, 4 to 5 cm. long and nearly as broad when expanded, flesh-colored, the segments spatulate, usually rounded at apex; fruit oblong, 15 mm. long; seeds brownish.

Type species: Near Saltillo, Mexico.

Distribution: States of Nuevo Leon, Coahuila, and Zacatecas, Mexico.

Two different plants have been passing under the name *Echinocactus saltillensis*. The one now in the trade, called *E. ingens* var. *saltillensis* by Schumann, is a very large plant and is a true *Echinocactus* which we have already elsewhere described as *E. palmeri*;† the other, which is the one originally described by Poselger, is a small globular *Coryphantha* and has usually been taken for *Mammillaria scheeri*, more recently described as *M. valida*.

The clusters of connivent weak spines, so characteristic of this species, are not always shown in young plants and this may account for certain seeming discrepancies in the original descriptions. The nascent spines are sometimes red, bleaching white; the gland in the groove of the tubercle is bright red.

Illustrations: De Laet, Cat. Gén. f. 44; Schelle, Handb. Kakteenk. 239. f. 157; Tribune Hort. 4: pl. 139; Rev. Hort. Belg. 40: after 196, as *Mammillaria radians*; Monatsschr. Kakteenk. 21: 99, as *Mammillaria valida*; (?) Blanc, Cacti 50. No. 599; (?) Cact. Journ. 2: 55, as *Echinocactus poselgerianus*; Monatsschr. Kakteenk. 18: 107, as *Mammillaria difficilis*; Rother, Praktischer Leitfaden Kakteen 31, as *Echinocactus scheeri*.

6. Coryphantha muehlenpfordtii (Poselger).

Mammillaria scheeri Mühlenpfordt, Allg. Gartenz. 15: 97. 1847. Not Mühlenpfordt, 1845.
Echinocactus muehlenpfordtii Poselger, Allg. Gartenz. 21: 102. 1853.
Mammillaria scheeri valida Engelmann, Proc. Amer. Acad. 3: 265. 1856.
Coryphantha scheeri Lemaire, Cactées 35. 1868.
Cactus scheeri Kuntze, Rev. Gen. Pl. 1: 261. 1891.*

Plants nearly globular, usually simple, short-oblong, 20 cm. long, 7.5 to 15 cm. in diameter; tubercles large, 1 to 2.5 cm. long; axils of young tubercles grooved and young spine-areoles very woolly; grooves bearing large dark-colored glands; spines variable, reddish to yellow with brown to black tips; radials 6 to 16, usually about 2 cm. long, straight; central spines 1 to 4, subulate, stouter than the radials, 3 to 3.5 cm. long, from nearly straight to curved at tip or even strongly hooked; flowers yellow, 6 cm. long; scales on flower-tube and outer perianth-segments more or less lacerated; inner perianth-segments oblong, entire, acute; fruit greenish, oblong, 3 to 3.5 cm. long, naked; seeds large, 3 mm. long, brown, shining, smooth.

Type locality: Mexico.

Distribution: Northern Chihuahua, western Texas, and southern New Mexico.

---

* We have not seen the type of this species but Bödeker has sent us a copy of the photograph of it left by Poselger.
† See Contr. U. S. Nat. Herb. 12: 290. 1909; Britton and Rose, Cactaceae 3: 172. 1922.

There has been considerable confusion regarding this species, which was first described as *Mammillaria scheeri* by Mühlenpfordt in 1847, but this proved to be a homonym. This led Poselger in 1853, when he transferred the species to *Echinocactus*, to publish it as *E. muehlenpfordtii*.

Dr. Engelmann in 1856 described a variety of *Mammillaria scheeri*, calling it *valida*. Some time afterwards he compared this variety with the type of the species and decided that they were the same. We have examined several specimens from near the type locality of the variety *valida*, which is near El Paso, Texas.

It is possible that Scheer's plant was a very young one, which might account for the differences in form and spines. The *Mammillaria scheeri* of Schumann's Monograph is a complex of 4 or 5 distinct species.

Illustrations: Allg. Gartenz. **15**: 97. pl. 2; Förster, Handb. Cact. ed. 2. 406. f. 44; Schumann, Gesamtb. Kakteen 485. f. 80; Monatsschr. Kakteenk. **8**: 23; **10**: 127; Schelle, Handb. Kakteenk. 237. f. 153, as *Mammillaria scheeri*.

FIG. 28.—Coryphantha muehlenpfordtii.   FIG. 29.—Coryphantha bumamma.

Figure 28 is from a photograph of a plant collected in western Texas by Mrs. S. L. Pattison in 1920.

7. Coryphantha guerkeana (Bödeker).

Mammillaria guerkeana Bödeker, Monatsschr. Kakteenk. **24**: 52. 1914.

Plant-body globular, 6 to 7 cm. in diameter; tubercles bluish green, somewhat broader than thick, bearing a large red gland at base of groove and sometimes at top; radial spines 9 to 12, yellow when young, spreading, bulbose at base, rather stout; central spines 3 or 4, rarely one of them stouter, often bent slightly at tip; flowering areoles very woolly; ovary oblong, naked; flower and fruit not seen.

Type locality: Mexico.
Distribution: Durango, Mexico.

This species is near *Coryphantha poselgeriana*, but is smaller and has different spines.

We have seen photographs of the type and have spine-clusters, all obtained from L. Quehl of Halle. We would also refer here specimens obtained by Dr. E. Palmer near Durango City in 1906 (No. 456).

Illustrations: Monatsschr. Kakteenk. **24**: 53, as *Mammillaria guerkeana*; Alianza Cientifica Universal **3**: pl. opp. 119, as *Mammillaria valida*.

8. Coryphantha echinoidea (Quehl).

> *Mammillaria echinoidea* Quehl, Monatsschr. Kakteenk. **23**: 42. 1913.

Plant solitary, globular or a little broader than high, 5 to 6 cm. in diameter, very woolly at apex; tubercles conic, 1.5 cm. high, 1.2 cm. broad at base; groove with 1 to 3 small, grayish glands; areoles elliptic, woolly when young, glabrate in age; radial spines 20 to 25, 1.5 cm. long, white with darker tips; central spines 1 to 3, a little stouter than the radials, one of them porrect, horn-colored; flowers rose-colored, 6 to 8 cm. broad; perianth-segments oblong, broad at apex, denticulate, sometimes mucronate; filaments numerous, red; fruit and seed unknown.

Type locality: Durango.
Distribution: Durango, Mexico.
Illustration: Monatsschr. Kakteenk. **23**: 42, as *Mammillaria echinoidea*.

9. Coryphantha clava (Pfeiffer) Lemaire, Cactées 34. 1868.

> *Mammillaria clava* Pfeiffer, Allg. Gartenz. **8**: 282. 1840.
> *Mammillaria schlechtendalii* Ehrenberg, Linnaea **14**: 377. 1840.
> *Mammillaria schlechtendalii levior* Salm-Dyck, Cact. Hort. Dyck. 1849. 127. 1850.
> *Echinocactus clavus* Poselger, Allg. Gartenz. **21**: 125. 1853.
> *Echinocactus schlechtendalii* Poselger, Allg. Gartenz. **21**: 125. 1853.
> *Cactus clavus* Kuntze, Rev. Gen. Pl. **1**: 260. 1891.
> *Cactus schlechtendalii* Kuntze, Rev. Gen. Pl. **1**: 261. 1891.

Plant-body club-shaped, deep green; axils of tubercles with white wool and with a red gland at base of groove; tubercles erect, elongated, somewhat 4-sided; spine-areoles white-villous; radial spines usually 7, straight, horn-colored, about equal; central spine 1, a little longer and stouter than the others; flowers very large, sometimes 9 cm. broad, pale yellow, with the outer segments tinged with red; perianth-segments glossy, linear-oblong to spatulate, outer ones entire, inner ones serrate and mucronate at apex; filaments orange; stigma-lobes 6, linear, yellow.

Type locality: Mexico.
Distribution: Mexico.
Coryphantha schlechtendalii Lemaire (Cactées 34. 1868) is usually given as a synonym of this species.
Illustrations: Curtis's Bot. Mag. **74**: pl. 4358; Loudon, Encycl. Pl. ed. 3. 1379. f. 19390, as *Mammillaria clava*.

10. Coryphantha octacantha (De Candolle).

> *Mammillaria octacantha* De Candolle, Mém. Mus. Hist. Nat. Paris **17**: 113. 1828.
> *Mammillaria leucacantha* De Candolle, Mém. Mus. Hist. Nat. Paris **17**: 113. 1828.
> *Mammillaria lehmanni* Otto in Pfeiffer, Enum. Cact. 23.' 1837.
> *Mammillaria macrothele* Martius in Pfeiffer, Enum. Cact. 24. 1837.
> *Mammillaria plaschnickii* Otto in Pfeiffer, Enum. Cact. 24. 1837.
> *Mammillaria aulacothele* Lemaire, Cact. Aliq. Nov. 8. 1838.
> *Mammillaria biglandulosa* Pfeiffer, Allg. Gartenz. **6**: 274. 1838.
> *Mammillaria sulcimamma* Pfeiffer, Allg. Gartenz. **6**: 274. 1838.
> *Mammillaria lehmannii sulcimamma* Miquel, Linnaea **12**: 9. 1838.
> *Mammillaria martiana* Pfeiffer, Linnaea **12**: 140. 1838.
> ? *Mammillaria thelocamptos* Lehmann, Linnaea **13**: Litt. 101. 1839.
> *Mammillaria aulacothele multispina* Scheidweiler, Bull. Acad. Aci. Brux. **6**: 92. 1839.
> *Mammillaria aulacothele spinosior* Monville in Lemaire, Cact. Gen. Nov. Sp. 93. 1839.
> *Mammillaria aulacothele sulcimamma* Pfeiffer in Walpers, Bot. Repert. **2**: 302. 1843.
> *Mammillaria aulacothele flavispina* Salm-Dyck, Cact. Hort. Dyck. 1844. 13. 1845.
> *Mammillaria polymorpha* Scheer in Mühlenpfordt, Allg. Gartenz. **14**: 373. 1846.
> *Mammillaria macrothele lehmanni* Salm-Dyck, Cact. Hort. Dyck. 1849. 19. 1850.
> *Mammillaria macrothele biglandulosa* Salm-Dyck, Cact. Hort. Dyck. 1849. 19. 1850.
> *Mammillaria plaschnickii straminea* Salm-Dyck, Cact. Hort. Dyck. 1849. 19. 1850.
> *Echinocactus macrothele* Poselger, Allg. Gartenz. **21**: 125. 1853.
> *Echinocactus plaschnickii* Poselger, Allg. Gartenz. **21**: 125. 1853.
> *Echinocactus macrothele lehmanni* Poselger, Allg. Gartenz. **21**: 125. 1853.
> *Echinocactus macrothele biglandulosus* Poselger, Allg. Gartenz. **21**: 125. 1853.
> *Coryphantha lehmanni* Lemaire, Cactées. 34. 1868.
> *Coryphantha aulacothele* Lemaire, Cactées. 34. 1868.
> *Cactus macrothele* Kuntze, Rev. Gen. Pl. **1**: 260. 1891.
> *Cactus aulacothele* Kuntze, Rev. Gen. Pl. **1**: 260. 1891.

Cactus biglandulosus Kuntze, Rev. Gen. Pl. 1: 260. 1891.
Cactus lehmannii Kuntze, Rev. Gen. Pl. 1: 260. 1891.
Cactus plaschnickii Kuntze, Rev. Gen. Pl. 1: 261. 1891.
Cactus octacanthus Kuntze, Rev. Gen. Pl. 1: 261. 1891.
Cactus martianus Kuntze, Rev. Gen. Pl. 1: 261. 1891.

Plant-body simple, cylindric, 3 dm. high, 12 to 15 cm. in diameter; axils of tubercles bearing white wool, the groove with 1 or 2 red glands; tubercles elongated, up to 25 mm. long, spreading, somewhat 4-angled but with broad bases; radial spines 8, spreading, rigid, horn-colored with black tips, 10 to 12 mm. long; central spines 1 or 2, stouter than the radials, brownish, 25 mm. long; flowers about 6 cm. broad, straw-colored; perianth-segments linear-oblong, obtuse; filaments reddish; style red; stigma-lobes yellow.

Type locality: Mexico.
Distribution: Central Mexico.
Mammillaria polymorpha Scheer (Mühlenpfordt, Allg. Gartenz. **14**: 373. 1846) is probably only an abnormal greenhouse form of this species.

Coryphantha aulacothele and *C. lehmanni* (Lemaire, Cactées 34. 1868) and *M. macrothele nigrispina* (Schelle, Handb. Kakteenk, 243. 1907) are only names but are usually referred here.

Mammillaria leucantha is credited to De Candolle by Steudel (Nom. ed. 2. **2**: 97. 1841), but we have not seen such a name used by De Candolle. It may be a misspelling for *M. leucacantha.* Steudel refers the name to *M. lehmannii*, while the Index Kewensis states that it equals *M. recurva.*

Cereus lehmannii Hortus is cited by Förster (Handb. Cact. 245. 1846) as a synonym of *M. lehmannii.*

Illustrations: Loudon, Encycl. Pl. ed. 2 and 3. 1201. f. 17362; Curtis's Bot. Mag. **65**: pl. 3634, as *Mammillaria lehmannii*; Monatsschr. Kakteenk. **20**: 85; Krook, Handb. Cact. 38, as *Mammillaria aulacothele;* Schelle, Handb. Kakteenk. 242. f. 161; Förster, Handb. Cact. ed. 2. 391. f. 39, as *Mammillaria macrothele.*

11. Coryphantha exsudans (Zuccarini) Lemaire.*

Mammillaria exsudans Zuccarini in Pfeiffer, Enum. Cact. 15. 1837.
Mammillaria brevimamma Zuccarini in Pfeiffer, Enum. Cact. 34. 1837.
Mammillaria glanduligera Otto and Dietrich, Allg. Gartenz. **16**: 298. 1848.
Mammillaria brevimamma exsudans Salm-Dyck, Cact. Hort. Dyck. 1849. 19. 1850.
Mammillaria asterias Cels in Salm-Dyck, Cact. Hort. Dyck. 1849. 129. 1850.
Echinocactus glanduligerus Poselger, Allg. Gartenz. **21**: 102. 1853.
Echinocactus brevimammus Poselger, Allg. Gartenz. **21**: 102. 1853.
Coryphantha glanduligera Lemaire, Cactées 34. 1868.
Coryphantha brevimamma Lemaire in Förster, Handb. Cact. ed. 2. 394. 1885.
Cactus brevimamma Kuntze, Rev. Gen. Pl. 1: 260. 1891.
Cactus exsudans Kuntze, Rev. Gen. Pl. 1: 260. 1891.
Cactus glanduliger Kuntze, Rev. Gen. Pl. 1: 260. 1891.

Subcylindric, 4 cm. in diameter; tubercles dull green, thick, ovate; glands in the axils of the tubercles pale yellow; spine-areoles somewhat tomentose, becoming naked; radial spines 6 or 7, 6 to 10 mm. long, slender, straight, spreading, yellow; central spine 1, erect, yellow but brown at tip, perhaps hooked; flowers yellow.

Type locality: Between Ixmiquilpan and Zimapán.
Distribution: Central Mexico.
All the synonyms cited above may or may not belong here. Our description is compiled mostly from Pfeiffer's.

Mammillaria curvata (Pfeiffer, Enum. Cact. 15. 1837) was given as a synonym of *Mammillaria exsudans.*

Illustrations: Monatsschr. Kakteenk. **23**: 147; Möllers Deutsche Gärt. Zeit. **25**: 475. f. 8, No. 30, as *Mammillaria glanduligera.*

---

* This binomial is credited to Lemaire by Rümpler (Förster, Handb. Cact. ed. 2. 395. 1885), but as a synonym of *Mammillaria brevimamma exsudans.*

12. Coryphantha erecta Lemaire, Cactées 34. 1868.

> *Mammillaria erecta* Lemaire in Pfeiffer, Allg. Gartenz. 5: 370. 1837.
> *Mammillaria ceratocentra* Berg, Allg. Gartenz. 8: 130. 1840.
> *Echincocactus erectus* Poselger, Allg. Gartenz. 21: 126. 1853.
> *Cactus erectus* Kuntze, Rev. Gen. Pl. 1: 260. 1891.
> *Cactus ceratocentrus* Kuntze, Rev. Gen. Pl. 1: 260. 1891.

Plant-body cylindric, yellowish green; axils of young tubercles white-woolly; tubercles obliquely conic, somewhat rhombiform at base; radial spines 8 to 14, subulate, ascending, yellowish; central spines 2, upper one short, lower one curved; flowers large, yellow; perianth-segments very narrow.

Type locality: Mexico.

Distribution: State of Hidalgo.

The plant described by Schumann has four central spines and may not belong to this species; his illustration answers it fairly well but does not show 4 centrals. We have recently examined specimens labeled *Mammillaria erecta* which were sent by Carl Ackerman, employed at the Huntington estate near Los Angeles, California; his plants grow in clumps

FIGS. 30 and 31.—Coryphantha erecta.

one meter in diameter; the larger branches are prostrate below, ascending or erect above, 3 dm. long; the spine-areoles are circular, white-felted when young; the spines are glossy yellow, the radials widely spreading; central spines often wanting or sometimes solitary, porrect, and shorter than the radials.

Mammillaria evarescentis, according to Lemaire (Cact. Aliq. Nov. 4. 1838), was a garden name improperly applied to this species.

The three names *Mammillaria evanescens, M. evarescens,* and *M. evarascens* were listed as synonyms of *M. erecta* by Förster (Handb. Cact. 243. 1846).

Illustrations: Schumann, Gesamtb. Kakteen 504. f. 82; Möllers Deutsche Gärt. Zeit. 25: 475. f. 8, No. 7; Lemaire, Icon. Cact. pl. 10, as *Mammilaria erecta.*

Figure 30 is from a photograph of a plant collected by Dr. Rose in Mexico in 1906 (No. 1072a) and figure 31 is from a photograph of a plant growing in the Huntington collection in southern California which was made by Ernest Braunton.

13. Coryphantha elephantidens Lemaire, Cactées 35. 1868.

> *Mammillaria elephantidens* Lemaire, Cact. Aliq. Nov. 1. 1838.
> *Echinocactus elephantidens* Poselger, Allg. Gartenz. 21: 102. 1853.
> *Cactus elephantidens* Kuntze, Rev. Gen. Pl. 1: 260. 1891.

Simple, subglobose, up to 14 cm. high and 19 cm. broad; tubercles very large, somewhat flattened, obtuse, 4 to 5 cm. long, densely woolly in the axils; areoles elliptic, when young woolly, in age naked; spines 8, all radial, somewhat unequal, subulate, the longest about 2 cm. long, spreading, when young brownish with yellowish bases, black at apex; flowers large, rose-colored, 11 cm. broad; perianth-segments numerous, narrowly oblong, apiculate.

Type locality: Not cited.

Distribution: Central Mexico, but Nicholson's Dictionary of Gardening says Paraguay in error.

This is a very characteristic plant but we know it only from illustrations. Walter Mundt once offered it for sale but his supply has been exhausted; he gives a good illustration of it in a group of cacti printed on his letter heads and he writes us that this plant has a large carmine flower.

Schelle (Handb. Kakteenk. 238. 1907) gives *M. elephantidens spinosissima* Rebut, without synonymy or description.

Illustrations: Dict. Gard. Nicholson 4: 563. f. 33; Suppl. 516. f. 550; Förster, Handb. Cact. ed. 2. 397. f. 40; Hort. Univ. 1: pl. 33; Pfeiffer, Abbild. Beschr. Cact. 2 pl. 20; Rümpler, Sukkulenten 206. f. 117; Garden 1: 396; Lemaire, Icon. Cact. pl. 2 [not pl. 3]; Herb. Génér. Amat. II. 2: pl. 17; Palmer. Cult. Cact. 111; Ann. Rep. Smiths. Inst. **1908**: pl. 14, f. 3; Goebel, Pflanz. Schild. 1: f. 34; Blanc, Cacti 68. No. 1224; Watson, Cact. Cult. 159. f. 60; Bergen in Rother, Praktischer Leitfaden Kakteen 5 ed. 1. 65; ed. 3. f. 38, as *Mammillaria elephantidens*.

14. Coryphantha bumamma (Ehrenberg).

    Mammillaria bumamma Ehrenberg, Allg. Gartenz. **17**: 243. 1849.
    Mammillaria elephantidens bumamma Schumann, Keys Monogr. Cact. 43. 1903.

Globular or somewhat depressed; tubercles few, very large, rounded at apex, bluish green, very woolly in their axils when young but glabrate in age; spines 5 to 8, subulate, grayish brown, more or less recurved, 2 cm. long or more, all radial; flower large, yellow, 5 to 6 cm. broad; inner perianth-segments narrowly oblong, obtuse or retuse.

Type locality: Mexico.

Distribution: Mexico.

This plant is perhaps nearest *Coryphantha elephantidens*, to which it was referred as a variety, but the flowers are much smaller and nearly yellow. Mundt states that the flowers are smaller but bright rose with a dark stripe. His plant, however, is not now in his possession.

The plants are often much depressed, arising only a little above the surface of the ground, and are firmly anchored in the soil by a thick root, almost equal in diameter to that of the stem itself.

Dr. Rose made two collections in Mexico which we would refer here, one on the pedregal near Yautepec, Morelos (No. 8530), and the other at Iguala, Guerrero (No. 9320).

Illustration: Engler and Prantl, Pflanzenfam. 3[6a]: 194. f. 67, as *Mammillaria bumamma*.

Plate v, figure 6, shows a plant collected by H. H. Rusby at Lemon Mountain, Guerrero, altitude 800 meters, July 28, 1910 (No. 4), which flowered in the New York Botanical Garden, September 11, 1911. Figure 29 is from a photograph showing a top view of a plant collected by Dr. C. Reiche at Iguala, Mexico, in 1921.

15. Coryphantha robustispina (Schott).

    Mammillaria robustispina Schott in Engelmann, Proc. Amer. Acad. **3**: 265. 1856.
    Cactus robustispinus Kuntze, Rev. Gen. Pl. **1**: 261. 1891.
    Mammillaria brownii Toumey, Bot. Gaz. **22**: 253. 1896.
    Cactus brownii Toumey, Bot. Gaz. **22**: 253. 1896.

Stems solitary or clustered, globular or a little longer than thick, 5 to 15 cm. high, densely armed and almost hidden by the spines; tubercles large, 2.5 to 2.8 cm. long, arranged in 13 somewhat spiraled rows, fleshy, in age thickly set one against the other, becoming more or less dorsally flattened, pale, grayish green, narrowly grooved; radial spines 12 to 15, the 3 lower very stout, brownish, the upper generally weaker, the 2 or 3 uppermost ones much weaker, clustered closely together and very pale, some of them sometimes crowded towards the center, the central spine solitary, very stout and erect or sometimes curved or even hooked, yellow, 3.5 cm. long; all the larger spines somewhat bulbous at base; flowers 5 to 6 cm. long, salmon-colored; ovary 20 to 25 mm. long, bearing 4 to 7 minute caducous scales; fruit narrowly oblong, 6 cm. long; seeds large, 3 mm. long, shining.

Type locality: Cited as Sonora in first publication of species; afterwards as south side of the Baboquivari Mountains in northern Sonora.

Distribution: Mountains of southern Arizona, southwestern New Mexico, and northern Sonora.

We have followed Mrs. K. Brandegee in referring *Mammillaria brownii* here, for not only do the original descriptions read much alike but the type localities for the two are in the same mountain range. *M. brownii* was described from a very small plant and differs considerably from mature individuals. Engelmann calls attention to the very large seeds, which he says are "larger than those of any other *Mammillaria* examined." He also states, "embryo with some albumen, curved; cotyledon foliaceous, approaching the structure of the seed of most *Echinocacti.*"

Dr. Shreve reports that the flowers appear in the summer and the fruits, which follow, hold over the following winter, gradually drying up. The fruits do not open by a basal pore as in other related species.

We would refer here specimens from Lordsburg, New Mexico, and Bowie, Arizona, which, have heretofore been referred to *Mammillaria valida*, now *Coryphantha muehlenpfordtii.*

Illustrations: Bot. Gaz. **22:** 254, as *Mammillaria brownii*; Cact. Journ. **1:** 85; Cact. Mex. Bound. pl. 74, f. 8, as *Mammillaria robustispina.*

16. Coryphantha connivens sp. nov.

Globular or somewhat depressed, 8 to 10 cm. broad, somewhat woolly at the crown at flowering time but becoming glabrate; spines all radial but of two kinds; one kind 5 or 6, spreading or curved backward, subulate, horn-colored, the other 8 to 10, from upper part of spine-areole, clustered, erect, or toward top connivent, acicular, black at tip; flowers yellow, 6 to 7 cm. broad; perianth-segments narrowly oblong, acuminate; fruit greenish, oblong, 3 cm. long; seeds brown, oblong, 2 mm. long.

This species is common in the Valley of Mexico, especially on the pedregal. Dr. Rose collected it first in 1901 and again in 1905 and 1906; the type is his No. 8372 from near Tlalpam, collected in 1905. Dr. C. Reiche also collected it between Tacubaya and Santa Fé in 1922, and according to him the plant from this locality is the one referred to *Mammillaria pycnacantha* by Schumann (Gesamtb. Kakteen 489. 1898).

The species is characterized by the peculiar clusters of spines in the upper angle of the areoles. A small plant was sent by O. Solis from Tlalpam in 1907, but it has fewer acicular spines than described above.

17. Coryphantha pectinata (Engelmann).

Mammillaria pectinata Engelmann, Proc. Amer. Acad. **3:** 266. 1856.
Mammillaria pectinata cristata Hortus in Förster. Handb. Cact. ed. 2. 403. 1885.
Cactus pectinatus Kuntze, Rev. Gen. Pl. **1:** 259. 1891.

Usually simple, globose, 3 to 6 cm. in diameter; tubercles usually arranged in 13 spirals; upper tubercles 10 to 12 mm. long, about twice as long as lower ones; areoles a little longer than broad; spines 16 to 24, all radial, those on lower areoles appressed and often a little recurved, those from upper part of upper areoles 12 to 18 mm. long, connivent over apex, yellowish white with black tips; flowers yellow, 5 cm. long; ovary 6 to 8 mm. long; fruit 12 mm. long.

Type locality: On the Pecos River in western Texas.

Distribution: Southern Texas and adjacent parts of Mexico.

Coulter and Schumann refer it to *Mammillaria radians* De Candolle, but it doubtless is a distinct species.

This plant is well illustrated by Engelmann and should be easily recognized. It appears to have been collected only rarely. The only representatives we have of

1. *Coryphantha nickelsae*, from Monterey, Mexico.
2. *Neobesseya similis*, from Texas.

it are flowers and a spine-cluster from the herbarium of J. W. Toumey, collected in his cactus garden at Tucson, June 12, 1896, and a small specimen from near the type locality obtained by Vernon Bailey, March 22, 1890, and more recently by Fisher at Langtry, Texas.

Illustrations: Cact. Journ. **1:** 114; **2:** 6; Dict. Gard. Nicholson Suppl. 514. f. 546; Förster, Handb. Cact. ed. 2. 402. f. 42; Rümpler, Sukkulenten 204. f. 115; Journ. Hort. Home Gard. III. **46:** 379; Cact. Mex. Bound. pl. 11; Watson, Cact. Cult. 169. f. 67; ed. 3. f. 44; West Amer. Sci. **13:** 40; Blanc, Cacti 73. No. 1459; Cassell's Dict. Gard. **2:** 48; Remark, Kakteenfreund 15, as *Mammillaria pectinata*; Schelle, Handb. Kakteenk. 240. f. 158, as *M. radians impexicoma* [Schelle's illustration is the same as Engelmann's].

Figure 31a is from a photograph of a plant obtained by George L. Fisher near Langtry, Texas, in 1922.

Fig. 31a.—Coryphantha pectinata.

Fig. 31b.—Coryphantha echinus.

18. Coryphantha nickelsae (K. Brandegee).

Mammillaria nickelsae K. Brandegee, Zoe **5:** 31. 1900.

Described as globular, densely cespitose, often 7 cm. high, pale green and glaucous; older plants becoming purplish; tubercles almost hidden by the overlapping spines, rather broad at base, low, not densely arranged; spines 14 to 16, all radial (a few forming a small fascicle at top of groove), slender, at first simply spreading but afterward bent back and interlaced with those of adjoining tubercles, 8 to 10 mm. long, at first yellowish at base with dark tips, but afterwards bleaching; flowers described as bright yellow, with a red center, 5 to 7 cm. broad; fruit nearly globular, 5 to 7 mm. long, green; seeds small, brown.

Type locality: Mexico, southward from Laredo, Texas.
Distribution: Northern Nuevo León, Mexico.

Plants collected by Robert Runyon in March 1921, on Mount La Mitra, near Monterey, which we believe should be referred here, deserve some detailed description. They grow in clusters of 4 to 12. From the axils of the lower tubercles near the surface of the ground numerous young plants or buds originate; the young spines are pale yellow, with reddish-brown tips, in age some bleaching white, others brownish to nearly black throughout; many of the first areoles have only radial spines but old plants often have one central spine 1.5 to 2 cm. long, from all the upper areoles; flowers large, light yellow; inner perianth-segments spreading, linear-lanceolate, acuminate; anthers bright yellow.

Plate III, figure 1, is from a photograph of the plant collected by Mr. Runyon, which was made at his home in Brownsville, Texas, September 15, 1921. Figure 32 is from a photograph of a specimen sent us by Dr. Richard E. Kunze in 1911.

19. Coryphantha compacta (Engelmann).

Mammillaria compacta Engelmann in Wislizenus, Mem. Tour North. Mex. 105. 1848.
Cactus compactus Kuntze, Rev. Gen. Pl. 1: 260. 1891.

Plants solitary, somewhat depressed, 3 to 6 cm. high, 5 to 8 cm. broad; tubercles in 13 rows, much crowded, 8 mm. long, sulcate above; radial spines 14 to 16, rigid, appressed, interwoven with adjacent ones, whitish, 10 to 20 mm. long; central spines usually wanting; flowers 2 cm. long and broad, yellow; fruit oval; seeds smooth and yellow.

Type locality: Cosihuiriachi, Chihuahua.
Distribution: Mountains of Chihuahua.

This species had long been known only from the original plant collected by Wislizenus, but in 1908 Dr. Rose visited the type locality, where he re-collected the plant, which later flowered at Washington.

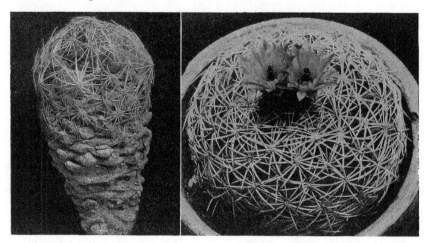

FIG. 32.—Coryphantha nickelsae.   FIG. 33.—Coryphantha compacta.

The name *Coryphantha compacta* occurs in C. R. Orcutt's Circular to Cactus Fanciers, 1922.

Illustrations: Cact. Mex. Bound. pl. 74, f. 2 (seeds); Dict. Gard. Nicholson Suppl. 515. f. 548; Bull. U. S. Dept. Agr. Bur. Pl. Ind. **262**: pl. 2, f. 1; Watson, Cact. Cult. ed. 2. 254. f. 95; ed. 3. 76. f. 35, as *Mammillaria compacta*.

Figure 33 is from a photograph of the plant collected by Dr. Rose.

20. Coryphantha radians (De Candolle).

Mammillaria radians De Candolle, Mém. Mus. Hist. Nat. Paris **17**: 111. 1828.
Mammillaria impexicoma Lemaire, Cact. Aliq. Nov. 5. 1838.
Mammillaria daimonoceras Lemaire, Cact. Aliq. Nov. 5. 1838.
Mammillaria radians globosa Scheidweiler, Bull. Acad. Sci. Brux. **5**: 494. 1838.
Mammillaria cornifera impexicoma Salm-Dyck, Cact. Hort. Dyck. 1849. 20. 1850.
Echinocactus corniferus impexicomus Poselger, Allg. Gartenz. **21**: 102. 1853.
Echinocactus radicans Poselger, Allg. Gartenz. **21**: 107. 1853.
Coryphantha daimonoceras Lemaire, Cactées 35. 1868.
Cactus radians Kuntze, Rev. Gen. Pl. 1: 261. 1891.
Cactus radians pectinoides Coulter, Contr. U. S. Nat. Herb. **3**: 114. 1894.
Mammillaria radians impexicoma Schumann, Gesamtb. Kakteen 495. 1898.
Mammillaria radians daemonoceras Schumann, Gesamtb. Kakteen 496. 1898.

Solitary, globose, either obtuse or depressed at the top, 7.5 cm. in diameter; tubercles ovoid, lar e; axils of tubercles naked; areoles glabrate; spines all radial, 16 to 18, white or sometimes yellowish, 10 to 12 mm. long, rigid, tomentose when young; flowers lemon-yellow, with outer segments tinged with red, about 10 cm. broad, the segments narrowly oblong to spatulate, acute, somewhat toothed toward the apex.

Type locality: Mexico.

Distribution: Central Mexico.

It is difficult to ascertain what the true *Mammillaria radians* of De Candolle really is. The type plant was described from specimens collected by Thomas Coulter, probably in eastern Mexico. We believe that specimens collected by Dr. Edward Palmer near San Luis Potosí, Mexico, represent the species as well as any plants we have yet seen; these, however, are cespitose as well as solitary. The species seems nearest *Coryphantha compacta*.

Cactus radians pectinoides Coulter, based on Eschanzier's plant from San Luis Potosí (1891), we have not seen but suspect that it belongs here.

Fig. 34.—Coryphantha radians.　　　　Fig. 35.—Coryphantha sulcolanata.

Mammillaria monoclova is only a garden name cited by Schumann (Gesamtb. Kakteen 495. 1898) as a synonym of this species.

Coryphantha impexicoma, credited to Lemaire, is given as a synonym of *Mammillaria cornifera impexicoma* Salm-Dyck by Rümpler (Förster, Handb. Cact. ed. 2. 414. 1885).

Illustrations: Blühende Kakteen **2**: pl. 102; Tribune Hort. **4**: pl. 139; Succulenta **5**: 57, as *Mammillaria radians;* Monatsschr. Kakteenk. **15**: 7, as *M. radians impexicoma.*

Figure 34 is from a photograph of the plant collected at San Rafael by Dr. Chaffey in 1910.

21. Coryphantha sulcolanata Lemaire, Cactées 35. 1868.

> *Mammillaria sulcolanata* Lemaire, Cact. Aliq. Nov. 2. 1838.
> *Echinocactus sulcolanatus* Poselger, Allg. Gartenz. 21: 102. 1853.
> *Mammillaria conimamma* Linke, Allg. Gartenz. 25: 239. 1857.
> *Mammillaria cornimamma* N. E. Brown, Gard. Chron. III. 2: 186. 1887.
> *Cactus sulcolanatus* Kuntze, Rev. Gen. Pl. 1: 261. 1891.

Subglobose, somewhat depressed, cespitose, 5 cm. high, 6 cm. thick or more; tubercles somewhat 5-angled at base, subconic above, their axils very woolly when young; spines 9 or 10, all radial, unequal, 12 to 16 mm. long, the lower and upper weaker and shorter than the lateral ones, brownish with black tips, but when young whitish yellow with purple tips; flowers large, 4 cm. long or more, widely spreading, 6 cm. broad or more; perianth-segments oblong, acute.

Type locality: Not cited, but Rümpler states that the plant was collected by Galeotti near Mineral del Monte, Hidalgo, in 1836.

Distribution: Mexico, perhaps Hidalgo, but definite range unknown.

Aulacothele sulcolanatum Monville (Lemaire, Icon. Cact. pl. 10. 1841–1847), referred here as a synonym, seems never to have been published.

Mammillaria retusa Scheidweiler is sometimes referred here also and the name has priority over *M. sulcolanata,* but we are treating it as distinct.

Echinocactus conimamma Linke was cited by Schumann (Monatsschr. Kakteenk. **5:** 75. 1895) by mistake for *Mammillaria conimamma* Linke. *M. conimamma major* is listed by Haage (Cact. Kultur ed. 2. 179. 1900).

The name *Mammillaria sulcolanata macracantha* (Walpers, Repert. Bot. **2:** 273. 1843) was without description.

Illustrations: Haage, Cact. Kultur ed. 2. 178, as *Mammillaria bumamma*; Blanc, Hints on Cacti 68. No. 1224, as *Mammillaria elephantidens*; Lemaire, Icon. Cact. pl. 10; Förster, Handb. Cact. ed. 2. 408. f. 45; Schelle, Handb. Kakteenk. 238. f. 154; Watson, Cact. Cult. 178. f. 72; ed. 3. f. 49; Deutsche Gärt. Zeit. **6:** 65; Dict. Gard. Nicholson **4:** 565. f. 40; Suppl. 518. f. 558, as *Mammillaria sulcolanata*; Möllers Deutsche Gärt. Zeit. **25:** 475. f. 8, No. 2, as *Mammillaria conimamma*; Lemaire, Cactées 35. f. 2.

Figure 35 is from a photograph of the plant collected by Dr. Rose near Pachuca in 1905.

FIG. 36.—Coryphantha retusa.   FIG. 37.—Coryphantha salm-dyckiana.

22. Coryphantha retusa (Pfeiffer).

Mammillaria retusa Pfeiffer, Allg. Gartenz. **5:** 369. 1837.
Cactus retusus Kuntze, Rev. Gen. Pl. **1:** 261. 1891.

Plants depressed-globose, 5 to 10 cm. in diameter, the top very woolly; tubercles rather large; areoles elliptic; spines 6 to 12, all radial, appressed, or even curved backward, yellowish to brownish, subulate, except 2 or 3 acicular ones at upper part of areoles; flowers central, yellow, about 3 cm. long; inner perianth-segments oblong, acute.

Type locality: Mexico.

Distribution: Oaxaca, Mexico.

We have referred to this species a plant common in Oaxaca, which answers the original description very well. It was collected by Pringle in 1894 (No. 5706) and by Conzatti in

1907, 1909, and 1920. It has also been sent us from the same region by O. Solis and B. P. Reko.

Figure 36 is from a photograph of a plant sent from Oaxaca by O. Solís in 1920.

23. Coryphantha palmeri sp. nov.

Plant-body globular; tubercles closely set in about 13 rows but not very regularly arranged, pale green, not very flaccid; radial spines 11 to 14, rather stout, spreading nearly at right angles to central one, yellowish; tips often blackish; central spine one, stout, terete, hooked at apex; young areoles very woolly; flowers central, pale yellow to nearly white, about 3 cm. long; outer perianth-segments linear-oblong, acute, brownish on broad mid-rib, entire, the inner yellow throughout, acuminate; stamens numerous; stigma-lobes 9, linear, cream-colored.

Collected by Dr. Edward Palmer on stony ridge near Durango, Mexico, and flowered in Washington, July 1906 (No. 557, type). Here seem to belong plants collected by Dr. Palmer at Agua Nueva, April 1905 (No. 561), and at Saltillo, October 1904 (No. 438), and July 1905 (No. 703), and also by F. E. Lloyd in Zacatecas, 1908 (No. 9).

24. Coryphantha cornifera (De Candolle) Lemaire, Cactées 35. 1868.

Mammillaria cornifera De Candolle, Mém. Mus. Hist. Nat. Paris **17**: 112. 1828.
Mammillaria pfeifferana De Vriese, Tydschr. Nat. Geschr. **6**: 51. 1839.
Mammillaria scolymoides Scheidweiler, Allg. Gartenz. **9**: 44. 1841.
Mammillaria scolymoides longiseta Salm-Dyck, Cact. Hort. Dyck. 1849. 132. 1850.
Mammillaria scolymoides nigricans Salm-Dyck, Cact. Hort. Dyck. 1849. 132. 1850.
Echinocactus corniferus Poselger, Allg. Gartenz. **21**: 102. 1853.
Echinocactus corniferus longisetus Poselger, Allg. Gartenz. **21**: 102. 1853.
Echinocactus corniferus nigricans Poselger, Allg. Gartenz. **21**: 102. 1853.
Echinocactus corniferus scolymoides Poselger, Allg. Gartenz. **21**: 102. 1853.
Cactus corniferus Kuntze, Rev. Gen. Pl. **1**: 260. 1891.
Cactus pfeifferanus Kuntze, Rev. Gen. Pl. **1**: 261. 1891.
Cactus scolymoides Kuntze, Rev. Gen. Pl. **1**: 261. 1891.

Plant solitary, globose, pale green; tubercles short, broad, somewhat imbricated, 12 cm. high; radial spines 16 or 17, grayish, 10 to 12 mm. long; central spine 1, stout, erect or subincurved, generally dark colored, 14 to 16 mm. long; flowers yellow, tinged with red, 7 cm. broad; inner perianth-segments oblanceolate, acuminate; fruit not seen.

Type locality: Mexico.
Distribution: Central Mexico.

We refer here a plant collected by Dr. Rose near San Juan del Rio, August 17, 1905.

Schumann referred *Mammillaria scolymoides* to *Mammillaria radians*, but its relationship is rather with *M. cornifera* as suggested by Schumann.

Mammillaria cornifera mutica Salm-Dyck (Cact. Hort. Dyck. 1849. 20. 1850), taken up afterwards as *Echinocactus corniferus muticus* by Poselger (Allg. Gartenz. **21**: 102. 1853), was without description and to it was referred *Mammillaria radians* Hortus.

Illustrations: Schumann, Gesamtb. Kakteen 492. f. 81; Thomas, Zimmerkultur Kakteen 55; Bull. U. S. Dept. Agr. Bur. Pl. Ind. **262**: pl. 1; Blühende Kakteen **3**: pl. 125; Monatsschr. Kakteenk. **14**: 73, as *Mammillaris cornifera*; Karsten and Schenck, Vegetationsbilder **2**: pl. 20e, as *Mammillaria scolymoides*; Tydschr. Nat. Geschr. **6**: pl. 1, f. 2, as *Mammillaria pfeifferana*.

Plate 11, figure 4, shows a plant collected by Dr. C. A. Purpus in Coahuila in 1905 which flowered in the New York Botanical Garden.

25. Coryphantha salm-dyckiana (Scheer).

Mammillaria salm-dyckiana Scheer in Salm-Dyck, Cact. Hort. Dyck. 1849. 134. 1850.
Mammillaria salm-dyckiana brunnea Salm-Dyck, Allg. Gartenz. **18**: 394. 1850.
Echinocactus salm-dyckianus Poselger, Allg. Gartenz. **21**: 102. 1853.
Cactus salm-dyckianus Kuntze, Rev. Gen. Pl. **1**: 261. 1891.
Mammillaria delaetiana Quehl, Monatsschr. Kakteenk. **18**: 59. 1908.

Plants either solitary or in clusters, nearly globular or sometimes club-shaped, 10 'o 15 cm. in diameter, light green; tubercles rather short, closely set; radial spines about 15, spreading, slender, 10 to 15 mm. long, grayish or whitish; central spines 1 to 4, reddish to black, the 3 upper ones when present ascending and those near top of plant connivent, the lowest central stouter than others, 2 to 2.5 cm. long, porrect or curved downward; flowers large, 4 cm. long; outer perianth-segments greenish or tinged with red, the inner pale yellow; filaments greenish yellow; stigma-lobes 7.

Type locality: Near Chihuahua, Mexico.

Distribution: Common in the state of Chihuahua, Mexico.

Mammillaria salm-dyckiana was originally collected by John Potts near Chihuahua City and sent to Kew; its flowers and fruit were unknown. Schumann referred it as a synonym of *M. scheeri*, but we believe that it must be distinct and that *M. delaetiana* is the same. It was described from plants distributed by de Laet, who probably obtained them from C. R. Orcutt.

In 1908 Dr. E. Palmer collected some fine plants near Chihuahua City, from which our flower characters have been drawn.

Illustrations: Monatsschr. Kakteenk. **18**: 59; **20**: 92, as *Mammillaria delaetiana*.

Figure 37 is from a photograph of the plant collected by Dr. E. Palmer near Chihuahua City in 1908.

26. Coryphantha pallida sp. nov.

Plants either solitary or in clusters of about 10 or more, globular, 12 cm. in diameter or less, bluish green; tubercles in 13 rows, short and thick, closely set; radial spines 20 or more, white, ap-

FIG. 38.—Coryphantha pallida.   FIG. 39.—Coryphantha pycnacantha.

pressed; centrals usually 3, but sometimes more, the two upper more or less ascending, the lower porrect or curved downward, with tip black, or sometimes black throughout; flowers very large, often 7 cm. long and nearly as broad; outer perianth-segments narrow, greenish yellow, with a reddish stripe on back; inner perianth-segments pale lemon-yellow, broader than outermost, acuminate; ovary bearing a few narrow scales; stamens deep red, numerous; style yellow, longer than stamens; stigma-lobes 9; fruit greenish brown, 2 cm. long; seeds brown, shining, broader at apex than below.

Common in calcareous soil about Tehuacán, Mexico. Collected by J. N. Rose in 1901 (No. 5583, type), in 1905 (Nos. 99972 and 10001), and in 1906. Living specimens were also obtained and these have flowered repeatedly in cultivation. It was also collected by C. G. Pringle in 1901 (No. 8573) and distributed as *Mammillaria pycnacantha?*.

In young plants the spines are not so numerous, the central spine is single, porrect, slightly curved, with black tips.

Figure 38 is from a photograph of the plant collected at the type locality in 1906.

27. Coryphantha pycnacantha (Martius) Lemaire, Cactées 35. 1868.

> ? *Mammillaria latimamma* De Candolle, Mém. Mus. Hist. Nat. Paris **17:** 114. 1828.
> *Mammillaria pycnacantha* Martius, Nov. Act. Nat. Cur. **16:** 325. 1832.
> ? *Mammillaria acanthostephes* Lehmann, Allg. Gartenz. **3:** 228. 1835.
> *Mammillaria arietina* Lemaire * Cact. Aliq. Nov. 10. 1838.
> *Mammillaria scepontocentra* Lemaire, Cact. Gen. Nov. Sp. 43. 1839.
> *Mammillaria arietina spinosior* Lemaire, Cact. Gen. Nov. Sp. 94. 1839.
> *Mammillaria pycnacantha spinosior* Monville in Salm-Dyck, Cact. Hort. Dyck. 1844. 14. 1845.
> *Mammillaria magnimamma arietina* Salm-Dyck in Förster, Handb. Cact. 235. 1846.
> *Mammillaria winkleri* Förster, Allg. Gartenz. **15:** 50. 1847.
> *Mammillaria magnimamma lutescens* Salm-Dyck, Cact. Hort. Dyck. 1849. 17, 121. 1850.
> *Echinocactus winkleri* Poselger, Allg. Gartenz. **21:** 102. 1853.
> ? *Echinocactus acanthostephes* Poselger, Allg. Gartenz. **21:** 102. 1853.
> *Echinocactus pycnacanthus* Poselger, Allg. Gartenz. **21:** 102. 1853.
> *Mammillaria acanthostephes recta* Hortus in Labouret, Monogr. Cact. 138. 1853.
> ? *Coryphantha acanthostephes* Lemaire, Cactées 35. 1868.
> *Cactus acanthostephes* Kuntze, Rev. Gen. Pl. **1:** 260. 1891.
> *Cactus latimamma* Kuntze, Rev. Gen. Pl. **1:** 260. 1891.
> *Cactus pycnacanthus* Kuntze, Rev. Gen. Pl. **1:** 261. 1891.
> *Cactus scepontocentrus* Kuntze, Rev. Gen. Pl. **1:** 261. 1891.
> *Cactus winkleri* Kuntze, Rev. Gen. Pl. **1:** 261. 1891.

Plant solitary, globular to cylindric, about 8 cm. high; tubercles broad, grooved above, glaucous-green; radial spines 10 to 12, slender, 10 to 16 mm. long; central spines about 4, stouter than the radials, about 25 mm. long, more or less curved backward, usually black; flowers from near center of plant, 25 mm. in diameter, yellowish; perianth-segments numerous, very narrow; stigma-lobes 5 or 6, white.

Type locality: Near the city of Oaxaca, Mexico.

Distribution: Oaxaca, Mexico.

The skeleton of the type of this species is preserved in the Munich Museum and Dr. Rose obtained a cluster of spines from this specimen in 1912.

Coryphantha pycnacantha has long been a desideratum. In September 1920 Professor Conzatti sent several small plants from near the type locality. In these, the radial spines are white, the centrals (3) are nearly black, and all more or less curved backward. In the center of the plant a quantity of white wool is developed, so abundant that it can be gathered for commercial use. With the specimens of Professor Conzatti are samples of the wool with an inquiry as to its value as a fiber.

Mammillaria magnimamma spinosior Lemaire (Salm-Dyck, Cact. Hort. Dyck. 1844. 12. 1845) was not described at the place here cited. Labouret afterwards refers it as a synonym of *M. magnimamma lutescens*.

Mammillaria cephalophora Salm-Dyck (Cact. Hort. Dyck. 1849. 137. 1850; *Echinocactus cephalophorus* Poselger, Allg. Gartenz. **21:** 102. 1853; *Cactus cephalophorus* Kuntze, Rev. Gen. Pl. **1:** 260. 1891) was a new name for *Melocactus mammillariaeformis* † Salm-Dyck (Allg. Gartenz. **4:** 148. 1836). It was first described as a *Melocactus* (because of its woolly crown), but it seems to be more like a *Coryphantha*. Its exact origin in Mexico seems to be unknown and the flowers had not been described up to 1850. Pfeiffer stated that the seeds obtained from a dead plant were similar to those of *Mammillaria coronaria*. Schumann discussed it under *M. pycnacantha* in a note. Hemsley (Biol. Centr. Amer. Bot. **1:** 502) listed it as a *Melocactus*.

Mammillaria pycnacantha scepontocentra Monville (Labouret, Monogr. Cact. 136. 1853) belongs here by implication.

Mammillaria magnimamma Otto was referred by De Candolle (Mém. Cact. 17. 1834) to his *M. latimamma*, now referred to this species.

---

* Schumann refers this name as a synonym of *Mammillaria centricirrha*.
† Schumann spells this name *Melocactus mamillariiformis*.

Echinocactus radiatus Hortus Belg. was referred as a synonym of *Mammillaria pycnacantha* by Pfeiffer (Enum. Cact. 180. 1837).

Illustrations: Nov. Act. Nat. Cur. **16**: pl. 17; Loudon, Encycl. Pl. ed. 3. 1379. f. 19387; Abh. Bayer. Akad. Wiss. München **2**: pl. 3; Pfeiffer and Otto, Abbild. Beschr. Cact. **1**: pl. 26; Curtis's Bot. Mag. **69**: pl. 3972, as *Mammillaria pycnacantha*.

Figure 39 is reproduced from the first illustration cited above; a spine-cluster is also shown.

28. Coryphantha echinus (Engelmann).

Mammillaria echinus Engelmann, Proc. Amer. Acad. **3**: 267. 1856.
Cactus echinus Kuntze, Rev. Gen. Pl. **1**: 260. 1891.
Mammillaria radians echinus Schumann, Gesamtb. Kakteen 496. 1898.

Solitary, globose to subconic, 3 to 5 cm. in diameter, almost hidden under the closely appressed spines; areoles orbicular or a little longer than broad; radial spines numerous, white, 10 to 16 mm. long; central spines 3 or 4, the 3 upper erect or connivent over the apex, the lower one porrect on side of plant, erect near top, subulate, straight, 1.5 to 2.5 cm. long, often blackish; flowers 2.5 to 5 cm. long, yellow; outer perianth-segments linear-lanceolate; inner perianth-segments 20 to 30, narrow; stigma-lobes about 12; fruit oblong, 12 mm. long.

Type locality: On the Pecos River, Texas.
Distribution: Western Texas.

The flowers with the type plant seem to have been shriveled, for Engelmann describes them as large, apparently about 1½ or 2 inches long; in a later description he states that they are yellow. This species is very rare in collections and we have seen no flowers of it. All the illustrations cited below are based on the figure in the Mexican Boundary Survey.

The name *Coryphantha echinus* occurs in C. R. Orcutt's Circular to Cactus Fanciers, 1922.

Illustrations: Cact. Mex. Bound. pl. 10; Dict. Gard. Nicholson **4**: 562. f. 32; Suppl. 515. f. 549; Watson, Cact. Cult. 157. f. 59; ed. 3. f. 37; Förster, Handb. Cact. ed. 2. 404. f. 43; Blanc, Cacti 68. f. 1228, as *Mammillaria echinus*; Schelle, Handb. Kakteenk. 240. f. 159, as *M. radians echinus*.

Figure 31*b* is from a photograph of a plant obtained by George L. Fisher near Langtry, Texas, in 1922.

29. Coryphantha durangensis (Rünge).

Mammillaria durangensis Rünge in Schumann, Gesamtb. Kakteen 478. 1898.

Plants solitary or in small clusters, short-cylindric, 10 cm. long or less, somewhat glaucous; tubercles rather prominent, in 5 or 8 series, somewhat compressed dorsally, very woolly in the axils; radial spines 6 to 8, acicular, spreading, 1 cm. long or less; central spine solitary, often erect, those of uppermost areoles connivent, black; flowers very small, about 2 cm. long, when fully expanded 2.5 to 4 cm. broad; outer perianth-segments dark purple or with only a purple stripe down center; inner perianth-segments cream-colored to pale lemon-yellow; filaments cream-colored, about length of style; style and stigma-lobes cream-colored, the latter 5, linear and curved backward; fruit globular, 5 to 8 mm. in diameter, naked, greenish; seeds brown, about 1 mm. broad.

Type locality: Villa Lerdo, Durango, Mexico.
Distribution: Northern Mexico.

Dr. E. Chaffey has collected this plant for us several times at the type locality, but it does not survive long under glass. In 1911 he found a cristate form with the lobes flattened like the joints of an *Opuntia*, bearing flowers along the edges.

This is *Mammillaria compressa* of Hildmann's Catalogue, according to Schumann (Gesamtb. Kakteen 479. 1898).

Illustration: Wiener Ill. Gart. Zeit. **29**: 411. f. 105, as *Mammillaria radians*.

Plate v, figure 4, shows a plant sent by Dr. Chaffey from the type locality in 1918, which flowered in the New York Botanical Garden, April 8, 1918. Figure 40 is from a photograph of a potted plant sent by Dr. Chaffey in 1910 which flowered in Washington; figure 41 is from a photograph of another plant sent by Dr. Chaffey in 1910.

30. Coryphantha chlorantha (Engelmann).

Mammillaria chlorantha Engelmann in Rothrock, Rep. U. S. Geogr. Surv. **6**: 127. 1878.
Cactus radiosus chloranthus Coulter, Contr. U. S. Nat. Herb. **3**: 121. 1894.
Mammillaria radiosa chlorantha * Schumann, Gesamtb. Kakteen 481. 1898.

Plant cylindric, sometimes 20 to 25 cm. high, 8 cm. in diameter; tubercles closely set and entirely hidden by the densely matted spines; flowers small, 35 mm. broad; outer perianth-segments ciliate; inner perianth-segments yellow or greenish yellow, linear-lanceolate, acute; stigma-lobes white; fruit central, green, 2.5 cm. long, juicy, bearing 5 or 6 scales near top; seeds brown, flattened, 1.5 mm. long, reticulated.

Type locality: Southern Utah, east of Saint George.

Distribution: Southern Utah, western Arizona, central Nevada, and eastern southern California.

Mammillaria utahensis Hildmann, cited by Schumann (Gesamtb. Kakteen 481. 1898) as a synonym of *M. radiosa*, may have been based on this plant.

Illustrations: Förster, Handb. Cact. ed. 2. 328. f. 33; Gartenflora **32**: 87; Deutsche Gärt. Zeit. **7**: 53, as *Mammillaria chlorantha*; Schelle, Handb. Kakteenk. 236. f. 151, as *M. radiosa chlorantha*.

Plate v, figure 7, is from a plant collected by I. Tidestrom at the type locality in 1919, which flowered in the New York Botanical Garden, May 27, 1919. Figure 42 is from a photograph of a plant collected by Major E. A. Goldman in Prospect Valley, Arizona.

FIGS. 40 and 41.—Coryphantha durangensis.

31. Coryphantha vivipara (Nuttall) Britton and Rose in Britton and Brown, Illustr. Fl. ed. 2. **2**: 571. 1913.

Cactus viviparus Nuttall, Fraser's Cat. No. 22. 1813.
Mammillaria vivipara Haworth, Suppl. Pl. Succ. 72. 1819.
Mammillaria radiosa Engelmann, Bost. Journ. Nat. Hist. **6**: 196. 1850.
Echinocactus radiosus Poselger, Allg. Gartenz. **21**: 107. 1853.
Echinocactus viviparus Poselger, Allg. Gartenz. **21**: 107. 1853.
Mammillaria vivipara vera Engelmann, Proc. Amer. Acad. **3**: 269. 1856.
Mammillaria vivipara radiosa Engelmann, Proc. Amer. Acad. **3**: 269. 1856.
Mammillaria vivipara radiosa Engelmann, Cact. Mex. Bound. 15. 1859, as subspecies.
Cactus radiosus Coulter, Contr. U. S. Nat. Herb. **3**: 120. 1894.
Mammillaria hirschliana Haage, Monatsschr. Kakteenk. **6**: 127. 1896.
Coryphantha radiosa Rydberg, Fl. Rocky Mountains 581. 1917.

---

* Schumann credits this trinomial to Engelmann at the place here cited, although we believe that Engelmann never used it.

Plants solitary or in clusters forming mounds 3 to 5 dm. in diameter, globular, with prominent tubercles; areoles large, woolly; radial spines about 16, rather delicate, radiating, white; centrals 4 to 6, divergent, much stouter, brownish, swollen at base; ovary green, naked; outer perianth-segments greenish; inner ones somewhat pinkish, long-ciliate; innermost perianth-segments pinkish purple, narrow, acuminate, entire, spreading; filaments much shorter than the segments, pinkish, but paler below; style greenish to purple above, longer than the stamens; stigma-lobes linear, purple, about 8, apiculate; fruit green when mature, juicy, nearly globular, 1.5 cm. in diameter, with several (sometimes 5 or 6) small ciliate scales scattered over its surface; seeds light brown, 1.5 mm. long.

Type locality: "Near the Mandan towns on the Missouri, lat. near 49°."

Distribution: Manitoba to Alberta, Kansas, south to northern Texas and Colorado.

The group to which *Coryphantha vivipara* belongs has always been very puzzling. Dr. Engelmann, our greatest authority on this group, was sometimes of one opinion and

FIG. 42.—Coryphantha chlorantha.   FIG. 43.—Coryphantha neo-mexicana.

sometimes of another. Schumann rejected the specific name *vivipara* of Haworth for this plant since he thought that it was not the same as the *vivipara* of Engelmann, but in this he must be wrong, for *Mammillaria vivipara* Haworth was based upon *Cactus viviparus* Pursh, a name previously used by Nuttall, and both Pursh's and Nuttall's descriptions were based on the specimens collected by Nuttall in "Upper Louisiana" in 1812. This is undoubtedly the plant which Engelmann had in mind and which he called variety *vera*. We have not seen the type, but Pursh stated that he had seen flowers in Lambert's Garden.

Engelmann's remarks regarding the variability of the species are interesting. In the Proceedings of the American Academy (**3**: 269) he says:

"The extreme forms are certainly very unlike one another, but the transitions are so gradual that I can not draw strict limits between them."

Coryphantha vivipara and the three following species are closely related.

This plant is a day bloomer, and according to Engelmann the flowers become fully expanded about one o'clock in the afternoon.

Hooker in Curtis's Botanical Magazine (pl. 7718) figures and describes a plant purchased from D. M. Andrews of Boulder, Colorado, in which all the spines are brown, the flower is rose-red, and the stigma-lobes are linear and white.

Mammillaria montana is described briefly and figured (f. 1399) by Blanc in Hints on Cacti, p. 72. It is also described and figured by Darel (Illustr. Handb. Kakteen 96. f. 81), who says that it comes from Montana and Utah. It is illustrated by Haage (Cact. Kultur ed. 2. 187). It is apparently the same as *Coryphantha vivipara.*

Illustrations: Cycl. Amer. Hort. Bailey **2:** f. 1356; Stand. Cycl. Hort. Bailey **4:** f. 2315; Tribune Hort. **4:** pl. 140; Curtis's Bot. Mag. **126:** pl. 7718; De Laet, Cat. Gén. f. 43; Cact. Mex. Bound. pl. 74, f. 3 (seed); Meehan's Monthly **9:** pl. 9, as *Mammillaria vivipara;* Clements, Rocky Mountain Flow. pl. 32, f. 7; Clements, Fl. Mount. Plain pl. 32, f. 7; Britton and Brown, Illustr. Fl. **2:** 462. f. 2526, as *Cactus viviparus;* Monatsschr. Kakteenk. **3:** 132; Schelle, Handb. Kakteenk. 236. f. 150; Floralia **42:** 375, as *Mammillaria radiosa;* Cact. Mex. Bound. pl. 74, f. 5 (seed), as *Mammillaria radiosa texana;* Cact. Mex. Bound. pl. 74. f. 4 (seed), as *M. radiosa borealis;* Britton and Brown, Illustr. Fl. ed. 2. **2:** f. 2985.

32. Coryphantha neo-mexicana (Engelmann).

Mammillaria vivipara radiosa neo-mexicana Engelmann, Proc. Amer. Acad. **3:** 269. 1856.
Mammillaria radiosa neo-mexicana Engelmann, Cact. Mex. Bound. 64. 1859.
Mammillaria radiosa borealis Engelmann, Cact. Mex. Bound. 68. 1859.
Mammillaria radiosa texana Engelmann, Cact. Mex. Bound. 68. 1859.
Cactus radiosus neo-mexicanus Coulter, Contr. U. S. Nat. Herb. **3:** 120. 1894.
Cactus neo-mexicanus Small, Fl. Southeast. U. S. 812. 1903.
Mammillaria neo-mexicana A. Nelson in Coulter and Nelson, Man. Bot. Rocky Mountains 327. 1909.

Plants usually solitary, globular to short-oblong, 8 to 12 cm. long, the whole body usually hidden under a mass of spines; radial spines numerous, acicular, usually white; central spines several, much stouter than the radials, pale below, brown or black towards top; flowers 4 to 5 cm. broad when fully expanded; outer perianth-segments greenish or the ones nearer center purplish, ciliate; inner perianth-segments broadly linear, acuminate and apiculate, more or less serrate above; filaments greenish, much shorter than perianth-segments; stigma-lobes extending beyond filaments, white, obtuse, not apiculate as in *Coryphantha vivipara;* fruit 2.5 cm. long, green, juicy, naked except a few hairy scales near top, capped by withered perianth, depressed at apex.

Type locality: Western Texas to New Mexico, doubtless at El Paso.
Distribution: Western Texas, New Mexico, and northern Chihuahua.

The distribution of this species can not be stated at present very definitely. It may be that some of the plants from northern New Mexico, especially those found in the mountains, may better be referred to *C. vivipara,* and the same is true of some of the plants from Texas. It is probable that the plants from central Texas and perhaps northwestern Texas may all be referred to *C. vivipara.* We have no Mexican plants before us but we have plants from El Paso, just over the Mexican Boundary line. Just how far south the species extends we do not know. We have greatly restricted the range from that given by Coulter in the Contributions from the U. S. National Herbarium (**3:** 120. 1894).

Illustrations: Gartenwelt **4:** 159; Cact. Mex. Bound. pl. 13; Förster, Handb. Cact. ed. 2. 304. f. 30, as *Mammillaria radiosa neo-mexicana;* Watson, Cact. Cult. 181. f. 73; ed. 3. f. 50; Dict. Gard. Nicholson **4:** 566. f. 41, as *Mammillaria vivipara radiosa;* Dict. Gard. Nicholson Suppl. 517. f. 554, as *Mammillaria radiosa;* Cact. Mex. Bound. pl. 74 (seed), as *Mammillaria borealis.*

Plate 11, figure 1, shows a plant sent from Canutillo, Texas, by Mrs. S. L. Pattison in 1920; figure 1a shows the fruit. Figure 43 is from a photograph of a plant collected by Dr. Rose near Albuquerque, New Mexico, in 1908.

33. Coryphantha arizonica (Engelmann).

Mammillaria arizonica Engelmann, Bot. Calif. **1:** 124. 1876.
Cactus radiosus arizonicus Coulter, Contr. U. S. Nat. Herb. **3:** 121. 1894.
Mammillaria radiosa arizonica Schumann, Gesamtb. Kakteen 481. 1898.

Sometimes cespitose, forming large clumps a meter broad; each head globose to ovoid, 7.5 to 10 cm. in diameter; tubercles about 2.5 cm. long, cylindric, ascending, deeply grooved; spines numerous, straight, rigid; radial spines 15 to 20, 10 to 30 mm. long, whitish; inner spines 3 to 6, stouter than the radial ones, deep brown above; flowers large, 5 to 7 cm. broad, rose-colored; outer perianth-segments 30 to 40, linear-subulate, with fimbriate margin; inner perianth-segments 40 to 50, lanceolate-linear, attenuate; stigma-lobes 8 to 10, white; fruit oval, green; seeds compressed, light brown, pitted.

Type locality: Northern Arizona.

Distribution: Northern Arizona, especially along the Upper River of the Grand Canyon, and perhaps also in southern Utah.*

Mammillaria arizonica Engelmann, when first described, was a complex. Engelmann states that it was found "on rocky and sandy soil in northern Arizona from the Colorado eastward (Coues, Palmer, F. Bischoff) and into southern Utah (J. E. Johnson); probably in southeastern California." Engelmann afterwards described Johnson's plant from Utah as *M. chlorantha* and the California plant is doubtless his *M. deserti*. We have in the U. S. National Herbarium Palmer's specimen from Arizona but we have not seen the plant of Coues nor of Bischoff.

FIGS. 44 and 45.—Coryphantha deserti.

The northern range of this species is very uncertain. Engelmann extended it into southern Utah.

Plate v, figure 5, shows a plant sent by M. A. H. Spencer from the Grand Canyon, Arizona, in May 1907, which afterwards flowered in Washington.

34. Coryphantha deserti (Engelmann).

FIG. 46.—Flower of C. deserti.

> *Mammillaria deserti* Engelmann, Bot .Calif. **2**: 449. 1880.
> *Cactus radiosus deserti* Coulter, Contr. U. S. Nat. Herb. **3**: 121. 1894.
> *Cactus radiosus alversonii* Coulter, Contr. U. S. Nat. Herb. **3**: 122. 1894.
> *Mammillaria alversonii* Zeissold, Monatsschr. Kakteenk. **5**: 70. 1895.
> *Mammillaria radiosa alversonii* Schumann, Gesamtb. Kakteen 481. 1898.
> *Mammillaria radiosa deserti* Schumann, Gesamtb. Kakteen 481. 1898.

Solitary or cespitose, usually cylindric, sometimes 2 dm. high, 6 to 9 cm. in diameter, densely covered with spines; radial spines white except at tip,

---

* Our Utah reference is based on some detached flowers collected by M. E. Jones and a barren plant sent by Dr. C. D. Marsh in 1922. Both collections came from above Salina.

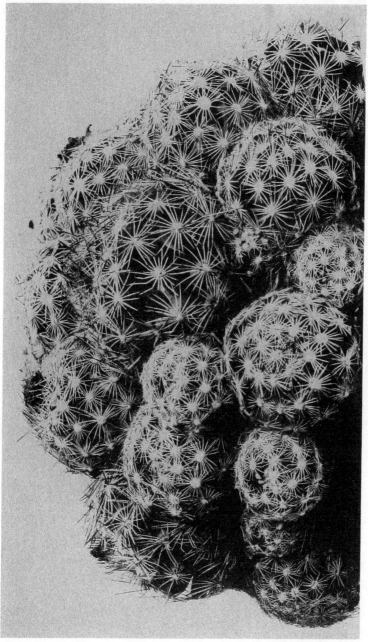

Coryphantha aggregata, from Arizona.

spreading; central spines several, sometimes as many as 14, much stouter than the radials, slightly spreading, those toward top of plant connivent, black or bluish black in their upper half, shading into red, nearly white at base; flowers 3 cm. long and nearly as broad when expanded, light pink, opening in bright sunlight; scales and outer perianth-segments ciliate; inner perianth-segments narrow, acute.

Type locality: Ivanpah, California.

Distribution: Deserts of southern California and southern Nevada.

This species is characterized by its stiff spines, with bluish-black tips shading into red, and is known in southern California as fox-tail cactus. The original description of *Mammillaria deserti* states that the flowers are straw-colored, tipped with pink, and this suggests *Coryphantha chlorantha* but we believe that it belongs with *Mammillaria alversonii*, which certainly has pinkish flowers, and since the name *deserti* is older than *alversonii* it is substituted for it.

Illustrations: Cact. Journ. **1**: pl. for February, in part; Alverson's Cat. pl. facing 8, as *Mammillaria alversonii*; Schumann, Gesamtb. Kakteen 480. f. 79, as *M. radiosa alversonii*.

Figure 44 is from a photograph of a single plant sent by E. C. Rost; figure 45 is from a photograph of a clump photographed by E. C. Rost in its natural surroundings; figure 46 shows a flower taken from Mr. Rost's plant.

 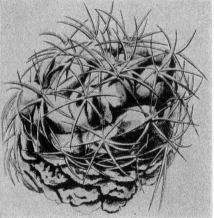

FIG. 47.—Coryphantha aggregata.   FIG. 48.—Mammillaria recurvispina.

35. Coryphantha aggregata (Engelmann).

Mammilaria aggregata Engelmann in Emory, Mil. Reconn. 157. 1848.
Cereus aggregatus Coulter, Contr. U. S. Nat. Herb. **3**: 396. 1896, as to name.
Echinocereus aggregatus Rydberg, Bull. Torr. Club **33**: 146. 1906, as to name.

Plants solitary or cespitose, globular to short-oblong, very spiny; radial spines numerous, stouter than those of *Coryphantha vivipara*, white, often with brown tips, appressed; central spines several, stout, all erect and appressed or one often porrect, those towards top of plant connivent; flowers very large and showy, purplish, 5 to 7 cm. broad; outer perianth-segments ciliate; inner perianth-segments narrowly oblanceolate, often 6 mm. broad, acute, apiculate; stigma-lobes 8 to 10, elongated, white; fruit green, oblong, 2 to 2.5 cm. long, naked or occasionally bearing a small scale on the side, juicy; seeds dark brown, 2 mm. long.

Type locality: Head waters of the Gila.

Distribution: Western New Mexico, southeastern Arizona, and northern Sonora.

Mammillaria aggregata came from the headwaters of the Gila. The type was not preserved and is known only from a drawing reproduced in Emory's report. There has been much discussion about the identity of the plant; Coulter transferred it to *Cereus*, referring to it *Cereus coccineus* and *C. phoeniceus* and assigning to it a wide range, Colorado to San Luis Potosí. Rydberg transferred the name to *Echinocereus* but applied it to the same group of plants described by Coulter. A careful restudy of the original illustration and Engelmann's description and a restudy of all the cacti of similar habit in the southwest leads us to a different conclusion from that reached by Dr. Coulter and Dr. Rydberg. Engelmann, who described it as a *Mammillaria*, says that it appears to be allied to *M. vivipara*, and this we believe is its true relationship. A *Mammillaria* from the region about Flagstaff often forms the great clusters mentioned by Engelmann, and while we believe that it differs from the one found in northern Arizona it is certainly a near ally, probably representing the closely related species from southeastern Arizona and south-western New Mexico which has often passed as *M. arizonica*.

Engelmann referred a specimen which he had from Sonora to his variety *Mammillaria vivipara neo-mexicana* with the remark that it was "a form with more spines than any other."

Plate IV shows a clump sent by Mrs. Ruth C. Ross from near Aravaipa, Arizona, in July 1922. Figure 47 is from a photograph of a single plant obtained by Dr. Rose near Benson May 1, 1908, which afterwards flowered in Washington.

36. Coryphantha cubensis Britton and Rose, Torreya 12: 15. 1912.

>   *Mammillaria urbaniana* Vaupel, Monatsschr. Kakteenk. 22: 65. 1912.

Plants depressed-globose, tufted, 2 to 3 cm. broad, pale green; tubercles numerous, vertically compressed, 6 to 7 mm. long, 4 to 5 mm. wide, about 3 mm. thick, grooved on upper side from apex to below middle, the groove very distinct; spines about 10, whitish, radiating, acicular but weak, 3 to 4 mm. long, those of young tubercles subtended by a tuft of silvery white hairs, 1.5 mm. long; flowers pale yellowish green, 16 mm. high, the segments acute; filaments, style, and stigma-lobes yellowish; fruit red, less than 1 cm. long, naked; seeds black, somewhat angled.

Type locality: Among stones in barren savanna, southeast of Holguin, Oriente, Cuba. *Distribution:* Type locality and vicinity.

This species is very inconspicuous and perhaps for that reason is rare in collections. It has only twice, to our knowledge, been collected, both times by Dr. J. A. Shafer, once in 1909 (No. 2946) and again in 1912 (No. 12432), who gave a short account of its discovery in the Journal of the New York Botanical Garden (No. 155). He states that it barely pro-trudes through the layer of broken stones that filled the interstices between the larger rocks; that the largest plants were scarcely an inch in diameter, one of them bearing a small yellowish flower. It lives only a short time in greenhouse cultivation.

On account of the name *Mammillaria cubensis* Zuccarini (Labouret, Monogr. Cact. 59. 1853) Vaupel gave a new specific name to the plant when he transferred it from *Cory-phantha*.

Plate V, figure 1, shows the plant collected by Dr. Shafer in 1912 which flowered in the New York Botanical Garden in July of the same year; figure 1a shows the fruit and figure 1b shows a tubercle from the same plant.

37. Coryphantha sulcata (Engelmann).

>   *Mammillaria sulcata*[*] Engelmann, Bost. Journ. Nat. Hist. 5: 246. 1845.
>   *Mammillaria strobiliformis* Mühlenpfordt, Allg. Gartenz. 16: 19. 1848. Not Engelmann, 1848.
>   *Mammillaria calcarata* Engelmann, Bost. Journ. Nat. Hist. 6: 195. 1850.
>   *Coryphantha calcarata* Lemaire, Cactées 35. 1868.
>   *Cactus calcaratus* Kuntze, Rev. Gen. Pl. 1: 259. 1891.
>   *Cactus scolymoides sulcatus* Coulter, Contr. U. S. Nat. Herb. 3: 116. 1894.
>   *Mammillaria radians sulcata* Schumann, Gesamtb. Kakteen 496. 1898.
>   *Cactus sulcatus* Small, Fl. Southeast. U. S. 812. 1903.

Cespitose, 8 to 12 cm. in diameter; tubercles rather large, 10 to 12 mm. long, somewhat flat-ened, soft; radial spines acicular, straight, white; central spines several, one somewhat stouter

---

> [*] Förster (Handb. Cact. 255. 1846) credits such a name to Pfeiffer but it is without description.

M. E. Eaton del. 1 to 4, 6, 7
A. A. Newton del. 5

1. Flowering plant of *Coryphantha cubensis*.
1a. Fruit of same.
1b. Tubercle of same.
2. Flowering plant of *Neomammillaria confusa*.
3. Flowering plant of *Neomammillaria geminispina*.

4. Top of flowering plant of *Coryphantha durangensis*.
5. Flowering plant of *Coryphantha arizonica*.
6. Flowering plant of *Coryphantha bumamma*.
7. Flowering plant of *Coryphantha chlorantha*.

than the others, porrect or slightly curved outward, others erect; flowers several, from near center of plant, 5 cm. in diameter or more, yellow, with a red center; inner perianth-segments lanceolate, apiculate; filaments reddish; style greenish yellow, exserted beyond stamens; stigma-lobes 7 to 10, yellow, notched at apex;* fruit oblong, greenish; seeds oblong, shining, dark brown.

Type locality: Industry, Texas.
Distribution: Southern Texas.

The herbarium sheets of this plant, sent us from the Missouri Botanical Garden, contain seeds, fruit, and style. Dr. Coulter speaks of seeing the spines of the type.

The name *Mammillaria sulcata*, first given by Engelmann, was changed by him to *M. calcarata* on account of *M. sulcata* Pfeiffer, but this was a later name and hence can not replace Engelmann's first one.

This species was collected by Lindheimer at Industry, Texas, growing with *Mammillaria similis*, but while the two are similar in habit, this plant differs from *M. similis* in having green fruit and brown oblong seeds instead of red fruit and black globose seeds, as well as in other ways. It has not been collected much in recent years and its characters

Fig. 49.—Coryphantha sulcata.

and range have been involved with other species. Miss Ellen D. Schulz sent us plants from San Antonio, Texas, in June 1921, and Robert Runyon sent us plants and photographs in 1922, which have enabled us to restudy the species in connection with its type now kept in the Engelmann Herbarium in the Missouri Botanical Garden.

Mammillaria goerngii was given by Haage (Cact. Kultur ed. 2. 183. 1900) as a new name for *M. calcarata*.

Illustrations: Cact. Mex. Bound. pl. 74. f. 1, as *Mammillaria calcarata*; Monatsschr. Kakteenk. **27**: 65, as *Mammillaria radians sulcata*.

Plate x, figure 1, shows a plant photographed by Robert Runyon at Sabinal, Texas, April 28, 1922. Figure 49 is from a photograph of four fruits sent by Professor Albert Ruth, of Polytechnic, Texas, in 1922.

---

*Whether this is a constant character we do not know, but we have observed it in three flowers, all from the same plant. It has not been noted before in any other species of *Coryphantha*.

PUBLISHED SPECIES, PERHAPS OF THIS GENUS.

MAMMILLARIA CALOCHLORA Hortus, Monatsschr. Kakteenk. **26**: 167. 1916; **2** : 133. 1917.

This seems undoubtedly a species of *Coryphantha*, but we have not been able to identify it. There is considerable confusion regarding this plant, as the following note from Meyer would indicate:

"I have gotten Mr. Quehl to send me the flower of *Mammillaria calochlora* Hort. and I see that this also agrees exactly with the flower of Grässner's *M. delaetiana*. As third and last I have now gotten Mr. de Laet to send me also a little plant of equal size of his genuine *M. delaetiana* Quehl and this one is entirely different from the two others in form and color of the body, areoles, and spines."

We have a small specimen and a photograph sent us by L. Quehl in 1921.

MAMMILLARIA CORDIGERA Heese, Gartenflora **59**: 445. 1910.

Short-cylindric, 6 cm. high, 4.5 cm. in diameter; tubercles 4-angled, broader than long, grooved above; spine-areoles longer than broad; radial spines 4 to 15, white, spreading; central spines 4, erect, curved if not hooked at apex, 15 mm. long; flowers and fruit unknown.

Type locality: Not cited.
Distribution: Doubtless Mexico.

This species we know only from descriptions and illustration. The illustration is so much like that of *Mammillaria bombycina* that we at first were inclined to combine them. From the observations of others there seem to be important technical differences which separate them, not only specifically but also generically. It may prove to be a synonym of *C. sulcolanata*, for we have recently examined a skeleton sent us by Bödeker which resembles very much the plants collected by Rose in Hidalgo, Mexico, which we have already referred to that species.

Illustration: Gartenflora **59**: f. 50, as *Mammillaria cordigera*.

MAMMILLARIA CORNUTA Hildmann in Schumann, Gesamtb. Kakteen 496. 1898.

Simple, grayish green, somewhat depressed, 4 to 5 cm. high, 6 to 8 cm. in diameter; tubercles spiraled, in 5 to 8 series; radial spines 5 to 7, subulate, straight or somewhat curved, white, 4 to 8 mm. long; central spine solitary, horn-colored; flowers said to be rose-colored; fruit unknown.

Type locality: Mexico.

From the description it is difficult to identify this species; its rose-colored flowers suggest a relationship with *Coryphantha elephantidens* but its spine-clusters are differently described.

MAMMILLARIA POTOSIANA Jacobi, Allg. Gartenz. **24**: 92. 1856.

Erect, cylindric, light green; tubercles conical, triangular at base, bearing 2 yellow glands in their axils; radial spines 15 or 16, subulate, equal or nearly so, 6 mm. long; central spine solitary, porrect but somewhat incurved at apex, subulate, 10 to 12 mm. long; flowers yellow.

Type locality: San Luis Potosí, Mexico.

Jacobi comments on the species as follows:

"Comptroller Shäfer in Münster received this beautiful plant in a shipment of plants from San Luis Potosí in Mexico, under the name of *Mammillaria raphidacantha*. From the given description it is adequately clear that the plant considered is another and undescribed one. The form of the tubercles as well as the number and form of the spines is other evidence, also the grooves upon the upper sides of the tubercles which are always present in the case of *M. raphidacantha* are here lacking throughout.

"The stem of the plant is cylindrical, dark green, finely punctate with white dots; tubercles conical, 3-angled at the base, gradually flattened above; axils sinuate with 2 yellow glands, inclosed by a ring of yellowish-white tomentum; areoles terminal, oval, the younger ones whitish tomentose, later naked; radial spines 15 or 16, radiating, somewhat recurved, needle-formed, two-colored. In older plants there appears here and there a longer and stronger central spine with the tip slightly bent downward. All the spines are awl-shaped and stiff.

"The radial spines when young are white with brownish (burnt) tips, later amber-colored above and below, grayish in the middle. The plant described is 3″ high and a little more than an inch in diameter; radials 3, centrals 5 or 6 lines long. The plant in my possession did indeed bloom last summer but I was hindered unfortunately in describing the flowers in detail. They are smaller than those of *M. raphidacantha*, very similar in form, but the petals are yellow with saffron-yellow central stripes on the outer side."

Although Jacobi states definitely that the tubercles are not grooved on the upper surface, yet the presence of glands would indicate that the plant is not a *Mammillaria* but, more likely, a *Coryphantha* of the Series *Recurvatae* and perhaps one of the species already described. We have never seen glands in the axils of tubercles, except in genera having grooved tubercles. In cultivated specimens growing under abnormal conditions tubercles are sometimes produced without a groove and with glands in their axils.

MAMMILLARIA RAMOSISSIMA Quehl, Monatsschr. Kakteenk. **18:** 127. 1908.

Globose to short-cylindric, dull grayish green; radial spines about 12, about 1 cm. long; central spines usually 1, sometimes 2 or 3; flowers and fruit unknown.

Type locality: Not cited.
Illustration: Monatsschr. Kakteenk. **18:** 127.

MAMMILLARIA RECURVISPINA De Vriese, Tijdschr. Nat. Geschr. **6:** 53. 1839.

Cactus recurvispinus Kuntze, Rev. Gen. Pl. 1: 261. 1891.

Solitary, somewhat depressed, about 16 cm. in diameter, glaucous; tubercles few, large, somewhat compressed, obtuse; areoles and axils of tubercles described as naked; spines all radial, 8, subulate, more or less incurved; flowers and fruit unknown.

Type locality: Mexico.

This plant was referred by Labouret to *Mammillaria sulcolanata* but was discussed by Schumann under *M. scheeri*; judging from the illustration, it is not close to either of these species but it is much nearer *Coryphantha bumamma*.

Illustration: Tijdschr. Nat. Geschr. **6:** pl. 1. f. 1.

Figure 48 is a reproduction of the illustration cited above.

Mammillaria speciosa De Vriese (Tijdschr. Nat. Geschr. **6:** 52. 1839. Not Don, 1830) is listed by Schumann among the species not known to him. It probably belongs to some species of *Coryphantha*.

The following names are without descriptions and can not be referred to any known species: *Coryphantha conspicua* Lemaire, Cactées 34. 1868; *Coryphantha engelmannii* Lemaire, Cactées 34. 1868; *Coryphantha hookeri* Lemaire, Cactées 34. 1868; *Coryphantha sublanata* Lemaire, Cactées 35. 1868.

7. NEOBESSEYA gen. nov.

Simple or tufted cacti, globose or somewhat depressed; tubercles irregular or somewhat spiraled, most of them grooved on upper side; flowers borne near top of plant, large, yellow or pink, probably always day-blooming; fruit globose, bright red, indehiscent; seeds black, globose, pitted, with a prominent white aril.

Type species: *Mammillaria missouriensis* Sweet.

Four species are recognized, all from the Great Plains of the United States.

The generic name commemorates Dr. Charles Edwin Bessey (1845–1915), professor in the University of Nebraska and for many years one of our eminent botanical teachers.

The genus is nearest *Coryphantha*, but it has very different fruit and seeds.

KEY TO SPECIES.

Flowers yellow.	
Outer perianth-segments naked	1. *N. wissmannii*
Outer perianth-segments ciliate.	
Inner perianth-segments long-acuminate	2. *N. similis*
Inner perianth-segments at most acute	3. *N. missouriensis*
Flowers grayish pink	4. *N. notesteinii*

1. Neobesseya wissmannii (Hildmann).

Mammillaria similis robustior Engelmann, Bost. Journ. Nat. Hist. **6**: 200. 1850.
Mammillaria nuttallii robustior Engelmann, Proc. Amer. Acad. **3**: 265. 1856.
Mammillaria missouriensis robustior S. Watson,
    Bibl. Index **1**: 403. 1878.
Cactus missouriensis robustior Coulter, Contr. U. S.
    Nat. Herb. **3**: 111. 1894.
Mammillaria wissmannii Hildmann in Schumann,
    Gesamtb. Kakteen 498. 1898.
Cactus robustior Small, Fl. Southeast. U. S. 812.
    1903.

Plant solitary, or forming mounds 2 to 3 dm. in
diameter and 1 dm. high with 25 heads or more;
areoles elliptic when young, conspicuously white-
woolly, the head usually globose, tubercles rather
large, spreading, somewhat narrowed towards apex;
spines 7 to 14, when young white to brownish, in age
gray with yellow swollen base, acicular, 1.5 to 2 cm.
long, sometimes all radial and spreading, rarely 1 or
2 centrals and these porrect; flowers large, 4 to 5 cm.
long, dark yellow; scales on flower-tube strongly
nerved; margin of perianth-segments naked; inner
segments abruptly long-apiculate; fruit globose, 8
mm. in diameter.

FIG. 50.—Neobesseya wissmannii.

Type locality: Not cited, presumably Texas.
Distribution: Central Texas.
Illustration: Blühende Kakteen **1**: pl. 5, as
Mammillaria wissmannii.

Figure 50 is a reproduction of the illustration cited above.

2. Neobesseya similis (Engelmann).

Mammillaria similis Engelmann, Bost. Journ. Nat. Hist. **5**: 246. 1845.
Mammillaria similis caespitosa Engelmann, Bost. Journ. Nat. Hist. **6**: 200. 1850.
Echinocactus similis Poselger, Allg. Gartenz. **21**: 107. 1853.
Mammillaria nuttallii caespitosa Engelmann, Proc. Amer. Acad. **3**: 265. 1856.
Mammillaria missouriensis caespitosa S. Watson, Bibl. Index **1**: 403. 1878.
Cactus missouriensis similis Coulter, Contr. U. S. Nat. Herb. **3**: 111. 1894.
Mammillaria missouriensis similis Schumann, Gesamtb. Kakteen 498. 1898.
Cactus similis Small, Fl. Southeast. U. S. 812. 1903.
Coryphantha similis Britton and Rose in Britton and Brown, Illustr. Fl. ed. 2. **2**: 571. 1913.

Plants sometimes growing in large clumps 1 to 1.5 dm. high by 2 to 3 dm. in diameter, containing
25 individuals or more; larger plants 6 to 10 cm. in diameter; tubercles deep green, cylindric, sometimes 2 cm. long, when young the groove filled with white wool; spines all puberulent; radial spines
12 to 15, spreading, dirty white with brownish tips; central spine solitary or often wanting, similar to
but stouter and longer than the radials; flowers 5 to 6 cm. long, light yellow, the outer lobes tinged
with brown and green; inner perianth-segments long, narrow, acuminate; flower-tube definite,
covered nearly to its base with short greenish stamens; style green; stigma-lobes 4 to 6, linear;
fruit globular or short-oblong, 10 to 20 mm. in diameter; seeds large, globose, 2 mm. in diameter.

Type locality: Near Industry, Texas.
Distribution: Eastern Texas.

Engelmann says that the flowers and fruits are larger than in *Mammillaria nuttallii.*
The inner perianth-segments gradually taper to the apex.

S. Watson and others refer here *Mammillaria caespitosa* Gray (Struct. Bot. 421. f. 838),
but the plant illustrated by Gray is *Echinocereus reichenbachii.* The Index Kewensis
refers *Mammillaria caespitosa* Gray, as they also do *Mammillaria similis*, to *Mammillaria
missouriensis.* (See Cactaceae **3**: 26).

Illustration: Cact. Mex. Bound. pl. 74, f. 7, as *Mammillaria nuttallii caespitosa* (seed).

Plate III, figure 2, shows a plant collected by F. E. Upham at Fort Worth, Texas, which
flowered in Washington.

3. Neobesseya missouriensis (Sweet).

Cactus mammillaris Nuttall, Gen. Pl. 1: 295. 1818. Not Linnaeus, 1753.
Mammillaria missouriensis Sweet, Hort. Brit. 171. 1826.
Mammillaria simplex Torrey and Gray, Fl. N. Amer. 1: 553. 1840.
Mammillaria nuttallii Engelmann, Pl. Fendl. 49. 1849.
Mammillaria nuttallii borealis Engelmann, Proc. Amer. Acad. 3: 264. 1856.
Cactus missouriensis Kuntze, Rev. Gen. Pl. 1: 259. 1891.
Mammillaria missouriensis nuttallii Schelle, Handb. Kakteenk. 241. 1907.
Coryphantha missouriensis Britton and Rose in Britton and Brown, Illustr. Fl. ed. 2. 2: 570. 1913.

Plants solitary or cespitose, globose, 2.5 to 5 cm. in diameter; tubercles more or less spiraled, 10 to 15 mm. long; spines 10 to 20, acicular, gray, pubescent, all radial or sometimes 1 central; flowers greenish yellow; outer perianth-segments narrowly oblong, gradually tapering to an acute apex, ciliate; inner segments linear-lanceolate, attenuate; fruit globose, scarlet, about 1 cm. in diameter; seeds 1 mm. in diameter.

Type locality: On the high hills of the Missouri, probably to the mountains.

Distribution: North Dakota to Montana, Colorado to Kansas, Oklahoma, and perhaps northern Texas.

This little cactus has a wide distribution on the Great Plains; both its conspicuous yellow flowers and its round red fruits are very attractive.

Coryphantha nuttallii, credited to Engelmann, is cited as a synonym of *Mammillaria nuttallii* by Rümpler (Förster, Handb. Cact. ed. 2. 407. 1885).

Illustrations: Meehan's Monthly 10: pl. 3; Gartenwelt 1: 85, as *Mammillaria missouriensis*; Gartenwelt 1: 89, as *M. missouriensis viridescens*; Britton and Brown, Illustr. Fl. 2: f. 2525, as *Cactus missouriensis*; Schelle, Handb. Kakteenk. 241. f. 160, as *M. missouriensis nuttalliii*; Cact. Mex. Bound. pl. 74, f. 6, as *M. nuttallii borealis*; Blanc, Cacti 72. No. 1426; Blühende Kakteen 3: pl. 145, as *M. nuttallii*; Britton and Brown, Illustr. Fl. ed. 2. 2: f. 2984, as *Coryphantha missouriensis*.

Plate XI, figure 4, shows a plant from a large clump sent by Professor C. O. Chambers in 1921 from Stillwater, Oklahoma.

4. Neobesseya notesteinii (Britton).

Mammillaria notesteinii Britton, Bull. Torr. Club 18: 367. 1891.
Cactus notesteinii Rydberg, Mem. N. Y. Bot. Gard. 1: 272. 1900.

Oval, solitary or cespitose, about 3 cm. in diameter; tubercles nearly terete, about 6 mm. high; spines 12 to 18, white, turning gray, weak, slender, 8 to 12 mm. long, pubescent throughout, a central one usually present and frequently pink-tipped; flowers 15 to 25 mm. broad, ash-gray, tinged and penciled with pink, the segments broadly linear-oblong, mucronate; fruit obovoid; seeds black, globose, pitted.

Type locality: Near Deer Lodge, Montana.

Distribution: Known only from the type locality.

Professor F. N. Notestein, who first collected and observed this little cactus, found it in gravelly soil near a small creek; it differs from the other species of the genus in the color of the flowers and the more pubescent spines.

8. ESCOBARIA gen. nov.

Globose or cylindric, usually cespitose cacti, never milky; tubercles grooved above, persisting as knobs at the base of old plants after the spines have fallen; spines both central and radial, never hooked; flowers small, regular, appearing from top of plant at bottom of groove of young tubercles; stamens and style included; fruit red, naked (or with one scale), indehiscent, globular to oblong, crowned by the withering perianth; seeds brown to black; aril basal or subventral, oval.

Type species: *Mammillaria tuberculosa* Engelmann.

The two species of this genus known to Schumann were placed by him in the subgenus *Coryphantha* of *Mammillaria*; they are like the *Coryphanthae* in having grooved flower-bearing tubercles, but are otherwise different, especially in the flowers, fruit, and seeds.

Eight species are known from northern Mexico and southern Texas.

The genus commemorates the work of two distinguished Mexicans, the Escobar brothers, Rómulo and Numa, of Mexico City and Juárez.

54 THE CACTACEAE.

KEY TO SPECIES.

Outer perianth-segments ciliate.
  Groove of tubercles without glands.
    Flowers large for the genus, 2 to 2.5 cm. long.
      Plants elongated; seeds very small, brown, with ventral hilum.............. 1. *E. tuberculosa*
      Plants usually globose; seeds larger than in *E. tuberculosa*, black, with a sub-
        basal hilum.. 2. *E. dasyacantha*
    Flowers small, about 1.5 cm. long.
      Plants globose to stout-cylindric.
        Inner perianth-segments pointed.
          Inner perianth-segments broad............................... 3. *E. chihuahuensis*
          Inner perianth-segments narrow.............................. 4. *E. runyonii*
        Inner perianth-segments obtuse................................. 5. *E. chaffeyi*
      Plants slender-cylindric.. 6. *E. sneedii*
  Groove of tubercles with glands... 7. *E. bella*
Outer perianth-segments eciliate.. 8. *E. lloydii*

1. Escobaria tuberculosa (Engelmann).

    Mammillaria strobiliformis Scheer in Salm-Dyck, Cact. Hort. Dyck. 1849. 104. 1850. Not Engelmann,
      1848.
    Echinocactus strobiliformis Poselger, Allg. Gartenz. 21: 107. 1853.
    Mammillaria tuberculosa Engelmann, Proc. Amer. Acad. 3: 268. 1856.
    Cactus tuberculosus Kuntze, Rev. Gen. Pl. 1: 261. 1891.
    Cactus strobiliformis Kuntze, Rev. Gen. Pl. 1: 261. 1891.
    Mammillaria strobiliformis pubescens Quehl, Monatsschr. Kakteenk. 17: 87. 1907.
    Mammillaria strobiliformis durispina Quehl, Monatsschr. Kakteenk. 17: 87. 1907.
    Mammillaria strobiliformis rufispina Quehl, Monatsschr. Kakteenk. 17: 87. 1907.
    Mammillaria strobiliformis caespititia Quehl, Monatsschr. Kakteenk. 19: 173. 1909.

FIG. 51.—Escobaria tuberculosa.   FIG. 52.—Escobaria dasyacantha.

    Usually growing in clumps, cylindric or becoming so, 5 to 18 cm. high, 2 to 6 cm. in diameter; tubercles more or less regularly arranged in spirals, 6 mm. long; radial spines numerous, white, sometimes as many as 30, acicular, 4 to 15 mm. long; central spines several, stouter than radials, brown to blackish or colored only at tips, one of them usually porrect; flowers 2.5 cm. in diameter when fully expanded, light pink; outer perianth-segments acute, ciliate; inner perianth-segments narrowly pointed; fruit oblong, up to 20 mm. long, red; seeds pitted, with a small ventral hilum.

    Type locality: Mountains near El Paso and eastward.
    Distribution: Southwestern Texas, southern New Mexico, and adjacent Mexico.
    Flowers appear in the afternoon and last for two days at least.
    The name *Coryphantha tuberculosa* occurs in C. R. Orcutt's Circular to Cactus Fanciers, 1922.

Illustrations: Cact. Mex. Bound. pl. 12, f. 1 to 16, as *Mammillaria tuberculosa*; Förster, Handb. Cact. ed. 2. 417. f. 46; Schelle, Handb. Kakteenk. 235. f. 149, as *M. strobiliformis.*

Figure 51 is from a photograph of the plant sent by Dr. Shreve from near El Paso, Texas, in 1920.

2. Escobaria dasyacantha (Engelmann).

Mammillaria dasyacantha Engelmann, Proc. Amer. Acad. 3: 268. 1856.
Cactus dasyacanthus Kuntze, Rev. Gen. Pl. 1: 259. 1891.

Globose to short-oblong, usually 4 to 7 cm. in diameter but sometimes 20 cm. long; radial spines 20 or more, white, bristle-like; central spines about 9, stouter and longer than the radials, upper half usually reddish or brownish, often 2 cm. long; flowers pinkish; perianth-segments narrowly oblong, ciliate, apiculate; stigma-lobes green; fruit clavate, scarlet, 15 to 20 mm. long; seeds black, 1 mm. in diameter, slightly flattened, pitted, with a narrow white subbasal hilum.

Type locality: El Paso and eastward.

Distribution: Western Texas, southern New Mexico, and northern Chihuahua.

We have examined the type of this species which was collected by Charles Wright at El Paso in 1852.

Escobaria dasyacantha is sometimes mistaken for *Escobaria tuberculosa,* but the stems are usually globose and the seeds larger and of a different shape. Engelmann speaks of its resemblance to *Echinocactus intertextus dasyacanthus,* now *Echinomastus dasyacanthus,* but this is only superficial, for the flowers, fruit, and seeds of the two species are very different. The name *Coryphantha dasyacantha* occurs in C. R. Orcutt's Circular to Cactus Fanciers, 1922. We had never seen this plant in cultivation until it was recently sent by Mrs. S. L. Pattison from western Texas.

Illustrations: Cact. Mex. Bound. pl. 12, f. 17 to 22, as *Mammillaria dasyacantha.*

Plate VII, figure 1, shows a plant sent by Mrs. S. L. Pattison from near El Paso, Texas, in 1921 which flowered in the New York Botanical Garden. Figure 52 is from a photograph of another plant sent by Mrs. Pattison from the same region.

3. Escobaria chihuahuensis sp. nov.

Plants often solitary, perhaps also cespitose, globose to short-cylindric, very spiny; tubercles short, usually hidden by the spines; radial spines numerous, spreading; central spines several, longer than radials, usually brown or black in upper part; flowers small, 1 to 1.5 cm. long, purple; outer perianth-segments broad, often rounded at apex with ciliate margins; inner perianth-segments pointed.

Common in the mountains near Chihuahua, where it was collected by Palmer (No. 72, type) in 1908 and by Pringle (Nos. 250, 251) in 1885.

This plant should be compared with *Mammillaria grusonii* Rünge (Gartenflora **38:** 105. f. 20. 1889). L. Quehl believed that *M. grusonii* was closely related to *M. scheeri,* but he apparently knew it only from the original illustration and description. It does not suggest any of the species of *Coryphantha* to us.

4. Escobaria runyonii sp. nov.

Cespitose, with numerous (sometimes 100) globose to short-oblong heads, grayish green, 3 to 5 cm. long with fibrous roots; tubercles 5 mm. long, terete in section with very narrow groove above; groove at first white-woolly, not glandular; radial spines numerous, acicular, white, 4 to 5 mm. long; central spines stouter than radials, 5 to 7, slightly spreading with brown or black tips, 6 to 8 mm. long; flowers 1.5 cm. long, pale purple; segments with a dark purple stripe down the middle and pale margins; outer perianth-segments narrow-oblong, with thin ciliate margins; inner perianth-segments narrower than the outer, with margins entire, acute; filaments purplish; style very pale; stigma-lobes 6, green; fruit scarlet, globose to short-oblong, 6 to 9 mm. long, juicy.

Collected by Robert Runyon in July 1921 and again in October of the same year near Reynosa, Mexico, about 75 miles up the Rio Grande from Brownsville, Texas, and on

August 10, 1921, near Rio Grande, Starr County, Texas. The plant flowered in Washington March 13, 1922.

Plate VI, figure 1, is from a photograph of the type plant taken by Robert Runyon. Figure 53 is from a photograph taken by Robert Runyon.

5. Escobaria chaffeyi sp. nov.

Short-cylindric, 6 to 12 cm. long by 5 to 6 cm. in diameter, almost covered by the numerous white spines; tubercles rather short, light green, with a narrow groove above; radial spines numerous, spreading, bristly; central spines several, a little shorter than the radials and brown or black-tipped; flowers 15 mm. long, cream-colored or sometimes purplish; outer perianth-segments ciliate; inner perianth-segments oblong, obtuse, entire; style white; stigma-lobes very short, yellowish green; fruit crimson, 2 cm. long.

Collected by Dr. Elswood Chaffey near Cedros, Zacatecas, Mexico, in June 1910 (No. 5, type), and by F. E. Lloyd near the same locality in 1908 (No. 29).

FIG. 53.—Escobaria runyonii.   FIG. 54.—Escobaria sneedii.

6. Escobaria sneedii sp. nov.

Densely cespitose, sometimes with as many as 50 joints, creeping or spreading; joints cylindric, up to 6 cm. long, 1 to 2 cm. in diameter; tubercles numerous, hidden under the many spines, terete, 2 to 3 mm. long, in age naked; groove narrow, hairy throughout its length; axils of tubercles not setose; spines 20 in a cluster or more, nearly white, or the larger ones brown at tip, longest one 6 mm. long, all usually appressed, but the longer ones near top, connivent; flowers small, 10 mm. long or less when dry, the outer segments long-ciliate; fruit (immature) a little longer than thick, 5 to 7 cm. long, green (?), at first juicy, naked; seeds globose, brown, nearly 1 mm. in diameter, pitted.

This curious little plant was sent us in February 1921 by Mrs. S. L. Pattison from southwestern Texas; it was collected by J. R. Sneed, who at first found only three clumps, but afterwards a fourth clump was discovered and again it was found in June 1921 just after it had flowered. It is known from a single station on the Franklin Mountains, Texas. According to Mrs. Slater the flowers are pink to saffron.

Figure 54 is from a photograph of a single plant sent by Mrs. Pattison in 1921.

7. Escobaria bella sp. nov.

Cespitose, cylindric, 6 to 8 cm. long; tubercles nearly terete, 1.5 to 2 cm. long, the groove white-hairy, with a narrow brownish gland near center; radial spines several, whitish, 1 cm. long or less; central spines 3 to 5, brown, unequal, the largest 2 cm. long or more, ascending; flowers central,

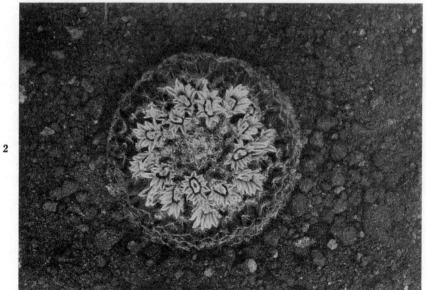

1. *Escobaria runyonii*, from Texas.
2. *Neomammillaria hemisphaerica*, from Texas.

small, rotate, nearly 2 cm. broad; perianth-segments pinkish with pale margins, linear-oblong, acute, the outer ones ciliate; filaments reddish; upper part of style and stigma-lobes green.

Collected by J. N. Rose and Wm. R. Fitch on hills of Devil's River, Texas (No. 17991).

Plate VII, figure 4, shows the type, which flowered in the New York Botanical Garden, March 31, 1914; figure 4a shows a tubercle with its gland-bearing groove.

8. Escobaria lloydii sp. nov.

Plant growing in clumps and resembling a small species of *Echinocereus*; old plants bearing naked corky tubercles; radial spines about 20, spreading, slender, white; central spines several, stout, with black or with brownish tips, 2 cm. long; flowers greenish with a central stripe on outside, 2.5 cm. long; filaments, style, and stigma-lobes green; fruit red, globose to short-oblong, 6 to 12 mm. long; seeds black, pitted, globose, 1 mm. in diameter.

Collected by F. E. Lloyd in foothills of Sierra Zuluaga, Zacatecas, Mexico, March 29, 1908 (No. 5).

This species is near *Escobaria tuberculosa*, but it has much stouter central spines and greenish white, eciliate inner perianth-segments.

SPECIES PERHAPS OF THIS RELATIONSHIP.

MAMMILLARIA EMSKOETTERIANA Quehl, Monatsschr. Kakteenk. **20**: 139. 1910.

Cespitose, globose to short-cylindric, 5 cm. high; tubercles conic, their axils naked; radial spines 20 to 25; central spines 6 to 8, setaceous, white with black tips; flowers brownish yellow, 3 cm. long.

Type locality: Not cited.

We obtained a specimen of this plant from Quehl in 1913, but it has not done well nor has it flowered and we have not been able to refer it to any described species, but believe that it may be near *Escobaria tuberculosa*. Mr. Quehl believed that it was near *Mammillaria dasyacantha*, but if it came from San Luis Potosí, as Mr. Quehl supposed, it is doubtless specifically distinct from both. The following note is a translation of some remarks by Mr. Quehl:

"Our illustration shows a grafted specimen which has naturally grown more corpulent and consequently permits one to see better its general structure and the arrangement of the spines. Ungrafted specimens are thicker, lower, and, without other characteristics, can not be distinguished from a red-spined *Mammillaria pusilla* var. *multiceps*. Only a closer inspection reveals the wart-furrows and consequently the *Coryphantha*. The similarity is so great that I suspect that the new species is already more disseminated though not correctly recognized and the plants are either set aside or ignored as a form of *Mammillaria pusilla*. The plants before me were raised by Mr. Robert Emskötter, fancy and commercial gardener, of Magdeburg, after whom I have named the species, from mixed seed which he received from San Luis Potosí, so that Mexico may be regarded as its home."

Illustration: Monatsschr. Kakteenk. **20**: 139.

9. BARTSCHELLA gen. nov.

Usually cespitose, globose to short-oblong cactus; tubercles large, somewhat united with the adjacent ones as in certain species of *Echinocactanae*, terete, not grooved, juicy, not milky; spines both radial and central, the latter usually hooked; flowers borne near top of plant, large, light purple or lavender; fruit short, hidden among the tubercles, seemingly dry, circumscissile; seeds dull black, pitted, with a narrow cylindric base, slightly constricted above; hilum large, slightly depressed, triangular.

Type species: *Mammillaria schumannii* Hildmann.

While this genus is probably to be referred to the *Coryphanthanae*, it possesses some characters of certain species of *Echinocactanae*, but the origin of the flower is quite different from any of them. The flower is large, like that of some species of *Coryphantha*, but the tubercles are not grooved and the seeds are not brown and reticulated. It differs from the typical species of the so-called *Mammillaria* in its large flowers and black seeds, while from all of these genera it differs in its circumscissile fruit.

This monotypic genus is named for Dr. Paul Bartsch, curator in the United States National Museum, distinguished in conchology, who has sent us cacti from many out of the way places.

1. Bartschella schumannii (Hildmann).

Mammillaria schumannii Hildmann, Monatsschr. Kakteenk. 1: 125. 1891.
Mammillaria venusta K. Brandegee, Zoe 5: 8. 1900.

More or less cespitose (as many as 40 stems have been reported in a single cluster), 6 cm. high or less; axils slightly woolly, without bristles; radial spines 9 to 15, stout, 6 to 12 mm. long, brownish above, glabrous; central spines usually 1, sometimes 2 or 3, one of these usually hooked; in seedlings 10 or 11 radial spines developing, these spreading, feather-like with long spreading hairs; in one-year-old plants the spines simply puberulent, all white with brown tips and one central much longer than the others and strongly hooked; flower 3 to 4 cm. in diameter, the segments about 10, lanceolate, acuminate; stamens numerous, erect, shorter than the style; style slender, erect, pale; stigma-lobes 6, linear, green; fruit short, dull in color; seeds usually found in a cup between the tubercles, less than 1 mm. long.

FIG. 55.—Bartschella schumannii.   FIG. 56.—Pelecyphora aselliformis.

Type locality: Not cited.*
Distribution: Southern Lower California.

This species has been rare in collections, but considerable material was collected by Dr. Rose at Cape San Lucas, Lower California, in March 1911 (No. 16375). Living specimens were sent us from Lower California by Ivan M. Johnston in 1921.

Dr. C. H. Thompson writes under date of September 15, 1911, as follows:

"Your No. 16375, *Mammillaria venusta*, puzzles me. We received three plants from the New York Botanical Garden. Two are considerably shriveled but are reviving. The third is more plump and shows the vegetation characters better. In these it would readily be taken for *Mammillaria*, yet there are some appearances of the mamillate *Echinocacti*. You will observe how commonly adjacent tubercles cohere as in that group of *Echinocactus*, quite distinct from any *Mammillaria* that I know. Yet the position of the flower excludes if from *Echinocactus*. With flower and fruit characters you have observed it strikes me as being distinct from either genus."

---

* *Mammillaria schumannii* was described from a cultivated plant, but *M. venusta* came from San José del Cabo, Lower California.

M. E. Eaton del.

1. Flowering plant of *Escobaria dasyacantha*.
2. Fruit of *Dolichothele sphaerica*.
2a. Seed of same.
3. Flowering plant of *Neomammillaria arida*.
4. Flowering plant of *Escobaria bella*.

4a. Tubercle of same.
5. Flowering plant of *Neomammillaria crocidata*.
6. Flowering plant of *Bartschella schumannii*.
7. Flowering plant of *Neomammillaria carnea*.

Mammillaria schumanniana (Monatsschr. Kakteenk. **12:** 178. 1902) was evidently intended for *M. schumannii.*

Illustrations: Monatsschr. Kakteenk. **1:** facing 89; Thomas, Zimmerkultur Kakteen 51, as *Mammillaria schumannii.*

Plate VII, figure 6, shows a plant collected by Dr. Rose at Cape San Lucas, Lower California, in March 1911 (No. 16375), while a member of the scientific staff of the U. S. Steamer *Albatross.* Figure 55 is from a photograph of another plant from the same collection.

10. PELECYPHORA Ehrenberg, Bot. Zeit. 1: 737. 1843.

Plants small, cespitose, cylindric or globose, tuberculate, watery; tubercles not arranged on ribs, strongly flattened, crowned with an elliptic areole bearing a pectinate spine, never grooved; flowers borne near center, broad, campanulate, purplish, the segments in definite series; flower-tube very short, slender; stamens short; fruit small, naked; seeds black, smooth.

Only one species, native of Mexico, is here recognized, *Pelecyphora aselliformis* Ehrenberg, the type. A second species has generally been referred here but it differs so widely from the other that we have no hesitancy in segregating it generically (see genus No. 13, p. 64).

The generic name is from πέλεκυς hatchet, and φορός bearing, referring to the shape of the tubercles.

The plant has usually been regarded as a near relative of *Mammillaria*, but it has little in common with that genus. The flowers are central, borne in a mass of wool or hairs; the tubercles are not grooved and the seeds are black and smooth. It has been difficult for us, with the material at hand, to make out definitely the origin and position of the flower, but it seems to originate on the central sunken disk. This disk at first bears only clusters of hairs in the center of which the flower is produced. In time the flower opens and the tubercle, with its peculiar spiny crown, is developed, leaving in its axil the tuft of hairs about the flower.

1. Pelecyphora aselliformis Ehrenberg, Bot. Zeit. 1: 737. 1843.

    Pelecyphora aselliformis concolor Hooker in Curtis's Bot. Mag. **99:** pl. 6061. 1873.
    Pelecyphora aselliformis grandiflora Haage jr., Cact. Kultur ed. 2. 206. 1900.

Tufted, cylindric, 5 to 10 cm. high, 2.5 to 5 cm. in diameter, covered with tubercles arranged in spirals; tubercles strongly flattened laterally, somewhat stalked at base; areoles at top of tubercles very long and narrow, crowned by an elongated, scale-like spine with numerous lateral ridges, usually free at tip, giving a peculiar pectinate appearance; flowers 3 cm. broad or more, campanulate; perianth-segments in 4 rows, the outer ones sometimes white, oblong, acute; stamens borne at top of flower-tube, much shorter than perianth-segments; stigma-lobes 4, erect; seeds 1 mm. broad, kidney-shaped.

Type locality: Mexico.

Distribution: About San Luis Potosí, Mexico.

This plant does not do well in cultivation. It is known generally as the hatchet cactus, and is also called peote and peyote, also peyotillo and peotillo; it is said by the Mexicans to possess medicinal properties.

Mammillaria aselliformis, according to Watson (Cact. Cult. 188. 1889), was described in 1843, but we have found no other reference to it, except that Dr. A. Weber gives it as a synonym, crediting it to Monville. The name *Anhalonium aselliforme* Weber and *Ariocarpus aselliformis* Weber (Dict. Hort. Bois 931. 1898), quoted by Schumann as synonyms, were not formally published. *Pelecyphora fimbriata* Hildmann (Monatsschr. Kakteenk. **3:** 68. 1893), simply a name, may or may not belong here.

Illustrations: Haage, Cact. Kultur ed. 2. 206, as *Pelecyphora aselliformis grandiflora;* Amer. Gard. **11:** 474; Curtis's Bot. Mag. **99:** pl. 6061, as *Pelecyphora aselliformis concolor;*

Rümpler, Sukkulenten 208. f. 118; Gartenflora **34:** 25; Watson, Cact. Cult. 189. f. 75; ed. 3. f. 52; Cycl. Amer. Hort. Bailey **1:** 203. f. 303; Stand. Cycl. Hort. Bailey **2:** f. 718; Illustr. Hort. **5:** pl. 186; Förster, Handb. Cact. ed. 2. 237. f. 21; Cact. Journ. **1:** 107, 149; Krook, Handb. Cact. 34; Ann. Rep. Smiths. Inst. **1908:** pl. 14, f. 6; Palmer, Cult. Cact. 117; Schelle, Handb. Kakteenk. 275. f. 197; Monatsschr. Kakteenk. **29:** 81; Weinberg, Cacti 23; Knippel, Kakteen pl. 28; Möllers Deutsche Gärt. Zeit. **25:** 477. f. 11, No. 3; Garten-Zeitung **4:** 218. f. 50; Blanc, Cacti 78. No. 1710; West Amer. Sci. **11:** 8; Balt. Cact. Journ. **1:** 89; **2:** 164; Floralia **42:** 369; Remark, Kakteenfreund 22; Haage, Cact. Kultur ed. 2. 206.

Figure 56 is reproduced from a painting made by Miss E. I. Schutt in 1907, of a plant sent from San Luis Potosí in 1905 by Dr. E. Palmer.

11. PHELLOSPERMA gen. nov.

A globular to cylindric, usually cespitose cactus with a large, fleshy, branched root; tubercles not grooved above, not milky; flowers borne in axils of old tubercles, funnel-shaped; fruit globular to cylindric, red, depressed at apex; seeds large (for this group), dull black, not pitted but rugose, with a thick corky base nearly as large as the body.

Type species: *Mammillaria tetrancistra* Engelmann.

This genus differs from all its relatives in its very peculiar seeds. The flower, in its shape and origin, suggests the following genus, but in its color and size resembles *Coryphantha radiosa*. A single species is known, native of the western United States.

The generic name is from φελλός cork, and σπέρμα seed, referring to the corky base of the seed.

1. Phellosperma tetrancistra (Engelmann).

> *Mammillaria tetrancistra* Engelmann, Amer. Journ. Sci. II. **14:** 337. 1852.
> *Mammillaria phellosperma* Engelmann, Proc. Amer. Acad. **3:** 262. 1856.
> *Cactus phellospermus* Kuntze, Rev. Gen. Pl. **1:** 261. 1891.
> *Cactus tetrancistrus* Coulter, Contr. U. S. Nat. Herb. **3:** 104. 1894.

Solitary or cespitose, cylindric; sometimes becoming very large and then 3 dm. long, usually very spiny; root elongated, carrot-shaped or sometimes branched; tubercles terete, often elongated, their axils naked; radial spines numerous, acicular, white or sometimes with a brown tip, not pungent; central spines 1 to 4, stouter and longer than the radials, often brown or black, one or all strongly hooked; flower 3.5 to 4 cm. long, purple; base of tube slender, greenish, naked; scales and outer perianth-segments ciliate; style and stigma-lobes cream-colored; fruit rather variable in size, sometimes 3.7 cm. long, becoming dry in age, with a depressed umbilicus; seeds black, dull, 2 mm. long.

Fig. 57.—Seed of Phellosperma tetrancistra.

Type locality: San Felipe, California.

Distribution: Western Arizona, southeastern California, southern Utah, and southern Nevada; probably northern Lower California.

Mr. C. R. Orcutt, under date of March 5, 1922, comments on the distribution of this plant as follows:

"It reaches its greatest development on sandy and gravelly slopes near the White Water River east of Banning, California. It no doubt enters Lower California, for I believe that I have found it within a mile of the boundary line. It is comparatively rare in Arizona."

We have seen no specimens from Utah, but suspect that the plants from that state which have been referred to *Mammillaria grahamii* probably belong here. The species should be looked for in northern Lower California and Sonora.

Illustrations: Cact. Mex. Bound. pl. 7; Engler and Prantl, Pflanzenfam. **3**[6a]**:** 162. f. 56, B; Cact. Journ. **1:** pl. for February; Bol. Direccion Estudios Biol. **2:** f. 3; Monatsschr. Kakteenk. **20:** 167, as *Mammillaria phellosperma.*

Figure 58 is from a photograph of a plant sent from California in 1921 by E. C. Rost; figure 57 shows a seed taken from a plant sent by Loren G. Polhamus in 1921 from Bard, California.

12. DOLICHOTHELE (Schumann) gen. nov.

Plant-body globose, more or less cespitose, soft in texture, never milky; tubercles elongated, not grooved above; flowers borne in axils of old tubercles, very large, with a definite funnel-shaped tube; inner perianth-segments yellow, spatulate, tapering into a claw and borne on top of tube; stamens forming a spiral about style and borne on whole face of throat, but forming a definite ring at top of throat; style slender; stigma-lobes linear; ovary exserted, naked; fruit smooth, greenish, purplish, or red, globose, ellipsoid or short-oblong; seeds black or brownish.

Type species: *Mammillaria longimamma* De Candolle.

The generic name is from δολιχός long, and θηλή nipple, referring to the elongated tubercles.

The fruit is not often collected and is not well known. Dr. Rose obtained a single fruit of one of the species, the only one we had then seen, in a private collection in Rome in 1915; this is nearly globular, red, thin-walled, many-seeded; the seeds are brownish, pitted, slightly flattened, pointed at base, with a small sub-basal hilum. In October 1921, Robert Runyon sent us a number of fruits which were greenish white to purplish, with black seeds, these somewhat flattened and pitted.

Fig. 58.—Phellosperma tetrancistra.   Fig. 59.—Dolichothele longimamma.

Mammillaria camptotricha Dams (Gartenwelt **10**: 14. 1905) is usually considered as a close relative of this group, but it differs widely from it in the flowers as well as in other ways, and we believe that it is not congeneric with it (see page 126).

Three species, natives of southern Texas and northern and central Mexico, are recognized.

KEY TO SPECIES.

Spines glabrous, even when very young; species of Texas and northern Mexico.......... 1. *D. sphaerica*
Spines puberulent; species of central Mexico.
    Tubercles very long (sometimes 5 cm. long), pale green, glaucous; radials 6 or more;
        central spines usually present.. 2. *D. longimamma*
    Tubercles much shorter, bright green; radial spines 4 or 5; central spines none....... 3. *D. uberiformis*

1. Dolichothele sphaerica (Dietrich).

    Mammillaria sphaerica Dietrich in Poselger, Allg. Gartenz. **21**: 94. 1853.
    Cactus sphaericus Kuntze, Rev. Gen. Pl. **1**: 261. 1891.
    Mammillaria longimamma sphaerica K. Brandegee, Cycl. Amer. Hort. Bailey **2**: 975. 1900.

Low and depressed, often growing in large cespitose masses 2 dm. in diameter, with a large thickened root; tubercles soft and turgid, resembling those of the following species (*D. longimamma*) but shorter, 12 to 16 mm. long; areoles small, circular, at first short-lanate; spines 12 to 15, glabrous, generally pale yellow, a little darker at base at first, in age darker, often reddish, 7 to 9 mm. long, spreading or a little curved backward; central spine 1, straight; flowers appearing toward top of plant but not from axils of younger tubercles, with a rotate limb 6 to 7 cm. broad; inner perianth-segments widely spreading, oblanceolate, acute to apiculate, tapering at base into a slender claw; stigma-lobes 8, yellow, narrow; fruit greenish white to purplish, short-oblong, 10 to 15 mm. long, juicy, very fragrant; seeds black, flattened, with a straight ventral face, rounded on the back, pitted; hilum subventral.

Type locality: Near Corpus Christi, Texas.

Distribution: Southern Texas and northern Mexico, especially along the Rio Grande from Eagle Pass to the sea.

Mr. R. D. Camp and Mr. Robert Runyon have recently found this species in abundance about Brownsville. With the aid of their material and the excellent photograph made by Mr. Runyon we have been able to present a detailed description of this plant.

FIG. 60.—Dolichothele sphaerica.   FIG. 61.—Dolichothele longimamma.

According to Mr. Runyon, the flowers are very large and handsome. The fruit does not ripen until about the middle of October, and in one plant a single fruit continued to grow until the 27th of March and had a pronounced pleasing odor. This is the first case which has come under our notice in which any of the *Coryphanthanae* develop any odor in the fruits.

Illustration: Haage and Schmidt, Haupt-Verz. **1912**: 36, as *Mammillaria sphaerica*.

Plate 1, figure 2, is from a photograph sent us by Robert Runyon from Brownsville, Texas; plate VII, figure 2, shows a fruit and figure 2*a* shows a seed from a plant collected by Mr. Runyon at Brownsville in 1921. Figure 60 is from a photograph of a flowering plant made by Mr. Runyon at Brownsville in 1921.

2. Dolichothele longimamma (De Candolle).

> *Mammillaria longimamma* De Candolle, Mém. Mus. Hist. Nat. Paris 17: 113. 1828.
> *Mammillaria longimamma hexacentra* Berg, Allg. Gartenz. 8: 130. 1840.
> *Mammillaria longimamma gigantothele* Berg in Förster, Handb. Cact. 183. 1846.
> *Mammillaria longimamma congesta* Hortus in Förster, Handb. Cact. 183. 1846.
> *Mammillaria uberiformis hexacentra* Salm-Dyck, Cact. Hort. Dyck. 1849. 6. 1850.

Mammillaria melaleuca Karwinsky in Salm-Dyck, Cact. Hort. Dyck. 1849. 108.   1850.
Mammillaria globosa Link, Allg. Gartenz. **25**: 240.   1857.
Mammillaria uberiformis gracilior Meinshausen, Wöchenschr. Gärtn. Pflanz. **1** : 26.   1858.
Mammillaria longimamma luteola Hortus in Förster, Handb. Cact. ed. 2. 246.   1885.
Cactus longimamma Kuntze, Rev. Gen. Pl. **1** : 260.   1891.
Cactus melaleucus Kuntze, Rev. Gen. Pl. **1** : 260.   1891.
Mammillaria longimamma globosa Schumann, Gesamtb. Kakteen 508.   1898.

Solitary or cespitose, about 10 cm. high; tubercles elongated, 5 cm. long, somewhat glaucous, their axils hairy or naked; spine-areoles with white hairs when young, in age naked; radial spines 6 to 12, widely spreading, acicular, 2.5 mm. long, white to pale yellow, swollen and darker at base, puberulent; central spines 1 to 3, usually solitary, porrect, similar to the radials but usually darker with a blackish tip; flowers citron-yellow, 4 to 6 cm. long.

Type locality: Mexico.
Distribution: Central Mexico.

F. Haage jr. in his Choice Cacti lists ten varieties under this species; those not accounted for elsewhere are *ludwigii* and *melaleuca*.

Grässner in his Kakteen 1912 and also 1914 listed *Mammillaria longimamma* var. *ludwigii*. This may be a printer's error.

Mammillaria longimamma melaleuca is in the trade (Grässner). *Mammillaria longimamma pseudo-melaleuca* is advertised by Haage and Schmidt in their 1922 Catalogue.

Mammillaria longimamma spinosior (Wöchenschr. Gärtn. Pflanz. **1**: 26. 1858), credited o Link's Catalogue, but without description, is of this relationship.

Mammillaria hexacentra Otto and *Mammillaria gigantothele* (Förster, Handb. Cact. 183. 1846) were never described.

Krook (Handb. Cact. 41. 1855) mentions the variety *congesta* Hortus but gives no description. Several varieties of *Mammillaria longimamma* are in gardens; the following are mentioned by Schelle: *cristata*, *compacta* (the name cited by Rümpler in 1885), *major*, *laeta*, and *malaena*.

Mammillaria centricirrha flaviflora is referred by Schumann as a synonym of *M. melaleuca* which we have listed among the synonyms of *Dolichothele longimamma*. *M. alpina* Martius, mentioned elsewhere, may be of this relationship.

Illustrations: Monatsschr. Kakteenk. **29**: 81, as *Mammillaria longimamma globosa*; Möllers Deutsche Gärt. Zeit. **25**: 475. f. 8, No. 23, as *M. longimamma gigantothele*; Blühende Kakteen **2**: pl. 73; De Candolle, Mém. Cact. pl. 5; Schumann, Gesamtb. Kakteen 792. f. 114; Monatsschr. Kakteenk. **8**: 149; Schelle, Handb. Kakteenk. 244. f. 162; Förster, Handb. Cact. ed. 2. f. 22, a and b; Ann. Rep. Smiths. Inst. **1908**: pl. 14, f. 2; Watson, Cact. Cult. 164. f. 63; ed. 3. f. 40; Dict. Gard. Nicholson **4**: 564. f. 35; Suppl. 516. f. 551; De Laet, Cat. Gén. f. 89, as *Mammillaria longimamma*.

Figure 61 is from a photograph obtained from L. Quehl; figure 59 is from a photograph of the plant collected by Dr. E. Palmer near Victoria, Mexico, in 1907.

3. Dolichothele uberiformis (Zuccarini).

Mammillaria uberiformis Zuccarini in Pfeiffer, Enum. Cact. 23.   1837.
Mammillaria uberiformis major Hortus in Förster, Handb. Cact. ed. 2. 244.   1885.
Mammillaria uberiformis variegata Hortus in Förster, Handb. Cact. ed. 2. 244.   1885.
Mammillaria laeta Rümpler in Förster, Handb. Cact. ed. 2. 247.   1885.
Cactus uberiformis Kuntze, Rev. Gen. Pl. **1**: 261.   1891.
Mammillaria longimamma uberiformis Schumann, Gesamtb. Kakteen 508.   1898.

Globose, about 7.5 cm. high and 10 cm. in diameter; tubercles elongated, 2.5 to 3 cm. long, 12 to 15 mm. in diameter, bright green, shining, their axils naked; spine-areoles nearly naked; spines 4 or 5, all radial, puberulent, horn-colored to reddish, nearly equal; flowers yellow, 3 cm. broad; outer perianth-segments reddish; inner perianth-segments in 2 series, oblong, acute, acuminate; filaments white; style yellow; stigma-lobes 5 or 6, reflexed.

Type locality: Near Pachuca, Mexico.
Distribution: Central Mexico.

Illustrations: Pfeiffer and Otto, Abbild. Beschr. Cact. 1: pl. 13; Abh. Bayer. Akad. Wiss. München 2: pl. 1, VII. f. 6; Rümpler, Sukkulenten 196. f. 109, as *Mammillaria uberiformis.*

Figure 62 is reproduced from the first illustration cited above.

13. SOLISIA gen. nov.

Plants very small, solitary, globular, tuberculate, milky; tubercles not arranged in ribs, small covered by broad pectinate spines; areoles very narrow and long; flowers lateral, yellow, small borne in axils of old tubercles; axils of tubercles neither hairy nor woolly; fruit naked, small, oblong; seeds black, smooth, dome-shaped with a broad basal hilum.

The type species, *Pelecyphora pectinata* B. Stein, is here segregated from *Pelecyphora,* with which it has little in common; it differs in being solitary, not cespitose, and in having the juice milky, not watery; the flowers small, lateral and yellow, not large, central and purple; the axils of the tubercles naked, not woolly; and the hilum of the seed broad and large, not small.

FIG. 62.—Dolichothele uberiformis.   FIG. 63.—Solisia pectinata.

The genus is named in honor of Octavio Solís of the City of Mexico, an earnest student of the cacti. Only one species is known.

1. Solisia pectinata (B. Stein).

Pelecyphora pectinata B. Stein, Gartenflora 34: 25. 1885.
Pelecyphora aselliformis pectinifera Rümpler in Förster, Handb. Cact. ed. 2. 238. 1885.
Pelecyphora aselliformis pectinata Nicholson,* Dict. Gard. 4: 585. 1888.
Pelecyphora aselliformis cristata Watson, Cact. Cult. 190. 1899.
Mammillaria pectinifera Weber, Dict. Hort. Bois 804. 1898.

Plants 1 to 3 cm. in diameter, fibrous-rooted, entirely hidden by the large overlapping spine-clusters; areoles narrow and long; spines 20 to 40, all radial, 1.5 to 2 mm. long, white, appressed; flowers small; fruit 6 mm. long; seed 1 mm. long.

Type locality: Mexico.
Distribution: Tehuacán, Mexico.

The cristate form of this species, when grown as a graft on some of the *Cereus* allies, becomes much larger than the normal form.

---

* Haage (Cact. Kultur ed. 2. 206. 1900) credits the variety to Ehrenberg.

This plant is very rare in living collections and is known only from a few localities near Tehuacán; one of these is near El Riego Hotel, where Dr. Rose obtained some 50 plants in 1905 but all have since died. We have been endeavoring since to obtain additional plants but Dr. Reko reports that this hill has been burned over and that no plants can now be found. Dr. Rose found it scattered over the top and side of a rounded hill, growing here and there among the stones and stunted plant life, looking not unlike the dull earth and pebbles.

Illustrations: Gartenflora **34**: 25; Garten-Zeitung **4**: 182. f. 42, No. 14; 217. f. 48; Grässner, Kakteen **1912**: 29; Ann. Rep. Smiths. Inst. **1908**: pl. 14, f. 5; Möllers Deutsche Gärt. Zeit. **25**: 477. f. 11, No. 4; **29**: 88. f. 10 (abnormal form), as *Pelecyphora pectinata*; Monatsschr. Kakteenk. **29**: 81; Schelle, Handb. Kakteenk. 275. f. 198; Garten-Zeitung **4**: 217. f. 49, as *P. pectinata cristata*; Monatsschr. Kakteenk. **3**: 172. f. 5, as *P. aselliformis pectinata*.

Figure 63 is from an enlarged photograph showing the top of a plant, collected by Dr. Rose at Tehuacán in 1905.

14. NEOMAMMILLARIA nom. nov.

Mammillaria* Haworth, Syn. Pl. Succ. 177. 1812. Not Stackhouse, 1809.

Plants globose, depressed-globose, or short-cylindric, occasionally much elongated, some with milky, others with watery juice; tubercles arranged in more or less spiraled rows, never on vertical ribs, terete, angled or sometimes flattened, never grooved on upper surface, usually bearing wool or hairs and sometimes bristles, but without glands in their axils and crowned by the spine-areoles; spines in clusters on top of tubercles, sometimes all alike, sometimes with central ones very different from the radial, all straight or sometimes one or more of central spines hooked; flowers, so far as known, diurnal, all from axils of old tubercles, much alike as to size and shape, more or less campanulate, comparatively small, variously colored, commonly red, yellowish or white to pinkish; perianth-segments rather narrow, spreading; stamens numerous, borne on base of perianth-tube, short, included; style about length of stamens; stigma-lobes linear; fruit usually clavate, rarely if ever globose, usually ripening rapidly, naked, scarlet (*Mammillaria brandegeei* with some scales and white fruit, according to Schumann) or white or greenish in a few species; seeds brown in some species, black in others.

The type is *Mammillaria simplex* Haworth, based on *Cactus mammillaris* Linnaeus.

We have given much time in attempting to group the species into definite series but have not succeeded, since many of the species are little known and incompletely described.

The name, *Neomammillaria*, as here used, replaces the name *Mammillaria* of Haworth (1812), which is a homonym of the *Mammillaria* of Stackhouse (1809), a genus of *Algae*.

The genus, as here treated, differs from Schumann's treatment (Gesamtb. Kakteen 472–601, 1898) in that we exclude three of his four subgenera, *Coryphantha*, *Dolichothele*, and *Cochemiea*, giving them generic rank. From his fourth subgenus we have excluded *Mammillaria micromeris* as the type of the genus *Epithelantha*† and *M. phellosperma* to the genus *Phellosperma* (see page 60).

The species, of which we recognize 150, are native chiefly of Mexico, extending northward into the southwestern United States; one species is reported as far north as Utah and Nevada. Two species are known from the West Indies (none is found in Jamaica or in the Lesser Antilles south of Antigua). Several species are known from Central America (none has been reported from Costa Rica, El Salvador, or Panama). One species is found in Venezuela and neighboring islands and one is described from Colombia, perhaps in error.

During the period of our investigation political conditions in Mexico have prevented our obtaining much original information concerning many of the species and have made it necessary for us to depend largely upon published descriptions and illustrations.

---

* The name was also spelled *Mammilaria* by Torrey and Gray (Flora **1**: 553) and *Mamillar a* by Reichenbach (Mössler, Handb. ed. 2. **1**: 1. 1827) and by Schumann (Gesamtb. Kakteen 472 and elsewhere).

† See Cactaceae, **3**: 92. 1922.

KEY TO SPECIES.

A. None of spines hooked (1–104, 150).
  B. Seeds brown (1–80).
    C. Tubercles giving off milk freely when pricked or cut (1–53).
      D. Axils of tubercles without bristles (1–33).
        E. Tubercles more or less elongated.
          F. Tubercles terete throughout.
            Spines yellow or reddish.
              Spines red... 1. *N. mammillaris*
              Spines yellow...................................... 2. *N. nivosa*
            Spines usually white except tips, at least not red or yellow.
              Central spines 1 or 2.
                Central spines about length of radials................ 3. *N. gaumeri*
                Central spines much longer than radials.............. 4. *N. petrophila*
              Central spines 4 to 7.
                Outer perianth-segments entire; central spines long, slender 5. *N. arida*
                Outer perianth-segments erose; central spines not elon-
                      gated, stouter than in preceding species. 6. *N. brandegeei*
          FF. Tubercles more or less angled.
            G. Tubercles nearly terete towards apex.
              Outer perianth-segments and scales more or less fim-
                briate.
                Flowers reddish................................. 7. *N. gummifera*
                Flowers light yellow............................. 8. *N. macdougalii*
               Outer perianth-segments and scales entire.
                Radial spines white; flowers pinkish................ 9. *N. heyderi*
                Radial spines brownish; flowers white to cream-
                colored.
                Plant hemispheric; radial spines 9 to 13; perianth-
                  segments acute...................... 10. *N. hemisphaerica*
                Plant much flattened; radial spines 18 or fewer;
                  perianth-segments acuminate.......... 11. *N. applanata*
          GG. Tubercles angled to top.
            H. Spines very unequal, some much elongated.
              Spines whitish................................. 12. *N. phymatothele*
              Spines horn-colored, reddish or black.
                No definite central spine.
                Spines horn-colored, short, curved........... 13. *N. magnimamma*
                Spines reddish, long....................... 14. *N. macracantha*
                Central spines definite.
                Central spines 2........................... 15. *N. johnstonii*
                Central spines solitary.
                  Central spine 2 to 3 cm. long.............. 16. *N. melanocentra*
                  Central spine 1 cm. long; oblong .... 17. *N. runyonii*
           HH. Spines nearly equal, at least none much elongated.
            Flowers red to pinkish.
              Outer perianth-segments ciliate.................. 18. *N. sartorii*
              Outer perianth-segments not ciliate (so far as known).
              Central spines none.
                Spines pinkish with black tips.............. 19. *N. seitziana*
                Spines straw-colored throughout.............. 20. *N. ortegae*
              Central spines 1 or more.
                Central spines solitary; radial spines nearly equal 21. *N. meiacantha*
                Central spines 2; some of radials very short.... 22. *N. scrippsiana*
            Flowers yellowish.
             Central spines 4 to 6.......................... 23. *N. gigantea*
             Central spines usually wanting.................. 24. *N. peninsularis*
   EE. Tubercles very short, symmetric.
      F. Plants globose or depressed.
        G. Axils of tubercles naked............................. 25. *N. flavovirens*
        GG. Axils of tubercles laniferous.
           Some of spines deciduous......................... 26. *N. sempervivi*
           None of spines deciduous.
             Central spines present.......................... 27. *N obscura*
            Central spines wanting.......................... 28. *N. crocidata*
      FF. Stems cylindric or ovoid.
        Central spines wanting.
           Tubercles nearly terete............................ 29. *N. polythele*
           Tubercles 4-angled.
             Tubercles pointed; axils very woolly................. 30. *N. carnea*
             Tubercles not pointed; axils not very woolly............. 31. *N. lloydii*
        Central spines several.
           Radial spines reduced to short bristles................... 32. *N. zuccariniana*
           Radial spines more elongated than in last species.......... 33. *N. formosa*

KEY TO SPECIES—continued.

DD. Axils of tubercles with bristles as well as wool (33–53).
  E. Some of spines much elongated, curved, and flexuous.
    Definite central spines wanting............................ 34. *N. compressa*
    Central spines present.
      Central spines weak..................................... 35. *N. mystax*
      Central spines stiff..................................... 36. *N. petterssonii*
  EE. None of spines elongated, or if elongated, not flexuous.
    F. Tubercles terete or nearly so.
      G. Axils of tubercles bearing yellow wool.................. 37. *N. eichlamii*
      GG. Axils of tubercles bearing white wool.
        H. Spines all radial.
          Spines 5 or 6, in young plants sometimes only 4....... 38. *N. karwinskiana*
          Spines always 4................................. 39. *N. praelii*
        HH. Spines both radial and central.
          I. Radial spines numerous, 12 or more.
            Central spines reddish, not much longer than the
              radials.
                Outer perianth-segments ciliate.............. 40. *N. standleyi*
                Outer perianth-segments setose.............. 41. *N. evermanniana*
            Central spines elongated, usually white except at tip.
                Flowers yellow........................... 42. *N. parkinsonii*
                Flowers dark red......................... 43. *N. geminispina*
          II. Radial spines few, 5 to 9.
            Spines black when young.................... 44. *N. pyrrhocephala*
            Spines at most brownish.
               Flowers yellow........................... 45. *N. woburnensis*
               Flowers pinkish.......................... 46. *N. collinsii*
    FF. Tubercles strongly angled.
      G. Spines both radial and central.
        Radial spines numerous........................... 47. *N. chinocephala*
        Radial spines few, bristle-like.
          Central spines 4 to 6......................... 48. *N. tenampensis*
          Central spines 2............................. 49. *N. polygona*
      GG. Spines few, all of one kind.
        Flowers yellow................................. 50. *N. confusa*
        Flowers rose-colored or white.
          Flowers rose-colored.
            Plants globose; stigma-lobes 4 or 5.............. 51. *N. villifera*
            Plants cylindric; stigma-lobes 8.................. 52. *N. polyedra*
          Flowers white.............................. 53. *N. conzattii*
CC. Milk-tubes developed, if at all, only in stem; tubercles not milky (54–80).
  D. Central spines wanting.
    Spines subulate; areoles elliptic.............................. 54. *N. napina*
    Spines mostly acicular; areoles circular.
      Spines numerous...................................... 55. *N. lanata*
      Spines few (4 to 6).
        Spines 5 or 6, short, straight....................... 56. *N. kewensis*
        Spines 4, elongated, curved.
          Flowers large (2.5 cm. broad)................... 57. *N. subpolyedra*
          Flowers small.
            Spines long and weak....................... 58. *N. galeottii*
            Spines subulate............................ 59. *N. tetracantha*
  DD. Central spines present.
    E. Central spines usually 2, sometimes solitary.
      F. Radial spines 20 or more.
        Central spines stout and not very long; stigma-lobes white.
          Plant round or nearly so at apex; central spine often 1....... 60. *N. elegans*
          Plant strongly umbilicate; central spines always 2............ 61. *N. pseudoperbella*
        Central spines long............................. 62. *N. dealbata*
      FF. Radial spines 20 or fewer.
        Radial spines white, bristle-like.
          Stigma-lobes red.
            Globose or somewhat elongated...................... 63. *N. haageana*
            Depressed-globose................................. 64. *N. perbella*
          Stigma-lobes white.
            Radial spines appressed............................. 65. *N. collina*
            Radial spines not appressed........................ 66. *N. donatii*
        Radial spines brownish when young, stouter than in the last..... 67. *N. mundtii*
    EE. Central spines usually 4, sometimes more.
      F. Central spines white or yellow.
        Radial spines white.
          Plant globose.
            Axils of tubercles not setose; central spines usually 4...... 68. *N. celsiana*
            Axils of tubercles setose; central spines usually 9........ 69. *N. aureiceps*
          Plant cylindric.
            Plants from Yucatan............................... 70. *N. yucatanensis*
            Plants from Central America........................ 71. *N. ruestii*

KEY TO SPECIES—continued.

Radial spines yellow.
Plants globular .... 72. *N. pringlei*
Plants slender-cylindric............................. 73. *N. cerralboa*
FF. Central spines brown or black.
Central spines black................................. 74. *N. phaeacantha*
Central spines brown.
Axils of tubercles not setose........................... 75. *N. graessneriana*
Axils of tubercles setose.
Tubercles closely set.
Central spines not very different from radial.
Plant body elongated; spines brownish or reddish.... 76. *N. spinosissima*
Plant body globose; radial spines whitish........... 77. *N. densispina*
Central spines very different from the radial.......... 78. *N. nunezii*
Tubercles spreading.
Central spines unequal; stigma-lobes green............ 79. *N. amoena*
Central spines nearly equal; stigma-lobes rose-colored... 80. *N. rhodantha*
BB. Seeds black; neither tubercles nor stems milky (81–104).
C. Spines plumose... 81. *N. plumosa*
CC. Spines not plumose.
D. Radial spines weak and hair-like.
Central spines with yellow tips............................... 82. *N. prolifera*
Central spines with brown tips................................ 83. *N. multiceps*
DD. Radial spines not hair-like.
E. Spines yellow.
Spines 2 to 8, glabrous, more or less twisted or bent.............. 84. *N. camptotricha*
Spines about 20, pubescent, straight......................... 85. *N. eriacantha*
EE. Spines not yellow.
F. Spines 25 to 80.
Spines pubescent or lanate.
Spines lanate, 25 to 30............................. 86. *N. schiedeana*
Spines pubescent or puberulent....................... 87. *N. lasiacantha*
Spines not pubescent.
Spines all very much alike.
Perianth-segments obtuse........................... 88. *N. denudata*
Perianth-segments pointed.
Flowers about 7 mm. long....................... 89. *N. lenta*
Flowers about 2 cm. long....................... 90. *N. candida*
Central spines 1 to 6, very unlike others................. 91. *N. vetula*
FF. Spines 20 or fewer but sometimes more in *N. oliviae* and *N. pottsii*.
Plant globose.
Flowers red... 92. *N. fertilis*
Flowers white.
Central spines solitary; radials 7 to 9................. 93. *N. decipiens*
Central spines 5 to 8; radials 16 to 20................. 94. *N. discolor*
Plant cylindric.
Joints very fragile, breaking loose when touched or jarred. 95. *N. fragilis*
Joints not fragile.
Spines all radial, recurved, sometimes with one central.. 96. *N. elongata*
Spines both radial and central.
Plants globose to short-cylindric.................... 97. *N. oliviae*
Plants slender-cylindric.
Axils of tubercles not bristly.
Spines all yellow........................... 98. *N. echinaria*
Spines not yellow.
Upper central spines more or less connivent
over top of plant...................... 99. *N. pottsii*
Upper central spines not connivent........... 100. *N. mazatlanensis*
Axils of tubercles bristly.
Stems slender-cylindric; central Mexican species.. 101. *N. sphacelata*
Stems short-cylindric or globose (sometimes globose
in *N. palmeri*); Lower Californian species.
Spines nearly white or at least becoming so; seeds
minute.
Spines all white or nearly so; spine-areoles at
first lanate........................... 102. *N. albicans*
Spines tan with dark tips; spine-areoles not
lanate................................. 103. *N. slevinii*
Spines not white; seeds 3 mm. long........... 104. *N. palmeri*
AA. Some of central spines hooked; radial spines never hooked (105–149).
B. Tubercles milky; seeds brown.
Plants globose... 105. *N. uncinata*
Plants cylindric.. 106. *N. hamata*
BB. Tubercles not milky except sometimes in *N. rekoi*; seeds mostly black.
C. Seeds brown.
Fruit red; flowers from side of plant................................. 107. *N. rekoi*
Fruit green; flowers from near base of plant......................... 108. *N. solisii*

KEY TO SPECIES—continued.

CC. Seeds black.
  D. Fruit elongated, clavate, ripening quickly.
    E. Seeds not rugose.
      F. Plants usually small; spines setaceous to delicately acicular.
        Central spines yellow.
          Central spines glabrous............................... 109. *N. pygmaea*
          Central spines pubescent.
            Flowers white.. 110. *N. wildii*
            Flowers yellowish................................... 111. *N. seideliana*
        Central spines red to brown.
          Outer perianth-segments ciliate.
            Central spines shorter than flower; perianth-segments acute 112. *N. barbata*
            Central spines longer than flower; perianth-segments obtuse 113. *N. mercadensis*
          Outer perianth-segments entire.
            Axils of tubercles setose.
              Inner perianth-segments white to yellowish.
                Central spines 3 or 4.
                  Radial spines about 25; flowers 2 cm. long........ 114. *N. kunzeana*
                  Radial spines about 20; flowers 1 cm. long........ 115. *N. hirsuta*
                Central spines 7 to 9............................. 116. *N. multihamata*
              Inner perianth-segments red or reddish.
                Radial spines weak and hair-like.
                  Central spines several......................... 117. *N. longicoma*
                  Central spine solitary.......................... 118. *N. bocasana*
                Radial spines stiff.
                  Radial spines glabrous......................... 119. *N. multiformis*
                  Radial spines pubescent....................... 120. *N. scheidweileriana*
             Axils of tubercles not setose.
              Flowers 2.2 cm. long or more.
                Central spines solitary........................... 121. *N. saffordii*
                Central spines 3................................ 122. *N. schelhasei*
              Flowers 1 to 1.5 cm. long.
              Plants cespitose............................... 123. *N. glochidiata*
              Plants solitary.
                Inner perianth-segments acuminate.............. 124. *N. trichacantha*
                Inner perianth-segments acute only.............. 125. *N. painteri*
      FF. Plants stout; central spines at least stout-acicular to subulate.
        G. Outer perianth-segments ciliate.
          Fruit purple, ovoid to globular.
            Radial spines 15 or less; fruit large (25 mm. long)..... 126. *N. wrightii*
            Radial spines 20 to 30; fruit 10 to 15 mm. long....... 127. *N. viridiflora*
           Fruit scarlet, clavate.
            Outer perianth-segments long-ciliate................. 128. *N. wilcoxii*
            Other perianth-segments and upper scales short-ciliate.
              Perianth rotate; stigma-lobes red................. 129. *N. mainae*
              Perianth campanulate; stigma-lobes green.
                Flowers white............................... 130. *N. boedekeriana*
                Flowers purple to pinkish.
                  Radial spines often as many as 30.
                    Inner perianth-segments acuminate......... 131. *N. microcarpa*
                    Inner perianth-segments usually obtuse or
                      rounded...................... 132. *N. milleri*
                  Radial spines often as few as 12; inner perianth-
                  segments acute................... 133. *N. sheldonii*
        GG. Perianth-segments not ciliate.
          H. Setae in axils of tubercles.
            I. Seeds constricted above base.
              J. Flowers greenish or pink, and small.
                K. Flowers greenish, 10 to 12 mm. long; cen-
                  tral spines yellowish to reddish...... 134. *N. armillata*
                KK. Flowers pink, 20 mm. long or more; cen-
                  tral spines dark brown............. 135. *N. fraileana*
               JJ. Flowers nearly white, seeds much larger...... 136. *N. swinglei*
            II. Seeds not constricted above base.
              Central spines several; flowers yellowish....... 137. *N. dioica*
              Central spines usually solitary; flowers rose-colored. 138. *N. goodridgei*
        HH. Setae wanting in axils of tubercles.
          I. Flowers rotate................................... 139. *N. zephyranthoides*
          II. Flowers campanulate.
            Plants globose.
              Flowers white.............................. 140. *N. carretii*
              Flowers pink to purplish.
                Inner perianth-segments obtuse............ 141. *N. jaliscana*
                Inner perianth-segments acute to acuminate. 142. *N. bombycina*
            Plants slender, elongated and cylindric.
              Flowers small, pinkish..................... 143. *N. occidentalis*
              Flowers large, purplish.................... 144. *N. fasciculata*
      EE. Seeds rugose (perhaps a generic type)............................. 145. *N. nelsonii*
    DD. Fruit depressed, long-persisting (perhaps a generic type).............. 146. *N. longiflora*
AAA. Species not grouped: 147, *N. tacubayensis*; 148, *N. umbrina*; 149, *N. verhaertiana*; 150, *N. xanthina*.

1. Neomammillaria mammillaris (Linnaeus).

Cactus mammillaris Linnaeus, Sp. Pl. 1: 466. 1753.
Cactus mammillaris glaber De Candolle, Pl. Succ. 137. 1799.
Mammillaria simplex Haworth, Syn. Pl. Succ. 177. 1812.
? Mammillaria conica * Haworth, Suppl. Pl. Succ. 71. 1819.
Mammillaria parvimamma Haworth, Suppl. Pl. Succ. 72. 1819.
Cactus microthele Sprengel, Syst. 2: 494. 1825.
Mammillaria simplex parvimamma Lemaire, Cact. Gen. Nov. Sp. 98. 1839.
Mammillaria caracassana † Otto in Salm-Dyck, Cact. Hort. Dyck. 1849. 107. 1850.
Mammillaria mammillaris Karsten, Deutsche Fl. 888. 1882.
? Cactus conicus Kuntze, Rev. Gen. Pl. 1: 259. 1891.
Cactus parvimammus Kuntze, Rev. Gen. Pl. 1: 259. 1891.

Globose to short-cylindric, 4 to 6 cm. high; tubercles short, 5 to 7 mm. long, conic, nearly terete, pale green, only slightly woolly in their axils; spine-areoles bearing a dense mass of white wool when young; spines reddish brown, acicular; radial spines 10 to 12, spreading, 5 to 7 mm. long; central spines 3 or 4, stouter and a little longer than the radials; flowers 8 to 10 mm. long, cream-colored; outer perianth-segments narrow, bearing long mucronate tips; fruit 15 to 20 mm. long, red; seeds minute, brown.

Fig. 64.—Neomammillaria mammillaris.

Fig. 65.—Neomammillaria macdougalii.

Type locality: Tropical America.
Distribution: Northern Venezuela and neighboring Dutch Islands.

This plant was the first-known species of the genus and the only one known to Linnaeus; it was described and illustrated by Commelin in 1697 and by Hermann in 1698. It was one of the first cacti discovered; Aiton states that it was cultivated by Bishop Compton before 1688. The cited distribution of the species has usually been inexact or erroneous; Linnaeus gave no definite locality but restricted it to the warm parts of America.

Nuttall assigns it also to the hills of the Missouri River, and De Candolle's range covers that of both Linnaeus and of Nuttall. Nuttall's plant was subsequently found to be different from the one of the Caribbean region. Schumann gives the range as the West Indies but his description covers two or three species. A number of his references are erroneous, for neither Wright's plant (No. 2619, as Mammillaria pusilla) from Cuba nor Haworth's plant (Syn. Pl. Succ. 177, as Mammillaria prolifera) from the West Indies belongs

---

* Tubercles large, conic; spines less than 10, all radial, red, but paler at base; flowers and fruit unknown. Neither Pfeiffer nor Schumann knew this species or its origin. The Index Kewensis refers it to South America. If from that region it might be a species of Discocactus, near D. placentiformis, but it may belong here.
† This is the original spelling, but Schumann wrote it M. caracasana.

here. The name *Cactus prolifer* Willdenow (Pfeiffer, Enum. Cact. 9. 1837) is doubtless to be referred here. Fawcett lists the plant from Jamaica (as *Mammillaria simplex*) but no specimens are known to us from that island, which was searched by Dr. Britton and the late Mr. William Harris.

De Tussac (Fl. Antill. 2: 216, pl. 32) refers it to Santo Domingo and he describes and figures it, mentioning a locality in the desert near Gonaives which, however, is in Haiti; his illustration, while undoubtedly of this species, is not an original but copied from that of De Candolle (Pl. Succ. pl. 111). The only similar plant we know from his locality is *Mammillaria pusilla*, described as *M. pusilla haitiensis* by Schumann, which has been collected by Buch at this locality, and we have specimens from other collectors. We now believe that *Neomammillaria mammillaris* is confined to the coast of Venezuela and the adjacent islands, among which is Curaçao. In 1913 Dr. Britton and Dr. Shafer found it common on the top of a limestone hill in Curaçao (No. 3085) and in the same year Mr. Pittier obtained living plants near Cabo Blanco, Venezuela (No. 6471). These two are the only collections which have been made in recent years.

Steudel (1821), under *Mammillaria simplex*, compares this species with *Cereus flavescens* and *C. lanuginosus*, but he must have meant *Cactus* instead of *Cereus*.

Mammillaria microthele Monville and *M. micrantha* Hortus are names which Rümpler (Förster, Handb. Cact. ed. 2. 335. 1885) refers to *M. caracassana*; Salm-Dyck (Cact. Hort. Dyck. 1844. 9. 1845) also referred to it *M. micracantha* Monville.

Mammillaria simplex affinis Otto is mentioned by Förster (Handb. Cact. 217. 1846), but is not described.

Mammillaria karstenii Poselger (Allg. Gartenz. 21: 95. 1853) is listed by Schumann among his little-known species. The Index Kewensis states that it comes from Argentina, which is doubtless a mistake. The type locality is given as "La Canada," a common Spanish locality name. If collected by Karsten, it probably was obtained in Venezuela, in which case it would probably be referable to *Neomammillaria mammillaris*.

Mammillaria fuliginosa Salm-Dyck (Cact. Hort. Dyck. 1849. 93. 1850) we do not know, but if it came from Venezuela, where it is referred doubtfully by the Index Kewensis, it would belong here.

Illustrations: Hermann, Parad. 132. pl. 137, as *Echinomelocactus minor*, etc.; Commelin, Hort. Amst. 1: 105. f. 55; Plukenet, Opera Bot. 1: 148. pl. 29, f. 1, as *Ficoides*, etc.; Bradley, Hist. Pl. Succ. 3: 11. pl. 29, as melon-thistle; Loudon, Encycl. Pl. ed. 2 and 3. 410. f. 6839; De Candolle, Pl. Succ. 137. pl. 111; Fl. Antill. 2: pl. 32, as *Cactus mammillaris*; De Candolle, Mém. Cact. pl. 7, as *Mammillaria simplex*.

Figure 64 is reproduced from a colored drawing by Miss M. E. Eaton of a plant obtained by Dr. Britton and Dr. Shafer on Curaçao in 1913, which fruited the same year in the New York Botanical Garden.

2. Neomammillaria nivosa (Link).

Mammillaria nivosa Link in Pfeiffer, Enum. Cact. 11. 1837.
Cactus nivosus Kuntze, Rev. Gen. Pl. 1: 259. 1891.
Coryphantha nivosa Britton, Ann. Mo. Bot. Gard. 2: 45. 1915.

Often forming large clusters 8 dm. in diameter, of 25 heads or more; separate specimens usually globose but sometimes cylindric, the largest ones 18 cm. in diameter, very spiny; tubercles milky, 10 mm. long, their axils filled with white wool; spines usually 14, bright yellow, acicular, the longer ones 1.5 cm. long; spine-areoles when young woolly, in age naked; flowers cream-colored, 1.5 cm. long; fruit clavate, 12 mm. long, red; seeds brown.

Type locality: Tortola Island, Virgin Islands.

Distribution: Southern Bahamas, Mona, Desecheo, Culebra, Buck Island, St. Thomas, Little St. James Island, Tortola, and Antigua.

Known as the snowy cactus in the Virgin Islands and as the woolly nipple-cactus in the Bahamas.

The plant inhabits crevices of rocks and locally is very abundant. On Mona Island, between Porto Rico and Santo Domingo in the Mona Passage, it exists in immense numbers on the limestone plateau.

Mammillaria tortolensis (Pfeiffer, Enum. Cact. 11. 1837) was published by Pfeiffer as a synonym of *M. nivosa*. The same or similar plant was briefly described by Forbes (Journ. Hort. Tour 148, 1837).

Illustrations: Förster, Handb. Cact. ed. 2. 331. f. 34; Schelle, Handb. Kakteenk. 264. f. 186; De Laet, Cat. Gén. f. 46; Blühende Kakteen **3**: pl. 165, as *Mammillaria nivosa*.

Figure 66 is from a photograph of a plant collected on Turks Island, British West Indies, in July 1916 and sent us by the Director of the New York Aquarium; figure 243 (Britton and Rose, Cactaceae **3**: p. 231) shows the plant on Mona Island, Porto Rico.

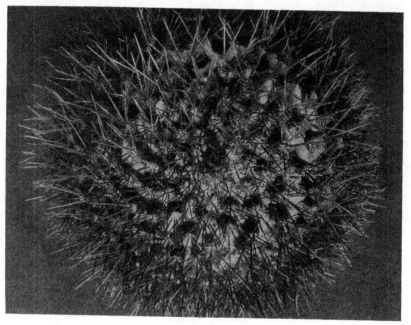

Fig. 66.—Neomammillaria nivosa.

3. Neomammillaria gaumeri sp. nov.

Cespitose, the branches short, globose to short-cylindric, up to 15 cm. long, growing half hidden in the sand; tubercles dark green, short, nearly terete, obtuse, 5 to 7 mm. long, very milky; axils naked even when young; spine-areoles conspicuously white-woolly at first, soon naked; radial spines 10 to 12, spreading, acicular, white with brown tips or lower ones in cluster darker, 5 to 7 mm. long; central spine solitary, porrect, usually brown; flowers very abundant from near top of plant but not from axils of young areoles, creamy white, small, 10 to 14 mm. long; outer perianth-segments greenish, brown-tipped; scales on flower-tube broadly ovate, scarious; fruit crimson, clavate, 18 to 20 mm. long, naked.

Common in the sand dunes of Progreso, Yucatan; collected first by George P. Gaumer and sons, April 1916 (No. 23349, type); re-collected in 1918 and again in 1921.

This species is remarkable for its unusual habitat and was the first of the genus reported from Yucatan. A second species has since been collected by Dr. Gaumer (see p. 114).

PLATE VIII

M. E. Eaton del.

1. Fruiting plant of *Neomammillaria gaumeri*.
2. Flowering plant of *Neomammillaria heyderi*.
2a. Fruit of same.
3. Flowering and fruiting plant of *Neomammillaria hemisphaerica*.

4. Flowering plant of *Neomammillaria compressa*.
5. Flowering plant of *Neomammillaria geminispina*.
6. Flowering plant of *Neomammillaria hemisphaerica*.

It is perhaps nearest some of the species from Texas, such as *N. hemisphaerica* and *N. heyderi*, but when growing it is easily distinguished by the peculiar white mats of wool on the young spine-areoles.

The following interesting note has been contributed by Dr. Gaumer, in whose honor the plant is named:

"The flowers begin to open at 8 a. m., are fully open at noon, close at dawn, and shrink the next morning, leaving the ovary wholly imbedded in the mass of the plant at the base of the tubercles; it remains dormant from 3 to 6 months, then suddenly develops to an inch in length in 48 hours. If put away in a dry place the bright crimson berries last from 3 to 6 months without decaying or changing their color. When thoroughly ripe they have a rather pleasant sweetish taste and are said to be edible.

"The plant multiplies by seed and by segmentation; this latter is accomplished by the plant putting out numerous shoots from its upper surface; these send out roots; the old plant decays and the little ones are often rolled about by the cattle or by the winds, and later send out stronger roots that finally anchor them to the sand, generally under a clump of brush."

Plate VIII, figure 1, shows the type plant which flowered in the New York Botanical Garden, July 24, 1918, soon after its arrival from Yucatan; plate XIII, figure 2, is from a photograph of the plant showing the large masses of white wool at the young spine-areoles.

4. Neomammillaria petrophila (Brandegee).

Mammillaria petrophila Brandegee, Zoe 5: 193. 1904.

Sometimes cespitose, milky, globular, 15 cm. in diameter or less; tubercles short, broad at base; spines at first chestnut-colored, becoming pale in age; radial spines 10, about 1 cm. long, a little spreading; central spine 1 (rarely 2), 2 cm. long, darker and stouter than the radials; flowers bright greenish yellow, 18 to 20 mm. long; perianth-segments hardly acute, sometimes slightly erose; stamens and style yellow; stigma-lobes 6; fruit small, roundish; seeds reddish brown, smooth, less than 1 mm. long.

Type locality: Sierra de la Laguna, Lower California.

Distribution: Mountains of southern Lower California.

We know this species only from description and illustration.

Illustration: Monatsschr. Kakteenk. **17:** 57, as *Mammillaria petrophila*.

5. Neomammillaria arida (Rose).

Mammillaria arida Rose in Quehl, Monatsschr. Kakteenk. **23:** 181. 1913.

Plants usually single, globular, deeply seated in the ground, 3 to 6 cm. in diameter, containing much milk and giving it off freely when injured; tubercles nearly terete; radial spines about 15, pale, ascending, the bases sometimes yellowish and the tips dark; central spines 4 to 7, 12 to 16 mm. long, much longer than the radials, dark brown, erect; flowers 1 cm. long; outer perianth-segments dark purple with lighter margins, entire; inner perianth-segments cream-colored to almost pale yellow; stamens pale; stigma-lobes green; fruit clavate, red, 15 cm. long; seeds brown.

Type locality: Hills near Pichilinque Island near La Paz, Lower California.

Distribution: Known only from the type locality.

Plate VII, figure 3, shows one of the plants collected by Dr. Rose in 1911 which flowered in the New York Botanical Garden, July 2, 1912.

6. Neomammillaria brandegeei (Coulter).

Cactus brandegeei Coulter, Contr. U. S. Nat. Herb. **3:** 96. 1894.
Cactus gabbii Coulter, Contr. U. S. Nat. Herb. **3:**109. 1894.
Mammillaria gabbii Engelmann in K. Brandegee, Erythea **5:** 116. 1897.
Mammillaria brandegeei K. Brandegee, Erythea **5:** 116. 1897.

Cylindric to globular, flattened, solitary or in clusters of 2 to 8; tubercles angled; axils woolly; radial spines 9 to 16, 8 to 10 mm. long, yellowish brown; central spines 3 to 6, a little longer and darker than the radials; flowers 15 mm. long; outer perianth-segments ovate, striate, ciliate; inner perianth-segments greenish yellow, narrower than the outer, entire; fruit white (according to Schumann), bearing a few narrow scales.

Type locality: San Jorge, Lower California.

Distribution: Lower California, San Quintin, and southward.

If we are right in referring *Mammillaria gabbii* here, this species was first collected by W. M. Gabb in southern Lower California in 1867 and was described by Dr. Engelmann as a new species but was not published. In 1894 Dr. Coulter published Engelmann's description, but used the name of *Cactus gabbii*. On a previous page, however, he published *Cactus brandegeei* which, if the same, takes precedence.

We have placed this species next to *Neomammillaria arida*, which is known to have nearly terete tubercles, while *N. brandegeei* is described as having angled tubercles, as they certainly are in herbarium specimens; whether this species has angled or terete tubercles in life we are in doubt.

We have not seen fresh fruit of this plant but Schumann describes it as white, which is unusual in this genus; it is also peculiar in bearing several small scales.

Illustrations: Blühende Kakteen **2**: pl. 119; Schumann, Gesamtb. Kakteen Nachtr. 137. f. 34; Monatsschr. Kakteenk. **11**: 153, as *Mammillaria brandegeei*.

7. Neomammillaria gummifera (Engelmann).

> *Mammillaria gummifera* Engelmann in Wislizenus, Mem. Tour North. Mex. 105. 1848.
> *Cactus gummifer** Kuntze, Rev. Gen. Pl. **1**: 260. 1891.

Depressed-globose, 8 to 12 cm. in diameter; tubercles light green, milky, somewhat 4-angled; axils of tubercles and spine-areoles white-tomentose when young; radial spines 10 to 12, ascending, white with brownish or even blackish tips, the lower ones stouter and longer than the others, often 2 to 2.5 cm. long and somewhat recurved; central spines 1 or 2, sometimes 4; flowers 3 cm. long, 12 to 25 mm. wide when fully open, brownish red outside; inner perianth-segments reddish white with dark red band in middle.

Type locality: Cosihuiriachi, Chihuahua.

Distribution: Northern Mexico.

This species was collected by Dr. A. Wislizenus in the state of Chihuahua, Mexico, about 1846. Specimens were sent to Dr. Engelmann at St. Louis, who described it in 1848 but without seeing flowers or fruit; two years afterward he described the flowers but the fruit is yet unknown. In 1894 Dr. J. M. Coulter redescribed the species, stating that it had never been re-collected. Professor Schumann in his Monograph does not recognize it, but refers it to his list of doubtful species. In 1908 Dr. Rose visited the type locality and obtained a single living specimen.

Illustrations: Cact. Mex. Bound. pl. 9, f. 18 to 20, as *Mammillaria gummifera*.

8. Neomammillaria macdougalii (Rose).

> *Mammillaria macdougalii* Rose, Stand. Cycl. Hort. Bailey **4**: 1982. 1916.

Usually low and flattened on top, but very old plants sometimes nearly globular and then 12 to 15 cm. in diameter with a carrot-shaped root; tubercles flattened dorsally, strongly angled, deep green; young areoles bearing white wool, but becoming naked in age; axils of tubercles often bearing long white wool; radial spines 10 to 12, white or somewhat yellowish, the lower ones a little stouter, brown or black at top or sometimes throughout; central spines 1 or 2, stout, yellowish, brown-tipped, similar to the radials; flowers 3.5 cm. long, cream-colored; outer perianth-segments short-fimbriate; fruit red, clavate, 3 cm. long.

Type locality: Near Tucson, Arizona.

Distribution: Southeastern Arizona.

Figure 65 is from a photograph of a plant collected by Dr. MacDougal in the Santa Catalina Mountains; figure 67 is from a photogragh of another plant sent by Dr. MacDougal from the same region in November 1909.

---

*Coulter writes this name *Cactus gummiferus* (Contr. U. S. Nat. Herb. **3**: 98. 1894).

9. Neomammillaria heyderi (Mühlenpfordt).

Mammillaria heyderi Mühlenpfordt, Allg. Gartenz. **16**: 20. 1848.
Cactus heyderi Kuntze, Rev. Gen. Pl. **1**: 260. 1891.
? Mammillaria buchheimeana Quehl, Monatsschr. Kakteenk. **27**: 97. 1917.

Plant globose or somewhat flattened at apex; tubercles conic, 12 mm. long, when young bearing wool in their axils; young spine-areoles white-woolly; radial spines 20 to 22, white, setaceous, the lower ones stouter and longer; central spine solitary, brown at base and apex, 5 to 6 mm. long; flowers pinkish, the segments linear-oblong; fruit oblong, red.

Type locality: Not cited.
Distribution: Texas and northern Mexico.
Illustration: Schulz, Wild Fl. San Antonio pl. 13 in part, as *M. heyderi.*

Plate VIII, figure 2, shows a plant sent to Dr. Rose by Mrs. S. L. Pattison in 1921 which flowered in the New York Botanical Garden on April 21 of that year; figure 2a shows the fruit.

Fig. 67.—Neomammillaria macdougalii.   Fig. 68.—Neomammillaria phymatothele.

10. Neomammillaria hemisphaerica (Engelmann).

Mammillaria hemisphaerica Engelmann in Wislizenus, Mem. Tour North. Mex. 105. 1848.
Mammillaria heyderi hemisphaerica Engelmann, Proc. Amer. Acad. **3**: 263. 1856.
Cactus heyderi hemisphaericus Coulter, Contr. U. S. Nat. Herb. **3**: 97. 1894.
Cactus hemisphaericus Small, Fl. Southeast. U. S. 811. 1903.

Deep-seated in the soil, hemispheric, 8 to 12 cm. broad, dark green; tubercles only slightly angled, not very closely set, 1 to 1.5 cm. long, somewhat pointed, their axils nearly naked in the dormant stages; spine-areoles woolly when young, becoming glabrate in age; radial spines 9 to 13, widely spreading, acicular, the upper ones more delicate, 4 to 8 mm. long, brownish or smoky, often with black tips; central spine solitary, porrect, brown; flowers small, cream-colored, 1 to 1.5 cm. long; inner perianth-segments acute; filaments pinkish; style pinkish; stigma-lobes 6 to 10, greenish yellow; fruit slender, clavate, red, 1 to 1.5 cm. long.

Type locality: Below Matamoros on the Rio Grande.
Distribution: Southeastern Texas and northeastern Mexico.

This species was collected in 1846 by the St. Louis Volunteers in the Mexican War and taken back to Dr. George Engelmann; it flowered and he described it briefly in 1848 and in more detail in 1850. It was recently re-collected near Brownsville, Texas, just across the river from Matamoros by Robert Runyon and sent to us with a photograph taken *in situ,* here reproduced (plate VI, figure 2).

This species differs from *Neomammillaria applanata* in being less flattened and in having fewer spines and white flowers.

Cactus heyderi hemisphaericus, as treated by Coulter, must be a composite, the western and southern forms probably representing different species.

Illustrations: Cact. Mex. Bound. pl. 9, f. 15 to 17, as *Mammillaria heyderi hemisphaerica*.

Plate VIII, figure 6, shows a flowering plant from near Brownsville, Texas, collected by Robert Runyon; figure 3 shows a flowering and fruiting plant obtained by Dr. Rose at Laredo, Texas, in 1913, which flowered in the New York Botanical Garden, March 23, 1914; plate VI, figure 2, is from a photograph taken near Brownsville, Texas, by Robert Runyon in 1920.

11. Neomammillaria applanata (Engelmann).

Mammillaria applanata Engelmann in Wislizenus, Mem. Tour North. Mex. 105. 1848.
Mammillaria declivis Dietrich, Allg. Gartenz. 18: 235. 1850.
Mammillaria texensis Labouret, Monogr. Cact. 89. 1853.
Mammillaria heyderi applanata Engelmann, Proc. Amer. Acad. 3: 263. 1856.
Cactus texensis Kuntze, Rev. Gen. Pl. 1: 261. 1891.

Plants much flattened; tubercles somewhat angled, their axils naked; radial spines 10 to 18, the radials widely spreading, lower ones darker brown than upper; central spine one, porrect, dark brown; young spine-areoles very woolly; flower-buds pointed, greenish; outer perianth-segments greenish, lanceolate, acuminate margins not ciliate; inner segments 2.5 cm. long, cream-colored, lanceolate, acuminate, with a broad green stripe down the middle; filaments white, shorter than the style; stigma-lobes green; fruit scarlet, naked, 2.5 to 3.5 cm. long; seeds brown.

Type locality: Rocky plains on the Pierdenales, Texas.

Distribution: Central and southern Texas.

The description is based on plants flowering in cultivation. It is one of the earliest species to flower in the spring, beginning soon after the first of March; the fruit requires a full year to mature.

Mammillaria lindheimeri Engelmann, given by Hemsley (Biol. Centr. Amer. Bot. 1: 525. 1880) and by the Index Kewensis as a synonym of *M. texensis*, belongs here.

Neomammilaria applanata, *N. heyderi*, and *N. hemisphaerica* are closely related and may represent races of the same species.

Illustrations: Blanc, Cacti 66. No. 1116; Gartenflora 30: 412; Cact. Journ. 1: pl. for March; Meehan's Monthly 1: 4; Balt. Cact. Journ. 1: 138; 2: 259; Förster, Handb. Cact. ed. 2. 333. f. 35, as *Mammillaria applanata*; Ann. Rep. Smiths. Inst. 1908: pl. 9, f. 1; Gartenflora 29: 52, as *Mammillaria heyderi*; Schelle, Handb. Kakteenk. 263. f. 185; Blühende Kakteen 1: pl. 43; Cact. Mex. Bound. pl. 9, f. 4 to 14, as *Mammillaria heyderi applanata*.

Plate IX, figure 1, shows a plant in flower and fruit, collected by Dr. Rose on hills above Devil's River, Texas, in 1913, which flowered in the New York Botanical Garden, February 2, 1914.

12. Neomammillaria phymatothele (Berg).

Mammillaria phymatothele Berg, Allg. Gartenz. 8: 129. 1840.
Mammillaria ludwigii Ehrenberg, Linnaea 14: 376. 1840.
Cactus ludwigii Kuntze, Rev. Gen. Pl. 1: 260. 1891.
Cactus phymatothele Kuntze, Rev. Gen. Pl. 1: 261. 1891.

Simple, subglobose, glaucous-green; axils of young tubercles bearing white wool, becoming naked; tubercles large, 4-sided; areoles when young white-woolly, in age naked; radial spines 7 to 10, grayish white, the three upper smaller, the central (Schumann says 1 or 2) recurved; flowers described by Schumann as carmine-colored.

Type locality: Mexico.

Distribution: Central Mexico.

We know this species only from the description and illustration.

M. E. Eaton del. 1 to 4
D. G. Passmore del. 5

1. Flowering and fruiting plant of *Neomammillaria applanata*.
2. Top of fruiting plant of *Neomammillaria karwinskiana*.
3. Top of flowering plant of *Neomammillaria aureiceps*.
4. Flowering plant of *Neomammillaria macracantha*.
5. Fruiting plant of *Neomammillaria mystax*.

Illustration: Blühende Kakteen 1: pl. 32, as *Mammillaria centricirrha* var.

Figure 69 is from a photograph sent us by L. Quehl; figure 68 is a reproduction of the illustration cited above; figure 70 shows a plant grown in the Missouri Botanical Garden in 1905 as *Cactus neumannianus.*

FIGS. 69 and 70.—Neomammillaria phymatothele.

13. Neomammillaria magnimamma (Haworth).

Mammillaria magnimamma Haworth, Phil. Mag. 63: 41. 1824.
Mammillaria divergens De Candolle, Mém. Mus. Hist. Nat. Paris 17: 113. 1828.
Mammillaria gladiata Martius, Nov. Act. Nat. Cur. 16: 336. 1832.
Mammillaria ceratophora Lehmann, Allg. Gartenz. 3: 228. 1835.
Mammillaria recurva Lehmann in Pfeiffer, Enum. Cact. 15. 1837.
Mammillaria hystrix Martius in Pfeiffer, Enum. Cact. 21. 1837.
Mammillaria ehrenbergii Pfeiffer, Allg. Gartenz. 6: 274. 1838.
Mammillaria microceras Lemaire, Cact. Aliq. Nov. 6. 1838.
Mammillaria deflexispina Lemaire, Cact. Aliq. Nov. 6. 1838.
Mammillaria versicolor Scheidweiler, Bull. Acad. Sci. Brux. 5: 494. 1838.
? *Mammillaria conopsea* Scheidweiler, Bull. Acad. Sci. Brux. 5: 496. 1838.
Mammillaria centricirrha Lemaire, Cact. Gen. Nov. Sp. 42. 1839.
Mammillaria centricirrha macrothele Lemaire, Cact. Gen. Nov. Sp. 42. 1839.
Mammillaria neumanniana Lemaire, Cact. Gen. Nov. Sp. 53. 1839.
Mammillaria conopsea longispina Scheidweiler, Bull. Acad. Sci. Brux. 6: 92. 1839.
Mammillaria pentacantha Pfeiffer, Allg. Gartenz. 8: 406. 1840.
Cactus magnimamma Salm-Dyck in Steudel, Nom. ed. 2. 1: 246. 1840.
Mammillaria subcurvata Dietrich, Allg. Gartenz. 12: 232. 1844.
Mammillaria diadema Mühlenpfordt, Allg. Gartenz. 13: 346. 1845.
Mammillaria krameri Mühlenpfordt, Allg. Gartenz. 13: 347. 1845.
Mammillaria foersteri Mühlenpfordt, Allg. Gartenz. 14: 371. 1846.
? *Mammillaria tetracentra* Otto in Förster, Handb. Cact. 214. 1846.
Mammillaria bockii Förster, Allg. Gartenz. 15: 50. 1847.
Mammillaria pazzanii Stieber, Bot. Zeit. 5: 491. 1847.
Mammillaria divaricata Dietrich, Allg. Gartenz. 16: 210. 1848.
Mammillaria hopferiana Linke, Allg. Gartenz. 16: 329. 1848.
Mammillaria glauca Dietrich in Linke, Allg. Gartenz. 16: 330. 1848.
Mammillaria centricirrha hopferiana Salm-Dyck, Cact. Hort. Dyck. 1849. 17, 123. 1850.
Mammillaria megacantha Salm-Dyck, Cact. Hort. Dyck. 1849. 123. 1850.
Mammillaria megacantha rigidior Salm-Dyck, Cact. Hort. Dyck. 1849. 18, 124. 1850.

Mammillaria uberimamma Monville in Labouret, Monogr. Cact. 120. 1853.
? *Mammillaria cirrosa* * Poselger, Allg. Gartenz. **21**: 94. 1853.
Mammillaria pachytele Poselger, Allg. Gartenz. **23**: 17. 1855.
Mammillaria lactescens † Meinshausen, Wöchenschr. Gärtn. Pflanz. **2**: 117. 1859.
Mammillaria falcata Hortus in Förster, Handb. Cact. ed. 2. 345. 1885.
Mammillaria gebweileriana Haage in Förster, Handb. Cact. ed. 2. 358. 1885.
Mammillaria schmidtii Sencke in Förster, Handb. Cact. ed. 2. 376. 1885.
Mammillaria krameri viridis Haage in Förster, Handb. Cact. ed. 2. 372. 1885.
Cactus bockii Kuntze, Rev. Gen. Pl. **1**: 260. 1891.
Cactus centricirrhus Kuntze, Rev. Gen. Pl. **1**: 260. 1891.
Cactus conopseus Kuntze, Rev. Gen. Pl. **1**: 260. 1891.
Cactus diadema Kuntze, Rev. Gen. Pl. **1**: 260. 1891.
Cactus divergens Kuntze, Rev. Gen. Pl. **1**: 260. 1891.
Cactus ehrenbergii Kuntze, Rev. Gen. Pl. **1**: 260. 1891.
Cactus foersteri Kuntze, Rev. Gen. Pl. **1**: 260. 1891.
Cactus gladiatus Kuntze, Rev. Gen. Pl. **1**: 260. 1891.
Cactus glaucus Kuntze, Rev. Gen. Pl. **1**: 260. 1891.
Cactus krameri Kuntze, Rev. Gen. Pl. **1**: 260. 1891.
Cactus lactescens Kuntze, Rev. Gen. Pl. **1**: 260. 1891.
Cactus megacanthus Kuntze, Rev. Gen. Pl. **1**: 260. 1891.
Cactus microceras Kuntze, Rev. Gen. Pl. **1**: 260. 1891.
Cactus hystrix Kuntze, Rev. Gen. Pl. **1**: 260. 1891.
Cactus divaricatus Kuntze, Rev. Gen. Pl. **1**: 261. 1891. Not Lamarck, 1783.
Cactus neumannianus Kuntze, Rev. Gen. Pl. **1**: 261. 1891.
Cactus pazzanii Kuntze, Rev. Gen. Pl. **1**: 261. 1891.
Cactus pentacanthus Kuntze, Rev. Gen. Pl. **1**: 261. 1891.
Cactus recurvus Kuntze, Rev. Gen. Pl. **1**: 261. 1891. Not Miller, 1768.
Cactus versicolor Kuntze, Rev. Gen. Pl. **1**: 261. 1891.
Cactus tetracentrus Kuntze, Rev. Gen. Pl. **1**: 261. 1891.
Cactus subcurvatus Kuntze, Rev. Gen. Pl. **1**: 261. 1891.
Mammillaria centricirrha magnimamma Schumann, Gesamtb. Kakteen 582. 1898.
Mammillaria centricirrha divergens Schumann, Gesamtb. Kakteen 582. 1898.
Mammillaria centricirrha bockii Schumann, Gesamtb. Kakteen 582. 1898.
Mammillaria centricirrha recurva Schumann, Gesamtb. Kakteen 582. 1898.
Mammillaria centricirrha krameri Schumann, Gesamtb. Kakteen 582. 1898.

Fig. 71.—Neomammillaria magnimamma.   Fig. 72.—Neomammillaria macracantha.

Globose, the larger plants 10 cm. in diameter, sometimes solitary but oftener cespitose with 25 in a cluster or more, very milky throughout; tubercles conic or somewhat flattened or faintly 4-angled, 1 cm. long, the axils when young densely woolly; spines 3 to 5, very unequal in length, the upper ones short and straight, the lower one or two 1.5 to 4.5 cm. long, recurved or incurved, all horn-colored, with black tips; flowers cream-colored; fruit clavate, 2 cm. long, crimson; seeds brownish.

---

* Schumann refers *Mammillaria cirrosa* (he spells it *M. cirrhosa*) doubtfully to *M. centricirrha,* but judging from the description it may belong elsewhere.
† Here was referred *M. neumanni glabrescens* Regel (Förster, Handb. Cact. ed. 2. 370. 1885).

Type locality: Not cited.

Distribution: Central Mexico.

This plant is very common in central Mexico, especially in the Valley of Mexico, about Tula, farther north, and also east of the City of Mexico. It makes large cespitose mounds, sometimes with many-headed branches, and has peculiar incurved spines and small flowers. It is frequently collected and has been shipped abundantly to Europe, where it has been much named, often from single joints. Our synonymy shows 34 specific names under *Mammillaria* and nearly as many under *Cactus*. Some writers have given these names varietal rank, so that this species now has about 100 names. It is a very characteristic plant and, while it may easily be confused with other species, yet, when clearly understood, its distinctness is evident.

Mammillaria zooderi was referred by Schumann (Gesamtb. Kakteen 582. 1898) as a synonym of *M. centricirrha* but the Index Kewensis Suppl. 5. cites Schelle (Handb. Kakteenk. 268. 1907), who gives it as a synonym of *M. centricirrha zooderi*. Neither the specific nor the varietal name can be considered published.

Schelle (Handb. Kakteenk. 266 to 268. 1907) lists 62 varietal names of *Mammillaria centricirrha*, all but one or two of which are based on species of the same name. Some of these perhaps are to be referred elsewhere, but we have listed them here as follows:

amoena	falcata	lactescens	posteriana
arietina	foersteri	lehmannii	pulchra
bockii	gebweileriana	longispina	recurva
boucheana	gladiata	macracantha	schiedeana
ceratophora	glauca	magnimamma	schmidtii
cirrhosa	globosa	megacantha	spinosior
conopsea	grandidens	microceras	subcurvata
cristata	guilleminiana	montsii	tetracantha
deflexispina	hopfferiana	moritziana	uberimamma
destorum	hystrix	neumanniana	valida
de tampico	hystrix grandicornis	nordmannii	versicolor
diacantha	hystrix longispina	obconella	viridis
diadema	jorderi	pachythele	zooderi
divaricata	krameri	pazzanii	zuccariniana
divergens	krameri longispina	pentacantha	
ehrenbergii	krausei	polygona	

The following garden names are listed by Schumann (Gesamtb. Kakteen 582. 1898) as belonging to this species:

boucheana	hystrix	moritziana	tetracantha
destorum	jorderi	nordmannii	viridis
de tampico	lehmannii	obconella	zooderi
grandicornis	longispina	posteriana	
grandidens	montsii	spinosior	

Illustrations: Hort. Belge **5**: pl. 6, as *Mammillaria conopsea*; Reiche, Elem. Bot. f. 166, as *M. centricirrha*. Schelle's figure (Handb. Kakteenk. 268. f. 189) we are not able to place. The illustration in Blühende Kakteen (**1**: pl. 32) as *M. centricirrha* var. does not seem to be of this relationship.

Plate XI, figure 1, shows a small potted plant which flowered in the New York Botanical Garden, May 6, 1913. Figure 71 is from a photograph of a plant obtained on the pedregal near San Angel, Valley of Mexico, by O. Solis in 1919.

14. Neomammillaria macracantha (De Candolle).

Mammillaria macracantha De Candolle, Mém. Mus. Hist. Nat. Paris **17**: 113. 1828.
Cactus macracanthus Kuntze, Rev. Gen. Pl. **1**: 260. 1891.
Cactus alternatus Coulter, Contr. U. S. Nat. Herb. **3**: 95. 1894.
Mammillaria centricirrha macracantha Schumann, Gesamtb. Kakteen 582. 1898.

Depressed-globose, 2 to 3 cm. high, 6 to 15 cm. in diameter; axils of old tubercles naked, of young ones densely lanate; tubercles ovoid, somewhat 4-sided; young spine-areoles somewhat tomentose; spines 1 or 2, somewhat angled, elongated, the longest 5 cm. long (but not elongated in greenhouse specimens), porrect or more or less reflexed, reddish in age; flowers dark pink, a little longer than the tubercles; perianth-segments linear, spreading; stigma-lobes 5 to 7, rose-colored.

Type locality: Mexico.

Distribution: San Luis Potosí.

Our description is based on plants from San Luis Potosí, Mexico, especially those collected by Mrs. Vera in 1912.

Schumann refers *Mammillaria macracantha* to *M. centricirrha* but it must be different. Rümpler refers to it also *M. zuccarinii*, but this has different flowers and we have recognized it as a species. *M. macrantha* (Förster, Handb. Cact. 189. 1894) is referred here.

Illustrations: De Candolle, Mém. Cact. pl. 9; Förster, Handb. Cact. ed. 2. 378. f. 38, as *Mammillaria macracantha;* Schumann, Gesamtb. Kakteen f. 93; Thomas, Zimmerkultur Kakteen 57; Möllers Deutsche Gärt. Zeit. **25**: 475. f. 8, No. 13, as *M. centricirrha macracantha;* (?) Engler and Prantl, Pflanzenfam. **3**[6a]: 170. f. 57, E, as *M. centricirrha.*

FIG. 72a.—Neomammillaria macracantha.

Plate IX, figure 4, shows a plant received from Kew in 1902, which flowered in the New York Botanical Garden on April 27, 1912. Figure 72a is a reproduction of the first illustration cited above; figure 72 is from a photograph of the plant distributed by the Kew Gardens in 1902 which flowered in the New York Botanical Garden in 1905.

15. Neomammillaria johnstonii sp. nov.

Plants large for the genus, globular to short-oblong, 15 to 20 cm. high, slightly depressed at apex; tubercles 1 to 1.5 cm. long, 4-angled throughout, somewhat bluish, naked in their axils, milky; spine-areoles when young short-floccose, in age glabrate, circular; radial spines 10 to 14, white, but with brown tips, somewhat spreading, stiff-acicular; central spines 2, much longer and stouter than the radials, slightly diverging, bluish brown; flowers from near top of plant but from axils of old tubercles, campanulate, 2 cm. long; outer perianth-segments ovate-lanceolate, greenish white with a reddish-brown mid-rib; inner perianth-segments narrow, acuminate, white; filaments short, pinkish; style pinkish; stigma-lobes linear, 6 or 7, green.

1. *Coryphantha sulcata*, from Sabinal, Texas.
2. *Neomammillaria runyonii*, from Monterey, Mexico.

Collected at San Carlos Bay, Sonora, Mexico, by Ivan M. Johnston in 1921 (No. 4373) and flowered in Washington in April 1922 and April 1923.

Figure 72*b* is from a photograph of the type specimen.

16. Neomammillaria melanocentra (Poselger).

Mammillaria melanocentra Poselger, Allg. Gartenz. **23**: 17. 1855.
Mammillaria erinacea Poselger, Allg. Gartenz. **23**: 18. 1855.
Mammillaria valida Weber, Dict. Hort. Bois 806. 1898.

Short-cylindric, glaucous-green; tubercles in 8 and 13 spirals, strongly angled; radial spines 6, stout-subulate, 1.5 to 2 cm. long, brownish; central spines solitary, black, 2 to 3 cm. long, greatly overtopping the stem; flowers pinkish red, the segments linear, acute.

Fig. 72*b*.—Neomammillaria johnstonii.

Type locality: Near Monterey, Mexico.
Distribution: Mexico, but range unknown.
Illustration: Blühende Kakteen **3**: pl. 129, as *Mammillaria melanocentra*.

Figure 73 is a reproduction of the illustration cited above.

17. Neomammillaria runyonii sp. nov.

Plants deep-seated, depressed; tubercles milky, elongated, 1.5 cm. long, strongly 4-angled, their tips widely separated from each other, their axils long-woolly (never setose), especially when young, sometimes permanently so; young spine-areoles long-woolly, but in age glabrate; radial spines 6 to 8, slightly ascending, the outer ones stouter and often dark brown in color, the inner ones about half the length of the outer and nearly white; central spine solitary, brown to black, erect, 10 to 14 mm. long; flowers about 2 cm. long, purple; perianth-segments oblong; fruit red, clavate, 12 to 16 mm. long; seeds brown.

Collected on El Mirador, near Monterey, Mexico, by Robert Runyon in 1921.

Plate x, figure 2, is from a photograph of one of the plants Mr. Runyon originally brought from El Mirador.

18. Neomammillaria sartorii (J. A. Purpus).

Mammillaria sartorii J. A. Purpus, Monatsschr. Kakteenk. 21: 50. 1911.

Globose to short-cylindric, 5 to 13 cm. in diameter, cespitose, very milky, bluish green; tubercles strongly 4-angled, pointed, 8 to 12 mm. long, their axils without bristles and in time without wool; spine-areoles circular when young, densely white-woolly but in age glabrate; spines 4 to 6, very unequal, 5 to 8 mm. long, whitish or sometimes brownish, the central spine solitary; flowers 1.5 to about 2 cm. long, deep carmine; perianth-segments oblong, apiculate, the tip dry, the outer ciliate, the inner serrulate; stamens and style purplish above; stigma-lobes 4, purple, short; fruit carmine; seeds brown.

Type locality: Barranca de Panoaya, Vera Cruz, Mexico.

Distribution: Mountains of Vera Cruz, 300 to 600 meters altitude.

Our description of this interesting and variable little plant is drawn from specimens sent to us by Dr. C. A. Purpus in 1920, collected at the type locality. There the plant

FIG. 73.—Neomammillaria melanocentra.　　FIG. 74.—Neomammillaria seitziana.

grows among rocks in rich humus of the decaying leaves in half shade or in the sun. It is very different from any other *Neomammillaria* which we have seen; the tubercles are copiously milky and the slightest bruise causes the white milk to ooze out. It flowered in Washington in April 1923.

Dr. C. A. Purpus writes that this species is common in many of the barrancas of Vera Cruz and that it is very variable. When first described two forms (*brevispina* and *longispina*) were characterized.

The species was named for Florantino Sartorius (1837–1908) who assisted Dr. Purpus for many years in his botanical expeditions. He was a son of Carlos Sartorius (1795–1872), a distinguished scientist who went to Mexico about 1825, where he made large collections of plants. Mr. W. Botting Hemsley (Biol. Centr. Amer. Bot. 4: 123) states that his herbarium was left to the Smithsonian Institution, but no record of this gift can now be found nor can any of his plants be found in the U. S. National Herbarium.

Here may or may not belong *Mammillaria rebsamiana* (Cact. Journ. 2: 176), advertised as a new discovery by Louis Murillo, who lived at Jalapa, Mexico.

Illustration: Monatsschr. Kakteenk. 21: 51, as *Mammillaria sartorii.*

Figure 75 is from a photograph showing two plants sent from the type locality of *Mammillaria sartorii* by Dr. Purpus in 1920.

19. Neomammillaria seitziana (Martius).

Mammillaria seitziana Martius in Pfeiffer, Enum. Cact. 18. 1837.
Mammillaria foveolata Mühlenpfordt, Allg. Gartenz. 14: 372. 1846.
Cactus foveolatus Kuntze, Rev. Gen. Pl. 1: 260. 1891.
Cactus seitzianus Kuntze, Rev. Gen. Pl. 1: 261. 1891.

Solitary or somewhat proliferous at base, cylindric, 12 cm. high; tubercles green, conic, somewhat angled; axils of tubercles woolly; areoles at first white-woolly, becoming glabrate; spines 4,* the upper and lower longer than the lateral; flowers rose-colored, about 25 mm. long; outer perianth-segments olive colored; inner perianth-segments linear, lanceolate, white, nerved with red; stamens white; stigma-lobes 6.

Type locality: Ixmiquilpan, Mexico.
Distribution: State of Hidalgo.
We have not seen this species and hence our description is compiled.

FIG. 75.—Neomammillaria sartorii.

Mammillaria senckena and *M. senckei* are two names listed as synonyms of this species, but we do not find that they have ever been published.
Illustration: Pfeiffer and Otto, Abbild. Beschr. Cact. 1: pl. 8, as *Mammillaria seitziana.* Figure 74 is reproduced from the illustration cited above.

20. Neomammillaria ortegae sp. nov.

Simple to short-clavate, 5 to 8 cm. in diameter, light green, lactiferous; tubercles rather short (8 to 10 mm. long), broader at base, obscurely 4-angled, somewhat pointed, very woolly but not setose in their axils; spines all radial, 3 or 4, more commonly 4 (sometimes with 1 or 2 small additional spines or bristles, perhaps deciduous), spreading, straw-colored, 6 to 10 mm. long; flowers small; fruit clavate, 1 cm. long; seeds numerous, small, angled, brown.

Collected by J. G. Ortega in Sinaloa, Mexico, in 1921 and 1922.

Figure 76 shows the type specimens as photographed in the U. S. National Museum under the direction of A. J. Olmstead.

---

* Schumann says central spines yellow.

21. Neomammillaria meiacantha (Engelmann).

Mammillaria meiacantha Engelmann, Proc. Amer. Acad. **3**: 263. 1856.
Cactus meiacanthus Kuntze, Rev. Gen. Pl. **1**: 260. 1891.

Somewhat depressed, 12 cm. broad or more; tubercles milky, bluish green, more or less angled, somewhat flattened dorsally, their axils naked; spines 5 to 9, ascending, pale flesh-colored, the tips darker, the lower a little stouter than the upper; central spines porrect, similar to but a little stouter than radials and often subradial; spine-areoles short-woolly at first; flowers not very abundant, at least on cultivated plants; inner perianth-segments white with a pink stripe along inside of midrib, one-fourth its width, greenish brown on outside; filaments white; style pink; stigma-lobes yellow; fruit scarlet, 22 mm. long; seeds brownish.

Type locality: Western Texas and New Mexico.

Distribution: Texas, New Mexico, and northern Mexico.

According to Dr. Engelmann, this species was first obtained in New Mexico by the Missouri Volunteers in 1847 and it has frequently been collected since that time. In Mexico it extends as far south as Zacatecas, but develops into some unusual forms. It was repeatedly collected in Zacatecas by F. E. Lloyd in 1908.

Fig. 76.—Neomammillaria ortegae.

Illustrations: Blühende Kakteen **1**: pl. 47 *; Blanc, Cacti 71. No. 1388; Cycl. Amer. Hort. Bailey **2**: f. 1357; Stand. Cycl. Hort. Bailey **4**: f. 2316; West Amer. Sci. **13**: 39; Schelle, Handb. Kakteenk. 258. f. 190; Cact. Mex. Bound. pl. 9, f. 1 to 3; Cact. Journ. **1**: pl. for October, as *Mammillaria meiacantha*.

Figure 77 shows the plant illustrated in the Mexican Boundary Report as cited above.

22. Neomammillaria scrippsiana sp. nov.

Globose or becoming short-cylindric, 6 cm. high; tubercles milky, in 26 rows, bluish green, very woolly in axils when young; spine-areoles very woolly at first; radial spines 8 to 10, slender, pale with reddish tips; central spines generally 2, a little longer than radials, brown throughout, slightly divergent; flowers borne near top of plant but not in axils of youngest tubercles, about 1 cm. long, pinkish, with margins of perianth-segments paler; anthers pinkish; stigma-lobes about 6, recurved, cream-colored.

Collected by Dr. Rose in the barranca of Guadalajara, Jalisco, in September 1903 (No. 871, type). The plant has flowered repeatedly in Washington since April 1906. Specimens were afterward collected near the same place by C. R. Orcutt. It is named in honor of E. W. Scripps, the founder of Science Service and The Scripps Institution for Biological Research of the University of California.

Figure 78 is from a photograph of the type specimen.

* This plate is labeled *Mammillaria meionacantha*, but described under *M. meonacantha*.

23. Neomammillaria gigantea (Hildmann).

Mammillaria gigantea Hildmann in Schumann, Gesamtb. Kakteen 578. 1898.

Solitary or cespitose, depressed-globose, 10 cm. high, 15 to 17 cm. in diameter; axils of tubercles lanate; radial spines 12, subulate, white, 3 mm. long; central spines 4 to 6, stout, 2 cm. long, curved, yellowish brown; flowers yellowish green.

Type locality: Guanajuato, Mexico.
Distribution: Known only from the type locality.

Mammillaria macdowellii Heese and *M. guanajuatensis* Rünge are two names referred here by Schumann (Gesamtb. Kakteen 578. 1898), but they were not published.

Plate XI, figure 3, shows a plant in fruit, collected by Dr. Safford at the type locality.

24. Neomammillaria peninsularis sp. nov.

Plants solitary or in clusters, deeply seated in the ground, more or less flat-topped, bluish green, the stems and tubercles very milky; tubercles erect, pointed, 4-angled, pale green; radial spines 4 to 8,

FIG. 77.—Neomammillaria meiacantha.

FIG. 78.—Neomammillaria scrippsiana.

nearly erect, short and pale with brown tips, one sometimes nearly central; axils of tubercles bearing long wool but in age naked; flowers 1.5 cm. long, arising from old tubercles but near the center; outer perianth-segments narrow, reddish; inner perianth-segments narrow, acuminate, green or light yellow with erose margins; stamens pale; style longer than stamens; stigma-lobes green, linear.

Collected by Dr. Rose at Cape San Lucas, Lower California, March 23, 1911 (No. 16377).

25. Neomammillaria flavovirens (Salm-Dyck).

Mammillaria flavovirens Salm-Dyck, Cact. Hort. Dyck. 1849. 117. 1850.

Either solitary or somewhat cespitose, globose or short-cylindric, 6 to 8 cm. high, light or yellowish green; tubercles somewhat 4-angled; axils naked; radial spines 5, slender, subulate; central spines solitary, porrect; flowers white, streaked with rose.

Type locality: Not cited.
Distribution: Mexico.

The above description is compiled, since the species is not otherwise known to us.

Mammillaria flavovirens cristata Salm-Dyck (Cact. Hort. Dyck. 1849. 16. 1850) is only a name.

The name *Mammillaria daedalea viridis* Fennel is given by Labouret (Monogr. Cact. 100. 1853) as a synonym of *M. flavovirens*.

26. Neomammillaria sempervivi (De Candolle).

Mammillaria sempervivi De Candolle, Mém. Mus. Hist. Nat. Paris **17**: 114. 1828.
Mammillaria sempervivi tetracantha De Candolle, Mém. Mus. Hist. Nat. Paris **17**: 114. 1828.
Mammillaria caput-medusae Otto in Pfeiffer, Enum. Cact. 22. 1837.
Mammillaria diacantha Lemaire, Cact. Aliq. Nov. 2. 1838.
Mammillaria sempervivi laeteviridis Salm-Dyck, Cact. Hort. Dyck. 1849. 113. 1850.
Mammillaria caput-medusae centrispina Salm-Dyck in Labouret, Monogr. Cact. 91. 1853.
Mammillaria caput-medusae crassior Salm-Dyck in Labouret, Monogr. Cact. 91. 1853.
Mammillaria caput-medusae tetracantha Salm-Dyck in Labouret, Monogr. Cact. 91. 1853.
Cactus sempervivi Kuntze, Rev. Gen. Pl. **1**: 261. 1891.

FIG. 79.—Neomammillaria sempervivi.   FIG. 80.—Neomammillaria polythele.

Solitary or somewhat cespitose, flattened above, narrowed below; axils of tubercles very woolly; tubercles short, milky, angled; spine-areoles very woolly when young, but glabrate in age; radial spines 3 to 7, short, white, caducous; central spines 2, ascending, brownish, stoutish; flowers dull white with reddish lines; inner perianth-segments acute, spreading.

Type locality: Mexico.
Distribution: Central Mexico.

Dr. Rose collected what he took to be this species in the Barranca Sierra de la Mesa, Hidalgo, Mexico, in 1905, but this plant differs somewhat from De Candolle's illustration. The central spines, while generally 2, are sometimes 3 and are not so stout; the radial spines are deciduous, as they should be in this species. It flowered once at Washington.

An examination of the original description of *Mammillaria caput-medusae* suggests the probability that this species is identical with *Mammillaria sempervivi*. The two names appeared in collections in 1829 and may have come from a common source. Indeed, Schumann credits T. Coulter with having obtained *M. caput-medusae*, while we know that *M. sempervivi* was based on Coulter's plant and, then, too, Pfeiffer refers *M. sempervivi* as a synonym of *M. caput-medusae*. Knippel's illustration of *M. caput-medusae* (pl. 19) seems to be referable here. Nicholson states that *M. caput-medusae* is only a form of this species.

Mammillaria staurotypa (Förster, Handb. Cact. 221. 1846), credited to Scheidweiler by Schumann and referred by him as a synonym of *M. caput-medusae*, seems never to have been described but may belong here.

The two varieties of *Mammillaria caput-medusae, tetracantha* and *hexacantha*, given by Salm-Dyck (Cact. Hort. Dyck. 1844. 10. 1845) are without description. The first was afterwards described by Labouret.

Illustrations: De Candolle, Mém. Cact. pl. 8; Förster, Handb. Cact. ed. 2. 344. f. 36; Schumann, Gesamtb. Kakteen 589. f. 95; Dict. Gard. Nicholson **4:** 565. f. 38; Suppl. 518. f. 556; Watson, Cact. Cult. 175. f. 70, as *Mammillaria sempervivi*; Schelle, Handb. Kakteenk. 270. f. 192; Succulenta **5:** 51, as *M. caput-medusae.*

Figure 79 is a reproduction of the first illustration cited above.

27. Neomammillaria obscura (Hildmann).

Mammillaria obscura Hildmann, Monatsschr. Kakteenk. I: 52. 1891.

Solitary, depressed-globose, blackish green; axils woolly; tubercles arranged in 13 and 21 spirals, angled, stout, woolly in their axils but not setose; radial spines 6 to 8, subulate, white, unequal, the upper ones shorter than the lower; central spines 2 to 4, the lower one slightly curved, black; flowers small, yellowish white.

Type locality: Mexico.
Distribution: Mexico, but range unknown.

The plant is known to us only from description and illustration.

Seeds of this species were introduced into Germany from Mexico about 1885 by Mr. Droege and flowers were obtained in 1891.

The earlier name, *Mammillaria obscura* Scheidweiler (Förster, Handb. Cact. 213. 1846), but used only as a synonym and for some other plant, does not interfere with our present use of the name.

Illustration: Monatsschr. Kakteenk. I: facing 52, as *Mammillaria obscura.*

28. Neomammillaria crocidata (Lemaire).

Mammillaria crocidata Lemaire, Cact. Aliq. Nov. 9. 1838.
Mammillaria webbiana Lemaire, Cact. Gen. Nov. Sp. 45. 1839.
Cactus crocidatus Kuntze, Rev. Gen. Pl. I: 260. 1891.
Cactus webbianus Kuntze, Rev. Gen. Pl. I: 261. 1891.

Plant globose or a little depressed, 5 to 6 cm. in diameter; radial spines 6 or 7, dark brown or nearly black; central spines none; axils of tubercles in young plant densely woolly; flowers from axils of old tubercles near the top of plant, small, reddish purple, 12 to 14 mm. long; outer perianth-segments ciliate; inner perianth-segments acuminate; filaments, style, and stigma-lobes reddish; stigma-lobes 3 or 4; fruit not seen.

Type locality: Mexico.
Distribution: Central Mexico.

Described here from plants collected by Dr. Rose near Querétaro, Mexico, in 1906, which flowered in August and September 1908, and again in April 1909 (No. 1072). Our specimen has more spines than the original *M. crocidata*; it is also near *M. carnea* but with different colored stigma-lobes; its tubercles are about 6 mm. high.

Schumann places this species near *M. carnea* and among the cylindric species, but it was originally described as depressed.

Mammillaria crocidata quadrispina Pfeiffer and Salm-Dyck, mentioned by Förster (Handb. Cact. 220. 1846) as a rare form and afterwards briefly described by Labouret (Monogr. Cact. 93. 1853), may or may not belong here.

Plate VII, figure 5, shows a flowering plant collected by Dr. Rose in Querétaro in 1906 and painted in the New York Botanical Garden, September 5, 1911.

29. Neomammillaria polythele (Martius).

Mammillaria polythele Martius, Nov. Act. Nat. Cur. **16**: 328. 1832.
Mammillaria quadrispina Martius, Nov. Act. Nat. Cur. **16**: 329. 1832.
Mammillaria columnaris Martius, Nov. Act. Nat. Cur. **16**: 330. 1832.
Mammillaria affinis De Candolle, Mém. Cact. 11. 1834.
Mammillaria setosa Pfeiffer, Allg. Gartenz. **3**: 379. 1835.
Mammillaria polythele quadrispina Salm-Dyck in Walpers, Repert. Bot. **2**: 271. 1843.
Mammillaria polythele columnaris Salm-Dyck in Walpers, Repert. Bot. **2**: 271. 1843.
Mammillaria polythele setosa Salm-Dyck, Cact. Hort. Dyck. 1844. 9. 1845.
Mammillaria polythele hexacantha Salm-Dyck, Cact. Hort. Dyck. 1849. 15. 1850.
Mammillaria polythele latimamma Salm-Dyck, Cact. Hort. Dyck. 1849. 112. 1850.
Cactus affinis Kuntze, Rev. Gen. Pl. **1**: 260. 1891.
Cactus quadrispinus Kuntze, Rev. Gen. Pl. **1**: 261. 1891.
Cactus setosus Kuntze, Rev. Gen. Pl. **1**: 261. 1891.
Cactus polythele Kuntze, Rev. Gen. Pl. **1**: 261. 1891.
? *Mammillaria hidalgensis* Purpus, Monatsschr. Kakteenk. **17**: 118. 1907.

Elongated, cylindric, often 3 to 5 dm. high, 7 to 10 cm. in diameter; tubercles milky, in about 21 spirals, 10 to 12 mm. long, nearly terete, somewhat narrowed toward apex, dull green; axils of young tubercles densely long-woolly, the wool nearly covering the top of the plant, in age becoming naked; spines 2 to 4, sometimes 6, all radial, somewhat spreading, 1 to 2.5 cm. long, reddish, straight or a little curved; flowers from near top of plant, reddish, 8 to 10 mm. long; perianth-segments narrow, acuminate; fruit red, clavate; seeds small, brownish.

Type locality: Mexico.

Distribution: State of Hidalgo.

In 1905 Dr. Rose collected living plants of this species near Ixmiquilpan. It is a rather striking plant, growing very tall and flowering near the top.

Schumann places this species in the Section *Hydrochylus*, in which the sap is watery, but Martius in his original description says definitely that it is milky.

Mammillaria aciculata Otto (Pfeiffer, Enum. Cact. 29. 1837; *M. polythele aciculata* Salm-Dyck, Cact. Hort. Dyck. 1844. 9. 1845) is referred here by Schumann but should be excluded; it came from the cold regions of Mexico and was described as having 20 white slender radial spines.

Mammillaria columnaris minor Martius and *M. quadrispina major*, mentioned by Förster (Handb. Cact. 214, 215. 1846), probably belong here.

Mammillaria cataphracta Martius was given by Pfeiffer (Enum. Cact. 11. 1837) as a synonym of *M. affinis* and by Salm-Dyck (Hort. Dyck. 155. 1834) as a synonym of *M. angularis*.

Illustrations: Nov. Act. Nat. Cur. **16**: pl. 19, as *Mammillaria polythele*; Monatsschr. Kakteenk. **17**: 119; Möllers Deutsche Gärt. Zeit. **25**: 475. f. 8, No. 10, as *M. hidalgensis*; De Candolle, Mém. Cact. pl. 6, as *M. affinis*; Abh. Bayer. Akad. Wiss. München **2**: pl. 1, I. f. 2, as *M. columnaris*.

Figure 80 is from a photograph of a plant collected in the state of Hidalgo in 1905 which has heretofore passed as *Mammillaria hidalgensis*.

30. Neomammillaria carnea (Zuccarini).

Mammillaria carnea Zuccarini in Pfeiffer, Enum. Cact. 19. 1837.
Mammillaria subtetragona Dietrich, Allg. Gartenz. **8**: 169. 1840.
Mammillaria aeruginosa Scheidweiler, Allg. Gartenz. **8**: 338. 1840.
Mammillaria pallescens Scheidweiler, Allg. Gartenz. **9**: 42. 1841.
Mammillaria villifera carnea Salm-Dyck, Cact. Hort. Dyck. 1849. 16. 1850.
Mammillaria villifera aeruginosa Salm-Dyck, Cact. Hort. Dyck. 1849. 16. 1850.
Mammillaria villifera cirrosa* Salm-Dyck, Cact. Hort. Dyck. 1849. 115. 1850.
Cactus aeruginosus Kuntze, Rev. Gen. Pl. **1**: 260. 1891.
Cactus carneus Kuntze, Rev. Gen. Pl. **1**: 260. 1891.
Cactus pallescens Kuntze, Rev. Gen. Pl. **1**: 261. 1891.
Cactus subtetragonus Kuntze, Rev. Gen. Pl. **1**: 261. 1891.
Mammillaria carnea cirrosa Gürke, Blühende Kakteen **1**: under pl. 60. 1905.
Mammillaria carnea aeruginosa Gürke, Blühende Kakteen **1**: under pl. 60. 1905.

Plants solitary, cylindric, 8 to 9 cm. high; tubercles 4-angled, milky, their axils woolly, the upper ones erect; spines 4, straight, reddish, the lower one 10 mm. long, twice as long as the other 3; flowers

---

* Förster (Handb. Cact. ed. 2. 342. 1885) spells this name, *cirrhosa*.

borne in the old axils; outer perianth-segments nearly 2 cm. long, nearly erect, flesh-colored; fruit pear-shaped, obtuse, bright red.

Type locality: Ixmiquilpan, Mexico.

Distribution: Central and southern Mexico.

Mammillaria villifera Otto, referred here by Schumann, must belong elsewhere, since the axils of the tubercles bear setae.

Illustrations: Blühende Kakteen **1**: pl. 60; Monatsschr. Kakteenk. **28**: 59; Schelle, Handbk. Kakteenk. 271. f. 193, as *Mammillaria carnea.*

Plate VII, figure 7, shows a plant collected by Dr. Rose at Tehuacán in 1906, which flowered in the New York Botanical Garden, May 4, 1912. Figure 81 is from a photograph of a plant collected by Dr. Rose at Tehuacán in 1905.

FIG. 81.—Neomammillaria carnea.   FIG. 82.—Neomammillaria lloydii.

31. Neomammillaria lloydii sp. nov.

Plant-body at first flattened but in cultivation becoming elongated, sometimes 10 cm. long, 6 to 7 cm. in diameter; axils of young tubercles only slightly woolly; tubercles milky, small, numerous, 4-angled, woolly when quite young; radial spines 3 or 4, ascending, glabrous, the uppermost one red or dark brown, the others whitish, 2 to 5 mm. long; central spines none; flowers in a ring near center of plant; outer perianth-segments dark red with light or colored margins; inner perianth-segments white with a tinge of red, and dark-red central stripes, not ciliate, apiculate, spreading above; filaments pale below, pinkish above; style pinkish above.

Collected by F. E. Lloyd in the State of Zacatecas, Mexico, in 1909, and flowered in Washington in 1911, 1912 (March), and 1915 (April).

Figure 82 is from a photograph of the type plant (Lloyd, No. 55).

32. Neomammillaria zuccariniana (Martius).

Mammillaria zuccariniana Martius, Nov. Act. Nat. Cur. **16**: 331. 1832.

Globose to elongated-cylindric, 8 to 20 cm. long, bluish green, milky; areoles and axils of young tubercles filled with white wool; radial spines wanting or represented by very stout bristles; central spines 2 to 4, black, unequal, 2 to 12 mm. long, spreading; flowers about 1 cm. long, with a broad open throat; outer perianth-segments brownish, acute; inner perianth-segments lanceolate, acute, entire, magenta-colored; filaments purplish; stigma-lobes 3 or 4, purplish, broad, truncate; fruit red, 10 mm. long; seeds brownish.

Type locality: Mexico.

Distribution: San Luis Potosí, Mexico.

We have had this plant in cultivation for a number of years; Dr. E. Palmer obtained it near San Luis Potosí in 1905 (No. 590); it was also collected by Mrs. Vera from the same locality in 1912.

Illustration: Martius, Nov. Act. Nat. Cur. **16:** pl. 20, as *Mammillaria zuccariniana.*

Figure 83 is from a photograph of a plant collected by Dr. E. Palmer near Alvarez, San Luis Potosí, May 1905, which afterwards flowered in Washington, D. C.

33. Neomammillaria formosa (Galeotti).

Mammillaria formosa Galeotti in Scheidweiler, Bull. Acad. Sci. Brux. **5:** 497. 1838.
Mammillaria formosa microthele Salm-Dyck, Cact. Hort. Dyck. 1849. 87. 1850.
Mammillaria formosa dispicula Monville in Labouret, Monogr. Cact. 60. 1853.
Mammillaria formosa gracilispina Monville in Labouret, Monogr. Cact. 60. 1853.
Mammillaria formosa laevior Monville in Labouret, Monogr. Cact. 60. 1853.
Cactus formosus Kuntze, Rev. Gen. Pl. **1:** 260. 1891.

Somewhat clavate, sunken at the apex; axils lanate; tubercles spirally arranged, obtusely 4-angled, light green; areoles naked; radial spines 20 to 22, white, rigid, radiating; central spines 6, spreading, thickened at base, at first flesh-colored at base, black at tip, becoming black throughout or grayish; flowers red.

Type locality: Near San Felipe.

Distribution: San Luis Potosí, Mexico, according to Hemsley.

Dr. Safford has referred here a plant collected by Dr. E. Palmer at San Luis Potosí which may be the plant which is passing under this name, but it does not seem to answer the original descriptions.

Illustration: Garten-Zeitung **4:** 182. f. 42, No. 11, as *Mammillaria formosa.*

Fig. 83.—Neomammillaria zuccariniana.　　　Fig. 84.—Neomammillaria compressa.

34. Neomammillaria compressa (De Candolle).

Mammillaria compressa De Candolle, Mém. Mus. Hist. Nat. Paris **17:** 112. 1828.
Mammillaria subangularis De Candolle, Mém. Mus. Hist. Nat. Paris **17:** 112. 1828.
Mammillaria triacantha De Candolle, Mém. Mus. Hist. Nat. Paris **17:** 113. 1828.
Mammillaria cirrhifera Martius, Nov. Act. Nat. Cur. **16:** 334. 1832.
Mammillaria angularis Link and Otto in Pfeiffer, Enum. Cact. 12. 1837.
Mammillaria cirrhifera angulosior Lemaire, Cact. Gen. Nov. Sp. 95. 1839.
Mammillaria longiseta Mühlenpfordt, Allg. Gartenz. **13:** 346. 1845.

Mammillaria cirrifera longiseta Salm-Dyck, Cact. Hort. Dyck. 1849. 18. 1850.
Mammillaria squarrosa Meinshausen Wochenschr. Gärten. Pflanz. 2: 116. 1859.
Cactus cirrhifer Kuntze, Rev. Gen. Pl. 1: 260. 1891.
Cactus compressus Kuntze, Rev. Gen. Pl. 1: 260. 1891. Not Salisbury, 1796.
Cactus longisetus Kuntze, Rev. Gen. Pl. 1: 260. 1891.
Cactus squarrosus Kuntze, Rev. Gen. Pl. 1: 261. 1891.
Cactus subangularis Kuntze, Rev. Gen. Pl. 1: 261. 1891.
Cactus triacanthus Kuntze, Rev. Gen. Pl. 1: 261. 1891.
? *Mammillaria angularis fulvispina* Schumann, Gesamtb. Kakteen 576. 1898.
Mammillaria angularis longiseta Salm-Dyck in Schumann, Gesamtb. Kakteen 576. 1898.
Mammillaria angularis triacantha Salm-Dyck in Schumann, Gesamtb. Kakteen 576. 1898.
Mammillaria angularis compressa Schumann, Gesamtb. Kakteen 577. 1898.
Mammillaria oettingenii Zeissold, Monatsschr. Kakteenk. 8: 10. 1898.
Mammillaria kleinschmidtiana Zeissold, Monatsschr. Kakteenk. 8: 21. 1898.

Growing in large clumps, cylindric, pale bluish green; axils of tubercles white-woolly, setose; tubercles short, compressed laterally, keeled below, more rounded above; young spine-areoles white-woolly; principal spines 4, sometimes with 1 to 3 very short accessory ones from the lower part of the areole; lower spine much longer, spreading or recurved, 5 to 6 cm. long, somewhat angled; all spines pale, more or less tinged with brown, with dark tips; flower small, pinkish, 10 to 12 mm. long; outer perianth-segments acute, somewhat ciliate; inner perianth-segments narrow, acuminate, with spreading tips; stamens and style pale; stigma-lobes 5, linear; fruit clavate, red; seeds brown.

FIGS. 85 and 86.—Neomammillaria compressa.

Type locality: Mexico.
Distribution: Central Mexico.

Our description is drawn largely from specimens which flowered in March 1908 and which were collected by Dr. Rose at Higuerillas, Querétaro, in 1905. Dr. Rose also found this species very abundant in the deserts of Querétaro and living specimens brought back by him have frequently flowered both in New York and Washington. These are identical with plants sent from Berlin, labeled *Mammillaria angularis longiseta*. The species as here treated is variable and more exhaustive field work might require some modifications in the description.

The varieties *Mammillaria cirrhifera major* and *M. cirrhifera fulvispina* (Salm-Dyck, Cact. Hort. Dyck. 1844. 11. 1845) are without descriptions. The two varieties *M. cirrhifera albispina* and *M. centricirrha macrothele* were listed as synonyms of *M. subangularis* by Walpers (Repert. Bot. 2: 272. 1843). To *M. subangularis* is also referred *M. subcirrhifera* by Förster (Handb. Cact. 234. 1846).

Here doubtless belongs *Mammillaria angularis fulvescens* (Salm-Dyck, Cact. Hort. Dyck. 1849. 18. 1850).

Illustrations: Schelle, Handb. Kakteenk. 264. f. 187 (?); Möllers Deutsche Gärt. Zeit. **25:** 486. f. 20; Krook, Handb. Cact. 38 (?); Gartenwelt 15: 410; Contr. U. S. Nat. Herb. **10:** pl. 16, f. B, as *Mammillaria angularis*; Möllers Deutsche Gärt. Zeit. **25:** 475. f. 8, No. 8, as *M. angularis compressa*; Pfeiffer and Otto, Abbild. Beschr. Cact. **1:** pl. 7, as *M. cirrhifera*; Möllers Deutsche Gärt. Zeit. **25:** 475. f. 8, No. 28, as *M. angularis rufispina*; Blanc, Cacti 67. f. 1170; Cact. Journ. **1:** pl. for March; **2:** 7, 93, as *M. cirrhifera longispina*.

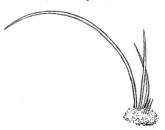

FIG. 87.—Spine-cluster of N. compressa.

Plate VIII, figure 4, shows a plant collected by Dr. Rose in Querétaro in 1905, which flowered in the New York Botanical Garden, March 17, 1913. Figure 84 is from a photograph of a plant sent by Dr. E. Palmer from San Luis Potosí in 1905; figure 85 is from a photograph of a plant collected by Dr. Rose near Higuerillas, Mexico, in 1905; figure 86 is from a photograph of this plant growing in the open in the Huntington Collection, southern California; figure 88 is from a photograph of a plant sent by Dr. C. A. Purpus from Minas de San Rafael in 1910; figure 87 shows spines from the plant collected by Dr. Rose at Ixmiquilpan in 1905.

FIG. 88.—Neomammillaria compressa.

35. Neomammillaria mystax (Martius).

Mammillaria mystax Martius, Nov. Act. Nat. Cur. **16:** 332. 1832.
Mammillaria leucotricha Scheidweiler, Allg. Gartenz. 8: 338. 1840.
Mammillaria zanthotricha Scheidweiler, Allg. Gartenz. 8: 338. 1840.
Mammillaria mutabilis Scheidweiler, Allg. Gartenz. 9: 43. 1841.
Mammillaria funkii Scheidweiler, Allg. Gartenz. 9: 43. 1841.
Mammillaria autumnalis Dietrich, Allg. Gartenz. 16: 297. 1848.
Mammillaria mutabilis xanthotricha Salm-Dyck, Cact. Hort. Dyck. 1849. 17, 120. 1850.
Mammillaria maschalacantha Monville in Labouret, Monogr. Cact. 106. 1853.
Mammillaria maschalacantha leucotricha Monville in Labouret, Monogr. Cact. 106. 1853.

Mammillaria maschalacantha xantotricha Monville in Labouret, Monogr. Cact. 106. 1853.
Cactus funckii Kuntze, Rev. Gen. Pl. 1: 260. 1891.
Cactus maschalacanthus Kuntze, Rev. Gen. Pl. 1: 260. 1891.
Cactus leucotrichus Kuntze, Rev. Gen. Pl. 1: 261. 1891.
Cactus mutabilis Kuntze, Rev. Gen. Pl. 1: 261. 1891.
Cactus mystax Kuntze, Rev. Gen. Pl. 1: 261. 1891.
Cactus xanthotrichus Kuntze, Rev. Gen. Pl. 1: 261. 1891.

Globose to short-cylindric, 7 to 15 cm. high, flat-topped; tubercles in as many as 34 rows, thickly set, full of milk which freely flows when pricked or cut; radial spines 8 to 10, small, white; central spines 4, 3 about twice as long as the radial ones, the other much elongated, 6 to 7 cm. long; flowers 1.5 to 2 cm. long, appearing in 2 or 3 rows, very abundant; inner perianth-segments dark red, 12 mm. long; stigma-lobes 4 or 5, greenish; fruit red, 2 to 2.5 cm. long.

Type locality: Mexico. According to Hemsley, Karwinsky's plant, which is the type, came from Ixmiquilpan and San Pedro Nolasco at about 6,000 feet altitude.

Distribution: Highlands of southern central Mexico.

This species is characterized by the long, erect, central spines which overtop the plant in the wild state; in cultivation these elongated spines do not always occur. The species is common in cultivation; in collections it is usually known as *Mammillaria mutabilis.*

FIG. 89.—Neomammillaria mystax.   90.—Neomammillaria petterssonii.

Mammillaria krauseana, a name from Gruson's Catalogue, is cited by Schumann (Gesamtb. Kakteen 595. 1898) as a synonym of *M. mutabilis.*

Mammillaria meschalacantha Hortus (Salm-Dyck, Cact. Hort. Dyck. 1844. 10. 1845), according to the Index Kewensis, is a misspelling for *M. maschalacantha.*

Mammillaria maschalacantha dolichacantha Monville was given as a doubtful synonym of *M. maschalacantha* by Labouret (Monogr. Cact. 106. 1853).

Mammillaria mutabilis autumnalis (Monatsschr. Kakteenk. **30:** February 1920) is offered for sale by Grässner.

Mammillaria mutabilis laevior Salm-Dyck (Cact. Hort. Dyck. 1849. 17, 120. 1850), with *M. leucocarpa* Scheidweiler as a synonym, was given as a variety of *M. mutabilis,* but it was not described. *M. xanthotricha laevior* Salm-Dyck (Cact. Hort. Dyck. 1844. 11. 1845), also undescribed, seems to be the same.

Schumann refers here *Mammillaria cirrhifera*, but certainly Pfeiffer's illustration (Abbild. Beschr. Cact. 1: pl. 7) with its long, curved, radial spines and no centrals is very different; we have referred it to *Neomammillaria magnimamma*.

Illustrations: Nov. Act. Nat. Cur. 16: pl. 21; Möllers Deutsche Gärt. Zeit. 25: 475. f. 8, No. 17, as *Mammillaria mystax*; Möllers Deutsche Gärt. Zeit. 25: 475. f. 8, No. 1; Karsten and Schenck, Vegetationsbilder 1: pl. 44; Schelle, Handb. Kakteenk. 272. f. 195, as *M. mutabilis*; Schelle, Handb. Kakteenk. 273. f. 196, as *M. mutabilis longispina*.

Plate IX, figure 5, shows a plant collected by Dr. Rose at Tehuacan which flowered and fruited in Washington in 1907. Figure 89 is from a photograph of a potted plant obtained by Dr. Rose at Tehuacan in 1905.

36. Neomammillaria petterssonii (Hildmann).

Mammillaria petterssonii Hildmann, Deutsche Garten-Zeitung 1886: 185.   1886.
Mammillaria heeseana McDowell, Monatsschr. Kakteenk. 6: 125.   1896.

Plants rather large for this genus, cylindric, 2 dm. high or more, very spiny; tubercles arranged in 13 or 21 spirals, terete, setose in their axils; radial spines 10 to 12, white, with black tips; central spines 4, the longer ones 4.5 cm. long; flowers unknown; fruit small, naked, oblong.

Type locality: Mexico.
Distribution: Guanajuato, Mexico.

We have followed Schumann in uniting *Mammillaria petterssonii* and *M. heeseana* but have selected the older name.

Dr. Rose collected this plant in Guanajuato in 1889 (No. 4846) and Dr. Safford obtained it there a few years later.

Mammillaria heeseana brevispina and *M. heeseana longispina* are two varieties listed by Schelle.

Illustrations: Ann. Rep. Smiths. Inst. 1908: pl. 7; Schelle, Handb. Kakteenk. 265. f. 188, as *Mammillaria heeseana*; Blanc, Cacti 70. f. 1350, as *M. krameri* (this is the same figure as that used by Schelle as *M. heeseana*); Cact. Journ. 1: pl. for March; Blanc, Cacti 73. No. 1460; Deutsche Garten-Zeitung 1886: 186. f. 45, as M. *petterssonii*.*

Figure 90 is a reproduction of the first illustration cited above.

37. Neomammillaria eichlamii (Quehl).

Mammillaria eichlamii Quehl, Monatsschr. Kakteenk. 18: 65.   1898.

Solitary or growing in large clumps of 25 or more, but loosely held together; plant-body cylindric, 6 to 15 cm. long; tubercles yellowish green, very milky, only slightly angled; axils filled with dense yellow (sometimes whitish) wool and longer white bristles; radial spines 7 or 8, ascending, whitish with brown tips; central spines usually 1, rarely 2, stouter, darker colored than the radials; spine-areoles when young filled with short yellow wool, in age glabrate; flower-buds covered with long wool; outer perianth-segments narrow, acuminate, with a dark red stripe down the center, otherwise cream-colored, slightly ciliate; inner perianth-segments narrowly lanceolate, acuminate, entire, cream-colored to light lemon-yellow; style longer than the stamens, pale; stigma-lobes linear, 4 to 6, yellow, obtuse.

Type locality: Guatemala.
Distribution: Guatemala and Honduras.

This plant differs from the other Guatemalan species in the yellow wool in the axils of the tubercles and in the areoles.

Our first knowledge of this species came from a photograph and living and herbarium material collected by Dr. William R. Maxon in Guatemala in 1905. In 1908 Quehl described it as new from specimens sent by F. Eichlam; the plant since then has been common in cultivation. It flowered first in Washington, December 1909.

---

* This name appears as *M. petersonii* in Blanc and Schumann.

The plant is named for Federico Eichlam (1862–1911), an enthusiastic cactus collector who made very valuable discoveries in Guatemala. He published a cactus list in 1911 (Kakteen-Verzeichnis Abgeschlosen Ende 1910).

Illustrations: Monatsschr. Kakteenk. **19**: 7; Möllers Deutsche Gärt. Zeit. **25:** 475. f. 8, No. 14, as *Mammillaria eichlamii.*

Figure 91 is from a photograph of a plant collected in Guatemala by F. Eichlam in 1908.

38. Neomammillaria karwinskiana (Martius).

> *Mammillaria karwinskiana* Martius, Nov. Act. Nat. Cur. **16:** 335. 1832.
> (?) *Mammillaria fischeri* Pfeiffer, Allg. Gartenz. **4:** 257. 1836.
> *Mammillaria centrispina* Pfeiffer, Allg. Gartenz. **4:** 258. 1836.
> *Mammillaria karwinskiana flavescens* Zuccarini in Pfeiffer, Enum. Cact. 19. 1837.
> (?) *Mammillaria virens* Scheidweiler, Allg. Gartenz. **9:** 43. 1841.
> *Mammillaria karwinskiana virens* Salm-Dyck, Cact. Hort. Dyck. 1844. 10. 1845.
> *Mammillaria karwinskiana centrispina* Salm-Dyck, Cact. Hort. Dyck. 1844. 10. 1845.
> *Cactus centrispinus* Kuntze, Rev. Gen. Pl. **1:** 260. 1891.
> *Cactus fischeri* Kuntze, Rev. Gen. Pl. **1:** 260. 1891.
> *Cactus karwinskianus* Kuntze, Rev. Gen. Pl. **1:** 260. 1891.
> *Cactus virens* Kuntze, Rev. Gen. Pl. **1:** 261. 1891.

FIG. 91.—Neomammillaria eichlamii.   FIG. 92.—Neomammillaria karwinskiana.

Globose to cylindric, somewhat flattened above; tubercles terete, yielding milk when pricked; axils very woolly and with long conspicuous white or brown-tipped bristles, much longer than the tubercles; spines 4, 5, or 6, all radial, sometimes one nearer the center than the others, nearly equal, short, brown or blackish at the tips or throughout; flowers nearly 2 cm. long, the scales and outer perianth-segments narrow, reddish except at the margins, ciliate; inner perianth-segments broader, cream-colored, not ciliate, mucronate-tipped; stamens cream-colored, much shorter than the inner perianth-segments; style a little longer than the stamens; stigma-lobes 5, cream-colored; fruit 15 mm. long, red; seeds brown.

Type locality: Mexico.

Distribution: Oaxaca, Mexico.

This species is near *Neomammillaria mystax* but the spines are usually radial, short, and nearly equal. Specimens sent to Washington in 1918 had some of the lowermost spines much elongated and curved backward, sometimes 2.5 cm. long.

The plant flowers readily in cultivation. Professor C. Conzatti has repeatedly sent it to us from Oaxaca.

Illustrations: Nov. Act. Nat. Cur. **16:** pl. 22; Ann. Rep. Smiths. Inst. **1908:** pl. 14, f. 4, as *Mammillaria karwinskiana.*

Plate XI, figure 2, shows a plant collected by Dr. Rose in Oaxaca in 1906, which flowered in Washington, April 16, 1907; plate IX, figure 2, shows a plant collected by B. P. Reko also in Oaxaca, which fruited in the New York Botanical Garden in 1918. Figure 92 is from a photograph of a plant collected by Dr. Rose in Oaxaca in 1906.

Related to this species is the following:

MAMMILLARIA KNIPPELIANA Quehl, Monatsschr. Kakteenk. **17:** 59. 1907.

Stem solitary, about 7 cm. high by 6 cm. in diameter, slightly depressed at apex; tubercles when young pyramidal, 4-sided, 8 mm. long, their axils setose; areoles circular, at first white-woolly, soon glabrate; spines usually 6, up to 6 cm. long, whitish with blood-red or brown tips, sometimes accompanied with smaller spines; flowers and native country unknown.

FIG. 93.—Neomammillaria standleyi.   FIG. 94.—Neomammillaria parkinsonii.

39. Neomammillaria praelii (Mühlenpfordt).

Mammillaria praelii Mühlenpfordt, Allg. Gartenz. **14:** 372. 1846.
Mammillaria viridis praelii Salm-Dyck, Cact. Hort. Dyck. 1849. 16. 1850.
Mammillaria viridis Salm-Dyck, Cact. Hort. Dyck. 1849. 116. 1850.
Mammillaria inclinis Lemaire, Illustr. Hort. **5:** Misc. 9. 1858.
Cactus praelii Kuntze, Rev. Gen. Pl. **1:** 261. 1891.
Cactus viridis Kuntze, Rev. Gen. Pl. **1:** 261. 1891.

Globose, light green, sunken at the apex; axils of the tubercles lanate and setose; tubercles somewhat 4-angled; spine-areoles villous; spines 4, radial, forming a cross, the uppermost and lowermost elongated; flowers and fruit unknown.

Type locality: Guatemala.
Distribution: Guatemala.

We do not know this species but we are following previous authors in our classification of it. When flowers and fruit become known this may be subject to modification. Until recently it and *Neomammillaria woburnensis* were the only species of this genus known from Guatemala; neither was known in cultivation. Through the efforts of Dr. William R. Maxon, Mr. F. Eichlam, Professor Kellermann, and others, much material has been collected, new species discovered, and *N. woburnensis* rediscovered, but not *N. praelii.*

M. E. Eaton del. 1, 4
E. I. Schutt del. 2
D. G. Passmore del. 8

1. Flowering plant of *Neomammillaria magnimamma*.
2. Flowering plant of *Neomammillaria karwinskiana*.
3. Fruiting plant of *Neomammillaria gigantea*.
4. Fruiting plant of *Neobesseya missouriensis*.

Schumann described the plant in some detail, but apparently confused it with another species, possibly *Mammillaria karwinskiana*, inasmuch as he reported it from Oaxaca as well as from Guatemala. He referred here as a synonym *M. viridis* Salm-Dyck (Cact. Hort. Dyck. 1849. 16. 1850), which may be the Mexican element.

40. Neomammillaria standleyi sp. nov.

Plants usually solitary, nearly globular, often 10 cm. in diameter, pale green, densely covered with spines; axils of tubercles containing white bristles, the flowering and fruiting ones filled with dense white wool; radial spines about 16, slightly spreading, white except the dark tips; central spines 4, longer and stouter than the radials, porrect, reddish brown; flowers rather small, about 12 mm. long, purplish; inner perianth-segments oblong, entire; filaments pale; stigma-lobes green; fruit scarlet, 12 to 16 mm. long; seeds brownish.

Collected by Rose, Standley, and Russell on rocks in the Sierra de Alamos, Sonora, Mexico, March 14, 1910 (No. 12849).

It is common in dry stony places above Alamos, where both living and herbarium specimens were obtained, and is an attractive plant flowering freely in cultivation.

The plant is named for Paul C. Standley of the U. S. National Museum.

Figure 93 is from a photograph of the type specimen which flowered in Washington.

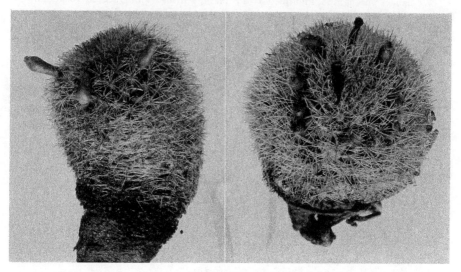

Fig. 95.—Neomammillaria evermanniana.

41. Neomammillaria evermanniana sp. nov.

Globose to elongate-turbinate, 5 to 7 cm. in diameter, lactiferous; tubercles closely set, terete, nearly hidden under the numerous slender spines; axils of tubercles at first very woolly and setose; spines white except at tip and there brown; radial spines 12 to 15; central spines 3, erect or nearly so; fruit red, about 1 cm. long; seeds brown.

Collected by Ivan M. Johnston on Cerralbo Island, Gulf of California, 1921 (No. 4058)· Mr. Johnston writes of it as follows:

"I found it growing wedged in narrow dirt-filled cracks on the canyon side of the island. It is quite common on this island, usually growing singly, but one cespitose mass with 19 unequal heads was observed."

The species is named for Dr. Barton W. Evermann, Director of the Museum of the California Academy of Sciences, who organized the scientific expedition to the Gulf of California in 1921, which obtained this as well as many other new and rare plants.

Related to this species, but perhaps distinct from it, is Johnston's No. 3121 from Nolasco Island, Gulf of California. It has fewer spines (about 10 radials and 1 or 2 centrals).

Figure 95 is from a photograph of plants from the type collection.

42. Neomammillaria parkinsonii (Ehrenberg).

Mammillaria parkinsonii Ehrenberg, Linnaea 14: 375. 1840.
Cactus parkinsonii Kuntze, Rev. Gen. Pl. 1: 261. 1891.

Cespitose, somewhat depressed to cylindric, 15 cm. high, 7.5 cm. in diameter, globose, glaucous, green; axils of tubercles lanate and setose; tubercles milky, short, conic; radial spines numerous (20 or more), setaceous, short, white; central spines 2 or sometimes 4 or 5, brownish at tip; flowers surrounded by a mass of wool, small, yellowish; inner perianth-segments apiculate; stigma-lobes elongated; fruit clavate, scarlet, 1 cm. long; seeds brown.

Type locality: At San Onofre in the Mineral del Doctor, Mexico.

Distribution: Central Mexico.

We have a photograph, identified as this plant, sent us by L. Quehl in 1921, and also specimens which are like this photograph, collected by Dr. Rose near Higuerillas, Querétaro, Mexico, in 1905 (No. 9798).

The plant was named for John Parkinson, at one time British Consul-General in Mexico, who died in Paris, April 3, 1847.

Mammillaria parkinsonii rubra (Förster, Handb. Cact. 196. 1846) is only a name.

Mammillaria parkinsonii waltonii we do not know, although it is frequently referred to in cactus literature. Haage and Schmidt offer it for sale in their catalogue (1920) under the name of *M. waltonii* Quehl.

Illustrations: Cact. Journ. 1: pl. for March, as *Mammillaria waltonii*; Gartenwelt 14: 232; Möllers Deutsche Gärt. Zeit. 25: 475. f. 8, No. 15; Rother, Praktischer Leitfaden Kakteen 39, as *M. parkinsonii*.

Figure 94 is from a photograph sent by L. Quehl.

43. Neomammillaria geminispina (Haworth).

Mammillaria geminispina Haworth in Gillies, Phil. Mag. 63: 42. 1824.
Mammillaria bicolor Lehmann, Samen. Hamb. Gartz. 7. 1830.
Mammillaria nivea Wendland in Pfeiffer, Enum. Cact. 27. 1837.
Mammillaria daedalea Scheidweiler, Hort. Belge 4: 16. 1837.
Mammillaria toaldoae Lehmann, Linnaea 12: 13. 1838.
Mammillaria eburnea Miquel, Linnaea 12: 14. 1838.
Mammillaria nivea daedalea Lemaire, Cact. Gen. Nov. Sp. 101. 1839.
Mammillaria nobilis Pfeiffer, Allg. Gartenz. 8: 282. 1840.
Mammillaria bicolor longispina Salm-Dyck, Cact. Hort. Dyck. 1844. 6. 1845.
Mammillaria bicolor cristata Salm-Dyck, Cact. Hort. Dyck. 1844. 6. 1845.
Mammillaria bicolor nobilis Förster, Handb. Cact. 198. 1846.
Cactus geminispinus Kuntze, Rev. Gen. Pl. 1: 260. 1891.
Cactus niveus Kuntze, Rev. Gen. Pl. 1: 261. 1891.
Cactus nobilis Kuntze, Rev. Gen. Pl. 1: 261. 1891. Not Lamarck, 1783.
Mammillaria bicolor nivea Schumann, Gesamtb. Kakteen 569. 1898.

Cespitose, or single in cultivation, cylindric, somewhat glaucous; axils woolly; tubercles terete, conic; radial spines numerous (16 to 20), very short, setaceous, white; central spines 2 to 4, stouter and longer than the radials, about 25 mm. long, black-tipped; flowers dark red; inner perianth-segments oblong, obtuse, serrate.

Type locality: Mexico.

Distribution: North-central Mexico.

Mammillaria daedalea, which is referred here by Schumann, is based on an abnormal specimen which has elongated, contorted stems and looks very unlike the typical plant. Scheidweiler illustrated his species.

Mammillaria nivea cristata Salm-Dyck (Walpers, Repert. Bot. **2**: 270. 1843) is only a name. *M. nivea wendlei* Pfeiffer (Labouret, Monogr. Cact. 57. 1853) was given as a synonym of *M. bicolor*.

To this relationship we would refer the plant which has long been known in collections under the name of *Mammillaria potosina** and *M. potosina* var. *longispina*. It resembles *M. celsiana* in the spines, but the tubercles are milky and the stem is more elongated. We have seen the following illustration: Möllers Deutsche Gärt. Zeit. **25**: 475. f. 8, No. 9, as *M. potosina*.

De Candolle (Prodr. **3**: 459. 1828) referred here *Cactus columnaris* Mociño and Sessé.

FIG. 96.—Neomammillaria collinsii.   FIG. 97.—Neomammillaria geminispina.

Illustrations: Wiener Ill. Gart. Zeit. **11**: pl. 3, in part, as *Mammillaria nobilis*; Hort. Belge **4**: pl. 1, as *M. daedalea*; Möllers Deutsche Gärt. Zeit. **25**: 475. f. 8, No. 4, as *M. bicolor nobilis*; Cact. Journ. **1**: pl. for March, as *M. nivea cristata*; Cact. Journ. **1**: pl. for March, as *M. nivea longispina*; Pfeiffer and Otto, Abbild. Beschr. Cact. **1**: pl. 3; De Laet, Cat. Gén. f. 50, No. 8; Wiener Ill. Gart. Zeit. **29**: f. 22, No. 8; Knippel, Kakteen pl. 19, as *M. bicolor*.

Plate v, figure 3, shows a flowering plant sent by Carl Ackerman which flowered in the New York Botanical Garden, October 9, 1920; plate VIII, figure 5, shows a plant which flowered in the New York Botanical Garden, November 11, 1911. Figure 97 is from a photograph by Ernest Braunton showing a plant grown in southern California.

44. Neomammillaria pyrrhocephala (Scheidweiler).

Mammillaria pyrrhocephala Scheidweiler, Allg. Gartenz. **9**: 42. 1841.
Mammillaria mallettiana Cels, Portef. Hort. **2**: 222.
Mammillaria senckei† Förster, Handb. Cact. 227. 1846.
Mammillaria pyrrhocephala donkelaeri Salm-Dyck, Cact. Hort. Dyck. 1849. 17, 121. 1850.
Cactus pyrrhocephalus Kuntze, Rev. Gen. Pl. **1**: 261. 1891.

Cylindric; axils lanate and setose; tubercles angled, green or subglaucous; areoles bearing yellowish wool; spines all black when young, when old becoming gray below; radial spines 6, spreading, the upper ones a little longer; central spines single, erect; flowers red.

Type locality: Real del Monte, Mexico.
Distribution: Hidalgo and, perhaps, Oaxaca.

---

* This name is sometimes credited to Rebut (Möllers Deutsche Gärt. Zeit. **25**: 475: 1910) but if he published it we are unaware of it.
† This was originally written *M. senkii*, although the plant was named for F. Senke of Leipzig.

We have followed Schumann and others who refer this species also to Oaxaca but the plants from that state may represent more than one species. In fact, the plant figured in Blühende Kakteen we have described as new (see No. 50), while the one illustrated by Mr. H. H. Thompson is like others sent by Dr. Reko and Professor Conzatti, which we have referred here.

Illustration: Thompson, U. S. Dept. Agr. Bur. Pl. Ind. Bull. **262:** pl. 2, f. 2, as *Mammillaria pyrrhocephala.*

Figure 100 is from a photograph of the plant sent to Washington by Dr. Reko from Oaxaca in 1919.

45. Neomammillaria woburnensis (Scheer).

Mammillaria woburnensis* Scheer, Lond. Journ. Bot. **4:** 136. 1845.
Cactus woburnensis Kuntze, Rev. Gen. Pl. **1:** 261. 1891.
Mammillaria chapinensis Eichlam and Quehl, Monatsschr. Kakteenk. **19:** 1. 1909.

FIG. 98.—Neomammillaria woburnensis.　　　FIG. 99.—Neomammillaria chinocephala.

Growing in clumps, giving off new plants from all parts of the body, globose to cylindric, dull green, milky; tubercles angled, setose and woolly in their axils; radial spines 5 to 9, yellowish or white; central spines 1 to 8, often long, reddish or yellow; flowers yellow, small, about 1 cm. long; fruit red, clavate, 18 to 25 mm. long; seeds minute, brown.

Type locality: Guatemala.

Distribution: Guatemala.

For a long time little was known about this plant, but a few years ago it was discovered in abundance by Wm. R. Maxon (1905) and by F. Eichlam (1908). It was given a new name, *Mammillaria chapinensis*, under which it is to be found in most collections.

The plant was described by Frederick Scheer from a barren specimen in the Royal Botanical Garden at Kew, sent from Guatemala. It was named for Woburn Abbey, where there was once a very large collection of cacti under the care of James Forbes.

Illustrations: Monatsschr. Kakteenk. **24:** 87; Succulenta **4:** 40; Möllers Deutsche Gärt. Zeit. **25:** 475. f. 8, No. 12, as *Mammillaria chapinensis.*

Figure 98 is from a photograph of a plant sent to Washington by F. Eichlam in 1908.

---

* This name was originally printed by Scheer as *Mamillaria voburnensis.*

46. Neomammillaria collinsii sp. nov.

Plants becoming large clumps, the individuals globose, 4 cm. in diameter; tubercles terete, milky, green, but becoming bronzed or even a deep purple; axils of tubercles both lanate and setose; radial spines usually 7, pale yellowish below, with dark brown or blackish tips, subequal, 5 to 7 mm. long; central spine 1, similar to or a little longer and usually darker than the radials; flowers 12 to 15 mm. long; outer perianth-segments reddish with a yellowish margin, ciliate; inner perianth-segments lighter, entire, acuminate; fruit clavate, 1.5 to 2 cm. long, deep red; seeds brownish.

Collected by G. N. Collins at San Gerónimo, near Tehuantepec, Mexico, December 1906, and flowered in Washington, July and August 1909, type, and near the same locality by A. Groeschner, February 1923.

Figures 96 and 103 are from photographs showing the type plant in flower and fruit.

47. Neomammillaria chinocephala (J. A. Purpus).

Mammillaria chinocephala J. A. Purpus, Monatsschr. Kakteenk. 16: 41. 1906.

Plant-body globose, sometimes 8 cm. in diameter, almost hidden by the white spines; tubercles very milky; axils of tubercles densely filled with white wool and numerous hair-like bristles; tubercles low; radial spines 35 to 40, somewhat pectinate, spreading; central spines 2 to 7, more or less divergent, much stouter than the radials, rigid, white with brownish tips; flowers 1 cm. long, rose-red; fruit clavate, red; seeds small, brown.

Type locality: Sierra de Parras, Coahuila, Mexico.

Distribution: Highlands of central Mexico.

This species is common in collections, both living and dried, and it is surprising that it remained so long undescribed. It was distributed by Pringle in 1890 as *Mammillaria acanthophlegma*. It resembles very much a large plant of *Mammillaria elegans*, but the tubercles are milky and bear setae in their axils.

Illustrations: Monatsschr. Kakteenk. 16: 43; 20: 46, as *Mammillaria chinocephala*.

Figure 99 is from a photograph of a plant collected by Dr. Purpus at Minas de San Rafael, Mexico, in 1910.

48. Neomammillaria tenampensis sp. nov.

Globose, light green, 5 to 6 cm. in diameter; tubercles 6 to 7 mm. long, 4-sided, milky, pointed; axils of upper tubercles naked, but those producing flowers filled with yellow wool and numerous yellow bristles, while in the older axils the wool disappears and the bristles become white; spines 4 to 6, brownish with dark tips, ascending, surrounded at base by 8 to 10 small white bristles; wool in young spine-areoles yellowish; outermost perianth-segments small, brownish, the outer ones lanceolate, acuminate, similar to the inner ones, all ciliate; inner perianth-segments reddish purple, 8 to 10 mm. long, lanceolate, apiculate, denticulate; stamens much shorter than the perianth-segments; filaments pale below, purplish above; style reddish; stigma-lobes 4 or 5.

Collected by C. A. Purpus in the Barranca de Tenampa, Mexico, in 1909 and flowered in Washington in November 1910.

Figure 102 is from a photograph of the type specimen.

49. Neomammillaria polygona (Salm-Dyck).

Mammillaria polygona Salm-Dyck, Cact. Hort. Dyck, 1849. 120. 1850.
Cactus polygonus Kuntze, Rev. Gen. Pl. 1: 261. 1891. Not Lamarck, 1783.

Subclavate, 10 cm. high, simple; axils of tubercles lanate and setose; tubercles 4-angled; radial spines about 8, 2 or 3 upper ones minute, the 4 lateral ones and the lowermost one longer; central spines 2, stout, brownish at tip, often long and recurved; flowers pale rose-colored; stigma-lobes 5 or 6, linear.

Type locality: Not cited.

Distribution: Mexico, according to Labouret.

Schumann lists this species among those unknown to him. Rümpler refers it to *Mammillaria subpolyedra*, but it must be related more nearly to *M. polyedra*, with which it was compared by Salm-Dyck. We know it only from descriptions.

Mammillaria polyedra spinosior Salm-Dyck (Cact. Hort. Dyck. 1849. 17. 1850) is usually referred here, but was never described.

Related to this species is the following:

MAMMILLARIA ECHINOPS Scheidweiler, Hort. Belge **5**: 95. 1838.

Simple, globose or a little broader than high, 8 cm. in diameter, lactiferous; tubercles ovoid, light green, somewhat 4-angled, lanate and setose in their axils; radial spines 12 or 13, the upper three much shorter, setose, the others about equal; central spines 4, stout when young, white, with rosy brown tips, these black in age; flowers not known; fruit red, clavate, 8 mm. long.

Type locality: Mexico.

We have not been able to associate this description or illustration with any species which we know. The author believed that it was related to *Mammillaria polyedra*. The setae in the axils of the tubercles suggest this relationship, but we believe that it is very distinct from that species.

The original description seems to have been unknown to the compilers of the Index Kewensis and to Schumann, for they refer the name to Förster's Handbuch, where it is used as a synonym of another species. Förster, followed by the Index Kewensis, refers it as a synonym of *Mammillaria oothele*, which is a very different plant if we can judge from the description.

Illustration: Hort. Belge **5**: pl. 5.

50. Neomammillaria confusa sp. nov.

At first solitary, becoming cespitose, globose to short-cylindric, deep green; axils densely white-woolly and setose; tubercles short, a little flattened, 4-angled, pointed; spines 4 to 6, all radial, ascending, at first yellowish with brown tips, in age white below, 2 to 3 mm. long; flowers yellow, small, about 8 mm. long, opening for 2 or 3 successive days; outer perianth-segments ovate, ciliate, with a black tip; inner perianth-segments spreading, acute; filaments and style yellowish white; stigma-lobes 6, greenish yellow.

In 1912 Dr. Rose obtained a plant from W. Mundt near Berlin which flowered in the New York Botanical Garden in April 1914 and in 1918 and which we have designated as the type. It is not known in the wild state, but is doubtless from Mexico.

This is the plant which Schumann described and figured as *Mammillaria pyrrhocephala*, but it does not accord with the original description.

Illustration: Blühende Kakteen **1**: pl. 20, as *Mammillaria pyrrhocephala*.

Plate v, figure 2, shows the type plant.

51. Neomammillaria villifera (Otto).

Mammillaria villifera Otto in Pfeiffer, Enum. Cact. 18. 1837.
Cactus villifer Kuntze, Rev. Gen. Pl. **1**: 261. 1891.
Mammillaria carnea villifera Gürke, Blühende Kakteen **1**: under pl. 60. 1905.

Subglobose, proliferous; axils lanate and setose; tubercles angled; areoles at first lanate, in age naked; spines 4, rigid, straight, the lowest one longer (8 mm. long), at first purplish, in age black; flowers pale rose-colored; inner perianth-sgements 14, acute; stigma-lobes 4 or 5.

Type locality: Mexico.
Distribution: Mexico, but range not known.

The species is often referred to *Mammillaria carnea*, but the axils are setose.

52. Neomammillaria polyedra (Martius).

Mammillaria polyedra Martius, Nov. Act. Nat. Cur. **16**: 326. 1832.
Mammillaria polytricha Salm-Dyck, Allg. Gartenz. **10**: 289. 1842.
Mammillaria polytricha hexacantha Salm-Dyck, Allg. Gartenz. **10**: 289. 1842.
Mammillaria polytricha tetracantha Salm-Dyck, Allg. Gartenz. **10**: 290. 1842.
Mammillaria polyedra laevior Salm-Dyck in Labouret, Monogr. Cact. 105. 1853.
Mammillaria polyedra scleracantha Labouret, Monogr. Cact. 105. 1853.
Cactus polyedrus Kuntze, Rev. Gen. Pl. **1**: 261. 1891.
Cactus polytrichus Kuntze, Rev. Gen. Pl. **1**: 261. 1891.

Solitary, cylindric or somewhat thicker above; axils of tubercles setose; tubercles 12 mm. long, flattened dorsally, angled, pointed; spines 4, ascending, short, grayish with purplish tips; flowers inconspicuous, reddish; inner perianth-segments short-acuminate; anthers white; style white, longer than the stamens; stigma-lobes 8, greenish; fruit unknown.

Type locality: Near Oaxaca, Mexico.

Distribution: Southern Mexico.

This species was collected by Baron Karwinsky near Oaxaca City, about 1832. It has been reported over a large area of central Mexico, but is doubtless much more restricted in range. One small specimen from near the type locality was sent to Washington in 1909.

FIG. 100.—Neomammillaria pyrrhocephala.   FIG. 101.—Neomammillaria polyedra.

Mammillaria anisacantha Hortus first appeared as a synonym of *M. polyedra anisacantha* Salm-Dyck (Cact. Hort. Dyck. 1844. 11. 1845) and then as a synonym of *M. polyedra laevior* Salm-Dyck (Cact. Hort. Dyck. 1849. 17. 1850); neither of the varieties was here described, but the latter was briefly characterized by Labouret. *Mammillaria scleracantha* is cited from Monville's Catalogue of 1846 but we have not seen this publication; it does occur as a synonym of *M. polyedra scleracantha* in Labouret's Monograph, p. 105.

Illustrations: Martius, Nov. Act. Nat. Cur. **16:** pl. 18; Blühende Kakteen **2:** pl. 112; Schelle, Handb. Kakteenk. 271. f. 194, as *Mammillaria polyedra*.

Plate XII, figure 5, shows the plant sent from the Berlin Botanical Garden in 1914 which flowered in the New York Botanical Garden on April 1, 1918. Figure 101 shows the type plant, being a reproduction of the first illustration cited above.

53. Neomammillaria conzattii sp. nov.

Short-cylindric, 8 cm. high, sometimes branched at apex, dark green, very milky; axils of young tubercles bearing abundant white wool and conspicuous white bristles; tubercles short, 4 to 5 mm. long, somewhat angled; young spine-areoles woolly; spines 4 or 5, all radial, somewhat spreading, brownish, the tips usually darker than the bases; flowers opening in bright sunlight, white, campanulate, sometimes tinged with red, about 2 cm. long, the segments somewhat spreading, narrowly oblong, the outer ones serrulate, apiculate; style pale green; stigma-lobes 3, white.

Collected by C. Conzatti on Cerro San Felipe, Oaxaca, in 1907 and flowered in 1913 (type); collected again in 1921 (No. 4140) and flowered in April 1922.

Figure 104 is from a photograph of the plant collected by C. Conzatti in 1921.

54. Neomammillaria napina (Purpus).

Mammillaria napina Purpus, Monatsschr. Kakteenk. **22**: 161.   1912.

Roots thick, but when in a cluster of 3 or 4 somewhat spindle-shaped; plants globose, 4 to 6 cm. in diameter; tubercles low, terete in section, not at all milky; spines all radial, 10 to 12, pectinate, white or yellowish, spreading and interlacing; flowers unknown.

Type locality: Mountains west of Tehuacán, Mexico.
Distribution: Southern Mexico.

The plant was collected by C. A. Purpus in 1911.   In 1901 Dr. Rose collected near Tehuacán three small plants which we now believe are to be referred here; these differ from the type plant chiefly in having usually one porrect central spine 5 to 8 mm. long. Some of the spine-clusters have no central spines and then they look very much like those of *Neomammillaria napina*.   Dr. Rose's plants were globose when collected but now are cylindric, and after 20 years are less than 6 cm. high; they have never flowered.

Illustration: Monatsschr. Kakteenk. **23**: 123, as *Mammillaria napina*.

FIG. 102.—Neomammillaria tenampensis.   FIG. 103.—Neomammillaria collinsii.

55. Neomammillaria lanata sp. nov.

Small, short-cylindric; tubercles short, 2 to 4 mm. long; spine-areoles short-elliptic; spines 12 to 14, all radial, widely spreading, white except the brown bases; flowering areoles very woolly, the young flowers surrounded by a mass of long white hairs; flowers very small, 6 to 7 mm. long, red; inner perianth-segments about 15, oblong, obtuse or acutish, spreading above; stigma-lobes 3, short, obtuse.

Collected by C. A. Purpus near Río de Santa Luisa, Mexico, in 1907 and since grown in Washington.

Figure 105 is from a photograph of the type specimen.

56. Neomammillaria kewensis (Salm-Dyck).

Mammillaria kewensis Salm-Dyck, Cact. Hort. Dyck. 1849. 112.   1850.

Globose to cylindric, 3 to 4 cm. in diameter; tubercles short, terete, when young short-woolly in the axils and at the areoles; spines 5 or 6, all radial, 4 or 5 mm. long, brown with dark tips; axils of tubercles bearing crisp hairs; flowers about 15 mm. long, reddish purple; perianth-segments lanceolate, acute; stigma-lobes 5, reddish.

Type locality: Not cited.
Distribution: Doubtless Mexico.

We have had a living plant from Haage and Schmidt and one from Quehl which we have used in our description.

Salm-Dyck (Cact. Hort. Dyck. 1849. 15. 1850) mentions *Mammillaria kewensis* var. *albispina* and also *M. spectabilis* Hortus as synonyms.

This plant was named for the Royal Botanic Gardens, Kew.

Illustration: Möllers Deutsche Gärt. Zeit. **25**: 475. f. 8, No. 3, as *Mammillaria kewensis.*

Figure 106 is reproduced from a photograph sent us by L. Quehl in 1921.

57. Neomammillaria subpolyedra (Salm-Dyck).

Mammillaria subpolyedra Salm-Dyck, Hort. Dyck. 343. 1834.
Cactus subpolyedrus Kuntze, Rev. Gen. Pl. 1: 261. 1891.

Solitary, subcylindric, 10 cm. high, 6 cm. in diameter; tubercles pointed, strongly angled; axils and spine-areoles white-woolly; spines 4, at first blackish purple, becoming paler but the tips remaining purplish, the lowest one the largest; flowers 2.5 cm. broad; perianth-segments obtuse, erose, with a darker midrib; fruit red, 2.5 cm. long, pyriform, 12 mm. in diameter at apex.

FIG. 104.—Neomammillaria conzattii.　　　　FIG. 105.—Neomammillaria lanata.

Type locality: Not cited.

Distribution: According to Rümpler, Zimapán and Ixmiquilpan, Mexico.

Some of the illustrations here cited do not correspond very well with the original description. This species is listed by Schumann with those unknown to him, and it is known to us only from descriptions and illustrations.

Mammillaria polygona Zuccarini (Pfeiffer, Enum. Cact. 17. 1837) is referred here but it was never described. Salm-Dyck afterwards used the name for a very different plant.

Mammillaria jalappensis and *M. anisacantha* are referred by Pfeiffer (Enum. Cact. 17. 1837) as synonyms of *M. subpolyedra.*

Illustrations: (?) Förster, Handb. Cact. ed. 2. 357. f. 37; (?) Dict. Gard. Nicholson 4: 565. f. 39; Suppl. 518. f. 557; Watson, Cact. Cult. 176. f. 71; ed. 3. f. 48, as *Mammillaria subpolyedra.*

Figure 107 is reproduced from the illustration used in Nicholson's Dictionary.

58. Neomammillaria galeottii (Scheidweiler).

Mammillaria galeottii Scheidweiler, Hort. Belge 4: 93. 1837.
Mammillaria obconella galeottii Scheidweiler, Hort. Belge 4: 93. 1837.
Mammillaria dolichocentra galeottii Salm-Dyck in Förster, Handb. Cact. 213. 1846.
Mammillaria dolichocentra phaeacantha Labouret, Monogr. Cact. 50. 1853.

Solitary or cespitose, globose; tubercles pointed; spines 4, elongated, the upper ones erect and connivent over apex of plant, on older tubercles weak and spreading, 2.5 cm. long, pale rose to crimson.

Type locality: Mexico.

Distribution: Mexico.

We have not seen this plant, but we have examined the illustration which accompanies the original description. L. Quehl has had it in cultivation, and sent us a photograph.

This must be a very distinct species and not at all closely related to *Mammillaria dolichocentra,* to which Schumann referred it as a variety, crediting himself as the authority; the name, however, had been used by Förster in 1846. The illustrations in Förster's Handbuch der Cacteenkunde and in Nicholson's Dictionary cited below probably are not to be referred here and they certainly should not be referred to *Mammillaria dolichocentra.*

Mammillaria obscura galeottii Salm-Dyck (Förster, Handb. Cact. 213. 1846) is mentioned as a synonym of this species, but so far as we can learn it was never described.

Illustrations: Hort. Belge **4:** pl. 6; Rother, Praktischer Leitfaden Kakteen 37, as *Mammillaria galeottii;* Förster, Handb. Cact. ed. 2. 323. f. 32; Dict. Gard. Nicholson **2:** 321. f. 508, as *Mammillaria dolichocentra.*

Fig. 106.—Neomammillaria kewensis.   Fig. 107.—Neomammillaria subpolyedra.

59. Neomammillaria tetracantha (Salm-Dyck).

Mammillaria tetracantha Salm-Dyck in Pfeiffer, Enum. Cact. 18. 1837.
Mammillaria obconella Scheidweiler, Hort. Belge **4:** 93. 1837.
Mammillaria dolichocentra Lemaire, Cact. Aliq. Nov. 3. 1838.
Mammillaria dolichocentra staminea Labouret, Monogr. Cact. 50. 1853.
Cactus obconella Kuntze, Rev. Gen. Pl. **1:** 259. 1891.
Cactus dolichocentrus Kuntze, Rev. Gen. Pl. **1:** 260. 1891.
Cactus tetracanthus Kuntze, Rev. Gen. Pl. **1:** 261. 1891.
Mammillaria rigidispina Hildmann, Monatsschr. Kakteenk. **3:** 112. 1893.
Mammillaria dolichocentra brevispina Rünge, Monatsschr. Kakteenk. **3:** 112. 1893.

Nearly globular, 6 to 8 cm. in diameter; axils of tubercles with scanty persistent wool; tubercles 8 to 10 mm. long, obscurely 4-angled; areoles small, at first lanate, somewhat 4-angled; spines 4, all radial, slender, the 3 lower equal, the upper one incurved, longer, 25 mm. long, when young all yellowish white, in age grayish yellow or brown; flowers numerous from towards top of plant, small, pinkish to rose-colored; inner perianth-segments narrowly lanceolate, acuminate.

Type locality: Mexico, but no definite locality cited.

Distribution: Mexico, but range unknown.

Schumann refers here *Mammillaria longispina* Reichenbach (Suppl. Terscheck Cact. Verz.; see also Walpers, Repert. Bot. **2**: 301. 1843) and *M. obconella* Scheidweiler (Hort. Belge **4**:93.f.6.1837), but we are uncertain as to their relationship. To the former Walpers refers as a synonym *M. galeottii* Otto.

Mammillaria dolichacantha Lemaire (Förster, Handb. Cact. 213. 1846) and *M. dolichocentra picta* (Salm-Dyck, Cact. Hort. Dyck. 1844. 9. 1845) were never described.

Illustrations: Curtis's Bot. Mag. **70**: pl. 4060, as *Mammillaria tetracantha*; Cassell's Dict. Gard. **2**: 48; Karsten, Deutsche Fl. 887. f. 501, No. 2; ed. 2. **2**: 456. f. 605, No. 2; Schelle, Handb. Kakteenk. 260. f. 182; Watson, Cact. Cult. 155. f. 58; ed. 3. f. 36; Lemaire, Icon. Cact. pl. 5; Schumann, Gesamtb. Kakteen 558. f. 91; Förster, Handb. Cact. ed. 2. 322. f. 31; Gartenwelt **9**: 265; Lemaire, Cactées 37. f. 3, as *Mammillaria dolichocentra*; Rev. Hort. **1861**: 270. f. 72, as *Mammillaria*; Thomas, Zimmerkultur Kakteen 54; Monatsschr. Kakteenk. **3**: 113, as *Mammillaria rigidispina.*

Figure 108 is a reproduction of the first illustration cited above.

60. Neomammillaria elegans (De Candolle).

Mammillaria geminispina * De Candolle, Mém. Mus. Hist. Nat. Paris **17**: 30.   1828.   Not Haworth, 1824.
Mammillaria elegans De Candolle, Mém. Mus. Hist. Nat. Paris **17**: 111.   1828.
Mammillaria elegans minor De Candolle, Mém. Mus. Hist. Nat. Paris **17**: 111.   1828.
? *Mammillaria elegans globosa* De Candolle, Mém. Mus. Hist. Nat. Paris **17**: 111.   1828.
Mammillaria acanthophlegma Lehmann, Del. Sem. Hamb.   1832.
Mammillaria supertexta Martius in Pfeiffer, Enum. Cact. 25.   1837.
Mammillaria dyckiana Zuccarini in Pfeiffer, Enum. Cact. 26.   1837.
Mammillaria elegans micrantha Lemaire, Cact. Gen. Nov. Sp. 100.   1839.
Mammillaria geminispina tetracantha Lemaire, Cact. Gen. Nov. Sp. 100.   1839.
Mammillaria klugii Ehrenberg, Bot. Zeit. **2**: 834.   1844.
Mammillaria meisneri Ehrenberg, Bot. Zeit. **2**: 834.   1844.
Mammillaria kunthii Ehrenberg, Bot. Zeit. **2**: 835.   1844.
Mammillaria splendens Ehrenberg, Allg. Gartenz. **17**: 242.   1849.
Mammillaria acanthophlegma decandollii Salm-Dyck, Cact. Hort. Dyck. 1849. 9.   1850.
Mammillaria elegans klugii Salm-Dyck, Cact. Hort. Dyck. 1849. 9.   1850.
Mammillaria acanthophlegma meisneri Salm-Dyck, Cact. Hort. Dyck. 1849. 9.   1850.
Mammillaria supertexta tetracantha Salm-Dyck in Labouret, Monogr. Cact. 61.   1853.
Mammillaria acanthophlegma elegans Monville in Labouret, Monogr. Cact. 63.   1853.
Mammillaria acanthophlegma monacantha Monville in Labouret, Monogr. Cact. 63.   1853.
Mammillaria acanthophlegma leucocephala Monville in Labouret, Monogr. Cact. 63.   1853.
Mammillaria acanthophlegma abducta Monville in Labouret, Monogr. Cact. 64.   1853.
Cactus acanthophlegma Kuntze, Rev. Gen. Pl. **1**: 260.   1891.
Cactus dyckianus Kuntze, Rev. Gen. Pl. **1**: 260.   1891.
Cactus elegans Kuntze, Rev. Gen. Pl. **1**: 260.   1891.   Not Link, 1822.
Cactus kunthii Kuntze, Rev. Gen. Pl. **1**: 260.   1891.
Cactus klugii Kuntze, Rev. Gen. Pl. **1**: 260.   1891.
Cactus meissneri Kuntze, Rev. Gen. Pl. **1**: 261.   1891.
Cactus supertextus Kuntze, Rev. Gen. Pl. **1**: 261.   1891.

Simple, obovate to globose, 5 cm. in diameter, somewhat umbilicate at apex; tubercles ovate, naked in their axils, not lactiferous; spine-areoles tomentose when young; radial spines stiff, bristle-like, 25 to 30, white, spreading; central spines 1 (sometimes 2 or 3), rigid.

Type locality: Mexico.
Distribution: Central Mexico.

This species was based on Thomas Coulter's No. 48 from Mexico but no definite locality was cited.   The type was not preserved nor is there any illustration extant of the original.   De Candolle may have had more than one species before him when he drew up his description, for he described two varieties, one of which has bristles in the axils of the tubercles, which are never found in *Neomammillaria elegans* as we have treated it here.

Plants named *Mammillaria elegans* are to be found in most collections of cacti, but the name is often applied to several closely allied species.   A plant from northern Mexico, *Mammillaria chinocephala*, resembles it very much but has milky tubercles.   Other species

---

* Here De Candolle referred *Cactus columnaris* Mociño and Sessé (De Candolle, Prodr. **3**: 459. 1828), which Schumann has inadvertently taken up as *Mammillaria columnaris* Mociño and Sessé (Gesamtb. Kakteen 565. 1898).

108   THE CACTACEAE.

which have passed as *M. elegans* have recently been described as *Mammillaria pseudo-perbella* and *M. perbella*.

Mammillaria supertexta caespitosa* Monville (Salm-Dyck, Cact. Hort. Dyck. 1844. 6. 1845) is only a name; *M. supertexta compacta* Scheidweiler (Labouret, Monogr. Cact. 61. 1853) was given as a synonym of *M. supertexta tetracantha* but may not belong here.

The name *Mammillaria leucocephala* Hortus is given by Pfeiffer as a synonym of *M. acanthophlegma. M. recta* Miquel (Labouret, Monogr. Cact. 63. 1853) occurs only as a synonym for the same species.

Illustrations: Blühende Kakteen **3**: pl. 139; Cact. Journ. **1**: pl. for February; Schelle, Handb. Kakteenk. 261. f. 183; Schumann, Gesamtb. Kakteen 564. f. 92, as *Mammillaria elegans*; Mém. Mus. Hist. Nat. Paris **17**: pl. 3, as *Mammillaria geminispina*; Cact. Journ. **1**: pl. for February in part, as *Mammillaria supertexta*; Möllers Deutsche Gärt. Zeit. **25**: 475. f. 8, No. 24, as *Mammillaria dyckiana*.

Fig. 108.—Neomammillaria tetracantha.   Fig. 109.—Neomammillaria elegans.

Figure 109 is from a photograph of the plant grown in the Huntington Collection near Los Angeles, California, as this species.

Of this relationship are the following:

MAMMILLARIA CONSPICUA J. A. Purpus, Monatsschr. Kakteenk. **22**: 163. 1912.

Simple, cylindric to globose, not milky; spine-areoles small, short-elliptic, when young a little woolly, in age glabrate; radial spines 10 to 25, rigid; central spines 2, a little curved; fruit red; seeds 1 mm. long.

Type locality: Near Zapotitlán, Puebla, Mexico.
Illustration: Monatsschr. Kakteenk. **24**: 37.

MAMMILLARIA MICROTHELE Mühlenpfordt, Allg. Gartenz. **16**: 11.   1848.

Cactus bispinus* Coulter, Contr. U. S. Nat. Herb. **3**: 101. 1894.

Cespitose, many-headed; joints globose, small; tubercles when dry 6 mm. long, naked or woolly in their axils; radial spines 22 to 24, white-setiform, spreading, 2 to 4 mm. long; central spines 2, much stouter than the radials, 2 mm. long or less; flowers flesh-colored without, white within, small, only 3 to 4 mm. long when dried; fruit clavate, 10 mm. long; seeds rather large, probably black.

Type locality: Not known but supposed to be Mexico.
Distribution: Mexico.

Our description is drawn from the original, supplemented by specimens in the Engelmann Herbarium obtained from Salm-Dyck's garden in January 1857, which consist cf two packets, one containing a few spine-clusters and the other several withered flowers and nearly ripe fruits; these latter are labeled "Baumann 857." Engelmann and Coulter compare this species with *Mammillaria micromeris* but we believe that it is related to *M. elegans* and its allies.

It seems to have been described from specimens of Haage of unknown origin but supposed to be from Mexico; Coulter's reference, on the statement of Budd, that it occurs within the southern border of Pecos County, Texas, is to be doubted.

Coulter renamed *Mammillaria microthele* because of an older *Cactus microthele*. Martius used the name *M. microthele* in 1829 (Hort. Reg. Monac. 127) but without description. The names *M. brongniartii* Hortus, *M. microthele brongniartii*, and *M. compacta* Hortus (not Engelmann, 1848) have been used (Salm-Dyck, Cact. Hort. Dyck. 1849. 9. 1850) but without descriptions.

FIG. 110.—Neomammillaria pseudoperbella.　　　FIG. 111.—Neomammillaria dealbata.

61. Neomammillaria pseudoperbella (Quehl).

Mammillaria pseudoperbella Quehl, Monatsschr. Kakteenk. **19**: 188. 1909.
Mammillaria pseudoperbella rufispina Quehl, Monatsschr. Kakteenk. **26**: 94. 1916.

Solitary, or few together, globose to short-cylindric, very spiny, depressed at apex; tubercles short-cylindric; radial spines 20 to 30, setaceous, white, short; central spines 2, one erect, the other turned backwards; flowers small, purple; perianth-segments narrow-oblong, with an ovate acute tip; style longer than the filaments, pinkish; stigma-lobes 3, obtuse.

Type locality: Mexico.
Distribution: Central Mexico.
The flowers of this plant were not known when first described nor was its exact origin known. An illustration of it was given. We have also received a dead plant from Bödeker. This illustration and specimen seem to point to a species which has been frequently sent to us from Oaxaca by Conzatti, Reko, and Solís. These plants from Oaxaca normally have 2 short, stout, divergent, central spines. In one specimen sent by Professor Conzatti in 1922 the central spines are often 2 and 4, with one of the centrals more elongated and those near the top of the plant connivent.

Illustration: Monatsschr. Kakteenk. **19**: 189, as *Mammillaria pseudoperbella*.

Plate XII, figure 1, shows a plant sent by C. Conzatti from Oaxaca, in 1921. Figure 110 is from a photograph of the type specimen.

62. Neomammillaria dealbata (Dietrich).

Mammillaria dealbata Dietrich, Allg. Gartenz. **14**: 309. 1846.
Cactus dealbatus Kuntze, Rev. Gen. Pl. **1**: 260. 1891.

Globose to short-cylindric, glaucous, more or less depressed at apex but almost hidden by the many closely appressed spine-clusters; axils of tubercles and young spine-areoles densely lanate but in age glabrate; radial spines about 20, white, short, appressed; central spines 2, much stouter and longer than the radials, sometimes 1 cm. long, the upper ones often erect, white below, brown or black at tip; flowers small, carmine; fruit clavate, red; seeds brown.

Type locality: Mexico.
Distribution: Central Mexico, especially on the pedregal about the City of Mexico.

We have referred to this species a plant which is very common in the Valley of Mexico and which is known in collections as *Mammillaria peacockii*. The name, first used by Rümpler (Förster, Handb. Cact. ed. 2. 286. 1885), was given as a synonym of *Mammillaria dealbata*. It was offered for sale by Grässner as *M. elegans dealbata* (Monatsschr. Kakteenk. February 1920).

Illustration: Grässner, Haupt-Verz. Kakteen **1912**: 23, as *Mammillaria peacockii*.

Plate XII, figure 3, shows a plant from Mexico, sent to the New York Botanical Garden in 1911. Figure 111 is from a photograph of a plant sent by Dr. Reiche from the Valley of Mexico in 1922.

FIG. 112.—Neomammillaria haageana.   FIG. 113.—Neomammillaria mundtii.

63. Neomammillaria haageana (Pfeiffer).

Mammillaria haageana Pfeiffer, Allg. Gartenz. **4**: 257. 1836.
Mammillaria diacantha Haage in Steudel, Nom. ed. 2. **2**: 96. 1841. Not Lemaire, 1838.
Mammillaria haageana validior Monville in Labouret, Monogr. Cact. 54. 1853.
Cactus haageanus Kuntze, Rev. Gen. Pl. **1**: 260. 1891.

Somewhat cespitose, the individual plants globose or somewhat elongated in age; axils slightly woolly; radial spines about 20, radiating, white; central spines 2, a little longer than the radials, black; flowers small, carmine-rose.

Type locality: Mexico.
Distribution: Mexico, but range unknown.

Pfeiffer (Enum. Cact. 26. 1837) refers here *Mammillaria diacantha nigra* which Haage had listed in his Catalogue of 1836. Here Pfeiffer also refers *M. perote* (Allg. Gartenz. **4**: 257. 1836) of gardens.

Illustrations: Dict. Gard. Nicholson **2**: 321. f. 509; Cact. Journ. **1**: 165; Knippel, Kakteen f. 21; Förster, Handb. Cact. ed. 2. 284. f. 29; Watson, Cact. Cult. 163. f. 62; ed. 3. f. 39; Schelle, Handb. Kakteenk. 262. f. 184; Rümpler, Sukkulenten 201. f. 114, as *Mammillaria haageana*.

M. E. Eaton del.

1. Flowering plant of *Neomammillaria pseudoperbella*.
2. Top of flowering plant of *Neomammillaria spinosissima*.
3. Flowering plant of *Neomammillaria dealbata*.
4. Flowering plant of *Neomammillaria amoena*.
5. Flowering plant of *Neomammillaria polyedra*.
6. Flowering plant of *Neomammillaria celsiana*.

Figure 112 is reproduced from the first illustration cited above. Nicholson recorded the receipt of the plant figured by him from Haage.

64. Neomammillaria perbella (Hildmann).

Mammillaria perbella Hildmann in Schumann, Gesamtb. Kakteen 567. 1898.

Solitary or somewhat cespitose, depressed-globose, glaucous-green; tubercles short-conic, their axils lanate; radial spines 14 to 18, 1 to 1.5 mm. long, setaceous, white; central spines 2, very short (4 to 6 mm. long); flowers 9 to 10 mm. long, reddish; stigma-lobes red.

Type locality: Mexico.
Distribution: Mexico, but range unknown.
We know this species from description only; Schumann places it near *Mammillaria donatii.*

FIG. 114.—Neomammillaria donatii.   FIG. 115.—Neomammillaria collina.

65. Neomammillaria collina (J. A. Purpus).

Mammillaria collina J. A. Purpus, Monatsschr. Kakteenk. **22**: 162. 1912.

Solitary, globose, 12 to 13 cm. in diameter, somewhat depressed at apex; tubercles cylindric, 1 cm. long or less, woolly in their axils; radial spines 16 to 18, white, 4 mm. long; central spines 1 or 2, longer than the radials; flowers rose-colored, 1.5 to 2 cm. long; fruit 2 cm. long, red.

Type locality: Esperanza, Puebla, Mexico.
Distribution: Puebla, Mexico.
We refer here specimens collected near the type locality in 1912 by Dr. C. A. Purpus.
Illustrations: Monatsschr. Kakteenk. **23**: 99; Grässner, Haupt-Verz. Kakteen **1914**: 28, as *Mammillaria collina.*
Figure 115 shows a plant sent by Dr. Purpus to Washington.

66. Neomammillaria donatii (Berge).

Mammillaria donatii Berge in Schumann, Gesamtb. Kakteen Nachtr. 135. 1903.

Usually simple, stout and globose, but sometimes cespitose, glaucous-green; tubercles small, conic, naked in their axils; radial spines 16 to 18, 8 mm. long, glassy; central spines 2, yellowish black, 10 mm. long; flowers reddish, 15 mm. long; style and stigma-lobes white.

Type locality: Mexico.
Distribution: Mexico.
We do not know the exact type locality or distribution of this plant. It is now in the trade and we recently obtained a specimen from Haage and Schmidt.

Figure 114 is from a photograph of the plant received from Haage and Schmidt in 1920, referred to above.

67. Neomammillaria mundtii (Schumann).

Mammillaria mundtii Schumann, Monatsschr. Kakteenk. **13**: 141. 1903.

Solitary, so far as known, globose, 6 to 7 cm. in diameter; tubercles not milky, nearly terete, dark green, rather short and stubby, their axils naked; spine-areoles circular, somewhat lanate when young; radial spines 8 to 19, swollen at base, spreading or somewhat curved backward, 6 to 8 mm. long, brownish when young, the tips usually darker; central spines 2, a little stouter and longer than the radials, porrect; flower from toward the center of the plant, 2 cm. long.

Type locality: Not cited.

Distribution: Mexico, but known only from cultivated plants.

We know this plant from a specimen sent to Washington in 1921 by W. Mundt, in whose honor the species had been named.

Illustration: Monatsschr. Kakteenk. **13**: 142, as *Mammillaria mundtii.*

Figure 113 is a reproduction of a photograph sent us by L. Quehl in 1921.

FIG. 116.—Neomammillaria celsiana.

68. Neomammillaria celsiana (Lemaire).

Mammillaria celsiana Lemaire, Cact. Gen. Nov. Sp. 41. 1839.
Mammillaria muehlenpfordtii Förster, Allg. Gartenz. **15**: 49. 1847.
Mammillaria schaeferi Fennel, Allg. Gartenz. **15**: 66. 1847.
Mammillaria schaeferi longispina Haage, Hamb. Gartenz. **17**: 160. 1861.
Cactus muehlenpfordtii Kuntze, Rev. Gen. Pl. **1**: 260. 1891.
Cactus celsianus Kuntze, Rev. Gen. Pl. **1**: 261. 1891.
Cactus schaeferi Kuntze, Rev. Gen. Pl. **1**: 261. 1891.
(?) *Mammillaria perringii* Hildmann, Gartenwelt **10**: 250. 1906.

Plant-body subglobose, becoming cylindric, 10 to 12.5 cm. high, 7.5 cm. in diameter, deep green; axils of tubercles woolly; tubercles conic, compact; spine-areoles small, round, woolly when young; radial spines 24 to 26, about equal, white, setaceous; central spines 4 to 6, rarely 7, somewhat longer than the radials, terete, rigid, pale yellow, more or less recurved and unequal, 8 to 16 mm. long; flowers red; fruit described as green.

Type locality: Not cited.

Distribution: Southern Mexico.

In 1920 Professor Conzatti sent us two specimens from the District of Cuicatlán, Oaxaca, which we refer here; these are the only plants of this species we have seen.

According to Salm-Dyck, *Mammillaria celsiana* differs from *M. rutila* in its columnar stem and in its spines.

Schumann refers *Mammillaria perringii* to *M. celsiana*, while Hildmann claims that it is possible that the two may be distinct, but we do not have the material at hand to decide definitely.

Mammillaria lanifera Haworth (Phil. Mag. **63**: 41. 1824; *Cactus lanifer* Kuntze, Rev. Gen. Pl. **1**: 260. 1891) is referred here by Schumann; it is probably different but, if not, the name has priority over *M. celsiana*. To *M. lanifera* De Candolle (Prodr. **3**: 459. 1828) refers *Cactus canescens* Mociño and Sessé. *M. geminispina monacantha* Lemaire (Cact. Gen. Nov. Sp. 100. 1839) was supposed to be the same as *M. lanifera*. *Mammillaria polycephala* Mühlenpfordt (Allg. Gartenz. **13**: 347. 1845; *Cactus polycephalus* Kuntze, Rev. Gen. Pl. **1**: 261. 1891) was referred by Schumann to *M. elegans*, but it was described with 4 central spines. It seems to be related to *M. crucigera*, which we have tentatively referred to *M. celsiana*, which has yellow central spines, while both *M. polycephala* and *M. elegans* have white centrals.

FIGS. 117 and 118.—Neomammillaria aureiceps.

Mammillaria supertexta dichotoma (Salm-Dyck, Cact. Hort. Dyck. 1849. 9. 1850) is based on *M. polycephala*.

Mammillaria crucigera Martius (Nov. Act. Nat. Cur. **16**: 340. pl. 25, f. 2. 1832; *Cactus cruciger* Kuntze, Rev. Gen. Pl. **1**: 260. 1891) is related to this species, judging from the description, but the illustration suggests that it is a distinct species. It was collected by Karwinsky in Mexico, but he does not give a definite locality. It was unknown to Schumann.

Illustrations: Gartenwelt **10**: 250; Möllers Deutsche Gärt. Zeit. **25**: 475. f. 8, No. 29, as *Mammillaria celsiana*; Gartenwelt **10**: 250, as *Mammillaria perringii*; ? Mém. Mus. Hist. Nat. Paris **17**: pl. 4, as *Mammillaria lanifera*; ? Martius, Nov. Act. Nat. Cur. **16**: pl. 25, f. 2, as *Mammillaria crucigera*.

Plate XII, figure 6, shows a plant in the New York Botanical Garden which flowered October 16, 1911. Figure 116 is from a photograph of two plants sent by Professor Conzatti in 1920.

69. Neomammillaria aureiceps (Lemaire).

Mammillaria aureiceps Lemaire, Cact. Aliq. Nov. 8. 1838.
Mammillaria rhodantha aureiceps Salm-Dyck, Cact. Hort. Dyck. 1844. 7. 1845.
Cactus aureiceps Kuntze, Rev. Gen. Pl. 1: 260. 1891.

Globose to short-oblong, 8 to 10 cm. in diameter; tubercles short, terete in section, woolly and setose in their axils; radial spines about 20, bristle-like, white, 5 to 8 mm. long, spreading; central spines several, sometimes as many as 9, yellow, stouter and longer than the radials, 10 to 14 mm. long, somewhat spreading and a little curved inward; flowers small, dark red.

Type locality: Mexico.
Distribution: Valley of Mexico.

Our description is based on specimens recently sent us by Dr. Karl Reiche as *Mammillaria rhodantha*, under which name it usually passes. *M. rhodantha*, however, has different spines and is more strictly a mountain species.

Plate IX, figure 3, shows a plant sent from the Edinburgh Botanical Garden in 1902 as *Mammillaria rhodantha* which flowered in the New York Botanical Garden, October 15, 1912. Figures 117 and 118 give two views of this plant sent us by Dr. Reiche from the Valley of Mexico.

FIG. 119.—Neomammillaria yucatanensis.   FIG. 120.—Neomammillaria ruestii.

70. Neomammillaria yucatanensis sp. nov.

Plants in clumps of 4, erect, cylindric, not milky, 10 to 15 cm. long, 3 to 6 cm. in diameter, very spiny; tubercles conic, woolly in their axils but not setose; radial spines about 20, white, spreading, acicular; central spines 4 or rarely 5, much stouter than the radials, 6 to 8 mm. long, slightly spreading above, yellowish brown; "flowers very small, rose; fruit oblong, bright red."

Collected by George F. Gaumer at Progreso, Yucatan, Mexico, in 1918 (No. 23939) and again in 1921 (No. 24367, type).

We have not seen this species in flower or fruit but Dr. Gaumer has described them as above. He says that the plant is rare on the land side of the coastal marshes.

Figure 119 is from a photograph of the plant sent in 1921 by Dr. Gaumer.

71. Neomammillaria ruestii (Quehl).

Mammillaria ruestii Quehl, Monatsschr. Kakteenk. **15**: 173. 1905.
Mammillaria celsiana guatemalensis Eichlam, Monatsschr. Kakteenk. **19**: 59. 1909.

Cylindric, 6 to 7 cm. high, 4 to 5 cm. in diameter, light green, almost hidden by the spines; axils of tubercles more or less woolly, at least when young; flowering areoles at first quite woolly; radial spines 20 or more, white, glossy, 5 to 6 mm. long, spreading; central spines usually 4, sometimes 5, much stouter than the radials, yellow, swollen at base, ascending, 7 to 8 mm. long; flowers small, sometimes almost hidden by the spines, 8 mm. long; inner perianth-segments about 25, lanceolate, acute, pale purple, the margins almost colorless; filaments colorless below, purplish above; style pale; stigma-lobes 4, linear, elongated, reflexed; fruit clavate, red; seeds brown.

Type locality: Honduras.
Distribution: Honduras and Guatemala.

We have had the Guatemala plant under observation for 14 years and it has both flowered and fruited.

Figure 120 is from a photograph of a plant sent by Dr. A. W. Kellermann from Guatemala in 1908.

Fig. 121.—Neomammillaria pringlei.

72. Neomammillaria pringlei (Coulter).

Cactus pringlei Coulter, Contr. U. S. Nat. Herb. **3**: 109. 1894.
Mammillaria pringlei K. Brandegee, Zoe **5**: 7. 1900.

Solitary, with long fibrous roots, usually globose, but sometimes depressed or short-cylindric, 6 to 16 cm. high, 6 to 7 cm. in diameter; tubercles dull green, terete, conic, 6 to 10 mm. long; axils of tubercles woolly and setose; spines all yellow; radial spines 18 to 20, setaceous, spreading, 5 to 8 mm. long; central spines 5 to 7, much stouter and longer than the radials, more or less recurved, 2 to 2.5 cm. long, those from the upper areoles curved over the apex of the plant; flowers deep red, 8 to 10 mm. long; fruit borne in a circle near the middle of the plant, oblong, 12 to 15 mm. long; seeds small, brown.

Type locality: Cited as San Luis Potosí, but doubtless Tultenango Canyon, state of Mexico, according to Pringle, who collected the type.
Distribution: Known only from the type locality.

Dr. Rose collected living specimens from the type locality some years ago but these never flowered. In April 1921 we sent Dr. Reiche to the type locality and he obtained thirteen beautiful specimens, one of which was in fruit.

Coulter (Contr. U. S. Nat. Herb. **3**: 109. 1894) states that *Cactus pringlei* was near *Cactus rhodanthus sulphureospinus*, which was based on *M. sulphurea* Förster.

Figure 121 is from a photograph of the plants collected at Tultenango Canyon in 1921.

73. Neomammillaria cerralboa sp. nov.

Cylindric, solitary, 1 to 1.5 dm. high, 5 to 6 cm. in diameter; tubercles not milky, yellowish, terete, obtuse, closely set; spines all yellow, very much alike, about 11, one usually more central, the longer ones nearly 2 cm. long; flowers small, 1 cm. long or less, forming a circle around the plant about 3 cm. below the top.

Collected by Ivan M. Johnston on Cerralbo Island, Gulf of California, June 6, 1921 (No. 4038). The next day on the same island he collected three more plants (No. 4053) which seem to be referable here, except that two of them have hooked spines; Dr. Rose also collected on this same island (No. 16877) in 1911 specimens with hooked spines which are like Mr. Johnston's plant. Whether this plant has normally these two forms or whether the hooked-spined one is a hybrid we are unable to determine.

Figure 121*a* is a photograph of the type plant, collected by Johnston (No. 4038).

74. Neomammillaria phaeacantha (Lemaire).

Mammillaria phaeacantha Lemaire, Cact. Gen. Nov. Sp. 47. 1839.
Mammillaria nigricans Fennel, Allg. Gartenz. **15**: 66. 1847.
Cactus nigricans Kuntze, Rev. Gen. Pl. **1**: 261. 1891. Not Haworth, 1803.
Cactus phaeacanthus Kuntze, Rev. Gen. Pl. **1**: 261. 1891.

FIG. 121*a*. Neomammillaria cerralboa.

FIG. 122.—Neomammillaria phaeacantha.   FIG. 123.—Neomammillaria graessneriana.

Globose or somewhat depressed, green; axils of tubercles woolly; tubercles conic, hardly, if at all, angled; spine-areoles small, yellowish tomentose (probably so only when young); radial spines

16 to 20, white, setaceous; central spines 4, black, subulate, spreading or reflexed, the lowest one longest; flowers from upper part of plant, dark red; perianth-segments oblong, acuminate.

Type locality: Mexico.

Distribution: Mexico, but range unknown.

This species has not been recognized by recent writers, and while we have seen no specimens we believe it deserves specific rank.

Schumann refers *Mammillaria nigricans* definitely to *M. rhodantha* but, it appears to us, without justification; the Index Kewensis has referred it, we believe properly, to *M. phaeacantha.*

Mammillaria phaeacantha rigidior (Salm-Dyck, Cact. Hort. Dyck. 1844. 8. 1845) is only a name.

Illustration: Pfeiffer, Abbild. Beschr. Cact. **2**: pl. 23, as *Mammillaria nigricans.*

Figure 122 is reproduced from the illustration cited above.

75. Neomammillaria graessneriana (Bödeker).

Mammillaria graessneriana Bödeker, Monatsschr. Kakteenk. **30**: 84. 1920.

Solitary, or becoming cespitose, globose, 6 to 8 cm. in diameter, dark bluish green, somewhat depressed at apex; tubercles 4-angled, 8 mm. long, not milky, obtuse or truncate at apex, not setose in their axils; spine-areoles circular, white-woolly when young, nearly naked in age; radial spines 18 to 20, acicular, 6 to 8 mm. long, white; central spines 2 to 4, stouter than the radials, spreading, 8 mm. long, reddish brown; flowers small, somewhat distant from the apex of the plant.

Type locality: Mexico.

Distribution: Mexico, but range unknown.

Illustration: Monatsschr. Kakteenk. **30**: 85, as *Mammillaria graessneriana.*

Figure 123 is reproduced from the illustration cited above.

FIGS. 124 and 125.—Neomammillaria spinosissima.

76. Neomammillaria spinosissima (Lemaire).

Mammillaria spinosissima Lemaire, Cact. Aliq. Nov. 4. 1838.
Mammillaria polycentra Berg, Allg. Gartenz. **8**: 130. 1840.
Mammillaria auricoma Dietrich, Allg. Gartenz. **14**: 308. 1846.
Mammillaria polyacantha Ehrenberg, Allg. Gartenz. **16**: 265. 1848.
Mammillaria polyactina Ehrenberg, Allg. Gartenz. **16**: 266. 1848.
Mammillaria hepatica Ehrenberg, Allg. Gartenz. **16**: 267. 1848.
Mammillaria pomacea Ehrenberg, Allg. Gartenz. **16**: 267. 1848.
Mammillaria pulcherrima Ehrenberg, Allg. Gartenz. **17**: 249. 1849.
Mammillaria pretiosa Ehrenberg, Allg. Gartenz. **17**: 250. 1849.

Mammillaria caesia Ehrenberg, Allg. Gartenz. 17: 251. 1849.
Mammillaria mirabilis Ehrenberg, Allg. Gartenz. 17: 251. 1849.
Mammillaria pruinosa Ehrenberg, Allg. Gartenz. 17: 261. 1849.
Mammillaria seegeri Ehrenberg, Allg. Gartenz. 17: 261. 1849.
Mammillaria haseloffii Ehrenberg, Allg. Gartenz. 17: 303. 1849.
Mammillaria herrmannii Ehrenberg, Allg. Gartenz. 17: 303. 1849.
Mammillaria aurorea Ehrenberg, Allg. Gartenz. 17: 303. 1849.
Mammillaria linkeana Ehrenberg, Allg. Gartenz. 17: 308. 1849.
Mammillaria vulpina Ehrenberg, Allg. Gartenz. 17: 308. 1849.
Mammillaria eximia Ehrenberg, Allg. Gartenz. 17: 309. 1849.
Mammillaria isabellina Ehrenberg, Allg. Gartenz. 17: 309. 1849.
Mammillaria spinosissima brunnea Salm-Dyck, Cact. Hort. Dyck. 1849. 8.   1850.
Mammillaria spinosissima flavida Salm-Dyck, Cact. Hort. Dyck. 1849. 8.   1850.
Mammillaria spinosissima rubens Salm-Dyck, Cact. Hort. Dyck. 1849. 8.   1850.
Mammillaria herrmanni flavicans Salm-Dyck, Cact. Hort. Dyck. 1849. 8.   1850.
Mammillaria seegeri gracilispina Salm-Dyck, Cact. Hort. Dyck. 1849. 8.   1850.
Mammillaria seegeri pruinosa Salm-Dyck, Cact. Hort. Dyck. 1849. 8.   1850.
Mammillaria uhdeana Salm-Dyck, Cact. Hort. Dyck. 1849. 83.   1850.
Mammillaria spinosissima hepatica Labouret, Monogr. Cact. 37.   1853.
Mammillaria castaneoides Lemaire in Labouret, Monogr. Cact. 37.   1853.
Mammillaria seegeri mirabilis Labouret, Monogr. Cact. 37.   1853.
Mammillaria sanguinea Haage jr. in Regel, Act. Hort. Petrop. 8: 276.   1883.
Mammillaria poselgeriana Haage in Förster, Handb. Cact. ed. 2. 269.   1885.
Mammillaria pretiosa cristata Hildmann in Förster, Handb. Cact. ed. 2. 273.   1885.
Cactus auricomus Kuntze, Rev. Gen. Pl. 1: 260.   1891.
Cactus auroreus Kuntze, Rev. Gen. Pl. 1: 260.   1891.
Cactus eximius Kuntze, Rev. Gen. Pl. 1: 260.   1891.
Cactus isabellinus Kuntze, Rev. Gen. Pl. 1: 260.   1891.
Cactus linkeanus Kuntze, Rev. Gen. Pl. 1: 260.   1891.
Cactus mirabilis Kuntze, Rev. Gen. Pl. 1: 260.   1891.
Cactus polycentrus Kuntze, Rev. Gen. Pl. 1: 261.   1891.
Cactus pomaceus Kuntze, Rev. Gen. Pl. 1: 261.   1891.
Cactus pretiosus Kuntze, Rev. Gen. Pl. 1: 261.   1891.
Cactus pulcherrimus Kuntze, Rev. Gen. Pl. 1: 261.   1891.
Cactus spinosissimus Kuntze, Rev. Gen. Pl. 1: 261.   1891.
Cactus vulpinus Kuntze, Rev. Gen. Pl. 1: 261.   1891.
Mammillaria spinosissima sanguinea Haage in Brandegee, Cycl. Amer. Hort. Bailey 2: 976.   1900.
Mammillaria spinosissima aurorea Gürke, Blühende Kakteen 2: under pl. 71.   1905.
Mammillaria spinosissima auricoma Gürke, Blühende Kakteen 2: under pl. 71.   1905.
Mammillaria spinosissima eximia Gürke, Blühende Kakteen 2: under pl. 71.   1905.
Mammillaria spinosissima haseloffii Gürke, Blühende Kakteen 2: under pl. 71.   1905.
Mammillaria spinosissima herrmannii Gürke, Blühende Kakteen 2: under pl. 71.   1905.
Mammillaria spinosissima isabellina Gürke, Blühende Kakteen 2: under pl. 71.   1905.
Mammillaria spinosissima linkeana Gürke, Blühende Kakteen 2: under pl. 71.   1905.
Mammillaria spinosissima mirabilis Gürke, Blühende Kakteen 2: under pl. 71.   1905.
Mammillaria spinosissima pruinosa Gürke, Blühende Kakteen 2: under pl. 71.   1905.
Mammillaria spinosissima pulcherrima Gürke, Blühende Kakteen 2: under pl. 71.   1905.
Mammillaria spinosissima seegeri Gürke, Blühende Kakteen 2: under pl. 71.   1905.
Mammillaria spinosissima vulpina Gürke, Blühende Kakteen 2: under pl. 71.   1905.

Cylindric, 7 to 30 cm. long, 2.5 to 10 cm. in diameter, almost hidden under a dense covering of spines; axils of tubercles setose; tubercles very short, 2 to 3 mm. long; spines brownish to red, usually weak, hardly pungent; radial spines about 20, 1 cm. long or less; central spines 7 or 8, 2 cm. long or more; flowers from the upper part of the plant, purplish, 12 mm. long; inner perianth-segments acute; filaments much shorter than the perianth-segments, purple.

Type locality: Not cited.
Distribution: Mountains of central Mexico.

The above description is drawn from collections obtained in the high mountains between the City of Mexico and Cuernavaca. There seems to be little doubt but that they are the *M. sanguinea* Haage which Schumann refers to *M. spinosissima*.

We are disposed to refer here *Echinocactus spinosissimus* (Forbes, Journ. Hort. Tour Germ. 152. 1837). Forbes did not have much knowledge of the cacti but was the gardener of the Duke of Bedford, who sent him to the Continent of Europe in 1835, where he obtained many cacti and on his return to England published a list of them, sometimes with brief descriptions. The names had been given to him by Pfeiffer and others who were studying this family. As he published his list very promptly after his return to England many names appear there first or in the same year as in Pfeiffer's Enumeratio. *Mammillaria spinosissima* may have been in cultivation at the time of Forbes's visit to Germany, for it was published in 1838.

Illustrations: Möllers Deutsche Gärt. Zeit. **25**: 475. f. 8. No. 26, as *Mammillaria poselgeriana*; Gartenflora **32**: pl. 1111; Dict. Gard. Nicholson **2**: 322. f. 510; Förster, Handb. Cact. ed. 2. 271. f. 28; Watson, Cact. Cult. 172. f. 68; ed. 3. f. 46, as *Mammillaria sanguinea*; Möllers Deutsche Gärt. Zeit. **25**: 475. f. 8, No. 11, as *Mammillaria eximia*; Möllers Deutsche Gärt. Zeit. **25**: 475. f. 8, No. 18, as *Mammillaria spinosissima auricoma*; Balt. Cact. Journ. **2**: 150, as *M. spinosissima brunnea*; Möllers Deutsche Gärt. Zeit. **25**: 487. f. 21; Cact. Journ. **2**: 93; Blanc, Cacti 74. No. 1580; Schelle, Handb. Kakteenk. 253. f. 174; Blühende Kakteen **2**: pl. 71, as *Mammillaria spinosissima.*

Plate XII, figure 2, shows a plant collected by Dr. Rose at El Parque, Mexico, in 1906. Figure 124 is from a photograph of a plant sent to the New York Botanical Garden by Frank Weinberg in 1906 as *Cactus spinosissimus*; figure 125 is from a photograph of a plant sent by William Brockway from the mountains above the City of Mexico.

FIG. 126.—Neomammillaria densispina.   FIG. 127.—Neomammillaria nunezii.

77. Neomammillaria densispina (Coulter).

Cactus densispinus Coulter, Contr. U. S. Nat. Herb. **3**: 96. 1894.
Mammillaria pseudofuscata Quehl, Monatsschr. Kakteenk. **24**: 114. 1914.

Globose, 6 to 10 cm. in diameter, entirely hidden by the dense covering of spines; tubercles short and thick, green, not milky; radial spines 25 or more, slightly spreading, about 1 cm. long, whitish or pale yellow; central spines 5 or 6, longer than the radials, 10 to 12 mm. long, the upper half or third dark brown; flowers purple without, yellowish within, 1.5 cm. long; seeds obovate, reddish brown, 1 mm. in diameter.

Type locality: San Luis Potosí, Mexico.

Distribution: San Luis Potosí, Mexico.

We have had this plant in cultivation since 1912, specimens having been sent to Washington by Mrs. Irene Vera from San Luis Potosí. Our plant is probably a part of the type collection of Quehl's *Mammillaria pseudofuscata*, as Mrs. Vera wrote us that she had sent specimens to Germany which had been identified as *M. fuscata.* Our plant has been compared with Eschanzier's specimen from the same locality which is the type of Coulter's

Cactus densispinus and we are convinced that they are the same; Coulter's type is now in the Field Museum of Natural History.

Illustration: Monatsschr. Kakteenk. **24:** 115, as *Mammillaria pseudofuscata*.

Figure 126 shows the plant sent by Mrs. Vera from San Luis Potosí.

78. Neomammillaria nunezii sp. nov.

Globose to cylindric, 1.5 cm. long, 6 to 8 cm. in diameter; tubercles closely set, short, terete in section, setose in their axils; radial spines white, stiff, about 30, widely spreading; central spines 2 to 4, stout, 10 to 15 mm. long, brown to nearly blackish at tips; fruit 2.5 cm. long, clavate, white or tinged with pink; seeds small, brown.

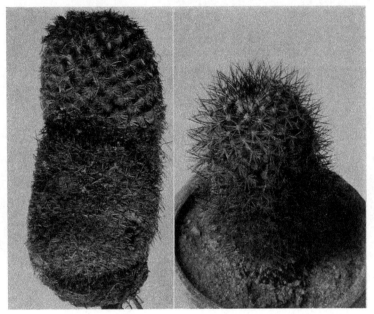

FIG. 128.—Neomammillaria nunezii.　　　　FIG. 129.—Neomammillaria rhodantha.

Collected by Professor C. Núñez at Buenavista de Cuellar, Guerrero, Mexico, in 1921 (Nos. 1, 2 and 3), and communicated to us by Octavio Solís.　This species is rather variable in habit and spines and is very unlike anything that we have heretofore studied.

Figures 127 and 128 are from photographs of the top and side of two plants of this collection.

79. Neomammillaria amoena (Hoppfer).

Mammillaria amoena Hoppfer in Salm-Dyck, Cact. Hort. Dyck. 1849. 99.　1850.

Stems robust, columnar; tubercles green, ovoid, obtuse, subglaucous; radial spines 16, slender, radiating, white; central spines 2, rigid, yellowish brown, 8 to 10 mm. long, the upper one longer and recurved; flowers appearing from axils above middle of plant, 2 cm. long; tube cone-shaped, green; outer perianth-segments somewhat brownish; inner perianth-segments with a pale-brown central stripe; margins nearly white, obtuse, entire; stamens short; filaments pale; anthers red; style pale green; stigma-lobes green, linear.

Type locality: Not cited.
Distribution: Central Mexico.

Förster's Handbuch (254. 1846) is often given as the place of publication, but while the name is found in the place cited it is without description.

Plate XII, figure 4, shows a plant which flowered in the New York Botanical Garden in 1912, sent from Cuernavaca, Mexico, by Wm. Brockway the preceding year. Figure 130 is from a photograph of a plant from the same collection which flowered in Washington.

FIG. 130.—Neomammillaria amoena.　　　　FIG. 131.—Neomammillaria plumosa.

80. Neomammillaria rhodantha (Link and Otto).

Mammillaria rhodantha Link and Otto, Icon. Pl. Rar. 51. 1829.
Mammillaria pulchra Haworth in Edwards's Bot. Reg. 16: pl. 1329. 1830.
Mammillaria fulvispina Haworth, Phil. Mag. 7: 108. 1830.
? *Mammillaria inuncta* Hoffmannsegg, Preiss-Verz. ed. 7. 23. 1833.
Mammillaria erinacea Wendland, Cact. Herrenh. 1835.
Mammillaria chrysacantha Otto in Pfeiffer, Enum. Cact. 28. 1837.
Mammillaria fuscata Pfeiffer, Enum. Cact. 28. 1837.
Mammillaria tentaculata Otto in Pfeiffer, Enum. Cact. 29. 1837.
Mammillaria rhodantha rubens Pfeiffer, Enum. Cact. 31. 1837.
Mammillaria rhodantha andreae * Otto in Pfeiffer, Enum. Cact. 31. 1837.
Mammillaria rhodantha prolifera Pfeiffer, Enum. Cact. 31. 1837.
Mammillaria rhodantha neglecta Pfeiffer, Enum. Cact. 31. 1837.
Mammillaria rhodantha wendlandii Pfeiffer, Enum. Cact. 31. 1837.
Mammillaria ruficeps Lemaire, Cact. Gen. Nov. Sp. 37. 1839.
Mammillaria odieriana Lemaire, Cact. Gen. Nov. Sp. 46. 1839.
Mammillaria pyrrhochracantha Lemaire, Cact. Gen. Nov. Sp. 51. 1839.
Mammillaria rhodantha major Monville in Lemaire, Cact. Gen. Nov. Sp. 98. 1839.
Mammillaria pfeifferi Booth in Scheidweiler, Bull. Acad. Sci. Brux. 6: 93. 1839.
Mammillaria pfeifferi altissima Scheidweiler, Bull. Acad. Sci. Brux. 6: 93. 1839.
Mammillaria pfeifferi dichotoma Scheidweiler, Bull. Acad. Sci. Brux. 6: 93. 1839.
Mammillaria pfeifferi flaviceps Scheidweiler, Bull. Acad. Sci. Brux. 6: 93. 1839.
Mammillaria pfeifferi fulvispina Scheidweiler, Bull. Acad. Sci. Brux. 6: 93. 1839.
Mammillaria pfeifferi variabilis Scheidweiler, Bull. Acad. Sci. Brux. 6: 93. 1839.
? *Mammillaria crassispina* Pfeiffer, Allg. Gartenz. 8: 406. 1840.
? *Mammillaria stenocephala* Scheidweiler, Allg. Gartenz. 9: 43. 1841.
? *Mammillaria imbricata* Wegener, Allg. Gartenz. 12: 66. 1844.
Mammillaria crassispina gracilior Salm-Dyck, Cact. Hort. Dyck. 1844. 8. 1845.
Mammillaria rhodantha centrispina Link in Förster, Handb. Cact. 198. 1846.
Mammillaria sulphurea Sencke in Förster, Handb. Cact. 200. 1846.
Mammillaria robusta Otto in Förster, Handb. Cact. 207. 1846.
Mammillaria tentaculata ruficeps Förster, Handb. Cact. 207. 1846.
Mammillaria stueberi Otto in Förster, Handb. Cact. 517. 1846.
Mammillaria fulvispina rubescens Salm-Dyck, Cact. Hort. Dyck. 1849. 10. 1850.

* *Mammillaria andreae* was used by Schumann (Gesamtb. Kakteen 598. 1898).

Mammillaria rhodantha sulphurea Salm-Dyck, Cact. Hort. Dyck. 1849. 11.   1850.
Mammillaria rhodantha ruficeps Salm-Dyck, Cact. Hort. Dyck. 1849. 11.   1850.
Mammillaria chrysacantha fuscata Salm-Dyck, Cact. Hort. Dyck. 1849. 12.   1850.
Mammillaria rhodantha rubescens Salm-Dyck, Cact. Hort. Dyck. 1849. 97.   1850.
Mammillaria odieriana rigidior Salm-Dyck, Cact. Hort. Dyck. 1849. 98.   1850.
Mammillaria lanifera Salm-Dyck, Cact. Hort. Dyck. 1849. 98.   1850.   Not Haworth, 1824.
? *Mammillaria russea* Dietrich, Allg. Gartenz. 19: 347.   1851.
Mammillaria odieriana rubra Sencke in Förster, Handb. Cact. ed. 2. 295.   1885.
Mammillaria odieriana cristata Hortus in Förster, Handb. Cact. ed. 2. 295.   1885.
Mammillaria tentaculata picta Förster, Handb. Cact. ed. 2. 309.   1885.
Mammillaria crassispina rufa Rümpler in Förster, Handb. Cact. ed. 2. 311.   1885.
Cactus chrysacanthus Kuntze, Rev. Gen. Pl. 1: 260.   1891.
Cactus crassispinus Kuntze, Rev. Gen. Pl. 1: 260.   1891.
Cactus fuscatus Kuntze, Rev. Gen. Pl. 1: 260.   1891.
Cactus odieranus Kuntze, Rev. Gen. Pl. 1: 261.   1891.
Cactus pyrrhochroacanthus Kuntze, Rev. Gen. Pl. 1: 261.   1891.
Cactus rhodanthus Kuntze, Rev. Gen. Pl. 1: 261.   1891.
Cactus ruficeps Kuntze, Rev. Gen. Pl. 1: 261.   1891.
Cactus stenocephalus Kuntze, Rev. Gen. Pl. 1: 261.   1891.
Cactus stueberi Kuntze, Rev. Gen. Pl. 1: 261.   1891.
Cactus tentaculatus Kuntze, Rev. Gen. Pl. 1: 261.   1891.
Cactus capillaris Coulter, Contr. U. S. Nat. Herb. 3: 107.   1894.
Cactus rhodanthus sulphureospinus Coulter, Contr. U. S. Nat. Herb. 3: 107.   1894.
Mammillaria rhodantha pfeifferi Schumann, Gesamtb. Kakteen 550.   1898.
Mammillaria rhodantha rubra Schumann, Gesamtb. Kakteen 550.   1898.
Mammillaria rhodantha ruberrima Schumann, Gesamtb. Kakteen 550.   1898.
Mammillaria rhodantha pyramidalis Schumann, Gesamtb. Kakteen 550.   1898.
Mammillaria rhodantha callaena Schumann, Gesamtb. Kakteen 550.   1898.
Mammillaria rhodantha crassispina Schumann, Gesamtb. Kakteen 550.   1898.
Mammillaria rhodantha droegeana Schumann, Gesamtb. Kakteen 550.   1898.
Mammillaria rhodantha chrysacantha Schumann, Gesamtb. Kakteen 550.   1898.
Mammillaria rhodantha stenocephala Schumann, Gesamtb. Kakteen 550.   1898.
Mammillaria rhodantha fuscata Schumann, Gesamtb. Kakteen 551.   1898.
Mammillaria rhodantha odieriana Schelle, Handb. Kakteenk. 257.   1907.
Mammillaria rhodantha fulvispina Schelle, Handb. Kakteenk. 257.   1907.
Mammillaria rhodantha tentaculata Hortus in Schelle, Handb. Kakteenk. 257.   1907.

Cylindric, 1 to 3 dm. long, erect, dull green; tubercles terete, somewhat narrowed toward the apex, 3 to 5 mm. long, not yielding milk when pricked; axils of tubercles sometimes bearing bristles, often naked; radial spines 15 to 20, white, 5 to 7 mm. long; central spines 4 to 6, reddish brown, straight, ascending, much stouter than the radials, 10 to 12 mm. long; flowers numerous, rose-colored, 12 mm. broad; inner perianth-segments linear, somewhat spreading, pointed; filaments red; stigma-lobes 4 or 5, rose-colored; fruit 2.5 cm. long, cylindric, lilac to red; seeds brownish.

Type locality: Mexico.

Distribution: Probably central Mexico.

We have had this plant in cultivation but it has never flowered with us; however, it is very distinct from anything else we know.

Mammillaria flaviceps is referred by Labouret to *M. crassispina*, now usually referred to *M. rhodantha*.

Mammillaria floccigera (and its variety *longispina*) Förster (Handb. Cact. 254. 1846), as well as *M. aurata* and *M. hybrida* (Pfeiffer, Enum. Cact. 31. 1837), are given by Schumann as synonyms of *M. rhodantha*, but none of them was described at the places cited.

Mammillaria erinacea Wendland is unknown to us; it is referred to *Mammillaria rhodantha* by the Index Kewensis, but whether it was described or not we do not know.

Mammillaria fulvispina was said by Haworth to come from Brazil, while the Index Kewensis refers it to Brazil and Mexico. If of this relationship, it is from Mexico. *Mammillaria radula* Scheidweiler (Förster, Handb. Cact. 208. 1846), referred by Schumann as a synonym of this species, was given by Förster as a synonym of *Mammillaria phaeacantha*.

Mammillaria pyramidalis Link and Otto (Verh. Ver. Beförd. Gartenb. 6: 429. 1830), given as a synonym of this species by Schumann, is only a name.

Mammillaria atrata Mackie (Curtis's Bot. Mag. 65: pl. 3642. 1839) is also referred here by Schumann. The plant is supposed to have come from Chile and is probably referable to *Neoporteria*, which see (Cactaceae 3: 97. 1922).

Mammillaria pyrrhocentra Otto, its var. *gracilior* (Salm-Dyck, Cact. Hort. Dyck. 1844. 8. 1845), and *M. fulvispina pyrrhocentra* Salm-Dyck (Cact. Hort. Dyck. 1849. 10. 1850) were referred as synonyms of *M. rhodantha* by Schumann, but were not described at the places cited.

Mammillaria aurea Pfeiffer (Förster, Handb. Cact. 200. 1846; *M. rhodantha aurea* Salm-Dyck, Cact. Hort. Dyck. 1849. 11. 1850) is referred here. We have found no description of it. *M. odieriana aurea* Salm-Dyck (Cact. Hort. Dyck. 1844. 7. 1845), also undescribed, may be the same.

Mammillaria rhodantha cristata (Förster, Handb. Cact. ed. 2. 292. 1885) is only an abnormal form.

Mammillaria recurvispina Hildmann (Schelle, Handb. Kakteenk. 257. 1907) is given without synonymy or description. *M. rhodantha schochiana* (*M. schochiana* Hortus) is also given at the same place, but so far as we can learn has not been published.

Mammillaria tentaculata conothele Monville is given by Labouret (Monogr. Cact. 55. 1853) as a synonym of *M. stueberi*, while he refers *M. tentaculata fulvispina* (Monogr. Cact. 44. 1853) to *M. fulvispina*.

Mammillaria tentaculata rubra (Förster, Handb. Cact. 207. 1846) was given as a synonym of *M. tentaculata ruficeps*.

Mammillaria olivacea was cited by Pfeiffer (Enum. Cact. 180. 1837) as a synonym of *M. tentaculata*.

Mammillaria neglecta was given as a synonym of *M. rhodantha neglecta* by Salm-Dyck (Cact. Hort. Dyck. 1849. 11. 1850).

Mammillaria rhodantha var. *inuncta* Hoffmannsegg was listed by Labouret (Monogr. Cact. 45. 1853) as one of the synonyms of *M. rhodantha*. *M. rhodantha rubra* was given by Rümpler (Förster, Handb. Cact. ed. 2. 292. 1885) as a synonym of *M. rhodantha ruficeps*, but afterwards was formally published by Schumann.

Mammillaria rhodantha celsii Lemaire (Labouret, Monogr. Cact. 48. 1853) was given as a synonym of *M. lanifera*. It probably belongs to *M. rhodantha*. *Cactus capillaris* was made by Coulter because of the older *Mammillaria lanifera* of Haworth. Palmer's plant from Saltillo (1880), preserved in the Missouri Botanical Garden, is very different and suggests that the labels have been mixed.

Illustrations: Knippel, Kakteen pl. 24; De Laet, Cat. Gén. f. 50, No. 9; Wiener Illustr. Gart. Zeit. **29**: f. 22, No. 9; Haage and Schmidt, Haupt-Verz. Cact. **1912**: 37; Link and Otto, Icon. Pl. Rar. pl. 26; Gard. Chron. III. **42**: 290. f. 116; Schelle, Handb. Kakteenk. 256. f. 179; Gartenwelt **12**: 200; Abh. Bayer. Akad. Wiss. München **2**: pl. 1, 1. f. 3; Möllers Deutsche Gärt. Zeit. **25**: 475. f. 8, No. 22, as *Mammillaria rhodantha*; Grässner, Haupt-Verz. Kakteen **1912**: 24; Schelle, Handb. Kakteenk. 258. f. 181, as *Mammillaria rhodantha pfeifferi*; Grässner, Haupt-Verz. Kakteen **1912**: 24, as *M. rhodantha fuscata*; Schelle, Handb. Kakteenk. 258. f. 180, as *M. rhodantha fulvispina*; Blanc, Cacti 73. No. 1434; Cact. Journ. **1**: 43, as *M. odieriana*; Cact. Journ. **1**: 43; pl. for February, as *M. pfeifferi*; Edwards's Bot. Reg. **16**: pl. 1329, as *M. pulchra*; Nov. Act. Nat. Cur. **19**: pl. 16, f. 8, as *Mammillaria tentaculata*.

Figure 129 is from a photograph of a plant obtained by Dr. Rose through W. Mundt in 1913, which is now growing at Washington.

81. Neomammillaria plumosa (Weber).

Mammillaria plumosa Weber, Dict. Hort. Bois 804. 1898.

Small, growing in dense clusters sometimes 15 cm. broad, entirely covered by the mass of white spines; tubercles small, somewhat woolly in their axils, 2 to 3 mm. long; spines about 40, all radial, weak, plumose, 3 to 7 mm. long; flowers white, small, 3 to 4 mm. long; perianth-segments with a red line running down the center; seeds black.

Type locality: Northern Mexico.

Distribution: Northern Mexico.

This plant for a long time passed in the trade under the name of *Mammillaria lasiacantha*, but it is, of course, very different. It is a very striking species and differs from all the others in its feather-like spines. We have had it under observation since 1907 and it has only once flowered (1921).

According to Walton, it is called the feather ball on account of the feather-like spines.

Illustrations: Möllers Deutsche Gärt. Zeit. **25**: 475. f. 8, No. 16; Schelle, Handb. Kakteenk. 252. f. 173; Ann. Rep. Smiths. Inst. **1908**: pl. 3, f. 6; Haage, Cact. Kultur ed. 2. 189; Journ. Hort. Home Farm. III. **60**: 7, as *Mammillaria plumosa*; Cact. Journ. **1**: pl. for February, in part; Darel, Illustr. Handb. Kakteenk. 94. f. 76; Blanc, Hints on Cacti 70. f. 1355; Blanc, Illustr. Price List Cacti 13, as *M. lasiacantha*.

Figure 131 is from a photograph, furnished by Dr. Safford, showing the spines.

82. Neomammillaria prolifera (Miller).

Cactus proliferus* Miller, Gard. Dict. ed. 8. No. 6. 1768.
Cactus glomeratus Lamarck, Encycl. 1: 537. 1783.
Cactus mammillaris prolifer Aiton, Hort. Kew. 2: 150. 1789.
Mammillaria prolifera Haworth, Syn. Pl. Succ. 177. 1812.
Cactus pusillus De Candolle, Cact. Hort. Monsp. 184. 1813. Not Haworth, 1803.
Cactus stellatus Willdenow, Enum. Pl. Suppl. 30. 1813.
Mammillaria stellaris Haworth, Suppl. Pl. Succ. 72. 1819.
Mammillaria pusilla Sweet, Hort. Brit. 171. 1826.
Mammillaria stellata Sweet, Hort. Brit. 171. 1826.
Mammillaria glomerata De Candolle, Prodr. 3: 459. 1828.
Mammillaria pusilla major Pfeiffer, Enum. Cact. 36. 1837.
Cactus haworthianus Kuntze, Rev. Gen. Pl. 1: 259. 1891.
Mammillaria pusilla haitiensis Schumann, Blühende Kakteen 1: under pl. 46. 1904.

Low, growing in colonies often 6 dm. in diameter, the individual plants globose or cylindric, 3 to 6 cm. in diameter, of soft texture; tubercles conic, about 8 mm. long, spreading; axils of tubercles with long, hair-like bristles; radial spines many, hair-like; central spines 5 to 12, much stouter than the radials, with bright yellow tips, puberulent; flowers borne in old axils but toward top of plant, small, yellowish white; inner perianth-segments erect, pale yellow, with brownish mid-rib, acute; filaments pale rose-colored; anthers at first deflexed inward; style shcrter than filaments; stigma-lobes 3, yellow; fruit crowned by persistent withering perianth, clavate, somewhat curved, 1.5 to 2 cm. long, scarlet; seeds black, pitted, a little depressed; aril white, triangular.

Type locality: West Indies.

Distribution: Cuba and Hispaniola. Loddiges reports it from South America, doubtless in error.

At the United States Naval Station, Guantánamo Bay, Cuba, the plant grows in low, dry thickets and is quite inconspicuous but abundant.

Dr. Shafer referred to this species (Bull. N. Y. Bot. Gard. **13**: 139) as *Mammariella*, without description or citation.

Burmann's plate (201, f. 1) of this plant shows most of the tubercles without spines or hairs but these have doubtless been omitted by the artist, for Plumier says (Cat. p. 19): "*Melocactus* minimus, lanuginosus et tuberosus."

Haworth (Phil. Mag. **7**: 114. 1830) would exclude *Mammillaria pusilla* (Mém. Mus. Hist. Nat. Paris **17**: pl. 2, f. 1) as figured by De Candolle. His illustration is evidently faulty, but his description seems to answer our plant.

The name *Mammillaria pusilla minor* occurred in the Index of the Cacti in the Botanical Garden of Berlin for 1829 (Verh. Ver. Beförd. **6**: 429. 1830), but it is without description. It is mentioned again by Salm-Dyck (Hort. Dyck. 156. 1834), who credits the name to Otto, but he does not describe it.

---

* Otto Kuntze (Rev. Gen. Pl. 1: 259. 1891) publishes this binomial as *Cactus prolifer*. Pfeiffer (Enum. Cact. 9. 1837) uses this later binomial for another species, crediting it to Willdenow, but we do not find it used elsewhere.

Mammillaria granulata Meinshausen (Wöchenschr. Gärtn. Pflanz. **1**: 264. 1858; *Cactus granulatus* Kuntze, Rev. Gen. Pl. **1**: 260. 1891) was described without the flowers and fruit being known and it has never been identified. Meinshausen says that it has the habit of *M. pusilla*, but he considered it different otherwise.

Cactus stellaris was given by Haworth (Suppl. Pl. Succ. 72. 1819) instead of *C. stellatus* Willdenow.

Mammillaria pusilla cristata (Schelle, Handb. Kakteenk. 249. 1907) is probably only a form.

Illustrations: Loudon, Encycl. Pl. 410. f. 6842, as *Cactus stellaris*; Loddiges, Bot. Cab. **1**: pl. 79, as *Cactus stellatus*; Plukenet, Opera Bot. **1**: pl. 29, f. 2, as *Ficoides* etc.; Dict. Hort. Nicholson Suppl. 514. f. 547; Abh. Bayer. Akad. Wiss. München **2**: pl. 1, VIII, f. 7; Rümpler, Sukkulenten 197. f. 110; Monatsschr. Kakteenk. **8**: 73; Mém. Mus. Hist. Nat. Paris **17**: pl. 2, f. 1; Ann. Rep. Smiths. Inst. **1908**: pl. 2, f. 4; Blanc, Cacti 74, No. 1500; Schumann, Gesamtb. Kakteen f. 87; Blühende Kakteen **1**: pl. 46; Ann. Inst. Roy. Hort. Fromont **2**: pl. 1, f. B; Watson, Cact. Cult. ed. 2. 255. f. 96; ed. 3. f. 45; Remark, Kakteenfreund 15; Cact. Journ. **2**: 6, as *Mammillaria pusilla*.

Figure 132 is from a photograph by Ernest Braunton of a clump of plants growing in the Huntington collection near Los Angeles, California.

FIG. 132.—Neomammillaria prolifera.   FIG 133.—Neomammillaria multiceps.

83. Neomammillaria multiceps (Salm-Dyck).

Mammillaria multiceps Salm-Dyck, Cact. Hort. Dyck. 1849. 81. 1850.
Mammillaria multiceps elongata Meinshausen, Wöchenschr. Gärtn. Pflanz. **1**: 27. 1858.
Mammillaria multiceps grisea Meinshausen, Wöchenschr. Gärtn. Pflanz. **1**: 27. 1858.
Mammillaria multiceps humilis Meinshausen, Wöchenschr. Gärtn. Pflanz. **1**: 27. 1858.
Mammillaria multiceps perpusilla Meinshausen, Wöchenschr. Gärtn. Pflanz. **1**: 27. 1858.
Mammillaria pusilla texana Engelmann, Cact. Mex. Bound. 5. 1859.
Mammillaria texana Poselger in Young, Fl. Texas. 279. 1873.
Cactus multiceps Kuntze, Rev. Gen. Pl. **1**: 260. 1891.
Cactus stellatus texanus Coulter, Contr. U. S. Nat. Herb. **3**: 108. 1894.
Cactus texanus Small, Fl. Southeast. U. S. 812. 1903.

Cespitose, often forming large clumps; separate plants globose to short-oblong, often only 1 to 2 cm. in diameter; tubercles small, terete, hairy in their axils; radial spines hair-like, white; central spines several, pubescent, yellowish at base, dark brown above; flowers about 12 mm. long, whitish to yellowish salmon, often becoming reddish on outside; fruit oblong, 8 to 12 mm. long, scarlet; seeds black, 1 mm. long, punctate.

Type locality: Not cited.

Distribution: Texas and northeastern Mexico.

It is sometimes classified as a variety of *Mammillaria prolifera*, from which it differs in having the central spines always brown-tipped instead of golden yellow; it is somewhat smaller, with slightly smaller seeds.

Mr. Robert Runyon says that this plant forms clumps usually about 10 cm. broad, but sometimes broader. It is never very plentiful but has a rather wide distribution, and seems to prefer mesquite thickets where the soil is very rich, but occasionally is found on rocky hillsides.

Mammillaria pusilla mexicana, offered for sale by Grässner (Monatsschr. Kakteenk. February 1920), probably belongs here.

Mammillaria caespititia Hortus was referred by Salm-Dyck as a synonym of *M. multiceps*. *M. pusilla caespititia* (Schelle, Handb. Kakteenk. 249. 1907) is the same.

Mammillaria parvissima Karwinsky (Wöchenschr. Gärtn. Pflanz. 1: 27. 1858) is sometimes credited to Meinshausen, but seems never to have been described. *M. perpusilla* Meinshausen, given only as a synonym, belongs here and the name occurs on the page mentioned above.

Fig. 134.—Neomammillaria multiceps.

Illustrations: Cact. Mex. Bound. pl. 5; Cact. Journ. 2: 93; Förster, Handb. Cact. ed. 2. 262. f. 25; Schelle, Handb. Kakteenk. 249. f. 168, as *Mammillaria pusilla texana*.

Plate XIV, figure 5, shows a very small plant in flower, collected by Robert Runyon near Brownsville, Texas, in 1921; figure 6 shows a plant received from the Missouri Botanical Garden in 1904 which flowered in the New York Botanical Garden in March 1912. Figure 134 is from a photograph of a plant collected near Victoria, Mexico, by Dr. Edward Palmer, which was grown for many years in Washington; figure 133 shows a small plant photographed by Robert Runyon on July 10, 1921.

84. Neomammillaria camptotricha (Dams).

Mammillaria camptotricha Dams, Gartenwelt 10: 14. 1905.

Plants globose, cespitose, deep green, 5 cm. in diameter; tubercles somewhat elongated, often curved, 2 cm. long, terete, not at all milky, bearing bristles in the axils; spines 2 to 4, described as up to as many as 8, yellowish, bristle-like, spreading and twisted or bent, often 3 cm. long; spine-areoles small, circular, a little woolly at first; axils of tubercles bristly; flowers small, about 1 cm. long; outer perianth-segments greenish; inner perianth-segments white, 10 mm. long, acute.

Type locality: Mexico.

Distribution: Deserts of eastern Querétaro, Mexico.

This plant was collected by Rose and Painter between Higuerillas and San Pablo, August 23, 1905 (No. 11536), and flowered in Washington on October 3, 1905. In 1913 L. Quehl of Halle sent us some flowers of this species.

Illustrations: Blühende Kakteen **3**: pl. 151; Möllers Deutsche Gärt. Zeit. **25**: 475. f. 8, No. 6, as *Mammillaria camptotricha.*

Figure 135 is from a photograph of the plant collected by Dr. Rose in 1905.

85. Neomammillaria eriacantha (Link and Otto).

> *Mammillaria eriacantha* Link and Otto in Pfeiffer, Enum. Cact. 32. 1837.
> *Cactus eriacanthus* Kuntze, Rev. Gen. Pl. **1**: 260. 1891.

Solitary or cespitose, 10 to 15 cm. high, cylindric, 5 cm. in diameter; tubercles spiraled, in 22 rows; radial spines about 20, delicate, spreading, pubescent; central spines 2, widely spreading, stouter than the radials, also pubescent, yellowish; flowers borne in a ring above the middle of the plant, yellow, 14 mm. broad; inner perianth-segments about 14, linear, acute; stigma-lobes 4; fruit at first greenish white, afterwards tinged with red, short-clavate.

FIG. 135.—Neomammillaria camptotricha.　　FIG. 136.—Neomammillaria schiedeana.

Type locality: Mexico.

Distribution: Central Mexico.

A plant, collected by McDowell, was seen in the collection of the Instituto Medico Nacional in the City of Mexico, but no specimen was obtained.

Mammillaria columbiana Salm-Dyck (Cact. Hort. Dyck. 1849. 99. 1850) is probably to be referred here. It is doubtless of Mexican rather than of Colombian origin.

Mammillaria eriantha (Pfeiffer, Enum. Cact. 32. 1837), referred here by Pfeiffer, was never described.

Mammillaria cylindracea De Candolle (Mém. Mus. Hist. Nat. Paris **17**: 111. 1828) is referred here by Schumann and also by Pfeiffer and Otto, but the description of it would suggest a different species. Kuntze changes the name to *Cactus cylindraceus* (Rev. Gen. Pl. **1**: 260. 1891). Here is also referred *Mammillaria cylindrica flavispina* (Labouret, Monogr. Cact. 88. 1853).

Illustrations: Pfeiffer and Otto, Abbild. Beschr. Cact. **1**: pl. 25; Schelle, Handb. Kakteenk. 256. f. 178, as *Mammillaria eriacantha.*

Figure 138 is a reproduction of the first illustration above cited.

86. Neomammillaria schiedeana (Ehrenberg).

Mammillaria schiedeana Ehrenberg in Schlechtendal, Allg. Gartenz. **6:** 249. 1838.
? *Mammillaria sericata* Lemaire, Cact. Gen. Nov. Sp. 44. 1839.
Cactus schiedianus Kuntze, Rev. Gen. Pl. **1:** 261. 1891.
Mammillaria dumetorum J. A. Purpus, Monatsschr. Kakteenk. **22:** 149. 1912.
? *Mammillaria cephalophora* Quehl, Monatsschr. Kakteenk. **24:** 158. 1914. Not Salm-Dyck, 1850.

Densely cespitose, somewhat soft in texture; axils of tubercles bearing long bristle-like white hairs; tubercles green, terete; radial spines about 30, white, spreading, bristle-like, puberulent; central spines 6 to 10, spreading and appressed against the radials, a little stouter, often tinged with yellow; flowers 15 mm. long; inner perianth-segments white; filaments white; style cream-colored; stigma-lobes 4, short, obtuse.

Type locality: Near Puente de Dios, Mexico.
Distribution: Central Mexico.
The Index Kewensis refers *Mammillaria sericata* Lemaire to *M. magnimamma*.

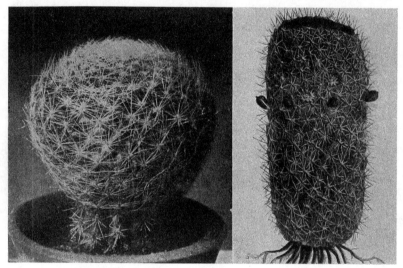

FIG. 137.—Neomammillaria lenta. FIG. 138.—Neomammillaria eriacantha.

Illustrations: Schumann, Gesamtb. Kakteen f. 113: Blühende Kakteen 1: pl. 13; Monatsschr. Kakteenk. **8:** 12; **13:** 92, f. A, as *Mammillaria schiedeana*; Monatsschr. Kakteenk. **23:** 89, as *Mammillaria dumetorum*; (?) Monatsschr. Kakteenk. **24:** 158, as *Mammillaria cephalophora*.

Figure 136 is from a photograph of a plant collected by Dr. C. A. Purpus at San Rafael, Mexico, in 1910.

87. Neomammillaria lasiacantha (Engelmann).

Mammillaria lasiacantha Engelmann, Proc. Amer. Acad. **3:** 261. 1856.
Mammillaria lasiacantha minor Engelmann, Cact. Mex. Bound. 5. 1859.
Cactus lasiacanthus Kuntze, Rev. Gen. Pl. **1:** 259. 1891.

Globose, 2 to 2.5 cm. in diameter; tubercles small, their axils naked; spines 40 to 60, in more than one series, white, puberulent, 2 to 4 mm. long; flowers 12 mm. long, whitish or pink; fruit 1 to 2 cm. long; seeds blackish, pitted.

Type locality: On the Pecos in western Texas.

Distribution: Western Texas and northern Chihuahua. Reported also from Arizona, but doubtless incorrectly.

We have seen no specimens of *N. lasiacantha*, except the type, but the following species, first described as a variety of *lasiacantha*, is very common in eastern Texas and northern Mexico. Possibly the two should be united, the typical form simply representing a juvenile phase.

Illustrations: Cact. Mex. Bound. pl. 3; Schumann, Gesamtb. Kakteen 522. f. 86; Engler and Prantl, Pflanzenfam. 3⁶ᵃ: f. 56, A; Blanc, Cacti 70. No. 1335; West Amer. Sci. **13:** 39, as *Mammillaria lasiacantha*.

88. Neomammillaria denudata (Engelmann).

Mammillaria lasiacantha denudata Engelmann, Cact. Mex. Bound. 5. 1859.
Cactus lasiacanthus denudatus Coulter, Contr. U. S. Nat. Herb. **3:** 100. 1894.
Mammillaria lasiandra denudata Quehl, Monatsschr. Kakteenk. **19:** 79. 1909.

Globose, 2.5 to 3.5 cm. in diameter; tubercles 5 to 6 mm. long; spines 50 to 80, glabrous or nearly so, 3 to 5 mm. long, the innermost usually much shorter; flowers and fruit from near the center but not from the axils of young tubercles; flowers 10 to 12 mm. long; perianth-segments few, about 12, oblong, obtuse, the margins white, the center light purple; stamens white; style and stigma-lobes green; fruit clavate, red, 1.5 to 2 cm. long; seeds black with basal hilum.

FIG. 139.—Neomammillaria denudata.   FIG. 140.—Neomammillaria lenta.

Type locality: Western Texas.
Distribution: Western Texas and northern Coahuila, Mexico.

The flowers open about mid-day and close at night; in one case which we recorded the flowers opened for six consecutive days.

Mammillaria rungii (Schumann, Gesamtb. Kakteen 522. 1898), an unpublished garden name, was supposed by Schumann to be referable to *M. lasiacantha denudata*.

Illustrations: Cact. Mex. Bound. pl. 4; Möllers Deutsche Gärt. Zeit. **25:** 475. f. 8, No. 21, as *Mammillaria lasiacantha denudata*.

Figure 139 is from a photograph of a plant collected by Elmer Stearns in 1909, which afterwards flowered in Washington.

89. Neomammillaria lenta (K. Brandegee).

Mammillaria lenta K. Brandegee, Zoe **5:** 194. 1904.

Described as cespitose; individuals globose to short-cylindric, almost hidden by the white delicate spines; tubercles very slender, light green; spine-areoles naked; spines about 40, very fragile; axils woolly and occasionally bearing a single bristle; flowers whitish, 7 mm. long; perianth-segments pointed; fruit red, clavate; seeds 1 mm. in diameter, dull black.

Type locality: Near Viesca, in Coahuila, Mexico.

Distribution: Coahuila, Mexico.

Illustration: Monatsschr. Kakteenk. **16**: 40, as *Mammillaria lenta*.

Figure 137 is from a photograph obtained from L. Quehl in 1921; figure 140 is from a photograph of a fruiting plant sent from Parras, Mexico, by C. A. Purpus in 1905.

90. Neomammillaria candida (Scheidweiler).

> *Mammillaria candida* Scheidweiler, Bull. Acad. Sci. Brux. **5**: 496. 1838.
> *Mammillaria sphaerotricha* Lemaire, Cact. Gen. Nov. Sp. 33. 1839.
> *Mammillaria humboldtii* Ehrenberg, Linnaea **14**: 378. 1840.
> *Mammillaria sphaerotricha rosea* Salm-Dyck, Cact. Hort. Dyck. 1849. 85. 1850.
> *Cactus humboldtii* Kuntze, Rev. Gen. Pl. **1**: 260. 1891. Not Humboldt, Bompland, and Kunth, 1823.
> *Cactus sphaerotrichus* Kuntze, Rev. Gen. Pl. **1**: 261. 1891.
> *Mammillaria candida rosea* Salm-Dyck in Schumann, Gesamtb. Kakteen 525. 1898.

Cespitose; individual plant globose, 5 to 7 cm. in diameter, almost hidden by the white spines; radial spines numerous, radiating; central spines 8 to 12, porrect, often brownish at tip, a little stouter than the radials; axils setose; flowers 2 cm. long, rose-colored; perianth-segments serrulate twards the apex; fruit red; seeds black.

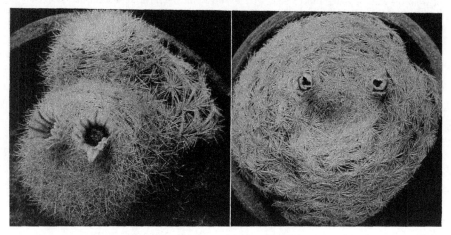

FIGS. 141 and 142.—Neomammillaria candida.

Type locality: Near San Luis Potosí.

Distribution: Central Mexico.

Illustrations: Monatsschr. Kakteenk. **29**: 141, as *Mammillaria candida rosea*; Hort. Belge **5**: pl. 117; Möllers Deutsche Gärt. Zeit. **25**: 475. f. 8, No. 27; Blühende Kakteen **3**: pl. 169, as *Mammillaria candida*.

Figure 141 is from a photograph of a plant obtained by Dr. Palmer near San Luis Potosí in 1905; figure 142 is from a photograph of a plant collected by C. A. Purpus from near the same locality in 1910.

91. Neomammillaria vetula (Martius).

> *Mammillaria vetula* Martius, Nov. Act. Nat. Cur. **16**: 338. 1832.
> *Cactus vetulus* Kuntze, Rev. Gen. Pl. **1**: 261. 1891.

Plant somewhat club-shaped, small, 4 to 5 cm. high; tubercles terete, light green, somewhat shining; axils of tubercles naked or sometimes with a small tuft of wool; radial spines about 25, spreading, white, bristle-like; central spines 1 to 6, stouter than the radials, brownish; flowers 12 to 15 mm. long, borne at upper part of plant; outer perianth-segments red with yellowish margins; inner perianth-segments cream-colored; filaments greenish; style green; stigma-lobes 5, white.

Type locality: San José del Oro, Hidalgo, Mexico.

Distribution: Hidalgo, Mexico.

The above description was drawn in part from a plant which flowered in Washington on November 8, 1912, and which had been sent to us by L. Buscationi from Catania, Italy. This plant gave off numerous young ones from the axils of the tubercles, but it has died.

Mammillaria vetula major Salm-Dyck (Walpers, Repert. Bot. **2**: 270. 1843) is said to be the same as *M. grandiflora* Hortus. If so, this must be different from *M. grandiflora* Otto, which we have referred to *Neolloydia conoidea.*

Illustration: Nov. Act. Nat. Cur. **16**: pl. 24, as *Mammillaria vetula.*

Figure 143 is reproduced from the illustration above cited.

FIG. 143.—Neomammillaria vetula.   FIG. 144.—Neomammillaria discolor.

92. Neomammillaria fertilis (Hildmann).

Mammillaria fertilis Hildmann in Schumann, Gesamtb. Kakteen 503. 1898.

Cespitose, the individual plant globose to short-cylindric, dark green; tubercles arranged in 8 or 13 rows, a little woolly in their axils; radial spines 7 to 10, acicular, 6 mm. long; central spines 1 or 2, straight, stouter than the radials, 10 mm. long; flowers deep crimson, 2 cm. long; inner perianth-segments linear-lanceolate, acute.

Type locality: Mexico, but definite station not given.

Distribution: Mexico, but range unknown.

We have not seen living specimens of this plant but L. Quehl of Halle had it growing in 1913 and sent us flowers which we have used in this description.

93. Neomammillaria decipiens (Scheidweiler).

Mammillaria decipiens Scheidweiler, Bull. Acad. Sci. Brux. **5**: 496. 1838.
Mammillaria anancistria* Lemaire, Cact. Gen. Nov. Sp. 39. 1839.
Mammillaria guilleminiana Lemaire, Cact. Gen. Nov. Sp. 48. 1839.
Mammillaria glochidiata inuncinata Lemaire, Cact. Gen. Nov. Sp. 102. 1839.
Cactus decipiens Kuntze, Rev. Gen. Pl. **1**: 260. 1891.
Cactus guilleminianus Kuntze, Rev. Gen. Pl. **1**: 261. 1891.
Cactus ancistrius Kuntze, Rev. Gen. Pl. **1**: 261. 1891.

Usually cespitose, deep green; tubercles soft, cylindric, about 1 cm. long, their axils bearing 2 or 3 bristles each; radial spines 7 to 9, spreading, slender, white, sometimes yellowish with brown

---

* Spelled *M. ancistria* by Walpers (Repert. Bot. **2**: 296. 1843.)

tips, puberulent when young; central spine 1, much longer than the radials, erect or ascending, 15 to 18 mm. long, dark brown; flower-buds pinkish, acute; flower 15 mm. long, broadly funnel-shaped; inner perianth-segments nearly white or faintly tinged with pink, acute; filaments white to pinkish; stigma-lobes 4, white or pinkish, slender, filiform.

Type locality: Not cited.

Distribution: San Luis Potosí.

The above description is drawn from plants growing in the top of *Calibanus caespitosus*, a curious, globose, lilliaceous plant of the desert of central Mexico, sent by Dr. E. Palmer from San Luis Potosí in 1905.

Schumann says that the axils of the tubercles are naked, while K. Brandegee describes them as bearing bristles as in our plant and so called for in the original description.

In some plants one or two of the upper radial spines are brown like the central spine; the flowers are delicately fragrant, remaining open during cloudy days. In cultivation this is one of the earliest species to flower; in 1918 it began to bloom early in January.

Mammillaria inuncinata (Lemaire, Cact. Gen. Nov. Sp. 39. 1839) was never described but belongs here.

Mammillaria ancistroides inuncinata Lemaire and *M. deficum* (Förster, Handb. Cact. 185. 1846), as synonyms, were referred here. *M. deficiens* Hortus (Salm-Dyck, Cact. Hort. Dyck. 1849. 7. 1850) is another name, used only as a synonym of this species.

Illustrations: Schumann, Gesamtb. Kakteen 528. f. 88; Knippel, Kakteen pl. 20; Schelle, Handb. Kakteenk. 249. f. 169; Blanc, Cacti 68. No. 1200, as *Mammillaria decipiens*.

Plate xIV, figure 3, is from a plant sent to the New York Botanical Garden by Weinberg in 1903, which flowered November 14, 1911.

94. Neomammillaria discolor (Haworth).

> *Mammillaria discolor* Haworth, Syn. Pl. Succ. 177. 1812.
> *Cactus depressus* De Candolle, Cact. Hort. Monsp. 84. 1813. Not Haworth, 1812.
> *Cactus pseudomammillaris* Salm-Dyck, Liste Pl. Gr. 1: 1. 1815.
> *Cactus spini* Colla, Mem. Accad. Sci. Torino 33: 133. 1826.
> *Mammillaria pseudomammarilaris* Pfeiffer, Allg. Gartenz. 3: 57. 1835.
> *Mammillaria discolor prolifera* Pfeiffer, Enum. Cact. 28. 1837.
> *Mammillaria albida* Haage in Pfeiffer, Enum. Cact. 28. 1837.
> *Mammillaria aciculata* Otto in Pfeiffer, Enum. Cact. 29. .1837.
> *Mammillaria discolor monstrosa* Monville in Lemaire, Cact. Gen. Nov. Sp. 99. 1839.
> *Mammillaria discolor albida* Salm-Dyck, Cact. Hort. Dyck. 1844. 7. 1845.
> ? *Mammillaria curvispina* Otto in Dietrich, Allg. Gartenz. 14: 204. 1846.
> ? *Mammillaria discolor pulchella* Otto in Förster, Handb. Cact. 206. 1846.
> *Mammillaria curvispina parviflora* A. Dietrich, Allg. Gartenz. 14: 204. 1846.
> *Mammillaria nitens* Otto in Linke, Allg. Gartenz. 16: 331. 1848.
> *Mammillaria pulchella* Otto in Linke, Allg. Gartenz. 16: 331. 1848.
> *Mammillaria discolor aciculata* Salm-Dyck, Cact. Hort. Dyck. 1849. 11. 1850.
> *Mammillaria discolor curvispina* Salm-Dyck, Cact. Hort. Dyck. 1849. 11. 1850.
> *Mammillaria discolor nitens* Salm-Dyck, Cact. Hort. Dyck. 1849. 11. 1850.
> *Mammillaria polythele aciculata* Salm-Dyck, Cact. Hort. Dyck. 1849. 15. 1850.
> *Mammillaria pulchella nigricans* Monville in Labouret, Monogr. Cact. 40. 1853.
> *Cactus aciculatus* Kuntze, Rev. Gen. Pl. 1: 260. 1891.
> *Cactus discolor* Kuntze, Rev. Gen. Pl. 1: 260. 1891.
> *Cactus pulchellus* Kuntze, Rev. Gen. Pl. 1: 261. 1891.

Globose or somewhat depressed, often solitary, about 7 cm. in diameter; tubercles ovoid-conic, arranged in 13 to 15 spirals, their axils naked; radial spines 16 to 20, white, setaceous, widely spreading; central spines about 6, stouter than the radials, straight, at first black with white bases; flowers 15 mm. broad when fully open; inner perianth-segments linear, white, with a violet-rose stripe; fruit red, 2.5 cm. long.

Type locality: Not cited.

Distribution: Puebla, according to Schumann.

We have been unable to identify definitely this species. As there seems to be no type preserved we must rely upon the short original description and the early illustrations. The illustration of Loddiges (Bot. Cab. 17: pl. 1871) shows a plant with yellowish-brown spines and must belong elsewhere.

Mammillaria depressa was credited by mistake to De Candolle by Pfeiffer in listing the synonyms of *M. discolor* (Enum. Cact. 28. 1837).

Mammillaria confinis Haage, according to Pfeiffer (Enum. Cact. 28. 1837), appeared in "Haage, Catal. Cact. 1836" and he lists it as a synonym of *M. albida.*

Mammillaria canescens Hortus (Pfeiffer, Enum. Cact. 28. 1837) was given as a synonym of *M. discolor.* This is different from *M. canescens* Jacobi (Allg. Gartenz. 24: 89. 1856) which Schumann lists among his unknown plants. (See also Lemaire, Cact. Gen. Nov. Sp. 99. 1839.)

Mammillaria coniflora Hortus and *M. discolor coniflora* Salm-Dyck (Cact. Hort. Dyck. 1849. 11. 1850) are only names which belong here.

Mammillaria discolor fulvescens Salm-Dyck (Cact. Hort. Dyck. 1844. 7. 1845) was not formally published at the place here cited.

Mammillaria discolor breviflora (Förster, Handb. Cact. 206. 1846), although not described at the place here cited, is usually referred here.

Cactus pseudomammillaris appeared simply as a name in 1815 (Desfontaines, Tab. Bot. ed. 2. 191), and again in Pfeiffer's Enumeratio (28. 1837) as a synonym of *Mammillaria discolor prolifera.* Pfeiffer credits the name to Salm-Dyck and gives the reference to Allgemeine Gartenzeitung (3: 57. 1835), but the name appeared there under *Mammillaria* along with *spinii* and *canescens. M. spinii,* credited to Colla, is given by Salm-Dyck (Cact. Hort. Dyck. 1849. 11. 1850) as a synonym of *M. discolor.*

Schumann lists *Mammillaria rhodacantha* Salm-Dyck (Cact. Hort. Dyck 1849. 96. 1850) among his unknown species. *M. rhodacantha pallidior* (Salm-Dyck, Cact. Hort. Dyck. 1844. 8. 1845) is only a name, while *M. discolor rhodacantha* (Walpers, Repert. Bot. 2: 271. 1843), although never described, seems to be the same as *M. rhodacantha.*

Illustrations: Mém. Mus. Hist. Nat. Paris 17: pl. 2, f. 2: Ann. Inst. Roy. Hort. Fromont 2: pl. 1, f. A; Loddiges, Bot. Cab. 17: pl. 1671 (?), as *Mammillaria discolor;* Mem. Accad. Sci. Torino 33: pl. 11, as *Cactus spini.*

Figure 144 is reproduced from the first illustration cited above.

FIG. 145.—Neomammillaria fragilis. FIG. 146.—Neomammillaria elongata.

95. Neomammillaria fragilis (Salm-Dyck).

Mammillaria fragilis Salm-Dyck, Cact. Hort. Dyck. 1849. 103. 1850.

Stems usually oblong or club-shaped, sprouting freely towards the top; branches globose and breaking off at the slightest touch; tubercles bright green, terete, their axils nearly naked; radial spines 12 to 14, white, naked, spreading; central spines usually wanting, especially on branches, if present 1 or 2, elongated, erect, brownish especially at tip; young spine-areoles with white wool; flowers from upper part of plant but not from center, small, lasting for several days; cream-colored with outer segments somewhat pinkish; petals broad with a mucronate tip; filaments and style pale.

Type locality: Not cited.

Distribution: Doubtless Mexico, but not known from wild plants.

Mrs. K. Brandegee, some years ago (Zoe **5:** 9. 1900), called attention to the fact that this fragile little plant did not answer Pfeiffer's description of *Mammillaria gracilis* and that Salm-Dyck had suggested the very appropriate name of *M. fragilis*, which we have adopted here. The plant is known in the trade as *Mammillaria gracilis pulchella*, under which designation we received plants from Haage and Schmidt in 1921.

Illustrations: Schumann, Gesamtb. Kakteen 552. f. 90 (?); Blühende Kakteen **2:** pl. 68; Monatsschr. Kakteenk. **6:** 2; Möllers Deutsche Gärt. Zeit. **25:** 475. f. 8, No. 19; Gartenwelt **12:** 333, as *Mammillaria gracilis*.

Figure 145 is from a photograph sent us by L. Quehl.

96. Neomammillaria elongata (De Candolle).

> *Mammillaria elongata* De Candolle, Mém. Mus. Hist. Nat. Paris **17:** 109. 1828.
> *Mammillaria subcrocea* De Candolle, Mém. Mus. Hist. Nat. Paris **17:** 110. 1828.
> *Mammillaria intertexta* De Candolle, Mém. Mus. Hist. Nat. Paris **17:** 110. 1828.
> *Mammillaria tenuis* De Candolle, Mém. Mus. Hist. Nat. Paris **17:** 110. 1828.
> *Mammillaria tenuis media* De Candolle, Mém. Mus. Hist. Nat. Paris **17:** 110. 1828.
> *? Mammillaria densa* Link and Otto, Icon. Pl. Rar. 69. 1830.
> *Mammillaria echinata densa* Pfeiffer, Enum. Cact. 6. 1837.
> *? Mammillaria stella-aurata* Martius in Zuccarini, Abh. Bayer. Akad. Wiss. München **2:** 101. 1837.
> *? Mammillaria minima* Reichenbach in Terscheck, Suppl. Cact. Verz. 1.
> *Echinocactus densus* Steudel, Nom. ed. 2. **1:** 536. 1840.
> *Mammillaria tenuis minima* Salm-Dyck in Walpers, Repert. Bot. **2:** 272. 1843.
> *Mammillaria subcrocea intertexta* Salm-Dyck, Cact. Hort. Dyck. 1844. 13. 1845.
> *Mammillaria elongata intertexta* Salm-Dyck, Cact. Hort. Dyck. 1849. 12. 1850.
> *Mammillaria elongata subcrocea* Salm-Dyck, Cact. Hort. Dyck. 1849. 12. 1850.
> *Mammillaria subcrocea rufescens* Salm-Dyck, Cact. Hort. Dyck. 1849. 100. 1850.
> *Mammillaria stella-aurata gracilispina* Salm-Dyck, Cact. Hort. Dyck. 1849. 101. 1850.
> *? Mammillaria anguinea* Otto in Salm-Dyck, Cact. Hort. Dyck. 1849. 101. 1850.
> *? Mammillaria subechinata* Salm-Dyck, Cact. Hort. Dyck. 1849. 101. 1850.
> *? Mammillaria rufocrocea* Salm-Dyck, Cact. Hort. Dyck. 1849. 102. 1850.
> *Cactus anguineus* Kuntze, Rev. Gen. Pl. **1:** 260. 1891.
> *Cactus densus* Kuntze, Rev. Gen. Pl. **1:** 260. 1891.
> *Cactus elongatus* Kuntze, Rev. Gen. Pl. **1:** 260. 1891. Not Willdenow, 1813.
> *Cactus intertextus* Kuntze, Rev. Gen. Pl. **1:** 260. 1891.
> *Cactus minimus* Kuntze, Rev. Gen. Pl. **1:** 260. 1891.
> *Cactus stella-auratus* Kuntze, Rev. Gen. Pl. **1:** 261. 1891.
> *Cactus subcroceus* Kuntze, Rev. Gen. Pl. **1:** 261. 1891.
> *Cactus subechinatus* Kuntze, Rev. Gen. Pl. **1:** 261. 1891.
> *Cactus tenuis* Kuntze, Rev. Gen. Pl. **1:** 261. 1891.
> *Mammillaria elongata tenuis* Schumann, Gesamtb. Kakteen 520. 1898.
> *Mammillaria elongata stella-aurata* Schumann, Gesamtb. Kakteen 520. 1898.
> *Mammillaria elongata anguinea* Schumann, Gesamtb. Kakteen 521. 1898.
> *Mammillaria elongata rufocrocea* Schumann, Gesamtb. Kakteen 521. 1898.

Densely cespitose, forming small clumps, erect, ascending or prostrate, 3 to 10 cm. long, 1 to 1.5 cm. in diameter, almost covered by a mass of interlocking spines; tubercles arranged in a few rows, usually in spirals, short, their axils naked; spines usually all radial but sometimes with 1 porrect central spine, yellow or with brown tips, more or less recurved, 8 to 12 mm. long; spine-areoles pubescent when young; flowers at the upper part of the plant, white or nearly so, 6 to 7 mm. long; perianth-segments about 12, rather broad, obtuse or sometimes apiculate.

Type locality: Mexico.

Distribution: Eastern Mexico.

Mammillaria supertexta rufa is referred to *M. elongata intertexta* by Labouret (Monogr. Cact. 68. 1853).

Mammillaria caespitosa was first listed by De Candolle (Prodr. **3:** 460. 1828). It next appears in 1830 as a synonym in a list of the cacti of the Botanical Garden of Berlin. In 1837 Pfeiffer (Enum. Cact. 6) gives it as a synonym of *M. echinata densa*.

The three varieties *Mammillaria tenuis arrecta*, *M. tenuis coerulescens*, and *M. tenuis derubescens* were garden names in the Botanical Garden at Berlin, listed by Förster (Handb. Cact. 240. 1846).

Walpers (Repert. Bot. **2:** 272. 1843) records *M. intertexta rufocrocea*, but without any description.

Labouret (Monogr. Cact. 67. 1853) records the variety *M. stella-aurata minima* Salm-Dyck.

The two varieties of *Mammillaria subcrocea, anguinea,* and *rutila* (Walpers, Repert. Bot. **2:** 272. 1843) are without descriptions.

Mammillaria elongata rufescens Salm-Dyck (Cact. Hort. Dyck. 1844. 12. 1845) was not described at the place here cited, while the variety *straminea* was a garden name (Förster, Handb. Cact. 240. 1846).

Illustrations: Schumann, Gesamtb. Kakteen 519. f. 85; Blühende Kakteen **3:** pl. 174, as *Mammillaria elongata*; Schelle, Handb. Kakteenk. 247. f. 165, as *M. elongata minima*; Blanc, Cacti 72. No. 1398, as *M. minima*; Link and Otto, Icon. Pl. Rar. pl. 35, as *M. densa*; Abh. Bayer. Akad. Wiss. München **2:** pl. 1. VIII. f. 5, as *M. stella-aurata*; Curtis's Bot. Mag. **65:** pl. 3646; Edwards's Bot. Reg. **18:** pl. 1523; De Candolle, Mém. Cact. pl. 1; Loudon, Encycl. Pl. ed. 2 and 3. 1201. f. 17359, as *M. tenuis.*

Figure 146 is from a photograph of the common form in cultivation.

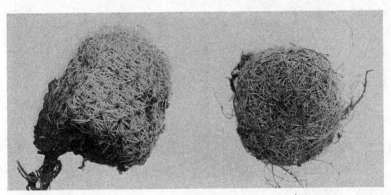

FIG. 147.—Neomammillaria oliviae.

97. Neomammillaria oliviae (Orcutt).

Mammillaria oliviae Orcutt, West Amer. Sci. **12:** 163. 1902.

Globose to short-cylindric, up to 10 cm. high, simple or becoming cespitose, sometimes as many as 8 together; tubercles ovoid, their axils naked; radial spines 25 to 36, snowy white or sometimes reddish brown, slender, rigid, 6 mm. long, the upper ones shorter; central spines 1 to 3, the lower one erect, rigid, white or tipped with chocolate brown; flowers about 3 cm. broad; perianth-segments lanceolate, acute, magenta, the upper part of the margins and tip with a narrow band of white; filaments deep magenta; style light pink; stigma-lobes olive-green; fruit scarlet, clavate, up to 2.5 cm. long; seeds small, black.

Type locality: West of Vail, a flag station on the Southern Pacific Railroad, near Tucson, Arizona.

Distribution: Mountains and deserts of Arizona.

Our description of the flowers is drawn from the notes and photograph of F. E. Lloyd's specimen sent us from Oro Blanco Mountains, Arizona. This is the only record we have had of this plant blooming, but fruiting plants were collected by C. R. Orcutt in 1922 (No. 802). It was first collected in considerable quantity by Mr. Orcutt, but his supply soon died out and most of the skeletons were sent to the U. S. National Herbarium, where they are now preserved.   In April 1921 Mr. Vernon Bailey rediscovered the species in Arizona and sent in a number of living specimens, but none has yet flowered.   Mr. Orcutt reports that he has collected specimens which have hooked spines.

Mr. Orcutt dedicated this species to his wife, Mrs. Olivia Orcutt.

Figure 147 is from a photograph of two plants sent by Mr. Vernon Bailey from Continental, Arizona, in 1920.

98. Neomammillaria echinaria (De Candolle).

Mammillaria echinaria De Candolle, Mém. Mus. Hist. Nat. Paris **17**: 110.   1828.
Mammillaria echinata De Candolle, Mém. Cact. 3.   1834.
Mammillaria gracilis Pfeiffer, Allg. Gartenz. **6**: 275.   1838.
Cactus echinaria Kuntze, Rev. Gen. Pl. **1**: 260.   1891.
Cactus gracilis Kuntze, Rev. Gen. Pl. **1**: 260.   1891.   Not Miller, 1770.
Mammillaria elongata echinata Schumann, Gesamtb. Kakteen 521.   1898.

Plants cespitose, often forming large clumps, ascending or spreading, about 1 dm. long, 1 to 1.5 cm. in diameter; tubercles short, terete, their axils naked; spines pale yellow to glassy white; radial spines about 15, spreading; central spine one, straight, acicular, about 1 cm. long; flowers and fruit not known.

FIG. 148.—Neomammillaria echinaria.   FIG. 149.—Neomammillaria rekoi.

Type locality: Mexico.

Distribution: Hidalgo, Mexico.

The above description is based on a plant collected by Dr. Rose in 1905 near Ixmiquilpan, and this we have had growing ever since.

The two varieties of *Mammillaria echinata*, *gracilior* Ehrenberg and *pallida*, published by Förster (Handb. Cact. 239. 1846), are probably only forms of the species.

The varieties of *Mammillaria gracilis* may or may not belong here.   They are as follows: var. *laetevirens* Salm-Dyck (as a synonym of var. *pulchella*), var. *pulchella* Hoppfer and *virens*, all given by Förster in 1846 (Handb. Cact. 242).   *Mammillaria elongata centrispina* (Förster, Handb. Cact. 240. 1846), which is only a name, may belong here.

Illustrations: Gartenflora **34**: pl. 1208, f. d, e, as *Mammillaria echinata*.

Figure 148 is from a photograph of the plant collected by Dr. Rose (No. 8990), mentioned above.

99. Neomammillaria pottsii (Scheer).

Mammillaria pottsii Scheer in Salm-Dyck, Cact. Hort. Dyck. 1849. 104.   1850.
Mammillaria leona Poselger, Allg. Gartenz. **21**: 94.   1853.
Echinocactus pottsianus Poselger, Allg. Gartenz. **21**: 107.   1853.
Cactus pottsii Kuntze, Rev. Gen. Pl. **1**: 261.   1891.

More or less cespitose, the individual plants cylindric, 12 cm. long or more; tubercles almost hidden by the spines; radial spines about 30, white, weak, short; central spines 6 to 12, much stouter and longer, more or less ascending, grayish with brown tips; axils of tubercles woolly; flowers borne in a circle about 2 cm. below top of plant, about 1 cm. long; inner perianth-segments light purple, somewhat spreading at tip, acute; stamens pale, much shorter than the style, purplish above; stigma-lobes narrow; fruit red, clavate; seeds blackish brown, the surface deeply pitted.

Type locality: Not cited.

Distribution: In the highlands of the Rio Grande, Texas; Nuevo León and Coahuila to Chihuahua and Zacatecas, Mexico.

This species is widely grown in collections but the flowers are inconspicuous.

In the Engelmann Collection, now in the Missouri Botanical Garden, is a specimen labeled "*Mammillaria pottsii vera*—original coll. Dyck. Jan. 1857." This proves to be identical with the plant well known in our collections as *M. leona*. With specimens of this plant in hand Salm-Dyck's description, which heretofore we had not understood, is clearly

FIG. 150.—Neomammillaria pottsii.

FIG. 151.—N. mazatlanensis.

interpreted, except that he states that the tubercle is slightly sulcate above. From the fact that Engelmann says that his specimen is "*M. pottsii vera*" we suspect that he may have had a plant like *M. tuberculosa* mixed with it. This seems to have been Poselger's idea, for he refers the plant to *Echinocactus*, doubtless on account of this supposed groove. The plant which Poselger describes under *Echinocactus pottsianus*, collected at Guerrero, south of the Rio Grande, is very different from Salm-Dyck's plant; his fragment, also deposited in the Missouri Botanical Garden, consists of a fruit, a few brownish seeds, and a spine-cluster, one attached to the top of a grooved tubercle, and is to be referred to *Escobaria tuberculosa*, or a related species. The specimen is too fragmentary to identify definitely. Poselger's misunderstanding of Salm-Dyck's plant left the way open for his species, *Mammillaria leona*, described shortly afterwards.

The description of the flower and fruit as given by Coulter is doubtless taken from Poselger but does not apply to the true *M. pottsii*. Our only Texas record is based on J. H. Ferriss's plant from the Big Bend of the Rio Grande, November 15, 1922.

Coryphantha pottsii occurs in C. R. Orcutt's Circular to Cactus Fanciers 1922 (unsigned and undated) to which he assigns *M. leona*.

Illustrations: Ann. Rep. Smiths. Inst. **1908**: pl. 2, f. 3; Blanc, Cacti 70. No. 1359, as *Mammillaria leona*.

Figure 150 is from a photograph of a cluster of plants obtained in Zacatecas by F. E. Lloyd in 1908.

100. Neomammillaria mazatlanensis (Schumann).

Mammillaria mazatlanensis Schumann, Monatsschr. Kakteenk. **11**: 154. 1901.
Mammillaria littoralis K. Brandegee, Kew Bull. Misc. Inf. **1908**: App. 91. 1908.

Plants cespitose, often forming broad clumps with many oblong heads, 4 to 10 cm. long, about 2 cm. in diameter; tubercles terete, 3 to 4 mm. long, their axils naked; radial spines 12 to 15, setaceous, spreading, white; central spines 4 to 6,* stouter than the radials, reddish, ascending, 8 to 10 mm. long; flowers from the axils of the old tubercles but towards the top of the plant, 3 cm. long or more, red; perianth-segments oblong, spreading; stigma-lobes 8, very long and slender.

Type locality: Mazatlán.

Distribution: On the hills near the sea, about Mazatlán, Mexico.

Dr. Rose collected this plant in 1897 and again in 1910. From this last collection we still have growing plants, but these have never flowered.

Mammillaria littoralis K. Brandegee, first mentioned in 1907 (Monatsschr. Kakteenk. **17**: 80), seems never to have been described by Mrs. Brandegee but was described in the Kew Bulletin as mentioned above, where it was stated to be from "California(?)." It was doubtless sent by Mrs. Brandegee from California but collected at Mazatlán.

Illustration: Monatsschr. Kakteenk. **15**: 155, as *Mammillaria mazatlanensis*.

Figure 151 is from a photograph sent by L. Quehl, showing a flowering plant.

101. Neomammillaria sphacelata (Martius).

Mammillaria sphacelata Martius, Nov. Act. Nat. Cur. **16**: 339. 1832.
Echinocactus sphacelatus Poselger, Allg. Gartenz. **21**: 107. 1853.
Cactus sphacelatus Kuntze, Rev. Gen. Pl. **1**: 261. 1891.

Usually densely cespitose, often grayish, forming clumps 3 to 4 dm. in diameter, the individual plants cylindric, more or less elongated, often 1 to 2 dm. high; radial spines 14 to 20, usually white with black tips; central spines 3 or 4, usually black or reddish throughout, sometimes becoming white in age; axils of tubercles often bearing tufts of short hairs and occasionally a few bristles; flowers about 15 mm. long, purplish; fruit red, clavate; seeds black, the surface deeply pitted.

Type locality: Mexico, possibly in Oaxaca or Puebla; it was collected by Karwinsky.

Distribution: Puebla and Oaxaca; Schumann reports it, but doubtless erroneously, from Hidalgo (Zimapán) and Sonora (Guaymas).

Illustrations: Nov. Act. Nat. Cur. **16**: pl. 25, f. 1; Monatsschr. Kakteenk. **28**: 74; Grässner, Haupt-Verz. Kakteen **1914**: 36, as *Mammillaria sphacelata*.

102. Neomammillaria albicans sp. nov.

Plants at first globose but becoming cylindric and then 10 to 20 cm. long, up to 6 cm. in diameter, often in clumps of 5 to 15; spines almost hiding the plant body and often pure white; radial

FIG. 152.—Neomammillaria albicans.

spines numerous, short, stiff, widely spreading; central spines several, straight, stiff, often brownish

---

* Sometimes one of the central spines is hooked, as is shown in plants from near the type locality collected by Señor J. G. Ortega in 1922.

or blackish at tip; spine-areoles when young densely white-woolly; fruit clavate, red, 10 to 18 mm. long; seeds black with basal hilum.

Collected on Santa Cruz Island, Gulf of California, by J. N. Rose, April 16, 1911 (No. 16842, type), and by Ivan M. Johnston in 1921 (No. 3912); also on the adjacent island of San Diego by Mr. Johnston (No. 3923).

This is a very beautiful plant which grows in small clusters and is covered with nearly pure white spines. A number of plants were brought back to the New York Botanical Garden in 1911 by Dr. Rose but they have all since died. We now have living plants sent in by Mr. Johnston from two localities.

Figure 152 is from a photograph of a plant sent by Mr. Johnston to Washington from the type locality.

Fig 153.—Neomammillaria slevinii.

103. Neomammillaria slevinii sp. nov.

Plants simple, cylindric, 1 dm. high or more, 5 to 6 cm. in diameter, entirely hidden under the many closely set spines; spines at top of plant pinkish below, with brown to blackish tips, on lower part of plant bleaching white; radial spines numerous, acicular, widely spreading; central spines about 6, a little longer and stouter than the radials, slightly spreading; flowers about 2 cm. broad; outer perianth-segments with a pinkish mid-rib; inner perianth-segments white; filaments pinkish; style nearly white; stigma-lobes nearly white; fruit red, about 1 cm. long; seeds black, nearly globular, with a projection at base and a large basal hilum.

Collected by J. N. Rose, March 31, 1911 (No. 16550, type), on San Josef Island, and by Ivan M. Johnston in 1921 (No. 3943) on San Francisco Island just off the southern end of San Josef Island.

This species is related to *Neomammillaria albicans*, but it has darker spines and the spine-areoles are not densely lanate.

The plant is named for J. R. Slevin, who was in charge of the scientific expedition of the California Academy of Sciences to the Gulf of California in 1921, at which time the plant was collected.

Figure 153 is from a photograph of one of the plants collected by Mr. Johnston and sent to Washington.

104. Neomammillaria palmeri (Coulter).

Cactus palmeri Coulter, Contr. U. S. Nat. Herb. **3**: 108. 1894.
Mammillaria dioica insularis K. Brandegee, Erythea **5**: 115. 1897.

Densely cespitose; individuals small; axils densely woolly and bristly; radial spines 25 to 30, slender, white, 5 mm. long, radiating; central spines 3 to 5, stouter and longer than the radials, brownish with black tips, straight, 7 to 8 mm. long; flowers cream-colored, sometimes tinged with pink; fruit clavate, scarlet; seeds black.

Type locality: "San Benito Island." *
Distribution: San Benito Islands and possibly Guadalupe Island off the west coast of Lower California.

Plate XIV, figure 7, shows the plant, collected on the San Benito Islands, which flowered in the New York Botanical Garden, April 1, 1912.

105. Neomammillaria uncinata (Zuccarini).

Mammillaria uncinata Zuccarini in Pfeiffer, Enum. Cact. 34. 1837.
Mammillaria bihamata Pfeiffer, Allg. Gartenz. **6**: 274. 1838.
Mammillaria depressa Scheidweiler, Bull. Acad. Sci. Brux. **5**: 494. 1838.
Mammillaria uncinata· biuncinata Lemaire, Cact. Gen. Nov. Sp. 96. 1839.
Mammillaria uncinata spinosior Lemaire, Cact. Gen. Nov. Sp. 96. 1839.
Mammillaria uncinata rhodacantha Hortus in Förster, Handb. Cact. ed. 2. 347. 1885.
Cactus bihamatus Kuntze, Rev. Gen. Pl. **1**: 260. 1891.
Cactus depressus Kuntze, Rev. Gen. Pl. **1**: 260. 1891. Not De Candolle, 1813.
Cactus uncinatus Kuntze, Rev. Gen. Pl. **1**: 261. 1891.

Globose or somewhat depressed, usually half-buried in the soil, 8 to 10 cm. in diameter; tubercles lactiferous, short, obtuse; axils of old tubercles naked, of young ones lanate, forming a mass of wool at top; young spine-areoles also lanate; radial spines 4 to 6, usually white, subulate, 4 to 5 mm. long; central spines usually solitary, sometimes 2 or 3, much stouter than the radials, 8 to 12 mm. long, brown, hooked at apex; flowers small, reddish white, about 2 cm. long; inner perianth-segments linear-oblong; stigma-lobes pinkish; fruit clavate, 10 to 18 mm. long, red; seeds small, brown.

Type locality: Mexico.
Distribution: Common in central Mexico, especially in Hidalgo and San Luis Potosí. Schumann reports it from Chihuahua, as collected by Wislizenus, but we suspect that there is an error. Pfeiffer does not give a definite locality for this species but Zuccarini, who redescribed the plant soon afterwards, says that Karwinsky obtained it in the mountains near Pachuca, Mexico.

This species and the following two are the only milk-bearing *Neomammillaria* which have hooked spines.

Mammillaria adunca Scheidweiler (Förster, Handb. Cact. *222*. 1846), referred here as a synonym, was never described.

Illustrations: Pfeiffer and Otto, Abbild. Beschr. Cact. **1**: pl. 19; Schumann, Gesamtb. Kakteen f. 94; Abh. Akad. Bayer. Wiss. München **2**: pl. 4, f. 3; Schelle, Handb. Kakteenk. 269. f. 191, as *Mammillaria uncinata.*

106. Neomammillaria hamata (Lehmann).

Cactus cylindricus Ortega, Nov. Rar. Pl. 128. 1800. Not Lamarck, 1783.
Mammillaria hamata Lehmann in Pfeiffer, Enum. Cact. 34. 1837.
Cactus hamatus Kuntze, Rev. Gen. Pl. **1**: 260. 1891.

Stem 6 dm. long, cylindric, somewhat branched at base, described as milky; tubercles conic or a little compressed; radial spines 15 to 20, white, spreading; central spines several, brownish, stouter than the radials, one of them hooked; flowers small, probably scarlet, from near top of plant but from

---

* Although San Benito Island is given as the type locality, San Benito is really a group of three small islands. Dr. Rose found this species on two of these islands in 1911 (No. 16042).

axils of old tubercles; inner perianth-segments lanceolate, acute; filaments half length of perianth-segments, white; stigma-lobes 4, yellowish; fruit slender, clavate, probably red; seeds minute, brown.

Type locality: Mexico.

Distribution: Mexico, but range not known.

Schumann referred both *Cactus cylindricus* and *Mammillaria hamata* to *M. coronaria,* but the last name must be excluded from this genus. The specific name, *cylindricus,* which has been used four times in the genus *Cactus,* can not be transferred to *Neomammillaria* on account of the earlier use of this specific name by Lamarck.

Mammillaria hamata was first mentioned in the Seed Catalogue of the Hamburg Garden in 1832.

FIG. 154.—Neomammillaria hamata.　　　　FIG. 155.—Neomammillaria wildii.

The following are usually referred as synonyms of *Mammillaria coronaria,* but probably belong here: *Mammillaria hamata brevispina* and *M. hamata principis* Salm-Dyck (Labouret, Monogr. Cact. 34. 1853) and *M. hamata longispina* Salm-Dyck (Cact. Hort. Dyck. 1844. 8. 1845). *Mammillaria principis* Monville (Labouret, Monogr. Cact. 34. 1853) was given as a synonym of the last variety here cited.

Illustration: Ortega, Nov. Rar. Pl. pl. 16, as *Cactus cylindricus.*

Figure 154 is reproduced from the illustration above cited.

107. Neomammillaria rekoi sp. nov.

Globular to short-cylindric, becoming 12 cm. long, 5 to 6 cm. in diameter, sometimes milky; tubercles green, terete, 8 to 10 mm. long, not very closely set, each bearing in its axil a tuft of short white wool and 1 to 8 long white bristles; radial spines spreading, about 20, white, delicately acicular, 4 to 6 mm. long; central spines 4, brown, much stouter than the radials, 10 to 15 mm. long, the lower one sometimes strongly hooked; flowers from axils of old tubercles, near top of plant; 1.5 cm. long, deep purple; inner perianth-segments narrowly oblong, apiculate; filaments and style purplish; stigma-lobes greenish; fruit clavate, red, 12 mm. long; seeds minute, brown.

FIG. 155a.—Neomammillaria rekoi.

This species has been sent to us repeatedly from Oaxaca, Mexico, by Dr. B. P. Reko and it has been named in his honor; we have selected as the type his specimen of 1921, which flowered in Washington.

This is a remarkable species, being the only one we know, except the following, which has the characters of watery tubercles, a hooked spine, and brown seeds, but some plants give out a very diluted milk and have no hooked spines.

Dr. Reko sent us a single plant in April 1922, which was about 12 cm. long and short-clavate; the central spines were mostly 4, but sometimes 5, and none of them hooked. In this specimen we obtained a diluted milky juice from the upper tubercles while the lower ones are entirely devoid of milk. It flowered in April 1923 and seemed to be referable here.

Figure 149 shows a plant sent by Dr. B. P. Reko from Oaxaca, Mexico, in 1919; figure 155a shows the plant collected by Dr. Reko in 1922, referred to above.

108. Neomammillaria solisii sp. nov.

Simple, globular or nearly so, 5 to 7 cm. in diameter, green or becoming purplish; tubercles 8 mm. long, terete in section, a little narrow towards the tip and thus separated above from the adjoining tubercles, their axils without wool even when quite young, and usually with 1 to many bristles; radial spines about 10 to 20, spreading, 6 to 7 mm. long, white, bristle-like; central spines 3 or 4, a little stouter than the radials, becoming brown, one of them strongly hooked (sometimes 2 cm. long).

Collected by Octavio Solís in Cerro de Buenavista de Cuellar, Guerrero, Mexico, in 1920 (No. 5) and in 1921, type, and at the same station by Professor C. Núñez in April and November 1921 (Nos. 4 and 6).

Figure 156 is from a photograph of a plant sent by Octavio Solís from Guerrero, Mexico, in 1920; figure 157 is from a photograph of a plant sent by Professor C. Núñez in 1922.

109. Neomammillaria pygmaea sp. nov.

Plant very small, globose to cylindric, 2 to 3 cm. in diameter; tubercles small, obtuse; radial spines about 15, white, stiff, hardly puberulent even under a lens; central spines 4, ascending, golden yellow, the lower one hooked, 5 to 6 mm. long; flowers about 1 cm. long, the outer segments tinged with red, apiculate; inner perianth-segments about 10, cream-colored; filaments greenish, much shorter than the perianth-segments; style greenish.

Collected by J. N. Rose near Cadereyta, Querétaro, Mexico, in 1905 (No. 9863). It has repeatedly flowered but was only 3 cm. high in 1921 when it died.

The species is known only from the single collection recorded above. It grows on stony hills in a very arid part of Querétaro. It is very inconspicuous and is easily overlooked in the field.

110. Neomammillaria wildii (Dietrich).

Mammillaria wildii Dietrich, Allg. Gartenz. 4: 137. 1836.
Mammillaria wildiana Otto in Pfeiffer, Enum. Cact. 37. 1837.
Mammillaria wildiana compacta Hortus in Förster, Handb. Cact. ed. 2. 258. 1885.
Mammillaria wildiana cristata Hortus in Förster, Handb. Cact. ed. 2. 258. 1885.
Cactus wildianus Kuntze, Rev. Gen. Pl. 1: 261. 1891.

Cylindric to globose, cespitose at base; axils of tubercles bearing rose-colored hairs and bristles; tubercles slender, elongated, 8 to 10 mm. long, obtuse, green or somewhat rose-colored at base; young areoles tomentose; spines all pubescent; radial spines 8 to 10, 8 mm. long, setiform, white; central spines 4, yellow, one of them hooked; flowers white, 12 mm. in diameter; inner perianth-segments acuminate; stigma-lobes 4 or 5, straw-colored; fruit clavate, red.

FIGS. 156 and 157.—Neomammillaria solisii.

Type locality: Mexico.

Distribution: State of Hidalgo, Mexico, according to Schumann.

We have had this plant growing for a number of years, obtained from other collectors, but we do not know its natural habitat. It sprouts freely and new plants are easily started. Dr. Rose examined a specimen, labeled *Mammillaria wildii*, in the Botanical Garden at Halle in 191; we have a cluster of spines and a flower of that plant.

Mammillaria glochidiata aurea (Pfeiffer, Enum. Cact. 37. 1837), although never described, is referred usually as a synonym of this species. The two varieties of *Mammallaria wildii*, *cristata* and *compacta*, are listed but not described by Schelle (Handb. Kakteenk. 251. 1907), the latter being offered for sale by Grässner in his Kakteen for 1914 as form *cristata*.

The two varieties, *Mammillaria wildiana major* and *M. wildiana spinosior*, were given by Walpers (Repert. Bot. 2: 270. 1843) as synonyms of *M. wildiana*. The variety *monstrosa* Cels was given by Rümpler (Förster, Handb. Cact. ed. 2. 258. 1885) as a synonym of *M. wildiana cristata*.

Illustrations: Blühende Kakteen 2: pl. 64; Monatsschr. Kakteenk. 32: 103, as *Mammillaria wildii*; Grässner, Haupt-Verz. Kakteen 1912: 27, as *M. wildii cristata*.

Plate xiv, figure 8, shows a plant from the Missouri Botanical Garden which flowered in the New York Botanical Garden, April 25, 1913. Figure 155 is reproduced from the first illustration cited above.

111. Neomammillaria seideliana (Quehl).

Mammillaria seideliana Quehl, Monatsschr. Kakteenk. **21**: 154. 1911.

Solitary, globose, becoming cespitose, 3 to 4 cm. in diameter; tubercles purplish, their axils naked; radial spines 20 to 25, white, long and slender, ascending, puberulent; central spines yellow, 3 or 4, puberulent when young, one hooked; flowers arising from near top of plant, about 15 to 18 mm. long, creamy yellow; outer perianth-segments brownish; inner perianth-segments oblong, acute; style cream-colored, much longer than stamens; stigma-lobes 5 or 6, cream-colored, obtuse; fruit persisting in axils of tubercles, apparently for a number of years; seeds black, with thick neck at base; the hilum basal, large.

Type locality: Zacatecas, Mexico.

Distribution: Known only from the state of Zacatecas.

Collected by F. E. Lloyd in Zacatecas, Mexico, in 1908 (No. 54), who states that he found but a single specimen, though he made diligent search for others.

Although the flowers appear to come from near the top of the plant they are all from axils of old tubercles. In the single specimen examined the flowers appeared before the plant began to form new tubercles. In *Mammillaria barbata*, a closely related species, the flowers occur at both the old and new tubercles, but so far as known no other species possesses that character, although there is no good reason for not finding it in closely related species.

FIGS. 157a and 158.—Neomammillaria seideliana.

We have also had a plant sent us by Haage and Schmidt; it is a profuse bloomer.

Illustration: Monatsschr. Kakteenk. **21**: 155, as *Mammillaria seideliana*.

Figure 157a is from a photograph of a plant sent us from Zacatecas, Mexico, by Professor Lloyd in 1908; figure 158 is from a photograph sent by L. Quehl.

112. Neomammillaria barbata (Engelmann).

Mammillaria barbata Engelmann in Wislizenus, Mem. Tour North. Mex. 105. 1848.
Cactus barbatus Kuntze, Rev. Gen. Pl. **1**: 260. 1891.

Often densely cespitose, globose, 3 to 4 cm. in diameter; radial spines 20 or more, acicular, spreading or ascending, white, sometimes with brown tips; central spines several, subulate, brown, puberulent, 1 or 2 hooked; flowers 15 mm. long; outer perianth-segments ovate to lanceolate, ciliate; inner perianth-segments erect or spreading at tip, light straw-colored or greenish, brown without, acute; filaments numerous, short, purplish; stigma-lobes 5 to 7, greenish.

Type locality: Cosihuirachi, Mexico.

Distribution: Western Chihuahua, Mexico.

This species was collected by Dr. Wislizenus in 1846 and rediscovered and collected at the type locality in 1908 by Dr. Rose, and upon this latter collection the above description is based. Schumann did not recognize the species, but thought that it might be near *Mammillaria grahamii*.

Illustrations: Cact. Mex. Bound. pl. 6, f. 9 to 12; Monatsschr. Kakteenk. **20**: 181; Gartenflora **34**: pl. 1208, f. a, b, c; **43**: pl. 1400, as *Mammillaria barbata*.

Figure 159 is from a photograph of the specimen collected by Dr. Rose in 1908 at the type locality.

113. Neomammillaria mercadensis (Patoni).

Mammillaria mercadensis Patoni, Alianza Cientifica Universal **1**: 54. 1910.
Mammillaria ocamponis Ochoterena, Bol. Direccion Estudios Biol. **2**: 355. 1918.

Solitary or cespitose, small, globose; radial spines numerous, sometimes 25, widely spreading, white; central spines 4 or 5, elongated, much longer than the flowers, one of them strongly hooked at apex; flowers small, pale rose-colored; perianth-segments oblong, obtuse.

Type locality: Cerro de Mercado, Durango.
Distribution: Durango, Mexico.
We know this plant only from descriptions and illustrations.

Fig. 159.—Neomammillaria barbata.

Illustrations: Alianza Cientifica Universal **3**: pl. facing 223, as *Mammillaria barbata*; Bol. Direccion Estudios Biol. **2**: facing 356, as *Mammillaria ocamponis*.

Figure 160 is from a photograph of the type plant, which has the same origin as the illustrations cited above.

114. Neomammillaria kunzeana (Bödeker and Quehl).

Mammillaria kunzeana Bödeker and Quehl, Monatsschr. Kakteenk. **22**: 177. 1912.
Mammillaria bocasana kunzeana Quehl, Monatsschr. Kakteenk. **26**: 46. 1916.

Cespitose, globose or sometimes becoming cylindric, light green; tubercles cylindric, setose in their axils; radial spines about 25, white, setaceous; central spines 3 or 4, brown, puberulent, one of them hooked; flowers white or yellowish white, rose-colored on the outside, 2 cm. long; inner perianth-segments acuminate; stigma-lobes 4, whitish yellow.

Type locality: Mexico.
Distribution: Mexico, but range unknown.
This species is dedicated to Dr. Richard Ernest Kunze (1838-1919), who was an enthusiastic student of cacti and for many years a resident of Phoenix, Arizona. He sent the plant to Germany in 1910.

Illustration: Monatsschr. Kakteenk. 22: 178, as *Mammillaria kunzeana*.

Plate XIV, figure 1, is of a plant obtained by Dr. Rose in 1912 from W. Mundt as *Mammillaria bocasana*, which flowered in the New York Botanical Garden, April 21, 1914.

115. Neomammillaria hirsuta (Bödeker).

Mammillaria hirsuta Bödeker, Monatsschr. Kakteenk. 29: 130. 1919.

Solitary or becoming cespitose, globose, about 6 cm. in diameter; tubercles 10 mm. long, in 8 or 13 spiraled rows, cylindric, their axils setose; spine-areoles naked; radial spines about 20, white, 10 to 15 mm. long; central spines 3 or 4, the lower one hooked; flowers small, 10 mm. long; fruit and seeds unknown.

Type locality: Mexico.

Distribution: Mexico, but range unknown.

The plant was exhibited by de Laet at Contich, Belgium, in 1914, as sent to him by Mrs. Nichols, presumably from northern Mexico.

Illustration: Monatsschr. Kakteenk. **29**: 131, as *Mammillaria hirsuta*.

FIG. 160.—Neomammillaria mercadensis.　　FIG. 161.—N. multihamata.

116. Neomammillaria multihamata (Bödeker).

Mammillaria multihamata Bödeker, Monatsschr. Kakteenk. **25**: 76. 1915.

Short-cylindric, about 5 cm. in diameter; tubercles cylindric, setose in their axils; spine-areoles white-lanate; radial spines 25, acicular, white, 8 mm. long; central spines 7 to 9, several of them hooked; flowers numerous from near top of plant, small, 1.5 cm. long; inner perianth-segments narrow, acute, spreading; seeds blackish brown.

Type locality: Mexico.

Distribution: Mexico, but range unknown.

This plant is in the trade. A specimen was sent us in 1914 by L. Quehl, but it never flowered and soon died.

Illustration: Monatsschr. Kakteenk. **25**: 77, as *Mammillaria multihamata*.

Figure 161 is reproduced from a photograph furnished by L. Quehl.

117. Neomammillaria longicoma sp. nov.

Cespitose, often forming broad clumps; individual specimens 3 to 5 cm. in diameter; tubercles conic, 4 to 5 mm. long, dark green, obtuse, bearing long white hairs in their axils; radial spines 25 or more, weak and hair-like, more or less interlocking; central spines 4, 10 to 12 mm. long, brown above, a little paler below, 1 or 2 hooked; flowers from axils of upper tubercles; outer perianth-segments pinkish, darker along the center; inner perianth-segments lanceolate, acute, nearly white or sometimes tinged with rose; stamens and style much shorter than the inner perianth-segments; stigma-lobes 3, cream-colored.

The plant is common about San Luis Potosí, Mexico, where it was collected by Dr. E. Palmer in 1905 (type) and by Mrs. Irene Vera in 1912. We have had it in cultivation since

1905. It differs considerably from *Neomammillaria kunzeana*, from the same region, in its hair-like radial spines. It is perhaps nearest *M. bocasana*, but that species has single central spines.

Illustration: Ann. Rep. Smiths. Inst. **1908**: pl. 4, f. 4, as *Mammillaria bocasana*.

Figure 162 is from a photograph of a plant (type) collected by Dr. E. Palmer near San Luis Potosí in 1905 and figure 165 shows a cluster of plants from the same colony.

118. Neomammillaria bocasana (Poselger).

Mammillaria bocasana Poselger, Allg. Gartenz. **21**: 94. 1853.
Cactus bocasanus Coulter, Contr. U. S. Nat. Herb. **3**: 104. 1894.

Cespitose, often forming large mounds; individual plants globose, 3 to 4 cm. in diameter, light green; tubercles slender, 6 to 8 mm. long, terete, their axils sometimes hairy or bristly; radial spines represented by numerous long white silky hairs; central spines solitary, 5 to 8 mm. long, brown, but paler at base, hooked, much shorter than the radial hairy ones; flower-buds rose-colored; flowers described as white; perianth-segments lanceolate-linear, acute, spreading; fruit "green, 4 mm. long; seeds cinnamon brown, oblique, broadly obovate, with narrowly basal hilum."

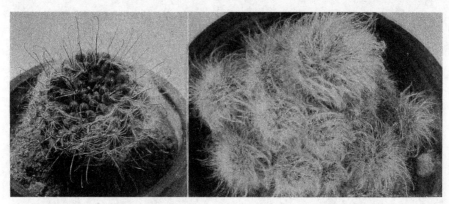

FIG. 162.—Neomammillaria longicoma.   FIG. 163.—Neomammillaria bocasana.

Type locality: Sierra de Bocas,* Mexico.

Distribution: Northern central Mexico, especially in San Luis Potosí.

This species has not been well understood and is usually misnamed in collections.

The two varieties of *Mamillaria bocasana*, *cristata* and *glochidiata*, are listed by Schelle (Handb. Kakteenk. 250. 1907), but not described. The former is offered for sale by Grässner in his Kakteen for 1914. We do not find that *M. bocasana splendens* Liebner and *M. bocasana sericata* Lemaire, mentioned by Quehl (Monatsschr. Kakteenk. **19**: 46. 1909), have ever been described.

Mammillaria schelhasei lanuginosior Hildmann (Schumann, Gesamtb. Kakteen 531. 1898) we have not seen but it may belong here.

Mammillaria bocasana splendens, credited to Schlechtendal, is offered for sale by Haage and Schmidt in their 1922 Catalogue.

Illustrations: Schelle, Handb. Kakteenk. 250. f. 170; Blanc, Cacti 67, No. 1148; West Amer. Sci. **13**: 40 (these three illustrations are from the same source); Blühende Kakteen **1**: pl. 35; Monatsschr. Kakteenk. **31**: 103; Schumann, Gesamtb. Kakteen f. 89, as *Mammillaria bocasana*; De Laet, Cat. Gén. 28. f. 42; Schelle, Handb. Kakteenk. 251. f.

*Coulter (Contr. U. S. Nat. Herb. **3**: 104) states that Poselger says the plant is from Texas "auf der Seira de Bocas," but in the original place of publication he does not give the state. Bocas, however, is in San Luis Potosí.

171; Rev. Hort. Belg. **40**: after 186; Tribune Hort. **4**: pl. 139 (these four illustrations are all from the same source); Möllers Deutsche Gärt. Zeit. **25**: 475. f. 8, No. 25; Monatsschr. Kakteenk. **29**: 81, as *Mammillaria bocasana cristata*.

Plate XIV, figure 2, shows a plant, collected by S. S. Hordes in 1915, which flowered in the New York Botanical Garden, May 11, 1916. Figure 163 shows a plant received from San Luis Potosí through Mrs. Irene Vera in 1912.

119. Neomammillaria multiformis sp. nov.

Cespitose, forming dense clumps, sometimes 25 or more from a single root, either globose or much elongated and 3 to 6 times as long as thick; tubercles terete, 6 to 8 mm. long, their axils bearing long white bristles and white wool; radial spines 30 or more, acicular, 8 mm. long, yellow

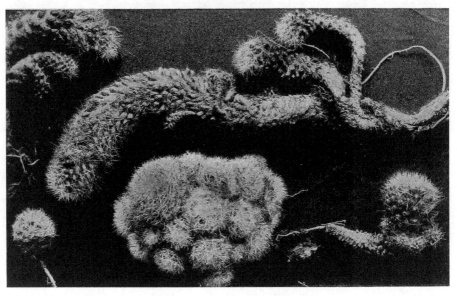

FIG. 164.—Neomammillaria multiformis.

or at least becoming so, ascending; central spines 4, a little longer and stouter than radials, nearly erect, reddish in upper part, one of them strongly hooked; flowers deep purplish red, 8 to 10 mm. long, usually broader than long; inner perianth-segments oblong, acute; filaments red; fruit nearly globose, at least when dry; seeds black.

Collected by Dr. E. Palmer at Alvarez, near San Luis Potosí, Mexico, in May 1905 (No. 591, type, and No. 592).

Figure 164 is from a photograph made from Dr. Palmer's specimen just after it was received in Washington.

120. Neomammillaria scheidweileriana (Otto).

Mammillaria glochidiata sericata Lemaire, Cact. Gen. Nov. Sp. 40. 1839.
Mammillaria scheidweileriana Otto in Dietrich, Allg. Gartenz. **9**: 179. 1841.
Mammillaria wildiana rosea Salm-Dyck, Cact. Hort. Dyck. 1849. 81. 1850.
Cactus scheidweilerianus Kuntze, Rev. Gen. Pl. **1**: 261. 1891.
Mammillaria monancistria* Berg in Schumann, Gesamtb. Kakteen 533. 1898.

* The publication of *Mammillaria monancistria* is usually referred to Förster's Handbuch (254. 1846), but the name occurs there without description.

Cespitose, globose to cylindric, light green; tubercles setose in their axils, in 8 and 13 spirals, cylindric; spines all puberulent; radial spines 9 to 11, setaceous, white, 1 cm. long; central spine, 1 to 4, brown, 1 or 2 hooked; flowers rose-colored, 12 to 13 mm. long.

Type locality: Mexico.

Distribution: Mexico, but range unknown.

The plant is known to us from description only.

121. Neomammillaria saffordii sp. nov.

Plants small, globose to short-cylindric, 3 to 4 cm. high, dull green, nearly hidden under the dense covering of spines; axils naked; spine-areoles when quite young slightly woolly, but early glabrate, circular; spines all puberulent under a lens when young; radial spines 12 to 14, somewhat ascending, but in age more or less curved outward, when just developing with bright red tips and white bases, later the lower part becoming yellowish; central spines single, stout, reddish, 1.5 cm. long, hooked at apex; flowers 2.5 cm. long, rose-colored; outer perianth-segments tipped by long bristles, the inner obtuse; stigma-lobes green.

This beautiful little species was collected by W. E. Safford, February 3, 1907, near Icamole, Nuevo León (No. 1250). Two plants, which were sent to Washington, flowered June 21, 1912; but they have not done well in cultivation. The plants sprout freely in cultivation and in this way we hope to distribute material to other collections. It is near *Mammillaria carretii* and was so figured by Dr. Safford, but it differs in several important respects from that species. It is named for Dr. Safford, the author of a very interesting paper, entitled Cactaceae of Northeastern and Central Mexico (Ann. Rep. Smiths. Inst. 1908), frequently referred to in these volumes.

Illustration: Ann. Rep. Smiths. Inst. **1908:** pl. 4, f. 2, as *Mammillaria carretii.*

Figure 168 is from a photograph of the type plant.

122. Neomammillaria schelhasei (Pfeiffer).

Mammillaria schelhasii Pfeiffer, Allg. Gartenz. **6:** 274. 1838.
Mammillaria glochidiata purpurea Scheidweiler, Bull. Acad. Sci. Brux. **5:** 495. 1838.
Cactus schelhasii Kuntze, Rev. Gen. Pl. **1:** 261. 1891.

Cespitose, forming a large hemispheric mound; individual plants globose to short-cylindric, olive-green; tubercles cylindric, their axils a little woolly, but not setose; radial spines 14 to 16, setaceous, white; central spines 3, brown, one hooked at apex; flowers large, 2.2 to 2.5 cm. long, salmon or rose-colored (Nicholson says white with line of rose down each petal); fruit 5 mm. long.

Type locality: Mineral del Monte, Mexico.

Distribution: Hidalgo, Mexico.

Salm-Dyck (Cact. Hort. Dyck. 1849. 7, 81. 1850) describes the three following varieties: *sericata, rosea,* and *triuncinata,* some of which may belong elsewhere. Of these Schumann recognizes only the last. The first Lemaire has referred to a different species, *Mammillaria glochidiata sericata* Lemaire (Cact. Gen. Nov. Sp. 40. 1839).

Illustrations: Schelle, Handb. Kakteenk. 252. f. 172; Dict. Gard. Nicholson **4:** 565. f. 37; Suppl. 518. f. 555; Förster, Handb. Cact. ed. 2. 254. f. 24 (32, in error); Rümpler, Sukkulenten 198. f. 111; Watson, Cact. Cult. 173. f. 69; ed. 3. f. 47; Knippel, Kakteen pl. 25; Blühende Kakteen **3:** pl. 170; Monatsschr. Kakteenk. **30:** 163, as *Mammillaria schelhasei;* Gartenflora **6:** pl. 207, as *M. schelhasei sericata.*

123. Neomammillaria glochidiata (Martius).

Mammillaria glochidiata Martius, Nov. Act. Nat. Cur. **16:** 337. 1832.
? *Mammillaria ancistroides* Lehmann, Del. Sem. Hort. Hamb. 1832.
Cactus glochidiatus Kuntze, Rev. Gen. Pl. **1:** 260. 1891.
Cactus ancistrodes Kuntze, Rev. Gen. Pl. **1:** 261. 1891.

Densely cespitose, forming clusters sometimes 15 cm. high; tubercles cylindric, green, shining, 8 to 15 mm. long, well separated from one another towards the tip, obtuse, terete; radial spines 12 to 15, widely spreading, puberulent, white, setiform, 10 to 12 mm. long; central spines 4, brownish, one of them hooked; flowers white; inner perianth-segments lanceolate, acuminate; style longer than the stamens; stigma-lobes 4 or 5, yellow; fruit clavate, scarlet, 16 mm. long; seeds black.

Type locality: Mexico.

Distribution: Southern Mexico.

Martius, who described this species, based it on a plant of Karwinsky, but did not cite a definite locality; Hemsley, however, records Karwinsky's plant as from near San Pedro Nolasco, Hidalgo, at 7,000 to 8,000 feet altitude.

As it is a high mountain species it would doubtless not remain long in cultivation. Pfeiffer refers here *Mammillaria criniformis* De Candolle (Mém. Cact. 8. pl. 4. 1834) and transfers his two varieties *rosea* and *albida* to *M. glochidiata* as variety *rosea* and *albida* (Enum. Cact. 37. 1837). *Mammillaria criniformis* must be very different, for it has only 8 to 10 radial spines and one central spine, and this yellow. The two varieties also may belong elsewhere; in fact, the variety *rosea* has been referred to *Mammillaria decipiens,*

Mammillaria ancistrata Schelhase (Salm-Dyck, Cact. Hort. Dyck. 1844. 8. 1845). given as a synonym of *M. ancistroides* Lemaire, is referred here by Schumann, perhaps wrongly.

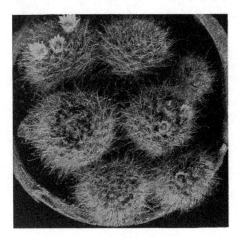

FIG. 165.—Neomammillaria longicoma.   FIG. 166.—Neomammillaria glochidiata.

Mammillaria ancistrina Hortus (Salm-Dyck, Cact. Hort. Dyck. 1849. 10. 1850) was given as a synonym of *M. ancistroides.*

To *Mammillaria ancistroides major* (Salm-Dyck, Cact. Hort. Dyck. 1844. 8. 1845) was referred *M. ancistrata* as a synonym. Afterwards it was briefly described in Förster's Handbuch.

Mammillaria bergeana, a name from Hildmann's Catalogue, is referred as a synonym of *M. glochidiata* (Schumann, Gesamtb. Kakteen 532. 1898), and so also is *M. glochidiata alba* (Förster, Handb. Cact. 188. 1846).

Mammillaria ancistroides Lehmann (Delect. Sem. Hort. Hamb. 1832) is usually referred to this species but it must go elsewhere; it has setae in the axils of the tubercles, the radial spines are 6 to 8, and the hooked spine is brown at tip.

Schumann (Gesamtb. Kakteen 532. 1898) describes two varieties, *crinita* and *prolifera.* The former is based on *Mammillaria crinita* De Candolle (Mém. Mus. Hist. Nat. Paris **17**: 112. 1828; *Cactus crinitus* Kuntze, Rev. Gen. Pl. **1**: 260. 1891), and has the central spines straight (at least so shown in the illustration, but described as hooked), and must be excluded from this species. *Mammillaria crinita pauciseta* De Candolle (Mém. Mus. Hist. Nat. Paris **17**: 112. 1828) may be of this relationship but we do not know it.

Other varietal names have been given, such as *M. glochidiata alba* (Förster, Handb. Cact. 188. 1846).

Illustrations: Blühende Kakteen **2**: pl. 82; Nov. Act. Nat. Cur. **16**: pl. 23, f. 1; Abh. Bayer. Akad. Wiss. München **2**: pl. 1, I. f. 4; Monatsschr. Kakteenk. **29**: 141, as *Mammillaria glochidiata*. The following illustrations we have not placed: De Candolle, Mém. Cact. pl. 3; Krook, Handb. Cact. 38, as *M. crinita*; De Candolle, Mém. Cact. pl. 4, as *M. criniformis*.

Figure 166 is reproduced from the original illustration of the type as shown in the second illustration cited above.

Fig. 167.—Neomammillaria trichacantha.   Fig. 168.—Neomammillaria saffordii.

124. Neomammillaria trichacantha (Schumann).

Mammillaria trichacantha Schumann, Gesamtb. Kakteen Nachtr. 133. 1903.

Solitary, globose to short-cylindric, small; tubercles small, clavate, 4 to 5 cm. high, slightly glaucous; radial spines 15 to 18, pubescent, acicular, white, 8 mm. long; central spines 2, brownish, 12 mm. long, one of them hooked; flowers red or yellow, 1.5 cm. long; inner perianth-segments lanceolate, widely spreading, acuminate; style pale green; stigma-lobes white.

Type locality: Not cited.

Distribution: Undoubtedly Mexico, but known only from cultivated plants.

The relationship of this species is somewhat uncertain. Schumann placed it next to *Mammillaria carretii* and described the flowers as red, while Quehl stated that the inner perianth-segments are pale yellow, and this is clearly shown by an unpublished study of Mrs. Gürke, made May 26, 1907, now in our possession. We have received such flowers from Quehl.

Quehl refers here *Mammillaria hamuligera* (sometimes written *M. lamuligera*) while Bödeker would keep it distinct. We have received flowers from Quehl which correspond with Mrs. Gürke's painting of *M. trichacantha*, but her plant may be different from Schumann's type, which had red flowers.

Illustrations: Schumann, Gesamtb. Kakteen Nachtr. 133. f. 33; Monatsschr. Kakteenk. **14**: 45, as *Mammillaria trichacantha*.

Figure 167 is reproduced from the first illustration cited above.

125. Neomammillaria painteri (Rose).

Mammillaria painteri Rose in Quehl, Monatsschr. Kakteenk. **27**: 22. 1917.
Mammillaria erythrosperma Bödeker, Monatsschr. Kakteenk. **28**: 101. 1918.
Mammillaria erythrosperma similis De Laet in Bödeker, Monatsschr. Kakteenk. **28**: 102. 1918.

Plant globose, small, 2 cm. in diameter, almost hidden by the spines; tubercles without bristles in their axils; radial spines about 20, stiff, white, puberulent under a hand lens; central spines 4 or 5, ascending, dark brown, one hooked, puberulent; flowers 15 mm. long, greenish white, the outer segments brownish; inner perianth-segments broad, with an ovate acute tip; stamens white; stigma-lobes cream-colored.

Type locality: Near San Juan del Rìo, Querétaro.

Distribution: Central Mexico.

Collected in Querétaro, Mexico, in 1905 by J. N. Rose. It has flowered repeatedly in cultivation (August 1909, June 1911, 1912, April 1915), and is nearest perhaps to *Neomammillaria kunzeana* and *N. multihamata*, but the axils of the tubercles are naked.

Illustrations: Monatsschr. Kakteenk. **27**: 23, as *Mammillaria painteri*; Monatsschr. Kakteenk. **28**: 103, as *M. erythrosperma* and var. *similis*.

Figure 169 is from a photograph of the type plant.

FIG. 169.—Neomammillaria painteri.   FIG. 170.—Neomammillaria microcarpa.

126. Neomammillaria wrightii (Engelmann).

Mammillaria wrightii Engelmann, Proc. Amer. Acad. **3**: 262. 1856.
Cactus wrightii Kuntze, Rev. Gen. Pl. **1**: 261. 1891.

Depressed-globose, simple; tubercles terete, 10 to 12 mm. long, with naked axils; radial spines 8 to 15, white, spreading, acicular; central spines 1 to 3, stouter than the radials, brown to black, 1 or sometimes 2 or 3 hooked at apex; flowers large, 25 mm. long and as broad as long when expanded; outer segments about 13, triangular-obtuse, fimbriate; inner perianth-segments bright purple; fruit obovoid, large, 25 mm. long, purple; seeds 1.5 mm. long, black, with a narrow ventral hilum.

Type locality: Anton Chico on the Pecos east of Santa Fé, New Mexico.

Distribution: Mountains of northeastern New Mexico.

Mammillaria wrightii as described by Dr. Engelmann is complex, his original description being based on two collections, one from the upper Pecos, the type, and one from the Santa Rita Copper mines in southwestern New Mexico. This latter specimen is referable to a new species described below. There has always existed much confusion regarding *M. wrightii*, and several species have been distributed under that name. It is very rare in collections. In the National Herbarium we have only a part of the type (3 clusters of spines) and spines and fruit collected by J. W. Toumey at White Oaks, New Mexico, October 20, 1896. Engelmann cites a specimen in Mexico (near Lake Santa Maria) which doubtless is to be referred elsewhere.

This species was named for Charles Wright (1811-1855), who explored extensively in Texas and Cuba.

Illustrations: Cact. Mex. Bound. pl. 8, f. 1 to 8; Monatsschr. Kakteenk. **14:** 9; Möllers Deutsche Gärt. Zeit. **25:** 475. f. 8, No. 5; West Amer. Sci. **13:** 40; Förster, Handb. Cact. ed. 2. 249. f. 23 (as f. 31, in error); Schelle, Handb. Kakteenk. 255. f. 177; Remark, Kakteenfreund 16, 17, as *M. wrightii*.

Figure 171 is a reproduction of the first illustration cited above.

Fig. 171.—Neomammillaria wrightii.   Fig. 172.—Neomammillaria mainae.

127. Neomammillaria viridiflora sp. nov.

Globular to short-oblong, 5 to 10 cm. long, the plant-body well hidden under the closely appressed radial spines; tubercles terete, small, naked in their axils; radial spines 20 to 30, widely spreading, white with brown tip, bristle-like, 10 to 12 mm. long; central spines much stouter than the radials, 1.5 to 2 cm. long, brown, one or more of them hooked; flowers greenish, narrowly campanulate, 1.5 cm. long; fruit globose to ovoid, 10 to 15 mm. long, purplish, very juicy; seeds |minute, 1 mm. long.

Collected by C. R. Orcutt on Superior-Miami Highway, near Boundary Monument, between Pinal and Gila counties, Arizona, 4,700 feet elevation, July, 1922 (No. 608, type), and by Mrs. Ruth C. Ross near Tula Spring, south of Aravaipa, Arizona, June 1922 (No. 14).

Here perhaps are to be referred plants collected in New Mexico by O. B. Metcalfe (Nos. 797, 803, and 820) and probably that part of *Mammillaria wrightii* which came from Santa Rita. Mr. Orcutt has repeatedly written to us about this green-flowered species, which we are now able to separate very distinctly from both *M. wrightii* and *M. wilcoxii*.

Dr. Forrest Shreve has also reported a green-flowered species from Arizona which he states is common in oak-woods.

128. Neomammillaria wilcoxii (Toumey).

Mammillaria wilcoxii Toumey in Schumann, Gesamtb. Kakteen 545. 1898.

Solitary, almost globose, flabby in texture, 10 cm. in diameter, almost covered by a mass of interlocking spines; axils of tubercles naked; radial spines 14 to 20, widely spreading, often 15 mm. long, bristle-like, white with colored tips; central spines 1 to 3, brown, 2 cm. long, 1 or more hooked; flowers pink to purple, large, 3 cm. long, 4 cm. broad when fully expanded; outer perianth-segments about 20, fringed with white hairs; inner perianth-segments about 40, in 2 rows.

Type locality: Arizona.
Distribution: Southeastern Arizona. It should be looked for in northern Sonora.

This species is very rare in living collections and in herbaria. When found in the field it is often associated with *Mammillaria grahamii* and *Coryphantha aggregata*, which has led to the suggestion that it might be a hybrid between these species.

The plant is named for General Timothy E. Wilcox, U. S. A., who collected extensively in Arizona, Oklahoma, Washington, and Alaska.

Illustration: Monatsschr. Kakteenk. **24**: 23, as *Mammillaria wilcoxii.*

Plate XIII, figure 1, is from a photograph of a plant collected at Calabasas, Arizona, by Dr. Rose in 1908 (No. 11955).

129. Neomammillaria mainae (K. Brandegee).

Mammillaria mainae K. Brandegee, Zoe **5**: 31. 1900.

Globose or somewhat depressed, 5 to 8 cm. broad; tubercles pale green, naked in their axils; spines all puberulent, at least when young; radial spines about 10, widely spreading, yellowish or white except the brownish tips; central spines usually stout, yellowish except the strongly hooked tip; flowers from upper part of plant but in old axils, about 2 cm. long, with a broad open throat; outer perianth-segments with a brownish stripe, inner ones with a reddish central stripe with broad nearly white margins; acute inner perianth-segments more or less spreading; stamens purplish; style also purplish, stout, much longer than stamens; stigma-lobes 5 or 6, purplish, elongated, linear; fruit red, globose to obovate, not projecting beyond the tubercles; seeds dull black, obovate, 1 mm. long, punctate, with a narrow basal hilum.

Type locality: South of Nogales, Sonora, Mexico.

Distribution: Northern Sonora.

For a long time it was known only from material collected by Mrs. F. M. Main, near Nogales, Mexico. It has been offered in the trade under the name of *Mammillaria galeottii*, to which, according to Mrs. K. Brandegee, it is not at all related. It was observed by Rose, Standley, and Russell in two localities near Hermosillo, Sonora, Mexico, and living plants were sent to Washington, which flowered in August 1910. This is not very close to any of the other species. It was collected again in Sonora by C. R. Orcutt in 1922.

Illustration: Monatsschr. Kakteenk. **22**: 19, as *Mammillaria mainae.*

Figure 172 is from a photograph of a specimen sent by Dr. Trelease from the Missouri Botanical Garden in 1910.

130. Neomammillaria boedekeriana (Quhel).

Mammillaria boedekeriana Quehl, Monatsschr. Kakteenk. **20**: 108. 1910.

Globose to ovoid, but in collections becoming cylindric, dull green; tubercles cylindric; radial spines about 20, white; central spines 3, brownish black, one hooked; axils naked; flowers white with brownish stripes.

Type locality: Not cited.

Distribution: Doubtless Mexico, but range unknown.

This plant, which was for a long time in cultivation in Europe, has, according to Mr. Bödeker, entirely disappeared. He writes that it is a prolific bloomer and that once he had a plant with 32 flowers open at the same time.

FIG. 172a.—Neomammillaria boedekeriana.

The species is named for Friederich Bödeker of Cologne, Germany. Quehl groups this species next to *Mammillaria wrightii.*

Illustration: Monatsschr. Kakteenk. **20**: 109, as *Mammillaria boedekeriana.*

Figure 172a is from a photograph of a plant which had been in cultivation 14 years by Bödeker. The photograph was sent to us in 1923.

1. *Neomammillaria wilcoxii*, from Calabasas, Arizona.
2. *Neomammillaria gaumeri*, from Yucatan, Mexico.

131. Neomammillaria microcarpa (Engelmann).

Mammillaria microcarpa Engelmann in Emory, Mil. Reconn. 157. 1848.
Mammillaria grahamii Engelmann, Proc. Amer. Acad. 3: 262. 1856.
Cactus grahamii Kuntze, Rev. Gen. Pl. 1: 260. 1891.
Mammillaria grahamii arizonica Quehl, Monatsschr. Kakteenk. 6: 44. 1896.
Coryphantha grahamii Rydberg, Fl. Rocky Mountains 581. 1917.

Globose to cylindric, simple or budding either at base or near middle, often cespitose, but in small clusters, sometimes 8 cm. high; tubercles small, corky when old; axils of tubercles naked; radial spines 15 to 30, spreading, white, sometimes with dark tips, slender, rigid, glabrous, 6 to 12 mm. long; central spines 1 to 3, dark, when more than one the lower stouter, often 18 mm. long, hooked; flowers from near top of plant, 2 to 2.5 cm. long, broadly funnel-shaped; outer perianth-segments ovate, obtuse, short-ciliate; inner perianth-segments purplish, sometimes with whitish margins, obovate, acuminate; style longer than stamens, purplish; stigma-lobes 7 or 8, linear, green; fruit clavate, 2 to 2.5 cm. long, scarlet; seeds black, shining, pitted, globose, 0.8 to 1 mm. in diameter.

Type locality: "On the Gila, 3,000 to 4,000 feet above the sea."

Distribution: Southwestern Texas and Chihuahua to Arizona and Sonora; recorded from southern California and southern Utah.

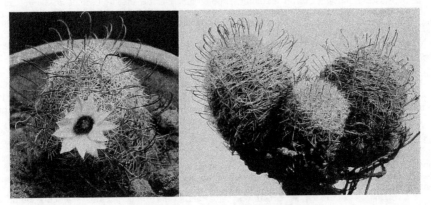

FIGS. 173 and 174.—Neomammillaria microcarpa.

Neomammillaria microcarpa has long been a favorite in living collections under the name of *Mammillaria grahamii*, but it does not do well in cultivation and soon dies out.

This plant is generally known under the name of *Mammillaria grahamii*. The specific name must now give place to an older one, *microcarpa*. *Mammillaria microcarpa* was based on a drawing made by J. M. Stanly, the artist on W. H. Emory's famous expedition across the continent. This drawing was sent to Dr. George Engelmann by Colonel Emory, early in 1848, with the following note: "November 4, 1846, abundant." From Emory's narrative map of his journey published later, in 1848, we know that on that date his camp was on the eastern side of the Gila and only one day's trip by pack train from the mouth of the San Pedro. His camp was "in a grove of cacti of all kinds; among them being the huge pitahaya [*Carnegiea gigantea*], one of which was 50 feet high." For years we have been striving to have this plant re-collected from the type locality; in 1908 Dr. Rose made an unsuccessful attempt to reach Emory's station.

Finally, at Dr. Rose's request, Mrs. Ruth C. Ross, on June 11, 1922, visited the locality at which Emory's party was camped on November 4, 1846, where he had said that the little *Mammillaria* was abundant. The *Mammillaria* which Mrs. Ross found there, also in some abundance, was the plant which has long passed as *M. grahamii*. Mrs. Ross deserves great credit for the enthusiasm which she has shown in visiting this remote locality and clearing up a botanical puzzle which had remained unsolved for 70 years.*

We have not seen any California or Utah plants and we suspect that the material so-named from those states may belong to the genus *Phellosperma*, which resembles this species in its hooked central spine. The plant is undoubtedly found in northern Mexico, but how far south it extends we are in doubt.

The variety *Mammillaria grahamii californica* has not been described.

Illustrations: Emory, Mil. Reconn. 157. No. 3, as *Mammillaria microcarpa*; Cact. Mex. Bound. pl. 6, f. 1 to 8; Bol. Direccion de Estudios Biol. **2**: f. 2; Rümpler, Sukkulenten 199. f. 112; Schelle, Handb. Kakteenk. 254. f. 176; Remark, Kakteenfreund 16, as *M. grahamii*; Cact. Journ. **1**: 171, as *M. grayhamii*.

Figures 170 and 173 are from photographs of the plants collected by Dr. Rose from the northern end of the Tucson Mountains, Arizona, April 22, 1908; figure 174 is from a photograph of a plant collected by Mrs. Ross at the type locality.

CACTUS ESCHANZIERI Coulter, Contr. U. S. Nat. Herb. **3**: 104. 1894.

"Depressed-globose, 3 cm. in diameter, simple; tubercles broader at base, 6 to 8 mm. long, with naked axils; spines all pubescent; radials 15 to 20, with dusky tips, the lateral 10 to 12 mm. long, the lower weaker, shorter, and curved, the upper shorter; solitary central spines reddish slender, somewhat twisted, usually hooked upwards, 15 to 25 mm. long; flowers red (?); fruit reddish (?), ovate, about 10 mm. long; seeds reddish, oblique-obovate, 1.2 mm. long, pitted, with subventral hilum."

It is stated at the original place of publication that the type collected by Eschanzier in 1901 was in the herbarium of Coulter, but it can not be found and is probably lost. Coulter says that it resembles *Cactus grahamii*, but judging from the description and its habitat it is not very near that species. It is evidently a *Neomammillaria*, possibly referable to one of the many species which have been described from San Luis Potosí.

132. Neomammillaria milleri sp. nov.

Globose to elongated cylindric, sometimes more than 2 dm. long and up to 8 cm. in diameter; tubercles closely set, rather thick, nearly 1 cm. long, the axils not bristly and seemingly always naked; radial spines about 20, widely spreading, 12 mm. long or less, white, with brownish tips; central spines 2 to 4, one or all hooked at apex, brown, about 2 mm. long; flowers campanulate, about 2 cm. long, the limb 2.5 cm. broad, purple to nearly pink; inner perianth-segments similar to the outer, oblong, the margins a little paler and somewhat undulate, the apex usually obtuse, often rounded, rarely acute; stamens pale purple; style white; stigma-lobes 7 to 9, linear, yellowish to cream-colored; fruit clavate, scarlet, 1.5 cm. long; seeds black.

Collected by Dr. Gerrit S. Miller jr., near Phoenix in 1921, and by Mrs. Bly near Kingman, June 29, 1921, and in 1922. It has been observed by C. R. Orcutt near Phoenix (No. 559a, type) and near Wickenburg (No. 559,) during the summer of 1922 and several fine specimens were sent in by him. He states that it has long been known as "*Mammillaria grahamii* var." and that it suggested at times *M. phellosperma*, *M. goodridgei*, and *M. grahamii*. It differs, however, from the first in its seeds, from the second in its naked axils, and from the last in its stouter habit and stronger central spines.

Figure 184a is from a photograph of the type, collected by Mr. Orcutt.

133. Neomammillaria sheldonii sp. nov.

Stems slender-cylindric, about 8 cm. high; axils of tubercles without setae; radial spines 12 to 15, pale with dark tips, the 3 or 4 upper ones darker, a little stouter and 1 or 2 of them subcentral, the true central erect or porrect, with upturned hook at end; outer perianth-segments ciliate; inner perianth-segments about 10, broad, acute, light purple with very pale margins; filaments and style light purple; stigma-lobes 6, green; fruit clavate, 2.5 to 3 cm. long, pale scarlet.

---

*Mrs. Ross's label bears this note: On upper terrace on right bank of Gila River in s. e. corner, section 15, t. 4 s. R. 16 E. (Christmas Triangle). From grove of cactus in which we believe Emory camped, Nov. 4, 1846.

PLATE XIV

M. E. Eaton del.

1. Flowering plant of *Neomammillaria kunzeana*.
2. Flowering plant of *Neomammillaria bocasana*.
3. Flowering plant of *Neomammillaria decipiens*.
4. Top of flowering plant of *Neomammillaria armillata*.
5. Flowering plant of *Neomammillaria multiceps*.
6. Flowering plant of *Neomammillaria multiceps*.
7. Flowering plant of *Neomammillaria palmeri*.
8. Flowering plant of *Neomammillaria wildii*.

This plant is described chiefly from the specimens collected by Rose, Standley, and Russell, near Hermosillo, Sonora, Mexico (No. 12366, type), but it has also been collected in Sonora by C. R. Orcutt and by Charles Sheldon, for whom it is named.

The plant differs from the *Neomammillaria microcarpa* in its stouter redder spines, in its heavier and shorter central spine with the hook more uniformly turned upward, and in its flowers, which appear to be smaller.

Figure 175 shows a plant collected by Rose, Standley, and Russell, in Hermosillo in 1910 (No. 12366), which flowered in Washington.

FIG. 175.—Neomammillaria sheldonii.   FIG. 176.—Neomammillaria carretii.

134. Neomammillaria armillata (K. Brandegee).

Mammillaris armillata K. Brandegee, Zoe 5: 7. 1900.

In clusters of 3 to 12, cylindric, sometimes 30 cm. high; tubercles bluish green, somewhat angled; axils setose and slightly woolly; radial spines 9 to 15, 7 to 12 mm. long, yellowish; central spines 1 to 4, but usually 2, brownish, the lowest one hooked and a little longer than the others; flowers 10 to 12 mm. long, greenish to flesh-colored; stigma-lobes greenish, short; fruit red, clavate, 15 to 30 mm. long; seeds black, punctate, constricted just above the base.

Type locality: San José del Cabo.

Distribution: Southern Lower California and on islands adjacent to it.

This species is very common in southern Lower California near the coast. Dr. Rose in 1911 collected it both at the type locality (No. 16455), and at Cape San Lucas (No. 16374). Similar to this is his plant (No. 16877) from Cerralbo Island off the coast of Lower California.

Illustration: Grässner, Haupt-Verz. Kakteen **1914**: 23, as *Mammillaria armillata*.

Plate XIV, figure 4, shows the top of a plant collected by Dr. Rose on Margarita Island, Lower California, in 1911 (No. 16302); plate XV, figure 2 shows a plant collected by Dr. Rose on Santa Maria Bay (No. 16276); figure 3 shows the top of a plant collected by Dr. Rose at San Esteban, Lower California; figure 4 shows another plant from the same island.

135. Neomammillaria fraileana sp. nov.

Stems elongated, cylindric, 1 to 1.5 dm. long; axils of tubercles naked or containing at most a single bristle; central spines dark brown, one of them strongly hooked; flowers rather large, pinkish;

inner perianth-segments acuminate, 2 to 2.5 cm. long, often lacerate towards the tip; filaments and style pinkish, the latter paler and much longer than the stamens; stigma-lobes 6, long and slender, rose-colored.

Collected by Dr. J. N. Rose on Pichilinque Island, March 27, 1911 (No. 16508, type); on Cerralbo Island, April 19, 1911 (No. 16895); and on Catalina Island, April 16, 1911 (No. 16831).

136. Neomammillaria swinglei sp. now.

Stems cylindric, 1 to 2 dm. long, 3 to 5 cm. in diameter; axils of tubercles more or less setose; radial spines rather stout for this group, spreading, dull white with dark tips; central spines 4, ascending, dark brown or black, the lowest one elongated (1 to 1.5 cm. long), hooked at apex or sometimes straight; outer perianth-segments greenish or sometimes pinkish; margins somewhat scarious; inner perianth-segments narrowly oblong, nearly white with a brown stripe down center; style pink, twice as long as the pink filaments; stigma-lobes 8, linear, pointed, green; fruit dark red, clavate, 14 to 18 mm. long; seeds 1 mm. in diameter, constricted below, black with a large elliptic basal hilum.

Common about Guaymas, Sonora; flowers and stems described from Rose's plant (No. 12568, type) and Johnston's plant (No. 3086), and the fruit and seeds from one collected by Swingle; also collected by Dr. W. S. W. Kew in 1920.

In cultivation the inodorous flowers remain open for several days (at least three).

Growing with this species (see Rose, No. 12569) were plants with all the central spines straight. This may be the plant from Guaymas which Scheer called "a very robust species of *Mammillaria sphaerica.*"* Neither flowers nor fruit were seen.

137. Neomammillaria dioica (K. Brandegee).

Mammillaria dioica K. Brandegee, Erythea **5**: 115. 1897.
Mammillaria fordii Orcutt, West Amer. Sci. **13**: 49. 1902.

Either solitary or clustered, cylindric, 5 to 25 cm. high or even higher;† axils of tubercles woolly and short-setose; radial spines 11 to 22, white, the tips often brownish to black or rose-colored throughout, 5 to 7 mm. long, spreading; central spines 3 or 4, brownish, the lower one a little longer than the others and hooked; flowers borne towards top of plant, yellowish white with purplish mid-rib, 10 to 22 mm. long, incompletely dioecious; outer and inner perianth-segments usually 6 each; outer perianth-segments reddish, especially along midrib, the inner ones oblong, pale cream-colored, notched or toothed near apex; style white; stigma-lobes 6, linear, bright yellow to brownish green; fruit scarlet, clavate, 10 to 25 mm. long; seeds black.

Type locality: West coast of Lower California.

Distribution: Southwestern California and northwestern Lower California. According to Mr. Orcutt, this plant extends east of the coastal mountains on the border of Imperial and San Diego Counties.

Although we have not seen the type of *Mammillaria fordii* we have referred it here on the advice of Mr. Orcutt, the author of this species.

Illustrations: Cact. Mex. Bound. pl. 8, f. 9 to 14, as *Mammillaria goodridgii.*

138. Neomammillaria goodridgei (Scheer).

Mammillaria goodridgei ‡ Scheer in Salm-Dyck, Cact. Hort. Dyck. 1849. 91.   1850.
Mammillaria goodridgii Scheer in Seemann, Bot. Herald 286.   1856.
Cactus goodridgii Kuntze, Rev. Gen. Pl. **1**: 260.   1891.

Stems clustered, erect, globose to cylindric, up to 10 cm. long, 3 to 4 cm. in diameter; axils of tubercles not setose; radial spines 12 to 15, spreading, white, sometimes with dark tips; central spines usually 1, white below, brown above, hooked; flowers perfect, rose-colored, 15 mm. long;

---

* Bot. Herald 286.

† In February 1922, Mr. C. R. Orcutt sent us a single plant from the Mason's Valley on the eastern side of the Coast Mountains in San Diego County, California, which was the largest solitary one we had ever seen, being more than 33 cm. long, 10 cm. in diameter, and weighed 3 lbs. 13 oz.   Three small buds were produced near the middle of the plant.

‡ Given as *Mammillaria goodrichii*, in error.

M. E. Eaton del.

1. Flowering plant of *Neomammillaria bombycina*.
2. Flowering plant of *Neomammillaria armillata*.
3. Top of flowering plant of *Neomammillaria armillata*.
4. Flowering plant of *Neomammillaria armillata*.
5. Flowering plant of *Neomammillaria goodridgei*.

segments oblong, obtuse or retuse; fruit clavate, 1.5 to 2 cm. long, scarlet, naked; seeds black, punctate, with a narrow basal hilum.

Type locality: Cedros Island, off Lower California.
Distribution: Cedros Island and the adjacent mainland of Lower California.

This species was originally collected on Cedros Island, by Mr. J. Goodridge, surgeon on the *Herald* during its memorable trip to the western coast of the Americas. The plant, which was sent to Scheer and named by him, was sent to Prince Salm-Dyck, who described it without knowing the flowers or fruit. The name has been associated with *N. dioica.*

Several collectors have visited Cedros Island, but all failed to find *Mammillaria goodridgei* until Dr. Rose collected it in 1911 (No. 16171); he also found it on the nearby mainland at Abreojos Point (No. 16248). Recently a plant was sent in from near Mulege by B. F. Hake.

Plate xv, figure 5, shows a plant collected by Dr. Rose at Mulegé, Lower California, in 1911, which flowered in the New York Botanical Garden, April 11, 1912.

Fig. 177.—Neomammillaria zephyranthoides.

139. Neomammillaria zephyranthoides (Scheidweiler).

Mammillaria zephyranthoides Scheidweiler, Allg. Gartenz. 9: 41. 1841.
Mammillaria fennelii Hopffer, Allg. Gartenz. 11: 3. 1843.
Cactus zephyranthodes Kuntze, Rev. Gen. Pl. 1: 261. 1891.

Depressed-globose to short-cylindric, up to 8 cm. high, 10 cm. in diameter; tubercles about 2 cm. long; radial spines 14 to 18, 8 to 10 mm. long, very slender, white; central spines 1 (sometimes 2), larger than the radials and hooked, at first purple, but in age yellowish at base; flowers 3 to 4 cm. broad with rotate limb; perianth-segments white with red stripes; fruit and seeds unknown.

Type locality: Oaxaca, altitude about 2,300 meters.
Distribution: Oaxaca, Mexico.

We have followed previous authors in referring here *Mammillaria fennelii* and Pfeiffer's illustration, based on his statement that the type plant was abnormal and much smaller than the one figured and with smaller tubercles.

The plant was in flower at Erfurt, Germany, where Dr. Rose studied it in 1912.

Illustrations: Pfeiffer, Abbild. Beschr. Cact. **2**: pl. 8, as *Mammillaria zephyranthiflora*; Schelle, Handb. Kakteenk. 254. f. 175, as *Mammillaria zephyranthoides*.

Figure 177 is reproduced from the first illustration cited above.

140. Neomammillaria carretii (Rebut).

Mammillaria carretii Rebut in Schumann, Gesamtb. Kakteen 542. 1898.

Solitary, dull green, globose, depressed, small, 5 to 6 cm. in diameter; tubercles cylindric; axils of tubercles naked; radial spines 14, subulate, spreading, recurved, nearly clothing the plant, long, yellowish; central spine 1, slender, chestnut-brown, hooked; flowers 2.5 cm. long; inner perianth-segments white, streaked with rose; fruit and seeds unknown.

FIG. 178.—Neomammillaria bombycina.   FIG. 179.—Neomammillaria occidentalis.

Type locality: Not cited.

Distribution: Doubtless Mexico, but no definite locality known.

We have not seen this species and know it only from descriptions and illustrations.

It is related to *Neomammillaria saffordii* but radial spines are yellow, flowers white with a streak of rose, and probably larger throughout.

Illustrations: Grässner, Haupt-Verz. Kakteen **1912**: 18; **1914**: 24, as *Mammillaria carretii*.

Figure 176 is reproduced from a photograph sent us by L. Quehl in 1921.

141. Neomammillaria jaliscana sp. nov.

Cespitose, globose, 5 cm. in diameter, bright green; tubercles in 13 rows, 4 to 5 mm. high; radial spines 30 or more, at right angles to the tubercles; central spines 4 to 6, reddish brown, darker toward the tips, one of them strongly hooked; axils naked; flowers pinkish to purplish, delicately fragrant, 1 cm. broad when fully expanded; outer segments ovate-oblong, acute or obtuse with a more or less serrulate margin; inner perianth-segments oblong, obtuse; filaments pinkish; stigma-lobes 3 or 4, white; fruit white, 8 mm. long; seeds black.

Collected by J. N. Rose at Río Blanco, near Guadalajara, Mexico, in September 1903 (No. 858, type), by C. R. Orcutt near Guadalajara and by B. P. Reko from the same locality in 1922 (No. 4410).

Dr. Rose introduced this species into cultivation but his plants all died. It flowered with us in March 1904 and again in 1923.

142. Neomammillaria bombycina (Quehl).

Mammillaria bombycina Quehl, Monatsschr. Kakteenk. **20**: 149. 1910.

Cylindric, 15 to 20 cm. long, 5 to 6 cm. in diameter; tubercles spiraled, obtuse; young areoles conspicuously white-woolly; radial spines numerous, acicular, widely spreading, short, 1 cm. long or less; central spines 4, elongated, a little spreading, those toward the top of plant erect, 2 cm. long, brown except at base, the lower one hooked; flowers from near top, light purple, about 1 cm. long; perianth-segments narrowly oblong; filaments and style pinkish; stigma-lobes 4, purplish.

Fig. 179a.—Neomammillaria occidentalis.

Type locality: Mexico.
Distribution: Mexico, but range unknown.

We have had this plant in cultivation for a number of years. It is a very attractive plant, the top being covered by a mass of white hairs which come from the closely set young tubercles.

Mammillaria cordigera Heese resembles this species very much in its spines and form, but is described as with grooved tubercles, which would exclude it from this genus (see page 50).

Illustration: Monatsschr. Kakteenk. **20**: 151, as *Mammillaria bombycina*.

Plate xv, figure 1, shows a plant received by Dr. Rose from M. de Laet in 1910 and probably from the type collection. Figure 178 is from a photograph of another plant from the same collection.

143. Neomammillaria occidentalis sp. nov.

Cespitose, the branches slender, cylindric, 10 cm. high, densely spiny; radial spines about 12, yellowish, spreading; central spines 4 or 5, reddish or brown, one of them longer and hooked; flowers small, 1 cm. long, pink; stigma-lobes 9, slender; fruit said to be red.

Collected by Dr. E. Palmer near Manzanillo, Colima, Mexico, December 1890 (No. 1053, type) and again from the same locality by Stephen E. Aguirre, American Vice-Consul-in-Charge, October 1922. Dr. Palmer's field notes say:

"A cactus quite plentiful among rocks in exposed places. Three flowers of a pink color and three red fruits were collected. The specimens of the plants collected were cut off close to the ground; they are a fair sample of plants of the average height and diameter, but in drying they shrink to three-fourths their original dimensions."

Figure 179 is from a photograph of a plant from the type collection; figure 179a is from a photograph of the plants referred to above, sent by Mr. Aguirre.

144. Neomammillaria fasciculata (Engelmann).

Mammillaria fasciculata Engelmann in Emory, Mil. Reconn. 157. 1848.
Cactus fasciculatus Kuntze, Rev. Gen. Pl. 1: 259. 1891.
Mammillaria thornberi Orcutt, West Amer. Sci. 12: 161. 1902.

Forming clumps, often containing many plants (as many as 110 have been noted), slender-cylindric, usually 5 to 8 cm., but sometimes 30 cm. high; axils of tubercles naked; radial spines 13 to 20, slender, 5 to 7 mm. long, white, with dark brown or nearly black tips; central spine usually 1, sometimes 2 or 3, often much elongated and 18 mm. long, brownish or black, one (sometimes all) strongly hooked; flowers broadly funnel-shaped, purplish; inner perianth-segments broad, acute; fruit short-clavate, scarlet, 8 mm. long; seeds black.

FIG. 180.—Neomammillaria fasciculata.   FIG. 181.—Neomammillaria longiflora.

Type locality: Along the Gila River.
Distribution: Southern Arizona.

This plant was found by Emory, October 20, 1846, on the Gila River, 3,000 or 4,000 feet above the sea, and was afterwards described by Engelmann from the sketch made in the field; for more that 50 years afterwards the plant remained otherwise unknown. About 1902 it was rediscovered by Professor Thornber and Mr. Orcutt near Tucson. On this latter collection Mr. Orcutt based *Mammillaria thornberi,* but he afterwards referred it to *M. fasciculata;* he is now inclined to question this reduction and thinks that *M. fasciculata* may be a species of *Echinocereus.* Engelmann, however, pointed out, when he described this species, that the spines were not arranged in vertical ribs as in *Echinocereus.* While we have not been able to prove beyond doubt the identity of the two names, as there is only one plant of this habit known from southeastern Arizona, we have admitted only one species and have used for it the older name; if a second species is afterwards found it may then be necessary to revise our conclusions. The plant has been collected several times since 1902 but it is still rare.

Illustration: Emory, Mil. Reconn. 157. f. 2, as *Mammillaria fasciculata.*

Figure 180 is from a photograph of a plant collected by F. E. Lloyd near Tucson in 1906.

145. Neomammillaria nelsonii sp. nov.

Globose, 5 cm. in diameter; tubercles numerous, small, terete, apparently not milky, 5 to 7 mm. long, their axils naked; radial spines about 15, acicular, white, 6 to 8 mm. long, spreading; central spines several, all like the radials; but one of them elongated, stouter and longer than the others, brown to black, strongly hooked, 12 to 15 mm. long; flowers unknown; fruit very slender, clavate, 3 cm. long or more, red, few-seeded; seeds globose, black, rugose, 2 mm. in diameter; hilum basal, triangular, white, depressed.

Collected by E. W. Nelson on cliffs at La Salada, Michoacán, Mexico, March 23, 1903 (No. 6932).

This plant in its form and in the color and shape of the fruit agrees with *Neomammillaria* but differs from all the species we know in its rather large rugose black seeds. It somewhat resembles *Neomammillaria zephyranthoides.*

Figure 182 shows the fruit, spine-cluster, and seed of the type.

146. Neomammillaria longiflora sp. nov.

Solitary or clustered, small, 3 cm. in diameter, apparently not at all milky; tubercles small, terete, not grooved on upper side, 5 to 7 mm. long, rather closely set and nearly hidden by the spines; radial spines about 30, acicular, 10 to 13 mm. long, yellow or straw-colored, somewhat spreading; central spines 4, reddish brown, much stouter than the radials, 3 of them straight, about length of radials, 1 of them hooked at apex, twice as long as others; flowers several, even on small plants, borne near top, 2 cm. long or more, with a distinct narrow tube; perianth-segments pinkish, oblong, acute; ovary very small, ovoid, more or less sunken in the axils, thin above and perhaps opening by an operculum, the lower part with the seeds persisting for years; seed nearly globose, minutely pitted, 1 to 1.5 mm. in diameter, black with a prominent white hilum.

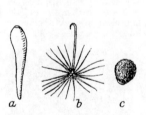

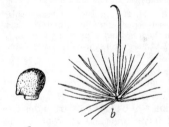

FIG. 182.—Fruit, spine-cluster, and seed of N. nelsonii.

FIG. 183.—Seed and spine-cluster of N. longiflora.

Collected at Santiago Papasquiaro, Durango, by Dr. Edward Palmer in 1897 (No. 89).

We have repeatedly studied this curious plant during the last 25 years, but have never been able to identify it or reach a definite conclusion as to its relationship. Our material consists of a single plant split down one side, bearing several withered flowers, and two detached flowers. Recently, we were sent a photograph of a cactus from Mexico, labeled *Mammillaria* n. sp., Sierra de Cacaria S. de Ulama, which seemed to be Dr. Palmer's plant and led us to make a detailed study of it. One of the peculiarities was the absence of an exserted ovary, so conspicuous in all the *Neomammillaria.* The cut stem showed an exposed sunken ovary, and by mere chance an old fruit with ripe seeds, probably several years old, was found in the axils of one of the oldest tubercles. As described above, the seeds are very unlike those of any species of *Neomammillaria.*

Figure 181 is a reproduction of the photograph mentioned above; figure 183 shows the seed and spine-cluster of the type.

147. Neomammillaria tacubayensis (Fedde).

Mammillaria tacubayensis Fedde, Nov. Gen. Sp. Ind. 1905. 443. 1905.

Globose, 3 to 5 cm. in diameter; radial spines 35 to 40, white, 3 to 5 mm. long; central spines 1, black, 5 to 6 mm. long, hooked; flower 1.5 cm. long.

Type locality: Near Tacubaya, Mexico.
Distribution: Mexico, but range unknown.
We know the plant only from the original description and illustration.
Illustration: Gartenflora **53**: 214. f. 33, as *Mammillaria* stella de Tacubaya (but legend placed under figure 32).

148. Neomammillaria umbrina (Ehrenberg).

Mammillaria umbrina Ehrenberg, Allg. Gartenz. **17**: 287. 1849.
Cactus umbrinus Kuntze, Rev. Gen. Pl. **1**: 261. 1891.

Simple or becoming cespitose, cylindric, 10 to 12.5 cm. high, dull green; tubercles conic; axils of tubercles naked; radial spines 22 to 25, spreading, white, 4 to 6 mm. long; central spines 4, 3 being 8 to 10 mm. long, one being 20 to 24 mm. long, hooked; flowers large, 2 cm. long; inner perianth-segments about 15, narrowly lanceolate, acute, purple; stamens numerous, described as connivent, white; style filiform, longer than the stamens; stigma-lobes 7, green.

Type locality: Mexico.
Distribution: Hidalgo, according to Schumann.
We know this species from description only; it is peculiar in having hooked spines and large flowers; it resembles somewhat *Neomammillaria zephyranthoides* but is undoubtedly distinct.

149. Neomammillaria verhaertiana (Bödeker).

Mammillaria verhaertiana Bödeker, Monatsschr. Kakteenk. **22**: 152. 1912.

Solitary, short-cylindric; tubercles subconic, their axils setose; radial spines 20 or more yellowish, setaceous, 1 cm. long, glabrous; central spines 4 to 8, stouter than the radials, brown at tip, one of them hooked at apex; flowers white, 2 cm. long, appearing in a circle below top of plant; outer perianth-segments broadly lanceolate, yellowish white; anthers rose-colored; style rose; stigma-lobes 8 or 9.

Type locality: Mexico.
Distribution: Known only from the type locality.

We know the plant only from descriptions and illustrations and a few-spine-clusters sent us by L. Quehl. Bödeker placed it next to *Mammillaria spinosissima*, but unlike that species one of the central spines is hooked.

The species is named for François Verhaert.

Illustration: Monatsschr. Kakteenk. **22**: 153, as *Mammillaria verhaertiana*.

150. Neomammillaria xanthina sp. nov.

Depressed-globose, 7 cm. high, 8 to 9 cm. broad, dull bluish green; axils of tubercles and spine-areoles densely white-woolly when young, glabrate in age; tubercles lactiferous, broader than high, the free part about 5 mm. long, somewhat flattened dorsally, arranged in 34 spiral

Fig. 184.—Neomammillaria xanthina.

rows; spine-areole circular, small; radial spines 10 to 12, spreading, acicular, white, 4 mm. long or less; central spines 2, stouter, but not much longer than the radials, somewhat brownish, more or less erect; flowers from the top of the plant but in the axils of old tubercles, the tube not exserted and the limb appressed against the adjacent tubercles; perianth rotate, 16 mm. broad, its segments, stamens, and style pale lemon-yellow; outer perianth-segments oblong, obtuse with ciliate margins, the inner a little longer than the outer, usually entire, oblong, usually retuse at apex, sometimes apiculate.

Sent by B. P. Reko (No. 4401) but collected by A. Groeschner from the vicinity of Monte Mercado, Durango, Mexico, in 1922 and flowered in Washington in May 1923.

Figure 184 is from a photograph of the type specimen.

LITTLE-KNOWN SPECIES PROBABLY OF THIS GENUS.

MAMMILLARIA ALPINA Martius in Salm-Dyck, Cact. Hort. Dyck. 1849. 79. 1850.

    Cactus alpinus Kuntze, Rev. Gen. Pl. 1: 260. 1891.

This plant has not been identified. Its large flowers, 2.5 cm. broad, suggest a species of *Coryphantha*. It was collected by Karwinsky in the state of Oaxaca.

MAMMILLARIA BELLATULA Förster, Allg. Gartenz. **15**: 51. 1847.

    Cactus bellatulus Kuntze, Rev. Gen. Pl. 1: 259. 1891.

Spherical, somewhat compressed, bright green; tubercles broadly cone-shaped, 4 mm. long, their axils naked; spine-areoles white-woolly when young; radial spines 12 to 16, whitish, bristle-like, spreading, 6 to 8 mm. long; central spines 2, straight, one pointing downward, the other upward, 12 to 16 mm. long, at first almost black, grayish brown in age; flowers and fruit unknown.

This species is said to have been grown from Brazilian seed; if this were true it would exclude it from this genus and for this reason Schumann questioned whether it might not be an *Echinocactus*. Judging from the description we believe that it is closely related to *Neomammillaria elegans* and is probably of Mexican origin.

MAMMILLARIA BERGII Miquel, Comment. Phytogr. 104. 1840.

Simple, subglobose, glaucous green; tubercles somewhat 4-angled at base, nearly terete above, woolly in the axils; spine-areoles woolly when young, becoming naked; spines 4, spreading, the uppermost one largest.

This plant is from Mexico.

MAMMILLARIA CAESPITITIA De Candolle, Mém. Mus. Hist. Nat. Paris **17**: 112. 1828.

    Mammillaria nitida Scheidweiler, Allg. Gartenz. 9: 42. 1841.
    Cactus caespititius Kuntze, Rev. Gen. Pl. 1: 260. 1891.

Densely cespitose, the clump 10 cm. in diameter; joints globose, 2.5 cm. in diameter; tubercles small, ovate; spines straight, rigid, when young whitish yellow, in age gray; radial spines 9 or 10; central spines 1 or 2, longer than the radials, erect; flowers and fruit unknown.

Both Pfeiffer and Schumann overlooked this species and it is doubtful if it can ever be identified. The plant was collected by Thomas Coulter in Mexico.

MAMMILLARIA CONICA Haworth, Suppl. Pl. Succ. 71. 1819.

Tubercles large, conic; spines less than 10, all radial, red but paler at base; flowers and fruit unknown.

Neither Pfeiffer nor Schumann knew this species or its origin. The Index Kewensis refers it to South America. If from that region it must be a species of *Discocactus*, near *D. placentiformis*.

MAMMILLARIA DIACENTRA Jacobi, Allg. Gartenz. **24**: 91. 1856.

Globose, about 7 cm. in diameter; tubercles milky, rhomboid at base, not setose in their axils; radial spines 5 or 6, white, with blackish tips; central spines 2, stouter and longer than the radials, grayish, with blackish tips, the lower centrals 2.5 cm. long or more; flowers small, reddish; style rose-colored: stigma-lobes 6.

This species was unknown to Schumann, and we are unable to group it; its origin is not recorded.

MAMMILLARIA FLAVESCENS Haworth, Suppl. Pl. Succ. 71. 1819.

Cactus mammillaris lanuginosus De Candolle, Pl. Succ. 111. 1799.
Cactus flavescens De Candolle, Cat. Hort. Monosp. 83. 1813.
Mammillaria straminea Haworth, Suppl. Pl. Succ. 71. 1819.
Cactus stramineus Sprengel, Syst. 2: 494. 1825, as to name.
Mammillaria simplex flavescens Schumann, Gesamtb. Kakteen 573. 1898.

This plant was first described in 1799 by De Candolle as "var. *β*" of *Cactus mammillaris* or *Cactus mammillaris lanuginosus* (Pl. Succ. pl. 111); at this time he referred to it certain citations of Plumier and Hermann which we now know belong to *Neomammillaria prolifera* and *N. mammillaris* respectively. This variety was raised to specific rank by De Candolle in 1813 as *Cactus flavescens* (Cact. Hort. Monosp. 83). From the more detailed description then given it is clear that *Cactus flavescens* can not be referred to either *N. prolifera* or *N. mammillaris*. It was transferred to the genus *Mammillaria* by Haworth in 1819, but he added little information except the statement that it had been in cultivation in the Chelsea Garden before 1811.

The question has been raised whether this plant is really West Indian. It is true that De Candolle does not state its origin, but it would be indicated that he believed that it was West Indian by his treating it as a variety of the common West Indian species and by his referring to it several West Indian descriptions when he later published it as a species. Pfeiffer states that it is tropical American. As *Neomammillaria mammillaris* is the only species known from South America it could not have come from that continent, and at that time no *Mammillaria* had been discovered in the United States or Mexico. Förster in 1846 says that it is West Indian, and this was Schumann's conclusion.

MAMMILLARIA FLAVICOMA Hortus in Förster, Handb. Cact. ed. 2. 298. 1885.

This species was described from garden plants of unknown origin. Schumann does not mention it in his monograph and it has remained unknown.

MAMMILLARIA GRISEA Salm-Dyck, Cact. Hort. Dyck. 1849. 110. 1850.

Cactus griseus Kuntze, Rev. Gen. Pl. **1**: 260. 1891.

Stout, short-cylindric, 10 to 12.5 cm. high, 7.5 cm. in diameter; tubercles glaucous-green, somewhat 4-angled, their axils woolly and setose; radial spines 10 to 12, spreading, short, rigid, white; central spines 4 to 6, white, with brown or blackish tips, on greenhouse plants 10 to 15 mm. long, but on wild plants 5 cm. long or more; flower and fruit unknown.

This is perhaps different from *Mammillaria grisea* Galeotti (Förster, Handb. Cact. 219. 1846), which was never described.

MAMMILLARIA HEINEI Ehrenberg, Bot. Zeit. **2**: 833. 1844.

Cactus heinei Kuntze, Rev. Gen. Pl. **1**: 260. 1891.

Schumann thought that this name was referable to *Mammillaria umbrina* but we have not been able to satisfy ourselves that the two are the same.

Much confusion is found in the spelling of the name; it sometimes appears as *M. haynii* and *M. haynei*. Salm-Dyck transfers two species of Ehrenberg to varieties of *M. haynii* but both are unknown to us. These varieties are as follows: var. *viridula* Salm-Dyck (Cact. Hort. Dyck. 1849. 10. 1850; *M. viridula* Ehrenberg, Allg. Gartenz. **16**: 267. 1848), and var. *minima* Salm-Dyck (Cact. Hort. Dyck 1849. 10. 1850; *M. digitalis* Ehrenberg, Allg. Gartenz. **16:** 267. 1848).

MAMMILLARIA HELICTERES De Candolle, Mém. Mus. Hist. Nat. Paris **17:** 31. pl. 5. 1828.

This name was based on Mociño and Sessé's drawing of a Mexican plant, which has never since been definitely identified. It was called by them *Cactus helicteres* (De Candolle, Prodr. **3:** 460. 1828), but it was renamed *Mammillaria convoluta* by St. Lager (Ann. Soc. Bot. Lyon **7:** 130. 1880). The published drawing indicates that the plant is of this genus.

MAMMILLARIA HEXACANTHA Salm-Dyck, Hort. Dyck. 344. 1834.

Cactus hexacanthus Kuntze, Rev. Gen. Pl. **1:** 260. 1891.

Solitary, short-cylindric; tubercles somewhat compressed, light green; areoles ovate to oblong when young, white-tomentose, glabrate in age; radial spines 25 to 30, white, 4 mm. long; central spines 6, stouter than the radials, brown, the 4 lateral ones 8 mm. long, the uppermost ones a little longer, the lowermost ones 18 mm. long, somewhat deflexed; flowers and fruit unknown.

This plant, which is of Mexican origin, is unknown to us except from description; Schumann referred it to *Mammillaria coronaria*, but it has nothing to do with that plant.

MAMMILLARIA IRREGULARIS De Candolle, Mém. Mus. Hist. Nat. Paris **17:** 111. 1828.

Cactus irregularis Kuntze, Rev. Gen. Pl. **1:** 260. 1891.

Cespitose, 5 cm. high, with a subtuberous base; joints ovoid, 2.5 cm. in diameter; spines all radial, 20 to 25, spreading or somewhat reflexed; flowers and fruit unknown.

This plant was collected by T. Coulter (No. 31). It has never been re-identified. It was grown at the Botanical Garden at Geneva, Switzerland, at the time the description was published but, unfortunately, no specimens were preserved; other types based on Coulter's plants are similarly lost and can never be certainly identified.

MAMMILLARIA JOOSSENSIANA Quehl, Monatsschr. Kakteenk. **18:** 95. 1908.

Simple, globose to cylindric, up to 5 cm. high, 3 cm. in diameter, pale green, slightly depressed at apex; young areoles white-woolly; radial spines 20, slender-subulate, straight, white, 12 mm. long; central spines 4, stouter than the radials, 15 mm. long or more, one of them often hooked; flowers small, yellow.

We know this plant, which is a native of Mexico, only from description and two small plants sent us by Frantz de Laet in 1922. Quehl places it in Schumann's classification just after *M. amoena*, although one of the central spines is hooked.

MAMMILLARIA LESAUNIERI Rebut in Schumann, Gesamtb. Kakteen 533. 1898.

Simple, globose, or a little longer than broad; tubercles conic, their axils naked; radial spines 11 to 13, slender, subulate, straight, white, 6 to 8 mm. long; central spines solitary, very short (5 mm. long or less), brownish, erect; flowers reddish, 2.5 cm. long.

Type locality: Described from cultivated plants.
Distribution: Supposed to be Mexico proper or Lower California.
This species is supposed to have the habit of *Mammillaria heyderi*.
Here probably belongs *Mammillaria lassonneriei* Rebut (Monatsschr. Kakteenk. **7:** 29. 1897), a garden name of which we have found no accompanying description. The dealer, Grässner, in his Catalogue of Cacti for 1912 (p. 21) and 1914 (p. 33) has illustrated *M. lassaunieri*.

MAMMILLARIA LEUCOCENTRA Berg, Allg. Gartenz. **8:** 130. 1840.

Cactus leucocentrus Kuntze, Rev. Gen. Pl. **1:** 260. 1891.

Ovoid, about 10 cm. high; tubercles ovoid, their axils very white-woolly; young spine-areoles white-tomentose at first, becoming naked; radial spines spreading, numerous, setose, white; central spines 4 to 6, stouter and longer than the radials, white throughout or with black.

Recorded from Oaxaca, but not identified.

MAMMILLARIA LORICATA Martius in Pfeiffer, Enum. Cact. 13. 1837.

> Echinocactus loricatus Poselger, Allg. Gartenz. 21: 107. 1853.
> Coryphantha loricata Lemaire, Cactées 35. 1868.
> Cactus loricatus Kuntze, Rev. Gen. Pl. 1: 260. 1891.

Solitary, simple, globose, 4 to 5 cm. in diameter, glaucous-green; tubercles short-ovate, 4-angled at base; radial spines 12, spreading, rigid, yellow, 6 to 8 mm. long; central spines 2, stouter than the radials, 8 to 10 mm. long, black at tip, the upper one straight, the lower one curved; flowers and fruit not described.

This plant is recorded as of Mexican origin, but we have found no description of it subsequent to the original and it may never be identified. Förster referred it to *Mammillaria polythele*, but Schumann did not know it.

Mammillaria heteracantha was referred here as a synonym by Pfeiffer (Enum. Cact. 13. 1837). This plant was mentioned by Martius (Verz. Konig. Bot. Gard. München 127. 1829), but so far as we can learn was never described.

MAMMILLARIA MONOCENTRA Jacobi, Allg. Gartenz. 24: 90. 1856.

Depressed-globose, up to 12 cm. high, about 8 cm. in diameter, umbilicate at apex; tubercles milky, somewhat rhomboid at•base, a little flattened, not setose in their axils; radial spines 6, white with black tips, a little spreading; central spine solitary, stouter and longer than the radials, about 2.5 cm. long; flowers rather large, rose-colored; style rose-colored; stigma-lobes 6, reddish yellow.

Jacobi referred this plant, presumably of Mexican origin, to the group *Angulosae-tetragonae* of Salm-Dyck.

Schumann placed it among his list of little-known species; we know it from description only.

MAMMILLARIA NERVOSA CRISTATA Journ. Hort. Home Farm. III. 60: (?) 7. 1910.

We know this plant only from a brief description and an illustration on pages 7 and 8 of the journal here cited:

" *Mammillaria nervosus cristatus* * grows in convoluted sinuous masses like a great brain-mass. The growths are covered with spiny mamillae (whence the name of the genus) and are of a dull olive-brownish hue. It, too, is Mexican."

We are not able to place this plant; it resembles the cristate form sometimes assumed by *Pediocactus simpsonii* and also resembles *Mammillaria bicolor* as shown by the illustration under *M. daedalea*.

Illustration: Journ. Hort. Home Farm. III. 60: 8 (or 7).

MAMMILLARIA NICHOLSONI Journ. Hort. Home Farm. III. 60: 7. 1910.

We know this species only from the illustration referred to below and the following brief note taken from the place of publication:

" *Mammillaria nicholsoni* resembles several of the Echinocactuses in external form. It was named we believe in honor of the late Mr. George Nicholson and came to Kew from the Swanley Collection. All our illustrations were secured at Kew where the collection is well cultivated. *M. nicholsoni* forms spherical masses with the typical protuberances or tubercles, these being tipped with sharp spines."

It is doubtless of Mexican origin.

Illustration: Journ. Hort. Home Farm. III. 60: 9.

MAMMILLARIA NUDA De Candolle, Prodr. 3: 460. 1828.

This is based on *Cactus nudus* (Mociño and Sessé, Pl. Mex. Sc. ined.), but has never been subsequently identified. It was also taken up by Otto Kuntze as *Cactus nudus* (Rev. Gen.

---

* This is the original spelling.

Pl. 1 : 261. 1891). The original description was based on a drawing and calls for a cylindric, unbranched plant, bearing unarmed tubercles and rose-colored flowers.

MAMMILLARIA PICTA Meinshausen, Wöchenschr. Gärtn. Pflanz. 1: 27. 1858.

Cactus pictus Kuntze, Rev. Gen. Pl. 1: 261. 1891.

Globose to ovoid, dull green; tubercles cylindric, somewhat oblique, obtuse, their axils setose; spines pubescent; radial spines 12, yellowish at base, white near middle, above dark purple; central spines 1 (rarely 2), erect; flowers greenish white; stigma-lobes 3.

This species is known from the description only. It was recorded as from Mexico.

MAMMILLARIA PLECOSTIGMA Meinshausen, Wöchenschr. Gärtn. Pflanz. 1: 27. 1858.

Mammillaria plecostigma major Meinshausen, Wöchenschr. Gärtn. Pflanz. 1: 27. 1858.
Mammillaria plecostigma minor Meinshausen, Wöchenschr. Gärtn. Pflanz. 1: 27. 1858.
Cactus plecostigma Kuntze, Rev. Gen. Pl. 1: 261. 1891.

Proliferous, the joints cylindric; tubercles cylindric, the apex oblique and rounded, with setae in their axils; radial spines 16 to 20, setaceous, white; central spines 3 or 4, at first yellow, becoming brown, one hooked at apex; flowers and fruit unknown.

Presumably of this genus and recorded as of Mexican origin; but not identified since it was described.

MAMMILLARIA PLINTHIMORPHA Jacobi, Allg. Gartenz. **24**: 92. 1856.

Cespitose, forming clumps 15 cm. in diameter or more; joints globose; tubercles 4-angled, obtuse, bearing yellowish white wool in their axils; spines 4, subulate, somewhat angled, flesh-colored with blackish tips, the upper one the longest and sometimes more than 2.5 cm. long; flowers not known.

This plant was collected by Galeotti in Mexico in 1847; we do not know it and it was listed by Schumann among his little-known species.

MAMMILLARIA PULCHRA Haworth in Edwards's Bot. Reg. **16**: pl. 1329. 1830.

Cactus pulcher Kuntze, Rev. Gen. Pl. 1: 261. 1891.

This species, which has yellow spines and dark-red flowers, was referred by Schumann, doubtfully, to *Mammillaria centricirrha*, and by Pfeiffer with doubt to *M. tentaculata*.

MAMMILLARIA RUTILA Zuccarini in Pfeiffer, Enum. Cact. 29. 1837.

Cactus rutilus Kuntze, Rev. Gen. Pl. 1: 261. 1891.

Simple, globose; axils of tubercles nearly naked; tubercles 1 cm. long, conic, dull green; areoles when young tomentose; radial spines 14 to 16, setiform, the upper ones smaller, 4 to 8 mm. long; central spines 4 to 6, spreading, rigid, 8 to 12 mm. long, curved, reddish brown, the lower one longest.

Type locality: Mexico.

This name is referred by Schumann to *M. coronaria*.

M. rutila pallidior Salm-Dyck (Cact. Hort. Dyck. 1849. 11. 1850) was never described, while *M. eugenia* (Salm-Dyck, Cact. Hort. Dyck. 1849. 11. 1850) is given as a synonym of *M. rutila*.

M. rutila octospina Scheidweiler (Bull. Acad. Sci. Brux. **6**: 91. 1839) is briefly described.

MAMMILLARIA SAXATILIS Scheer, Bot. Herald 286. 1856.

Cactus saxatilis Kuntze, Rev. Gen. Pl. 1: 261. 1891.

Plant small; spines brownish to straw-colored.

Only two plants were collected, somewhere in Mexico, by Potts and sent to Scheer; the flowers and fruit were not described. The species may never be identified.

MAMMILLARIA SCHMERWITZII Haage in Förster, Handb. Cact. ed. 2. 270. 1885.

Depressed-globose, 10 cm. in diameter, grassy green; radial spines 10 to 25, yellow, 4 to 5 mm. long; central spines 4 or 5, dark brown, 15 mm. long; flowers red.

This plant, recorded as of Mexican origin, was at one time offered for sale by A. Blanc and Company; we know it only from description and are unable to identify it.

MAMMMILLARIA SEEMANNII Scheer in Seemann, Bot. Herald 288. 1856.

Cactus seemannii Kuntze, Rev. Gen. Pl. 1: 261. 1891.

Hemispheric, stout, 10 cm. in diameter, 7.5 cm. high; tubercles somewhat ovoid, elongated, greenish, minutely punctate, their axils soon white-woolly; radial spines 11 to 13, nearly equal, less than 6 mm. long; central spines 1, shorter than the radials, subulate, straight, erect, blackish purple, becoming white.

This plant was sent to F. Scheer in 1850, who states that it came from Sonora or Durango. It is incompletely described and can not be identified. It may be a species of Coryphantha.

MAMMILLARIA SORORIA Meinshausen, Wochenschr. Gärtn. Pflanz. 1: 28. 1858.

Cactus sororius Kuntze, Rev. Gen. Pl. 1: 261. 1891.

Depressed-globose, 5 to 6 cm. high, 7.5 to 10 cm. in diameter, milky; tubercles angled, 12 mm. long, naked in their axils; radial spines 6, 4 mm. long; central spines 1, erect, stouter than the radials, blackish at apex; flowers greenish purple; stigma-lobes 4.

Recorded as of Mexican origin but otherwise unknown.

MAMMILLARIA SPINAUREA Salm-Dyck, Allg. Gartenz. 18: 594. 1850.

Cactus spinaureus Kuntze, Rev. Gen. Pl. 1: 261. 1891.

Globose or becoming depressed; tubercles light green, somewhat 4-angled, gibbous at base, obtuse and oblique at apex, their axils woolly; radial spines about 12, slender, rigid, spreading; central spines 5 or 6, twice as long and stouter than the radials, recurved or reflexed, yellow.

The above was sent by John Potts from Chihuahua in 1850; Scheer thought that it might have been collected in Durango or Sonora. We have not been able to identify it.

MAMMILLARIA SUAVEOLENS Rümpler in Förster, Handb. Cact. ed. 2. 297. 1885.

About 4 cm. high; radial spines 13 to 15; central spines 4, brown; flowers and fruit unknown.

The above is unidentifiable from the brief description. It was grown in Germany from Mexican seed.

MAMMILLARIA TROHARTII Hildmann in Schumann, Gesamtb. Kakteen 586. 1898.

Simple or proliferous and densely cespitose, globose or somewhat depressed, glaucous-green, small (6 cm. in diameter); axils naked; areoles at first woolly, afterwards naked; tubercles very small, scarcely angled; radial spines 5, with brown tips; central spines solitary, dark brown, subulate; flowers and fruit unknown.

M. trohartii is of Mexican origin.

MAMMILLARIA UNISETA Quehl, Monatsschr. Kakteenk. 14: 128. 1904.

Solitary, globose, about 5 cm. in diameter, somewhat depressed at apex; tubercles dark green, 4-angled; spines 6, about 3 mm. long, at first black, changing to gray; flowers and fruit unknown.

This plant was described from a specimen in the Botanical Garden at Halle of unknown origin, but doubtless from Mexico.

MAMMILLARIA VIPERINA J. A. Purpus, Monatsschr. Kakteenk. 22: 148. 1912.

Cespitose, decumbent, cylindric, 1.5 to 2 cm. in diameter; tubercles very short, sometimes nearly globular; spines numerous, 5 mm. long, whitish brown to brownish black; flowers and fruit unknown.

This plant came from Río de Zapotitlán, Puebla; we know it only from description and the very characteristic published illustration. Quehl, who had seen it, said that it was a form of Mammillaria elongata. We believe that it is near M. sphacelata and perhaps a distinct species. The plant figured by Grässner (Haupt-Verz. Kakteen 38. 1914) shows nearly upright branches.

Illustration: Monatsschr. Kakteenk. **23:** 21, as *Mammillaria viperina.*

MAMMILLARIA ZEYERIANA Haage jr. in Schumann, Gesamtb. Kakteen 574. 1898.

Simple, hemispheric to short-cylindric, up to 10 cm. high, pale glaucous-green; tubercles in 13 or 21 spirals, terete, 10 to 12 mm. long, their axils naked; spine-areoles elliptic, 3 mm. long; radial spines 10, white; central spines 4, the uppermost one curved, 15 mm. long, brownish; flowers and fruit unknown.

Described from Mexican plants; supposed to be of Mexican origin.

PLANTS KNOWN BY NAME ONLY.

Mammillaria acicularis Lemaire (Cact. Gen. Nov. Sp. 34. 1839) was described without the flowers, fruit, or native country being known and has not been identified; here belongs *Cactus acicularis* (Kuntze, Rev. Gen. Pl. **1:** 261. 1891), but *C. acicularis* (Kuntze, Rev. Gen. Pl. **1:** 260. 1891) based on some name of Lehmann we have not been able to find.

Mammillaria aulacantha, referred by Schumann and the Index Kewensis to De Candolle's Revision (Mém. Mus. Hist. Nat. Paris **17:** 113. 1828), is not to be found at the place cited by them; here probably belongs *Cactus aulacanthus* Kuntze (Rev. Gen. Pl. **1:** 260. 1891).

Mammillaria beneckei Ehrenberg (Förster, Handb. Cact. 210. 1846; *Cactus beneckei* Kuntze, Rev. Gen. Pl. **1:** 260. 1891) was referred to *M. coronaria* by Schumann.

Mammillaria brandi is described in Blanc, Hints on Cacti, p. 67, as "a rare Mexican sort, with very long straw-colored spines deflecting from the plant. Flowers cream-colored and very fragrant."

Mammillaria centa is mentioned by C. A. Purpus in a short article in Die Gartenwelt (**9:** 249. 1905).

Mammillaria chrysantha is listed by De Candolle (Prodr. **3:** 460. 1828) among species little known but not described. It is said to have been in the Berlin Botanic Garden.

Mammillaria circumtexta Martius (Hort. Reg. Monac. 127. 1829) seems never to have been described.

Mammillaria hochderferi is mentioned by C. A. Purpus in a short article in Die Gartenwelt (**9:** 249. 1905).

Mammillaria multiradiata (Martius, Hort. Reg. Monac. 127. 1829) is only a name.

Mammillaria nigra Ehrenberg (Allg. Gartenz. **17:** 287. 1849) was referred to *M. coronaria* by Schumann; *Cactus niger* Kuntze (Rev. Gen. Pl. **1:** 261. 1891) is a synonym of it.

Mammillaria parmentieri Link and Otto (Verh. Ver. Beförd. Gartenb. **6:** 429. 1830), without description, was doubtfully referred to *M. flavescens.* It was supposed, however, to have come from Mexico.

The following species, briefly described by F. Schlumberger (Rev. Hort. IV. **5:** 404. 1856), we do not know, nor do we find them mentioned elsewhere:

Mammillaria albiseta, with flowers like those of *M. spinosissima.*
Mammillaria bocasiana, with clear yellow flowers.
Mammillaria cunendstiana, with flowers like those of *M. clillifera.*
Mammillaria decholara, with very small red flowers.
Mammillaria klenneirii, with rose-colored flowers.
Mammillaria roematactina, with abundant small rose-red flowers.
Mammillaria saluciana, flowers 1.5 cm. long and of the same diameter, flesh-colored.

The following names, without descriptions, appear in Förster's Handbuch (254, 255, 1846). Some of the names have been used subsequently, but so far as our observation goes they are all still *nomen nudum*. *Mammillaria asteriflora* Cels, *M. binops* Haage, *M. cantera* Haage, *M. citrina* Scheidweiler, *M. contacta* Wendland, *M. coryphides* Forbes, *M. crinigera* Otto, *M. daedalea viridis* Fennel, *M. echinops* Fennel, *M. enneacantha* Otto, *M. heteracentra* Otto, *M. intricata* Otto, *M. miqueliana* Pfeiffer, *M. palmeri* Fennel, *M. pyrrhacantha* Pfeiffer, *M. pyrrhacantha pallida* Pfeiffer, *M. salmiana* Fennel, *M. stephani* Hortus, *M. suberecta* Pfeiffer, and *M. villosa* Fennel.

The following names appeared first, published by Forbes (Journ. Hort. Tour Germ. 147. 1837), but are so briefly described that they can not be identified: *Mammillaria cuneiflora* Hitchen, *M. cylindraca* Hitchen, *M. divaricata*, *M. flavescens* Hitchen, *M. grandis* Hitchen, *M. lutescens*, and *M. pulcherrima*. Some of these names were afterwards used, but whether they were applied to the same plants we can not tell.

The following names of *Mammillaria* listed by Haage (Cact. Kult. ed. 2. 1900) are without description: *brandtii* Haage jr., *bruennowii*, *celsiana longispina*, *de grandii*, *deleuili* Rebut, *desertorum*, *donkelaari*, *dubia* Hildmann, *fulvolanata*, *geminiflora*, *glabrescens*, *goeringii*, *grusonii similis*, *guebwilleriana* Haage jr., *hermantiana* Monville, *hevernickii* Senke, *lapaixi* Rebut, *microdasys*, *monothele*, *morini* Rebut, *multicolor*, *nickelsi*, *nigerrima*, *numina*, *polia* Sieber, *quehlii*, *rebuti*, *roii* Rebut, *roessingii* Gruson, *semilonia*, *simonis*, *tellii*, *variimamma* Ehrenberg, *villa-lerdo*, *wegeneri cristata*, and *xanthispina*.

Schumann, at the close of his treatment of the genus *Mammillaria* (Gesamtb. Kakteen 599. 1898), lists 158 names which he had not been able to refer. Later, Otto Kuntze referred many of the names to the genus *Cactus*, thus making many useless synonyms. Some of these names of Schumann we have been able to refer more or less definitely to other species, but there still remain many which we can not place. Most of them were described without flower and fruit, and since the types were not preserved it is doubtful if many more can be ever identified. The residue is as follows:

MAMMILLARIA ACTINOPLEA Ehrenberg, Allg. Gartenz. **16**: 266. 1848.

> *Mammillaria amabilis* Ehrenberg, Allg. Gartenz. **17**: 326. 1849.
> *Mammillaria albiseta* Hortus in Förster, Handb. Cact. ed. 2. 354. 1885.
> *Cactus actinopleus* Kuntze, Rev. Gen. Pl. **1**: 260. 1891.
> *Cactus amabilis* Kuntze, Rev. Gen. Pl. **1**: 260. 1891.

Mammillaria crebrispina nitida Monville (Labouret, Monogr. Cact. 75. 1853) is known only as a synonym.

MAMMILLARIA ARGENTEA Fennel, Allg. Gartenz. **15**: 66. 1847.

> *Cactus argenteus* Kuntze, Rev. Gen. Pl. **1**: 260. 1891.

MAMMILLARIA ATRORUBRA Ehrenberg, Allg. Gartenz. **17**: 327. 1849.

> *Cactus atroruber* Kuntze, Rev. Gen. Pl. **1**: 260. 1891.

MAMMILLARIA ATROSANGUINEA Ehrenberg, Allg. Gartenz. **17**: 270. 1849.

> *Cactus atrosanguineus* Kuntze, Rev. Gen. Pl. **1**: 260. 1891.

MAMMILLARIA BADISPINA Förster, Hamb. Gartenz. **17**: 159. 1861.

MAMMILLARIA BARLOWII Regel and Klein, Ind. Sem. Hort. Petrop. **1860**: 46. 1860.

> *Cactus barlowii* Kuntze, Rev. Gen. Pl. **1**: 260. 1891.

MAMMILARIA BERGENII Ehrenberg, Allg. Gartenz. **17**: 326. 1849.

MAMMILLARIA BIFURCA A. Dietrich, Allg. Gartenz. **18**: 186. 1850.

MAMMILLARIA BREVISETA Ehrenberg, Allg. Gartenz. **17**: 251. 1849.

> *Cactus brevisetus* Kuntze, Rev. Gen. Pl. **1**: 260. 1891.

MAMMILLARIA CLOSIANA Roumey, Bull. Soc. Bot. France **2**: 372. 1855.

MAMMILLARIA COROLLARIA Ehrenberg, Allg. Gartenz. **17**: 294. 1849.

> *Cactus corollarius* Kuntze, Rev. Gen. Pl. **1**: 260. 1891.

MAMMILLARIA CORONATA Scheidweiler, Allg. Gartenz. **8**: 338. 1840.

> *Cactus coronatus* Kuntze, Rev. Gen. Pl. **1**: 261. 1891.

MAMMILLARIA CURVISPINA Otto in Dietrich, Allg. Gartenz. **14**: 204. 1846.

> *Cactus curvispinus* Kuntze, Rev. Gen. Pl. **1**: 260. 1891. Not Bertero, 1829.

MAMMILLARIA DECORA Förster, Hamb. Gartenz. **17**: 159. 1861.
Mammillaria decora obscura Förster, Hamb. Gartenz. **17**: 159. 1861.
MAMMILLARIA EBORINA Ehrenberg, Allg. Gartenz. **17**: 309. 1849.
Cactus eborinus Kuntze, Rev. Gen. Pl. 1: 260. 1891.
MAMMILLARIA EMUNDTSIANA Hortus in Förster, Handb. Cact. ed. 2. 341. 1885.
MAMMILLARIA ERECTACANTHA Förster, Allg. Gartenz. **15**: 50. 1847.
Cactus erectacanthus Kuntze, Rev. Gen. Pl. 1: 260. 1891.
MAMMILLARIA EUCHLORA Ehrenberg, Allg. Gartenz. **16**: 266. 1848.
Cactus euchlorus Kuntze, Rev. Gen. Pl. 1: 260. 1891.
MAMMILLARIA FELLNERII Ehrenberg, Allg. Gartenz. **17**: 261. 1849.
Cactus fellneri Kuntze, Rev. Gen. Pl. 1: 260. 1891.
MAMMILLARIA FLAVA Ehrenberg, Allg. Gartenz. **17**: 261. 1849.
Mammillaria tomentosa flava Salm-Dyck, Cact. Hort. Dyck. 1849. 12. 1850.
MAMMILLARIA GEMINATA Scheidweiler, Allg. Gartenz. **9**: 42. 1841.
Cactus geminatus Kuntze, Rev. Gen. Pl. 1: 260. 1891.

Illustration: Möllers Deutsche Gärt. Zeit. **25**: 475. f. 8, No. 20.

MAMMILLARIA GIBBOSA Salm-Dyck, Hort. Dyck. 343. 1834.
Cactus gibbosus Kuntze, Rev. Gen. Pl. 1: 261. 1891. Not Haworth, 1812.
MAMMILLARIA GLABRATA Salm-Dyck, Cact. Hort. Dyck. 1849. 109. 1850.
Cactus glabratus Kuntze, Rev. Gen. Pl. 1: 260. 1891.
MAMMILLARIA GRANDICORNIS Mühlenpfordt, Allg. Gartenz. **14**: 372. 1846.
Cactus grandicornis Kuntze, Rev. Gen. Pl. 1: 260. 1891.
MAMMILLARIA HAEMATACTINA Ehrenberg, Allg. Gartenz. **16**: 266. 1848.
Cactus haematactinus Kuntze, Rev. Gen. Pl. 1: 260. 1891.
MAMMILLARIA INCURVA Scheidweiler, Bull. Acad. Sci. Brux. **6**: 92. 1839.
Cactus incurvus Kuntze, Rev. Gen. Pl. 1: 260. 1891.
MAMMILLARIA JUCUNDA Ehrenberg, Allg. Gartenz. **17**: 250. 1849.
Cactus jucundus Kuntze, Rev. Gen. Pl. 1: 260. 1891.
MAMMILLARIA KLEINII Regel, Ind. Sem. Hort. Petrop. **1860**: 47. 1860.
Cactus kleinii Kuntze, Rev. Gen. Pl. 1: 260. 1891.
MAMMILLARIA LAMPROCHAETA Jacobi, Allg. Gartenz. **24**: 82. 1856.
MAMMILLARIA LEUCODASYS Salm-Dyck in Scheer, Seemann, Bot. Herald 286. 1856.
Cactus leucodasys Kuntze, Rev. Gen. Pl. 1: 260. 1891.

Mexico. Probably *M. micromeris* Engelmann (*fide* Schumann).

MAMMILLARIA LEUCODICTIA Linke, Allg. Gartenz. **16**: 330. 1848.
Cactus leucodictyus Kuntze, Rev. Gen. Pl. 1: 260. 1891.
MAMMILLARIA LIVIDA Fennel, Allg. Gartenz. **15**: 66. 1847.
Cactus lividus Kuntze, Rev. Gen. Pl. 1: 260. 1891.

Mammillaria farinosa (Fennel, Allg. Gartenz. **15**: 66. 1847) is referred to *M. livida* by the Index Kewensis.

MAMMILLARIA MELANACANTHA Hortus in Förster, Handb. Cact. ed. 2. 386. 1885.
MAMMILLARIA MICANS Dietrich in Linke, Allg. Gartenz. **16**: 330. 1848.
Cactus micans Kuntze, Rev. Gen. Pl. 1: 260. 1891.
MAMMILLARIA MICRACANTHA Miquel, Linnaea **12**: 16. 1838.
Cactus micracanthus Kuntze, Rev. Gen. Pl. 1: 261. 1891.
MAMMILLARIA MUCRONATA Ehrenberg, Allg. Gartenz. **17**: 294. 1849.
Cactus mucronatus Kuntze, Rev. Gen. Pl. 1: 260. 1891.
MAMMILLARIA MULTISETA Ehrenberg, Allg. Gartenz. **17**: 242. 1849.
Cactus multisectus * Kuntze, Rev. Gen. Pl. 1: 261. 1891.
MAMMILLARIA OBLIQUA Ehrenberg, Allg. Gartenz. **17**: 250. 1849.
Cactus obliquus Kuntze, Rev. Gen. Pl. 1: 261. 1891.

---

* Kuntze's specific name is credited to Scheidweiler, but we do not find it.

MAMMILLARIA OBVALLATA Otto in Dietrich, Allg. Gartenz. **14**: 308.   1846.
  Cactus obvallatus Kuntze, Rev. Gen. Pl. **1**: 261.   1891.
MAMMILLARIA OLORINA Ehrenberg, Allg. Gartenz. **17**: 326.   1849.
  Cactus olorinus Kuntze, Rev. Gen. Pl. **1**: 261.   1891.
MAMMILLARIA OOTHELE Lemaire, Cact. Gen. Nov. Sp. 37.   1839.
  Mammillaria ovimamma Lemaire, Cact. Gen. Nov. Sp. 49.   1839.
  Mammillaria ovimamma brevispina Salm-Dyck, Cact. Hort. Dyck. 1849. 108.   1850.
  Mammillaria ovimamma oothele Labouret, Monogr. Cact. 85.   1853.
  Cactus oothele Kuntze, Rev. Gen. Pl. **1**: 261.   1891.
  Cactus ovimamma Kuntze, Rev. Gen. Pl. **1**: 261.   1891.
MAMMILLARIA PERSICINA Ehrenberg, Allg. Gartenz. **17**: 250.   1849.
  Cactus persicanus Kuntze, Rev. Gen. Pl. **1**: 261.   1891.
MAMMILLARIA PHAEOTRICA Monville in Labouret, Monogr. Cact. 39.   1853.
  Cactus phaeotrichus Kuntze, Rev. Gen. Pl. **1**: 261.   1891.
MAMMILLARIA PLEIOCEPHALA Regel and Klein, Ind. Sem. Hort. Petrop. **1860**: 47.   1860.
  Cactus pleiocephalus Kuntze, Rev. Gen. Pl. **1**: 261.   1891.
MAMMILLARIA POLYMORPHA Scheer in Mühlenpfordt, Allg. Gartenz. **14**: 373.   1846.
  Cactus polymorphus Kuntze, Rev. Gen. Pl. **1**: 261.   1891.
MAMMILLARIA PORPHYRACANTHA Jacobi, Allg. Gartenz. **24**: 81.   1856.
MAMMILLARIA PROCERA Ehrenberg, Allg. Gartenz. **17**: 241.   1849.
  Cactus procerus Kuntze, Rev. Gen. Pl. **1**: 261.   1891.
MAMMILLARIA PUGIONACANTHA Förster, Allg. Gartenz. **15**: 50.   1847.
  Cactus pugionacanthus Kuntze, Rev. Gen. Pl. **1**: 261.   1891.
MAMMILLARIA PUNCTATA Labouret in Förster, Handb. Cact. ed. 2. 293.   1885.
MAMMILLARIA PURPURASCENS Ehrenberg, Allg. Gartenz. **17**: 260.   1849.
MAMMILLARIA PURPUREA Ehrenberg, Allg. Gartenz. **17**: 270.   1849.
  Cactus purpureus Kuntze, Rev. Gen. Pl. **1**: 261.   1891.
MAMMILLARIA REGIA Ehrenberg, Allg. Gartenz. **17**: 269.   1849.
  Cactus regius Kuntze, Rev. Gen. Pl. **1**: 261.   1891.
MAMMILLARIA ROSEA Scheidweiler, Hort. Belge **5**: 118.   1838.
  Mammillaria rhodeocentra Lemaire, Cact. Gen. Nov. Sp. 52.   1839.
  Mammillaria discolor nigricans Salm-Dyck in Walpers, Repert. Bot. **2**: 271.   1843.
  Mammillaria rhodeocentra gracilispina Salm-Dyck, Cact. Hort. Dyck. 1849. 14.   1850.
  Cactus roseus Kuntze, Rev. Gen. Pl. **1**: 261.   1891.
  Cactus rhodeocentrus Kuntze, Rev. Gen. Pl. **1**: 261.   1891.

Salm-Dyck referred *Mammillaria rosea* to *M. rhodeocentra*, but the former is the older name.

Illustration: Hort. Belge **5**: pl. 7. as *Mammillaria rosea*.

MAMMILLARIA RUFIDULA Ehrenberg, Allg. Gartenz. **17**: 295.   1849.
  Cactus rufidulus Kuntze, Rev. Gen. Pl. **1**: 261.   1891.
MAMMILLARIA RUFO-CROCEA Salm-Dyck, Cact. Hort. Dyck. 1849. 102.   1850.
  Cactus rufo-croceus Kuntze, Rev. Gen. Pl. **1**: 261.   1891.
MAMMILLARIA RUSCHIANA Regel, Ind. Sem. Hort. Turic. 4.   1830, in adnot.
  Cactus rueschianus Kuntze, Rev. Gen. Pl. **1**: 261.   1891.
MAMMILLARIA SEIDELII Terscheck, Suppl. Cact. Verz. 1.
  Cactus seidelii Kuntze, Rev. Gen. Pl. **1**: 261.   1891.
MAMMILLARIA SEVERINI Regel and Klein, Ind. Sem. Hort. Petrop. **1860**: 46.   1860.
  Cactus severinii Kuntze, Rev. Gen. Pl. **1**: 261.   1891.
MAMMILLARIA SPECIOSA De Vriese, Tijdschr. Nat. Geschr. **6**: 52.   1839.   Not Gillies, 1830.
  Cactus vrieseanus Kuntze, Rev. Gen. Pl. **1**: 260.   1891.
MAMMILLARIA SPECTABILIS Mühlenpfordt, Allg. Gartenz. **13**: 346.   1845.
  Cactus spectabilis Kuntze, Rev. Gen. Pl. **1**: 261.   1891.
MAMMILLARIA SUBULIFERA Ehrenberg, Allg. Gartenz. **17**: 242.   1849.
  Cactus subulifer Kuntze, Rev. Gen. Pl. **1**: 261.   1891.
MAMMILLARIA TECTA Miquel, Linnaea **12**: 12.   1838.
  Cactus tectus Kuntze, Rev. Gen. Pl. **1**: 261.   1891.

MAMMILLARIA TOMENTOSA Ehrenberg, Allg. Gartenz. **17:** 262. 1849.
Cactus tomentosus Kuntze, Rev. Gen. Pl. 1: 261. 1891.
MAMMILLARIA VARIMAMMA Ehrenberg, Allg. Gartenz. **17:** 242. 1849.
Cactus varimamma Kuntze, Rev. Gen. Pl. 1: 261. 1891.
MAMMILLARIA WEGENERI Ehrenberg, Bot. Zeit. **1:** 738. 1843.
Cactus wegeneri Kuntze, Rev. Gen. Pl. 1: 261. 1891.
MAMMILLARIA ZEGSCHWITZII Terscheck, Suppl. Cact. Verz. 1.
Cactus zegschwitzii Kuntze, Rev. Gen. Pl. 1: 261. 1891.
MAMMILLARIA ZEPNICKII Ehrenberg, Bot. Zeit. **2:** 835. 1844.
Cactus zepnickii Kuntze, Rev. Gen. Pl. 1: 261. 1891.

NAMES TO BE EXCLUDED FROM THIS GENUS.

The names *Mammillaria solitaria*, *M. spinosa*, *M. caudata*, *M. ambigua*, and *M. quadrata*, credited to G. Don, with the synonyms *Cactus solitarius*, *C. spinosus*, *C. caudatus*, *C. ambiguus* [Not Bonpland, 1813], and *C. quadratus*, credited to Gillies, respectively, each with a single word description, viz., solitary, spiny, tailed, ambiguous, quadrate, appeared in 1830 (Loudon, Hort. Brit. 194). As they all are said to come from Chile they can not be of this alliance.

Mammillaria brachydelphys Schumann (Just, Bot. Jahresb. **26:** 343. 1898) seems to have been intended for *Maihuenia brachydelphys*.

Cereus caudatus Gillies (Sweet, Hort. Brit. ed. 3. 285. 1839) is probably the same as *M. caudata*.

Mammillaria corioides Bosch (Sweet, Hort. Brit. ed. 3. 281. 1839) was described as leather-like and native of South America. It can not be identified, but it is not of this relationship if it comes from South America. Schumann referred it to *Echinocactus*, but it does not belong to that genus as we now define it.

Mammillaria dichotoma (Sweet, Hort. Brit. ed. 3. 281. 1839), described only as forked, can not be identified.

Mammillaria mitis (De Candolle, Prodr. **3:** 460. 1828), without description, is credited to Miller (Dict. Gard.), but Miller never used the generic name *Mammillaria*. Pfeiffer and Förster also refer this name to Miller. Steudel states that it is from South America. Kuntze also refers to the same as *Cactus mitis* (Rev. Gen. Pl. **1:** 259. 1891). Schumann thought that it might be an *Echinocactus* and, if it really came from South America, as stated by the Index Kewensis, it is probably of the *Echinocactanae*.

Mammillaria speciosa Gillies (Sweet, Hort. Brit. ed. 2. 235. 1830), to which *Cactus speciosus* Gillies is referred as a synonym, is based upon some Chilean plant.

Mammillaria subulata Mühlenpfordt is listed both by Schumann and the Index Kewensis but the name intended was *Pereskia subulata!*

MAMMILLARIA CHILDSI Blanc, Illustr. Cat. 14. 1894.

"This fine *Mammillaria* was sent out by us as *M. pectinata* before we bloomed it, from the fact that small plants answered the description exactly. After blooming, however, we discovered that it was a valuable new variety and named it as above. When small, the spines are regular, short and white; as the plant becomes older the spines also increase in size and assume a beautiful purple color. Flowers very numerous, even on small plants; color a clear pink."

We have not been able to identify this plant definitely. From the illustration, which shows large flowers from the center of the plant, we judge that it can not be referred to *Neomammillaria* nor to any of its near relatives. It may be a *Coryphantha*; in fact, at first it was taken for *C. pectinata*. The spines, however, are shown as arranged on vertical ribs, while the central spine is shown as erect; these two characters along with the central purple flowers suggest *Echinomastus erectocentrus*.

Illustration: Blanc, Illustr. Cat. 14.

MAMMILLARIA CORONARIA Haworth, Rev. Pl. Succ. 69. 1821, as to name.

Cactus coronatus Willdenow, Enum. Pl. Hort. Berol. Suppl. 30. 1813. Not Lamarck, 1783.

Judging from Willdenow's original descriptions of this plant it is not of this genus. He says that it is 5 feet long and a foot in diameter and that the central spine of the areole is hooked. Its geographical origin was not recorded and its flowers were not described. It was grown at Berlin prior to 1813 and later at the Chelsea Garden, London. Descriptions of this species are based largely on *Cactus cylindricus* Ortega, a very different plant.

Through the courtesy of N. E. Brown we have a photograph of *Mammillaria coronaria* from Haworth's collection with the date "Feb. 20, 1846." This photograph answers Haworth's brief description and differs from Willdenow's in having the spines all straight. Haworth's plants we would refer to *Neomammillaria*.

Cactus coronarius Willdenow, given by Haworth as a synonym of *Mammillaria coronaria*, is a mistake for *C. coronatus*.

The variety *Mammillaria coronaria minor* was briefly described by Förster (Handb. Cact. 212. 1846).

MAMMILLARIA FULVISPINA Haworth, Phil. Mag. 7: 108. 1830.

Cactus fulvispinus Kuntze, Rev. Gen. Pl. 1: 260. 1891.
Mammillaria rhodantha fulvispina Schelle, Handb. Kakteenk. 257. 1907.

This plant was said by Haworth to come from Brazil and if so it is to be excluded from this relationship. Pfeiffer associated the name with a Mexican specimen which has led to its being referred by later writers to *M. rhodantha*. The varieties *M. fulvispina media* and *M. fulvispina minor* (Salm-Dyck, Cact. Hort. Dyck. 1844. 8. 1845) were not described.

MAMMILLARIA PICTURATA Labouret, Rev. Hort. IV. 4: 28. 1855.

Simple, cylindric, 8 cm. high, 5 cm. in diameter; radial spines 20, white, setiform, 4 mm. long; central spines 6, yellowish; flowers and fruit unknown.

Although Labouret stated that this plant came from Mendoza, Argentina, the Index Kewensis says Chile. If it is in southern South America, it does not belong to *Neomammillaria*.

The illustration (figure 184a) at the bottom of this page is of *Neomammillaria milleri* described on page 156.

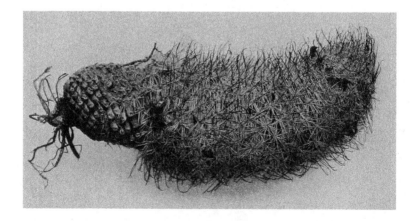

Subtribe 7. EPIPHYLLANAE.

Mostly epiphytic and night-blooming cacti, generally growing on trees, but sometimes on the earth when this is rich in humus, rarely in the crevices of rocks, much branched, spineless (except *Eccremocactus* and some species of *Epiphyllanthus*); joints several or many, usually flat except at base, often thin, with the areoles borne along the margin (except in *Epiphyllanthus*); flowers regular (except in *Zygocactus* and *Epiphyllanthus*); perianth various; filaments usually long and slender; style long and slender; fruit spineless, usually red or purple, either naked or bearing a few scales (rarely many), these usually with naked axils; seeds small, black.

We recognize 9 genera, diverse both in the plant-body and in the flowers. While apparently not closely related among themselves, the genera forming this subtribe are not any more closely related to other genera, either in the *Cereanae* or in the *Rhipsalidanae*.

KEY TO GENERA.

Plants branching dichotomously.
 Perianth irregular.
  Joints thin and leaf-like with toothed margin; areoles all marginal............ 1. *Zygocactus* (p. 177)
  Joints thick, without teeth, bearing areoles all around..................... 2. *Epiphyllanthus* (p. 180)
 Perianth regular or nearly so; joints thin................................... 3. *Schlumbergera* (p. 182)
Plants branching irregularly.
 Perianth-segments spreading or reflexed; flowers mostly large.
  Tube of flower definitely longer than limb.............................. 4. *Epiphyllum* (p. 185)
  Tube of flower not longer than limb.
   Perianth campanulate, its segments few.
    Stamens few; flowers small..................................... 5. *Disocactus* (p. 201)
    Stamens many; flowers large.................................... 6. *Chiapasia* (p. 203)
   Perianth short-funnelform, its segments many.
    Outer perianth-segments short, obtuse or rounded, the inner white... 7. *Eccremocactus* (p. 204)
    Outer perianth-segments acute or acuminate, the inner rose or red.... 8. *Nopalxochia* (p. 204)
 Perianth-segments erect; flowers small....................................... 9. *Wittia*. (p. 206)

1. ZYGOCACTUS Schumann in Martius, Fl. Bras. 4²: 223. 1890.

Stems dichotomously much branched, flattened, divided into short joints; flowers terminal, polychromic, irregular; ovary terete, smooth, gradually broadening from base, bearing minute scales at top; flower-tube abruptly bent just above the ovary, ending in a serrate mouth, bearing petaloid spreading scales scattered along its sides; stamens slender, white, arranged in 2 clusters; outer stamens borne along inside of flower-tube from near base to near middle; inner clusters of stamens about 20, arising from center and forming a short tube about base of style with an inner deflexed toothed membrane, upper part free, and all appressed against upper side of flower-tube and upper perianth-segments; style purple, slender, as long as stamens and usually not surrounded by them; stigma-lobes linear, purple, erect and adhering (so far as we have seen); fruit purple, turgid, not at all angled; skin thin; seeds dark brown to nearly black, shining.

Type species: *Epiphyllum truncatum* Haworth.

This genus has passed for many years under the name of *Epiphyllum* but that name was wrongly applied to it. One species is here recognized, although several have been proposed by previous authors.

The generic name is from ζυγόν yoke and κάκτος cactus, referring, doubtless, to the peculiarly jointed stems.

1. Zygocactus truncatus (Haworth) Schumann in Martius, Fl. Bras. 4²: 224. 1890.

 Epiphyllum truncatum Haworth, Suppl. Pl. Succ. 85. 1819.
 Cactus truncatus Link, Enum. Pl. 2: 24. 1822.
 Cereus truncatus Sweet, Hort. Brit. 272. 1826.
 Epiphyllum altensteinii Pfeiffer, Enum. Cact. 128. 1837.
 Epiphyllum truncatum altensteinii Lemaire, Cact. Gen. Nov. Sp. 76. 1839.
 Epiphyllum purpurascens Lemaire, Hort. Univ. 2: 349. 1841.
 Epiphyllum truncatum violaceum Morren, Belg. Hort. 16: 260. 1866.
 Epiphyllum truncatum spectabile Morren, Belg. Hort. 16: 260. 1866.
 Zygocactus altensteinii Schumann in Martius, Fl. Bras. 4²: 225. 1890.
 Epiphyllum delicatum N. E. Brown, Gard. Chron. III. 32: 411. 1902.
 Epiphyllum delicatulum Schumann, Monatsschr. Kakteenk. 13: 9. 1903.
 Zygocactus delicatus Britton and Rose, Contr. U. S. Nat. Herb. 16: 260. 1913.

Joints dark glossy green, about 3 cm. long, sharply serrate, with two prominent teeth at otherwise truncate apex; terminal areole broad and thin, filled with brown wool and bristles; flowers 6 to 7 cm. long; tube 2 cm. long; inner perianth-segments scarlet to white, oblong, obtuse to acute, reflexed; filaments white; style purple throughout; fruit obovoid, 1.5 to 2 cm. long.

Type locality: Brazil.

Distribution: Mountains, state of Rio de Janeiro, Brazil.

This species has been cultivated widely for many years under various names. It was introduced into cultivation about 1818 and, according to Edwards, flowered first in England in 1822 and has since been a great favorite as a household plant, blooming freely about the end of the year, hence the name Christmas cactus. It is also called crab cactus and ringent-flowered cactus.

Schumann gives as synonyms of this species *Epiphyllum salmoneum* and *E. spectabile*, referring them to Cels's Catalogue, which, however, we have not seen.

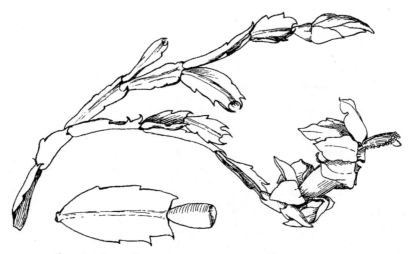

FIGS. 185 and 186.—Flowering branch and fruiting joint of Zygocactus truncatus.

Cereus truncatus altensteinii (Salm-Dyck, Hort. Dyck. 65. 1834) occurs in literature, sometimes attributed to Otto, but we have seen no description. We follow Löfgren, who refers *Zygocactus altensteinii* to *Z. truncatus.* The type came from the Organ Mountains near Rio de Janeiro; in 1915, Dr. Rose visited these mountains, where he found the true *Z. truncatus.*

There are many garden varieties, most of which are very beautiful. Among these are *Epiphyllum gibsonii*, introduced in 1886, with dark orange-red flowers, and *Epiphyllum guedeneyi*, of unknown origin, with large flowers, the outer segments white, tinged with sulphur, and the inner ones creamy white; the variety is referred by some to *Phyllocactus guedeneyi.* Nicholson (Dict. Gard. 1: 517) describes some of the best as follows:

"*Bicolor*, white, edged with rose; *coccineum*, rich deep scarlet; *elegans*, bright orange-red, centre rich purple; *magnificum*, flowers large, white, tips bright rose-colored; *roseum*, bright rose; *ruckerianum*, deep reddish purple, with a rich violet centre; *salmoneum*, reddish salmon; *spectabile*, white, with delicate purple margin; *violaceum superbum*, pure white, rich deep purple edge."

Rümpler (Förster, Handb. Cact. ed. 2. 870, 871. 1885) described nine varieties, among which are *cruentum* and *tricolor; E. truncatum cruentum* was also briefly described by Morren (Belg, Hort. 16: 260. 1866). Among other varieties are *albiflorum, aurantiacum, grandidens, minus, purpuraceum,* and *vanhoutteanum.*

Epiphyllum ruckeri Paxton (Mag. Bot. **12**: 46. 1846) was described from cultivated plants of unknown origin as an improved variety of *Epiphyllum truncatum*. It may have been a hybrid.

Epiphyllum truncatum multiflorum was given as a synonym of *Epiphyllum altensteinii* by Pfeiffer (Enum. Cact. 128. 1837).

Epiphyllum elegans Cels and *E. violaceum* Cels (Förster, Handb. Cact. 446. 1846) were supposed to be only varieties of *Epiphyllum truncatum*.

Schelle (Handb. Kakteenk. 223. 1907) lists more than fifty forms of *Epiphyllum truncatum*; the following not hitherto mentioned by us under *Epiphyllum* have the regular Latin form:

amabile roseum	maximum	salmoneum brasiliense	spectabile superbum
carmineum	morellianum	salmoneum flavum	splendens
gracile	pallidum roseum	salmoneum marginatum	translucens
grandiflorum rubrum	purpureum	salmoneum rubrum	violaceum album
harrisonii	rubrum violaceum	snowi	violaceum grandiflorum
lateritium album	salmoneum aurantiacum	spectabile carmineum	superbum
makoyanum			

Illustrations: Nov. Herb. Amat. pl. 83; Loudon, Encycl. Pl. 413. f. 6903; Loddiges, Bot. Cab. **13**: pl. 1207; Curtis's Bot. Mag. **52**: pl. 2562; Edwards's Bot. Reg. **9**: pl. 696;

Fig. 187.—Zygocactus truncatus.

Reichenbach, Fl. Exot. pl. 325; Hooker, Exot. Fl. **1**: pl. 20, as *Cactus truncatus*; Wiener Ill. Gart. Zeit. **18**: 265. f. 55, as *Phyllocactus delicatus*; Blühende Kakteen **1**: pl. 25; Cact. Journ. **1**: 34, 114; Cycl. Amer. Hort. Bailey **2**: f. 765; Engler and Prantl, Pflanzenfam. **3**[6a]: f. 61, A, B, C; Schumann, Gesamtb. Kakteen f. 9, 43; Hort. Univ. **7**: facing 132; Karsten, Deutsche Fl. 887. f. 501, No. 3; ed. 2. **2**: 456. f. 605, No. 3; Förster, Handb. Cact. ed. 2. 129. f. 5; Rümpler, Sukkulenten f. 87; Hort. Franc. **11. 4**: pl. 3; Schelle, Handb. Kakteenk. 223. f. 145; Balt. Cact. Journ. **1**: 49; Floralia **42**: 375; Gard. Chron. **1847**: 324; **11. 6**: 808. f. 148; **111. 7**: 173. f. 29; **111. 19**: f. 1; West Amer. Sci. **7**: 172; Amer. Gard. **11**: 534; Schelle, Handb. Kakteenk. 224. f. 146; 225. f. 147; Rother, Praktischer Leitfaden Kakteen 104; Belg. Hort. **16**: pl. opp. 257; also the vars. *spectabile, cruentum,* and *violaceum*; Deutsches Mag. Gart. Blumen. **1852**: pl. opp. 176. f. 2; Gartenwelt **4**: 230; Goebel, Pflanz. Schild. **1**: f. 55 (seedling); Garten-Zeitung **4**: 182. f. 42, No. 2; Jacquin, Ecl. Pl. Rar. **2**: pl. 142, as *Epiphyllum truncatum*; Pfeiffer and Otto, Abbild. Beschr. Cact. **1**: pl. 28, as *E. altensteinii;* Gard. Chron. **111. 32**: f. 140, as *E. delicatum*; Schumann, Gesamtb. Kakteen Nachtr. f. 9; Monatsschr. Kakteenk. **13**: 7, as *E. delicatulum*; Deutsches

Mag. Gart. Blumen. **1852**: pl. opp. 176, f. 1, as *E. truncatum elegans*; Arch. Jard. Bot. Rio de Janeiro **2**: pl. 3, as *Zygocactus delicatus*; Van Géel, Sert. Bot. **1**: pl. 117 as *Cactus truncatus*; Martius, Fl. Bras. 4²: pl. 46; Contr. U. S. Nat. Herb. **16**: pl. 80; Stand. Cycl. Hort. Bailey **6**: f. 4055.

Figure 185 shows a plant in the New York Botanical Garden which flowered December 15, 1911; figure 186 shows a fruiting joint collected by Dr. Rose in the Organ Mountains of Brazil in 1915 (No. 20819); figure 187 is from a photograph of a cultivated plant obtained by Dr. Rose in the Botanical Garden at Rio de Janeiro in 1915 (No. 20855).

2. EPIPHYLLANTHUS Berger, Rep. Mo. Bot. Gard. 16: 84. 1905.

Plants either epiphytic or growing in shade of rock in rich humus, often in clumps, more or less branched; joints globular, cylindric or much flattened; areoles scattered over surface of joints, circular, tomentose, either with or without spines; flowers zygomorphic, slender, purple to white; stamens somewhat exserted, arranged in 2 series, those forming the inner series united at base; style slender, a little longer than stamens; ovary angled, bearing a few small scales; fruit small.

Type species: *Epiphyllanthus obtusangulus* Berger.

The type of this genus has long been treated as a species of *Cereus*, although its dissimilarity to *Cereus* proper or to any of its immediate relatives must have been observed. It was left to Alwin Berger to call attention to its true alliance and to propose for it a new generic name; his statement regarding it is so clear that we quote from it as follows:

"This very strange little plant, still rare in cultivation, can not be considered either a *Cereus* or an *Epiphyllum*. But no doubt is it much more nearly allied to the latter than to the former genus. Schumann brought it into *Cereus* on account of its round and ribbed stems, but there exists no *Cereus* of a similar articulated growth; only with *Rhipsalis* and *Epiphyllum* can it be compared. The plant resembles somewhat a minute *Platyopuntia*. The joints are slightly flattened and have numerous little prominent areoles distributed spirally all over the surface. In this it differs greatly from *Epiphyllum* with which it agrees in all the characters of the flowers, the angular, nearly alate ovary, and especially in the inner stamens being united at the base into a small incurved membrane. Also, the fruit resembles more that of an *Epiphyllum* than that of a *Cereus*. The flowers rise from the top of the joints as in *Epiphyllum*. The plant is best considered as generically different from both, but must be placed with *Epiphyllum* and *Rhipsalis* among the *Inarmatae* of K. Schumann."

We recognize 3 species, all from central Brazil. All occur on the high mountain Itatiaya, province of Rio de Janeiro; what their actual relationships may be can be determined only by further field observations. They may all be referable to one variable species.

The generic name was given because of the resemblance of the flowers of the type species to those of *Epiphyllum truncatum* (*Zygocactus*).

KEY TO SPECIES.

Joints or some of them flattened, *Opuntia*-like.. 1. *E. obovatus*
Joints terete or obtusely angled.
    Flower purple to rose.. 2. *E. microsphaericus*
    Flowers white.. 3. *E. candidus*

1. Epiphyllanthus obovatus (Engelmann).

Epiphyllum obovatum Engelmann in Schumann, Gesamtb. Kakteen 224. 1897.
Epiphyllum opuntioides Löfgren and Dusén, Arch. Mus. Nac. Rio de Janeiro 13: 49. 1905.
Zygocactus opuntioides Löfgren, Arch. Jard. Bot. Rio de Janeiro 2: 26. 1918.

Usually growing in shade of rocks, at first erect, becoming more or less decumbent, very much branched; joints usually 5 to 7 cm. long, obovate to oblong, more or less flattened, often suggesting small joints of some *Opuntia*, bearing scattered areoles and these often spinescent; old and lower joints often nearly terete, bearing large areoles with numerous short yellow spines; flowers 5 cm. long, purple; ovary naked.

Type locality: Brazil.

Distribution: Central Brazil.

Dr. Rose collected this species on Itatiaya, altitude about 2,300 meters, in July 1915 (No. 20495); the plant did not do well in cultivation with us and his specimens died.

Illustrations: Arch. Jard. Bot. Rio de Janeiro **2**: pl. 4, as *Zygocactus opuntioides*; Arkiv Bot. Stockholm **8**: pt. 7. 10, as *Epiphyllum opuntioides.*

Figure 188 is reproduced from the first illustration cited above; figure 189 is reproduced from the second illustration cited above.

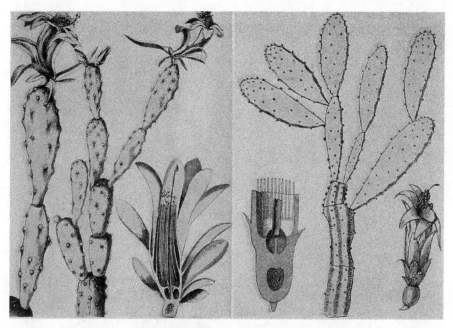

FIGS. 188 and 189.—Epiphyllanthus obovatus.

2. Epiphyllanthus microsphaericus (Schumann).

Cereus microsphaericus Schumann in Martius, Fl. Bras. 4²: 197.   1890.
Cereus parvulus Schumann in Martius, Fl. Bras. 4²: 197.   1890.
Cereus obtusangulus Schumann in Martius, Fl. Bras. 4²: 198.   1890.
Cereus anomalus Schumann, Keys Monogr. Cact. 16.   1903.
Epiphyllanthus obtusangulus Berger, Rep. Mo. Bot. Gard. **16**: 84.   1905.
Zygocactus obtusangulus Löfgren, Arch. Jard. Bot. Rio de Janeiro **2**: 28.   1918.

Low, at first erect, much branched and more or less prostrate, growing under rocks and perhaps epiphytic on trees; joints slender, terete or obtusely angled, somewhat spiny or often naked; flowers all terminal, purple to rose.

Type locality. Province of Rio de Janeiro, Brazil.
Distribution: Central Brazil.

Dr. Rose collected this species on Itatiaya, Brazil, in 1915 (No. 20494), growing at higher altitudes than *E. obovatus.*

Epiphyllum obtusangulum Lindberg (Martius, Fl. Bras. 4²: 198. 1890), usually referred here as a synonym, has not been published.

Illustrations: Arch. Jard. Bot. Rio de Janeiro **2**: pl. 5, as *Zygocactus obtusangulus*; Schumann, Gesamtb. Kakteen f. 30, as *Cereus obtusangulus.*

Figure 190 is reproduced from the first illustration cited above.

3. Epiphyllanthus candidus (Löfgren).

Zygocactus candidus Löfgren, Arch. Jard. Bot. Rio de Janeiro **2**: 30. 1918.

Usually epiphytic on shrubs, but sometimes growing in the shade of large boulders; joints usually terete or nearly so, 2 to 4 cm. long, naked or sometimes bristly; flowers solitary, terminal, white; fruit globose, red.

Type locality: On Itatiaya, Brazil.
Distribution: Known only from the type locality.

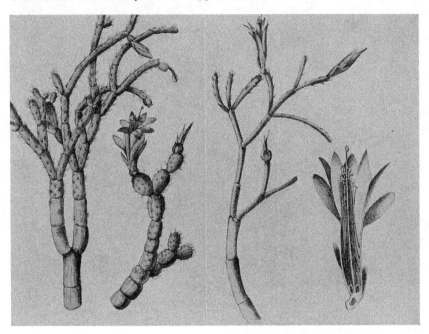

FIG. 190.—Epiphyllanthus microsphaericus.   FIG. 191.—Epiphyllanthus candidus.

Dr. Rose collected this species on the very top of Itatiaya, growing in the shade of rocks (No. 20610) and in the deep cleft of the rock cap through which the ascent to the top is made.

Epiphyllum candidum Barboso-Rodrigues (Arch. Jard. Bot. Rio de Janeiro **2**: 30. 1918) is only a name.

Illustration: Arch. Jard. Bot. Rio de Janeiro **2**: pl. 6, as *Zygocactus candidus*.

Figure 191 is reproduced from the illustration cited above.

3. SCHLUMBERGERA Lemaire, Rev. Hort. IV. **7** 253. 1858.

Similar in habit to *Zygocactus*; stems much branched; joints short, crenate or serrate, mostly flattened; flowers purple to scarlet, regular; tube very short; stamens in 2 clusters, one scattered over the throat, the other forming a short tube at base of flower and surrounding style or free at base; ovary and fruit strongly 5-angled, naked or rarely bearing areole on one of the ribs and crowned by 5 more or less persistent, sepal-like scales; fruit hard, often remaining on plant for a long time.

Type species: *Epiphyllum russellianum* Hooker.

The taxonomic history of the two species here recognized is interesting. *Schlumbergera gaertneri* was at first supposed to be conspecific with *S. russelliana* and was made a variety of that species by Regel. In 1890 Schumann considered them distinct species but congeneric; in 1897 he referred them to different genera. Both species are native of Brazil.

These plants have usually been associated with *Zygocactus truncatus* and all included in *Epiphyllum*. Although resembling *Zygocactus* very much in habit, they differ from it in flower and fruit characters. The flowers are nearly regular, not strongly oblique; are nearly rotate, not elongated; the stamens are of equal length and in a cylindric cluster shorter than the style, not of unequal lengths and in a flattened cluster, not extending beyond the style; the ovary and fruit are strongly angled, not terete.

Lemaire named the genus for Frederick Schlumberger, an amateur student of plants and a collector of cacti, begonias, and bromelias.

KEY TO SPECIES.

Flowers scarlet.. 1. *S. gaertneri*
Flowers purplish.. 2. *S. russelliana*

1. Schlumbergera gaertneri (Regel) Britton and Rose, Contr. U. S. Nat. Herb. **16**: 260. 1913.

> *Epiphyllum russellianum gaertneri* Regel, Gartenflora **33**: 323. 1884.
> *Epiphyllum makoyanum* W. Watson, Gard. and For. **2**: 243. 1889.
> *Epiphyllum gaertneri* Schumann in Martius, Fl. Bras. **4**²: 218. 1890.
> *Phyllocactus gaertneri* Schumann in Rümpler, Sukkulenten 147. 1892.

FIG. 192.—Schlumbergera gaertneri.

Branches spreading, the terminal ones often pendent; joints usually flattened, but sometimes 3 to 6-angled, fleshy, 5 cm. long or more by 2 cm. broad, dull green except the purplish crenate margins; areoles small, with short white wool and a few yellowish bristles; flowers 1 to 3, usually all at distal end of the terminal branches, 4 cm. long, dark scarlet; outermost perianth-segments usually 5, short, thick, triangular, drying separately from the others; outer perianth-segments spreading; innermost perianth-segments more erect, nearly distinct, acute; all of the segments, except the 5 outer ones, more or less coalesce and withering, remaining on top of ovary; ovary crowned by a slightly depressed disk or umbilicus with upturned margin, which passes into the flower-tube; on the margin are borne the free stamens; style slender, 1.5 cm. long, red; stigma-lobes 6, linear, cream-colored; ovary dark red, angled, 12 mm. long; fruit red, oblong, 15 mm. long, depressed at apex, in cultivation ripening in July.

Type locality: Minas Geraes, Brazil.

Distribution: Brazil.

While the joints are usually much flattened, yet they are sometimes strongly angled. In some cases too the juvenile growth is peculiar, forming short stubby joints with 6 ribs, with closely set areoles, each bearing a cluster of 7 or more bristly spines.

The plant flowers abundantly in Washington in April.

The two varieties *Epiphyllum gaertneri coccineum* and *E. gaertneri mackoyanum* (Monatsschr. Kakteenk. **7**: 101. 1897) are doubtless forms of this species.

Illustrations: Wiener Ill. Gart. Zeit. **10**: 136. f. 60; Rev. Hort. **59**: pl. opp. 516; Blanc, Cacti 64. 1002; Cact. Journ. **1**: 9, 114; Gartenflora **33**: pl. 1172; **39**: f. 96; Rev. Hort. Belg. **15**: f. 23; pl. [19.] f. 2, opp. 229, as *Epiphyllum russellianum gaertneri*; Schelle, Handb. Kakteenk. 213. f. 141; Curtis's Bot. Mag. **117**: pl. 7201; Gartenwelt **10**: 559, as *Epiphyllum gaertneri*; Blühende Kakteen **1**: pl. 21; Thomas, Zimmerkultur Kakteen

FIG. 193.—Schlumbergera gaertneri.

19; Monatsschr. Kakteenk. **4**: 107; Rümpler, Sukkulenten 148. f. 80, as *Phyllocactus gaertneri*; Rev. Hort. Belg. **15**: pl. [19.] f. 1, opp. 229; Journ. Hort. Home Farm. III. **18**: 362. f. 58, as *Epiphyllum makoyanum*.

Figure 192 is from a photograph of a plant which flowered in the New York Botanical Garden in 1912; figure 193 shows a fruiting plant in the collections of the U. S. Department of Agriculture at Washington, D. C.

2. Schlumbergera russelliana (Gardner) Britton and Rose, Contr. U. S. Nat. Herb. **16**: 261. 1913.

    Cereus russellianus Gardner in Lemaire, Hort. Univ. **1**: 31. 1839.
    Epiphyllum russellianum Hooker in Curtis's Bot. Mag. **66**: pl. 3717. 1840.
    Phyllocactus russellianus Salm-Dyck, Cact. Hort. Dyck. 1844. 37. 1845.
    Epiphyllum truncatum russellianum G. Don in Loudon, Encycl. Pl. ed. 3. 1378. 1855.
    Schlumbergera epiphylloides Lemaire, Rev. Hort. IV. **7**: 253. 1858.

Epiphytic, growing on trees, rocks, or in humus, often found in dark crevices, 1 to 3 dm. long, either hanging or erect, much branched, divided into short joints; joints 1 to 2.5 cm. long; lower joints usually terete, covered with a brown epidermis; young joints green, flat, usually thin, with 1 or 2 small teeth on a side, 8 mm. broad or less, usually truncate at apex; areoles in axils of teeth, small, naked or bearing 1 or 2 bristles; flowers terminal, 4 to 5 cm. long, reddish purple; style slender, purple; ovary glabrous, sharply 4-angled, 1-celled; ovules numerous, arranged in 4 or 5 vertical double rows along walls of ovary; fruit described as red, 4-angled, or narrowly winded.

Type locality: Organ Mountains, Brazil.
Distribution: Brazil.

This plant was introduced into England in 1839. It was named by G. Gardner for the Duke of Bedford, who had sent him to Brazil to collect plants. The Duke of Bedford brought together at Woburn Abbey a very large and choice collection of cacti which became one of the finest in England. His gardener, Mr. James A. Forbes, published a catalogue of this collection in 1837.

Two varieties of this species are mentioned in horticultural works, namely, var. *rubra* and var. *superbum* under *Epiphyllum russellianum*.

Illustrations: Curtis's Bot. Mag. **66**: pl. 3717; Watson, Cact. Cult. 42. f. 9; Dict. Gard. Nicholson Suppl. 346. f. 370: Gartenflora **33**: pl. 1172; Förster, Handb. Cact. ed. 2. 873. f. 119; Rother, Praktischer Leitfaden Kakteen 106; Paxton's Mag. Bot. **10**: facing 245, as *Epiphyllum russellianum*; Hort. Univ. **1**: pl. 5, as *Cereus russellianus*; Rümpler, Sukkulenten 146. f. 79, as *Phyllocactus russellianus*; Cycl. Amer. Hort. Bailey **2**: f. 766, as *Epiphyllum truncatum russellianum* (perhaps a hybrid); Contr. U. S. Nat. Herb. **16**: pl. 81.

SPECIES OF THIS RELATIONSHIP.

EPIPHYLLUM BRIDGESII Lemaire, Illustr. Hort. **8**: Misc. 5. 1861.
Epiphyllum truncatum bridgesii Rümpler in Förster, Handb. Cact. ed. 2. 870. 1885.

Epiphytic; joints green, flattened with 2 or more crenations on the side; areoles more or less setose, the setae yellowish brown; flowers terminal, 6 cm. long, nearly regular, purplish to crimson; perianth-segments oblong, acute; stamens long-exserted; style about as much exserted as stamens, purplish; ovary angled, angles sometimes bearing setose areoles.

Type locality: Not cited. Described from garden plants of unknown origin.
Distribution: Brazil cr Bolivia or both.

This plant was described by Lemaire from a vegetative specimen seen in the collection of L. Desmet and from one in the collection of Schlumberger. He associated it with *Epiphyllum russellianum*, with which it must be allied, rather than with *E. truncatum*, to which it is referred as a synonym by the Index Kewensis.

Schlumberger had named the plant *Epiphyllum rueckerianum*, and here this name, often referred to in horticultural literature, should be referred.

It was briefly described by W. Watson (Gard. For. **2**: 243. 1889), who writes of its being awarded a first-class certificate at a flower show.

Schumann unfortunately describes the flower as zygomorphic, which may be an error; specimens recently sent to us from A. Berger have a regular flower. The ovary was originally described as angled and this is one of the differences between *Zygocactus* and *Schlumbergera*.

We do not know the origin of this plant. As it seems to have been introduced by Bridges it may have come from Bolivia, where he did much of his work.

This plant is sometimes called *Epiphyllum truncatum rueckerianum*.

Illustration: Dict. Hort. Bois 497. f. 347, as *Epiphyllum ruckerianum*.

4. EPIPHYLLUM (Hermann) Haworth, Syn. Pl. Succ. 197. 1812.

Phyllocactus Link, Handb. Erkenn. Gewächse **2**: 10. 1831.
Phyllocereus Miquel, Bull. Sci. Phys. Nat. Néerl. 112. 1839.

Plants mostly epiphytic, the main stem often terete and woody; branches usually much flattened, often thin and leaf-like, sometimes 3-winged; areoles small, borne along the margins of the flattened branches; spines usually wanting in mature plants, but often represented in seedlings and juvenile forms by slender bristles; true leaves wanting; cotyledons rather large, sometimes persisting for a long time; flowers usually large, in some species nocturnal, in others diurnal, either odorless or very fragrant; flower-tube longer than the limb, in some species greatly elongated; filaments usually long, borne at top of tube or scattered over surface of throat; style elongated, white or colored; stigma-lobes several, linear; perianth soon dropping from the ovary; fruit globular or short-oblong to narrowly oblong, often with low ridges, sometimes tubercled, red or purple, edible or insipid, when mature splitting down one side and exposing the white or crimson pulpy interior; seeds black, shining.

Type species: *Cactus phyllanthus* Linnaeus.

The generic name is from ἐπί upon, and φύλλον leaf, as it was supposed that the flowers were borne on leaves; it is a misnomer, for the flowers are not borne on leaves but on stems as in all other cacti.

In 1890 K. Schumann recognized 15 species; but, as a number of new ones were described soon afterward, he increased this number to 21 in his Keys of the Monograph published in 1903. In our treatment 16 species are recognized.

The name *Epiphyllum* is often used for a different group of cacti, that is, the crab cactus; the type species of *Epiphyllum* is, however, in the genus as we have here limited it. When Haworth published the genus he referred to it but one species, *Epiphyllum phyllanthus*, but he later added another species, *E. truncatum*, which, when it was found to belong to a different generic type, was erroneously allowed to retain the name *Epiphyllum*, while *Epiphyllum phyllanthus* became the type of the genus *Phyllocactus*, which, when first described in 1831, contained but a single species, so that *Epiphyllum* and *Phyllocactus* were based on the same type and *Phyllocactus* is a synonym of *Epiphyllum*. This is also true of *Phyllocereus*, which was based on the *Epiphyllum* of Haworth (Syn. Pl. Succ. 197. 1812), where only *E. phyllanthus* is described.

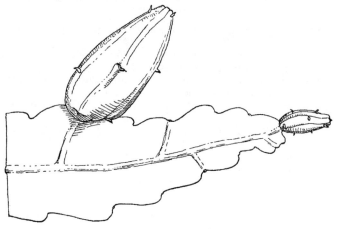

FIG. 194.—Top of fruiting branch of Epiphyllum phyllanthus.   × 0.66.

The pre-Linnaean species of this genus were usually referred to *Cereus* and, for it, the section *Alati* in *Cereus* was proposed by De Candolle (Prodr. **3**: 469. 1828). Linnaeus, however, referred the only species which he recognized to *Cactus*, and Philip Miller referred the same species to *Opuntia*, but neither have had many followers.

Haworth (Phil. Mag. **6**: 108, 109. 1829) followed by Don (Hist. Dichl. Pl. **3**: 170. 1834) divides the genus into two sections, the *Nocturna* and the *Diurna*.

Phyllarthus Nicker (Elam. **2**: 85. 1790) is generally supposed to be a synonym of this group but the genus is not typified; the Index Kewensis refers it to *Cereus* (?); Dr. E. L. Greene (Leaflets **1**: 52) says that it applied to *Phyllanthus* and *Opuntia* of earlier authors; the *Phyllanthus* here referred to was *Cactus phyllanthus* Linnaeus.

Hermann (Par. Botavus Prodr. Add. 2. 1689) first used the name *Epiphyllum* when he listed the name *Epiphyllum americanum*. Haworth credited the name also to Hermann in 1812 when he established the genus.

KEY TO SPECIES.

A. Perianth-tube 7 to 9 times as long as the limb................................... 1. *E. phyllanthus*
AA. Perianth-tube 1½ to 3 times as long as the limb.
  B. Ultimate joints acuminate.
    Flowers 25 to 30 cm. long.. 2. *E. oxypetalum*
    Flowers 10 to 15 cm. long.
      Margins of joints crenate.. 3. *E. pumilum*
      Margins of joints undulate... 4. *E. caudatum*
  BB. Ultimate joints acute, obtuse or rounded.
    C. Joints deeply lobed.
      Joints 2 to 7 cm. broad.
        Lobes of joints spreading; outer perianth-segments lemon-yellow......... 5. *E. darrahii*
        Lobes of joints pointing forward; outer perianth-segments reddish yellow.. 6. *E. anguliger*
      Joints very large, up to 25 cm. broad................................. 7. *E. grandilobum*
    CC. Joints crenate or nearly entire.
      D. Joints deeply crenate, thick; perianth-tube bearing foliaceous scales...... 8. *E. crenatum*
      DD. Joints low-crenate to nearly entire; perianth-tube without foliaceous scales.
        E. Sinui of the joint-margins very narrow; flowers up to 20 cm. broad;
                  stamens yellow.................... 9. *E. mac opterum*
        EE. Sinui of the joint-margins open; flowers 15 cm. wide or less; stamens
           not yellow.
           Ovary and fruit bearing linear scales........................... 10. *E. lepidocarpum*
           Ovary and fruit without linear scales.
             Flowers 10 to 13 cm. long................................. 11. *E. pittieri*
             Flowers 15 to 28 cm. long.
                Flowers about 28 cm. long; style orange.................... 12. *E. guatemalense*
                Flowers 15 to 25 cm. long; style white or pink.
                  Joints very stiff.. 13. *E. strictum*
                Joints flexible or moderately stiff.
                  Joints very large, up to 1 meter long and 12 cm. wide..... 14. *E. stenopetalum*
                  Joints smaller, rarely ever 7 cm. wide.
                    Joints shallowly crenate or subdentate; species of Costa
                      Rica....................................... 15. *E. cartagense*
                    Joints deeply crenate; species of Tobago, Trinidad, and
                      Venezuela................................... 16. *E. hookeri*

1. **Epiphyllum phyllanthus** (Linnaeus) Haworth, Syn. Pl. Succ. 197. 1812.

    Cactus phyllanthus Linnaeus, Sp. Pl. 469. 1753.
    Opuntia phyllanthus Miller, Gard. Dict. ed. 8. No. 9. 1768.
    Cereus phyllanthus De Candolle, Prodr. 3: 469. 1828.
    Phyllocactus phyllanthus Link, Handb. Erkenn. Gewächse 2: 11. 1831.
    Rhipsalis macrocarpa Miquel, Bull. Sci. Phys. Nat. Néerl. **1838**: 49. 1838 (in most part).
    Rhipsalis phyllanthus Schumann in Martius, Fl. Bras. 4²: 298. 1890 (in part).
    Hariota macrocarpa Kuntze, Rev. Gen. Pl. 1: 263. 1891.
    Phyllocactus phyllanthus paraguayensis Weber, Dict. Hort. Bois 957. 1898.
    Phyllocactus phyllanthus boliviensis Weber, Dict. Hort. Bois 957. 1898.
    Phyllocactus phyllanthus columbiensis Weber, Dict. Hort. Bois 957. 1898.
    Epiphyllum gaillardae Britton and Rose, Contr. U. S. Nat. Herb. **16**: 240. 1913.
    Phyllocactus gaillardae Vaupel, Monatsschr. Kakteenk. **23**: 87. 1913.

Elongated and much branched; main branches narrow, terete or 3 or 4-angled, woody; terminal joints elongated, terete or 3-angled below, usually flat or thin, rarely 3-winged, bright green with a purple margin, sometimes 7 cm. broad, obtuse, the margin coarsely serrate, the teeth obtuse; flower slender, 25 to 30 cm. long, the slender tube very much longer than the limb, green, the limb greenish or white, its segments narrow, 2 to 2.5 cm. long; scales on flower-tube few, minute, spreading; style long, slender, pinkish (Schumann says white); filaments short; stigma-lobes short, white; fruit oblong, 7 to 9 cm. long, somewhat 8-ribbed, bright red; pulp white; seeds large, black, numerous.

According to De Candolle, the flowers are nocturnal and odoriferous.

Type locality: Brazil.

Distribution: Panama to British Guiana, Bolivia, Peru, and Brazil.   Recorded from Paraguay.

The species has been recorded from the West Indies, apparently erroneously.

Our description is based on field notes made by Dr. Rose in Brazil in 1915 which differ slightly from published descriptions. This plant is common in the woods along the coast of eastern Brazil, often growing in inaccessible places high up in the great trees. Open flowers were not seen, but buds, fruit, and seeds were obtained. Living plants were collected and these have done well; one flower appeared in the collection of the Department of Agriculture during Dr. Rose's absence in Ecuador in 1918. The caretaker, Mr. Fraile, describes the flower as long and slender and very unlike other species of *Epiphyllum*, of which he has seen many (Rose, No. 19627). It fruited in the New York Botanical Garden in 1920. The plant is called flor de baile or flower of the ball.

An *Epiphyllum* grows in the lowlands of Ecuador which we have tentatively referred here, although we have never seen its flowers or fruits. Dr. Rose collected it below Huigra, September 8, 1918 (No. 22614), and again above Santa Rosa near Limón Playo, October 17 (No. 23493).

Cactus phyllanthus of Linnaeus (Sp. Pl. 469. 1753) and *Epiphyllum, phyllanthus* Haworth (Syn. Pl. Succ. 197. 1812) both contain references not only to this species but to *Epiphyllum phyllanthoides* also.

The variety *columbiensis* was described by both Weber and Schumann with a flower-tube only 6 cm. long.

FIG. 195.—Seedling of Epiphyllum phyllanthus. × o.6.

Cereus phyllanthus marginatus Parmentier is mentioned by Lemaire (Cact. Gen. Nov. Sp. 76. 1839) but not described.

Illustrations: Contr. U. S. Nat. Herb. **16**: pl. 68, as *Epiphyllum gaillardae*; Pfeiffer and Otto, Abbild. Beschr. Cact. **1**: pl. 10, f. 1, as *Cereus phyllanthus*; Petiver, Gazoph. Dec. pl. 59, f. 10. 1709, as *Heliotropium*, etc.; Dillenius, Hort. Elth. pl. 64, as *Cereus scolopendrii*, etc.; De Candolle, Pl. Succ. Hist. pl. 145; Vellozo, Fl. Flum. **5**: pl. 33 (except flower), as *Cactus phyllanthus*; Monatsschr. Kakteenk. **2**: 73, as *Phyllocactus phyllanthus*; Martius, Fl. Bras. **4²**: pl. 44.

Figure 194 is from a photograph of a fruiting branch borne on the specimens obtained by Dr. Rose in Brazil in 1915; figure 195 shows a seedling with its two large cotyledons, grown from seeds sent by Mrs. D. D. Gaillard from Panama.

2. Epiphyllum oxypetalum (De Candolle) Haworth, Phil. Mag. **6**: 109. 1829.

> *Cereus oxypetalus* De Candolle, Prodr. **3**: 470. 1828.
> *Cereus latifrons* Pfeiffer, Enum. Cact. 125. 1837.
> *Phyllocactus oxypetalus* Link in Walpers, Repert. Bot. **2**: 341. 1843.
> *Phyllocactus latifrons* Link in Walpers, Repert. Bot. **2**: 341. 1843.
> *Phyllocactus grandis* Lemaire, Fl. Serr. **3**: 255b. 1847.
> *Phyllocactus guyanensis* Brongnart in Labouret, Monogr. Cact. 416. 1853.
> *Epiphyllum acuminatum* Schumann in Martius, Fl. Bras. **4²**: 222. 1890.
> *Phyllocactus acuminatus* Schumann, Gesamtb. Kakteen 213. 1897.
> *Phyllocactus purpusii* Weingart, Monatsschr. Kakteenk. **17**: 34. 1907.
> *Epiphyllum grande* Britton and Rose, Contr. U. S. Nat. Herb. **16**: 257. 1913.

Plants stout, 3 meters long or more, much branched; branches flat and thin, 10 to 12 cm. broad, long-acuminate, deeply crenate; flowers opening in the evening, drooping and limp after anthesis, fragrant; tube of flower 13 to 15 cm. long, rather stout, red, about 1 cm. thick, bearing distant narrow scales about 10 mm. long; outer perianth-segments narrow, reddish to amber, 8 to 10 cm. long; inner perianth-segments oblong, white; stamens numerous, white; style white, thick, 20 cm. long; stigma-lobes numerous, cream-colored, entire.

Type locality: Mexico.

Distribution: Mexico and Guatemala, Venezuela, and Brazil. Widely cultivated in the tropics and doubtless an escape in many places.

This species has long been cultivated and has always been a great favorite on account of the ease with which it is grown and the abundance of large flowers it furnishes. These begin to open in the early evening and are perfect about midnight.

According to Mr. Pittier, this plant is known as flor de baile in Venezuela.

Epiphyllum latifrons Zuccarini (Pfeiffer, Enum. Cact. 125. 1837) was given as a synonym of *Cereus latifrons* when that name was first published.

The name *Cactus oxypetalus* Mociño and Sessé was the first one given to this plant, but De Candolle (Prodr. **3**: 470. 1828) published the species as a *Cereus*, citing the above name as a synonym.

The following hybrids were listed by Labouret (Monogr. Cact. 429. 1853) between *Phyllocactus latifrons* and some other species of *Epiphyllum* or related genera; *Phyllocactus longipes, P. lothii, P. londonii, P. macquianus, P. maelenii, P. maurantianus, P. mexicanus, P. roseus albus, P. roseus superbus, P. selloi, P. smoli,* and *P. smithii.*

Illustrations: Monatsschr. Kakteenk. **17**: 35, as *Phyllocactus purpusii*; Meehans' Monthly **12**: 188, as *Epiphyllum latifrons*; Mém. Mus. Hist. Nat. Paris **17**: pl. 14, as *Cereus oxypetalus*; Förster, Handb. Cact. ed. 2. 849. f. 112, as *Phyllocactus oxypetalus*; Rother, Praktischer Leitfaden Kakteen 93; Monatsschr. Kakteenk. **20**: 123, as *Phyllocactus grandis*; Martius, Fl. Bras. **4**²: pl. 45, as *Epiphyllum acuminatum*; Engler and Prantl, Pflanzenfam. **3**⁶ᵃ: f. 59, D, as *Phyllocactus acuminatus*; Gard. Chron. **1849**: 788; Pfeiffer and Otto, Abbild. Beschr. Cact. **1**: pl. 10, f. 2, 3; Curtis's Bot. Mag. **67**: pl. 3813, as *Cereus latifrons*; Gartenwelt **10**: 560; Cact. Journ. **1**: 55; Goebel, Pflanz. Schild. **1**: pl. 2, f. 6; Schelle, Handb. Kakteenk. 209. f. 139; 210. f. 140, as *Phyllocactus latifrons.*

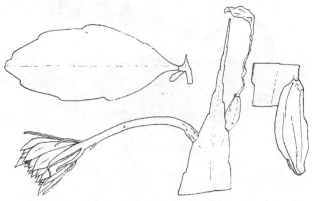

FIG. 196.—Epiphyllum pumilum. × 0.5.

3. Epiphyllum pumilum Britton and Rose, Contr. U. S. Nat. Herb. **16**: 258. 1913.

Phyllocactus pumilus Vaupel, Monatsschr. Kakteenk. **23**: 117. 1893.

At first erect or ascending but often becoming pendent, sometimes 5 meters long; main stems terete; branches of two types; some of them elongated, 8 to 15 dm. long, terete, whip-like, sometimes becoming flattened at tip; some broad and flattened, rarely 3-winged, except at base, usually acute or acuminate, 1 to 6 dm. long, 3 to 8.5 cm. broad, becoming thick when old, the margin remotely toothed; flowers small for the genus; tube 5 to 6 cm. long, greenish white to reddish, bearing a few very small ascending and appressed reddish scales; outer perianth-segments linear, greenish or reddish, acute; inner perianth-segments white, lanceolate, acuminate, 3 to 4 cm. long; stamens in two groups; style slender, white, oblong, 4 to 7 cm. long, 2 to 2.5 cm. in diameter; fruit brilliant cerise when ripe, 5 to 7-ridged, bearing a few very small reddish ascending scales; pulp of fruit white, edible, sweet; seeds minute, jet-black.

Type locality: Guatemala.

Distribution: Lowlands of Guatemala.

This species has frequently been collected in Guatemala and is usually called *Epiphyllum pittieri*, which it somewhat resembles in the size of the flower, but the style is always white.

The flowers are night-blooming and sweet-scented. The fruit is much sought after by the Guatemalan Indians, who call it pitahaya.

The above description is based on living specimens, full notes, and drawings, furnished by Harry Johnson, a very keen observer, at one time stationed in Guatemala.

Figure 196 is copied from pencil sketches made by Mr. Harry Johnson at Chamá, Alta Verapaz, Guatemala, in 1920.

4. Epiphyllum caudatum Britton and Rose, Contr. U. S. Nat. Herb. **16**: 256.   1913.

   Phyllocactus caudatus Vaupel, Monatsschr. Kakteenk. **23**: 116.   1913.

Old stems terete and slender; lateral branches elongated-lanceolate, cuneately narrowed at base into a terete stalk, long-acuminate, 15 to 20 cm. long, 3 to 4 cm. wide, the margins low-undulate; flowers white, the tube slender, about 7 cm. long; inner perianth-segments about 6 cm. long; ovary and most of the flower-tube quite naked.

Fig. 197.—Epiphyllum caudatum.

Type locality: Near Comaltepec, Oaxaca, Mexico, altitude 540 to 900 meters.

Distribution: Known only from the type locality.

We have seen no specimens of this species except the type, but Dr. B. P. Reko, under date of June 28, 1919, wrote that he had seen the plant not only at Comaltepec, but at other places in the Sierra Juárez.

A plant sent from Chiapas, Mexico, by Dr. C. A. Purpus in 1920 has joints with similar acuminate tips, but the margins are indented. We do not know its flowers.

Figure 197 is from a photograph of the type specimen.

5. Epiphyllum darrahii (Schumann) Britton and Rose, Contr. U. S. Nat. Herb. **16**: 256.   1913.

   Phyllocactus darrahii Schumann, Gesamtb. Kakteen Nachtr. 69.   1903.

Stems much branched, often terete and woody below; joints rather thick, 2 to 3 dm. long, 3 to 5 cm. wide, deeply lobed, sometimes nearly to the midrib, the lobes usually obtuse; tube of flower 9 cm. long, somewhat curved, greenish; scales on tube and ovary small, linear, green, appressed; outer perianth-segments 10, linear, spreading or reflexed, acute, 4 cm. long, lemon-yellow; inner perianth-segments pure white, nearly as long as the outer, broader and more erect, short-

M. E. Eaton del.

1. Flowering plant of *Epiphyllum darrahii*.
2. Top of flowering plant of *Epiphyllum pittieri*.

acuminate; filaments white, nearly as long as the perianth-segments; style overtopping the stamens, pure white; stigma-lobes 8, linear.

Type locality: Mexico.

Distribution: Mexico, but range unknown.

This species was named for Charles Darrah of Heaton Mersey near Manchester, England (1844-1903). His large and valuable collection of succulents, especially cacti, was presented to the Corporation of Manchester by his widow and family and is now housed in specially constructed houses in Alexander Park. In 1908 the late Robert Lamb published a catalogue of 129 pages of this collection.

The plant is cultivated in Mexico and is much prized as a potted plant for the patio; one of these was obtained by Dr. Rose in Ixmiquilpan in 1905 (No. 9091). Living specimens were sent home and these have repeatedly flowered in Washington and New York. It flowers abundantly, its blossoms giving off a most delicious honeysuckle-like fragrance; we have seen no specimens of wild plants.

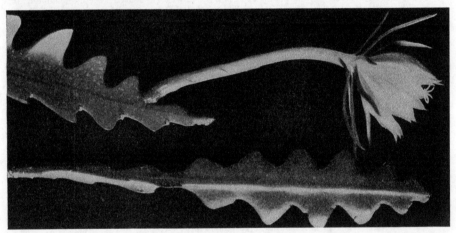

FIG. 198.—Epiphyllum darrahii.

Illustration: Blühende Kakteen **2**: pl. 91, as *Phyllocactus darrahii.*

Plate XVI, figure 1, shows the plant in flower, collected by Dr. Rose in Mexico in 1905, which flowered in the New York Botanical Garden in September 1917. Figure 198 is from a photograph of the plant collected by Dr. Rose at Ixmiquilpan, Mexico, in 1905 which afterwards flowered in Washington.

6. Epiphyllum anguliger (Lemaire) Don in Loudon, Encycl. Pl. ed. 3. 1380. 1855.

Phyllocactus anguliger Lemaire, Jard. Fleur. 1: pl. 92. 1851.
Phyllocactus serratus Brongnart in Labouret, Monogr. Cact. 417. 1853.

Much branched; stems and lower branches terete; upper branches flattened with deeply toothed margins, rather fleshy; areoles small, usually felted and sometimes bearing 1 or 2 white bristles; flower-tube stout, without scales, about 8 cm. long; outer perianth-segments brownish yellow, inner perianth-segments white, oblong, acuminate, about 5 cm. long; style slender, white.

Type locality: Near Matanejo, Mexico.

Distribution: Central and southern Mexico.

We know the species only from cultivated plants. When not in flower it is difficult to distinguish it from *Epiphyllum darrahii.*

This plant was first distributed by the Horticultural Society of London, which obtained it from the collector, T. Hartweg, in 1846, from southern Mexico, where it was found growing on oak trees.

Phyllocactus angularis occurs in the index of Labouret's Monograph (511), credited to Lemaire, and also is listed in the Index Kewensis. It may have been a manuscript name for this species.

Illustrations: Lemaire, Jard. Fleur. **1**: pl. 92; Lindley and Paxton, Fl. Gard. **1**: pl. 34; Curtis's Bot. Mag. **85**: pl. 5100; Dict. Gard. Nicholson **3**: f. 134; Amer. Gard. **11**: 538; Möllers Deutsche Gärt. Zeit. **25**: 477. f. 11, No. 24; Cycl. Amer. Hort. Bailey **1**: f. 306; Palmer, Cult. Cact. 167; Watson, Cact. Cult. 48. f. 11; ed. 3. f. 9; Floralia **42**: 377, as *Phyllocactus anguliger*.

7. Epiphyllum grandilobum (Weber) Britton and Rose, Contr. U. S. Nat. Herb. **16**: 257.   1913.

  Phyllocactus grandilobus Weber, Bull. Mus. Hist. Nat. Paris **8**: 464.   1902.

Branches bright green, very large, up to 25 cm. broad with the margins deeply lobed and with a thick midvein and obtuse or rounded apex; lobes rounded, 3 to 5 cm. long; flowers described as large, white, opening at night; fruit red without.

Type locality: La Hondura, Costa Rica.

Distribution: Costa Rica.

Weber speaks of this as a very remarkable species of which he had not seen flowers or fruit. His description was based on specimens collected by Wercklé in 1900 and also by Pittier in 1905.

Specimens of the type collection were obtained by Mr. Wm. R. Maxon from A. Brade in Costa Rica in 1906 (No. 13), but these have never flowered. In the New York Botanical Garden is a small specimen received from Wercklé in 1902 as *Epiphyllum grandilobum*; this shows one very deep lobe; a young joint shows shallow crenations and suggests *E. macropterum.* A plant of this relationship was collected by Mr. Pittier in Panama in 1911 (No. 4229) and is now growing in Washington, but has not flowered.

We believe that *Phyllocactus macrolobus* of Schumann's Keys belongs here, the specific name in error for *grandilobus.*

8. Epiphyllum crenatum (Lindley) G. Don in Loudon, Encycl. Pl. ed. 3. 1378.   1855.

  Cereus crenatus Lindley in Edwards's Bot. Reg. **30**: pl. 31.   1844.
  *Phyllocactus crenatus** Lemaire, Hort. Univ. **6**: 87.   1845.
  Phyllocactus caulorrhizus Lemaire, Jard. Fleur. **1**: Misc. 6.   1851.
  Epiphyllum caulorhizum G. Don in Loudon, Encycl. Pl. ed. 3. 1380.   1855.

Old stems woody and terete; branches glaucous, often rooting at the tips, rather stiff, 2 to 3 cm. broad, obtuse, erect, at least at first, with large deep crenations, cuneate at base, the midrib thick; areoles at base of stem and branches often bearing hairs or small bristles; flowers very fragrant, rather large, the limb 10 to 12 cm. broad, cream-colored to greenish yellow, tube 10 to 12 cm. long, slender, bearing linear scales 2 to 3 cm. long; inner perianth-segments oblanceolate, 6 cm. long; filaments yellow; style white; stigma-lobes narrow; ovary scaly, some of the scales 2 cm. long, somewhat spreading.

Type locality: Honduras.

Distribution: Honduras and Guatemala.

This species has long been a favorite with gardeners, and many hybrids with it have been produced; the flowers, which are delicately fragrant, are diurnal and remain expanded for several days.

Among hybrids with other species are *Phyllocactus crenatus amaranthinus, P. elegans, erleri, haageanus, lateritius, roseus, splendens, superbus,* and *vogelii.*

Illustrations: Edwards's Bot. Reg. **30**: pl. 31, as *Cereus crenatus*; Blühende Kakteen **3**: pl. 180, as *Phyllocactus crenatus vogelii*; Gartenflora **40**: pl. 1347; Garten-Zeitung **4**: 182.

---

    * This name was also published by Walpers in 1843 (Repert. Bot. 2. 820).

M. E. Eaton del.

1. End of branch of *Epiphyllum macropterum*.
2. Base of branch of same.

f. 42, No. 4; Rother, Praktischer Leitfaden Kakteen 80, as *Phyllocactus crenatus*; Loudon, Encycl. Pl. ed. 3. 1379. f. 19401.

Figure 199 is from a photograph showing the base and tip of a branch of this species sent from Guatemala.

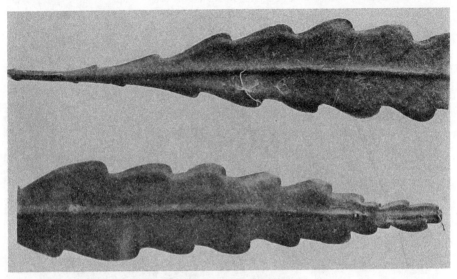

FIG. 199.—Epiphyllum crenatum.

9. Epiphyllum macropterum (Lemaire).

Phyllocactus macropterus Lemaire, Illustr. Hort. **11**: Misc. 73. 1864.
Phyllocactus thomasianus Schumann, Monatsschr. Kakteenk. **5**: 6. 1895.
Phyllocactus costaricensis Weber, Bull. Mus. Hist. Nat. Paris **8**: 463. 1902.
? *Phyllocactus macrocarpus* Weber, Bull. Mus. Hist. Nat. Paris **8**: 464. 1902.
Epiphyllum costaricense Britton and Rose, Contr. U. S. Nat. Herb. **16**: 256. 1913.
Epiphyllum thomasianum Britton and Rose, Contr. U. S. Nat. Herb. **16**: 259. 1913.

Plants up to 2 meters long, the joints weak, sometimes 10 cm. broad, thin, their margins horny; areoles distant (4 to 6 cm. apart) along the slightly indented margins; flower very large for genus, long, curved as in *Epiphyllum oxypetalum*; scales of ovary small, green, spreading, with long hairs in their axils; scales on tube longer (10 to 12 mm. long), less spreading but similar to those on ovary, acute; outer perianth-segments narrow, salmon-colored or with yellow tips, 10 cm. long; inner perianth-segments pure white, 8 to 9 cm. long, 2 to 3 cm. broad; tube of the flower 10 to 12 cm. long; throat 5 to 6 cm. long, funnelform, narrow below, 3 cm. broad at top; stamens lemon-yellow, slender, in 2 definite clusters, a single continuous row at top of throat, the second cluster scattered all over throat except for intervals of 2 cm. below upper one; style stout, 20 cm. long, pure white.

Type locality: Not cited.
Distribution: Costa Rica.

According to Mr. Fraile, the flower always comes out on the under side of the joint and lies appressed to it, instead of standing out free from it as in other species of the genus.

A vigorous plant in greenhouse cultivation but it flowers only sparingly.

Illustrations: Monatsschr. Kakteenk. **5**: pl. [1]; Blühende Kakteen **1**: pl. 41, as *Phyllocactus thomasianus*.

Plate XVII, figure 1, shows a branch of a plant sent by Dr. Wm. R. Maxon from San José, Costa Rica, which flowered in the New York Botanical Garden in 1912; figure 2 shows the base of the branch.   Figure 200 is from a photograph showing the top and base of a joint.

10. Epiphyllum lepidocarpum (Weber) Britton and Rose, Contr. U. S. Nat. Herb. **16**: 257.   1913.

    Phyllocactus lepidocarpus Weber, Bull. Mus. Hist. Nat. Paris 8: 462.   1902.

    Old and lower part of stems woody, cylindric; upper branches usually flattened, sometimes 3-winged, thickish, but not very stiff, 2 to 3 cm. wide; margins cut "stair-like," the areole closed by a small scale bearing in its axil short wool and a few bristles; flowers 20 cm. long, white and night-blooming; stamens in 2 rows; style white; fruit 9 cm. long by 4 cm. in diameter, violet-red, covered with long scales, at first erect, but finally becoming reflexed; flesh described as white,* acidulous, somewhat agreeable to the taste.

FIG. 200.—Epiphyllum macropterum.

Type locality: Near Cartago, Costa Rica.

Distribution: Known only from the type locality.

    Our description is based on that of M. Weber.   The very scaly fruit should be characteristic, but plants received from Costa Rica under the name *Phyllocactus lepidocarpus* produced smooth fruits at the New York Botanical Garden.

11. Epiphyllum pittieri (Weber) Britton and Rose, Contr. U. S. Nat. Herb. **16**: 258.   1913.

    Phyllocactus pittieri Weber, Dict. Hort. Bois 957.   1898.

    Stem usually terete below, much divided, 2 to 3 meters long; branches flat and thin, mostly 5 cm. wide or less, the margins coarsely toothed; flowers rather small, the tube about 8 cm. long, white to greenish white, bearing a few red, ascending scales; outer perianth-segments 4 to 4.5 cm. long, narrow, yellowish green, or some of the lower ones tinged with red, acute; inner perianth-segments white, a little shorter than the outer; stamens white, erect, in 2 series, longer than the style; style white above, red or purplish below; ovary with a few red scales; fruit dark red, 2 cm. long; seeds dull black.

---

*Mr. Wercklé, who discovered this species, states in a letter (September 22, 1921) that the flesh is crimson.

Epiphyllum pittieri, from Costa Rica.

Type locality: Costa Rica.

Distribution: Costa Rica.

This species is an abundant bloomer, flowering in cultivation usually in January but also at other times of the year; its flowers are the smallest of the genus.

Plate XVI, figure 2, shows a flowering branch from the specimen sent by Mr. Pittier from Zent, Costa Rica, in 1904; plate XVIII shows another plant of the same collection which flowered in Washington.

12. Epiphyllum guatemalense Britton and Rose, Contr. U. S. Nat. Herb. **16**: 257. 1913.

　　Phyllocactus guatemalensis Vaupel, Monatsschr. Kakteenk. **23**: 116. 1913.

Plant rather stout, in cultivation a meter long or longer; old stem woody, with gray bark, terete; branches green, flat, 4 to 8 cm. broad, narrowed at base and there terete, coarsely crenate, obtuse at apex; flower-bud pointed; flowers nocturnal, about 28 cm. long; tube about 15 cm. long,

FIG. 201.—Epiphyllum guatemalense.

straight or nearly so, green or yellowish green, somewhat angled, at least below, bearing only a few red-tipped scales; inner central part of tube densely pilose; outer perianth-segments scale-like with red reflexed tips; inner pure white, narrow, 8 to 9 cm. long, acuminate; stamens borne on whole surface of rather short throat and therefore in more than one series; filaments pure white; style 25 cm. long, somewhat glossy, orange; stigma-lobes orange; ovary pale, bearing a few spreading scales.

Type locality: Guatemala.

Distribution: Guatemala, but range unknown.

Two very distinct forms occur in this species which are hard to explain. They are so different that it seemed at first they must represent two distinct species, as they occur on separate plants. In one (it may be simply the juvenile form) the joints are rather thin and broad (5 to 8 cm. broad), the margins soft, with low broad undulations separated by a narrow, nearly closed sinus; in the other (it may perhaps be the adult form) the joints are stiff and narrow, the margins horny, the undulations with an open triangular sinus.

Illustration: Contr. U. S. Nat. Herb. **16**: pl. 78.

Figure 201 is from a photograph of the type plant.

13. Epiphyllum strictum (Lemaire) Britton and Rose, Contr. U. S. Nat. Herb. **16**: 259.   1913.

Phyllocactus strictus Lemaire, Illustr. Hort. 1: Misc. 107.   1854.

Plant up to 2 meters long; joints linear, green, 5 to 8 cm. broad, coarsely serrate, stiff; tube of flower 13 to 15 cm. long, slender, green, bearing a few distant scales 8 to 12 mm. long; outer perianth-segments pink, the inner white, narrow, acuminate, 6 to 8 cm. long; filaments white; style pink or red; stigma-lobes yellow; fruit globose, 4 to 5 cm. in diameter; seeds black.

Type locality: Cuba, but the plant was grown there from seed.

Distribution: Southern Mexico and Guatemala to Panama.

The plant was found in the wild state in Honduras by Mr. Percy Wilson in 1902.   All the other specimens studied by us are from cultivated plants.   The species is common in collections.

Illustrations: Schumann, Gesamtb. Kakteen f. 41; Monatsschr. Kakteenk. **6**: 183; Thomas, Zimmerkultur Kakteen 18, as *Phyllocactus strictus.*

FIG. 202.—Epiphyllum stenopetalum.

14. Epiphyllum stenopetalum (Förster) Britton and Rose, Contr. U. S. Nat. Herb. **16**: 259.   1913.

Phyllocactus stenopetalus Förster, Handb. Cact. 441.   1846.

Described as with the habit of *Epiphyllum latifrons* but with different flowers, these delicately fragrant; flower-tube 12 to 15 cm. long, bearing small, spreading, rose-colored scales; outer perianth-segments rose-colored to reddish green; inner perianth-segments white, elongated, linear (7 to 8 cm. long, very narrow, 4 to 7 mm. broad), spreading or recurved; stamens somewhat exserted; style slender, pink or purplish; stigma-lobes 12 to 14, yellow; fruit unknown.

Type locality: Not cited.

Distribution: Oaxaca, Mexico.

This plant is a night-bloomer but the flowers are late in closing, sometimes remaining partially open as late as 9 o'clock in the morning.

The above description is compiled from that of Salm-Dyck with reference to a plant at the New York Botanical Garden, received from Paris in 1909.

It resembles *E. strictum* but the joints are more flexible and broader and it has somewhat larger flowers than that species; we have a herbarium specimen identified by Schumann which was collected by P. Sintenis from a cultivated plant grown in Porto Rico.

In 1911 C. Conzatti sent us from Coyula, Cuicatlán, Oaxaca, cuttings of what we now take to be this species. These grew into vigorous plants 3 meters long and flowered in Washington in 1921 and 1922.

Illustration: Goebel, Pflanz. Schild. 1: f. 56, as *Phyllocactus stenopetalus* (seedling).

Figure 202 is from a photograph showing the top and base of a branch from Professor Conzatti's plant.

15. Epiphyllum cartagense (Weber) Britton and Rose, Contr. U. S. Nat. Herb. **16**: 256. 1913.

> *Phyllocactus cartagensis* Weber, Bull. Mus. Hist. Nat. Paris 8: 462. 1902.
> *Phyllocactus cartagensis refractus* Weber, Bull. Mus. Hist. Nat. Paris 8: 462. 1902.
> *Phyllocactus cartagensis robustus* Weber, Monatsschr. Kakteenk. 15: 180. 1905.

Plants 2 to 3 meters long, usually more or less flattened in the lower and older parts; joints short or elongated, 4 to 5 cm. broad, coarsely toothed or crenate, green; flowers opening at night, the slender tube 10 to 15 cm. long, reddish, bearing a few short distant scales; outer perianth-segments pink to yellowish; inner segments 5 to 7 cm. long, white; stamens in one series; filaments white; style pink to white; stigma-lobes yellow; fruit oblong, 7 to 8 cm. long, 3 cm. in diameter, red without, white within.

Type locality: Near Cartago, Costa Rica.

Distribution: Costa Rica.

A species apparently composed of several races, differing in margins of the joints, in size of flowers, and in color of style. It is called in Costa Rica platanillo de monte.

16. Epiphyllum hookeri Haworth, Phil. Mag. **6**: 108. 1829.

> *Cereus hookeri* Link and Otto, Cat. Sem. Hort. Berol. 1828.
> *Cereus marginatus* Salm-Dyck, Hort. Dyck. 340. 1834. Not De Candolle, 1828.
> *Phyllocactus hookeri* Salm-Dyck, Cact. Hort. Dyck. 1842.

Plants usually 2 to 3 meters long, but sometimes 7 meters long; joints 5 to 9 cm. broad, rather thin, light green, deeply crenate; flowers inodorous, the tube slender, 11 to 13 cm. long, greenish, bearing a few narrow, slightly spreading, rose-tipped scales; outer perianth-segments narrow, greenish pink, sometimes rose-colored at tip, the inner pure white, narrow, 5 cm. long; stamens in a single series, attached at top of throat; filaments white; style carmine, except yellowish base and pinkish top, smooth in upper half, papillose in lower half; stigma-lobes yellow; ovary green, somewhat angled, 2 cm. long, bearing a few small spreading scales; fruit oblong, 8 cm. long, red, somewhat angled, bearing a few scattered scales; seeds numerous, black, shining, reniform.

Type locality: Cited as Brazil, presumably in error.

Distribution: Tobago, Trinidad, and northern Venezuela.

This plant when it first flowered in cultivation in 1826 was taken for *Cactus phyllanthus* and was so figured and described in the Botanical Magazine, but it was soon discovered to be very different from that species.

While Brazil is cited as the type locality for this species we have seen no specimens from any point south of Venezuela. The plant is and has been widely cultivated in tropical America, commonly under the erroneous name, *Epiphyllum phyllanthus.* In Trinidad it forms great masses on trees and on coastal cliffs, ascending the trees to a length of 10 meters or more, branching profusely, and is very floriferous.

Phyllocactus marginatus Salm-Dyck (Cact. Hort. Dyck. 1844. 37. 1845) doubtless belongs here.

Illustrations: Pfeiffer and Otto, Abbild. Beschr. Cact. **1**: pl. 5, as *Cereus hookeri*; Curtis's Bot. Mag. **53**: pl. 2692; Loudon, Encycl. Pl. 413. f. 6901, as *Cactus phyllanthus*; Addisonia **5**: pl. 192.

Plate XIX shows a flowering branch from a specimen sent by W. E. Broadway from the Island of Tobago in 1909.

HYBRIDS

EPIPHYLLUM ACKERMANNII Haworth, Phil. Mag. **6**: 109. 1829.

> *Cactus ackermannii* Lindley in Edwards's Bot. Reg. **16**: pl. 1331. 1830.
> *Cereus ackermannii* Otto in Pfeiffer, Enum. Cact. 123. 1837.
> *Phyllocactus ackermannii* Salm-Dyck, Cact. Hort. Dyck. 1841. 38. 1842.

Branches weak, flat, and thin with crenate margins; areoles felted, often bristly or with weak spines, especially on the young growth; flowers day-blooming, very large, sometimes 1.5 to 2 dm. broad, crimson; inner perianth-segments oblong, acute; filaments long, weak, declined; style more or less declined, pinkish; stigma-lobes white; ovary more or less bristly.

Type locality: Mexico.

Distribution: Mexico.

This species was originally described as from Mexican plants sent to Haworth from Ackermann and, supposedly, from wild plants, but the general belief now is that the plant is of hybrid origin. The flowers are so much like those of *Heliocereus* that this genus probably furnished one of its parents (see Botanical Magazine, pl. 3598).

On the other hand, E. A. Goldman collected in Chiapas a series of specimens which seems to represent more than one species, but all the flowers are similar to those of *Epiphyllum ackermannii* and one of the specimens may represent the wild state of that species. The plants all have flat joints bearing clusters of spines in their areoles.

Many garden varieties and artificial hybrids have been obtained from this plant, some described under English and others under Latin names.

Illustrations: Edwards's Bot. Reg. **16**: pl. 1331, as *Cactus ackermannii*; Curtis's Bot. Mag. **64**: pl. 3598, as *Cereus ackermannii*; Blühende Kakteen **1**: pl. 49; Cycl. Amer. Hort. Bailey **3**: f. 1773; Dict. Gard. Nicholson **3**: f. 133; Karsten, Deutsche Fl. 887. f. 501, No. 6; ed. 2. **2**: 456. f. 605, No. 6; Förster, Handb. Cact. ed. 2. 841. f. 111; Rümpler, Sukkulenten 149. f. 81; Watson, Cact. Cult. 47. f. 10; Rother, Praktischer Leitfaden Kakteen 97; ed. 3. f. 8; Amer. Gard. **11**: pl. opp. 445; Gartenflora **32**: 374, as *Phyllocactus ackermannii*; Loudon, Encycl. Pl. 1202. f. 17368; Encycl. Brittanica ed. 11. **4**: 926. f. 3, as *Phyllocactus*; Rev. Hort. **1861**: 226. f. 44; Stand. Cycl. Hort. Bailey **2**: f. 1402.

Cactus hybridus was described and illustrated by P. C. Van Géel (Sert. Bot. **1**: pl. **115. 1832**). He states that it is known in Great Britain as *C. ackermannii*.

EPIPHYLLUM HYBRIDUM Hortus in Pfeiffer, Enum. Cact. 121. 1837.

This was given as a synonym of *Cereus speciosissimus lateritius*, which is briefly mentioned in volume 2 (p. 128) of this work.

EPIPHYLLUM JENKENSONII G. Don, Gen. Hist. Dichl. Pl. **3**: 170. 1834.

> *Epiphyllum speciosum jenkensonii* G. Don in Loudon, Encycl. Pl. ed. 2. 1202. 1841.

This plant is an artificial hybrid raised from *Heliocereus speciosissimus*, impregnated by the pollen of *Epiphyllum phyllanthoides*; it has branches 3-angled at base but flattened above, with areoles very prominent and spiny; flowers large, 10 cm. broad and deep scarlet; fruit nearly globular, purple, 2.5 cm. in diameter, its areoles bearing a few spines and bristles. We have had it to flower and fruit in cultivation.

EPIPHYLLUM SPLENDIDUM Paxton, Mag. Bot. **1**: 49. 1834.

> *Cereus splendidus* Steudel, Nom. ed. 2. **1**: 336. 1840.
> *Epiphyllum aitoni* Steudel, Nom. ed. 2. **1**: 561. 1840.
> *Epiphyllum hitchenii* Steudel, Nom. ed. 2. **1**: 561. 1840.

We know this plant only from a colored illustration (Paxton, Mag. Bot. **1**: pl. facing 49). The flower is very large, 10 inches broad, red, tinged with orange; flower-tube much shorter than limb, and suggests a relationship with *Epiphyllum ackermannii*. Branches flat and strongly crenate. It

Epiphyllum hookeri, from Tobago, West Indies.

is said to be a native of Mexico, but probably is of hybrid origin. It is described as having one of the largest flowers among cacti, rivaling *Selenicereus grandiflorus* and *Heliocereus speciosus*.

Epiphyllum splendens Hortus (Ann. Fl. Pom. 343, 1839) is referred here by the Index Kewensis. It is, however, described on page 345 and illustrated on plate 44. This illustration is very different from that of Paxton.

PHYLLOCACTUS ALBUS GRANDIFLORUS.

Illustration: Cact. Journ. 1: 37.

PHYLLOCACTUS ALBUS SUPERBUS.

Illustration: Blanc, Cacti 88, No. 2511.

PHYLLOCACTUS COOPERI Regel, Gartenflora 33: 218. 1884.

This is a hybrid between *Epiphyllum crenatum* and *Selenicereus grandiflorus*. It has large yellowish flowers.

Illustrations: Gartenflora 33: pl. 1176, as *Phyllocactus crenato grandiflorus*; Cassell's Dict. Gard. 2: 192.

PHYLLOCACTUS EREBUS.

This is a large red-flowered hybrid.
Illustration: Blühende Kakteen 3: pl. 160.

PHYLLOCACTUS HAAGEI.

This is doubtless a garden hybrid related to *Epiphyllum ackermannii*. It has large flowers, 12.5 cm. broad, at first flesh-colored, becoming carmine.
Illustrations: Dict. Gard. Nicholson 4: 590. f. 58; Watson, Cact. Cult. 54. f. 13.

PHYLLOCACTUS HIBRIDUS GORDONIANUS.

Illustration: Blühende Kakteen 1: pl. 36.

PHYLLOCACTUS HIBRIDUS WRAYI.

This is said to be a cross between *Selenicereus grandiflorus* and *Epiphyllum crenatum*.
Illustration: Blühende Kakteen 2: pl. 62.

PHYLLOCACTUS HILDMANNII.

Illustration: Gartenflora 44: pl. 1421, f. 2.

PHYLLOCACTUS MARSUS.

Illustration: Dict. Gard. Nicholson Suppl. 598. f. 631.

PHYLLOCACTUS PFERSDORFFII.

Illustrations: Cact. Journ. 1: 38; Rümpler, Sukkulenten f. 85; Schelle, Handb. Kakteenk. 221. f. 144.

PHYLLOCACTUS ROSEUS GRANDIFLORUS Watson, Cact. Cult. 55. 1889.

This was figured and described by Watson with flowers 15 cm. long and broad, nodding and white (!); doubtless of hybrid origin; it may be the same as *Phyllocactus roseus grandissimus* (Monatsschr. Kakteenk. 19: 182. 1909).
Illustrations: Dict. Gard. Nicholson 4: 591. f. 59; Förster, Handb. Cact. ed. 2. 857. f. 117; Watson, Cact. Cult. 55. f. 14.

PHYLLOCACTUS RUESTII Weingart, Monatsschr. Kakteenk. 24: 123. 1914.

We have not seen this plant and Mr. Weingart, who described it, says that he does not possess either living or herbarium material but that it is still growing at Halle, Germany.

THE CACTACEAE.

PHYLLOCACTUS TRIUMPHANS.

Illustration: Monatsschr. Kakteenk. **20:** 3.

Epiphyllum speciosum lateritium Henslow (Loudon, Encycl. Pl. ed. 2. 1202. 1841), an English hybrid, produced in 1828, is described as having brick-colored flowers.

Phyllocactus tonduzii Weber is mentioned by Schumann (Monatsschr. Kakteenk. **10:** 127. 1900).

Phyllocactus tuna is a name used by Wercklé (Monatsschr. Kakteenk. **15:** 180. 1905) for a Costa Rican plant, without description.

Phyllocactus weingartii Berger (Monatschr. Kakteenk. **30:** 33. 1920) is related to *Epiphyllum ackermannii.*

Charles Simon in 1893 published a list of 62 names of *Epiphyllum,* most of which are undoubtedly hybrids and some are referable to *Zygocactus.* The following binominals and trinominals are in the usual Latin form for specific and varietal names and are not recorded elsewhere:

album violaceum	grandiflorum	palidum roseum	translucens
amabile	grande superbum	purpureum	tricolor
aurantiacum	harrissoni	roseum	violaceum elegans
brasiliense	hercule	rubrum violaceum	violaceum grandiflorum
carminatum	latetium album	ruckerianum superbum	violaceum rubrum
cruentum	maximum	salmoneum marginatum	violaceum speciosum
gracilis	multiflorum	spectabile coccineum	violaceum superbum

There are many Latin names of *Phyllocactus* in catalogues, representing hybrids. We give below only those which have been used more or less in general botanical works, either as binominals or trinominals in regular Latin form:

acutifrons	buestii	germania	lunus
agatha	campmannii	guentneri	mexicanus *
alatus	caparti	hamburgiensis	purpureus
albus superbiens	capelleanus	hauffii	ruelcheri
alexandrinae	carolus magnus	helenus	speciosissimus ·feltonii
amarantinus	castneri	hempelii	superbus
arnoldi	chico	hibridus	ulbrechtii
aurantiacus superbus	coccineus	incomparabilis minuatus	victoria-regia
belgicus	demouline	jenkinsonii	vogelii
bergeri	dolores	kerthii	wippermannii
bleindlii	epirus	laarsenii	wrayi
boehmii	fuertii	lorenzii	zarka

In 1897 Charles Simon, of Saint-Ouen, Paris, published a list of 370 names of *Phyllocactus,* most of which were probably hybrids. The following binominals and trinominals are in the usual Latin form for specific and varietal names and are not recorded elsewhere:

ackermannii hybridus	crenatus hirsutis	funkii	lorentzii
ackermannii major	crenatus lateralis	gloriosus	ludmani
alatus major	crenatus latifolius	gordonianus	ludwigi
albus grandidissimus	crenatus luteus	grandidissimus	maigretii
albus perfectus	crenatus ruber	grandiflorus	magnificus
albus superbissimus	crispielsi	grandiflorus albiflorus	makoyi
amabilis	curtissi	grandiflorus ruber	mayanus
amabilis perfectus	dangeli	guebwillerianus	meyerianus
atrosanguineus	decumbens	guedeneyi	meuhlenpfordtii
aurantiacus	deveauxi	hansii	mulhousianus
bergei	dieffenbacchianus	havermansii	multiflorus
billiardieri	dumoulini	hitchensis	neubertii
binderi	edwarsii	ignescens	niedtii
blindii	elegans	jenkinsonii superbus	niger
boliviensis	erectus perfectus	johnsonii	nitens
bollwillerianus	erectus superbus	jordanis	nymphoea beata
bothii	ernesti	kampmannii	paraguayensis
brongnarti	erubescens	kermesimus magnus	pentneri
burmeisteri	fastuosus	kiardi	phyllantoides
colmariensis	feasti	kranzii	phyllantoides crenatus
colombiensis	felonis	krausei	potstachianus
courantii	feltoni	laetingii	poulletianus
crassuliefolius	floribundus	laloyi	pressleri
crenatus amarantinus	formosus	laudowi	
crenatus caulorhizus	franzii	leopoldii	

---

* A hybrid referred by Index Kewensis to *Cereus mexicanus.*

pulcherrimus	roseus miniatus	schmidtii	tettani
quilliardetti	roseus splendidus	sellowii	tricolor
raveaudii	roydii	specillimus	undiflorus
rebuti	ruber	speciosissimus	vandesii
reichei	ruber perfectus	speciosissimus grandiflorus	vonhoffini
reineckii	ruber violaceus	speciosus roseus	vitellinus
roseus carmineus	sarniensis	splendens	warscewiczii
roseus carneus	schaffieri	splendidus	wittmackianus
roseus floribundus	schallerianus	stenesi	
roseus grandidissimus	schlimmi	superbissimus	

CACTUS ENSIFORMIS Biden, Gard. Chron, II. **20**: 53. 1883.

This is evidently some *Epiphyllum* hybrid. It was sent to H. B. Biden from Manchester, England, in 1883 and flowered the same year. Its flowers were described as 6 inches across, white, richly scented, and remaining open for 3 days.

Cactus speciosus grandiflorus (Monatsschr. Kakteenk. **14**: 11. 1904) is supposed to be some hybrid *Epiphyllum*.

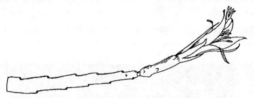

FIG. 203.—Tip of branch with flower of Disocactus biformis. × 0.8.

5. DISOCACTUS Lindley in Edwards's Bot. Reg. **31**: pl. 9. 1845.

Disisocactus Kunze, Bot. Zeit. **3**: 533. 1845.

Irregularly branching, spineless epiphytes, the stem terete; branches numerous, flattened; areoles marginal; flowers diurnal, borne near tips of branches, nearly regular; tube shorter than limb; perianth-segments few, elongated, spreading; ovary small, cylindric, elongated, bearing a few minute scales; fruit globular to ovoid, not at all angled.

FIG. 204.—Disocactus biformis.

Type species: *Cereus biformis* Lindley.

We recognize two species, both from Central America.

The name is from δίς twice, and κάκτος cactus, and was given because the perianth-segments of the inner and outer series were equal in the type specimens.

KEY TO SPECIES.

Slender with linear lateral branches, their margins slightly toothed; style and stamens about length of
　　　　　perianth-segments.. 1. *D. biformis*
Spreading with oblanceolate lateral branches, their margins crenate; style and stamens long-exserted.. 2. *D. eichlamii*

1. Disocactus biformis Lindley in Edwards's Bot. Reg. 31: pl. 9. 1845.

　　　Cereus biformis Lindley in Edwards's Bot. Reg. 29: Misc. 51. 1843.
　　　Disisocactus biformis Kunze, Bot. Zeit. 3: 533. 1845.
　　　Phyllocactus biformis Labouret, Monogr. Cact. 418. 1853.
　　　Epiphyllum biforme G. Don in Loudon, Encycl. Pl. ed. 3. 1378. 1855.

Plant 2 dm. long or longer; branches linear, 5 to 8 cm. long, 1 to 2 cm. broad, with serrate margins; flower-bud elongated, curved upward, pointed; tube of the flower about 1 cm. long, the segments 8 (rarely 9), magenta, about 3 cm. long, the outer 4 or 5 spreading or curved backward, linear, the inner 3 or 4 broader and more erect; stamens 10 to 12, slightly exserted, borne in 2 series at top of tube; style slender, purple; stigma-lobes 4, white; ovary short-oblong, green, somewhat tubercled, with a few areoles subtended by small ovate scales; fruit ovoid, 1.5 cm. long, turgid, wine-colored.

Type locality: Honduras. The species described from a garden specimen, introduced into England in 1839.

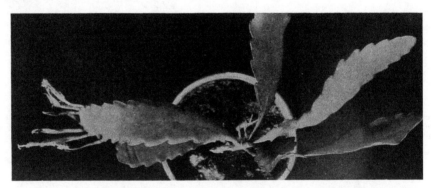

FIG. 205.—Disocactus eichlamii.

Distribution: Honduras and Guatemala.

We have had this plant under observation for a number of years. It is rather a shy bloomer with us, although we get one or two flowers each spring; the flowers open in the night or early morning and remain open all day; they begin to wither the second morning. The perianth-segments are more widely spreading in the morning than in the afternoon. The flower is almost horizontal and the tube proper is about the length of the ovary. The fruit matures very slowly. In 1920 we had a plant flower in April, but the fruit did not mature until July 8.

Illustrations: Förster, Handb. Cact. ed. 2. 876. f. 120; Rümpler, Sukkulenten f. 86, as *Disisocactus biformis*; Blühende Kakteen 1: pl. 54; Curtis's Bot. Mag. 101: pl. 6156; Dict. Gard. Nicholson 3: f. 135; Monatsschr. Kakteenk. 9: 141; Watson, Cact. Cult. 50. f. 12, as *Phyllocactus biformis*; Loudon, Encycl. Pl. ed. 3. 1379. f. 19403, as *Epiphyllum biforme*; Edwards's Bot. Reg. 31: pl. 9; Palmer, Cult. Cact. 175.

Plate XXXII, figure 2, shows a branch of a fruiting plant sent to Dr Rose by Robert Lamb of Manchester, England, in 1912. Figure 203 shows the flower of the same plant; figure 204 is from a photograph of the same plant in flower.

2. Disocactus eichlamii (Weingart) Britton and Rose, Contr. U. S. Nat. Herb. **16:** 259. 1913.

> *Phyllocactus eichlamii* Weingart, Monatsschr. Kakteenk. **21:** 5. 1911.

Branching near the base; branches oblong, 2 to 3 dm. long, 3 to 5 cm. broad, cuneate at base, obtuse, thickish, strongly crenate; flowers several at the uppermost areoles, slender, 4 cm. long; stamens and style exserted; stigma-lobes 5; fruit red, 1 cm. in diameter with white pulp; seeds 1.5 mm. long.

Type locality: Guatemala.

Distribution: Guatemala.

This plant we know only from the collection of F. Eichlam; living specimens were sent to Washington by him which soon afterward flowered, but these have since died. Eichlam wrote that the flowers were a brilliant red. The species was named for Federico Eichlam, who lived in Guatemala at the time of his death in 1911.

Illustration: Contr. U. S. Nat. Herb. **16:** pl. 79.

Figure 205 is from a photograph of the type plant in flower.

6. CHIAPASIA gen. nov.

An epiphytic spineless cactus, the branches flattened, crenate, with slender terete bases, the large flowers borne at upper areoles; perianth narrowly campanulate; tube about half as long as limb, bearing a few small triangular scales; segments about 8, linear, recurved, spreading; ovary ovoid, shorter than tube, also with a few small scales; filaments about 20, not longer than perianth-segments; stigma-lobes few.

Type species: *Epiphyllum nelsonii* Britton and Rose.

A monotypic genus, its name taken from that of the Mexican state in which it grows.

1. Chiapasia nelsonii Britton and Rose.

> *Epiphyllum nelsonii* Britton and Rose, Contr. U. S. Nat. Herb. **16:** 257. 1913.
> *Phyllocactus nelsonii* Vaupel, Monatsschr. Kakteenk. **23:** 116. 1913.
> *Phyllocactus chiapensis* J. A. Purpus, Monatsschr. Kakteenk. **28:** 118. 1918.

Fig. 206.—Chiapasia nelsonii.

Branches 6 to 12 dm. long, slender and terete below, flat and thin above, 3 to 4 cm. broad; margin low, crenate; flowers light rose-red; tube 2 to 3 cm. long; segments about 6 cm. long, narrow, acute.

Type locality: Near Chicharras, Chiapas, Mexico, altitude 900 to 1,800 meters.

Distribution: Known only from Chiapas.

Our first description of this plant was based on an herbarium specimen, but considerable additional information is now known regarding its habit. A very fine plant was grown at Darmstadt by J. A. Purpus, a photograph of which we have, showing that the main branches are long and terete while the lateral branches are broad and thin, often pendent, with 1 to 3 flowers near the end; the flowers are horizontal, with the perianth-segments more or less recurved; the stamens and style are slender and long-exserted.

Illustration: Monatsschr. Kakteenk. **28**: 119, as *Phyllocactus chiapensis.*

Figure 206 is from a photograph of the type specimen.

7. ECCREMOCACTUS Britton and Rose, Contr. U. S. Nat. Herb. **16**: 261. 1913.

Plants epiphytic, pendent (erect or ascending in cultivation), several-jointed, the joints flat and thickish with spine-bearing marginal areoles or in cultivation often spineless; flowers solitary at upper areoles, funnelform, the short, nearly cylindric tube bearing small somewhat spreading scales, but no spines; perianth withering-persistent, its segments obtuse, rounded, or the innermost acutish; stamens and style white, included, slender, declinate; fruit carmine-red, oblong, with a few spineless areoles; seeds numerous, minute, black.

Type species: *Eccremocactus bradei* Britton and Rose.

Only one species is known, a native of Costa Rica. We have had the plant in cultivation for a number of years; it is a shy bloomer.

The generic name is from ἐκκρεμής hanging from, and κακτος cactus.

1. Eccremocactus bradei Britton and Rose, Contr. U. S. Nat. Herb. **16**: 262. 1913.

Phyllocactus bradei Vaupel, Monatsschr. Kakteenk. **23**: 118. 1913.

Epiphytic on trees; joints 15 to 30 cm. long, 5 to 10 cm. broad, light dull green, flat, but the central axis somewhat elevated on both sides, the margins shallowly crenate, with small spine-bearing areoles in the sinuses; spines solitary or in 2's or 3's, dark brown, 6 mm. long or less; flowers developing very slowly, 6 to 7 cm. long, slightly asymmetrical; outermost perianth-segments thick, shining, pinkish; outer ones oblong, thinner, pinkish white; inner perianth-segments oblong, obtuse, 3 to 3.5 cm. long; flower-tube 1 cm. long; throat broad, short, covered with stamens; filaments very slender, delicate, white, strongly declined; style slender, nearly white, slightly pinkish above, elongated, glabrous; stigma-lobes 8; ovary angled by the elongated tubercles; its areoles bearing a line of short hairs, subtended by thick ovate purple scales; seeds 1.5 mm. long.

Type locality: Cerro Turriwares, near Orotina (formerly Santo Domingo de San Mateo), Costa Rica.

Distribution: In dense forests at low altitudes, Costa Rica.

Our attention was first called to this plant by Dr. Maxon, who obtained specimens from Mr. Alfredo Brade in 1906; these bloomed in June 1911, but good flowers were not obtained. In 1913 Otón Jiménez sent specimens to Dr. Rose which flowered in 1918.

The flowers open in the night and are closed on the following morning. The branches of wild plants bear clusters of spines at the areole, but our cultivated plants are spineless and in the vegetative state resemble those of a turgid *Epiphyllum.* When the plant sent by Mr. Jiménez from Costa Rica (No. 905) flowered in 1921 seven flower-buds were produced from the seven uppermost areoles.

Illustration: Contr. U. S. Nat. Herb. **16**: pl. 83.

Plate xx is from a photograph of a plant sent to Washington by Otón Jiménez in 1913, which flowered in May 1921.

8. NOPALXOCHIA gen. nov.

A flat-jointed, spineless epiphytic cactus; joints crenate, the rather large, short-funnelform, rose or red flowers, solitary at lateral marginal areoles; flower-tube about as long as limb, bearing several narrow scales; outer perianth-segments short, acute, reflexed or spreading; inner spreading or connivent, acute; stamens numerous.

Type species: *Cactus phyllanthoides* De Candolle.

A monotypic Mexican genus, the name taken from the Aztec of Hernández.

Eccremocactus bradei, from Costa Rica.

1. Nopalxochia phyllanthoides (De Candolle).

Cactus phyllanthoides De Candolle, Cat. Hort. Monsp. 84. 1813.
Cactus speciosus Bonpland, Descr. Pl. Rar. Malm. 8. 1813. Not Cavanilles, 1803.
Epiphyllum speciosum Haworth, Suppl. Pl. Succ. 84. 1819.
Cactus elegans Link, Enum. **2**: 25. 1822.
Epiphyllum phyllanthoides Sweet, Hort. Brit. 172. 1826.
Cereus phyllanthoides De Candolle, Prodr. **3**: 469. 1828.
Phyllocactus phyllanthoides Link, Handb. Gewächs. **2**: 11. 1831.
Opuntia speciosa Steudel, Nom. ed. 2. **2**: 222. 1841.

Stems somewhat woody, branching, the branches terete at base, flattened and thin above, sometimes 5 cm. broad, green; margin of branches coarsely crenate; flowers diurnal, the tube 2 cm. long; inner perianth-segments oblong, more or less spreading; filaments and style elongated, slender; stigma-lobes 5 to 7.

Type locality: Mexico.

Distribution: Mexico or Colombia, but known only from cultivated plants.

The distribution of this species is assigned to Mexico, but both Edwards and Sims state definitely that it was first observed by Humboldt and Bonpland near the village of Turbaco, which is a few leagues south of Cartagena, Colombia. From seeds collected at that time, plants were grown in the garden of La Malmaison; one of these flowered in 1811 and was described and illustrated as *Cactus speciosus* in 1813.

This is one of the oldest known species of cacti; it was figured by Hernández in 1651 and by Plukenet in 1691. It has long been in cultivation, perhaps in prehistoric times.

It is often hybridized with other species. The following hybrids with it are given: *Phyllocactus phyllanthoides albiflorus, striatus, striatus multiflorus.*

Salm-Dyck (Hort. Dyck. 65. 1834) lists four varieties as follows: *Cereus phyllanthoides curtisii, C. phyllanthoides guillardieri, C. phyllanthoides jenkinsonii,* and *C. phyllanthoides vandesii.* Pfeiffer (Enum. Cact. 124. 1837)

FIG. 207.—Nopalxochia phyllanthoides.

also mentions *Cereus phyllanthoides albiflorus.*.

Epiphyllum vandesii Don (Gen. Hist. D[1]chl. Pl. **3**: 170. 1834) is a hybrid produced by placing the pollen of *Epiphyllum phyllanthoides* on the stigmas of *Heliocereus elegantissimus.*

Illustrations: Plukenet, Phyt. pl. 247, f. 5, as *Phillanthos*; Loudon, Encycl. Pl. 413. f. 6902; Curtis's Bot. Mag. **46**: pl. 2092, as *Cactus phyllanthoides*; Schumann, Gesamtb. Kakteen 217. f. 42; Monatsschr. Kakteenk. **7**: 87; Wiener Ill. Gart. Zeit. **28**: f. 39; Gartenwelt **4**: 560; **5**: 6 and pl. facing 6; Ann. Rep. Smiths. Inst. **1908**: f. 24; Möllers Deutsche Gärt. Zeit. **11**: 61; Goebel, Pflanz. Schild. **1**: f. 13, 52, 54, as *Phyllocactus phyllanthoides*; Bonpland, Descr. Pl. Rar. pl. 3; Edwards's Bot. Reg. **4**: pl. 304; Herb. Génér. Amat. **4**: pl. 244, as *Cactus speciosus*; Loudon, Encycl. Pl. ed. 2. 1202. f. 17367, as *Epiphyllum speciosum*; Pfeiffer, Abbild. Beschr. Cact. **2**: pl. 17, as *Cereus phyllanthoides* var. *stricta*; Ann. Inst. Roy. Hort. Fromont **2**: pl. 1, f. E, as *E. phyllanthoides*; Hort. Ripul. pl. 10; Van Géel, Sert. Bot. pl. 111, as *Cactus alatus.*

Figure 207 is a reproduction of Bonpland's illustration as *Cactus speciosus.*

9. WITTIA Schumann, Monatsschr. Kakteenk. 13: 117. 1903.

Epiphytic, branching cacti, pendent from trees and rocks; joints elongated, flattened or some-what thickened, spineless, the margins more or less crenate; flowers small for this group, not fugacious, with a definite tube, much longer than limb; perianth-segments short, erect; style (so far as known) slender, white, a little exserted; fruit small, berry-like.

Type species: *Wittia amazonica* Schumann.

In vegetative characters this genus is similar to some of the *Rhipsalidanae*, but the flower has a tube longer than the limb.

The genus is named for N. H. Witt, who made valuable collections in Brazil.

We recognize two species, natives of Panama and northern South America.

KEY TO SPECIES.

Fruit roughened by small tubercles.. 1. *W. amazonica*
Fruit smooth... 2. *W. panamensis*

FIG. 208.—Wittia amazonica.

1. Wittia amazonica Schumann, Monatsschr. Kakteenk. 13: 117. 1903.

Branches flattened except at base, 15 to 40 cm. long, 4.5 to 9 cm. broad, often with constrictions, cuneate at base, coarsely crenate, obtuse or acute at apex; flowers 2.5 cm. long; perianth wine-colored, 2 cm. long, cylindric, somewhat curved; perianth-segments 10, erect, in 2 series; stamens included, in 2 series; style 18 mm. long; stigma-lobes 5; ovary strongly tuberculate; scales on ovary 3-angled; fruit 12 to 17 mm. long, deeply umbilicate at apex.

Type locality: Near Laeticia and Tarapoto, Peru.
Distribution: Northeastern Peru, not far from the Brazilian border.
We know the plant from description and illustration only.
Illustration: Monatsschr. Kakteenk. 13: 119.
Figure 208 is reproduced from the illustration cited above.

2. Wittia panamensis Britton and Rose, Contr. U. S. Nat. Herb. **16:** 241.   1913.

Branches much flattened, up to 1 meter long, 4 to 7 cm. wide, low-crenate; flowers sometimes 15 or more on a branch but solitary at areoles on upper half of joint, purple, becoming straight, 2.5 to 3.5 cm. long, 5-angled, stiff; outer perianth-segments 10, in 2 series, equal, obtuse; outermost ones angled on back; inner perianth-segments 5, similar to outer but thinner, not angled or only slightly so, a little longer, all erect; innermost segments 10 or 11, thinner, paler, and much smaller than outer, apiculate, sometimes toothed above; tube proper 5 to 6 mm. long, the throat 10 mm. long; stamens many, in 2 series, one on base of throat on long filaments, one on top of throat on short filaments, all included; stigma-lobes 4, white, but remaining in a close cluster, the top exserted beyond perianth-segments; ovary globular, purple, bearing a few scarious scales.

Type locality: Mountains above Chepo, Panama.

Distribution: Panama, Colombia, and perhaps Venezuela.

Mr. Henri Pittier has collected in Venezuela a plant which is closely related to this species (No. 7656).

Illustrations: Contr. U. S. Nat. Herb. **16:** pl. 73; Curtis's Bot. Mag. **145:** pl. 8799.

Figure 209, shown below, is from a photograph showing the plant collected by Mr. Pittier in 1912 which afterwards flowered in Washington.

Subtribe 8. RHIPSALIDANAE.

Mostly epiphytic cacti, generally growing on trees but sometimes clambering over rocks or pendent from them, much branched; branches alternate or often in whorls, slender, terete, angled or flat and thin, spineless, except in *Pfeiffera* and *Acanthorhipsalis*; flowers regular, mostly small, rotate, and without any tube or with a very short tube; stamens usually few, attached to disk or near base of flower-tube; style usually short; fruit a small juicy berry, white, red, or purple; seeds minute.

We have placed this subtribe at the end of our monograph because it appears to us to represent the most extreme differentiation within the family. It is indeed difficult to explain to most people that its species are really cacti.

We recognize eight closely related genera.

KEY TO GENERA.

Flowers with a short definite tube.
Joints terete. . . . . 1. *Erythrorhipsalis* (p. 208)
Joints flattened, ribbed, or angled.
Joints and flowers terminal. . . . . 2. *Rhipsalidopsis* (p. 209)
Joints and flowers normally lateral.
Joints with spiny areoles; ovary and fruit with areoles subtended by scales.
Joints ribbed; fruit-areoles spiny. . . . . 3. *Pfeiffera* (p. 210)
Joints flattened or 3-winged; fruit-areoles not spiny. . . . . 4. *Acanthorhipsalis*(p. 211)
Joints not spiny; fruit mostly without areoles.
Upper joints normally flattened; areoles not pilose . . . . 5. *Pseudorhipsalis* (p. 213)
Upper joints flattened or 3-angled; areoles long-pilose. . . . . 6. *Lepismium* (p. 215)
Flowers without tube.
Petals erect; ends of same joint unlike; flowers and branches always terminal. . . . . . . 7. *Hatiora* (p. 216)
Petals usually widely spreading; ends of same joint usually similar; flowers and
branches lateral or terminal. . . . . 8. *Rhipsalis* (p. 219)

1. ERYTHRORHIPSALIS Berger, Monatsschr. Kakteenk. 30: 4. 1920.

Epiphytic, with slender terete stem and branches, often pendent; branches dichotomous or sometimes in whorls of 3 to 6; areoles scattered, small, all bearing several bristles; flowers terminal, regular, diurnal, white to rose-colored with a short but definite tube; ovary and fruit bristly, the latter red; seeds much larger than in *Rhipsalis*.

Type species: *Rhipsalis pilocarpa* Löfgren.

The generic name is from ἐρυθρός red, and *Rhipsalis*, referring to the red fruit and to the resemblance of this genus to *Rhipsalis*. Only one species is known.

The genus resembles in habit some of the species of *Rhipsalis* with round stems but has a distinct flower-tube, on the top of which the stamens are borne. It also differs from *Rhipsalis* in having a long exserted style, exserted even in the bud; in its slowly opening flower (requiring several days to expand); in its very fragrant flower; in having its ovary and fruit bearing areoles, each with a cluster of bristles; and in its larger seeds.

Löfgren, when he described *Rhipsalis pilocarpa*, was inclined to think that it might belong to *Pfeiffera*. In his latest treatment of it (Arch. Jard. Bot. Rio de Janeiro 1: 68) he referred it and *Pfeiffera ianthothele* to *Rhipsalis* under the subgenus *Pfeiffera*.

At the place cited above, Berger proposed that *Rhipsalis pilocarpa* should be regarded as a new subgenus of *Rhipsalis* but at the same time he incidentally made it the type of a new genus. Mr. Berger, who has written most interestingly of it, says in part:

"In 1903, Löfgren made known *Rhipsalis pilocarpa* (Monatsschr. Kakteenk. 13: 52 to 57) which formerly had a fairly wide distribution in our collections. I received it from various sources, the finest specimens coming from the Botanic Garden in Bremen, from which place it was sent for naming. The plant grew well in my hothouse but appeared to prefer it cooler and sunnier. The habit picture in the Monatsschrift, above cited, is not exactly right. The plant is striking because of its beautiful bristles; it is very odd. In general it does not differ from the rest of the species of *Rhipsalis*. Because of its beautiful bristles one is persuaded to put it into *Ophiorhipsalis*. Meantime the habit, the flowers, and the fruit show themselves to be a fundamental obstacle.

"In all the species of *Rhipsalis* which I have had experience with, the ovary and later the fruit are entirely naked; at the most there is at times a little scale. In these plants, however, the ovaries, which in form remind one of those of *Cereus*, bear a mass of small tubercles with little scales, in whose

M. E. Eaton del.

1. Fruiting branch of *Rhipsalis grandiflora*.
2. Flowering branch of *Rhipsalis lindbergiana*.
3. Fruiting branch of *Rhipsalis shaferi*.
4. Flowering and fruiting branch of *Rhipsalis lindbergiana*.
5. Flowering plant of *Erythrorhipsalis pilocarpa*.
6. Flowering branch of *Rhipsalis grandiflora*.

axils are a large number of projecting white bristles. Still more different is the fruit, which Löfgren did not know. It is striking because of its size, about 10 to 12 mm. by 10 to 12 mm., and while the rest of the *Rhipsalis* fruits in size, form, and color resemble mistletoe berries or are rarely yellow or pale rose, these are strongly wine-red and beset with numerous bristles bearing small areoles, forming a wreath on the umbilicus of the fruit, like *Cereus* and especially *Opuntia*, only the bristles are white and not pricking. In cross-sections the fruit is also red but has a watery sap and a larger number of seeds, coiled on the placenta in the middle of the fruit. The seeds are about double the size of those of *Rhipsalis.*"

1. Erythrorhipsalis pilocarpa (Löfgren) Berger, Monatsschr. Kakteenk. **30**: 4. 1920.

Rhipsalis pilocarpa Löfgren, Monatsschr. Kakteenk. **13**: 52. 1903.

Stems dark green to purple, at first erect, sometimes 4 dm. long and unbranched, terminated by 2 to 4 branches in a whorl, the ultimate branches often only 1 cm. long, in time the whole plant becoming pendent; joints clustered, when withering somewhat angled, tipped by yellow bristles; areoles filled with long setose hairs or bristles subtended by ovate scarious bracts; flowers at ends of terminal branches, very fragrant, opening slowly, up to 2 cm. broad; flower-tube 2 mm. long, reddish on the inside; outer perianth-segments 5 or 6, triangular, rose-colored; inner perianth-segments 10 to 15, spreading or sometimes recurved, lanceolate, acuminate, 10 mm. long, white or cream-colored with pinkish tips; stamens numerous, red at bases; ovary with several areoles, bearing as many as 10 bristles, subtended by small scarious scales and surrounded by purple spots; style exserted in the bud; stigma-lobes 4 to 8, white, spreading apart the second day after the appearance of the style and before the stamens appear.

Type locality: Ytu and Ypanema, São Paulo, Brazil.

Distribution: States of São Paulo and Rio de Janeiro, Brazil.

Pfeiffera rhipsaloides Löfgren (Monatsschr. Kakteenk. **13**: 54. 1903) was another name suggested for this plant when it was first described.

Illustrations: Blühende Kakteen **2**: pl. 99; Monatsschr. Kakteenk. **13**: 55; Rev. Centr. Sci. Campinas No. 4, opp. 188; Möllers Deutsche Gärt. Zeit. **25**: 477. f. 11, No. 11, 20; Arch. Jard. Bot. Rio de Janeiro **1**: pl. 1, as *Rhipsalis pilocarpa.*

Plate XXI, figure 5, is of a plant in the New York Botanical Garden which flowered in April 1919 and was obtained by Dr. Shafer from Dr. Löfgren at Rio de Janeiro in 1917.

2. RHIPSALIDOPSIS gen. nov.

Somewhat shrubby, erect, reclining or pendulous, the joints 3 to 5-angled; branches usually several, terminal; areoles small, sometimes bearing setae; flowers terminal, with a broad rotate limb and a very short tube; stamens erect; style slender; fruit unknown.

Type species: *Rhipsalis rosea* Lagerheim.
One species is known, native of southern Brazil.

This plant was originally described as *Rhipsalis*, but it has a much larger flower and the perianth-segments are united into a short tube. In habit it resembles some of the species of *Epiphyllanthus* but has a regular flower. We have placed it near *Pfeiffera*, but we do not believe that it is close to that genus, for it has a rotate flower and the flowers and branches are terminal, as in *Zygocactus*.

The generic name is given on account of its resemblance to some of the species of *Rhipsalis*.

1. Rhipsalidopsis rosea (Lagerheim).

Rhipsalis rosea Lagerheim, Svensk Bot. Tidskr. **6**: 717. 1912.

Branches short, 1 to 3, strongly 4-angled or sometimes 3 or 5-angled, with concave sides; buds red; flowers 3.7 cm. broad, fragrant; perianth-segments few, rose-colored; stamens 11 mm. long, rose-colored; style 13 mm. long, rose-colored; stigma-lobes 3, white, 3 mm. long.

FIG. 210.—Rhipsalidopsis rosea.

Type locality: Woods near Caiguava, state of Parana, Brazil, altitude 1,100 to 1,300 meters.

Distribution: Southern Brazil.

Illustrations: Svensk Bot. Tidskr. **6:** pl. 28; Arch. Jard. Bot. Rio de Janeiro **2:** pl. 14, 15; Monatsschr. Kakteenk. **32:** 121, as *Rhipsalis rosea.*

Figure 210 is reproduced from the first illustration above cited.

3. PFEIFFERA Salm-Dyck, Cact. Hort. Dyck. 1844. 40. 1845.

Epiphytic, with a woody base; branches in wild state hanging, mostly 4-angled, not emitting aerial roots; spines several, acicular; flowers regular, diurnal, pale yellow to rose-colored (sometimes described as purple-red), small, the segments united at base into a very short tube; stamens included, some borne on flower-tube and some on disk; ovary and fruit spiny; seeds black, oblong.

Type species: *Cereus ianthothele* Monville.

Only one species is known, and this was first described as a *Cereus* and afterwards referred to *Rhipsalis.* We agree with the author in regarding it as a distinct genus.

The genus was named for Dr. Ludwig Pfeiffer, a physician by profession and one of the most distinguished authorities on the Cactaceae. He visited Cuba in 1838-1839. Dr. Pfeiffer was born July 4, 1805, at Kassel, Germany, and died in 1877.

Fig. 211.—Acanthorhipsalis micrantha.

1. Pfeiffera ianthothele (Monville) Weber, Dict. Hort. Bois 944. 1898.

Cereus ianthothele * Monville, Hort. Univ. **1:** 218.   1839.
Pfeiffera cereiformis Salm-Dyck, Cact. Hort. Dyck. 1844. 41.   1845.
Rhipsalis cereiformis Förster, Handb. Cact. 454.   1846.
Hariota cereiformis Kuntze, Rev. Gen. Pl. **1:** 262.   1891.
Rhipsalis ianthothele K. Brandegee, Cycl. Amer. Hort. Bailey **4:** 1514.   1902.

Stem weak, spreading or pendent, 3 to 6 dm. long, 2 cm. in diameter or less; joints 8 to 12 cm. long, 3 to 5-ribbed, 10 mm. in diameter, light green, spiny; ribs tuberculate; areoles 10 mm. apart; spines 6 or 7, 5 to 7 mm. long, yellowish; flowers including the ovary about 15 mm. long; inner perianth-segments 5, pale yellow to cream-colored, acute, erect or slightly spreading at tip; stamens numerous, shorter than the perianth-segments, included; style longer than stamens; stigma-lobes 8, linear, spreading; ovary strongly tuberculate, purplish, its areoles bearing white bristly spines; fruit globose, 12 to 16 mm. in diameter, rose-red, spiny; seeds numerous, black.

---

* The specific name is sometimes spelled *janthothele*; it was originally given as *ianthothelus.*

M. E. Eaton del.

1. Flowering branch of *Pfeiffera ianthothele*.
2. Flowering and fruiting branch of *Lepismium cruciforme*.
3. Top of fruiting branch of *Pfeiffera ianthothele*.
4. Flowering and fruiting branch of *Rhipsalis jamaicensis*.

5. Flowering branch of *Pseudorhipsalis alata*.
6. Flowering and fruiting branch of *Pseudorhipsalis himantoclada*.
7. Flowering branch of *Pfeiffera ianthothele*.

Type locality: Montevideo is cited in the original description, but this must be wrong.

Distribution: Northwestern Argentina, especially in the states of Salta, Tucuman, and Catamarca.

Illustrations: Goebel, Pflanz. Schild. **1**: 45, B; Palmer, Cult. Cact. 191; Förster, Handb. Cact. ed. 2. 895. f. 122; Pfeiffer, Abbild. Beschr. Cact. **2**: pl. 9; Garten-Zeitung **4**: 182. f. 42, No. 10, as *Pfeiffera cereiformis*; Schumann, Gesamtb. Kakteen 611. f. 97, A, B; Blühende Kakteen **3**: pl. 152.

Plate XXII, figures 1 and 7, shows flowering branches from a plant collected by Dr. Shafer in Argentina in 1917 (No. 71), which flowered in April 1919; figure 3 shows the mature fruit.

4. ACANTHORHIPSALIS (Schumann) gen. nov.

Small branching cacti, more or less epiphytic, growing on forest trees or creeping over rocks; joints flattened or sometimes 3-winged, short or elongated, their margins crenate or serrate; areoles spiny; flowers solitary from lateral areoles; perianth-segments united into a short tube; ovary bearing on its surface small scales with tufts of felt in their axils, at least in typical species; seeds small, black, narrowed at base.

The type is *Cereus micranthus* Vaupel and to this genus we have also referred two little-known species of *Rhipsalis*, both of which have flattened joints and spiny areoles. In their habit and armament they resemble *Acanthorhipsalis micrantha* more than they do the true species of *Rhipsalis*. The plants are native of Peru, Bolivia, and Argentina.

KEY TO SPECIES.

Joints crenate.
    Joints about 2 cm. broad; spines 5 to 15 mm. long.............................. 1. *A. micrantha*
    Joints usually 4 to 6 cm. broad; spines 4 mm. long or less........................ 2. *A. crenata*
Joints serrate... 3. *A. monacantha*

1. Acanthorhipsalis micrantha (Vaupel).

Cereus micranthus Vaupel, Bot. Jahrb. Engler **50**: Beibl. 111: 19. 1913.

Stems much branched; joints 2 or 3-winged, about 2 dm. long and 2 cm. broad, yellowish green, at least when dry; areoles 6 to 10 mm. apart; spines 3 to 10, 5 to 15 mm. long, brown to blackish, straight or a little curved; flower, including the ovary, 22 mm. long.

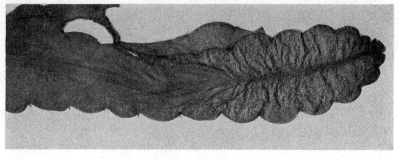

FIG. 212.—Acanthorhipsalis crenata.

Type locality: Sandía, southeastern Peru, altitude 2,100 meters.

Distribution: Known only from the type locality.

This plant was described by Dr. Vaupel as a species of *Cereus*, but as he writes us under date of October 20, 1920, it is of course not a *Cereus* in the stricter sense, but is more nearly related to *Rhipsalis*. This view was taken by Schumann, who had labeled it *Rhipsalis peruviana* Schumann (Vaupel, Bot. Jahrb. Engler **50**: Beibl. **111**: 19. 1913).

The plant was collected by A. Weberbauer July 31, 1902 (No. 1353); a fragment of the type, which is in the Berlin Herbarium, was sent us by Dr. Vaupel in 1920.

Figure 211 is from a photograph of a part of the type specimen now in the National Herbarium at Washington.

2. Acanthorhipsalis crenata (Britton).

Hariota crenata * Britton, Bull. Torr. Club **18**: 35. 1891.

Branches lateral, narrowly oblong, very flat, obtuse, 20 to 30 cm. long, 3 to 6 cm. broad, strongly crenate, with a stout central axis; areoles between crenations rather large, filled with wool and bearing 3 to 8 spines, these 2 to 4 mm. long; flowers red, lateral, small; berry 7 mm. in diameter.

Type locality: Yungas, Bolivia.
Distribution: Known only from the type locality.

When first described, this species was thought to be nearest the Brazilian *Rhipsalis platycarpa*, which it resembles, but that species has no spines.

Figure 212 is from a photograph of Dr. Rusby's herbarium specimen (No. 2047).

FIG. 213.—Acanthorhipsalis monacantha.

3. Acanthorhipsalis monacantha (Grisebach).

Rhipsalis monacantha Grisebach, Abh. Ges. Wiss. Göttingen **24**: 140.   1879.
Hartiota monacantha Kuntze, Rev. Gen. Pl. **1**: 263.   1891.

Epiphytic, branching; branches flat and thin, linear-oblong, 2 cm. broad, sometimes 8 dm. long, obtuse, cuneate at base; serrate (acuminate says Schumann, but figured by him as obtuse); areoles white-felted and spiny, spines 1 to 6, but usually only 1 or 2, 5 to 10 mm. long, yellow; flowers solitary at the areoles, lateral, white, 1 cm. long; fruit globular, 8 to 10 mm. in diameter, white; seeds blackish, pitted, obovoid.

Type locality: Oran, near San Andrés, Argentina.
Distribution: Northern Argentina.

---

* This name is printed *H. cinerea* in the Index Kewensis.

Illustration: Schumann, Gesamtb. Kakteen 633. f. 98, H, as *Rhipsalis monacantha.*
Figure 213 is from a photograph of a herbarium specimen collected at Calilegua, Jujuy, Argentina, by J. A. Shafer in 1917 (No. 56).

5. PSEUDORHIPSALIS gen. nov.

Epiphytic, much branched, and elongated cacti, at first erect, but soon prostrate or hanging; branches flattened, rather thin, serrate or crenate; flowers numerous, borne solitary at the lateral areoles, narrowly campanulate; segments united into a short but definite tube; ovary and fruit globular, bearing several scales; seeds black.

Two species are here included, of which *Cactus alatus* Swartz is made the generic type. These plants in their habit and branches resemble certain species of *Rhipsalis*, especially *R. ramulosa* and its relatives, but differ from all the species of *Rhipsalis* in having united perianth-segments and more scaly ovary and fruit.

KEY TO SPECIES.

Ovary and outer perianth-segments reddish.. 1. *P. himantoclada*
Ovary and outer perianth-segments greenish or yellowish green....................... 2. *P. alata*

1. Pseudorhipsalis himantoclada (Roland-Gosselin).

Rhipsalis himantoclada Roland-Gosselin, Bull. Soc. Bot. France **55**: 694. 1908.
Wittia costaricensis Britton and Rose, Contr. U. S. Nat. Herb. **16**: 261. 1913.

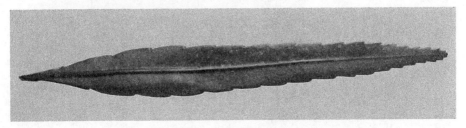

FIG. 214.—Pseudorhipsalis himantoclada.

Stems 4 to 5 dm. long, erect or curved, flat, 1 to 3 cm. broad, with horizontal branches narrowed at base, pointed, the margin low-serrate; areoles 12 to 15 mm. apart; ovary, tube, and sepals purplish; ovary 3 mm. long, bearing a few very short scales; tube of flower about 8 mm. long; inner perianth-segments white, obtuse, spreading; stamens erect; style white; stigma-lobes 4.

Type locality: Pozo Azul, Costa Rica.
Distribution: Costa Rica.

We are told by Mr. Otón Jiménez that Mr. Wercklé, who first collected the plant, would refer *Wittia costaricensis* here. He states also that it is very luxuriant and when growing wild becomes so large that one man can not carry a single plant.

Illustration: Contr. U. S. Nat. Herb. **16**: pl. 82, as *Wittia costaricensis.*

Plate XXII, figure 6, shows a flowering branch collected by Wercklé in 1907 which flowered in the New York Botanical Garden, December 20, 1911. Figure 214 is from a photograph of a terminal branch; figure 215 shows a flowering branch; figure 216 shows a flower cut longitudinally.

2. Pseudorhipsalis alata (Swartz).

Cactus alatus Swartz, Prodr. 77. 1788.
Cereus alatus De Candolle, Prodr. **3**: 470. 1828.
Epiphyllum alatum Haworth, Phil. Mag. **6**: 109. 1829. Not Haworth, 1819.
Rhipsalis swartziana Pfeiffer, Enum. Cact. 131. 1837.
Hariota swartziana Lemaire, Cact. Gen. Nov. Sp. 75. 1839.
Rhipsalis alata Schumann in Martius, Fl. Bras. **4²**: 288. 1890.
Hariota alata Kuntze, Rev. Gen. Pl. **1**: 262. 1891.
Rhipsalis harrisii Gürke, Monatsschr. Kakteenk. **18**: 180. 1809.

Pendent from trees and rocks, up to 5 meters long, branched; joints broadly linear to lanceolate or linear-oblong, 2 to 4 dm. long, 3 to 6 cm. broad, obtuse, the margin crenate-undulate; flowers yellowish white, 15 mm. long; flower-tube 4 mm. long; perianth-segments 10, lanceolate, acute; stamens numerous, about half as long as perianth; style slender; stigma-lobes 5; ovary somewhat tubercled, bearing several broad scales; fruit ovoid, 1 cm. long, yellowish green; seeds obovate, black, bearing depressed tubercles; hilum oblique.

Type locality: Jamaica.

Distribution: Mountains of Jamaica.

This plant has usually passed as a *Rhipsalis*, but its definite flower-tube and somewhat tubercled and scaly ovary exclude it from that genus. This species has long been known in Jamaica; it was mentioned by Sloane as a spineless *Opuntia* and it is also referred to by Patrick Browne.

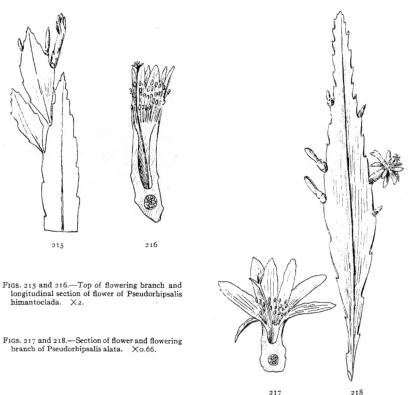

FIGS. 215 and 216.—Top of flowering branch and longitudinal section of flower of Pseudorhipsalis himantoclada. ×2.

FIGS. 217 and 218.—Section of flower and flowering branch of Pseudorhipsalis alata. ×0.66.

Cactus dentatus Ruiz (Martius, Fl. Bras. 4²: 288. 1890) was given as a synonym of *Rhipsalis alata* by Schumann, but better referred to *R. ramulosa* (see page 241).

Cereus alatus crassior Salm-Dyck (Hort. Dyck. 66. 1834) is only a name, which may or may not refer to the Jamaican plant.

Illustration: Torreya **9**: 157. f. 2, as *Rhipsalis alata*.

Plate XXII, figure 5, shows a plant collected by Dr. Britton in Jamaica in 1907, which flowered in the New York Botanical Garden, November 8, 1912. Figure 218 shows a flowering branch (natural size); figure 217 shows half of a flower with tube, perianth-segments, and stamens.

6. LEPISMIUM Pfeiffer, Allg. Gartenz. 3: 315, 380. 1835.

Saxicolous or epiphytic cacti, usually much branched and elongated, the branches flat, angled or 3-winged, the margins strongly crenate; areoles in the crenations producing a tuft of long white hairs; flowers 1 to 5 at an areole, white to pinkish; perianth-segments united at base into a short tube; filaments slender, adnate to flower-tube; stigma-lobes 4 or 5; fruit globose, smooth, turgid, purple; seeds minute; cotyledons broad, acuminate.

Type species: *Lepismium commune* Pfeiffer.

We recognize but one species, which has been described under many names. The generic name is from λεπίς, a scale, referring to the small scales subtending the areole.

1. Lepismium cruciforme (Vellozo) Miquel, Bull. Neerl. 49. 1838.

Cactus cruciformis Vellozo, Fl. Flum. 207. 1825.
Cereus tenuispinus Haworth, Phil. Mag. 1: 125. 1827.
Cereus myosurus Salm-Dyck in De Candolle, Prodr. 3: 469. 1828.
Cereus tenuis De Candolle, Prodr. 3: 469. 1828.
Cereus squamulosus Salm-Dyck in De Candolle, Prodr. 3: 469. 1828.
Cereus setosus Loddiges, Bot. Cab. 19: pl. 1887. 1832.
Lepismium tenue Pfeiffer, Allg. Gartenz. 3: 315. 1835.
Lepismium commune Pfeiffer, Allg. Gartenz. 3: 315. 1835.
Lepismium knightii Pfeiffer, Allg. Gartenz. 3: 380. 1835.
Lepismium myosurus Pfeiffer, Enum. Cact. 139. 1837.
Cereus cruciformis Steudel, Nom. ed. 2. 1: 333. 1840.
Rhipsalis myosurus Förster, Handb. Cact. 455. 1846.
Rhipsalis mittleri Förster, Handb. Cact. 455. 1846.
Rhipsalis knightii Förster, Handb. Cact. 456. 1846.
Lepismium myosurus knightii Salm-Dyck in Labouret, Monogr. Cact. 445. 1853.
Lepismium myosurus laevigatum Salm-Dyck in Labouret, Monogr. Cact. 446. 1853.
Lepismium radicans Vöchting, Jahrb. Wiss. Bot. Leipzig 9: 399. 1873.
Lepismium cavernosum Lindberg, Gartenflora 39: 151. 1890.
Rhipsalis brevibarbis Schumann in Martius, Fl. Bras. 4²: 268. 1890.
Rhipsalis squamulosa Schumann in Martius, Fl. Bras. 4²: 280. 1890.
Rhipsalis macropogon Schumann in Martius, Fl. Bras. 4²: 282. 1890.
Hariota cruciformis Kuntze, Rev. Gen. Pl. 1: 262. 1891.
Hariota squamulosa Kuntze, Rev. Gen. Pl. 1: 263. 1891.
Hariota knightii Kuntze, Rev. Gen. Pl. 1: 263. 1891.
Hariota knightii tenuispinis Kuntze, Rev. Gen. Pl. 1: 263. 1891.
Rhipsalis anceps Weber, Rev. Hort. 64: 427. 1892.
Rhipsalis cavernosa Schumann, Monatsschr. Kakteenk. 3: 24. 1893.
Rhipsalis radicans Weber, Dict. Hort. Bois 1047. 1898.
Rhipsalis radicans anceps Weber, Dict. Hort. Bois 1047. 1898.
Rhipsalis radicans ensiformis Weber, Dict. Hort. Bois 1047. 1898.
Lepismium cavernosum ensiforme Weber in Roland-Gosselin, Rev. Hort. 70: 108. 1899.

Usually creeping over rocks, freely rooting, appressed, somewhat branching; branches foliaceous, usually flat, sometimes 3, 4, or even 5-angled, linear-lanceolate, 2 cm. broad, narrowed at base, more or less purplish, especially on edges; margins somewhat repand; areoles sunken in margins; flowers white, 2 to 5 or even more from an areole, 12 to 13 mm. long; fruit globular, juicy, purplish to red, translucent, 6 to 12 mm. in diameter; seeds light brown to black, 1.8 mm. long.

Type locality: Coast of Brazil.
Distribution: States of Rio de Janeiro and Minas Geraes, Brazil.

Rhipsalis radicans rosea Weber (Dict. Hort. Bois 1047. 1898) has small rose-colored flowers which, according to Weber, resemble those of *R. myosurus*.

Schumann (Gesamtb. Kakteen 649. 1898) gives *Lepismium anceps* Weber (in Hort. Paris) as a synonym for *Rhipsalis anceps*. Here belongs also *R. ensiformis* Weber (Dict. Hort. Bois 1047. 1898).

Some of the plants now in cultivation are not so broadly winged as is shown in the illustration in Curtis's Botanical Magazine referred to below. This illustration was based upon specimens which were supposed to have come from Prince de Salm-Dyck and, therefore, presumably are typical.

Cereus knightii Parmentier (Pfeiffer, Enum. Cact. 139. 1837) is given as a synonym of *Lepismium knightii*.

Cactus tenuis Schott (De Candolle, Prodr. 3: 469. 1828) was cited as a synonym of *Cereus tenuis*.

Schumann cited *Lepismium mittleri* as a synonym of *Rhipsalis squamulosa*, referring it to Förster (Handb. Cact. 455. 1846), but the plant is there described as *Rhipsalis mittleri*.

Cereus elegans Hortus appeared first (Pfeiffer, Enum. Cact. 138. 1837) as a synonym of *Lepismium commune*, while the Index Kewensis refers it to *Rhipsalis mittleri*.

Cereus myosurus tenuior Salm-Dyck (Hort. Dyck. 65. 1834) is only a name.

Lepismium cavernosum minus Lindberg is a name mentioned by Roland-Gosselin (Rev. Hort. **70**: 108. 1899).

Lepismium duprei, the name mentioned by Salm-Dyck (Cact. Hort. Dyck. 1844. 41. 1845) and by Förster (Handb. Cact. 456. 1846) as in the collections at Paris, was never described.

Lepismium laevigatum Salm-Dyck (Cact. Hort. Dyck. 1844. 41. 1845) is without description, nor do we find it listed in the Index Kewensis.

Illustrations: Fl. Flum. **5**: pl. 29, as *Cactus cruciformis*; Loddiges, Bot. Cab. **19**: pl. 1887; Loudon, Encycl. Pl. ed. 2. 1202. f. 17365, as *Cereus setosus*; Palmer, Cult. Cact. 195, as *Lepismium*; Curtis's Bot. Mag. **66**: pl. 3755; Garten-Zeitung **4**: 182. f. 42, No. 3; Loudon's Encycl. Pl. ed. 3. 1380. f. 19411, as *Lepismium myosurum*; Monatsschr. Kakteenk. **3**: 41, as *L. knightii*; Abh. Bayer. Akad. Wiss. München **2**: pl. 7, f. 1; Curtis's Bot. Mag. **66**: pl. 3763; Förster, Handb. Cact. ed. 2. 898. f. 123 (in error 103); Loudon, Encycl. Pl. ed. 3. 1380. f. 19412; Nov. Act. Nat. Cur. **19**[1]: pl. 16, f. 12, as *L. commune*; Goebel, Pflanz. Schild. **1**: pl. 2, f. 3, 4, as *L. radicans* (seedling); Gartenwelt **16**: 633; Schumann, Gesamtb. Kakteen f. 98, C, D, as *Rhipsalis cavernosa*; Gartenflora **39**: f. 38, as *Lepismium cavernosum*; Martius, Fl. Bras. **4**[2]: pl. 55, f. 2, as *Rhipsalis macropogon*; Arch. Jard. Bot. Rio de Janeiro **1**: pl. 25, as *Rhipsalis radicans*; Arch. Jard. Bot. Rio de Janeiro **1**: pl. 24, as *R. myosura*; Möllers Deutsche Gärt. Zeit. **25**: 477. f. 11, No. 19, as *R. squamulosa*; Rev. Hort. 85: f. 152, as *R. anceps*.

Plate XXII, figure 2, shows the plant obtained by Dr. Rose in Brazil in 1915 which flowered November 18 of that year.

LEPISMIUM RAMOSISSIMUM Lemaire in Förster, Handb. Cact. ed. 2. 899. 1885.

> *Rhipsalis ramosissima* Schumann in Martius, Fl. Bras. **4**[2]: 299. 1890.
> *Hariota ramosissima* Kuntze, Rev. Gen. Pl. **1**: 263. 1891.

This is a very uncertain species which we know only from descriptions. It is from Brazil.

7. **HATIORA** Britton and Rose, Stand. Cycl. Hort. Bailey **3**: 1432. 1915.

> *Hariota* De Candolle, Mém. Cact. 23. 1834. Not Adanson, 1763.

Unarmed, slender, branched cacti; branches terete, short, arising in 2's or 3's from tops of older ones, smooth, leafless and spineless,* bearing several small areoles along their sides and each a large, woolly, terminal one from which the flower and succeeding branches arise; sepals usually in 2 series, outer ones broader and short, inner ones larger and more petal-like; petals distinct, narrowed toward base; stamens distinct, erect, borne on disk; stigma-lobes 4 or 5, erect or a little spreading, white; ovary globular, naked or nearly so.

Type species: *Rhipsalis salicornioides* Haworth.

Some six or seven species have been described; we recognize three.

The genus *Hariota* was named for Thomas Hariot, a botanist of the 16th century, *Hatiora* being an anagram. It is closely related to *Rhipsalis*, with which it is often united.

The flowers open only in bright sunlight and are rotate or nearly so. In the United States the plants flower under glass, usually in the winter from December to February, but sometimes as late as April.

---

*Sometimes peculiar lateral branches are produced which are made up of short, rounded joints with numerous areoles bearing several bristles or hairy spines. See illustrations of Schumann (Gesamtb. Kakteen f. 97, D) and Loddiges (Bot. Cab. 4: 369). In cases which we have observed these occur on stunted or starved plants, the areoles arranged in 6 rows forming low angles on the branchlets.

KEY TO SPECIES.

Lower part of joints slender, pedicel-like.. 1. *H. salicornioides*
Joints only slightly narrowed below or not narrowed.
    Joints clavate... 2. *H. bambusoides*
    Joints cylindric... 3. *H. cylindrica*

1. Hatiora salicornioides (Haworth) Britton and Rose, Stand. Cycl. Hort. Bailey **3**: 1433. 1915.

    Rhipsalis salicornoides Haworth, Suppl. Pl. Succ. 83. 1819.
    Cactus salicornioides* Link and Otto, Icon. Pl. Select. 49. 1822.
    Cactus lyratus Velozo, Fl. Flum. ed. 2. **4**: 205. 1825.
    Hariota salicornioides De Candolle, Mém. Cact. 23. 1834.
    Rhipsalis salicornioides strictior Salm-Dyck, Cact. Hort. Dyck. 1849. 230. 1850.
    Hariota salicornioides strictior Gürke, Blühende Kakteen **2**: under pl. 95. 1907.

Stems 1 to 2 meters long with a jointed cylindric trunk; branchlets club-shaped, the lower part very slender and pedicel-like, 1.5 to 3 cm. long, green or purplish; areoles of cultivated specimens without setae; flowers 8 to 10 mm. long, salmon-colored, the outer sepals short and obtuse; inner petals somewhat crenate, obtuse; filaments yellowish, at top appressed against style, shorter than petals; style yellowish; stigma-lobes 4 or 5, white.

Type locality: Recorded originally from the West Indies in error.

Distribution: Southeast Brazil.

These plants grow quite differently in the woods from the way they do in greenhouses. The following note was made by Dr. Rose in 1915 while collecting at Rio de Janeiro:

The plant grows on trunks of trees, its roots long and fibrous, 4 dm. long or more and wrapped about the trunk of the tree; at first it is erect, then spreading and finally pendent; it is then a meter long or more and very much branched; main stem and branches 5 to 10 mm. in diameter, made up of short terete joints (2 to 5 cm. long); branches in whorls of 2 to 6.

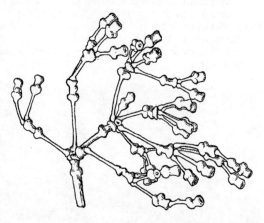

FIG. 219.—Unusual form of Hatiora salicornioides. × 0.8.

A very remarkable form, if not a distinct species, was obtained by Dr. Rose in the forest of Itatiaya, altitude 1,200 meters, in July 1915 (No. 20585). The terminal joints are 1 to 2 cm. long, the lower half slender, pedicel-like, the upper half twisted and contorted. This is well shown in our illustration (figure 219).

Rhipsalis salicornioides gracilior (Salm-Dyck, Cact. Hort. Dyck. 1844. 40. 1845; *Hariota salicornioides gracilior* Gürke, Blühende Kakteen. **2**: under pl. 95. 1907) is only a name.

The following varieties of *Rhipsalis salicornioides* of Weber are probably to be referred here: var. *gracilis* Weber (Dict. Hort. Bois 1048. 1898; *Rhipsalis gracilis* Weber and *Hariota gracilis* Weber, Dict. Hort. Bois 1048. 1898) and var. *stricta* Weber (Dict. Hort. Bois 1048. 1898; *Rhipsalis stricta* Cels, Dict. Hort. Bois 1048. 1898). The name *Rhipsalis stricta* seems never to have been published. Weber cited it as above, referring it to Cels as the author. Schumann uses the name earlier where he states that it was used in France for *Hariota salicornioides* (Monatsschr. Kakteenk. **4**: 74. 1894). Pfeiffer refers here as a synonym *Opuntia salicornioides* (Enum. Cact. 141. 1837), attributing the name to Sprengel, who, however, used it as *Cactus (Opuntia) salicornioides*. *Hariota sticta* has been used (Monatssch. Kakteenk. **5**: 22. 1895). The variety *ramosior* Salm-Dyck (Pfeiffer, Enum. Cact. 142. 1837) may or may not belong to this species.

---

* This name is often credited to Sprengel (Syst. **2**: 497. 1825).

Rhipsalis schottmuelleri Hortus is given by Schelle (Handb. Kakteenk. 227. 1907) as a synonym of *Hariota salicornioides schottmuelleri*, an unpublished variety.

Hariota villigera (Schumann in Martius, Fl. Bras. 4^2: 265. 1890; *Rhipsalis salicornioides villigera* Löfgren, Arch. Jard. Bot. Rio de Janeiro 1: 85. 1915) we know from description only; it seems to be stouter than *salicornioides* but may belong here. It was based on Sellow's specimen from São Paulo, but its flowers are unknown.

Illustrations: Loddiges, Bot. Cab. 4: pl. 369; Cact. Journ. 1: 180; Curtis's Bot. Mag. 51: pl. 2461; Blanc, Cacti 90. No. 1013; Balt. Cact. Journ. 1: 122; Gard. Chron. II. 6: 731. f. 134; Amer. Gard. 11: 463; Goebel, Pflanz. Schild. 1: pl. 4, f. 5, 6; Möllers Deutsche Gärt. Zeit. 25: 477. f. 11, No. 17; Arch. Jard. Bot. Rio de Janeiro 1: pl. 12; Gartenwelt 13: 117, as *Rhipsalis salicornioides*; Link and Otto, Icon. Pl. Select. pl. 21, as *Cactus salicornioides*; Schumann, Gesamtb. Kakteen f. 97, C, D; Martius, Fl. Bras. 4^2: pl. 52; Monatsschr. Kakteenk. 5: 23; Schelle, Handb. Kakteenk. 227. f. 148, as *Hariota salicornioides*; Rev. Hort. 1861: 110. f. 23, as *Rhipsalis salicorne*; Fl. Flum. 5: pl. 21, as *Cactus lyratus*.

Plate XXIII, figure 4, shows a plant in the New York Botanical Garden which flowered February 2, 1912. Figure 219 shows a peculiar form collected by Dr. Rose in Brazil in 1915.

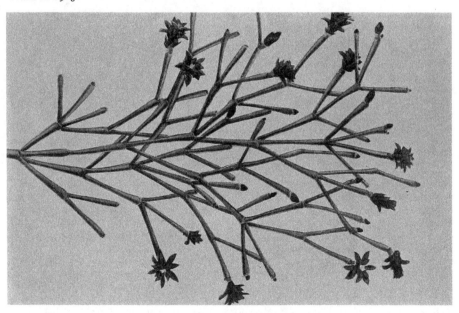

FIG. 220.—Hatiora bambusoides.

2. Hatiora bambusoides (Weber).

Rhipsalis salicornioides bambusoides Weber, Rev. Hort. 64: 429. 1892.
Hariota salicornioides bambusoides Schumann, Gesamtb. Kakteen 613. 1898.
Rhipsalis bambusoides Löfgren, Arch. Jard. Bot. Rio de Janeiro 2: 41. 1918.

Stems becoming 2 meters high and stouter than those of *H. salicornioides*; joints clavate, 3 to 5 cm. long, 4 mm. in diameter at the top; flowers orange; sepals obtuse; petals usually erect but sometimes spreading.

Type locality: Brazil.

Distribution: State of Rio de Janeiro, Brazil.

M. E. Eaton del.

1. Flowering branch of *Hatiora cylindrica*.
2. Fruiting branch of *Rhipsalis heteroclada*.
3. Fruiting branch of *Rhipsalis cribrata*.
4. Flowering branch of *Hatiora salicornioides*.

Introduced into Jardin des Plantes, Paris, France, from Brazil.

We have not seen this type material, but if plate 95 in the Blühende Kakteen is typical, our identification is correct. In the description accompanying this plate it is stated that the drawing was made from a plant sent by Mr. Weber to the Berlin Botanical Garden.

Hariota bambusoides Weber (Dict. Hort. Bois 1048. 1898) was given as a synonym but was never described.

Illustrations: Blühende Kakteen 2: pl. 95, as *Hariota salicornioides bambusoides*; Gartenwelt 13: 117, as *Rhipsalis salicornioides bambusoides*.

Figure 220 is reproduced from the first illustration cited above.

3. Hatiora cylindrica sp. nov.

Forming dense masses one meter in diameter or more; joints cylindric, 3 cm. long or less, pale green, becoming spotted or finally red throughout; flowers usually solitary, 12 mm. long; sepals ovate, short, red; petals orange to yellow, oblong, obtuse.

Collected by J. N. Rose in company with Dr. Löfgren and Señor Porto at Ilha Grande, Districto Federal, near Rio de Janeiro, July 22 to 24, 1915.

Illustration: Arch. Jard. Bot. Rio de Janeiro 2: pl. 13, ad *Rhipsalis bambusoides*.

Plate XXIII, figure 1, shows the plant collected by Dr. Rose on Ilha Grande, near Rio de Janeiro, which flowered in the New York Botanical Garden, December 18, 1918.

8. RHIPSALIS Gaertner, Fruct. Sem. 1 137. 1788.

*Hariota Adanson, Fam. Pl. 2: 243. 1763.
†Cassytha Miller, Gard. Dict. ed. 8. 1768. Not Linnaeus, 1753.

Cacti sometimes growing in humus, but usually epiphytic and hanging from trees, sometimes erect, sometimes clambering over rocks, more or less rooting or, when hanging, irregularly producing aërial roots; roots always fibrous; stems usually much branched (often heteromorphic), terete, angled or much flattened and leaf-like, very slender and thread-like or stout and stiff; leaves wanting or represented by minute bracts; areoles borne along margin of flat-branched forms, along ribs or scattered irregularly in other forms, usually small, bearing hairs, wool, bristles, and flowers; flowers usually solitary, but sometimes several from a single areole, opening night or day and remaining open for 1 to 8 days, small for the family; perianth-segments distinct, few, sometimes only 5, usually spreading, sometimes reflexed; filaments few or numerous, erect, slender, borne on outer margin of disk in one or two rows; style erect; stigma-lobes 3 or more, usually slender, spreading; ovary small, sometimes depressed or sunken in branch; fruit globular or oblong, sometimes angled when immature, but finally turgid, juicy, white or colored, usually naked (setose at areoles in 1 or 2 species) or sometimes bearing a few scales; seeds small, few to many.

Type species: *Rhipsalis cassutha* Gaertner.

The generic name is from ριψ wicker-work, referring to the slender, pliable branches of the typical species.

We recognize 57 species, although more than 115 names have been published.

The species range from Florida, Mexico, and the West Indies through continental America to Argentina; only 2 species are found in Mexico; 1 in Florida; 2 are known in the West Indies; a very few in northern South America; 3 or 4 only on the west coast of South America; and 5 or 6 in Argentina. The center of distribution is in the states of Rio de Janeiro, São Paulo, and Minas Geraes, in southern Brazil. In the little state of Rio de Janeiro and chiefly about the city of the same name, Dr. Rose collected 15 species in 1915.

The occurrence of species of *Rhipsalis*, in the wild state, in tropical Africa and in Ceylon, forms the only possible exception to the American natural distribution of cacti. Eight

---

*No species was cited by Adanson for his genus *Hariota* but it was based on Burmann's plate of Plumier (pl. 197, f. 2), which has been identified as *Cactus parasiticus* Lamarck, not Linnaeus. The type of *Cactus parasiticus* Linnaeus is a species of *Vanilla*, probably *V. claviculata* Swartz.

†Miller, in his Gardeners' Dictionary of 1768, described *Rhipsalis cassutha* under the name of *Cassytha filiformis*, a name which had already been published by Linnaeus for a wholly different plant. Miller's generic name. *Cassytha*, therefore, being a misidentification, should not be treated as a synonym proper of *Rhipsalis*, although usually so cited.

supposedly distinct species have been described by authors from tropical Africa, and *R. cassutha* has long been known to exist in Ceylon. M. Roland-Gosselin, a diligent French student of cacti, after an investigation of these Old World plants, published in 1912 a very interesting paper,* giving his conclusion that they are really all American species, their seeds having been transmitted to the Old World by migratory birds, and he referred them all to known American species. We have followed him in these reductions but we have not been able in all cases to study authentic specimens. It raises the interesting question if the Old World plants should be regarded as native or introduced.

In stem structure some of the species, such as *Rhipsalis elliptica*, approach very closely *Zygocactus truncatus*, while certain forms of *Epiphyllanthus* are easily mistaken for a *Rhipsalis*.

As we have treated the genus here, the flowers and fruits are fairly uniform. The stem structures are various and parallel in a way those of *Opuntia*, ranging from slender and terete to broad and thin; in some species they are leaf-like as in *Epiphyllum*, or 3-angled, suggesting *Hylocereus*. The areoles are usually small and bear only a small tuft of wool, but in some species they bear hairs or bristles. The flowers may open at any time of the day and in most species do not close at the approach of night; they are not readily affected by shade or direct sunlight and open but once.

KEY TO SPECIES.

A. Joints terete, ribbed or angled, none of them flat.
  B. Joints terete or young ones angled, smooth, or areoles bristly or hairy.
    C. Joints short, oblong, not more than 5 times as long as thick; areoles of young joints with a few long hairs.
      Lateral joints simple; flowers lateral. Series 1, *Mesembryanthemoides*....   1. *R. mesembryanthemoides*
      Lateral joints much branched; flowers terminal. Series 2, *Cereusculae*.....   2. *R. cereuscula*
    CC. Joints cylindric, rarely clavate, slender, short or elongated.
      D. Flowering areoles small, not very woolly, not depressed.
        E. Ultimate joints slender, about 2.5 mm. thick or less, relatively short.
          F. Young joints or some of them angled, their areoles bearing hairs. Series 3, *Prismaticae*.
            Species of Brazil or Madagascar.......................   3. *R. prismatica*
            Species of Costa Rica.................................   4. *R. simmleri*
        FF. All joints cylindric to clavate. Series 4, *Capilliformes*.
          Joints clavate...................................   5. *R. clavata*
          Joints cylindric.
            Ultimate joints up to 2.5 mm. thick; petals 9 mm. long..   6. *R. campos-portoana*
            Ultimate joints about 1.5 mm. thick; petals 6 mm. long or less.
              Plant stiff; areoles red; flowers white...............   7. *R. heteroclada*
              Plant weak; areoles not red.
                Flowers greenish white or yellowish, 5 to 6 mm. wide.   8. *R. capilliformis*
                Flowers white or nearly white, about 8 mm. wide.
                  Pendent; secondary branches 2 to 3-chotomous.....   9. *R. burchellii*
                  Spreading or diffuse; upper branches subverticillate. 10. *R. cribrata*
      EE. Ultimate joints stouter, mostly 3 to 10 mm. thick and elongated.
        F. Scale subtending the areoles inconspicuous or none.
          G. Ultimate joints 3 to 6 mm. thick. Series 5, *Cassuthae*.
            H. Ultimate joints definitely shorter than others, often verticillate.
              Plants weak, pendent..........................   11. *R. cassutha*
              Plants stiffer, not strictly pendent.
                 Ultimate joints slender......................   12. *R. virgata*
                 Ultimate joints stout.......................   13. *R. teres*
           HH. Ultimate joints not definitely shorter than others, simple or dichotomous, rarely verticillate.
            I. Areoles without bristles or with spreading bristles.
              Petals about 4 mm. long.
                Fruit naked
                  Petals pink..........................   14. *R. lindbergiana*
                    Petals white.........................   15. *R. shaferi*
                Fruit with scales and these setose in axils... 16. *R. fasciculata*
               Petals 8 to 12 mm. long.
                 Flowers purplish red; areoles not bristly.... 17. *R. pulchra*
                 Flowers white; areoles somewhat bristly.... 18. *R. lumbricoides*

---

* Bull. Soc. Bot. France **59**: 97-102. 1912. Translation in Torreya **13**: 151-156. 1913.

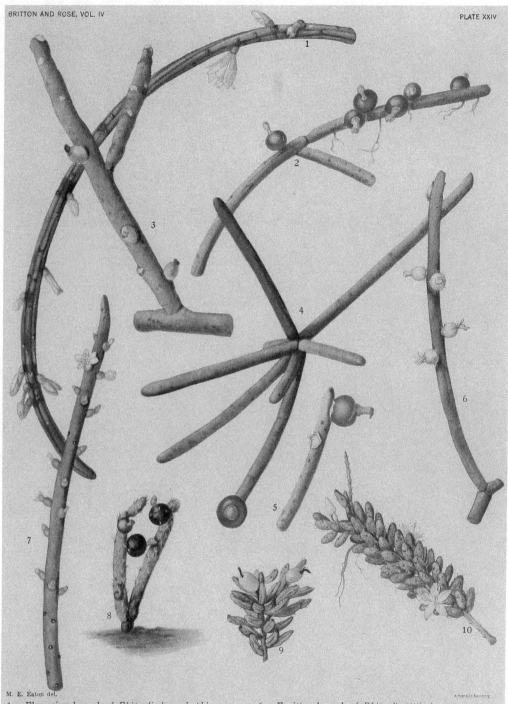

M. E. Eaton del.

1. Flowering branch of *Rhipsalis leucorhaphis*.
2. Fruiting branch of same.
3. Fruiting branch of *Rhipsalis megalantha*.
4. Fruiting branch of *Rhipsalis neves-armondii*.
5. Fruiting branch of same.

6. Fruiting branch of *Rhipsalis pittieri*.
7. Flowering and fruiting branch of *Rhipsalis shaferi*.
8. Fruiting branch of *Rhipsalis aculeata*.
9. Fruiting branch of *Rhipsalis mesembryanthemoides*.
10. Flowering branch of same.

KEY TO SPECIES—continued.

II. All areoles with appressed bristles............. 19. *R. aculeata*
GG. Ultimate joints 8 to 12 mm. thick. Series 6, *Grandiflorae.*
    Flowers 2 to 2.5 cm. broad........................ 20. *R. grandiflora*
    Flowers up to 4 cm. broad .... 21. *R. megalantha*
FF. Scale subtending the areole white, scarious, appressed, con-
    spicuous. Series 7, *Leucorhaphes.*
    Joints 5 to 8 mm. thick; areoles with deciduous bristles...... 22. *R. leucorhaphis*
    Joints 3 mm. thick; areoles without bristles.............. 23. *R. loefgrenii*
DD. Flowering areoles large, very woolly, depressed. Series 8, *Floccosae.*
    Ultimate joints much shorter than others; verticillate; plants stiff.... 24. *R. neves-armondii*
    Ultimate joints not much shorter than others; plants weak.
    Main branches stout, 8 to 10 mm. in diameter.
        Fruit pure white; Venezuelan species...................... 25. *R. pittieri*
        Fruit red or tinged with red; Brazilian and Argentine species.
            Fruit bright red.. 26. *R. pulvinigera*
            Fruit whitish but when mature tinged with red or purple.
                Fruit 5 mm. in diameter; flowers white, becoming yellowish.. 27. *R. floccosa*
                Fruit 8 to 10 mm. in diameter; flowers tinged with red, larger
                    than the last...................... 28. *R. tucumanensis*
    All branches slender, 3 to 6 mm. in diameter.
        Flowers white.. 29. *R. gibberula*
        Petals with red tips.................................... 30. *R. puniceo-discus*
BB. Joints ribbed or angled, at least when old.
    C. Some joints bristly, others unarmed, ribbed when old. Series 9, *Dissimiles.*.. 31. *R. dissimilis*
    CC. All joints unarmed, angled or winged.
        Joints 5-angled or 5-winged.
            Joints 5-winged; wings crenate. Series 10, *Pentapterae*.............. 32. *R. pentaptera*
            Joints 5-angled; ribs nearly continuous. Series 11, *Sulcatae*......... 33. *R. sulcata*
        Joints 3-angled.
            Angles of joints continuous, wingless. Series 12, *Trigonae*........... 34. *R. trigona*
            Angles of joints interrupted, winged. Series 13, *Paradoxae*.......... 35. *R. paradoxa*
AA. At least some of joints flat, on same plant or on different plants.
    B. Joints deeply serrate; flowers nodding. Series 14, *Houlletianae*............... 36. *R. houlletiana*
    BB. Joints repand, entire or crenate; flowers not nodding.
        C. Joints linear to linear-lanceolate. Series 15, *Lorentzianae.*
            D. Joints both flat and 3-angled or rarely 4-angled, mostly narrowly linear.
                Fruit black.
                    Joints 10 mm. wide or less; fruit 5 to 6 mm. in diameter......... 37. *R. warmingiana*
                    Joints 10 to 15 mm. wide; fruit about 10 mm. in diameter......... 38. *R. gonocarpa*
                Fruit white to reddish.
                    Joints long-acuminate................................. 39. *R. linearis*
                  Joints blunt.
                    Joints scarcely crenate; Ecuadorean and Peruvian species....... 40. *R. micrantha*
                    Joints definitely crenate; Costa Rican species.................. 41. *R. tonduzii*
            DD. All joints flat, linear-lanceolate to oblong.
                Joints deeply crenate, the lobes rounded....................... 42. *R. boliviana*
                Joints repand, low-crenate, or nearly entire.
                    Joints coriaceous, distinctly crenate......................... 43. *R. lorentziana*
                    Joints thin in texture, merely repand or low-crenate.
                      Bolivian species....................................... 44. *R. ramulosa*
                      West Indian and Central American species.
                      Larger joints 3 cm. wide............................... 45. *R. purpusii*
                      Joints 0.5 to 2.5 cm. wide.
                        Perianth-segments 7 to 8 mm. long, greenish white or pinkish 46. *R. coriacea*
                        Perianth-segments 5 to 6 mm. long, green............... 47. *R. jamaicensis*
    CC. Joints elliptic to oblong. Series 16, *Crispatae.*
        Joints thick, coriaceous.
            Joints oblong, cuneate at base.
                Flowers and fruit usually solitary at areoles...................... 48. *R. platycarpa*
                Flowers several at an areole.................................. 49. *R. russellii*
             Terminal joints short-oblong to elliptic.
                Fruit red... 50. *R. elliptica*
                Fruit white... 51. *R. pachyptera*
         Joints thin.
            Brazilian species.
                Joints purplish green, obovate............................... 52. *R. rhombea*
                Joints bright green or reddish.
                    Joints reddish green, the margins much crisped................. 53. *R. crispimarginata*
                  Joints bright green, the margins slightly crisped.
                      Ultimate joints broad, elliptic to obovate.................... 54. *R. crispata*
                      Ultimate joints narrow, oblong to narrowly obovate.......... 55. *R. oblonga*
            Bolivian species.. 56. *R. cuneata*
AAA. Species not grouped... 57. *R. roseana*

1. **Rhipsalis mesembryanthemoides** * Haworth, Rev. Pl. Succ. 71.   1821.

    Rhipsalis salicornioides * (variety B) Haworth, Suppl. Pl. Succ. 83.   1819.
    Hariota mesembrianthemoides † Lemaire, Cact. Gen. Nov. Sp. 74.   1839.

Branches very dissimilar; main branches elongated, slender, terete, more or less setose, often bearing aërial roots, covered with short stubby branchlets; these sometimes also bearing short joints, usually less than 1 cm. long, more or less angled, often with short setae from the small areoles; flower-buds small, pinkish; flowers solitary at areoles of the branchlets, opening in early morning, rather large, 1.5 cm. broad, white or light pink; petals 5, spreading, acute; stamens about 20, erect, white; style white; stigma-lobes 3, white; fruit short-oblong, 5 mm. long, white or tinged with red.

    Type locality: Not cited where published.
    Distribution: Rio de Janeiro, Brazil.
    The plant is common in cultivation; in nature it grows in dense masses on trunks of trees. It first flowered in cultivation in England in 1831. Its short joints have a fancied resemblance to species of *Mesembryanthemum.*
    A dried specimen of Haworth's plant is still preserved in London and through the kindness of N. E. Brown we have a photograph of it.
    Rhipsalis echinata was published as a synonym by Pfeiffer (Enum. Cact. 136. 1837).
    Illustrations: Cact. Journ. 1 : 180; Curtis's Bot. Mag. **58**: pl. 3078; Schumann, Gesamtb. Kakteen 633. f. 98, G; Monatsschr. Kakteenk. **2**: 9; **4**: 59; Arch. Jard. Bot. Rio de Janeiro 1 : pl. 11; Goebel, Pflanz. Schild. 1: pl. 4, f. 7; Loddiges, Bot. Cab. **20**: pl. 1920; Thomas, Zimmerkultur Kakteen 58.
    Plate XXIV, figure 9, shows a fruiting plant obtained by Dr. Rose in Rio de Janeiro in 1915 (No. 20246); figure 10 shows a flowering plant sent by Alwin Berger in 1908.

2. **Rhipsalis cereuscula** Haworth, Phil. Mag. **7**: 112.   1830.

    Hariota saglionis Lemaire, Cact. Aliq. 39.   1838.
    Rhipsalis saglionis Otto in Walpers, Repert. Bot. **2**: 936.   1843.
    Rhipsalis brachiata Hooker in Curtis's Bot. Mag. **69**: pl. 4039.   1843.
    Hariota cereuscula Kuntze, Rev. Gen. Pl. 1: 262.   1891.
    Rhipsalis saglionis rubrodiscus Löfgren, Arch. Jard. Bot. Rio de Janeiro 1: 80.   1915.

    Stems and branches terete; stem slender, usually elongate, often erect, sometimes 6 dm. high, crowned by a cluster of short branches; upper branches short, 2 to 6 times as long as thick, somewhat angled, the areoles bearing 2 to 4 short bristles; flowers terminal or near the ends of the branches, 16 mm. broad; petals about 12, spreading, pinkish to white with yellowish midrib; stigma-lobes 3 or 4; berries white.

    Type locality: Brazil.
    Distribution: Uruguay to central Brazil.
    Illustrations: Curtis's Bot. Mag. **69**; pl. 4039; Loudon, Encycl. Pl. ed. 3. 1380. f. 19408, as *Rhipsalis brachiata;* Cycl. Amer. Hort. Bailey **4**: f. 2101; Stand. Cycl. Hort. Bailey **5**: f. 3377; Cact. Journ. 1 : 180; Monatsschr. Kakteenk. **4**: 75, as *R. saglionis.*
    Plate XXVII, figure 3, is of a plant which flowered in the New York Botanical Garden in March 1912. Figure 221 is from a photograph of a flowering plant from Misiones, obtained by Dr. Rose in 1915 from Dr. Spegazzini.

3. **Rhipsalis prismatica** Rümpler in Förster, Handb. Cact. ed. 2. 884.   1885.

    Hariota prismatica Lemaire, Illustr. Hort. **10**: Misc. 84.   1863.
    Rhipsalis suareziana ‡ Weber, Rev. Hort. **64**: 425.   1892.
    Rhipsalis tetragona Weber, Rev. Hort. **64**: 428.   1892.

    Very much branched, prostrate; lower branches elongated and terete; upper branches short and somewhat angled; areoles more or less setose; flowers white; petals usually 5, obtuse; fruit small, pinkish to white, globose.

---

   * Haworth spelled this *R. mesembryanthoides* and also *R. salicornoides.*
   † The Index Kewensis gives the place of publication erroneously as Lemaire, Cact. Aliq. Nov. 39. 1838.
   ‡ According to the Index Kewensis, *Rhipsalis suarensis* Weber (Dict. Hort. Bois 1046. 1898) is the same.

M. E. Eaton del.

1. Fruiting branch of *Rhipsalis heteroclada*.
2. Fruiting branch of same.
3. Fruiting branch of *Rhipsalis capilliformis*.
4. Fruiting branch of *Rhipsalis virgata*.

Type locality: Not cited, but Förster and Weber state that it came from Brazil.

Distribution: Brazil and northern Madagascar, but range not known.

Weber thought that *Rhipsalis tetragona* was the same as *R. prismatica* Rümpler, but because he was not certain he described it as new.

Illustration: Gartenwelt **16**: 634, as *Rhipsalis suareziana*; Monatsschr. Kakteenk. **18**: 74, as *R. tetragona.*

Plate XXXII, figure 3, shows a plant from Berlin which flowered in the New York Botanical Garden on November 23, 1915.

FIG. 221.—Rhipsalis cereuscula.

4. Rhipsalis simmleri Beauverd, Bull. Herb. Boiss. II. **7**: 136. 1907.

Stems pendent, cylindric, 2 to 3 mm. in diameter, very much branched, the branches dichotomous or 3 or 4-verticillate, upper short and somewhat angled, quite unlike lower ones; flowers solitary, subterminal; petals white with pink tips, oblong, 6 to 8 mm. long; filaments 5 to 8 mm. long, white, filiform; style exserted, 9 mm. long; stigma-lobes ovate, reflexed, white; ovary obconic, 3 to 3.5 mm. in diameter; fruit white.

Type locality: Costa Rica.

Distribution: Costa Rica.

This species is named for Paul Simmler, chief gardener of the Boissier Collections at Geneva, Switzerland. The plant was introduced in a collection of orchids from Costa Rica and flowered in cultivation. Dr. Rose saw it when in Geneva in 1912 and obtained a small fragment, but he did not see it in flower.

Illustration: Bull, Herb. Boiss. II. **7**: 137.

5. Rhipsalis clavata Weber, Rev. Hort. **64**: 429. 1892.

    Rhipsalis clavata delicatula Löfgren, Arch. Jard. Bot. Rio de Janeiro **2**: 45. 1918.

Erect when young but soon hanging, often a meter long or more, much branched; joints all similar, narrowly clavate, sometimes 4-angled when young, short, 1 to 3 cm. long, deep green, becoming brown, produced in terminal whorls of 2 to 7; areoles few, sometimes bearing 1 to 5 white hairs; flowers near end of branches, white, 1.5 cm. long; petals hardly spreading; fruit spherical, 6 mm. in diameter, white or yellowish; seeds 1.5 cm. long.

Type locality: Petropolis, in the state of Rio de Janeiro, Brazil.
Distribution: State of Rio de Janeiro.

This species is much like *Hatiora* and it was really referred to *Hariota* at one time by Weber, himself. Schumann gives only one locality for it, but Dr. Rose found it on Corcobado in Rio de Janeiro, altitude 465 meters, growing on branches of trees, and on this plant the description has been partly based. Weber's manuscript name, *Hariota clavata*, has appeared only as a synonym of this species (Monatsschr. Kakteenk. **5**: 172. 1895).

Illustrations: Arch. Jard. Bot. Rio de Janeiro **2**: pl. 17, as *Rhipsalis clavata delicatula*; Arch. Jard. Bot. Rio de Janeiro **1**: pl. 13; Möllers Deutsche Gärt. Zeit. **25**: f. 11, No. 16.

6. Rhipsalis campos-portoana Löfgren, Arch. Jard. Bot. Rio de Janeiro **2**: 35.   1918.

Stem slender, terete, usually pendent, usually dichotomous; primary branches elongated; terminal branches in 2's or 4's, somewhat clavate, 3 to 5 cm. long; areoles few, naked; flowers terminal or usually so, white; petals about 8, slightly spreading, obtuse, up to 9 mm. long; fruit globose, 4 mm. in diameter, red.

Type locality: Serra de Itatiaya, Brazil
Distribution: Known only from the type locality.

This plant was collected by Dr. Rose and Campos Porto in July 1915 (No. 20612) and flowered in the Jardim Botanico do Rio de Janeiro in September of that year, and from this the description was drawn. Dr. Rose brought home living specimens but these have not yet flowered.

Illustration: Arch. Jard. Bot. Rio de Janeiro **2**: pl. 7.

7. Rhipsalis heteroclada nom. nov.

Stems stiff, dark green, but purple about areoles and tips of branches, often erect in cultivation, much branched toward top of plant; branches often in verticillate clusters, much more slender than the main stem, 1 to 2 mm. in diameter; areoles small, often bearing a single bristle; flowers small, white or greenish; petals 5, obtuse, spreading or recurved; filaments about 20, white, erect; style white, sunken at base into a little cup; stigma-lobes 3, white; ovary green, about 2 mm. long; fruit globose, 5 to 6 mm. in diameter, white.

This plant is very common in Brazilian collections, where it is planted on fruit trees. Dr. Rose found some beautiful examples in the Horto Bolanco Paulista, near São Paulo, and on Ilha Grande (Rose 20371, type).

Plate XXIII, figure 2, shows a fruiting branch obtained by Dr. Rose in Rio de Janeiro; plate XXV, figures 1 and 2, shows fruiting plants collected by Dr. Rose in Rio de Janeiro; plate XXXII, figure 1, shows a fruiting plant obtained by Dr. Rose in Rio de Janeiro.

8. Rhipsalis capilliformis Weber, Rev. Hort. **64**: 425.   1892.
   Rhipsalis gracilis N. E. Brown, Gard. Chron. III. **33**: 18.   1903.

Stems and branches very slender and weak, the main branches often much elongated, the branchlets short, spreading or drooping; flowers numerous, scattered along sides of branches, cream-colored, rotate, 5 to 6 mm. broad; petals few, sometimes only 5, short and obtuse; fruit globose, naked, white or pinkish, 4 to 5 mm. in diameter; seeds very numerous.

Type locality: Not cited.
Distribution: Eastern Brazil, but not known to us in the wild state.

PLATE XXVI

M. E. Eaton del.

1. Flowering branch of *Rhipsalis cribrata*.
2. Fruiting branch of *Rhipsalis capilliformis*.
3. Flowering branch of same.
4. Flowering branch of same.
5. Flowering branch of *Rhipsalis teres*.

This is a very attractive little plant, often forming a dense mass of delicate branches. It is a rather shy bloomer, but grows well in damp greenhouses.

Illustration: Gartenwelt **13**: 117.

Plate XXVI, figure 4, is from a plant obtained in the Botanical Garden at Brussels by Dr. Rose in 1912, which flowered and fruited in Washington in 1919; figure 3 shows a plant sent from Paris, France, which flowered in the New York Botanical Garden in 1911 (No. 14795); figure 2 is from a plant sent by R. Lamb, from Manchester, England; plate XXV, figure 3, shows a fruiting plant sent from Paris in 1901.

9. Rhipsalis burchellii nom. nov.

Rhipsalis cribrata Löfgren, Arch. Jard. Bot. Rio de Janeiro 1: 81. pl. 10. 1915. Not Rümpler, 1885.

Much branched, very weak, with long slender hanging branches, the branching usually dichotomous; ultimate branches usually 4 to 10 cm. long; flowers subterminal, campanulate, 10 to 12 mm. long, white; fruit turbinate, rose-colored.

This plant is very common in the forests about São Paulo. Dr. Rose collected it in the forest of Jabaquara, August 15, 1915 (No. 20857, type), and also in the Botanical Garden of Museu Paulista on August 14, 1915 (No. 20849).

This species is named for William John Burchell (1781-1863), who went to Brazil in 1825, where he made large and valuable collections.

Plate XXVII, figure 2, shows a fruiting branch taken from Dr. Rose's plant No. 20857.

10. Rhipsalis cribrata (Lemaire) Rümpler in Förster, Handb. Cact. ed. 2. 889. 1885.

Hariota cribrata Lemaire, Illustr. Hort. **4**: Misc. 12. 1857.
Rhipsalis pendula Vöchting, Jahrb. Wiss. Bot. Leipzig **9**: 371. 1873. Not Pfeiffer, 1837.
Rhipsalis penduliflora N. E. Brown, Gard. Chron. II. **7**: 716. 1877.
Hariota penduliflora Kuntze, Rev. Gen. Pl. 1: 263. 1891.
Rhipsalis cribrata filiformis Engelhardt in Möllers, Deutsche Gärt. Zeit. **18**: 585. 1903.

Woody at base, much branched; branches of two forms; stems terete, elongated, at first erect, then hanging, without aërial roots; terminal branches very short, 2 to 3 cm. long, usually in whorls of 2 to 20; areoles small, often with 1 or 2 small setae; flowers generally terminal, pendulous, white or cream-colored, 8 to 10 mm. long; petals usually 5 to 7, obtuse, drying yellow; filaments erect, numerous, white, salmon-colored at base; style white; stigma-lobes 3 or 4, spreading, white; ovary naked; fruit small, globose, 2 to 3 mm. in diameter, pinkish, terminated by the old perianth.

Type locality: Brazil.

Distribution: States of Minas Geraes, Rio de Janeiro, and São Paulo, Brazil.

This species was introduced into Europe in 1856 from Brazil, as some of the other species have been, through sendings of orchids, where it was discovered by Lemaire, and when it flowered the following year it was named and described by him.

Hariota penduliflora (Monatsschr. Kakteenk. **1**: 69. 1891) is listed but not described.

Rhipsalis penduliflora laxa, referred to by Schumann (Martius, Fl. Bras. **4**²: 276. 1890), comes from the gardens at Kew.

Illustrations: Möllers Deutsche Gärt. Zeit. **18**: 585, as *Rhipsalis cribrata filiformis*; Blühende Kakteen 1: pl. 27, A; Arch. Jard. Bot. Rio Janeiro 1: pl. 9, as *R. penduliflora*.

Plate XXIII, figure 3, shows a fruiting branch collected by Dr. Rose in Rio de Janeiro in 1915; plate XXVI, figure 1, shows a flowering branch obtained by Dr. Rose in Rio de Janeiro.

11. Rhipsalis cassutha* Gaertner, Fruct. Sem. **1**: 137. 1788.

Cassytha filiformis Miller, Gard. Dict. ed. 8. 1768. Not Linnaeus, 1753.
Cactus parasiticus Lamarck, Encycl. **1**: 541. 1783. Not Linnaeus, 1768.
Cactus pendulus Swartz, Prodr. 77. 1788.

---

* The original spelling given by Gaertner is as above. The usual spelling, however, is *R. cassytha*.

Rhipsalis parasitica Haworth, Syn. Pl. Succ. 187. 1812.
Cactus caripensis* Humboldt, Bonpland, and Kunth, Nov. Gen. et Sp. 6: 66.   1823.
Cereus caripensis De Candolle, Prodr. 3: 467.   1828.
Rhipsalis cassytha dichotoma De Candolle, Prodr. 3: 476.   1828.
Rhipsalis cassytha mauritiana† De Candolle, Prodr. 3: 476.   1828.
Rhipsalis cassytha mociniana De Candolle, Prodr. 3: 476.   1828.
Rhipsalis cassytha hookeriana De Candolle, Prodr. 3: 476.   1828.
Rhipsalis cassytha swartziana De Candolle, Mém. Mus. Hist. Nat. Paris 17: 80.   1828.
Rhipsalis dichotoma G. Don, Hist. Dichl. Pl. 3: 176.   1834.
Rhipsalis hookeriana G. Don, Hist. Dichl. Pl. 3: 176.   1834.
Rhipsalis cassythoides G. Don, Hist. Dichl. Pl. 3: 176.   1834.
Rhipsalis cassutha pendula Salm-Dyck in Pfeiffer, Enum. Cact. 134.   1837.
Rhipsalis undulata Pfeiffer, Enum. Cact. 136.   1837.
Hariota cassytha Lemaire, Cact, Gen. Nov. Sp. 75.   1839.
Cereus parasiticus Haworth in Steudel, Nom. ed. 2. 1: 335.   1840.
Rhipsalis aethiopica Welwitsch, Journ. Linn. Soc. Bot. 3: 152.   1859
Rhipsalis minutiflora Schumann in Martius, Fl. Bras. 4²: 271.   1890.
Hariota parasitica Kuntze, Rev. Gen. Pl. 1: 262.   1891.
Rhipsalis comorensis Weber, Rev. Hort. 64: 424.   1892.
Rhipsalis zanzibarica‡ Weber, Rev. Hort. 64: 425.   1892.

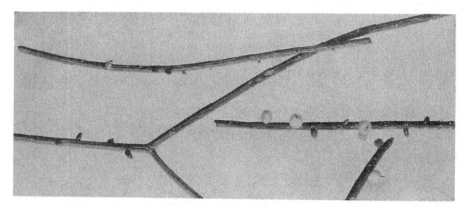

Fig. 222.—Rhipsalis cassutha.

Epiphytic or saxicolous, usually growing on trunk or branches of large trees, hanging in large clusters, 1 to 9 meters long, the branches weak and pendent; branches when young bearing 5 to 9 white bristles at the areoles, when old naked, terete, sometimes producing aërial roots, often only 3 mm. in diameter, light green, usually growing from tips of other branches, generally in pairs but sometimes in clusters of 6 or 8; flowers lateral, solitary, small, greenish in bud, sometimes subtended by a single bristle; petals 2 mm. long, cream-colored; stamens borne on disk; ovary exserted; fruit naked, white or pink, maturing a few days after flowering, globose, 5 mm. in diameter.

Type locality: Not cited.

Distribution: Florida, Mexico, Central America, West Indies, Panama to Dutch Guiana, eastern and southern Brazil, Colombia, Ecuador, Bolivia, and Peru, also in Ceylon and tropical Africa.

The fruit of *Rhipsalis cassutha*, while usually white, is sometimes described as red or pinkish.   Hooker, in his Exotic Flora, figured and described the fruit as flesh-colored. Weber, who received a red-fruited form from Costa Rica, has named it variety *rhodocarpa* (Dict. Hort. Bois 1046. 1898).   In the West Indies the plants inhabit moist districts and are most abundant in forests, but in the vicinity of Matanzas, Cuba, occur on cliffs.

---

* This name was written *Cactus garipensis* by Kunth (Syn. Pl. Aeq. 3: 370. 1824) and is so listed in the Index Kewensis.
† De Candolle gives *Cactus pendulinus* Sieber (Fl. Maur. 2. n. 259) as a synonym of this variety.
‡ Schumann (Gesamtb. Kakteen 623) spells the name *Rhipsalis sansibarica*.

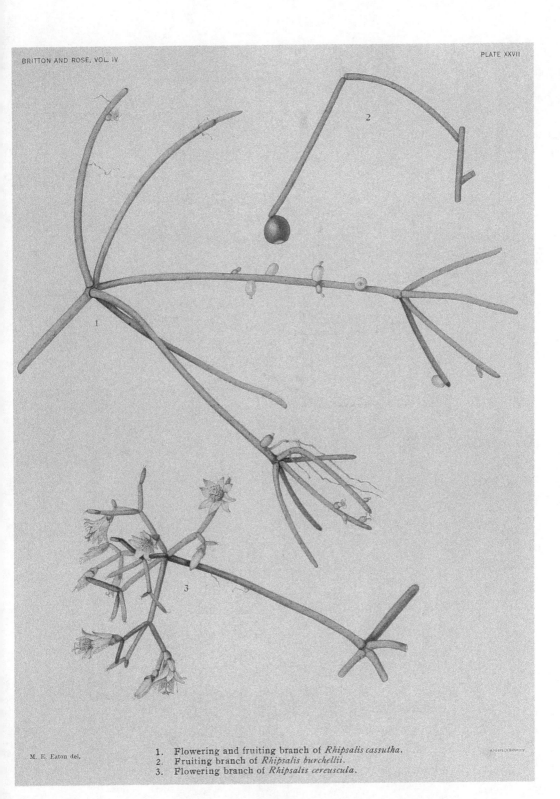

1. Flowering and fruiting branch of *Rhipsalis cassutha*.
2. Fruiting branch of *Rhipsalis burchellii*.
3. Flowering branch of *Rhipsalis cereuscula*.

Hitherto unknown wild within the continental United States, the plant was found on August 5, 1923, by C. A. Mosier on trees in Wallenstein's Hammock, Dade County, Florida.

Cactus cassythoides Mociño and Sessé was given by De Candolle (Prodr. **3**: 476. 1828) as a synonym of *R. cassytha mociniana.*

Löfgren (Arch. Jard. Bot. Rio de Janeiro **2**: 40. pl. 11. 1918) has figured and described as new a plant under the name of *Rhipsalis cassythoides* which may belong here. The name had already been used by Don and we have referred it as a synonym of *R. cassutha.*

Cactus epidendrum Linnaeus (Amoen. Acad. **8**: 257. 1785) is without description and has been referred to *Rhipsalis undulata.* It was from Surinam.

Cereus bacciferus (Hemsley, Biol. Centr. Amer. Bot. **1**: 548. 1880) appears only as a synonym of *Rhipsalis cassutha.*

Cassytha baccifera Miller (De Candolle, Prodr. **3**: 476. 1828) and *C. polysperma* Aiton (Gaertner, Fruct. Sem. Pl. **1**: 137. 1788) are known in synonymy only.

Rhipsalis pendula Hortus (Pfeiffer, Enum. Cact. 133. 1837) occurs only as a synonym. *Rhipsalis caripensis* Weber is listed as one of the synonyms of this species by Schumann (Gesamtb. Kakteen 622. 1898).

Rhipsalis cassytha vars. *major* and *pilosiuscula* (Salm-Dyck, Hort. Dyck. 228. 1834) and var. *tenuior* (Schumann, Monatsschr. Kakteenk. **1**: 78. 1891) are only names. The first has been referred to *R. floccosa,* while the second is sometimes referred to *R. pulvinigera.*

Illustrations: De Tussac, Fl. Antill. **3**: pl. 22, as *Cactus pendulus*; Plunkenet, Phyt. pl. 172, f. 2. 1692, as *Cuscuta baccifera,* etc.; De Candolle, Mém. Mus. Hist. Nat. Paris **17**: pl. 21; Förster, Handb. Cact. ed. 2. 888. f. 121. as *Rhipsalis cassytha mociniana*; Ann. Inst. Roy. Hort. Fromont **2**: pl. 3, as *R. parasitica*; Arch. Jard. Bot. Rio de Janeiro **2**: pl. 11, as *R. cassythoides*; Gartenwelt **13**: 117, as *R. minutiflora*; Möllers Deutsche Gärt. Zeit. **25**: 477. f. 11, No. 18, as *R. sansibarica*; (Hortus malabaricus pl. 7, *fide* Miller); Gaertner, Fruct. Sem. Pl. **1**: pl. 28, f. 1; Torreya **9**: 154. f. 1; Ann. Rep. Smiths. Inst. **1908**: 537. f. 1; Journ. N. Y. Bot. Gard. **11**: f. 23; Loudon, Encycl. Pl. 413. f. 6907; Loddiges, Bot. Cab. **9**: pl. 865; Goebel, Pflanz. Schild. **1**: pl. 4, f. 2; Karsten, Deutsche Fl. 887. f. 501, No. 5; ed. 2, **2**: 456. f. 605, No. 5; Stand. Cycl. Hort. Bailey **2**: f. 712; Nov. Act. Nat. Cur. **19**: pl. 16. f. 13; Hooker, Exot. Fl. **1**: pl. 2; Curtis's Bot. Mag. **58**: pl. 3080; Gartenwelt **16**: 633.

Plate XXVII, figure 1, shows a plant received from the Hope Botanical Garden in Jamaica. Figure 222 is from a photograph showing branches of a plant sent us from R. Lamb's collection at Manchester, England.

12. Rhipsalis virgata Weber, Rev. Hort. **64**: 425. 1892.

Main stem or branches 1 meter long or more, terete, about 5 mm. thick, erect or ascending but in time often pendent, often bearing aërial roots; upper branches short, 1 to 6 cm. long, terete; areoles small, a little hairy, often with a white or pinkish bristle, subtended by a minute bract; flowers borne along sides of the 2 and 3-year old branches, solitary at areoles, rotate, 8 to 10 mm. broad, open throughout day; outer perianth-segments few, ovate, greenish yellow, sometimes tinged with red; inner perianth-segments 4 to 6, oblong, cream-colored, obtuse; filaments erect, white; style white, about as long as stamens; stigma-lobes 3, white; ovary broader than high, crowned by a circle of scales and bearing one on the side.

Type locality: Described from a garden plant supposed to have come from Brazil.
Distribution: Eastern Brazil.
Illustration: Möllers Deutsche Gärt. Zeit. **25**: 477. f. 11, No. 12.

Plate XXV, figure 4, shows a plant, received from M. Simon of St. Ouen, Paris, in 1901, which flowered and fruited in the New York Botanical Garden in 1916.

13. Rhipsalis teres (Vellozo) Steudel, Nom. ed. 2. **2**: 449. 1841.

Cactus teres Vellozo, Fl. Flum. 207. 1825.
Rhipsalis conferta Salm-Dyck, Cact. Hort. Dyck. 1849. 229. 1850.
Hariota conferta Kuntze, Rev. Gen. Pl. **1**: 262. 1891.
Hariota teres Kuntze, Rev. Gen. Pl. **1**: 263. 1891.

Stems erect or spreading, woody at base, 10 to 12 mm. in diameter, much branched, especially above, with 5 to 12 short ultimate branches at top of main ones; old branches terete, green or blotched with red; flowers usually several at top of short terminal branches and scattered all along the primary ones, 10 to 12 mm. broad, pale yellow; petals widely spreading; filaments and style white, erect.

Type locality: Brazil.

Distribution: States of Minas Geraes, Rio de Janeiro, and São Paulo, Brazil.

Rhipsalis floribunda Schott was given by Schumann (Martius, Fl. Bras. 4²: 274. 1890) as a synonym of this species.

Illustrations: Vellozo, Fl. Flum. **5:** pl. 30, as *Cactus teres*; Möllers Deutsche Gärt. Zeit. **25:** 477. f. 11, No. 22; Garten-Zeitung **4:** 182. f. 42, No. 7, as *Rhipsalis conferta.*

Plate XXVI, figure 5, shows a plant received from Kew in 1902 which flowered in the New York Botanical Garden in 1917.

Fig. 223.—Rhipsalis shaferi.

14. Rhipsalis lindbergiana Schumann in Martius, Fl. Bras. 4²: 271. 1890.

> *Rhipsalis erythrocarpa* Schumann in Engler, Pflanzenw. Ost. Afrikas 282. 1895.
> *Hariota lindbergiana* Kuntze, Rev. Gen. Pl. 3²: 107. 1898.
> *Rhipsalis densiareolata* Löfgren, Arch. Jard. Bot. Rio de Janeiro 2: 41. 1918.

Very much branched, hanging from tree-trunks in great festoons, 1 to 2 meters long; joints elongated, 3 to 5 mm. in diameter; areoles filled with hairs and 2 bristles; flowers numerous, lateral, pinkish; ovary naked or nearly so; fruit light red, globose, 2 to 3 mm. in diameter, 16 to 20-seeded.

Type locality: Near the city of Rio de Janeiro.

Distribution: Mountainous regions in the state of Rio de Janeiro, Brazil, and Mount Kilman-Djaro, Africa.

Rhipsalis erythrocarpa Schumann was described from herbarium specimens, collected on Mount Kilman-Djaro, in tropical Africa. We refer it to *R. lindbergina* Schumann in deference to the opinion of Mr. Roland-Gosselin* but we have not had specimens for study.

Illustrations: Arch. Jard. Bot. Rio de Janeiro **2:** pl. 12, as *Rhipsalis densiareolata.* Rev. Hort. **85:** f. 152, in part, as *R. erythrolepis*; Martius, Fl. Bras. 4²: pl. 53; Arch. Jard; Bot. Rio de Janeiro **1:** pl. 4.

---

* Bull. Soc. Bot. France **59:** 100.

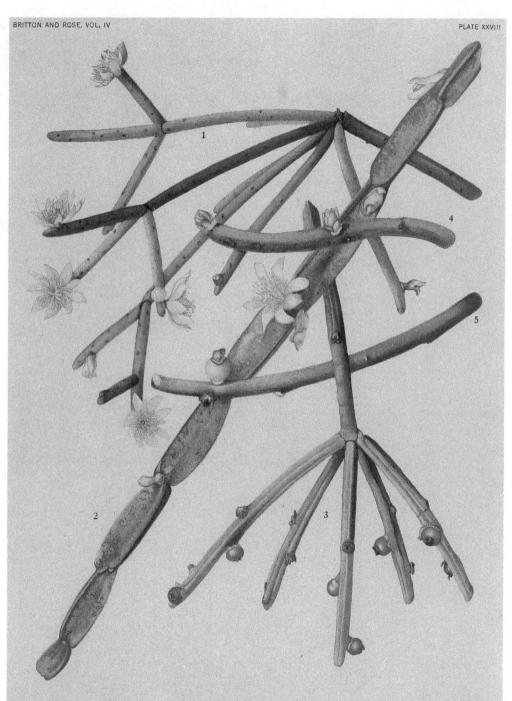

M. E. Eaton del.

1. Flowering branch of *Rhipsalis neves-armondii*.
2. Flowering branch of *Rhipsalis paradoxa*.
3. Fruiting branch of *Rhipsalis pulvinigera*.
4. Flowering branch of *Rhipsalis tucumanensis*.
5. Fruiting branch of same.

Plate xxi, figure 4, shows a branch from the plant collected by Dr. Rose on Tijuca in 1915 (No. 21174); figure 2 shows another plant collected by him in Rio de Janeiro, Brazil (No. 20309).

15. Rhipsalis shaferi sp. nov.

Stems at first stiff, erect or ascending, afterwards spreading or procumbent, 4 to 5 mm. thick, terete, green or more or less purplish at tips; juvenile and lower branches often bearing several bristles at areoles; upper branches without bristles or with a single appressed one; scales subtending the areoles small but broad; flowers numerous, scattered all along side of branch, solitary (rarely in pairs) at areoles, small, rotate, greenish white, 8 to 10 mm. broad; petals 5 or 6, short-oblong, obtuse; filaments greenish, erect; stigma-lobes 4, white; ovary not sunken in branch; fruit small, globose, 2 to 3 mm. in diameter, white or sometimes tinged with pink.

Collected by John A. Shafer on trees at Asunción, Paraguay, March 18, 1917 (No. 139), on trees at Trinidad, Paraguay, March 17, 1917 (No. 134, type), again in Paraguay (Nos. 145 and 147), and on trees at Posados, Misiones, Argentina (No. 131).

Plate xxiv, figure 7, shows a branch in flower; plate xxi, figure 3, shows a branch in fruit of the type which flowered in the New York Botanical Garden in 1921. Figure 223 is from a photograph of Shafer's No. 131, which flowered and fruited in Washington in 1921.

16. Rhipsalis fasciculata (Willdenow) Haworth, Suppl. Pl. Succ. 83. 1819.

Cactus fasciculatus Willdenow, Enum. Pl. Suppl. 33. 1813.
Rhipsalis horrida Baker, Journ. Linn. Soc. 21: 347. 1884.
Rhipsalis madagascarensis Weber, Ind. Sem. Hort. Paris 1889; Rev. Hort. 64: 424. 1892.
Hariota fasciculata Kuntze, Rev. Gen. Pl. 1: 262. 1891.
Hariota horrida Kuntze, Rev. Gen. Pl. 1: 263. 1891.

Stems woody, terete, much branched; branchlets clavate to cylindric, faintly ribbed when old, 4 mm. in diameter, with numerous areoles, each with a cluster of fragile hairs 3 to 4 mm. long; flowers lateral but not described; ovary not sunken in the branch; fruit globose, small, bearing a few areoles, these pubescent and setose.

Type locality: Not cited.
Distribution: Brazil and Madagascar.

We have studied Madagascan specimens of this plant sent from Kew and one sent from Bahia, Brazil, to Dr. Rose by L. Zehntner in 1920. De Candolle (Plantes Grasses 1: pl. 59) states that it occurred in Santo Domingo. Roland-Gosselin* says that it inhabits American Islands; our very extensive explorations in the West Indies have failed to discover it. The Brazilian plant differs only from the Madagascan by having fewer hairs at the areoles.

Rhipsalis pilosa Weber is listed by Schumann (Martius, Fl. Bras. 4²: 300. 1890) with the statement that it occurs in P. Rebut's Catalogue without description; A. Berger in a letter (dated March 7, 1920) states that this name is said to be a synonym of *R. madagascarensis*. It is illustrated (Möllers Deutsche Gärt. Zeit. 25: 477. f. 11, No. 20). *Rhipsalis madagascarensis dasycerca* Weber is listed by R. Lamb (Collection of Cacti 73. 1908.)

Illustrations: De Candolle, Pl. Succ. 1: pl. 59, as *Cactus parasiticus*; Curtis's Bot. Mag. 58: pl. 3079; Gartenwelt 13: 117, Ann. Inst. Roy. Hort. Fromont 2: pl. 1, f. G, as *R. fasciculata*; Loudon, Encycl. Pl. 413. f. 6908. as *R. parasitica*.

17. Rhipsalis pulchra Löfgren, Arch. Jard. Bot. Rio de Janeiro 1: 75. 1915.

Stems much branched, often pendent; branches often in whorls of 3's or 4's, 3 to 4 mm. in diameter, bright green; areoles minute, reddish; flowers few, usually from near the tips of terminal branches, purplish red, large, 12 to 14 mm. long; petals oblong, obtuse; stigma-lobes white; ovary purplish red.

Type locality: Serra da Mantiqueira, Brazil.
Distribution: State of Rio de Janeiro.

---

* Bull. Soc. Bot. France 59: 99.

Our living specimens came from the Organ Mountains, Rio de Janeiro, Brazil, obtained by J. N. Rose through Ph. Luetzelburg, September 21, 1915 (No. 21157).

Dr. Rose examined the type collected by A. O. Darby in 1915 in the Museu Paulista and obtained a fragment of it through the kindness of the Director.

Rhipsalis pulcherrima Löfgren (Monatsschr. Kakteenk. 9: 136. 1899) seems to have been the name first given to this plant.

Illustration: Arch. Jard. Bot. Rio de Janeiro 1: pl. 5.

Plate xxxi, figure 2, shows a flowering branch of the plant obtained by Dr. Rose in 1915 which flowered in the New York Botanical Garden in 1918 (No. 21151).

18. Rhipsalis lumbricoides Lemaire, Illustr. Hort. 6: Misc. 68.   1859.

> *Cereus lumbricoides* Lemaire, Cact. Gen. Nov. Sp. 60. 1839.
> *Rhipsalis sarmentacea* Otto and Dietrich, Allg. Gartenz. 9: 98. 1841.
> *Lepismium sarmentaceum* Vochting, Jahrb. Wiss. Bot. Leipzig 9: 399. 1873.
> *Hariota lumbricalis* Kuntze, Rev. Gen. Pl. 1: 263. 1891.
> *Hariota sarmentacea* Kuntze, Rev. Gen. Pl. 3²: 107. 1898.

FIG. 224.—Rhipsalis lumbricoides.

Stems terete when growing, but angled when dormant, 3 to 4 meters long, about 6 mm. thick rooting freely, much branched; young growth with 5 to 10 white bristles from each areole, usually spreading, but old branches naked; flowers white to cream-colored, sometimes tinged with green; petals few, often only 5, lanceolate, acute, 10 to 12 mm. long, acuminate; style slender, greenish, longer than the stamens; stigma-lobes 4, spreading, greenish; ovary naked; fruit white.

Type locality: Montevideo, Uruguay.

Distribution: Uruguay and Paraguay, also probably southern Brazil.   Hooker says that it is a native of Buenos Aires, but this is doubtless an error.

This plant flowered in Washington on March 16, 1915.   Schumann's drawing of the flower is not very good.

Rhipsalis sarmentosa (Monatsschr. Kakteenk. 4: 46. 1894) and *R. larmentacea* (Illustr. Hort. 6: 88. 1859) are misspellings for *R. sarmentacea.*

According to Lemaire (Cact. Gen. Nov. Sp. 60. 1839) *Cereus flagelliformis minor* Salm-Dyck (Hort. Dyck. 64. 1834) belongs here.   Grisebach (Symb. Fl. Argen. 139) referred *Cereus donkelaarii* here.

Illustrations: Martius, Fl. Bras. 4²: pl. 59; Curtis's Bot. Mag. 85: pl. 5136; Dict. Gard. Nicholson 4: 598. f. 60; Suppl. 635. f. 646; Engler and Prantl, Pflanzenfam. 3⁶ᵃ: f. 69, D, E; Gard. Chron. III. 2: 465. f. 95; Watson, Cact. Cult. 232. f. 90; ed. 3. f. 66, as *Rhipsalis sarmentacea*; Schumann, Gesamtb. Kakteen 633. f. 98, F; Arch. Jard. Bot. Rio de Janeiro 1: pl. 3; Gartenwelt 13: 117.

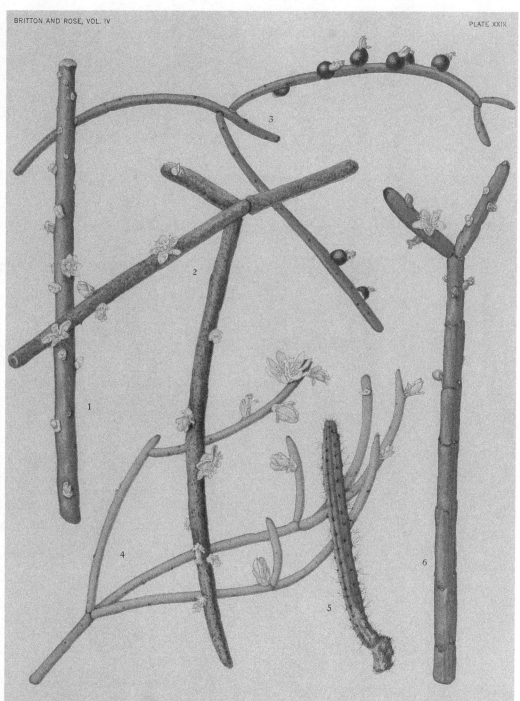

M. E. Eaton del.

A Hoen & Co. Baltimore.

1. Flowering branch of *Rhipsalis floccosa*.
2. Flowering branch of same.
3. Fruiting branch of *Rhipsalis puniceo-discus*.
4. Flowering branch of *Rhipsalis gibberula*.
5. Branch of *Rhipsalis dissimilis*.
6. Flowering and fruiting branch of same.

Figure 224 is from a photograph taken by H. Buch which was given to Dr. Rose when he was in La Plata, Argentina, in 1915.

19. Rhipsalis aculeata Weber, Rev. Hort. **64**: 428. 1892.

Stems terete, 3 to 4 mm. in diameter, somewhat angled and roughened in dried specimens; areoles close together, bearing wool and 8 to 10 appressed white bristles or spines; fruit not immersed, globose, 7 to 8 mm. in diameter, dark purple to nearly black, either naked or with 3 or 4 hairy areoles.

Type locality: Catamarca, Argentina.

Distribution: Northern Argentina, in the provinces of Catamarca and Tucuman.

A round-stemmed species collected by Otto Kuntze on the Sierra de Santa Cruz, Bolivia, and labeled *Hariota sarmentacea* may belong here.

This species is described by Schumann as 8 to 10-ribbed, but no ribs are shown in growing plants; in drying the branches are somewhat angled but one could hardly describe them as ribbed. Dr. Shafer made a single collection of this plant at Tucuman in 1917 (No. 92); part of this material is living in the New York Botanical Garden. Dr. Rose also obtained a specimen through one of his Argentina correspondents from Catamarca.

Plate XXIV, figure 8, is from Dr. Shafer's plant mentioned above.

20. Rhipsalis grandiflora Haworth, Suppl. Pl. Succ. 83. 1819.

> *Cactus funalis* Sprengel, Syst. **2**: 479. 1825.
> *Cactus cylindricus* Vellozo, Fl. Flum. 207. 1825. Not Lamarck, 1783. Not Ortega, 1800.
> *Rhipsalis funalis* Salm-Dyck in De Candolle, Prodr. **3**: 476. 1828.
> *Hariota funalis* Lemaire, Cact. Gen. Nov. Sp. 74. 1839.
> *Rhipsalis cylindrica* Steudel, Nom. ed. 2. **2**: 448. 1841.
> *Hariota cylindrica* Kuntze, Rev. Gen. Pl. **1**: 262. 1891.
> *Hariota grandiflora* Kuntze, Rev. Gen. Pl. **1**: 262. 1891.
> *Rhipsalis robusta* Lindberg, Monatsschr. Kakteenk. **6**: 53. 1896. Not Lemaire, 1860.
> *Rhipsalis hadrosoma* Lindberg, Monatsschr. Kakteenk. **6**: 96. 1896.

Branches divaricate, often reddish, especially about the areoles, stout, 8 to 10 mm. in diameter; flowers numerous, scattered all along branches, 12 mm. long, 2 cm. broad, light rose or cream-colored; sepals reddish; petals few, oblong, obtuse, widely spreading; anthers and style white; stigma-lobes 4, white; fruit naked, purplish, 6 to 7 mm. in diameter.

Type locality: Not cited.

Distribution: State of Rio de Janeiro, Brazil.

We have not seen the type specimen of this species, but through the kindness of Mr. N. E. Brown of Kew we have seen a photograph of Haworth's specimens, which are the same as the species here described. Haworth's plant was received from Brazil in 1816, sent by Messrs. Bowie and Cunningham.

Rhipsalis calamiformis (Pfeiffer, Enum. Cact. 135. 1837) was published as a synonym of *R. funalis*.

Walpers gives *Rhipsalis funalis gracilior* Pfeiffer (Repert. Bot. **2**: 279. 1843) as a synonym.

Illustrations: Gartenwelt **13**: 117; Watson, Cact. Cult. 228. f. 89; ed. 3. f. 65; Amer. Gard. **11**: 465; Dict. Gard. Nicholson **3**: 289. f. 365; Gartenflora **42**: 234. f. 48; Link and Otto, Icon. Pl. Rar. pl. 38, as *Rhipsalis funalis*; Vellozo, Fl. Flum. **5**: pl. 31, as *Cactus cylindricus*; Monatsschr. Kakteenk. **6**: 55, as *R. robusta*; Blühende Kakteen **3**: pl. 141; Monatsschr. Kakteenk. **7**: 151. f. 1 to 8; Arch. Jard. Bot. Rio de Janeiro **1**: pl. 7, as *R. hadrosoma*; Curtis's Bot. Mag. **54**: pl. 2740; Schumann, Gesamtb. Kakteen 633. f. 98, A; Martius, Fl. Bras. **4**²: pl. 54; Monatsschr. Kakteenk. **7**: 151. f. 9 to 11; Arch. Jard. Bot. Rio de Janeiro **1**: pl. 6.

Plate XXXI, figure 3, shows a plant collected by Dr. Rose near Rio de Janeiro in 1915 (No. 20746) which flowered in the New York Botanical Garden in 1918; figure 1 is of a plant which also flowered in the New York Botanical Garden, April 3, 1912; plate XXI, figures 1 and 6, shows the flowers and fruit of specimens sent by Alwin Berger in 1908.

21. Rhipsalis megalantha Löfgren, Monatsschr. Kakteenk. **9**: 134. 1899.

Rhipsalis novaesii Gürke, Monatsschr. Kakteenk. **19**: 12. 1909.

Plants stout, up to 1 cm. thick, at first erect but in time spreading or with pendent branches, dull green, often spotted with purple; areoles rather prominent, especially after flowering; flowers large, 4 cm. broad; petals 8 to 12, oblong, often shortly acuminate or obtuse, white; filaments erect, orange at base, rose-colored above; style thick, longer than the stamens; stigma-lobes 6 to 8; fruit surrounded with white hairs, rather small, 6 mm. in diameter, white or tinged with red; seeds nearly black.

Type locality: Island of São Sebastião, Brazil.

Distribution: Known only from the type locality, an island off the coast of Brazil, belonging to the state of São Paulo.

This plant is known wild only from the collection of Dr. Löfgren, but is now widely found in cultivation, sometimes under the names *Rhipsalis grandiflora* or *R. nevaesii*. It has the largest flower of any species of *Rhipsalis*.

Illustrations: Blühende Kakteen **2**: pl. 116; Monatsschr. Kakteenk. **19**: 13, as *Rhipsalis novaesii*; Monatsschr. Kakteenk. **9**: 137; Schumann, Gesamtb. Kakteen Nachtr. 147. f. 35; Arch. Jard. Bot. Rio de Janeiro **1**: pl. 8.

Plate XXIV, figure 3, shows a fruiting branch obtained by Dr. Rose in Rio de Janeiro, Brazil, in 1915 (No. 20400).

22. Rhipsalis leucorhaphis Schumann, Monatsschr. Kakteenk. **10**: 125. 1900.

Epiphytic, much branched, about 5 dm. long, rooting abundantly along the branches, jointed, 5 to 8 mm. in diameter, terete or showing 4 or 5 ribs in herbarium specimens; bristles 1 to 5, appressed, early deciduous; areoles subtended by an ovate papery bract; flowers white, nodding, large, 1.5 cm. long; petals only slightly spreading; filaments purplish or white with orange-colored base; stigma-lobes 3 or 4, greenish, spreading; ovary not sunken in the branch; fruit globose, bright red, 6 to 8 mm. in diameter; seeds numerous, brown.

Type locality: Estancia Tagatiya, Paraguay.

Distribution: Paraguay and northern Argentina.

We did not know this species until it was brought back by Dr. Shafer in 1917 from Paraguay, where he obtained good specimens; he also found it abundant in northern Argentina. Like many of the other species it grows in various situations, sometimes sprawling over rocks or growing on forest trees. One of his living plants fruited in the New York Botanical Garden and from this we have drawn part of our description.

Plate XXIV, figure 1, shows the plant in flower, and figure 2 shows it in fruit, collected by Dr. Shafer at Trinidad, Paraguay (No. 143).

23. Rhipsalis loefgrenii nom. nov.

Rhipsalis novaesii Löfgren, Arch. Jard. Bot. Rio de Janeiro **1**: 69. 1915. Not Gürke, 1909.

Stems long and slender, rooting freely all along stem, pale green to purple, terete, 3 mm. in diameter; areoles small, subtended by a large scarious bract with appressed hairs in axils when young; flowers very numerous, 12 to 15 mm. long, white, campanulate; filaments purplish at base; fruit purplish, 5 to 8 mm. in diameter.

Type locality: Near Campinas, Brazil.

Distribution: Brazil.

Dr. Rose saw the Löfgren type in the Botanical Garden at Rio de Janeiro and obtained living and herbarium specimens of the plant. Dr. Shafer also obtained living specimens from Löfgren in 1917.

Unfortunately, Löfgren's name was given to another plant by Gürke and for this reason we have renamed it in honor of Dr. Alberto Löfgren (1854-1918), who long studied this genus and published an excellent monograph of it in 1915.

Illustration: Arch. Jard. Bot. Rio de Janeiro **1**: pl. 2, as *Rhipsalis novaesii*.

M. E. Eaton del.

1. Fruiting branch of *Rhipsalis gonocarpa*.
2. Flowering branches of *Rhipsalis warmingiana*.
3. Fruiting branch of *Rhipsalis tonduzii*.
4. Fruiting branch of *Rhipsalis trigona*.
5. Flowering branch of *Rhipsalis pentaptera*.
6. Fruiting branch of same.

Figure 225a shows two branches with a single fruit 1.33 times natural size; figure 225b shows a branch twice natural size; figure 225c shows one of the bracts which subtend the areoles, 4 times natural size, all drawn from plants obtained by Dr. Shafer from Dr. Löfgren in 1917 and since grown in the New York Botanical Garden.

24. Rhipsalis neves-armondii
Schumann in Martius, Fl. Bras. 4²: 284. 1890.

> *? Rhipsalis rigida* Löfgren, Arch. Jard. Bot. Rio de Janeiro 1: 93. 1915.

Stems elongated, much branched, and hanging from trees in large clusters; branches arranged in whorls of 3 to 10, 4 to 5 mm. thick, terete, elongated, deep green; flowers widely spreading, 2 cm. broad, white to cream-colored; petals about 12,

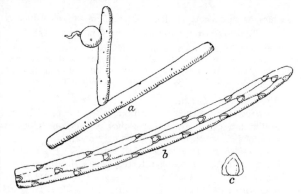

FIG. 225.—Rhipsalis loefgrenii. *a*, fruiting branch; *b*, tip of branch; *c*, bract.

acute; style erect, white; stigma-lobes 5, white; ovary sunken in the branch; fruit globose, red, 10 mm. in diameter; seeds brown.

Type locality: Mount Tijuca, Rio de Janeiro, Brazil.

Distribution: Rio de Janeiro, Brazil.

There has long been much uncertainty regarding this species and Dr. Rose, during his trip to South America, in 1915, endeavored to solve the problem. He first visited one of the three localities mentioned in the original description, namely Tijuca, a mountain near Rio de Janeiro. Here he found two species which belonged to the same group, *Rhipsalis grandiflora* and *R. pulvinigera*. He then visited the herbarium of the Museo Nacional, where he found specimens of *R. neves-armondii*. Unfortunately, they did not bear an original label but one doubtless written after the appearance of the description in the Flora Brasiliensis for the three localities mentioned therein. After studying this material carefully, he visited the mountain region just above Tijuca, namely Pica Popagaya, where he feels certain he has collected the true form, although the joints are more terete and the flowers are pure white instead of yellow; it is a singular *Rhipsalis* and a very shy bloomer. A second visit was then made to Tijuca, but lower down on the mountain, and here he again found this species.

Illustrations: Martius, Fl. Bras. 4²: pl. 56; Arch. Jard. Bot. Rio de Janeiro 1: pl. 19; Blühende Kakteen 2: pl. 80, A.

Plate XXVIII, figure 1, shows a flowering plant collected by Dr. Rose at the type locality in 1915 (No. 20673), which flowered in the New York Botanical Garden in 1916; Plate XXIV, figures 4 and 5, show branches from the same plant, in fruit.

25. Rhipsalis pittieri sp. nov.

Epiphytic, resembling in habit *Rhipsalis cassutha*; branches 5 to 6 mm. in diameter, dull green, terete; petals greenish yellow, 5 to 6 mm. long; ovary sunken in the stem, surrounded by white hairs; fruit maturing very slowly, white; seeds black.

Collected by H. Pittier near Hacienda Koster, Borburata, near Puerto Cabello, Venezuela, in 1913 (No. 6467), and flowered first in Washington in the fall of 1914 (October 16), the fruit maturing March 16, 1915. The plant has repeatedly flowered since. This species is perhaps nearest *Rhipsalis floccosa*, from Brazil, and is the most northern representative of the Series *Floccosae*.

Plate XXIV, figure 6, is of a fruiting specimen of the type plant.

26. Rhipsalis pulvinigera Lindberg, Gartenflora 38: 186. 1889.

Rhipsalis funalis minor Pfeiffer, Enum. Cact. 135. 1837.

Plant epiphytic, rather stout, at first erect but in time hanging, and then sometimes 3 to 5 meters long, the branches dull green with purple about the areoles, 5 to 7 mm. in diameter; terminal branches often in whorls of 3 to 5; flowers at first white, in age yellowish, 2 cm. broad; ovary sunken in the branch; fruit globose, red, 8 mm. in diameter.

Type locality: Brazil.

Distribution: In the coastal mountains of central Brazil.

Schumann gives *Rhipsalis grandiflora minor* (Gesamtb. Kakteen 644. 1898) as a synonym of this species, but he evidently meant *R. funalis minor*.

Rhipsalis cassytha pilosiuscula Salm-Dyck (Hort. Dyck. 228. 1834), although never described, probably is to be referred here.

Illustrations: Gartenflora 42: f. 48, as *Rhipsalis funalis*; Gartenflora 38: f. 33, 34; Rümpler, Sukkulenten 210. f. 119; 211. f. 120; Rev. Hort. 85: f. 152, in part.

Plate XXVIII, figure 3, is from a plant collected by Dr. Rose near Rio de Janeiro in 1915, which fruited in the New York Botanical Garden in 1915 (No. 43060).

27. Rhipsalis floccosa Salm-Dyck in Pfeiffer, Enum. Cact. 134. 1837.

Hariota floccosa Lemaire, Cact. Gen. Nov. Sp. 75. 1839.
Rhipsalis rugulosa Lemaire, Illustr. Hort. 8: after pl. 293. 1861.
Hariota rugosa Kuntze, Rev. Gen. Pl. 1: 263. 1891.

Stems slender, 5 to 8 mm. in diameter, much branched, at first erect, becoming pendent; branches alternate; flowers lateral, 2 cm. broad, white, tinged with yellow, surrounded by a tuft of wool; ovary sunken in the branch; fruit globose, 5 mm. in diameter, rose-colored or nearly white.

Type locality: Not cited.

Distribution: Brazil.

Rhipsalis cassytha major Salm-Dyck (Pfeiffer, Enum. Cact. 134. 1837), a synonym only, is referred here by Pfeiffer.

Hariota floccosa Cels was used as a synonym by Förster (Handb. Cact. 458. 1846), but was not technically published until 1891.

Illustration: Gartenflora 38: 185. f. 35.

Plate XXIX, figure 1, shows a flowering branch from a specimen sent by Mr. Lamb from Manchester, England, in 1914, and figure 2 shows a flowering branch collected by Dr. Rose in Brazil in 1915 which flowered in the New York Botanical Garden on February 24, 1922.

28. Rhipsalis tucumanensis Weber, Rev. Hort. 64: 426. 1892.

Hariota tucumanensis Kuntze, Rev. Gen. Pl. 3²: 107. 1898.

Epiphytic on forest trees, when young setose, but soon naked, much branched; branches often pendent, sometimes in whorls of 4, 4 to 10 mm. in diameter, when young nearly terete, bright green with a red spot at the areoles, when old angled, yellowish green; flowers one from an areole, 15 to 18 mm. in diameter, rosy white to cream-colored; sepals 4, white but rose-colored on the back; petals 8, ovate-lanceolate; stamens numerous, white, spreading, much shorter than petals; style white; stigma-lobes 4 or 5; ovary sunken in the branch, surrounded by a tuft of wool; fruit described as white tinged with red, but often red or pinkish, 8 to 10 mm. broad.

Type locality: Tucuman, Argentina.

Distribution: Tucuman to Catamarca, Argentina, and perhaps Bolivia and Paraguay.

PLATE XXXI

M. E. Eaton del.

1. Flowering branch of *Rhipsalis grandiflora*.
2. Flowering branch of *Rhipsalis pulchra*.
3. Flowering branch of *Rhipsalis grandiflora*.

This species is common in northern Argentina, where it was repeatedly collected by Dr. Shafer in 1917.

Of this relationship, but perhaps specifically distinct, is the plant sent by M. Bang (No. 2323) from Coripati, Yungas, Bolivia, distributed as *Rhipsalis salicornioides*. Here we have tentatively referred K. Fiebrig's plant (No. 5801) from the Upper Paraná, Paraguay.

Of plate XXVIII, figures 4 and 5 show flowering and fruiting branches from Dr. Shafer's collection from Calilegua, Argentina (Nos. 55 and 68), painted at the New York Botanical Garden, May 24, 1922.

FIG. 226.—Rhipsalis sulcata. Reduced.   FIG. 227.—Rhipsalis gibberula. ✕0.5.

29. Rhipsalis gibberula Weber, Rev. Hort. **64**: 426. 1892.

Stems 3 to 6 mm. thick, yellowish green, with dichotomous or trichotomous branches or sometimes with terminal whorls of 4 or 6; areole small; buds obtuse, pinkish, hairy when in flower; flowers scattered along branches toward tip, white to pale pink, 8 to 9 mm. long, 12 to 15 mm. broad; petals not widely spreading (at least in our specimen); stigma-lobes 3 to 6, white; fruit white, somewhat depressed, 8 to 10 mm. in diameter, 7 to 8 mm. high, the base sunken in the branch.

Type locality: Brazil.

Distribution: Organ Mountains, Brazil.

The species was described from plants brought to Paris from Brazil in 1887, their habitat not recorded, but Dr. Rose traced it to the Organ Mountains in 1915 and his plant flowered in the New York Botanical Garden in February 1921 (No. 21161). In 1902 a specimen was sent from Paris to the New York Botanical Garden and one specimen was obtained from R. Lamb, Superintendent of Parks at Manchester, England, in 1914, but neither has done well in cultivation.

Plate XXIX, figure 4, is from a plant collected by Dr. Rose in the Organ Mountains in 1915, which flowered in the New York Botanical Garden, February 17, 1921. Figure 227 shows the plant received from Paris in 1902 which flowered in the New York Botanical Garden on March 6, 1917.

30. Rhipsalis puniceo-discus G. A. Lindberg, Gartenflora **42**: 233. 1890.

Rhipsalis foveolata Weber, Dict. Hort. Bois 1047. 1898. According to Roland-Gosselin.
Rhipsalis chrysocarpa Löfgren, Arch. Jard. Bot. Rio de Janeiro 1: 94. 1915.
? Rhipsalis chrysantha Löfgren, Arch. Jard. Bot. Rio de Janeiro 1: 99. 1915.

Branches slender, almost filiform, hanging, pale green when young, freely rooting; branches in terminal whorls, often as many as 6; flowers large, 1.5 cm. long, white; perianth-segments widely spreading; stamens orange-colored, at least at base; fruit at first dark red but in age golden yellow.

Type locality: Not cited.

Distribution: Brazil.

This plant first passed in living collections as *R. funalis gracilis* (Gartenflora **42**: 233. 1893.)

Dr. Löfgren gave Dr. Rose a cutting of the original plant of *Rhipsalis chrysantha*.
Illustrations: Arch. Jard. Bot. Rio de Janeiro **1**: pl. 20, as *Rhipsalis chrysocarpa*; Rev.
Hort. **79**: 106. f. 33, as *R. foveolata*; Gartenflora **42**: 235. f. 49; Arch. Jard. Bot. Rio de
Janeiro **1**: pl. 21.

Plate xxix, figure 3, shows a plant also brought by Dr. Rose from Brazil (No. 20662)
which flowered and fruited in the New York Botanical Garden, March 7, 1921.

31. Rhipsalis dissimilis (G. A. Lindberg) Schumann in Martius, Fl. Bras. **4²**: 286.   1890.

> *Lepismium dissimile* G. A. Lindberg, Gartenflora **39**: 148.   1890.
> *Rhipsalis dissimilis setulosa* Weber, Rev. Hort. **64**: 428.   1892.
> *Rhipsalis pacheco-leonii* Löfgren, Arch. Jard. Bot. Rio de Janeiro **2**: 38.   1918.

In clumps on large limbs of trees and freely rooting; branches very diverse, some with numerous
bristly hairs from the areoles, others naked, erect, prostrate or even hanging; hairy branches with 9
very low ribs, the areoles close together, each with about 15 long white bristles; glabrous branches,
5-angled, with the areoles alternating as in *Rhipsalis paradoxa*; flower-buds red; flowers solitary,
about 6 mm. broad; petals few, oblong, obtuse, widely spreading, sometimes turned back, pinkish;
stamens erect, numerous, white; ovary sunken in the branch; style pinkish, erect; stigma-lobes 3 or
4, white.

Type locality: São Paulo, Brazil.

Distribution: States of São Paulo and Rio de Janeiro, Brazil.

We have referred *Rhipsalis pacheco-leonii* here after studying living specimens of
R. dissimilis and specimens from the type collection obtained by Dr. Rose in 1915 (No.
20707).

Rhipsalis setulosa Weber (Hort. Bois Paris) was published as a synonym of *R. dissimilis*
var. *setulosa*.

Illustrations: Gartenflora **39**: 148. f. 36, 37, as *Lepismium dissimile*; Arch. Jard. Bot.
Rio de Janeiro **2**: pl. 10, as *Rhipsalis pacheco-leonii*; Curtis's Bot. Mag. **131**: pl. 8013, as
R. dissimilis setulosa; Blühende Kakteen **2**: pl. 80, B; Gartenflora **40**: f. 121.

Plate xxix, figures 5 and 6, shows the two diverse forms which this plant takes, as does
also plate xxxii, figures 6 and 7.   The specimens were collected by Dr. Rose in the state of
Rio de Janeiro in 1915 and are a part of the type material of *R. pacheco-leonii*.

32. Rhipsalis pentaptera Pfeiffer in Dietrich, Allg. Gartenz. **4**: 105.   1836.

> *Hariota pentaptera* Lemaire, Cact. Gen. Nov. Sp. 75.   1839.

Branches stiff, bright green, 6 to 15 mm. in diameter, strongly 5 or 6-ribbed, the ribs indented at
areoles; areoles often 2 cm. apart, small, subtended by broad bracts, usually bearing 2 white bristles;
flowers usually scattered along whole length of branches, opening in daytime, 1 to 4 from an areole;
scales 4 or 5 at base of corolla, broad and obtuse; petals 5, reddish on back, cream-colored on face,
4 mm. long, obtuse; stamens numerous, about 25, free from petals, white, about as long as style;
style and stigma-lobes white; ovary truncate, naked; fruit 3 to 4 mm. in diameter, white, naked, or
with an occasional small scale.

Type locality: Not cited.   Otto says, in a note, probably Brazil.

Distribution: Southern Brazil and Uruguay.

A very common species in cultivation, flowering freely in March and April.

Hariota pentaptera Lemaire and *Rhipsalis pentagona* are given as synonyms of this
species by Förster (Handb. Cact. 453. 1846).

Illustrations: Pfeiffer and Otto, Abbild. Beschr. Cact. **1**: pl. 17, f. 1; Goebel, Pflanz.
Schild. **1**: pl. 4, f. 4; Gartenwelt **13**: 117; Möllers Deutsche Gärt. Zeit. **25**: 477. f. 11, No.
21; Rev. Hort. **85**: f. 152, in part.

Plate xxx, figures 5 and 6, shows a plant which flowered and fruited in the New
York Botanical Garden in 1912 and 1915, obtained from Paris, France, in 1902.

33. Rhipsalis sulcata Weber, Dict. Hort. Bois 1046.   1898.

Stems woody, sometimes 10 to 15 mm. in diameter, often long and pendent; branches elongated,
the joints 2 to 3 dm. long, 5-angled, light green; areoles remote (2.5 to 5 cm. apart), usually near the

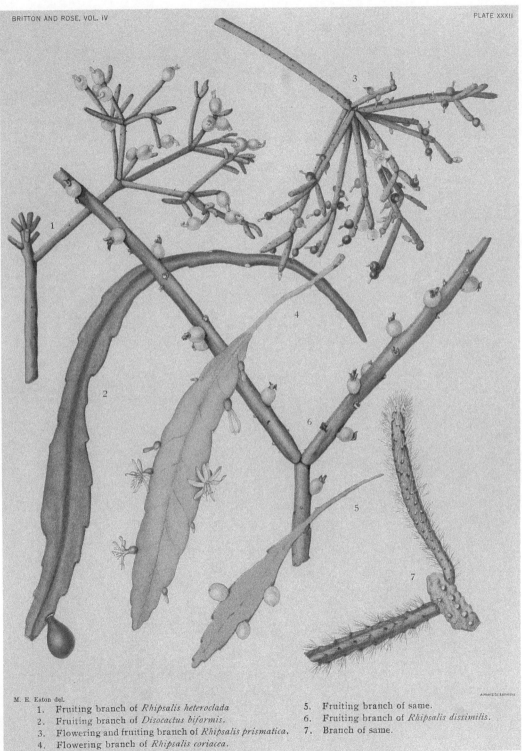

M. E. Eaton del.

1. Fruiting branch of *Rhipsalis heteroclada*
2. Fruiting branch of *Disocactus biformis*.
3. Flowering and fruiting branch of *Rhipsalis prismatica*.
4. Flowering branch of *Rhipsalis coriacea*.

5. Fruiting branch of same.
6. Fruiting branch of *Rhipsalis dissimilis*.
7. Branch of same.

center of a purple blotch; flowers solitary at the areoles, rather large, rotate, white to pinkish; ovary naked.

Type locality: Not cited.

Distribution: Not known in the wild state.

Weber found this plant in cultivation under the name of *Rhipsalis micrantha*, but it is very different from the true *R. micrantha* which comes from Ecuador.

Figure 226 shows a plant received from Paris in 1902 which flowered in the New York Botanical Garden on March 21, 1912.

34. Rhipsalis trigona Pfeiffer, Enum. Cact. 133. 1837.

> *Hariota trigona* Kuntze, Rev. Gen. Pl. 1: 263. 1891.

Stems stout, very much branched, 1.5 cm. in diameter, strongly 3-angled, the angles or ribs alternating with those of adjoining joints; flowers solitary, white to pinkish, widely spreading, sometimes 2 cm. broad; sepals usually 3, short, obtuse; petals generally 7, oblong, obtuse; filaments numerous, white; style white; stigma-lobes 4, white; ovary sunken in the branch; fruit globose, 8 to 10 cm. in diameter, red.

Type locality: Brazil.

Distribution: Brazil.

Wildeman states that the species is probably from the state of Rio de Janeiro.

Illustrations: Wildeman, Icon. Select. **5**: pl. 193; Arch. Jard. Bot. Rio de Janeiro **1**: pl. 23; Gartenflora **40**: 38. f. 15, 16; Gartenwelt **13**: 117.

Plate xxx, figure 4, shows a plant sent to Dr. Rose by R. Lamb of Manchester, England, in 1914, which flowered and fruited in the New York Botanical Garden in 1919.

35. Rhipsalis paradoxa Salm-Dyck, Cact. Hort. Dyck. 1844. 39. 1845.

> *Lepismium paradoxum* Salm-Dyck in Pfeiffer, Enum. Cact. 140. 1837.
> *Hariota alternata* Lemaire, Hort. Univ. **2**: 39. 1841.
> *Rhipsalis alternata* Lemaire, Cactées 80. 1868.
> *Hariota paradoxa* Kuntze, Rev. Gen. Pl. 1: 263. 1891.

Plants freely giving off aërial roots, branched, hanging in large clusters 1 meter long or more; branches in zigzag links, terminal, in pairs or in whorls of 3 to 8, more or less spreading, 3-winged, pale; flowering areoles very woolly, setose when young, borne at upper ends of ribs; flowers subterminal, large, 2 cm. long, white; ovary sunken in stem; fruit not seen.

Type locality: Brazil.

Distribution: Brazil, especially near the city of São Paulo, Brazil.

The young growth is glossy green, the areoles subtended by broad round bracts. Seedling plants are very different from the adult plant; they are strongly 4-angled, with each angle bearing closely-set areoles, filled with slender bristles and showing no resemblance to the typical form; gradually as the plants grow older their mature joints take on the normal form. This plant is a prolific bloomer and in the garden of the Museo Paulista it remains in flower for three weeks.

Pfeiffer (Enum. Cact. 140. 1837) gives *Cereus pterocaulis* Hortus as a synonym of *Lepismium paradoxum* while Förster (Handb. Cact. 453. 1846) gives *Rhipsalis pterocaulis* as a synonym of *R. paradoxa.*

Lepismium alternatum Hortus (Loudon, Hort. Brit. Suppl. **3**: 576. 1850) appeared as a questionable synonym of *Lepismium paradoxum.*

Illustrations: Herb. Génér. Amat. II. **2**: pl. 38; Hort. Univ. **2**: pl. 50, as *Hariota alternata*; Engler and Prantl, Pflanzenfam. **3**[6a]: f. 69, A, B; Schumann, Gesamtb. Kakteen 633. f. 98, B; Karsten, Deutsche Fl. 887. f. 501, No. 4; ed. 2. **2**: 456. f. 605, No. 4; Martius, Fl. Bras. **4**[2]: pl. 55, f. 1; Arch. Jard. Bot. Rio de Janeiro **1**: pl. 22; Goebel, Pflanz. Schild. **1**: pl. 1, f. 5; Rev. Hort. **85**: f. 152, in part; Karsten and Schenck, Vegetationsbilder **1**: pl. 6, f. c.

Plate xxviii, figure 2, is from a plant received from La Mortola in 1908 which flowered in the New York Botanical Garden in 1916.

36. Rhipsalis houlletiana Lemaire, Illustr. Hort. **5**: Misc. 64. 1858.

Rhipsalis houlletii Lemaire in Curtis's Bot. Mag. **100**: pl. 6089. 1874.
Rhipsalis regnellii Lindberg, Gartenflora **39**: 119. 1889.
Hariota houlletiana Kuntze, Rev. Gen. Pl. 1: 263. 1891.

Stems 1 to 2 meters long, slender, terete below but flat and broad above; branches flat and thin, 1 to 5 cm. broad, tapering into a petiole-like base; margin serrate; flowers numerous, bell-shaped with a red eye; petals cream-colored, turning pale yellow, lanceolate, acute; stamens numerous; ovary not sunken in the branch, strongly 4 to 5-angled; fruit not angled, globose, red, 5 to 6 mm. in diameter.

Type locality: Not cited.
Distribution: Brazil, in the states of Minas Geraes, Rio de Janeiro, and São Paulo. This species grows on trees in the mountains at an altitude of 1,000 meters.

Rhipsalis regnelliana appears in the general index for the Monatsschrift für Kakteenkunde (volumes 1–20) in place of *R. regnellii*.

Illustrations: Blühende Kakteen 1: pl. 56; Engler and Prantl, Pflanzenfam. **3**[6a]: f. 69, C; Gartenflora **39**: f. 29, 31 to 33; Schumann, Gesamtb. Kakteen f. 98, E; Martius, Fl. Bras. **4**[2]: pl. 58; Möllers Deutsche Gärt. Zeit. **25**: 477. f. 11, No. 14, as *Rhipsalis regnellii*; Curtis's Bot. Mag. **100**: pl. 6089; Gartenflora **39**: f. 30; Rümpler, Sukkulenten 212. f. 121, as *R. houlletii*; Rev. Hort. Belge **40**: after 186, as *R. kegnelli* (in error for *R. regnellii*); Blühende Kakteen 2: pl. 111; Arch. Jard. Bot. Rio de Janeiro 1: pl. 17.

Plate XXXIII, figure 1, shows a flowering plant collected by Dr. Rose in Rio de Janeiro, Brazil, in 1915 (No. 20307), which flowered in the New York Botanical Garden in 1918; figure 2 shows a plant obtained from M. Simon of St. Ouen, Paris, in 1901, as *Rhipsalis regnellii*, which flowered in the New York Botanical Garden December 16, 1916; figure 3 shows a dissected flower and figure 4 a fruiting branch; plate XXXIV, figure 1, shows a plant with flowers obtained in Paris in 1901; figure 2 shows a flower cut through the center.

37. Rhipsalis warmingiana Schumann in Martius, Fl. Bras. **4**[2]: 291. 1890.

At first erect, then spreading or hanging; branches elongated, jointed, 10 mm. wide or less, either flat or sharply 3 or 4-angled, more or less blotched or colored throughout with purple or red; flowers one at an areole, 20 mm. long, white, directed forward, the perianth-segments spreading, acute; stamens 25 to 30, white; ovary strongly angled; fruit globose, 5 to 6 mm. in diameter, dark purple to nearly black, capped by the withered flower.

Type locality: Near Lagoa Santa, Minas Geraes; two localities were cited when first described, this being the first.
Distribution: State of Minas Geraes, Brazil.

The plant has long been in cultivation, where it does well and blooms freely. Dr. Rose brought back a fresh supply from Brazil in 1915. According to Robert Lamb, the flowers have a perfume resembling that of a hyacinth.

Illustrations: Monatsschr. Kakteenk. **9**: 151; Arch. Jard. Bot. Rio de Janeiro 1: pl. 18; Gartenflora **41**: f. 5, 6, 7.

Plate XXX, figure 2, shows a plant from M. Simon which flowered in the New York Botanical Garden in 1912; plate XXXIV, figures 3 and 4, shows two fruiting branches received from the Berlin Botanical Garden in 1902.

38. Rhipsalis gonocarpa Weber, Rev. Hort. **64**: 427. 1892.

Very much branched; joints narrowly lanceolate to linear, crenate, 3-angled or flattened, becoming purplish; flowers lateral, white, 15 mm. long; petals 7 or 8, lanceolate; stamens 20 to 30, white; ovary strongly 3-angled; stigma-lobes 3 or 4; fruit terete, dark purple to black, globular to short-oblong, 10 to 12 mm. long.

Type locality: São Paulo, Brazil.
Distribution: State of São Paulo, Brazil.

M. E. Eaton del.

1. Flowering branch of *Rhipsalis houlletiana*.
2. Flowering branch of same.
3. Flower of same.
4. Fruiting branch of same.

Schumann (Gesamtb. Kakteen 641. 1898) refers here as a synonym *Rhipsalis ptero-carpa* Weber, a name which he had previously listed in the Flora Brasiliensis (4^2: 300. 1890).

Plate xxx, figure 1, is from a plant sent to Dr. Rose in 1914 by R. Lamb of Manchester, England, which flowered in the New York Botanical Garden in 1920.

39. Rhipsalis linearis Schumann in Martius, Fl. Bras. 4^2: 296. 1890.

Stems at first erect but afterwards spreading or prostrate, 6 to 8 dm. long, much branched; branches vary narrow, serrate, narrowed at base and woody; flowers white, 16 to 18 mm. long; fruit white, 5 mm. in diameter.

Type locality: Southern Brazil, but no definite locality cited. Localities in Paraguay and Argentina also cited in the original place of publication.

Distribution: Southern Brazil, Uruguay, Paraguay, and northern Argentina.

We know this species only from description.

40. Rhipsalis micrantha (HBK.) De Candolle, Prodr. 3: 476. 1828.

    Cactus micranthus Humboldt, Bonpland, and Kunth, Nov. Gen. et Sp. 6: 65. 1823.
    Hariota micrantha Kuntze, Rev. Gen. Pl. 1: 263. 1891.

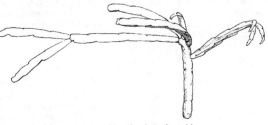

Either epiphytic and pendulous or clambering over rocks; branches 3 or 4-angled or flattened, 5 to 8 mm. broad; areoles small, remote, bearing often 1 to 4 bristles; flowers white, lateral, 7 mm. long including the ovary; petals cream-colored, spreading, obtuse; filaments, style, and stigma-lobes white; fruit 8 to 10 mm. long, naked, white to reddish, globose; seeds black.

FIG. 228.—Rhipsalis micrantha. ×0.5.

Type locality: Near Olleros, formerly in Ecuador, now in northern Peru.

Distribution: Ecuador and northern Peru.

Schumann describes this species as having 5 angles and cites only Humboldt's plant. The original description says 3 or 4-angled or compressed. The plant which he actually described is doubtless *Rhipsalis sulcata*, which has long passed in collections as *R. micrantha*.

Dr. Rose found this species quite common in southern Ecuador and brought back living specimens of it. The specimen in the New York Botanical Garden which came from Berlin agrees with Schumann's description.

Figure 228 shows a branch from the plant brought by Dr. Rose from southern Ecuador in 1918 (No. 23248).

41. Rhipsalis tonduzii Weber, Dict. Hort. Bois 1046. 1898.

Stems giving off aërial roots freely, at first erect but branches hanging, 1 cm. in diameter or less, normally 4 or 5-angled, sometimes 7-angled, but terminal branches often 3-angled or occasionally flattened and 2-angled; branches about 10 cm. long, usually terminal but always in clusters of 2 to 6, pale green; areoles close together, forming notches in the branch; flowers small, 12 mm. long, white; ovary exserted (Schumann says immersed); fruit globose, short-oblong, white, 7 to 10 mm. long, usually on upper half of terminal branches, resembling fruit of *Rhipsalis cassutha* but much longer, sometimes abortive and covered with hairs, thus resembling a small chestnut-bur, perhaps the result of insect stings; seeds oblong, numerous, black.

Type locality: Costa Rica.

Distribution: Costa Rica but range unknown.

This species flowered in Washington in March 1912, in June 1919, and again in April 1920; fruit was obtained July 31, 1919, and in April 1920.

Plate xxx, figure 3, shows a branch of a plant brought back from Costa Rica by Dr. Maxon in 1906.

Of this relationship is the following:

RHIPSALIS WERCKLEI Berger, Monatsschr. Kakteenk. **16**: 64. 1906.

Epiphytic, much branched, hanging, 3 to 6 dm. long; branches 2 to 4-angled, mostly 3, 8 to 10 cm. long, 10 mm. broad or less, without aërial roots; flowers borne singly along the whole branch, small; sepals 2; petals 4, creamy white; ovary not sunken in the branch; fruit globose, naked or with an occasional small scale, white, 5 mm. long; seeds numerous, brownish.

Type locality: Navarro, Costa Rica.

Distribution: Costa Rica.

The above description with regard to flowers and fruit has been copied. Our living specimens suggest that it may be different from *Rhipsalis tonduzii*, but whether specifically distinct will require further study to determine.

42. Rhipsalis boliviana (Britton) Lauterbach in Buchtien, Contr. Fl. Bolivia **1**: 145. 1910.

   Hariota boliviana Britton in Rusby, Mem. Torr. Club **3²**: 40. 1893.

Stems somewhat 4-angled and narrowly winged at base, setose at the areoles, the setae 5 to 10, yellowish white, about 2 mm. long; branches 1.5 to 30 cm. long, flattened and thin, 1 to 2 cm. broad, broadly crenate, the crenations 1.5 to 3 cm. long; flowers usually solitary but sometimes 2 or 3 at an areole, about 15 mm. long, one-half to two-thirds as broad, "yellow"; fruit globose, nearly 1 cm. in diameter, truncate at apex.

Type locality: Yungas, Bolivia.

Distribution: Wet forests of Bolivia.

43. Rhipsalis lorentziana Grisebach, Abh. Ges. Wiss. Göttingen **24**: 139. 1879.

Epiphytic on forest trees or clambering over rocks, freely rooting along stems; lower part of stem often terete; branches thin, flattened or sometimes 3-angled, usually elongated and narrow, sometimes more or less constricted near middle, 3 cm. broad or less, coarsely serrate, usually cuneate at base; flowers white, about 4 cm. long; ovary oblong, strongly angled, naked except a few scales at the top; fruit globose, purplish, 3 mm. in diameter.

Type locality: Oran, Argentina.

Distribution: Northwestern Argentina and to be expected in southern Bolivia.

Dr. Kurtz gave to Dr. Rose when he was in Córdoba, Argentina, in 1915, a part of the plant collected by Lorentz and Hieronymus in 1893 (No. 454), which proves to be the type.

44. Rhipsalis ramulosa (Salm-Dyck) Pfeiffer, Enum. Cact. 130. 1837.

   Cereus ramulosus Salm-Dyck, Hort. Dyck. 340. 1834.
   Hariota ramulosa Lemaire, Cact. Gen. Nov. Sp. 75. 1839.*

Stems woody, 3 dm. or more high, erect, terete; branches 7 to 12 cm. long, 1.2 to 2.5 cm. broad, pale green, with distant low crenations 12 to 20 mm. apart, when young often ciliate at areoles but in age naked; flowers solitary at the areoles, small, rotate, greenish white; sepals and petals 6 or 7, ovate-lanceolate, adhering to the base of the ovary, persistent; stamens 12 to 18; style filiform; stigma lobes inconspicuous; fruit glabrous, 5 to 6 mm. in diameter, white and subpellucid with 2 to 3 minute scales; seeds small, black.

Type locality: Not cited.

Distribution: Western Brazil and the adjacent borders of Bolivia and Peru (according to Vaupel).

Collected by R. S. Williams at Isapuri, Bolivia, altitude 1,550 feet, October 1, 1901 (No. 734). We have also referred here H. H. Rusby's No. 749 from trunk of trees near the cataracts of Bopi River, Bolivia, altitude 2,500 feet, September 8, 1921.

We know this plant from herbarium specimens; it is similar to *Rhipsalis lorentziana* but bearing scales on the ovary.

---

*Lemaire, in 1839 (Cact. Gen. Nov. Sp. 74, 75), combines *Rhipsalis* with *Hariota*, and 8 of the 10 species which he lists had not heretofore been referred to *Hariota*. They are, therefore, to be credited to Lemaire rather than to Otto Kuntze (Rev. Gen. Pl. **1**: 262. 1891), as has been done in the Index Kewensis.

M. E. Eaton del.

1. Flowering branch of *Rhipsalis houlletiana*.
2. Flower of same.
3. Fruiting branch of *Rhipsalis warmingiana*.
4. Fruiting branch of same.

Rhipsalis ramulosa has long been a doubtful species. Its origin was unknown at the time of its first publication, but Schumann in 1890 attributed it to Costa Rica, but this was evidently a mistake.

Vaupel has recently published an article (Zeitschrift für Sukkulentenkunde 1: 19. 1923) in which he states that the type was cultivated in the Botanical Garden of Berlin in 1833 and that specimens are now preserved in the herbarium there. He states that these are the same as the plant collected by Ule at Seringal, San Francisco, in the Upper Acre region of Brazil, about 10° south latitude, towards the border of Bolivia and Peru. He would also refer here a plant collected by Tafalla in 1790 at Pozugo in eastern Peru. *Cactus dentatus* Ruiz (Martius, Fl. Bras. 4²: 288. 1890), given as a synonym of *Rhipsalis alata* by Schumann, is based on Tafalla's plant and according to Vaupel should not have been credited to Ruiz.

Epiphyllum ramulosum, E. ciliare, and *E. ciliatum* were all given by Pfeiffer (Enum. Cact. 130. 1837) as synonyms of *Rhipsalis ramulosa.*

Figure 229 shows a drawing made from Mr. Williams's specimen.

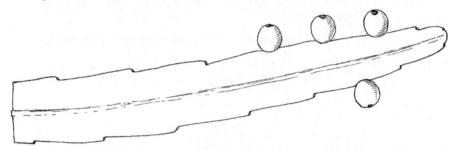

Fig. 229.—Top of fruiting branch of Rhipsalis ramulosa. × 0.75.

45. Rhipsalis purpusii Weingart, Monatsschr. Kakteenk. **28**: 78. 1918.

Plant epiphytic; stems 8 mm. in diameter, woody, terete, brown; branches weak, elongated, terete below, flattened above, thin, remotely crenate; flowers small, white, solitary.

Type locality: Cerro de Boqueron, Chiapas, Mexico.
Distribution: Known only from the type locality.

This must be related to the Costa Rican plant, *Rhipsalis coriacea,* and perhaps conspecific.

Illustrations: Monatsschr. Kakteenk. **28**: 79; Möllers Deutsche Gärt. Zeit. **35**: 117.

46. Rhipsalis coriacea Polakowsky, Linnaea **41**: 562. 1877.

  Hariota coriacea Kuntze, Rev. Gen. Pl. **1**: 262. 1891.
  Rhipsalis angustissima Weber Bull. Mus. Hist. Nat. Paris **8**: 465. 1902.
  Rhipsalis leiophloea Vaupel, Zeitschrift Sukkulentk. **1**: 20. 1923.

Stems 2 to 10 cm. high, woody and terete at base, with many lateral branches; branches often hanging, 1 to 3.5 cm. broad, thin, serrate, the teeth 1.5 to 2.5 cm. apart, bearing the small areoles; young branches purple, terete at first, but finally broad and flattened above; areoles at base of branch and sometimes but rarely on flattened part, bearing 2 to 7 long, hairy bristles; flowers rather narrow, including ovary 12 mm. long, each subtended by a shallow scale; sepals and petals erect below; sepals usually 3, cream-colored, tinged with red; petals greenish white to pinkish, usually 5 to 10, obtuse, 7 to 8 mm. long; stamens numerous, white; style white; stigma-lobes short, white; fruit white, 7 mm. in diameter, bearing several broad, rounded scales; seeds black.

Type locality: Near Cartago, Costa Rica.
Distribution: Widely distributed in Costa Rica.

This species flowers in March.

Rhipsalis coriacea, which originally came from Costa Rica, Schumann referred to *R. alata* of Jamaica, a plant of similar habit but yet very distinct.

Illustration: Bull. Mus. Hist. Nat. Paris **8**: 466, as *Rhipsalis angustissima*.

Of plate XXXII, figure 4 shows a flowering specimen and figure 5 a fruiting specimen from a plant collected by Wm. R. Maxon at Tunialba, Costa Rica, in April 1900, painted at the New York Botanical Garden on April 12, 1912.

47. Rhipsalis jamaicensis Britton and Harris, Torreya **9**: 159. 1909.

Pendent from trees, 3 to 10 dm. long, the main axis angular; joints 1 to 4 dm. long, 1 to 2.5 cm. broad, thin, dull green, bluntish at apex, narrowed into a short or elongated stipe at base, the margins low-crenate; flowers yellowish green, about 6 mm. long; perianth-segments about 7, oblong to oblanceolate, only a little spreading, obtusish; ovary oblong, bearing a few small scales; stamens 20 to 30; style longer than stamens; stigma-lobes 3; fruit globose, 6 to 8 mm. in diameter, white, the scales 3 mm. broad.

Type locality: Troy, Cockpit Country, Jamaica.
Distribution: Forests of Jamaica.
Illustration: Torreya **9**: 158. f. 3.

Plate XXII, figure 4, shows a plant with flowers and young fruit from Jamaica.

48. Rhipsalis platycarpa (Zuccarini) Pfeiffer, Enum. Cact. 131. 1837.

Epiphyllum platycarpum Zuccarini, Cat. Cact. Monac. 1836.
Cereus platycarpus Zuccarini, Abh. Bayer. Akad. Wiss. München 2: 736. 1857.
Hariota platycarpa Kuntze, Rev. Gen. Pl. 1: 263. 1891.

Branches broad and flat, 1 to 2 dm. long, 3 cm. broad or more, dull green becoming red when grown in sunlight, with broad deep crenations; flowers borne toward apex of branch, 1 to 3 from an areole, 16 to 18 mm. long, greenish yellow or dull white; petals 8 mm. long, ovate; stamens white; stigma-lobes 5, white; fruit (doubtless immature) naked, green, somewhat compressed, angled, truncate.

Type locality: Brazil.
Distribution: Organ Mountains, Brazil.

We have obtained plants of this species from Mr. Lamb at Manchester in 1904 and Dr. Rose found it wild in the Organ Mountains of Brazil in 1915 (No. 21159). It grows well in cultivation but it has never flowered with us.

Illustrations: Pfeiffer and Otto, Abbild. Beschr. Cact. **1**: pl. 17, f.2; Blühende Kakteen **2**: pl. 90.

Figure 230 shows a branch of the plant obtained by Dr. Rose in the Organ Mountains.

FIG. 230.—Rhipsalis platycarpa. ×0.4.

49. Rhipsalis russellii sp. nov.

Hanging in great clusters from the horizontal branches of trees; branches strongly flattened, 15 cm. long, 5 to 6 cm. broad, cuneate at base, strongly crenate, dark green or purplish along margins; flowers often 9 at an areole, minute; sepals few, obtuse, reddish at tips; petals usually 5, cream-colored, obtuse, 2 mm. long; fruit usually 1 at an areole, small, globose, 5 to 6 mm. in diameter, purple.

Collected by J. N. Rose and Paul G. Russell near Toca da Onca, Bahia, Brazil, June 27 to 29, 1915 (No. 20106). This species suggests *Rhipsalis elliptica*, but has very different flowers and fruit.

PLATE XXXV

M. E. Eaton del.

1. Fruiting branch of *Rhipsalis oblonga*.
2. Fruiting branch of *Rhipsalis elliptica*.
3. Flowering branch of *Rhipsalis crispata*.

Of plate XXXVII, figures 1 to 4 are from the type specimen which has repeatedly flowered and fruited in the New York Botanical Garden; figure 1 shows the tip of a flowering branch; figure 2 shows a cluster of six flowers; figure 3 shows a flower enlarged four diameters; figure 4 shows a fruiting branch.

50. Rhipsalis elliptica Lindberg in Martius, Fl. Bras. 4^2: 293.   1890.

> *Rhipsalis chloroptera* Weber, Dict. Hort. Bois 1045.   1898.
> *Rhipsalis elliptica helicoidea* Löfgren, Arch. Jard. Bot. Rio de Janeiro 2: 44.   1918.

Plants growing in clumps, at first ascending, often hanging from trees; joints flat and broad, oblong to elliptic, 3 to 20 cm. long, 2 to 7 cm. broad, the margin faintly to strongly crenate; flowers generally 1, sometimes 2 or 3 at an areole, 12 mm. broad; petals usually 5, yellowish, widely spreading, oblong, obtuse; filaments numerous, nearly erect, white; style white; stigma-lobes white, 5; ovary not sunken in the branch; fruit reddish, a little longer than broad, 6 to 7 mm. long.

Type locality: Near Sorocaba, south of Santos, São Paulo, Brazil.

Distribution: States of São Paulo and Santa Catharina, Brazil.

Illustrations: Arch. Jard. Bot. Rio de Janeiro 2: pl. 16, as *Rhipsalis elliptica helicoidea*; Blühende Kakteen 2: pl. 104, as *Rhipsalis chloroptera*; Arch. Jard. Bot. Rio de Janeiro 1: pl. 15.

Plate XXXV, figure 2, is from a plant collected by Dr. Rose at Jabaquara, near Rio de Janeiro, Brazil, in 1915, which flowered in the New York Botanical Garden in 1916.

51. Rhipsalis pachyptera Pfeiffer, Enum. Cact. 132.   1837.

> *Cactus alatus* Willdenow, Enum. Pl. Suppl. 35.   1813.   Not Swartz, 1788.
> *Epiphyllum alatum* Haworth, Suppl. Pl. Succ. 84.   1819.
> *Cactus triqueter* Vellozo, Fl. Flum. 206.   1825.   Not Willdenow, 1813.   Not Haworth, 1803.
> *Cereus alatus* Link and Otto, Icon. Pl. Rar. 77.   1830.
> *Lepismium fluminense* Miquel, Bull. Neerl. 48.   1838.
> *Rhipsalis robusta* Lemaire, Rev. Hort. IV. 9: 502.   1860.
> *Rhipsalis pachyptera purpurea* Corderoy, Gard. Chron. III. 2: 468.   1887.
> *Hariota triquetra* Kuntze, Rev. Gen. Pl. 1: 263.   1891.
> *Hariota pachyptera* Kuntze, Rev. Gen. Pl. 1: 263.   1891.
> *Hariota robusta* Kuntze, Rev. Gen. Pl. 1: 263.   1891.
> *Rhipsalis crassa* Schumann, Keys 54.   1903.

Stems much jointed, pendent; joints often 3 to 6 dm. long, 5 to 7 cm. broad, thickish, stiff, sometimes nearly orbicular, often purple, deeply crenate; flowers numerous, but solitary, rarely 2 to 4 from the areole, large; petals widely spreading, yellowish; stamens numerous; stigma-lobes 4 or 5, slender; fruit globose, white.

Type locality: Originally given as the West Indies,* but this is doubtless a mistake.

Distribution: States of Rio de Janeiro, Minas-Geraes, Santa Catherina, and São Paulo, Brazil.

The species grows in the high mountains on trunks of trees, altitude 1,000 meters, down to nearly sea-level.

A variety, *crassior* Salm-Dyck (Pfeiffer, Enum. Cact. 132. 1837), with thick green orbicular joints, has been described.

Steudel's name of *Rhipsalis alata* (Nom. ed. 2. 1: 333. 1840), given as a synonym of *Cereus alatus* De Candolle, is referred here by Schumann, but probably relates to *Pseudorhipsalis*.

This species was for a long time confused with *Rhipsalis alata*, a very distinct species from Jamaica, now referred by us to the genus *Pseudorhipsalis*.

Illustrations: Curtis's Bot. Mag. 55: pl. 2820,* as *Cactus alatus*; Vellozo, Fl. Flum. 5: pl. 25, as *Cactus triqueter*; pl. 33, as to flower only; Link and Otto, Icon. Pl. Rar. pl. 39, as *Cereus alatus*; Monatsschr. Kakteenk. 6: 55; 7: 151, in part, as *Rhipsalis robusta*;

---

*The plant, however, which Hooker described and figured (Curtis's Bot. Mag. 55: pl. 2820) as *Cactus alatus* and which Pfeiffer cited in his original description, came from the Organ Mountains near Rio de Janeiro, Brazil.

Blühende Kakteen 1: pl. 34; Martius, Fl. Bras. 4²: pl. 57; Paxton's Fl. Gard. 1: pl. 99; Arch. Jard. Bot. Rio de Janeiro 1: pl. 14; Möllers Deutsche Gärt. Zeit. 25: 477. f. 11, No. 15; Gartenwelt 13: 117; 16: 633, 635; Karsten and Schenck, Vegetationsbilder 1: pl. 5, f. d.

Plate XXXVII, figure 6, shows a fruiting branch from the plant obtained by Dr. Rose in Brazil in 1915 (No. 20346); plate XXXVI, figure 1, shows a flowering branch from a plant obtained from M. Simon, of Paris, in 1901.

52. Rhipsalis rhombea (Salm-Dyck) Pfeiffer, Enum. Cact. 130. 1837.

   Cereus rhombeus Salm-Dyck, Hort. Dyck. 341. 1834.
   Hariota rhombea Lemaire, Cact. Gen. Nov. Sp. 75. 1839.

Stems terete or angled; branches usually flat and thin, but sometimes 3-angled; joints oblong, 1 to 3 cm. broad, cuneate at base, strongly crenate, dark green or purple; flowers usually solitary at areoles but sometimes in 2's, small, cream-colored, with a red spot at base of stamens; sepals reddish; petals obtuse; fruit dark red.

FIG. 231.—Rhipsalis crispimarginata.

Type locality: Not cited.
Distribution: Brazil, but range unknown.

Cereus crispatus crenulatus, Epiphyllum crenulatum, and *E. rhombeum* were referred by Pfeiffer (Enum. Cact. 130. 1837) as synonyms of this species.

Here perhaps also belongs *Cereus crispatus latior* (Salm-Dyck, Hort. Dyck. 66. 1834), which is without description.

Illustrations: Gartenwelt 16: 635; Karsten and Schenck, Vegetationsbilder 1: pl. 6, f. e; Möllers Deutsche Gart. Zeit. 25: 477. f. 11, No. 13; Wildeman, Icon. Select. 2: pl. 67; Arch. Jard. Bot. Rio de Janeiro 1: pl. 16.

Plate XXXVI, figure 2, shows a flowering plant received from the Royal Botanic Garden at Kew in 1902 which flowered in the New York Botanical Garden in January 1912.

---

*According to the Index Kewensis this is *Cactus speciosus* Hooker, said to be equal to *Rhipsalis swartziana.*

M. E. Eaton del.

1.   Flowering branch of *Rhipsalis pachyptera*.
2.   Flowering branch of *Rhipsalis rhombea*.

53. Rhipsalis crispimarginata Löfgren, Arch. Jard. Bot. Rio de Janeiro 2: 37. 1918.

Plants pendulous, the main stem terete below, often 3-winged above; terminal branches in clusters, oblong, flat, obtuse, narrowed at base, shining green or tinged with purple, 4 to 6 cm. long; flowers usually solitary but sometimes 2 or 3 at an areole, white, drying pale yellow; sepals ovate-obtuse, reflexed; petals white, widely spreading, numerous; stigma-lobes white; fruit globose, white.

Type locality: Ilha Grande, near the city of Rio de Janeiro.
Distribution: State of Rio de Janeiro, Brazil.
Illustration: Arch. Jard. Bot. Rio de Janeiro 2: pl. 9.

Plate XXXVII, figure 5, shows a fruiting branch of the type collection obtained by Dr. Rose on Ilha Grande near Rio de Janeiro in 1915 (No. 20401). Figure 231 is from a photograph of Miss Eaton's painting of a plant given to Dr. Shafer by Dr. Löfgren in 1917 at Rio de Janeiro which flowered in the New York Botanical Garden in May 1922.

54. Rhipsalis crispata (Haworth) Pfeiffer, Enum. Cact. 130. 1837.

    Epiphyllum crispatum Haworth, Phil. Mag. 7: 111. 1830.
    Rhipsalis crispata latior Salm-Dyck in Pfeiffer, Enum. Cact. 130. 1837.
    Hariota crispata Lemaire, Cact. Gen. Nov. Sp. 75. 1839.
    Rhipsalis rhombea crispata Schumann, Gesamtb. Kakteen 638. 1898.

FIG. 232.—Rhipsalis crispata.

Branches divided into short flat joints 6 to 10 cm. long, broad both at base and apex, green, more or less crenate; flowers solitary or sometimes 2 to 4 at an areole, 10 to 12 mm. broad, cream-colored to light yellow; filaments numerous; fruit white, 7 mm. in diameter.

Type locality: Brazil.
Distribution: Brazil, but range unknown.

The synonyms *R. crispa* and its var. *major* (Förster, Handb. Cact. 450. 1846) and *R. crispa latior* Salm-Dyck (Walpers, Repert. Bot. 2: 279. 1843); *Hariota crispata latior* Lemaire (Cact. Gen. Nov. Sp. 75. 1839) doubtless belong here. Schumann, in Nachtrag, page 144, refers here *Rhipsalis swartziana* Pfeiffer.

Pfeiffer publishes as a synonym of the above *Cereus crispatus* (Enum. Cact. 130. 1837), as does also Förster. Schumann (Gesamtb. Kakteen 638. 1898) makes this species a variety of *Rhipsalis rhombea*, but later (Gesamtb. Kakteen Nachtr. 145. 1903) recognizes two species.

Cactus torquatus (Walpers, Repert. Bot. 2: 342. 1843), referred to *Rhipsalis rhombea* by Walpers, was only a garden name.

Illustrations: Arch. Jard. Bot. Rio de Janeiro 1: pl. 16, as *R. rhombea*; Gartenwelt 13: 117; Garten-Zeitung 1: 459. f. 109; Rev. Hort. 85: f. 152, in part.

Plate xxxv, figure 3, shows a flowering plant received from A. Berger in 1908. Figure 232 is from a photograph of Miss Eaton's painting of the plant obtained by Dr. Rose in Brazil in 1915 (No. 20708) which flowered and fruited in the New York Botanical Garden in 1922.

55. Rhipsalis oblonga Löfgren, Arch. Jard. Bot. Rio de Janeiro **2**: 36. 1918.

In cultivation bushy; main branches terete below, more or less flattened above; ultimate branches narrowly oblong, 5 to 15 cm. long, 1 to 2 cm. broad, shining green even in sunlight; flowers borne along the sides of the branches, solitary at the areoles; fruit globular to short-oblong, 3 to 4 mm. long, nearly white, naked, crowned by the withered perianth.

Type locality: On Ilha Grande, Brazil.
Distribution: Known only from the type locality.
Illustration: Arch. Jard. Bot. Rio de Janeiro **2**: pl. 8, as *Rhipsalis oblonga*.

Plate xxxv, figure 1, shows the plant grown by Dr. Löfgren at Rio de Janeiro and given to Dr. Shafer in 1917, which flowered and fruited in the New York Botanical Garden in May 1922.

56. Rhipsalis cuneata sp. nov.

Epiphytic on trees; joints oblong to spatulate, 8 to 12 cm. long, thin, obtuse, cuneate at base, strongly crenate, naked at the areoles or with a bristle or two; flowers so far as known solitary; fruit globose, 4 mm. in diameter, naked.

FIG. 233.—Rhipsalis cuneata.

Collected by R. S. Williams above San Juan, Bolivia, altitude 5,500 feet, April 2, 1902 (No. 2458). This species is known to us only from herbarium specimens.

Figure 233 is from a photograph of the specimen in the U. S. National Herbarium.

57. Rhipsalis roseana Berger, Zeitschrift für Sukkulentenkunde **1**: 22. 1923.

Lower joints flat, 15 to 20 mm. broad, distinctly alternately notched; areoles small, with a little tuft of white wool and a single short brown hair, 15 to 20 mm. apart, the upper ones more closely set; upper joints narrower and more linear or linear-lanceolate, 10 to 15 mm. broad and 6 to 12 cm. long or more, equally notched, smooth, bright green; some of the uppermost joints often narrower, 8 to 10 mm. broad and only shallowly notched, others triangular with prominent notched angles and excavated sides, others 1 cm. wide, with 3 or 4 prominent wing-like distinctly but remotely notched ribs and areoles about 4 cm. apart; flowers small, whitish yellow.

This species was described from cultivated plants of unknown origin. We believe that it may be from Colombia and we would refer here the following specimens: Wilson Popenoe's No. 518 from near San Miguel, Perdoma, Tolima, 1921, and Ellsworth P. Killip's No. 8203 from mountains west of Popayán, 1922.

Mr. Berger writes: "This new *Rhipsalis* is decidedly distinct from *R. wercklei;* its branches are shorter, broader, more deeply notched and of a firmer nature. Its growth too is far less quick and it does not form so promptly long and pendent shoots as *R. wercklei.*

M. E. Eaton del.

1. Flowering branch of *Rhipsalis russellii*.
2. Cluster of flowers of same.
3. Flower of same.
4. Fruiting branch of same.
5. Fruiting branch of *Rhipsalis crispimarginata*.
6. Fruiting branch of *Rhipsalis pachyptera*.

UNPUBLISHED OR INCOMPLETELY DESCRIBED SPECIES.

RHIPSALIS CHRYSANTHA Löfgren, Arch. Jard. Bot. Rio de Janeiro 1: 99. 1915.

We know this species only from description. Löfgren places it in his subgenus *Lepismium* near *Rhipsalis dissimilis*, but his descriptions suggest *R. rosea* (our *Rhipsalidopsis rosea*). Both names are based on Dr. P. Dusen's collections from Paramá, Brazil. It seems near *R. puniceo-discus*.

RHIPSALIS FRONDOSA Wercklé, Subregion Fitogeografica Costa Ricense 42. 1909.

The above name is given without description.

Weingart (Monatsschr. Kakteenk. 20: 185. 1910) refers to this plant as a new species represented in a sending from Costa Rica by Wercklé. Nothing further is known about it.

RHIPSALIS RIEDELIANA Regel, Ind. Sem. Hort. Petrop. 1860: 49. 1860.

   Hariota riedeliana Kuntze, Rev. Gen. Pl. 1: 263. 1891.

This plant was sent from Brazil by Riedel, but we do not know it. Schumann did not know it.

Rhipsalis bucheni Béhagnon (Rev. Hort. 85: 436. f. 152. 1913) we know only from the illustration of a poor potted plant and an incomplete description.

Rhipsalis carnosa and *R. lagenaria* are names mentioned by Vöchting (Jahrb. Wiss. Bot. Leipzig 9: 368, 372. 1873).

Rhipsalis erythrolepis Bénagnon (Rev. Hort. 85: f. 152, part) is known only from a potted plant of some species with broad, flat joints.

Rhipsalis filiformis seems to be only a garden name (Monatsschr. Kakteenk. 6: 47. 1896). It may be the same as *R. cribrata filiformis* Engelhardt (Möllers Deutsche Gärt. Zeit. 18: 585. 1903).

Rhipsalis itatiaiae Weber appears in Robert Lamb's Collection of Cacti, page 72, 1908, without description. In 1914 Mr. Lamb sent Dr. Rose a specimen under this name, but it has not bloomed. A part of this plant from Mount Itatiaya, Brazil, is now growing in the New York Botanical Garden (Rose, No. 888).

Rhipsalis macahensis Glaziou (Bull. Soc. Bot. France Mem. III. 326. 1909) is only a name. According to Glaziou he collected it on rocks and trees at Alto Macahé, Rio de Janeiro (No. 18262).

Rhipsalis microcarpa Steudel, is a name found only in Schumann's Index (Gesamtb. Kakteen 832. 1898).

Rhipsalis miquelii Lemaire (Cactées 80. 1868) is not described but it is grouped with *R. pachyptera*, *R. rhombea*, and other flat-jointed species. Lemaire also lists *R. turpinii* on the same page, associating it with *R. micrantha* and *R. trigona*.

Rhipsalis oligosperma Lindberg (Monatsschr. Kakteenk. 7: 21. 1897) is a name only.

Rhipsalis spathulata Otto (Sweet, Hort. Brit. ed. 3. 1839) Schumann thought might be a mistake for *Pereskia spathulata*. Kuntze takes it up, however, as *Hariota spathulata* (Rev. Gen. Pl. 1: 263. 1891).

The name *Rhipsalis taglionis* occurs in the Index Kewensis Supplement 1, by error for *R. saglionsis*.

Rhipsalis wettsteinii Schumann (Monatsschr. Kakteenk. 17: 48. 1907) is a name only.

The illustration (fig. 233a) given on next page is a reproduction of Plumier's plate 92 of Burmann Plantarum Americanum pubished in 1755 and now referred to *Neoabbattia paniculata*, discussed on page 280 of this volume.

CACTUS brachiatus & arboreus.

APPENDIX.

During the progress of our investigations much information has been received from numerous sources which could not be included in publication at the logical places. Some of this was taken up in the appendix to the first volume (Cactaceae 1: 216–225) and some in the appendix to the second (Cactaceae 2: 223-226); what remains is included in this appendix to the whole work.

Dr. David Griffiths, who studied the species of *Opuntia*, especially with relation to their economic possibilities, and grew many of them at experimental stations of the United States Department of Agriculture at Brownsville, Texas, and at Chico, California, has published and described many species as new. We have included these in our studies of the genus and have grouped them with the species known to us as accurately as has been possible from his published descriptions and illustrations and after examination of as many of his type specimens as we have been permitted to see; however, conditions were such that we have not been able to study a number of them. They have not been arranged for ready reference by students.

The preface to Volume I gives a list of volunteers who have made valuable contributions of specimens and data to this investigation. Many of these have continued to aid us.

Dr. Britton, in continuing his West Indian studies, investigated the cacti of Grenada and of Trinidad in 1920 and 1921 and published an account of the Trinidad species.

Dr. John K. Small has continued his investigation of the southeastern United States and of Florida in particular, in cooperation with Mr. Charles Deering, and has greatly increased our knowledge of the cacti existing there, including the discovery of many undescribed species of *Opuntia*.

Dr. Francis W. Pennell, Curator of Botany at the Academy of Natural Sciences of Philadelphia, and Mr. E. P. Killip, of the United States National Museum, made extensive botanical collections in Colombia in 1922, including some specimens of cacti, which we have studied.

Dr. Henry H. Rusby led the Mulford Biological Exploring Expedition to Bolivia in 1921–1922 and with the assistance of Dr. O. E. White obtained for us specimens of several little-known cacti.

Dr. Philip A. Munz has sent us cacti from the deserts of southern California.

Mr. C. Z. Nelson obtained cacti from southern Mexico, including a beautiful new species of *Selenicereus*.

Mr. Francis J. Dyer, while connected with the Consular Service in Honduras and at Nogales, Mexico, sent us many specimens from those stations.

Professor Harvey M. Hall, while making extensive explorations in the western United States in connection with his own work, has forwarded interesting cactus plants.

Dr. and Mrs. Charles D. Walcott have sent specimens from Alberta, Canada, some of them coming from near the most northern range of the family.

Dr. W. L. Abbott and Mr. E. C. Leonard made extensive collections in Haiti in 1920 and obtained a number of rare and little-known plants, including one which had been collected by Charles Plumier about 1698 and which proved to be a new genus; this was named by us for Dr. Abbott. More recently Dr. Abbott has sent us specimens from Santo Domingo.

Dr. George F. Gaumer, the veteran collector in Yucatan, has sent very important collections from his region, including a number of new species.

Mr. Ivan M. Johnston, who accompanied the scientific expedition sent out by the California Academy of Sciences in 1921 to explore the islands of the Gulf of California,

249

collected many cacti, especially species of *Neomammillaria*, duplicating many of the important discoveries made by Dr. Rose on the same islands in 1911. He has also sent us cacti from Colorado.

Professor Fortunato L. Herrera has sent some very interesting plants from eastern Peru, especially from about his home at Cuzco.

Mr. Robert Runyon has collected extensively in southern Texas and northern Mexico and has supplemented his specimens with some very beautiful photographs.

Dr. L. H. Bailey and his daughter, Miss Ethel Zoe Bailey, obtained valuable cacti from Venezuela, especially from the region about Ciudad Bolívar, on the Orinoco, in 1921.

Mr. W. B. Alexander was sent to Argentina by the Australian Government in 1920 and 1921 in search of enemies of the weed prickly pears and there made many important observations, especially on the genus *Opuntia*. He sent us two undescribed species.

Dr. B. P. Reko, a very diligent collector, has sent many cacti from Mexico, especially from Oaxaca, including several new to science.

Señor Octavio Solís, in charge of the cactus garden belonging to the Mexican Government in the City of Mexico, has sent many living plants from his country, especially of the genus *Neomammillaria*. To him we have dedicated the genus *Solisia*.

Señor J. G. Ortega has collected extensively on the west coast of Mexico, especially in the state of Sinaloa, and for him we have named *Neomammillaria ortegae*.

Mr. J. Francis Macbride and Mr. William Featherstone, who were in charge of the botanical expedition of the Field Museum to Peru in 1922 and 1923, made large and valuable collections of cacti in central and eastern Peru.

Mr. E. C. Rost has collected and photographed many interesting cacti for us in southern California and Lower California.

Dr. W. S. W. Kew explored extensively in Lower California in 1921 and sent not only many specimens but numerous habit photographs.

Mr. James H. Ferriss, while making various excursions through the western United States, has sent in many specimens. Among his interesting discoveries was the finding of *Neomammillaria pottsii* in southern Texas.

Mrs. S. L. Pattison, an enthusiastic collector in western Texas, has sent many valuable specimens, including new species collected by herself or for her by local collectors.

Mrs. Ruth C. Ross spent considerable time in eastern Arizona in 1921 and collected cacti along the route traversed by Emory in 1847, re-collecting certain species which he had discovered at that time.

Mr. Harry Johnson was located for about a year in Guatemala, during which time he sent a number of very interesting cacti, especially species of *Epiphyllum*. Some of these were accompanied by full notes and drawings.

Señor P. Campos-Porto has sent a number of interesting specimens from Brazil belonging to the genus *Rhipsalis*.

The following persons have contributed valuable specimens, usually from about their homes or while engaged in other work: G. W. Goldsmith, Colorado; B. C. Tharp, Texas; Charles O. Chambers, Oklahoma; James S. Holmes, Washington, D. C.; Joseph A. Holmes, Wyoming; William Hertrich, California; William Tell, Texas; Albert Ruth, Texas; D. C. Parman, Texas; Karl Reiche, Mexico; Gerold Stahel, Surinam; Rev. Louis Mille, Ecuador; H. M. Pilkington, Haiti; Percy L. Ports, Washington, D. C.; W. E. Broadway, Trinidad; A. F. Moeller, Mexico; W. E. Meyer, Bolivia; Stephen E. Aguirre, Mexico; Mrs. Elsie McElroy Slater, Texas; Paul C. Standley, Central America; R. D. Camp, Brownsville, Texas; George L. Fisher, Texas; A. V. Frič, Mexico; and Dana Lee, Wyoming.

As treated in this monograph the Cactus family is composed of 3 tribes. The first and second tribes are taken as units, but the third is composed of 8 subtribes. The number of genera recognized is 124 and the number of species is 1,235.

CORRECTIONS AND ADDITIONS TO VOLUME I.

On page 11, vol. 1, under *Pereskia pereskia*, add to illustrations: Garten-Zeitung **4**: 182. f. 42, No. 5; Gard. Chron. III. **20**: f. 108; Stand. Cycl. Hort. Bailey **2**: f. 714, as *Pereskia aculeata*; London, Encycl. Pl. ed. 3. 413, as *Cactus pereskia*; Möllers Deutsche Gärt. Zeit. **23**: 256. f. 15, as *Pereskia godseffiana*.

Also insert: *Pereskia longispina rubescens* Pfeiffer and *P. longispina rotundifolia* Salm-Dyck were given by Walpers (Repert. Bot. **2**: 283. 1843) as synonyms of *P. aculeata*, but they were not described.

On page 12, vol. 1, under *Pereskia autumnalis*, add to distribution: Common in Salvador where it is much planted for hedges.

Also add to illustrations: Monatsschr. Kakteenk. **25**: 35, as *Pereskiopsis autumnalis*; Engler and Drude, Veg. Erde **13**: f. 10, as *Pereskia guatemalensis*.

On page 14, vol. 1, under *Pereskia sacharosa*, add the synonym: *Pereskia amapola argentina* Weber in Weingart, Monatsschr. Kakteenk. **14**: 87. 1894.

On page 17, vol. 1, under *Pereskia guamacho*, insert: *Illustration:* Carnegie Inst. Wash. **269**: pl. 11, f. 92, 93.

On page 20, vol. 1, under *Pereskia grandifolia*, add to illustrations: Rümpler, Sukkulenten f. 128; Engler and Prantl, Pflanzenfam. **3**^{6a}: f. 57, J; Blühende Kakteen **3**: pl. 137; Watson, Cact. Cult. f. 6, in part; 222. f. 87; ed. 3. f. 63; Karsten, Deutsche Fl. ed. 2. **2**: 456. f. 605, No. 9; Loudon, Encycl. Pl. ed. 3. 1202. f. 1737↑; Van Géel, Sert. Bot. **4**: pl. 1, as *Pereskia bleo*; Dict. Gard. Nicholson **3**: 75. f. 81; Monatsschr. Kakteenk. **15**: 81.

Also add synonym: *Cactus grandiflorus* Link, Enum. Hort. Berol. **2**: 25. 1822.

On page 21, vol. 1, *Pereskia zinniaeflora*, add to illustrations: Watson, Cact. Cult. ed. 1 and 2. 223. f. 88; ed. 3. f. 64; Dict. Gard. Nicholson **4**: 586. f. 55.

On page 21, vol. 1, under *Pereskia horrida*, substitute for this name:

Pereskia humboldtii nom. nov.

Cactus horridus Humboldt, Bonpland, and Kunth, Nov. Gen. et Sp. **6**: 70. 1823. Not Salisbury, 1796.
Pereskia horrida De Candolle, Prodr. **3**: 475. 1828.

On page 24, vol. 1, at end of *Pereskia*, add: *Pereskia recurvifolia* and *P. galeottiana* are two names marked with an asterisk by Lemaire (Cactées 95. 1868), indicating that they are new. So far as we know they were never described.

On page 24, vol. 1, at end of *Pereskia* insert:

Pereskia pflanzii Vaupel, Zeitschrift Sukkulentenk. **1**: 56. 1923.

Tree about 15 meters high, with verticillate branches, not very spiny; leaves thick, ovoid, narrowed at base, 4 cm. long by 2 cm. broad; flowers solitary at apex of leafy branches; corolla 3 cm. long, rose-colored.

Type locality: Vicinity of Laguna Santa Isabel, Bolivia.
Distribution: Bolivia, but known only from type locality.

Pereskia verticillata Vaupel, Zeitschrift Sukkulentenk. **1**: 55. 1923.

Erect shrub, 2 meters high, very spiny, with verticillate branches; leaves thick, lanceolate, 5 cm. long by 1.5 cm. broad; flowers borne at apex of leafy branches; corolla 1.5 cm. long, rose-colored.

Type locality: Vicinity of Laguna Santa Isabel, Bolivia.
Distribution: Bolivia, but known only from type locality.

On page 27, vol. 1, under *Pereskiopsis chapistle*, add to illustration: Smiths. Misc. Coll. **50**: pl. 43.

On page 28, vol. 1, under *Pereskiopsis porteri*, add the synonym: *Opuntia rotundifolia* Brandegee, Zoe **2**: 21. 1891. Not *Pereskia rotundifolia* De Candolle, 1828.

On page 29, vol. 1, under *Pereskiopsis spathulata*, insert: *Illustration:* Möllers Deutsche Gärt. Zeit. **25**: 488. f. 22, No. 1, as *Pereskia spathulata*.

On page 29, vol. 1, under *Pereskiopsis pititache*, add to illustrations: Deutsche Gärt. Zeit. **8**: 33, as *Pereskia calandriniaefolia*.

On page 30, vol. 1, insert the following:

11. Pereskiopsis scandens sp. nov.

Slender, climbing or clambering over walls, up to 10 meters long; branches terete, grayish, smooth; areoles circular, white-woolly when young, gray in age, with a short spine (5 mm. long) and a bunch of brown glochids in the upper edge; leaves ovate, 1.5 to 2 cm. long, glabrous, acute; flowers yellow, from the areoles on old branches, appearing in June; fruit maturing slowly (perhaps requiring 2 to 3 years to ripen), very narrow, 5 to 7 cm. long, somewhat tubercled, with a deep umbilicus; seeds few.

Living specimens of *P. scandens* were sent by Dr. George F. Gaumer from Izamal, Yucatan, Mexico, in July 1921 (type). It was also collected by A. Schott at Mérida in 1865 (No. 409).

Withdraw the name *Pereskia zehntneri* from page 14, vol. 1, and substitute the following at the end of *Pereskiopsis* on page 30:

1a. QUIABENTIA gen. nov.

A low, leafy, much branched shrub with numerous horizontal branches, usually in whorls; leaves fleshy but flattened, stiff, borne at right angles to the branches; areoles large, white-felted, often with numerous spines, these acicular and white, the upper part of areole bearing glochids; flowers terminal, very large, bright red; ovary leafy, very narrow; stamens numerous, a little shorter than the style, much shorter than the petals; style short and stiff; stigma-lobes very short, obtuse; seeds white, a little flattened, covered with a hard bony aril as in *Opuntia*.

A monotypic genus, native of the semiarid region of Bahia, Brazil. The generic name is from quiabento, the native name of the plant.

1. Quiabentia zehntneri Britton and Rose.

    Pereskia zehntneri Britton and Rose, Cactaceae **1**: 14. 1919.

Flowers at ends of branches, large, 7 to 8 cm. broad, 3 to 4 cm. long, bright red, appearing in November; petals broad, retuse; ovary borne in the upper end of the branch, very narrow, 3 to 4 cm. long, bearing the usual leaves, areoles, and spines of the branches; fruit oblong to clavate, 6 to 7 cm. long, 1.5 cm. in diameter at the top, slightly angled by the low elongated tubercles running downward from the small scattered areoles, and finally without leaves, spines, or bristles, sterile below, with thick fleshy walls and with a small narrow seed-cavity; umbilicus broad, slightly depressed; seeds thick with flattened sides rounded on the back, 5 mm. in diameter.

In its large, red, rotate flowers this plant at once suggests a *Pereskia*. Its red flowers are so similar to those of *P. bahiensis* of the same region that at first we considered the two species congeneric. Now that we have studied the fruit and seed it is evident that *P. zehntneri* belongs to a very different genus. Then, too, the old areoles develop deciduous spines or bristles which are doubtless glochids; these occur on the upper part of the areoles but do not form the definite brush of the *Opuntiae*. These glochids would exclude it from the *Pereskieae*. It must therefore be referred to the *Opuntieae* and next to *Pereskiopsis*. In its broad, thick leaves it resembles that genus, but its flowers are terminal, very large, and rotate; its fruit is much elongated and the seeds are glabrous.

We are indebted to Dr. Leo Zehntner, a very keen observer, for many fine specimens and much information regarding it. He has found it only on a small calcareous mountain near the city of Bom Jesus da Lapa, Brazil, but it has been transplanted to the Horto Florestal of Joazeiro where it is well established and where it flowered three years after being replanted. In 1915 Dr. Rose brought living specimens to the New York Botanical Garden from this stock (No. 19722).

On page 32, vol. 1, under *Pterocactus tuberosus*, add the synonym: *Opuntia tuberosa albispina* Salm-Dyck in Förster, Handb. Cact. ed. 2. 911. 1885.

Also add to illustrations: Haage and Schmidt, Cat. Gen. 230. 1908; De Laet, Cat. Gén. f. 74, as *Pterocactus kuntzei*.

On page 34, vol. 1, under *Nopalea cochenillifera*, add the synonyms: *Cactus nopal* Thierry, Dict. Sci. Nat. **6:** 103. 1817; *Cactus splendidus* Thierry, Dict. Sci. Nat. **6:** 103. 1817; *Cactus campechianus* Thierry, Dict. Sci. Nat. **6:** 103. 1817; *Nopalea coccifera* Lemaire, Cactées 89. 1868.

Also add to illustrations: Loudon, Encycl. Pl. ed. 1 and 3. 412. f. 6888, as *Cactus cochenillifer*; Contr. U. S. Nat. Herb. **8:** pl. 48, as spineless opuntia; Knorr, Thesaurus pl. 0,1.

On page 37, vol. 1, under *Nopalea auberi*, insert:

Opuntia auberi was described as from Cuba, but as no *Nopalea* is known from Cuba we have been unable to account for this reference. The following incidents may explain it:

L. Pfeiffer described the plant in 1840 just after his return from Cuba, where he had gone with Otto in 1838. At Havana they visited the Botanical Garden, then in charge of Pedro Auber, for whom this plant was doubtless named. It is also stated that, although Pfeiffer made this trip especially to gather cacti, he saw only one species, *Opuntia horrida*. The probabilities, therefore, are that this plant was obtained from the Botanical Garden at Havana, perhaps with a statement from Auber that it was Cuban.

On page 37, vol. 1, under *Nopalea dejecta*, add the synonym: *Nopalea angustifrons*[*] Lindberg, Act. Soc. Sc. Fenn. **10:** 123. 1871.

Add to illustrations: Act. Soc. Sc. Fenn. **10:** pl. 2, as *Nopalea angustifrons*.

On page 41, vol. 1, under *Maihuenia poeppigii*, add to illustrations: Gartenflora **30:** 412, as *Pereskia poeppigii*.

On page 42, vol. 1, under *Maihuenia brachydelphys*, insert the synonym: *Opuntia brachydelphis* Schumann in Just, Bot. Jahresb. **26**[1]**:** 343. 1898.

Insert: *Mammillaria brachydelphis* is a clerical error for *Opuntia brachydelphis*.

On page 42, vol. 1, under *Opuntia*, add the synonym: *Cactus* Lemaire,[†] Cactées 86. 1868. Not Linnaeus, 1753.

On page 46, vol. 1, under *Opuntia ramosissima*, insert: *Opuntia tessellata denudata*, according to C. R. Orcutt, is only a form—spiny joints frequently occurring on the same plant with the spineless form; it is common in the Mojave Desert, California. It was mentioned by Alverson (Cact. Cat. 6) while *O. ramosissima denudata* is listed by Weinberg (Cacti 22). *O. ramosissima cristata* is mentioned by Schelle (Handb. Kakteenk. 41. 1907).

Also add to illustrations: Cact. Journ. **1:** pl. for February; Monatsschr. Kakteenk. **8:** 71, as *Opuntia tessellata cristata*; Stand. Cycl. Hort. Bailey **4:** f. 2596, 2610.

On page 47, vol. 1, under *Opuntia leptocaulis*, add the synonym: *Opuntia californica* Engelmann in Emory, Mil. Reconn. 158. 1848.

Also insert: *Opuntia stipata* (Schumann, Index Gesamtb. Kakteen 830. 1898) refers to *O. leptocaulis stipata*.

Also add to illustrations: Emory, Mil. Reconn. 158. No. 11, as *Opuntia californica*; Gartenwelt **11:** 75, as *O. vaginata*; Carnegie Inst. Wash. **269:** pl. 10, f. 89; pl. 11, f. 96; Stand. Cycl. Hort. Bailey **2:** f. 717; Schelle, Handb. Kakteenk. 41. f. 2; Möllers Deutsche Gärt. Zeit. **25:** 475. f. 9, No. 21.

On page 49, vol. 1, under *Opuntia caribaea*, insert: Dr. Britton endeavored to find this plant in Trinidad in 1920 and 1921 but failed and he could not learn anything about it. It appears probable that the drawing sent by Mr. Lockhart to Kew in 1825 was made from a Venezuelan plant.

On page 54, vol. 1, under *Opuntia clavellina*, add to illustration: Karsten and Schenck, Vegetationsbilder **13:** pl. 18, in part.

On page 56, vol. 1, under *Opuntia whipplei*, in last line of description read cm. as mm.

---

[*] The Index Kewensis refers this name to *Opuntia leucacantha*, but the illustration shows that it belongs to *Nopalea*.

[†] Lemaire in his Les Cactées, published in 1868, takes up the name *Cactus* for certain of the low, depressed, much branched or cespitose species of *Opuntia*. He lists a number of these on pages 87 and 88, but as they are not connected through published species their identification is made only by inference.

Add to illustration: Bull. Agr. Exper. Sta. N. Mex. **78**: pl. 11, 12; North Amer. Fauna **7**: pl. 9; Pac. R. Rep. **4**: pl. 17, f. 1 to 4; Stand. Cycl. Hort. Bailey **4**: f. 2609.

On page 57, vol. I, under *Opuntia acanthocarpa*, add to illustration: Stand. Cycl. Hort. Bailey **4**: f. 2606; Gartenwelt **11**: 75.

On page 57, vol. I, under *Opuntia echinocarpa* and *O. parryi*, respectively, add the synonyms: *Cactus echinocarpus* and *C. parryi* Lemaire, Cactées 88. 1868.

On page 58, vol. I, under *Opuntia bigelovii*, add to illustrations: MacDougal, Bot. N. Amer. Des. pl. 47: Shreve, Veg. Des. Mt. Range pl. 4; Contr. U. S. Nat. Herb. **16**: pl. 10; Stand. Cycl. Hort. Bailey **4**: f. 2607; Karsten and Schenck, Vegetationsbilder **4**: pl. 40, B.

On page 61, vol. I, under *Opuntia cholla*, insert: *Opuntia chella* (Index Kew. Suppl. 1: 302) is a typographical error for *O. cholla*.

On page 62, vol. I, under *Opuntia versicolor*, add to illustrations: Carnegie Inst. Wash. **269**: pl. 8, f. 81; pl. 9; MacDougal, Bot. N. Amer. Des. pl. 58; Plant World 9^{12}: f. 50.

On page 63, vol. I, under *Opuntia imbricata*, add the synonym: *Cactus imbricatus* Lemaire, Cactées 88. 1868. Also add to distribution: Oklahoma.

Insert: Rydberg (Fl. Rocky Mountains 576. 1917) reports this species from Utah under the name of *Opuntia arborescens*; we have seen no specimens of it from Utah.

Insert: *Cactus subquadriflorus* Mociño and Sessé (De Candolle, Prodr. **3**: 471. 1828), given as a synonym of *Opuntia rosea*, doubtless belongs here. Schumann's reference, *C. quadriflorus*, is incorrect. *C. subquadrifolius* (Cactaceae **3**: 65) is a clerical error.

Add to illustrations: Dict. Gard. Nicholson Suppl. 179. f. 195, as *Opuntia decipiens*; Dict. Gard. Nicholson **4**: 581. f. 52, as *O. rosea*; Stand. Cycl. Hort. Bailey **4**: f. 2608; Engler and Drude, Veg. Erde **13**: f. 28, in part; Gartenwelt **4**: 159, as *O. arborescens*; Bot. Jahrb. Engler **58**: Beibl. **129**: 33. f. 10.

On page 66, vol. I, under *Opuntia tunicata*, add to illustrations: Garden **13**: 107,* as *Opuntia exuviata*; Möllers Deutsche Gärt. Zeit. **25**: 476. f. 9, No. 7; Goebel, Pflanz. Schild. **1**: f. 36, as *O. stapeliae*; Contr. U. S. Nat. Herb. **10**: pl. 17, f. A.

On page 68, vol. I, under *Opuntia fulgida*, add to illustrations: MacDougal, Bot. N. Amer. Des. pl. 57, as *Opuntia mamillata*; MacDougal, Bot. N. Amer. Des. pl. 87.

On page 68, vol. I, under *Opuntia spinosior*, insert: This plant is sometimes found in the trade as *Opuntia arborescens spinosior* (see Grässner).

Add to illustrations: Emory, Mil. Reconn. App. 2. f. 10, as *Opuntia arborescens*; Shreve, Veg. Des. Mt. Range pl. 2, A.

On page 71, vol. I, under *Opuntia vestita*, insert: *Illustration:* Möllers Deutsche Gärt. Zeit. **25**: 476. f. 9, No. 8.

On page 73, vol. I, under *Opuntia clavarioides*, add to illustrations: Garden **13**: 107, as *Opuntia clavarioides cristata*; Rother, Praktischer Leitfaden Kakteen 106; Möllers Deutsche Gärt. Zeit. **15**: 67; **25**: 476. f. 9, No. 19; Thomas, Zimmerkultur Kakteen 59; Wiener Ill. Gärt. Zeit. **28**: f. 18; Monatsschr. Kakteenk. **32**: 131.

On page 73, vol. I, under *Opuntia salmiana*, insert: Extend range to central Argentina and habit to rocky hillsides (according to W. B. Alexander).

On page 75, vol. I, under *Opuntia subulata*, add to illustrations: Deutsche Gärt. Zeit. **8**: 32, as *Pereskia subulata*; Haage and Schmidt, Haupt-Verz. Kakteen **1919**: 169; Goebel, Pflanz. Schild. **1**: f. 35; Möllers Deutsche Gärt. Zeit. **25**: 476. f. 9, No. 15.

On page 78, vol. I, under *Opuntia cylindrica*, add to illustrations: Möllers Deutsche Gärt. Zeit. **25**: 476. f. 9, No. 12; Gartenwelt **15**: 539; Rother, Praktischer Leitfaden Kakteen 107; Cact. Journ. **1**: 100; Schelle, Handb. Kakteenk. 42. f. 4, as *Opuntia cylindrica cristata*; Wiener Illustr. Gartenz. **29**: f. 22, No. 10; De Laet, Cat. Gén. f. 88; Monatsschr. Kakteenk. **13**: 71; Schelle, Handb. Kakteenk. 42. f. 3.

---

*This illustration is very poor and is only tentatively referred here. If native to California, as one might infer from the account which accompanies the illustration, it may refer to a form of *Opuntia prolifera* or *O. echinocarpa*.

On page 80, vol. 1, under *Opuntia stanlyi*, add the synonym: *Cactus emoryi* Lemaire, Cactées 88. 1868.

Also add to illustrations: Schelle, Handb. Kakteenk. 38. f. 1, as *Opuntia emoryi*; Nat. Geogr. Mag. **21**: pl. on p. 716, as *O. kunzei*.

On page 80, vol. 1, under *Opuntia schottii*, insert: *Opuntia greggii* occurs only in Schumann's Index (Gesamtb. Kakteen 829) with page reference to *O. schottii greggii*.

On page 81, vol. 1, under *Opuntia clavata*, insert the synonym: *Cactus clavatus* Lemaire, Cactées 88. 1868.

Add to illustrations: Stand. Cycl. Hort. Bailey **4**: f. 2605.

On page 82, vol. 1, under *Opuntia pulchella*, add to illustration: MacDougal, Bot. N. Amer. Des. pl. 26, as *O. pusilla*.

On page 83, vol. 1, under *Opuntia bulbispina*, insert: *Cactus bulbispinus* Lemaire. (Cactées 88. 1868) was intended as a synonym of this species.

On page 89, vol. 1, under *Opuntia glomerata*, insert: Extend range to central and northern Argentina.

Insert: *Tephrocactus polyacanthus* (Index Kewensis Suppl. **1**: 421) was intended for *T. platyacanthus* Lemaire (Förster, Handb. Cact. ed. 2. 915. 1885).

Add to illustrations: Watson, Cact. Cult. ed. 1 and 2. 257. f. 97; ed. 3. f. 60, as *Opuntia papyracantha*; Dict. Gard. Nicholson **2**: 503. f. 755; Möllers Deutsche Gärt. Zeit. **25**: 476. f. 9, No. 1, as *O. platyacantha*; Schelle, Handb. Kakteenk. 45. f. 7, as *O. andicola*; De Laet, Cat. Gén. f. 60; Rev. Hort. Belg. **40**: after 186; Schelle, Handb. Kakteenk. 44. f. 6; Möllers Deutsche Gärt. Zeit. **25**: 476. f. 9, No. 2, as *O. diademata*.

On page 92, vol. 1, under *Opuntia aoracantha*, add to illustrations: Schelle, Handb Kakteenk. 44. f. 5.

On page 93, vol. 1, under *Opuntia hickenii*, insert: Mr. W. B. Alexander suggests that *Opuntia platyacantha* Spegazzini (not Salm-Dyck) is probably a synonym of this species.

On page 94, vol. 1, insert:

64a. Opuntia wetmorei sp. nov.

Forming low mounds of considerable extent with hundreds of branches; joints 4 to 10 cm. long, terete, turgid, 2. cm. in diameter or less, slightly tapering towards each end, dull green, but

FIG. 234.—Opuntia wetmorei, fruit, stem, and seeds.

dull purple around and especially below the areoles; leaves subtending the minute areoles, 1 to 2 mm. long, caducous; areoles circular, bearing tawny or white wool when young; glochids short, yellowish; spines numerous, very unequal, scarcely pungent, white to straw-colored or brownish, 3 or 4 of lower ones almost hair-like, reflexed or appressed to joints, 3 or 4 of uppermost erect or ascending, flattened, 2 to 3.5 cm. long; flowers not known; immature fruit glabrous at first, dull green, becoming reddish purple especially about the areoles, 3 cm. long, bearing long white bristly spines, especially from upper areoles, deeply umbilicate.

Collected by W. B. Alexander in the barranca of the Tunuyán River near Tunuyán, Mendoza, Argentina, March 22 and 23, 1921.

This species is perhaps nearest *Opuntia darwinii*. We are under great obligation to W. B. Alexander for sending us very fine living plants by Alexander Wetmore, who brought them to us directly from Argentina. Mr. Wetmore was with Mr. Alexander when the plant was collected and he has given us a word picture of the plant; we take pleasure in naming the species for him, not only in recognition of this service but also for obtaining other valuable specimens of cacti.

Figure 234 is from a photograph of the type plant, one-half natural size.

On page 95, vol. 1, under *Opuntia corrugata*, insert: *Tephrocactus rectrospinus* (Index Kewensis Suppl. 1: 421) is a misspelling for *T. rectrospinosus* Lemaire.

Also insert: *Illustrations:* Möllers Deutsche Gärt. Zeit. **25**: 476. f. 9, No. 11; 488. f. 22, No. 8.

On page 95, vol. 1, under *Opuntia ovata*, add:

Opuntia pusilla Salm-Dyck (Observ. Bot. **3**: 10. 1822. Not Haworth, 1812) was referred by Schumann to *O. corrugata*. We have seen a photograph of Haworth's specimen (bearing the date November 8, 1824) which seems to answer to Salm-Dyck's plant which we would refer here.

On page 96, vol. 1, under *Opuntia sphaerica*, add the synonym: *Opuntia ovata leonina* Schelle, Handb. Kakteenk. 46. 1907.

Also add to illustrations: Deutsche Gärt. Zeit. **7**: 313, as *Opuntia leonina*; Schelle, Handb. Kakteenk. 46. f. 8, as *O. grata leonina*.

On page 97, vol. 1, under *Opuntia pentlandii*, add the synonym: *Cactus bolivianus* Lemaire, Cactées 88. 1868.

Also add to illustrations: Watson, Cact. Cult. ed. 3. 106. f. 54; Deutsche Gärt. Zeit. **7**: 312; Schelle, Handb. Kakteenk. 58. f. 16, as *Opuntia boliviana*; Möllers Deutsche Gärt. Zeit. **25**: 476. f. 9, No. 14.

On page 99, vol. 1, insert:

76a. Opuntia alexanderi sp. nov.

Low, depressed, forming a small clump; joints readily detached, grayish green, strongly tubercled, globose, 2 to 3 cm. in diameter, nearly hidden by the numerous spines; areoles small, close together, circular; spines 4 to 12, up to 4 cm. long, flexible, white below, dark above or with black tips, scurfy-pubescent even in age; flowers not known; fruit red, dry, obovoid, 2 cm. long, lower areoles not spiny, but upper ones bearing 2 to 8 long, white, erect, weak spines overtopping the fruit; umbilicus of fruit depressed; seeds white, 5 to 6 mm. broad.

Collected by W. B. Alexander, between Chilecito and Fanatina, province of La Rioja, Argentina, February 19, 1921. Mr. Alexander studied this species in the field but could not identify it and sent it to us for study. It belongs to the subgenus *Tephrocactus*, but is not near any of the known species. We take great pleasure in naming it for Mr. Alexander, who has extensively studied the cacti in Argentina.

On page 100, vol. 1, under *Pumilae*, add to distribution: Venezuela.

On page 100, vol. 1, under *Opuntia pumila*, insert: *Illustration*: Möllers Deutsche Gärt. Zeit. **25**: 476. f. 9, No. 5.

On page 101, vol. 1, under *Opuntia pubescens*, add the synonym: *Cactus pubescens* Lemaire, Cactées 87. 1868.

On page 102, vol. 1, under *Opuntia curassavica*, add to illustrations: Dillenius, Hort. Elth. **2**: pl. 295, as tuna; Loudon, Encycl. Pl. 413. f. 6897, as *Cactus curassavicus*; Knorr, Thesaurus pl. o.2.

On page 102, vol. 1, insert:

80a. Opuntia abjecta Small, sp. nov.

Prostrate, often growing in large irregular patches on almost bare limestone or where some sand and humus has accumulated, irregularly branched; joints suborbicular, sometimes nearly subglobose, oval, or broadly obovate, mostly 4 to 8 cm. long, very thick, frequently turgid, light green, loosely attached to each other; leaves ovoid to conic-ovoid, 2 to 3 mm. long, ascending and slightly curved upward, green or purplish; glochids yellowish; spines setaceous-acicular, mostly solitary, brown, or reddish purple, mottled light and dark, becoming chalky gray when dry; the larger ones 2 to 6 cm. long; flowers usually solitary on a joint; berry urceolate, 1 to 1.5 cm. long, somewhat tuberculate, red or purple-red, rounded at base; umbilicus relatively broad, concave; seeds few, flattish, about 4 mm. wide.

On edge of hammock, southern end of Big Pine Key, Florida. Type collected in May 1921 by J. K. Small, preserved in the herbarium of the New York Botanical Garden.

Similar to *Opuntia drummondii* but with shorter joints, longer and more slender spines, and different fruit.

On page 105, vol. 1, under *Opuntia drummondii*, add to illustration: Journ. Elisha Mitchell Sci. Soc. **34**: pl. 13, 14.

On page 105, vol. 1, under *Opuntia tracyi*, insert:

Type Locality: Biloxi, Mississippi. *Distribution:* Southern Mississippi, southeastern Georgia to northern Florida.

On page 105, vol. 1, insert:

86a. Opuntia impedata Small, sp. nov.

Prostrate, ultimately copiously branched, the joints often piled several layers deep and forming viciously armed mats, elliptic or oblong, mostly 7 to 15 cm. long, rather thick, pale green; leaves

FIG. 235.—Opuntia impedata.

stout-subulate, 4 to 6 mm. long, erect or ascending, slightly curved upward, dark green; glochids brownish; spines subulate, usually numerous, solitary or 2 together, light gray, except the brown tip, salmon-colored when dry, and faintly banded when wet; flowers often several on a joint; ovary obconic, nearly terete; sepals green, outer lanceolate to ovate, 4 to 8 mm. long, acuminate, the inner much larger, with shoulders of very broad body narrowed into stoutish tip; corolla bright yellow, 4.5 to 5.5 cm. wide; petals about 12, 2.5 to 3 cm. long, broadly obovate to cuneate-obovate, broadly rounded at apex, mucronate; anthers nearly 2 mm. long; berry clavate, about 3 cm. long, narrowed at base; umbilicus rather small, somewhat concave; seeds rather few, 4 to 4.5 mm. in diameter.

Sand dunes, northeastern Florida. Type in the herbarium of the New York Botanical Garden; collected on dunes at Atlantic Beach, Florida, in April 1921, by J. K. Small.

Dr. Small notes that the stiff spines may penetrate leather shoes and that the plant is very prolific, both vegetatively and through its fruit.

Figure 235 is from a photograph taken by Dr. Small of the type plant.

On page 110, vol. I, insert:

Series 3a. PISCIFORMES.

Plants in dense colonies with turgid, very spiny, narrow, deep green joints, the spines conspicuously long and slender, salmon-colored in the first year, gray in the second; flowers numerous, bright yellow; berry turbinate-obovoid, 4 cm. long or less. The only species inhabits Florida.

96a. Opuntia pisciformis Small, sp. nov.

Prostrate, copiously branched, forming dense mats often 1 to 3 meters in diameter, with joints piled several layers deep, roots fibrous; joints narrowly elliptic, linear-elliptic, or spatulate, mostly

FIG. 236.—Opuntia pisciformis.

1 to 3 dm. long, very thick, deep green, readily detached; leaves stout-subulate, 2 to 4 mm. long, incurved; areoles rather prominent, mostly armed; spines solitary or 2 or 3 together, cream-colored, becoming salmon-colored and gray with a dark tip when dry, salmon when wet, the longer ones 5 to 6 cm. long; flowers numerous; ovary turbinate, angular and tuberculate; sepals green, the outer lanceolate to triangular-lanceolate, 9 to 12 mm. long, acuminate, the inner much larger, the broad ovate or suborbicular base broadly tapering into the very stout tip; corolla bright yellow, 6 to 7.5 cm. wide; petals about 12, 3 to 4 cm. long, broadly cuneate, mostly truncate or emarginate at apex, mucronate; anthers nearly 2 mm. long; berry broadly turbinate-obovoid, 3.5 to 4 cm. long, purple, narrowed at base, the umbilicus deeply concave; seeds rather numerous, 5 to 5.5 mm. in diameter.

Sand dunes, estuary of the Saint Johns River, Florida. Type in the herbarium of the New York Botanical Garden; collected on dunes at Atlantic Beach, Florida, in April 1921, by J. K. Small,
Figure 236 is from a photograph by Dr. Small of the type plant.

On page 113, vol. I, under *Opuntia tuna*, in first line read 1769 as 1768.
Add the synonyms: *Cactus horridus* Salisbury, Prodr. 348. 1796; *Opuntia tuna humilior* Salm-Dyck, Cact. Hort. Dyck. 1849. 66. 1860.
Insert: *Opuntia maidenii* Griffiths (Bull. Torr. Bot. Club **46**: 201. 1919) described from a cultivated plant sent from Australia and grown at Chico, California, seems referable to this species.
Add to illustrations: Loudon, Encycl. Pl. 411. f. 6880, as *Cactus polyanthos*; Monatsschr. Kakteenk. **6**: 25, as *Opuntia polyantha*; Deutsche Gärt. Zeit. **7**: 447, as *O. humilis*; Watson, Cact. Cult. ed. 3. f. 62; Cact. Journ. **2**: 169; Useful Wild Plants U. S. Canada, opp. 18, 108, 174; Stand Cycl. Hort. Bailey **4**: f. 2599; Schelle, Handb. Kakteenk. 51. f. 13; Remark, Kakteenfreund 24.

On page 115, vol. I, under *Opuntia antillana*, insert: *Opuntia domingensis* appears without description in Urban's Symbolae (**8**: 466. 1920). It was a manuscript name for which *O. antillana* was substituted.
On page 117, vol. I, under *Opuntia decumbens*, add to illustrations: Bull. U. S. Dept. Agr. **31**: pl. 7, f. 1, as *Opuntia puberula*; Möllers Deutsche Gärt. Zeit. **25**: 476. f. 9, No. 3.

On page 119, vol. I, under *Opuntia basilaris*, insert: *Opuntia dorffii* is advertised by Haage and Schmidt (Monatsschr. Kakteenk. **29**: September). We have had a cutting which we would refer to one of the forms of *O. basilaris*.
Also add to illustrations: Cact. Journ. **2**: 163, as *Opuntia basilaris albiflora*; Cact. Journ. **1**: pl. for October; Möllers Deutsche Gärt. Zeit. **25**: 476. f. 9, No. 13, as *O. basilaris cordata*; Möllers Deutsche Gärt. Zeit. **25**: f. 9, No. 9, as *O. basilaris minima*; Watson, Cact. Cult. ed. 3. f. 53; Deutsche Gärt. Zeit. **7**: 312; Remark, Kakteenfreund 23; Monatsschr. Kakteenk. **7**: 125; Stand. Cycl. Hort. Bailey **4**: f. 2597; Gartenflora **31**: 280; Schelle, Handb. Kakteenk. 47. f. 10.

On page 121, vol. I, under *Opuntia microdasys*, add to illustrations: Möllers Deutsche Gärt. Zeit. **25**: 488. f. 2, No. 4, as *Opuntia microdasys monstrosa*; Garden **13**: 107,* as *O. pubescens*; Schelle, Handb. Kakteenk. 47. f. 9; Möllers Deutsche Gärt. Zeit. **25**: 476. f. 9, No. 16; Karsten and Schenck, Vegetationsbilder **2**: pl. 22, B.
On page 123, vol. I, under *Opuntia pycnantha*, insert: *Opuntia pycnacantha* (Just's Jahresb. **24**²: 380. 18) seems to have been a misspelling for *O. pycnantha*.

On page 127, vol. I, under *Opuntia opuntia*, add the synonym: *Opuntia compressa* Macbride, Contr. Gray Herb. II. **65**: 41. 1922.
Also add to illustrations: Contr. U. S. Nat. Herb. **21**: pl. 23, B; Bailey, Sand Dunes Indiana 94; Ann. Inst. Roy. Hort. Fromont **2**: pl. 1, f. F; Deutsches Mag. Gart. Blumen. **1869**: pl. 17. opp. 257; Kraemer, Appl. Econ. Bot. f. 341, as *Opuntia vulgaris*; Watson, Cact. Cult. 212. f. 84; Ann. Rep. Bur. Amer. Ethn. **33**: pl. 20, A; Clements and Clements, Rocky Mt. Fl. pl. 32, f. 6; Clements, Fl. Mount. Plain pl. 32, f. 7, as *O. humifusa*; Wiener, Ill. Gart. Zeit. **2**: 40. f. 10, as *O. rafinesquiana*; Deutsche Gärt. Zeit. **7**: 447; Wiener Ill. Gärt. Zeit. **2**: f. 112, as *O. rafinesquiana arkansana*; Watson, Cact. Cult. ed. 3. f. 61;

---

* This illustration is very poor and the identification is based largely upon the description.

Schelle, Handb. Kakteenk. 50. f. 12; Belg. Hort. **26**: pl. 8; Illustr. Hort. **15**: pl. opp. 51; Deutsches Mag. Gart. Blumen. **1869**: pl. 17, opp. 257, as *O. rafinesquei*; Kraemer, Appl. Econ. Bot. f. 341.

On page 130, vol. 1, insert:

121*a*. **Opuntia eburnispina** Small, sp. nov.

Prostrate, widely branched and forming mats on dune sands, with tuberous roots; joints oval or suborbicular, varying to broadest above middle, thickish, 6 to 13 cm. long, pale green, somewhat shining, especially when young; leaves ovoid-subulate, 4 to 5 mm. long, pale green, recurved-spreading; spines relatively stout, 2 to 4 at an areole or sometimes solitary, 1 to 2 cm. long, ivory-white with yellowish tips when young, becoming dark gray, not spirally twisted, greenish when wet; flowers few; ovary obconic; sepals triangular, green, 5 to 7 mm. long; corolla clear yellow, 4 to 5 cm. wide; petals few, narrowly cuneate, often minutely pointed; berries obovoid, 2 cm. long or less.

Coastal sands, Cape Romano, Florida. Type specimens in the herbarium of the New York Botanical Garden; collected in May 1922, by J. K. Small.

Figure 237 is from a photograph by Dr. Small of the type plant.

Fig. 237.—Opuntia eburnispina.

On page 131, vol. 1, under *Opuntia macrorhiza*, add to illustrations: Watson, Cact. Cult. ed. 3. f. 59; Dict. Gard. Nicholson **4**: 580. f. 50, 51.

On page 131, vol. 1, under *Opuntia tortispina*, add the synonym: *Opuntia cymochila montana* Englemann, Proc. Amer. Acad. **3**: 296. 1856.

Also add to distribution: Southeastern Colorado. Established and slowly spreading east of Cincinnati, Ohio (E. T. Wherry).

Also add to illustrations: Watson, Cact. Cult. ed. 3. pl. opp. 102; Meehans' Monthly **11**: 57, as *Opuntia mesacantha*; Meehans' monthly **5**: 172, as *O. oplocarpa*.

On page 134, vol. 1, *Opuntia sulphurea*, insert: Mr. W. B. Alexander writes as follows concerning this species:

"This is by far the commonest species of *Opuntia* in the Argentine, where it is commonly known as 'penca,' i. e. the spiny plant, sometimes being distinguished from other larger species by the name 'penquilla' or 'penca chica.' The writer met with it in the provinces of Buenos Aires, Córdoba, San Luis, Mendoza, San Juan, La Rioja, Catamarca and Santiago del Estero."

Add to illustrations: Wiener Ill. Gärt. Zeit. **28**: f. 17, as *Opuntia maculacantha*; Möllers Deutsche Gärt. Zeit. **25**: 476. f. 9, No. 18.

On page 134, vol. 1, under *Opuntia soehrensii*, add the synonyms: *Cactus ayrampo* Azara, Voy. **2**: 526. 1809; *Opuntia haenquiana* Herrera, Rev. Univ. Cuzco **8**: 60. 1919.

Also insert: Azara's original description is interesting and a translation of it is given:

"A species of tunilla (cactus) which is found in the temperate gorges near the Cordillera produces the seed in question. The plant is found in arid and sterile soil where ordinarily this family of plants grows and thrives by creeping on the ground in such a way as to stifle all the others. From the seed confined within the round and spiny fruit is derived a color of a clear violet, brilliant and extremely agreeable to the eye but very superficial and very light, although it acquires a little stability and durability by the means of alum and some other chemicals."

On page 135, vol. 1, insert:

129a. Opuntia macbridei sp. nov.

A low bush, 6 dm. high, forming broad impenetrable thickets on gravelly river flats; joints obovate, 6 to 8 cm. broad, 8 to 15 cm. long, glabrous, at first light green, in age dark green; leaves minute, 1 to 2 mm. long, caducous; areoles on young joints hemispheric, brown-felted and with

FIG. 238.—Opuntia macbridei.

brown glochids, on old joints 2 to 3 cm. apart; spines 2 to 4, in age gray to horn-colored, with yellowish tips, very unequal, the longest up to 5 cm. long, stout-subulate; flowers very small, orange to orange-red; petals only 4 to 5 mm. long; ovary tuberculate, bearing many brown-felted tubercles but without spines, deeply umbilicate; fruit deeply umbilicate, red to purple.

Collected by Macbride and Featherstone at Huanuco, Peru, altitude 2,300 meters, August 28 to September 3, 1922 (No. 2365, type), and April 8, 1923 (No. 3250).

Mr. Macbride states that the seeds are brown. All the fruits we have seen were sterile; these sterile fruits on falling to the ground take root and form new plants.

This interesting plant, which proves to be undescribed, we have named for Mr. J. Francis Macbride, who led the Botanical Expedition of 1922 to South America, sent out by the Field Museum of Natural History, under the Captain Marshall Field fund.

Figure 238 is from a photograph showing the habit of this plant.

On page 135, vol. 1, under *Opuntia penicilligera*, insert: Mr. W. B. Alexander sends us the following account of this plant:

"This plant was met with close to the coast at Bahia Blanca, and near the foot of the Andes at Tunuyán. As remarked by Spegazzini, this species is very distinct from any other found in Argentina and there seems no reason for thinking that it may belong to the Series *Sulphureae* in which it is tentatively placed in the Cactaceae. It should surely either be the type of a separate series or be placed in the Series *Basilares*, to the members of which, judging by illustrations, it shows great resemblance.''

On page 138, vol. 1, under *Opuntia pottsii*, add to illustrations: Watson, Cact. Cult. ed. 3. f. 58; Dict. Gard. Nicholson **4**: 580. f. 49, as *Opuntia filipendula*.

On page 145, vol. 1, under *Opuntia phaeacantha*, add to illustrations: Deutsche Gärt. Zeit. **7**: 447, as *Opuntia camanchica*; Meehans' Monthly **11**: 57, as *O. phaeacantha major*; Shreve, Veg. Des. Mt. Range pl. 5, A, as *O. toumeyi*; De Laet, Cat. Gén. f. 58.

On page 147, vol. 1, under *Opuntia engelmannii*, add the synonym: *Opuntia engelmannii discata* C. Z. Nelson, Trans. Ill. State Acad. Sci. **12**: 124. 1919.

Also add to illustrations: Cact. Journ. **1**: pl. for February; **2**: 162, as *Opuntia engelmannii cristata*; Gard. Chron. III. **39**: 148. f. 58; Plant World **9**[12]: f. 49; Shreve, Veg. Des. Mt. Range pl. 5, B; Stand. Cycl. Hort. Bailey **4**: f. 2601; Scientific Month. **17**: 70, 71, 72.

On page 149, vol. 1, under *Opuntia discata*, add to illustrations: Carnegie Inst. Wash. **269**: pl. 10, f. 87.

On page 153, vol. 1, under *Opuntia bergeriana*, add to illustrations: Gartenwelt **11**: 75.

On page 153, vol. 1, under *Opuntia elatior*, add to illustrations: London, Encycl. Pl. ed. 3. 411. f. 6879, as *Cactus nigricans*.

On page 155, vol. 1, under *Opuntia boldinghii*, add to distribution: Chacachacare and Patos Islands, Trinidad.

On page 156, vol. 1, under *Opuntia vulgaris*, insert: *Opuntia gracilior* (Index Kewensis **3**: 357. 1894) is a mistake for *O. monacantha gracilior* Lemaire.

Add to illustrations: Möllers Deutsche Gärt. Zeit. **25**: 476. f. 9, No. 20, as *Opuntia monacantha variegata*; Pl. Utiles Madagascar 124. f. 39; 125. f. 39.

On page 158, vol. 1, under *Opuntia arechavaletai*, add to illustration: Karsten and Schenck, Vegetationsbilder **11**: pl. 17.

On page 158, vol. 1, under *Opuntia bonaerensis*, insert: Mr. W. B. Alexander writes of this species as follows:

This species was seen only on rocky slopes in the Sierra de la Ventana in the south of the province of Buenos Aires. It is known only from the few Sierras which rise from the pampas in the east of the province. There is little doubt that it is nearly related to *Opuntia vulgaris* Miller (*O. monacantha* Haworth) which was found by the writer at Rio de Janeiro and is familiar in Australia.

Add to illustration: Anal. Mus. Nac. Montevideo **5**: pl. 33, as *Opuntia chakensis*.

On page 159, vol. 1, insert after *Opuntia scheeri*: *Opuntia diversispina* Griffiths (Bull. Torr. Club **46**: 197. pl. 9. 1919) grown from seed of unknown origin at Brownsville, Texas, is described as similar to *O. scheeri* and in the accompanying illustration the joints resemble those of that species.

On page 160, vol. 1, *Opuntia chlorotica*, add to illustrations: Bull. N. Mex. Coll. Agr. No. 78. pl. [4]; Stand. Cycl. Hort. Bailey **4**: f. 2600.

On page 161, vol. 1, under *Opuntia laevis*, add to illustrations: MacDougal, Bot. N. Amer. Des. pl. 56.

On page 163, vol. 1, under *Opuntia dillenii*, add to illustrations: Garden **13**: 107,* as *Opuntia crassa;* Bull. Torr. Club **46**: pl. 10, as *O. maritima*; Lindley, Veg. King. ed. 3. 746. f. 498, No. 1, 2: Knorr, Thesaurus pl. o; Watson, Cact. Cult. ed. 3. f. 56.

On page 163, vol. 1, insert:

174a. Opuntia ochrocentra Small, sp. nov.

Erect, 1 meter tall or less, much branched or sometimes diffuse, with fibrous roots; joints elliptic to oval, varying to broadest above the middle, 1 to 3 dm. long, thickish, light green, not

---

* This illustration is not very good for this species. It is, however, the same one that Nicholson used (f. 757) and that W. Watson used (f. 86) as *Opuntia tuna*, which we have referred here.

repand; leaves ovoid, 2 to 4 mm. long, often purplish; areoles rather prominent; glochids yellowish brown; spines 5 to 6 together or sometimes fewer on new joints, yellow, stiff, subulate, reflexed, becoming gray when dry, yellowish green when wet, straight, the longer ones 4.6 to 5 cm. long; flowers rather few; ovary turbinate, even; sepals often purple-tinged, deltoid to rhombic-orbicular or rhombic-reniform, acute; corolla bright lemon-yellow, 7 to 8.5 cm. wide; petals few, cuneate, somewhat crisped; berry obovoid, red, about 2 cm. long.

On edge of hammock, southeastern end of Big Pine Key, Florida. Type specimens collected in December 1921, by J. K. Small, in the herbarium of the New York Botanical Garden.

Related to *O. dillenii*, differing in shape of the joints, which are not repand, and the strongly reflexed, scarcely flattened spines.

On page 166, vol. 1, under *Opuntia lindheimeri*, add to illustrations: Journ. Hered. Washington, 6^4: f. 19, as *Opuntia ellisiana*; Journ. Hered. Washington 6^4: f. 15, 16, as *O. cacanapa*; Journ. Hered. Washington 6^4: f. 17, 18; as *O. subarmata*; Journ. Hered. Washington 5: 223. f. 13; Schulz, 500 Wild Fl. San Antonio pl. 12.

Also insert: Dr. Small has found this plant established, after cultivation, in pine lands west of Halenville, Florida.

On page 167, vol. 1, under *Opuntia cantabrigiensis*, add to illustrations: Gartenwelt 10: 560; Gard. Chron. III. 33: 98. f. 42.

Also insert: Professor Duncan S. Johnson found this species naturalized on sand dunes at Beaufort, North Carolina, in 1899, and Doctor Small studied it there in 1922. At Cambridge, England, it has passed through many winters out of doors.

On page 168, vol. 1, under *Opuntia beckeriana*, insert: *Opuntia prostrata spinosior* (Schumann, Gesamtb. Kakteen 723. 1898) seems to have been a garden name which Schumann would refer to *O. beckeriana*.

On page 173, vol. 1, under *Opuntia tomentosa*, add to illustrations: Blanc, Cacti 82. No. 2200, as *Opuntia lurida*; Reiche, Elem. Bot. f. 165; Gartenwelt 11: 75.

On page 175, vol. 1, under *Opuntia leucotricha*, add to illustrations: Möllers Deutsche Gärt. Zeit. 25: 476. f. 9, No. 4, as *Opuntia leucacantha*; Cassell's Dict. Gard. 2: 138; Bull. U. S. Dept. Agr. 31: pl. 6, f. 2; pl. 7, f. 2; U. S. Dept. Agr. Bur. Pl. Ind. Bull. 262: pl. 4; pl. 5, f. 1.

Insert: Dr. John K. Small has found this plant naturalized in a hammock south of Fort Pierce, Florida, where it is reported as established during the Seminole wars.

On page 176, vol. 1, under *Opuntia orbiculata*, add to the illustrations: Schelle, Handb. Kakteenk. 48. f. 11, as *Opuntia crinifera*; Gartenwelt 11: 76, as *O. lanigera*.

In third line of description on page 177 read cm. as dm.

On page 178, vol. 1, under *Opuntia ficus-indica*, add to illustration: Engler and Prantl, Pflanzenfam. 3^{6a}: f. 57, H; Gard. Chron. III. 34: 89. f. 34; 92. f. 42; Karsten, Deutsche Fl. 887. f. 501. No. 10, 11; ed. 2. 2: 456. f. 605. No. 10, 11; Journ. Dept. Agr. S. Austr. 13: 764; Garten-Zeitung 4: 182. f. 42, No. 1; Stand. Cycl. Hort. Bailey 4: f. 2598; Watson, Cact. Cult. ed. 3. f. 57.

On page 180, vol. 1, under *Opuntia maxima*, add the synonym: *Cactus maximus* Colla, Mem. Accad. Sci. Torino 33: 140. 1826 (?).

Also insert: *Illustration:* Möllers Deutsche Gärt. Zeit. 25: 488. f. 22, No. 3, as *Opuntia labouretiana*.

On page 181, vol. 1, under *Opuntia hernandezii*, insert: *Opuntia hernandezii* first appeared in De Candolle's Prodromus (3: 474. 1828).

Also insert: *Nopal silvestre* Thierry (Förster, Handb. Cact. ed. 2. 929. 1885) is cited as a synonym of *Opuntia hernandezii*. This reference is given also in the Index Kewensis.

Also insert: *Illustration:* Förster, Handb. Cact. ed. 2. 930. f. 128.

On page 184, vol. 1, under *Opuntia streptacantha*, add to illustrations: Useful Wild Pl. U. S. Canada opp. 18, 108, 174, as *Opuntia tuna*.

On page 185, vol. 1, under *Opuntia megacantha*, insert: *Opuntia effulgia* Griffiths (Bull. Torr. Club **46**: 195. 1919) was obtained from San Luis Potosí, Mexico, and grown at Chico, California; *O. hispanica* Griffiths (Bull. Torr. Club **46**: 198. 1919) was described from a plant received from Spain and grown at Chico; *O. chata* Griffiths (Bull. Torr. Club **46**: 199. 1919), from Aguascalientes, Mexico, was grown at Brownsville, Texas, and at Chico; *O. obovata* Griffiths (Bull. Torr. Club **46**: 202. 1919) from Hepasote, Mexico, was also grown at Brownsville and at Chico; *O. amarilla* Griffiths (Bull. Torr. Club **46**: 205. 1919) was obtained in cultivation at Cardenas, Mexico, and grown at Chico. These are known to us only from descriptions and appear to be races of *O. megacantha* or of some of the related tall, white-spined species.

Add to illustrations: Ann. Rep. Smiths. Inst. **1917**: pl. 16, f. 2.

On page 191, vol. 1, under *Opuntia robusta*, insert: *Opuntia cyanea* Griffiths (Bull. Torr. Club **46**: 196. 1919) judging from the original description may be related to *O. robusta*.

Add to illustrations: Engler and Prantl, Pflanzenfam. **3**^{6a}: f. 56, G, as *Opuntia albicans*.

On page 194, vol. 1, *Opuntia fragilis*, add to illustrations: Watson, Cact. Cult. ed. 3. f. 55; Deutsche Gärt. Zeit. **7**: 313; Remark, Kakteenfreund **22**, as *Opuntia brachyarthra*; Schelle, Handb. Kakteenk. 56. f. 15, as *O. fragilis brachyarthra*; Meehans' Monthly **11**: 57.

On page 195, vol. 1, under *Opuntia arenaria*, add to illustration: Meehans' Monthly **11**: 57.

On page 195, vol. 1, *Opuntia erinacea*, add the synonym: *Opuntia ursus horribilis* Walton, Cact. Journ. **2**: 152. 1899.

Also add to illustrations: Cact. Journ. **1**: 93, as *Opuntia*; Möllers Deutsche Gärt. Zeit. **25**: 476. f. 9, No. 1c; Cycl. Amer. Hort. Bailey **3**: 1149. f. 1548; Stand. Cycl. Hort. Bailey **4**: 2363. f. 2603, as *O. ursina*; Meehans' Monthly **4**: 9; Monatsschr. Kakteenk. **14**: 105; N. Amer. Fauna **7**: pl. 11, as *O. rutila*.

On page 198, vol. 1, under *Opuntia rhodantha*, add to illustrations: Monatsschr. Kakteenk. **30**: 153, as *Opuntia xanthostemma*.

On page 199, vol. 1, under *Opuntia polyacantha*, add the synonym: *Opuntia missouriensis watsonii* Schumann, Gesamtb. Kakteen 735. 1898.

Also insert: Extend range to northwestern Oklahoma.

Add to illustrations: Rep. Mo. Bot. Gard. **13**: pl. opp. 13; Schelle, Handb. Kakteenk. 54. f. 14, as *Opuntia missouriensis*; Möllers Deutsche Gärt. Zeit. **25**: 476. f. 9, No. 6, as *O. schweriniana*; Scientific American **124**: 492; Meehans' Monthly **11**: 57; Stand. Cycl. Hort. Bailey **4**: f. 2604.

On page 201, vol. 1, under *Opuntia grandis*, add to illustration: The Garden **62**: 425; Möllers Deutsche Gärt. **25**: 476. f. 9, No. 17.

On page 203, vol. 1, under *Opuntia nashii*, insert: *Illustration*: Journ. N. Y. Bot. Gard. **6**: f. 3.

On page 204, vol. 1, *Opuntia spinosissima*, insert at end: A species of this series, *Spinosissimae*, occurs on Navassa Island off the southeastern point of Haiti; specimens were sent us by Mr. F. P. Dillan, Superintendent of Light Houses, San Juan, Porto Rico, but they are not complete enough to be specifically referred.

On page 206, vol. 1, under *Opuntia moniliformis*, add the synonyms: *Cactus reticulatus* Index Kewensis **1**: 369. 1893;* *Opuntia reticulata* Karsten, Deutsche Fl. ed. 2. **2**: 457. 1895; *Opuntia picardae* Urban, Repert. Sp. Nov. Fedde **16**: 35. 1919.

On page 208, vol. 1, under *Opuntia rubescens*, add to illustration: Carnegie Inst. Wash. **269**: pl. 10, f. 90. 91, as *Opuntia catacantha*.

---

* The Index Kewensis refers *Cactus reticulatus* to Descourtilz (Fl. Med. Antill. 1: pl. 68), but the formal name was not used by him.

On page 209, vol. 1, under *Opuntia brasiliensis*, add to distribution: Peru.

Insert: *Opuntia brasiliensis gracilior* Salm-Dyck was given by Förster (Handb. Cact. 500. 1846) as a synonym of *O. brasiliensis minor*.

Also insert: Dr. Small has found this plant established after planting on shell mounds and waste places in southern Florida.

Add to illustrations: Goebel, Pflanz. Schild. 1: f. 37, 38.

On page 211, vol. 1, under *Ammophilae*, substitute for characters of the series:

Erect species, sometimes with a definite continuous trunk, often much branched, the joints broad and flat, spiny or unarmed, the spines (when present) subulate or subulate-acicular, whitish, gray or reddish, the large flowers yellow.

The series now appears to be most nearly related to the Series *Tortispinae* (vol. 1: 126) and may be placed to follow it as series 7a. *Opuntia austrina* Small, of southern Florida, may be transferred from the *Tortispinae* to the *Ammophilae*.

On page 211, vol. 1, under *Opuntia ammophila*, insert: More recent collections of this plant by Dr. Small, show that its range extends south to Cape Romano, Florida, and that the definite trunk, at first taken as characteristic of it, is not always developed; his living plants from different stations show slight individual differences which do not appear to be specific. This species has been erroneously referred by Dr. Griffiths (Bull. Torr. Club 46: 201) to *Opuntia bartramii* Rafinesque.

On page 213, vol. 1, insert:

239a. Opuntia turgida Small, sp. nov.

Plant erect, more or less diffusely branched, 0.5 meter tall or less, with fibrous roots; joints elliptic to elliptic-obovate, 5 to 12 cm. long, thickish, deep green, sometimes slightly glaucous when young; leaves subulate, 6 to 10 mm. long, spreading and more or less recurved, green, sometimes accompanied by fine bristles, but without spines; areoles scattered, often prominent and densely bristly on the older joints; spines (as far as known) wanting; flowers often several on a joint; ovary obovoid or obconic-obovoid, 2 to 2.5 cm. long, slightly tubercled; sepals green or purple-tinged, the outer subulate to lanceolate, 4 to 10 mm. long, acute, the inner rhombic-ovate, fully 1.5 cm. long, stout-pointed; corolla bright yellow, 5.5 to 6.5 cm. wide; petals 10 to 12, about 3 cm. long, broadly cuneate, abruptly narrowed, rounded or subtruncate at the apex, mucronate; anthers 2 mm. long; berry obovoid, 2 to 2.5 cm. long, greenish purple, even, broadly rounded at the base, the umbilicus flat or a little depressed at the middle; seeds rather numerous, about 4 mm. in diameter, somewhat turgid.

Hammocks near Yulee and on the mainland along the Halifax River south of Daytona, Florida. Type collected about five miles south of Daytona, in December 1919, by J. K Small, preserved in the herbarium of the New York Botanical Garden.

This spineless, small-jointed species is tentatively referred to the Series *Ammophilae* on account of its fruit characters and erect habit. A plant sent from Kew to the New York Botanical Garden in 1902, under an unpublished name, very closely resembles this species.

On page 214, vol. 1, insert the following:

Opuntia napolea, offered for sale by Grässner (Monatsschr. Kakteenk. February 1920) we have not seen.

The name *Opuntia spirocentra* Engelmann and Bigelow (Haage, Verz. Cact. 30), found in the Index Kewensis, we have not been able to verify. As the name is credited to Engelmann and Bigelow and the habitat of the plant is said to be New Mexico it is doubtless an error and probably was intended for *O. macrocentra*.

Opuntia todari (Haage and Schmidt, Haupt-Verz. 230. 1912) is known only in the trade.

Cactus italicus referred by the Index Kewensis to Tenore (Steudel, Nom. ed. 2. 2: 246. 1840) occurs first in 1831 (Tenore, Syll. Pl. Neop. 241) where also occurs the name *Opuntia italica*. Both are unpublished but doubtless refer to some species of *Opuntia*.

CACTUS PARVIFOLIUS Ehrenberg in F. G. Dietrich, Vollst. Lex. Gaertn. 2:416. 1802.

An upright, cylindrical, almost articulate stem; the upper part bedecked with small, cylindrical, fleshy, pointed leaves; on lower part of the stem, at the place where the leaves are attached, stiff bristles are formed which are surrounded at the base by a whitish-gray, woolly substance; in old age the stem requires a support on account of its slender growth; if the stem is cut through in the middle and the wound well dried, young sprouts make their appearance at this place which serve to propagate the plant. South America is its home.

The above paragraph is a free translation of the description.

We have not been able to identify this plant, but it is probably some species of *Opuntia* or possibly *Tacinga funalis*.

Cereus vulnerator Cortes (Fl. Colombia 69. 1897) and *C. guasabara* Cortes (Fl. Colombia 208. 1897) are different names for the same plant. From the brief descriptions it is difficult to identify this plant but it certainly is not a *Cereus*. It suggests some sheathed-spined *Opuntia* such as *O. tunicata* which has been introduced into South America and is common in northern Ecuador. It is known as curuntilla or guasabara in Colombia.

CORRECTIONS AND ADDITIONS TO VOLUME II.

On page 4, vol. II, under *Cereus hexagonus*, add the synonyms: *Cereus regalis* Haworth, Suppl. Pl. Succ. 75. 1819; *Cactus regalis* Sprengel, Syst. 2: 476. 1825; *Cereus childsi* Blanc, Cacti 39, No. 375.

Insert: *Cereus cyaneus* Hortus is listed by Berger (Hort. Mortola 69. 1912) as a South American plant grown at La Mortola. From drawings sent by Berger it is probably to be referred to *C. hexagonus*.

Add to illustrations: Andrews, Bot. Rep. 8: pl. 513; Reichenbach, Fl. Exot. pl. 322; Van Géel, Sert. Bot. 1: pl. 114, as *Cactus hexagonus*; Blanc, Cacti 39. No. 375, as *Cereus childsi*.

On page 8, vol. II, under *Cereus jamacaru* insert: *Cereus caracore* (Gosselin, Bull. Soc. Acclim. France 51: 58. 1905) belongs to the group containing *C. jamacaru*, that is, it is a true *Cereus*, according to Gosselin. He does not claim that it is a good species. No species of *Cereus*, however, are natives of Chile, from which this plant is said to have come. If indigenous to that country it is more likely to be *Trichocereus chiloensis*.

On page 9, vol. II, under *Cereus jamacaru*, add to illustrations: Monatsschr. Kakteenk. 26: 181; Karsten, Deutsche Fl. ed. 2. 2: 456. f. 605, No. 8.

On page 11, vol. II, under *Cereus peruvianus*, add the synonyms: *Piptanthocereus peruvianus* Riccobono, Boll. R. Ort. Bot. Palermo 8: 232. 1909; *Piptanthocereus peruvianus monstruosus* Riccobono, Boll. Ort. Bot. Palermo 8: 233. 1909.

Also add to illustrations: Saint-Hilaire, Exp. Fam. Nat. 2: pl. 95, in part as f. 1(?); De Candolle, Pl. Succ. Hist. 1: pl. 58, as *Cactus peruvianus*; Blanc, Cacti 36. No. 252; Rother, Praktscher Leitfaden Kakteen 15, as *Cereus peruvianus monstrosus*; Karsten and Schenck, Vegetationsbilder 1: pl. 41; 42, f. b; Gartenwelt 6: 133; Mem. Acad. Roy. Sci. pl. 4, 5; Haage and Schmidt, Haupt-Verz. 1919: 134. f. 10737; Goebel, Pflanz. Schild. 1: f. 5, 53.

On page 14, vol. II, under *Cereus pernambucensis*, add to illustration: Remark, Kakteenfreund 7, as *Cereus formosus monstrosus*.

On page 17, vol. II, under *Cereus aethiops*, add to illustrations: Förster, Handb. Cact. ed. 2. 207. f. 15, as *Cereus landbeckii*; Blanc, Cacti 26. No. 27; Gartenwelt 16: 537, as *Cereus coerulescens*.

On page 19, vol. II, insert:

25. Cereus trigonodendron Schumann, Bot. Jahrb. Engler **40**: 413. 1908.

Simple, or in age with a much branched top, 15 meters high; trunk 5 meters long, smooth, 3 dm. in diameter or more; ribs 3 to 6, 2 to 3 cm. high; areoles in young growth 2 to 3 cm. apart, producing abundant white wool, 1 cm. long or more; spines 4 to 7, at first brown, subulate, 2 to 5.5 cm. long; flowers as in typical species of *Cereus*, 10 to 15 cm. long; fruit smooth, edible.

Type locality: Department of Loreto, Peru.
Distribution: Valleys of eastern Peru and Bolivia.

This species is briefly described on page 19 of volume II of The Cactaceae, but at that time we knew little about it and were disposed to exclude it from the genus *Cereus*. We have since had a photograph of the type specimen from Berlin. In December 1922 F. L. Herrera sent us flowers from the Santa Ana Valley, province of Convención, Peru, and in February 1923 we received herbarium specimens of branches and flowers from W. E. Meyer, collected in 1922 at Cachucla-Esperanza, Boni, Bolivia. It is found only in the Atlantic drainage of Peru and Bolivia and is therefore geographically within the range of the genus *Cereus* as limited by us.

Illustration: Bot. Jahrb. Engler **40**: pl. 10.

Figure 239 is from a photograph sent by Dr. Vaupel from Berlin.

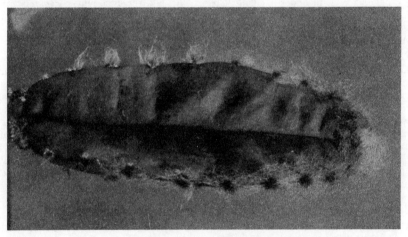

FIG. 239.—Cereus trigonodendron.

On page 20, vol. II, add at end of page: *Cereus amalonga* and its variety *cristata* are described in the Cactus Journal (**2**: 93, 104, 119) and both are illustrated in the plate for August of that volume. They are said to have been imported from Mexico. We are unable to identify these plants either from the descriptions or illustrations.

On page 21, vol. II, under *Cereus lormata*, insert: We listed *Cereus lormata* among the species unknown to us but we have since seen an illustration (Wiener Ill. Gart. Zeit. **11**: pl. 3, in part) of a barren plant. It has about 10 vertical ribs with clusters of subulate spines, some of them greatly elongated. It is probably not a true *Cereus*.

On page 21, vol. II, at end of *Cereus* add: *Cereus perviridis* Weingart is advertised by Haage & Schmidt (Cat. 1914). We have seen a cutting but do not know its relationship.

Cereus pitahaya variabilis Weingart (Monatschr. Kakteenk. **16**: 158. 1906) is only a form and is not described.

Cereus roezlii Haage jr. (Schumann, Gesamtb. Kakteen 64. f. 12. 1897) was described as columnar with 9 obtuse ribs, 9 to 12 radial spines, and one central spine much larger than the radials. Its flowers were unknown. It is said to come from the Andes of Peru or Ecuador. It is probably some species of *Lemaireocereus* or *Trichocereus*.

Cereus stolonifer Weber is listed by Schumann (Monatsschr. Kakteenk. 5: 43. 1895) as a plant grown in the Botanical Garden in Paris.

Cereus tripteris Salm-Dyck (De Candolle, Prodr. 3: 468. 1828) was described from barren plants of unknown origin and has never been identified.

Cereus uspenski Haage jr. is mentioned in a report by Karl Hirscht (Monatsschr. Kakteenk. 8: 109. 1898).

Cereus auratus Labouret (Rev. Hort. iv. 4: 27. 1855) is a tall *Cereus*-like plant, originally reported as from Peru, but the Index Kewensis says it is from Mexico. The four following varieties: *genuinus, intermedius, mollissimus,* and *pilosus* are briefly described by Regel & Klein (Ind. Sem. Hort. Petrop. 1860: 45. 1860); *Pilocereus auratus* (Rumpler in Förster, Handb. Cact. ed. 2. 650. 1885) is doubtless the same.

On page 22, vol. II, under *Monvillea cavendishii,* add to illustrations: Blühende Kakteen 3: pl. 171, as *Cereus euchlorus;* Blühende Kakteen 3: pl. 172, as *C. rhodoleucanthus;* Blühende Kakteen 3: pl. 178, as *C. cavendishii.*

On page 23, vol. II, under *Monvillea spegazzinii,* add the synonyms: *Piptanthocereus spegazzinii* Riccobono, Boll. R. Ort. Bot. Palermo 8: 233. 1909; *Cereus spegazzinii hassleri* Weingart, Monatsschr. Kakteenk. 32: 163. 1922.

Add to illustrations: De Laet, Cat. Gén. f. 28, as *Cereus spegazzinii.*

On page 27, vol. II, under *Cephalocereus senilis,* add to illustrations: Journ. Intern. Gard. Club 3: 640, as *Cephalocereus* sp.; Gard. Chron. III. 32: 35; Journ. Hort. Home Farm. III. 59: 625; Amer. Garden 11: 479; West Amer. Sci. 13: 16, as *Cereus senilis;* West Amer. Sci. 13: 23, as *C. hoppenstedtii;* Möllers Deutsche Gärt. Zeit. 25: 473. f. 5, No. 19; Remark, Kakteenfreund 20, as *Pilocereus hoppenstedtii;* Cact. Journ. 1: pl. 5; Gartenflora 27: 114; Deutsche Gärt. Zeit. 6: 64; Gard. Chron. 1873: f. 15; Garten-Zeitung 4: 182. f. 42, No. 6; Gartenwelt 2: 574; 16: 175; Watson, Cact. Cult. ed. 2. 260. f. 98; ed. 3. f. 34; West Amer. Sci. 9: 2; Journ. Intern. Gard. Club 3: 640; Blanc, Cacti 76. No. 1755; Weinberg, Cacti 26, as *Pilocereus senilis;* Palmer, Cult. Cact. 148, as *Pilocereus;* Engler and Drude, Veg. Erde 13: f. 30; Tribune Hort. 4: 283; Möllers Deutsche Gärt. Zeit. 25: 473. f. 5, No. 3; Schelle, Handb. Kakteenk. 108. f. 44, 45; Floralia 42: 370; Balt. Cact. Journ. 1: 116.

On page 30, vol. II, under *Cephalocereus fluminensis,* add to illustrations: Goebel, Pflanz. Schild. 11: pl. 3, f. 1 to 3, as *Pilocereus.*

On page 31, vol. II, under *Cephalocereus macrocephalus,* add to illustrations: Möllers Deutsche Gärt. Zeit. 25: 473. f. 5, No. 2.

On page 32, vol. II, under *Cephalocereus polylophus,* insert: *Pilocereus angulosus* Förster, according to Lemaire (Rev. Hort. 1862: 428. 1862) is little known; it is perhaps to be referred here.

Add to illustration: Gard. Chron. III. 50: 135. f. 64, c, as *Cereus polylophus.*

On page 42, vol. II, under *Cephalocereus arrabidae,* insert: The following names relate to this species and other names associated with it: *Pilocereus sublanatus* Förster (Haage, Verz. Cact. 22) is referred to *Cereus sublanatus* by the Index Kewensis. *Pilocereus tilophorus* (Index Kewensis) is evidently a mistake for *Cereus tilophorus. Pilocereus oligogonus* Lemaire (Rev. Hort. 1862: 428. 1862) is said to come from Mexico; the two varieties, *houlletianus* and *sublanatus,* given at this same place as synonyms, may or may not belong with it; they should doubtless be referred to the species bearing the same names respectively.

Add to illustrations: Möllers Deutsche Gärt. Zeit. 25: 473. f. 5, No. 9, as *Pilocereus exerens.*

On page 44, vol. II, under *Cephalocereus nobilis*, add the synonyms *Cereus polyptychus* Lemaire, Cact. Gen. Nov. Sp. 56. 1839; *Pilocereus polyptychus* Rümpler in Förster, Handb. Cact. ed. 2. 680. 1885.

Insert: The plant upon which this name was based was a small, barren one of unknown origin.

Insert: *Pilocereus houlletianus niger* (Förster, Handb. Cact. ed. 2. 676. 1885) is only a name given as a synonym of *P. niger*, while *P. niger aureus* is briefly described on the same page.

Add to illustrations: Möllers Deutsche Gärt. Zeit. **25**: 473. f. 5, No. 4, as *Pilocereus curtisi*.

On page 47, vol. II, under *Cephalocereus polygonus*, add the synonym: *Cephalocereus schlumbergeri* Urban, Symb. Antill. **8**: 464. 1920.

On page 49, vol. II, under *Cephalocereus lanuginosus*, add the synonym: *Pilocereus lanuginosus virens* Salm-Dyck in Förster, Handb. Cact. ed. 2. 672. 1885.

Insert: Curran reports that this fruit is edible (Inventory No. 50. p. 50. U. S. Dept. Agr. Bur. Plant Industry).

On page 51, vol. II, under *Cephalocereus royenii*, add to illustrations: Journ. N. Y. Bot. Gard. **15**: pl. 133, 134.

On page 52, vol. II, under *Cephalocereus leucocephalus*, add to illustrations: Watson, Cact. Cult. 145. f. 56; Deutsche Gärt. Zeit. **7**: 312, as *Pilocereus houlletianus*; Gard. Chron. III. **29**: f. 79, as *P. houlletianus leucocephalus*; Möllers Deutsche Gärt. Zeit. **25**: 473. f. 5, No. 14, as *P. cometes*; De Laet, Cat. Gén. No. 51, 52, 53; Möllers Deutsche Gärt. Zeit. **25**: 473. f. 5, No. 7; Blühende Kakteen **2**: pl. 79; West Amer. Sci. **13**: 24; Schelle, Handb. Kakteenk. 101. f. 40. as *P. houlletii*; Gard. Chron. III. **32**: 253, as *Cereus houlleti*.

On page 56, vol. II, under *Cephalocereus purpusii*, insert: Wilhelm Weingart, under date of June 18, 1921, wrote of this species as follows:

"*Cephalocereus purpusii* sp. nov. was collected by C. A. Purpus in 1902 near Mazatlán, was sent to me February 18, 1907, and bloomed in Darmstadt in 1918."

Figures 240 and 241 are reproduced from a drawing furnished by Wilhelm Weingart.

On page 56, vol. II, under *Cephalocereus catingicola*, add to illustrations: Vegetationsbilder **6**: pl. 14, as *Cereus catingicola*; Engler, Bot. Jahrb. **40**: Suppl. pl. 5, as *C. catingae*.

On page 58, vol. II, insert:

49. Cephalocereus collinsii sp. nov.

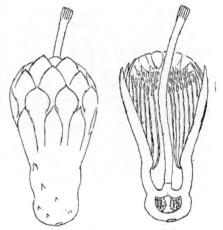

Figs. 240, 241.—Cephalocereus purpusii, flower.

About 3 meters high with few slender, elongated branches, these 3 to 4 cm. in diameter; ribs about 7, obtuse; tubercles about 1.5 cm. apart, circular, long-woolly as well as spiny; flowering areoles not much more woolly than the others; spines numerous, acicular, longer ones 3 to 4 cm. long; flowers borne near ends of branches, about 5 cm. long; fruit somewhat depressed, about 3 cm. broad; seed shining, black, 1.5 to 2 mm. broad.

Common in thickets near Tehuantepec, Oaxaca, Mexico. The type was collected by Dr. C. A. Purpus near Gerónimo in April 1923. It was reported by O. F. Cook and G. N. Collins from this region in 1902. The plant is named for Mr. Collins, who first brought it to our attention more than 20 years ago while carrying on field work in southern Mexico for the U. S. Department of Agriculture.

Figure 242 is from a photograph taken by Mr. Collins in 1902; it is three-fourths natural size.

On page 58, vol. II, under *Cephalocereus hermentianus*, add: *Illustration:* Möllers Deutsche Gärt. Zeit. **25**: 473. f. 5, No. 10, as *Pilocereus hermentianus.*

On page 58, vol. II, under *Pilocereus albisetosus*, add the synonyms: *Cactus albisetosus* Sprengel, Syst. **2**: 496. 1825; *Cactus albisetus* Steudel, Nom. ed. 2. **1**: 245. 1840.

On page 61, vol. II, under *Espostoa lanata*, add to illustrations: Schelle, Handb. Kakteenk. 105. f. 41, as *Pilocereus lanatus*; Schelle, Handb. Kakteenk. 105. f. 42, as *P. lanatus cristatus*; Wiener Ill. Gart. Zeit. **11**: pl. 3, in part, as *P. dautwitzii.*

On page 64, vol. II, under *Stetsonia coryne*, insert: W. B. Alexander wrote, under date of March 7, 1921, as follows:

"Noticing your statement that the fruit of *Stetsonia coryne* is unknown, I obtained a ripe specimen at La Rioja for you and am sending it by parcel post."

This we describe as follows:

Oblong, 6 cm. long, glabrous, bearing scattered scales, these 5 mm. broad, 1 mm. high, each with a cartaceous tip and a denticulate margin; seeds numerous, small, 1.5 mm. long, flattened, pitted; hilum large, basal.

On page 65, vol. II, under *Stetsonia coryne*, add to illustrations: Thomas, Zimmerkultur Kakteen 11, as *Cereus coryne.*

On page 66, vol. II, under *Escontria chiotilla*, add to illustrations: Möllers Deutsche Gärt. Zeit. **29**: 438. f. 13; Floralia **42**: 389, as *Cereus chiotilla.*

On page 69, vol. II, under *Pachycereus pringlei*, insert: The distribution of *Pachycereus pringlei* in northern Sonora is not well defined. Dr. MacDougal has recently visited northwestern Sonora and states that he saw it along the route between Altar and Port Libertad to within a hundred miles of the United States boundary. Prospectors and ranchers also speak of it as being abundant in the valley of the Asuncion or Altar River some miles to the northward. He writes of it as follows:

"On the whole, however, my chief interest was centered on the sowesa or *Pachycereus pringlei*. We began to get into this about 85 miles from the Gulf, and in the region below a thousand feet it attains perfectly tremendous size, as you will see from some photographic prints."

FIG. 242.—Cephalocereus collinsii.

Figure 243 is from a photograph obtained by Dr. MacDougal at Port Libertad, Sonora, May 4, 1923.

Also add to illustrations: Zeitschr. Ges. Erdk. **1916**: f. 6, in part; Contr. U. S. Nat. Herb. **16**: 131, A; 132, A; Karsten and Schenck, Vegetationsbilder **13**: pl. 13, as *Pachycereus calvus*; Contr. U. S. Nat. Herb. **16**: pl. 131, B, as *P. titan*; Ann. Rep. Smith. Inst. **1908**: 553. f. 17.

FIG. 243.—Pachycereus pringlei.

On page 71, vol. II, under *Pachycereus pecten-aboriginum*, add to illustrations: Engler and Drude, Veg. Erde **13**: 297. f. 9; Watson, Cact. Cult. ed. 2. 246. f. 92; ed. 3. f. 17; Karsten and Schenck, Vegetationsbilder **1**: pl. 48, as *Cereus pecten-aboriginum*; Karsten and Schenck, Vegetationsbilder **13**: pl. 14; Contr. U. S. Nat. Herb. **16**: pl. 132, B.

On page 71, vol. II, under *Pachycereus gaumeri*, add additional characters:

Ribs sometimes only 3, thin, 3 to 4 cm. high; areoles sometimes 2.5 cm. apart; fruit becoming dry, globose, 3 to 4 cm. in diameter, scales at base of fruit small, becoming long and foliaceous above, fleshy at base but tips thin and soon drying black; axils of scales felted, with a cluster of about 8 very short black spines; seeds numerous, brown, 4 mm. long.

The above description is drawn from specimens sent by Dr. Gaumer to Washington in June 1922.

On page 73, vol. II, under *Pachycereus chrysomallus*, add to illlustrations: Reiche, Elem. Bot. 226. f. 162, as *Pilocereus*; Möllers Deutsche Gärt. Zeit. **29**: 297. f. 5; U. S. Dept. Agr. Bur. Pl. Ind. Bull. **262**: pl. 13, f. 1, as *Pilocereus fulviceps*; Floralia **42**: 377; Belg. Hort. **3**: pl. 57, as *P. chrysomallus*; Möllers Deutsche Gärt. Zeit. **25**: 473. f. 5, No. 1.

On page 74, vol. II, under *Pachycereus marginatus*, insert: The two varieties, *Cereus marginatus monstrosus* (Monatsschr. Kakteenk. **19**: 62. 1909) and *C. marginatus cristatus* (Monatsschr. Kakteenk. **4**: 194. 1894) occur in the trade.

Also add to illustrations: West Amer. Sci. **13**: 6; Möllers Deutsche Gärt. Zeit. **25**: 472. f. 2, No. 15; De Laet, Cat. Gén. f. 23; Möllers Deutsche Gärt. Zeit. **29**: 355. f. 11; Bot. Jahrb. Engler **58**: Beibl. **129**: 27. f. 9, as *C. marginatus*; Remark, Kakteenfreund 7; Karsten and Schenck, Vegetationsbilder **1**: pl. 43; 48, as *Cereus gemmatus*; Ann. Rep. Smiths. Inst. **1908**: pl. 11, f. 2.

On page 76, vol. II, under *Pachycereus lepidanthus*, insert: Since the appearance of volume II, we have received flowers of this species from Wilhelm Weingart, which show a very close likeness to those of *Escontria chiotilla*. The fruit of the latter, however, is a juicy edible berry, while that of the former is described as dry. The illustrations here printed may lead to the rediscovery of this rare plant.

Figure 244 is from a photograph of a plant grown in Washington, showing a joint as it came from the field and also the young growth as developed in the greenhouse; figure 245 is from a photograph of two flowers and a spine-cluster.

Fig. 244.—Pachycereus lepidanthus.

On page 78, vol. II, under *Leptocereus leonii*, insert: Specimens collected in June 1923, by Brother León and Dr. Roig on Loma de Somorrostro, Jamaica, Havana Province, Cuba, show that the fruit of this species becomes 6.5 cm. long by 5.5 cm. thick, when fully mature. The fruits are borne near the ends of the branches, 3 or 4 close together.

On page 76, vol. II, under *Cereus tetazo*, insert: *Pilocereus tetetzo cristatus* Weber (Schumann, Gesamtb. Kakteen 176. 1897) is only a name and so is *Cephalocereus tetetzo* (Monatsschr. Kakteenk. **19**: 73. 1909) and *Cereus tetezo* and *C. tetetzo* (Monatsschr. Kakteenk. **17**: 79. 1907).

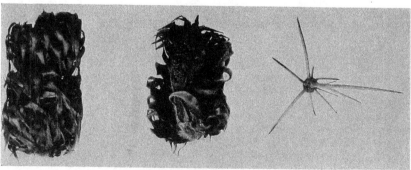

Fig. 245.—Pachycereus lepidanthus.

Also insert: *Illustration:* Bull. Soc. Nat. Acclim. **52**: 55. f. 14, as *Cereus tetezo*.

On page 82, vol. II, under *Eulychnia spinibarbis*, add to illustrations: Engler and Drude, Veg. Erde **8**: pl. 5, f. 11, as *Cereus coquimbanus*.

On page 86, vol. II, under *Lemaireocereus hollianus*, add to illustrations: Bull. Soc. Nat. Acclim. **52**: 45. f. 9, as *Cereus bavosus*.

On page 86, vol. II, under *Lemaireocereus hystrix*, add the synonym: *Cactus americanus* Vitman, Summa Pl. **3**: 209. 1789.

Insert: *Cactus americanus* is based on Bradley's illustration (Hist. Succ. Pl. 12) which De Candolle referred to *Cereus eburneus*, but as the plant came from the West Indies it is perhaps better referred to *Lemaireocereus hystrix*.

Also insert: We have recently obtained from N. E. Brown a photograph of Haworth's *Cereus hystrix*, with the date, "Oct. 24, 1824."

On page 87, vol. II, under *Lemaireocereus griseus*, add to illustrations: Monatsschr. Kakteenk. **24**: 5, as *Cereus eburneus*; Ann. Rep. Smiths. Inst. **1908**: pl. 9, f. 5.

On page 89, vol. II, under *Lemaireocereus eichlamii*, add to the description:

Fruit globular, about 5 cm. in diameter, becoming dry, not edible, thin-skinned, filled with numerous large seeds, the surface bearing scattered areoles, these densely short-felted with clusters of short spines subtended by small, ovate, acute scales; seeds black, 4 to 5 mm. long with a prominent hilum.

Insert: This plant is much used for hedges in Salvador and was obtained there by Mr. Paul C. Standley in the vicinity of Sonsonate, altitude 220 to 300 meters, March 1922 (No. 22328), but was not seen in the wild state. It is called there órgano. This species heretofore has been known only from Guatemala and was not known to us in fruit; this differs from that of the other species of *Lemaireocereus* in being rather dry with very large seeds.

Also insert: Figure 246 is from a photograph of the plant, sent by F. Eichlam in 1909 to Washington, which flowered in 1918.

FIG. 246.—Lamaireocereus eichlamii.

On page 91, vol. II, under *Lemaireocereus chende*, add to illustrations: Grässner, Haupt-Verz. Kakteen **1912**: 3, as *Cereus del moralii*.

On page 96, vol. II, under *Lemaireocereus weberi*, add to illustrations: Floralia **42**: 388, as *Cereus candelabrum*.

On page 96, vol. II, insert:

14a. Lemaireocereus beneckei (Ehrenberg).

> *Cereus beneckei* Ehrenberg, Bot. Zeit. **2**: 835. 1844.
> *Cereus farinosus* Haage in Salm-Dyck, Allg. Gartenz. **13**: 355. 1845.
> *Cereus beneckei farinosus* Salm-Dyck, Cact. Hort. Dyck. 1849. 49. 1850.
> *Piptanthocereus beneckei* Riccobono, Boll. R. Ort. Bot. Palermo **8**: 226. 1909.

Plants 4 to 5 meters high, much branched; branches 6 to 7 cm. in diameter, the growing tips very glaucous; ribs 7 or 8, strongly tuberculate, obtuse, separated by narrow intervals; areoles small, circular, borne on the upper side of the tubercle, brown to black-felted; spines 1 to 7, acicular, the longest sometimes 2.5 cm. long, brown to black; flowers night-blooming, small, 4 cm. long, greenish brown without; inner perianth-segments rose-colored to white (?); ovary globose, glaucous, tuberculate, its areoles brown-felted and bearing 3 to 7 acicular spines, the longest sometimes 2.5 cm. long and brown to black; fruit about 2 cm. in diameter, somewhat tubercled, bearing clusters of spines at the areoles, red; pericarp thick, somewhat fleshy; pulp disappearing, leaving the large seeds loose, these escaping by a basal pore as in *Oreocereus* and many of the *Echinocactanae*.

Type locality: Mexico on red lava beds.
Distribution: Central Mexico.

In volume II of The Cactaceae (p. 18), we described this plant under *Cereus* but with the statement that it was not a true *Cereus*; we were not then able to refer it to any known genus. At that time we knew little about the flowers and nothing accurate about the ovary and fruit. In 1921 Professor K. Reiche sent us some living plants from Iguala, the station from which Dr. Rose obtained his plants in 1905. These contained some old withered flowers and some well-developed ovaries which have enabled us to refer the plant to *Lemaireocereus*.

Figure 247 is from a photograph of K. Reiche's plant, slightly reduced, showing the top of a branch bearing an old flower and a half-ripe fruit.

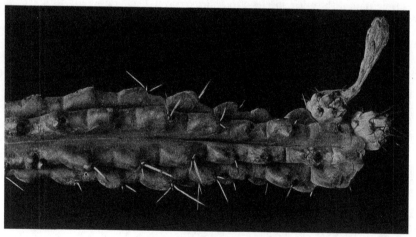

FIG. 247.—*Lemaireocereus beneckei.*

On page 98, vol. II, under *Lemaireocereus thurberi*, add to illustrations: Journ. N. Y. Bot. Gard. **3**: f. 13, as *Cereus thurberi*; Bull. U. S. Nat. Mus. **56**: pl. 8, f. 2; Karsten and Schenck, Vegetationsbilder **13**: pl. 15; 21, f. A; Contr. U. S. Nat. Herb. **16**: pl. 125, A; Amer. Bot. **20**: 88.

On page 108, vol. II, under *Bergerocactus emoryi*, add to illustrations: Cact. Journ. **1**: 59; Gartenwelt **11**: 498, as *Cereus emoryi*.

On page 111, vol. II, under *Wilcoxia poselgeri*, add to illustrations: Remark, Kakteenfreund 6; Deutsche Garten-Zeitung **1886**: f. 25, as *Cereus tuberosus*.

On page 111, vol. II, under *Wilcoxia striata*, insert: According to T. S. Brandegee (under date of June 8, 1921), the flowers of *Wilcoxia striata* are nocturnal.

On page 112, vol. II, under *Peniocereus greggii*, add to illustrations: Amer. Gard. **11**: 474, as *Cereus greggii*; Journ. Wash. Acad. **12**: 329. f. 1; Succulenta **4**: 71.

On page 113, vol. II, insert:

2. Peniocereus johnstonii Britton and Rose, Journ. Wash. Acad. **12**: 329.   1922.

A climbing or clambering plant, up to 3 meters long, with a very large fleshy root sometimes weighing 14 pounds; stems and branches 3 to 5-angled, the young growth not pubescent; spines 9 to 12, brown to black, glabrous; upper radial spines short, stubby, swollen at base, nearly black, the two lower light brown, elongated, bristle-like, reflexed; central spines 1 to 3, subulate, 4 to 8 mm. long; flower (only an old flower seen) about 15 cm. long; perianth-segments about 3 cm. long; the lower and outer ones bearing tawny hairs and long bristles; flower-tube slender, with prominent areoles on knobby projections and bearing tawny wool and bristly spines; fruit ovoid to oblong, about 6 cm. long, bearing prominent clusters of black spines, dry (?), many-seeded; seeds oblong, 3 mm. long or more, black, shining; seedling dark purple; cotyledons very thick, triangular.

Type locality: San Josef Island, off the east coast of southern Lower California.
Distribution: Southern Lower California.
This plant was always found growing up through bushes of *Olneya tesota.*
Illustrations: Journ. Wash. Acad. **12**: 330. f. 2; Succulenta **4**: 73.
Figure 248 shows a branch, old flowers, and seeds of the type specimen.

FIG. 248.—Peniocereus johnstonii, showing branch, old flower, and seeds.

On page 113, vol. II, under *Dendrocereus nudiflorus*, insert: In 1922 Dr. L. H. Bailey sent us two photographs and some stem-sections (No. 806) which he had obtained from the Botanic Garden at Roseau, Dominica. It grows as a low, rounded, much branched bush with the outer joints often pendent. Mr. Joseph Jones, curator of the Botanic Garden wrote that the group is made up of six plants which have not been cut back or interfered with in any way and have experienced two hurricanes without having a piece broken off. One of our colleagues, Dr. William R. Maxon, who had rediscovered this plant some years ago in Cuba, suggested that the plant grown in Dominica might be that species; a careful study of our material convinces us that he is correct. *Dendrocereus nudiflorus*, however, is naturally a large tree with a very definite trunk and a large, much branched top. An explanation of this inconsistency is that the Dominican plant was doubtless grown from cuttings, causing it to assume this bushy habit, a phenomenon also observed in other cacti.

Also insert: *Cereus undiflorus* is a misspelling, used by Sauvalle (Fl. Cuba 59. 1873) and reprinted in the Index Kewensis (**1**: 493).

Figure 249 is from one of the photographs sent us by Dr. Bailey.

On page 116, vol. II, under *Machaerocereus eruca*, add to illustratious: Journ. Intern. Gard. Club **3**: 641; Karsten and Schenck, Vegetationsbilder **13**: pl. 16, as *Cereus eruca*.

Fig. 249.—Dendrocereus nudiflorus.

On page 117, vol. II, under *Machaerocereus gummosus*, add to illustrations: Cact. Journ. **2**: 107, as *Cereus gummosus*; Zeitschr. Ges. Erdk. **1916**: f. 6, in part; Karsten and Schenck, Vegetationsbilder **13**: pl. 17, f. A.

On page 119, vol. II, under *Nyctocereus serpentinus*, add to illustrations: Watson, Cact. Cult. 67. f. 16; ed. 3. f. 12, as *Cereus serpentinus*.

On page 119, vol. II, under *Nyctocereus guatemalensis*, add to illustrations: Monatsschr. Kakteenk. **31**: 41, as *Cereus hirschtianus*.

On page 123, vol. II, under *Acanthocereus pentagonus*, add to illustrations: De Laet, Cat. Gén. f. 32, as *Cereus baxaniensis*; Monatsschr. Kakteenk. **32**: 21, as *C. princeps*.

On page 125, vol. II, insert the following:

3a. Acanthocereus floridanus Small, sp. nov.

Stems and branches diffusely spreading or reclining, 3 to 10 meters long, stout: joints prominently 3 to 5-angled, but mostly 3-angled, dark green, often forming impenetrable thickets: areoles remote, with mostly 4 to 7 slender or subulate spines, the central one often 1 to 2 cm. long: ovary stout-trumpet-shaped, 8 to 10 cm. long, with few large, separated tubercled areoles at the base, bearing mostly 3 to 5 diverging spines, those on the upper part usually with one spine each; outer perianth-segments deltoid to triangular-lanceolate or lanceolate-subulate and almost linear, the longer ones 3.5 to 4 cm. long, acuminate; inner perianth-segments broadly linear, 3.5 to 4.5 cm. long, about six times as long as wide, broadly acuminate; filaments adnate more than halfway up from the base of the hypanthium; anthers less than 2.5 mm. long.

Hammocks, along or near the coast, southern peninsular Florida, adjacent islands, and Florida Keys. Type collected by J. K. Small, on Key Largo, December 1917 and 1918; preserved in the herbarium of the New York Botanical Garden.

This Florida plant has been referred by us to *A. pentagonus*, but specimens recently collected by Dr. Small, including good flowers, which we had not seen before, indicate it to be a distinct species, characterized by its much shorter perianth and more spiny ovary.

Illustrations: Britton and Rose, Cactaceae **2**: 123. f. 182; 124. f. 184, as *Acanthocereus pentagonus*.

On page 129, vol. II, under *Heliocereus speciosus*, add to the illustrations: Herb. Génér. Amat. **4**: pl. 244; Colla, Hort. Ripul. pl. 10; Bonpl. Descr. Pl. Rar. pl. 3, as *Cactus* *speciosus*; Maund, Bot. **1**: pl. 12, as *C. speciosus lateritius*; Curtis's Bot. Mag. **49**: pl. 2306; Edwards's Bot. Reg. **6**: pl. 486, as *C. speciosissimus*; Edwards's Bot. Reg. **19**: pl. 1596, as *C. speciosissimus lateritius*; Newman, Illustr. Bot. 209; Abh. Bayer. Akad. Beschr. Cact. **2**: pl. 3, f. 5; De Laet, Cat. Gén. f. 24, as *Cereus speciosus*; Lindley, Veg. King. ed. 3. 746. f. 498; Curtis's Bot. Mag. **67**: pl. 3822; The Garden **53**: 153, as *C. speciosissimus*; Illustr. Hort. **32**: pl. 548, as *C. speciosissimus hoveyi*; Sci. Amer. **124**: 492, as *Heliocereus mallisoni*; Van Géel, Sert. Bot. **1**: 116, as *Cactus speciosissimus*.

On page 129, vol. II, under *Heliocereus cinnabarinus*, add: *Illustration:* Monatsschr. Kakteenk. **32**: 54, 55, as *Cereus cinnabarinus.*

On page 129, vol. II, under *Heliocereus amecamensis*, add to illustrations: Rother, Praktischer Leitfaden Kakteen 74, as *Cereus amecamensis.*

On page 132, vol. II, under *Trichocereus spachianus*, add to illustrations: Remark, Kakteenfreund 5, as *Cereus spachianus.*

On page 133, vol. II, under *Trichocereus pasacana*, insert: The name *Cephalocereus pasacana* (Engler and Prantl, Pflanzenfam. **3**⁶ᵃ: 182. 1894) has been used for this plant.

FIG. 250.—Borzicactus fieldianus.   FIG. 251.—Borzicactus fieldianus.

On page 136, vol. II, *Trichocereus macrogonus*, add: *Illustrations:* Garten-Zeitung **4**ᵃ: 182. f. 8, as *Cereus macrogonus.*

On page 140, vol. II, under *Trichocereus coquimbanus*, add to illustrations: Engler and Drude, Veg. Erde **8**: pl. 16, as *Cereus nigripilis.*

On page 140, vol. II, under *Trichocereus terscheckii*, insert: This cactus is the only timber found in the region of the Puna and in the western mountains of Argentina that can be utilized in any form. It is employed on a large scale in the mines for timbering the galleries, if these happen to be dry. It is called cardón.

Add: *Illustration:* Sci. Amer. **124**: 492.

On page 143, vol. II, under *Trichocereus candicans*, insert: The names *Cereus gladiatus vernaculatus* Monville (Labouret, Monogr. Cact. 327. 1853) and *C. gladiatus courantii* (Förster, Handb. Cact. ed. 2. 833. 1885) were given as synonyms of *C. candicans.*

Cereus candicans dumesnilianus is figured and briefly described in the Gardeners' Chronicle (III. **26**: 415. f. 132). It is an upright plant with long, straight spines; the flowers are large and pure white. It flowered in the collection of Justus Corderoy.

On page 144, vol. II, under *Trichocereus schickendantzii*, add to illustration: Möllers Deutsche Gärt. Zeit. **25**: 475. f. 7, No. 16, as *Echinopsis schickendantzii*.

On page 146, vol. II, under *Echinopsis catamarcensis*, add: *Illustration:* Möllers Deutsche Gärt. Zeit. **25**: 475. f. 7, No. 19.

On page 149, vol. II, under *Harrisia eriophora*, add to illustration: Journ. N. Y. Bot. Gard. **11**: 234. f. 34; Roig. Cact. Fl. Cub. pl. [5], as *Harrisia undata*.

Insert: *Cactus peruvianus jamaicensis* appears in Grisebach's Flora (Fl. Brit. W. Ind. 301. 1860) as a synonym of *Cereus eriophorus*, but refers to *Harrisia gracilis*.

On page 151, vol. II, under *Harrisia nashii*, add: *Illustration:* Descourtilz, Fl. Med. Antill. **1**: pl. 66, as *Cactus divaricatus*.

On page 151, vol. II, under *Harrisia gracilis*, add the synonym: *Cactus subrepandus* Sprengel, Syst. **2**: 495. 1825.

Add to illustrations: Förster, Handb. Cact. ed. 2. f. 139; Blühende Kakteen **2**: pl. 84; Watson, Cact. Cult. 85. f. 28; ed. 3. f. 19; Dict. Gard. Nicholson Suppl. 220. f. 255, as *Cereus repandus*; Addisonia **2**: pl. 61.

On page 154, vol. II, under *Harrisia aboriginum*, add: *Illustration:* Journ. N. Y. Bot. Gard. **22**: pl. 253.

On page 155, vol. II, under *Harrisia martinii*, add to illustrations: Addisonia **2**: pl. 68.

On page 157, vol. II, under *Harrisia bonplandii*, add to illustrations: Montasschr. Kakteenk. **25**: 3, as *Cereus bonplandii*.

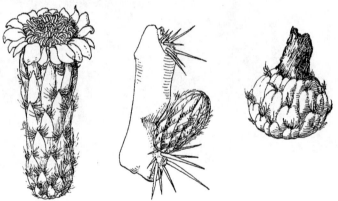

FIGS. 252, 253, 254.—Borzicactus fieldianus.

On page 163, vol. II, insert:

9. Borzicactus fieldianus sp. nov.

Forming thickets 3 to 6 meters high, the branches elongated, at first erect or ascending but sometimes becoming pendent or even prostrate; ribs few, perhaps only 6 or 7, stout, broad, 1 to 2 cm. high, depressed between the areoles and on young shoots and appearing as tubercled; areoles large, circular, short-lanate and spiny, with a depression extending upward from its upper side to constriction of rib; spines 6 to 10, white, subulate, very unequal, the longest ones 5 cm. long or longer; flowers several, from near tip of branches, but with only one from an areole, with a cylindric tube 6 to 7 cm. long and a very narrow limb; ovary and flower-tube bearing ovate, acute scales, 1 to 3 mm. long, these with long brown hairs in their axils; flower-tube within glabrous below its throat, bearing many stamens 4 cm. long; perianth-segments red, 1 cm. long; stamens exserted only beyond the perianth-segments, if at all; ovary globular, perhaps somewhat tuberculate, with scattered, long-hairy areoles; fruit probably fleshy, globular to ovoid, 2 cm. in diameter.

Collected by Macbride and Featherstone on gravelly river bluffs, eastern exposure at Huaraz, Peru, altitude about 2,600 meters, October 6, 1922 (No. 2519).

This very interesting plant we have named in honor of Captain Marshall Field, a patron of science, who financed the Botanical Expedition of 1922 to South America, sent out by the Field Museum of Natural History.

Figure 250 shows the habit of the plant, 251 a flowering branch, and figures 252 to 254 show flower, rib, and fruit.

On page 164, vol. II, under *Carnegiea gigantea*, add to illustrations:* Remark, Kakteenfreund 19, as *Pilocereus giganteus*; Nat. Geogr. Mag. 41: 373, as giant cactus; Tribune Hort. 4: 243; Journ. N. Y. Bot. Gard. 3: f. 15, 16, 17; 5: 173. f. 27; 6: f. 31, 32; Gartenwelt 8: 485; 11: 498; Schelle, Handb. Kakteenk. f. 20, 21; Cact. Mex. Bound. frontispiece; Bull. U. S. Nat. Mus. 56: pl. 8, f. 1; Useful Wild Pl. U. S. Canada opp. 112; Gartenflora 54: 589. f. 70; Gard. Chron. II. 20: 265. f. 39; Rev. Hort. IV. 3: 343. f. 20; Wiener Ill. Gart. Zeit. 11: 216. f. 47; Watson, Cact. Cult. 76. f. 22; Balt. Cact. Journ. 1: 67; Blanc, Cacti 30. No. 120; Carnegie Institution of Washington 6: pl. 1; De Laet, Cat. Gén. f. 26; Monatsschr. Kakteenk. 32: 87, as *Cereus giganteus*; Contr. U. S. Nat. Herb. 16: pl. 7; Amer. Bot. 26: 136; Stand. Cycl. Hort. Bailey; 1: pl. 3; 2: f. 819; Nat. Geogr. Mag. 44: 171.

On page 167, vol. II, for *Binghamia melanostele*, substitute for this name:

Binghamia multangularis (Willdenow).

Cactus multangularis Willdenow, Enum. Pl. Suppl. 33. 1813.
Cereus multangularis Haworth, Suppl. Pl. Succ. 75. 1819.
Echinocereus multangularis Rumpler in Förster, Handb. Cact. ed. 2. 825. 1885.
Cephalocereus melanostele Vaupel, Bot. Jahrb. Engler 50: Beibl. 111: 12. 1913.
Binghamia melanostele Britton and Rose, Cactaceae 2: 167. 1921.

Insert: We have recently obtained a photograph of Haworth's plant bearing the date "Oct. 29, 1824." A careful comparison of this photograph with photographs and specimens obtained by Dr. Rose in Peru in 1914 convinces us that this is the same plant as *Cephalocereus melanostele* which we referred to *Binghamia*.

Figure 255 is from a photograph of Haworth's plant, from N. E. Brown of Kew.

FIG. 255.—Binghamia multangularis.

On page 171, vol. II, under *Oreocereus celsianus*, add to illustrations: Karsten and Schenck, Vegetationsbilder 7: pl. 42; Möllers Deutsche Gärt. Zeit. 25: 473. f. 5, No. 11, as *Pilocereus celsianus*; Möllers Deutsche Gärt. Zeit. 25: 473. f. 5, No. 6, as *P. kranzleri*;† Watson, Cact. Cult. 146. f. 57, as *P. bruennowii*; Balt. Cact. Journ. 1: 133, as *P. fossulatus*; Monatsschr. Kakteenk. 31: 123; 32: 9; Gartenflora 62: f. 55, as *Cereus straussii*; Möllers Deutsche Gärt. Zeit. 25: 475. f. 5, No. 15, as *P. williamsii*; Amer. Mus. Journ. 16: 39; Bull. Pan Amer. Union 42: 408.

On page 174, vol. II, under *Cleistocactus baumannii*, add to illustrations: Deutsches Mag. Gart. Blumen. 1851: pl. opp. 48, as *Cereus tweediei*; Jard. Fleur. 1: pl. 48; De Laet, Cat. Gén. f. 25; Blanc, Cacti 24. f. 2;. West Amer. Sci. 13: 8, as *Cereus colubrinus*.

On page 177, vol. II, under *Lophocereus schottii*, add to illustrations: Karsten and Schenck, Vegetationsbilder 13: pl. 18, in part;

*Some of the illustrations cited here and on pages 166 and 167 do not have the technical name of the plants.
†This name is credited to Rümpler but he gives the spelling as *Pilocereus kanzleri*.

280 THE CACTACEAE.

Contr. U. S. Nat. Herb. **16**: pl. 126, B, as *Lophocereus australis*; Möllers Deutsche Gärt. Zeit. **25**: 473. f. 5, No. 8; Cycl. Amer. Hort. Bailey **3**: f. 1803; Schumann, Gesamtb. Kakteen f. 7, 8; Nachtr. f. 8; Monatsschr. Kakteenk. **11**: 10; **18**: 101, as *Pilocereus schottii*; West Amer. Sci. **13**: 16, as *Cereus sargentianus*; Rep. Mo. Bot. Gard. **16**: pl. 4, 5, 6, 7, 8, as *Cereus schottii*; Contr. U. S. Nat. Herb. **16**: pl. 125, B, as *Lophocereus schottii*; Thomas, Zimmerkultur Kakteen 17, as *Pilocereus sargentianus*.

On page 180, vol. II, under *Myrtillocactus geometrizans*, add to illustrations: Reiche, Elem. Bot. 228. f. 164; Engler and Drude, Veg. Erde **13**: f. 31; Möllers Deutsche Gärt. Zeit. **25**: 482. f. 13; West Amer. Sci. **13**: 15; Zeitschrift Sukkulenk. **1**: 31, as *Cereus geometrizans*.

On page 180, vol. II, under *Myrtillocactus cochal*, add to illustrations: Thomas, Zimmerkultur Kakteen 13, as *Cereus cochal*.

On page 183, vol. II, insert the following:

39. NEOABBOTTIA Britton and Rose, Smiths. Misc. Coll. **72**[9]: 2. 1921.

A tree-like cactus with a smooth, upright, terete trunk and a much branched top, the branches strongly winged or ribbed, normally from distal end of preceding branch, but sometimes from below tip and usually in the same plane; ribs thin and high, very spiny; flowers nocturnal, small, tubular, with a narrow limb, borne several together at distal end of a terminal branch from a small, felted cephalium; perianth persisting on the ovary; perianth-tube and ovary bearing small scales with short wool and an occasional bristle in their axils; perianth-segments very small; throat of flower a little broadened at top, bearing many stamens; style slender; fruit oblong, turgid, nearly naked, deeply umbilicate; seed minute, black, muricate.

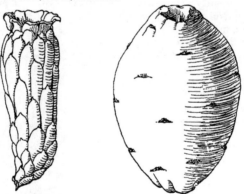

FIGS. 256, 257.—Flower and fruit of Neoabbottia paniculata. Natural size.

Type species: *Cactus paniculatus* Lamarck.

A monotypic genus of Hispaniola, dedicated to Dr. W. L. Abbott, a patron of natural history.

1. Neoabbottia paniculata (Lamarck) Britton and Rose, Smiths. Misc. Coll. **72**[9]: 3. 1921.

Cactus paniculatus Lamarck, Encycl. **1**: 540. 1783.
Cereus paniculatus De Candolle, Prodr. **3**: 466. 1828.

Plant 6 to 10 meters high or higher; trunk woody, 30 cm. in diameter, the wood close-grained, yellowish white; bark of trunk 1.5 cm. thick, brown, not spiny in age, smooth; branches 4 to 6 cm. broad, strongly 4-ribbed, occasionally 6-ribbed or winged; ribs thin, 1.5 to 2.5 cm. high, their margins somewhat crenate, areoles borne at base of sinuses, 1.5 to 2 cm. apart; spines 12 to 20, acicular, brownish to gray, 2 cm. long or less; cephalium 1 to 1.5 cm. in diameter, becoming elongated and angled; flowers straight, 5 cm. long, with a limb about 3 cm. broad, tube 6 to 7 mm. long, about 18 mm. in diameter, with walls 5 to 6 mm. thick; inner perianth-segments greenish white, short-oblong, about 1 cm. long, obtuse; throat 18 mm. long, covered with numerous filaments, these with a knee near base and pressing against style; stamens and style included; ovary and flower-tube tubercled, the former with short tubercles, the latter with oblong ones (sometimes 1.5 cm. long), each ending in a depressed areole subtended by a minute scale; areoles bearing a tuft of brown felt and an occasional brown bristle; fruit oblong in outline, 6 to 7 cm. long, 4 to 5.5 cm. in diameter, turgid, nearly naked; rind green, thick, hard; seeds rounded above, cuneate at base, with a large lateral depressed hilum.

Fig. 258.—Neoabbottia paniculata.

Type locality: Haiti.

Distribution: Hispaniola.

This plant was described by Plumier as follows: "*Melocactus* arborescens, tetragonus, flore ex albido." This description was repeated by Tournefort, with the addition of a single word, in 1719. Plumier's drawing of this plant was published long after his death by Burmann as plate 192 of the Plantarum Americanum, and upon this plate Lamarck based his *Cactus paniculatus*, which De Candolle a little later took up as *Cereus paniculatus*. Ever since, the plant has usually passed under the latter name, with an occasional reversal to the earlier one.

Until recently, the species has been known only from this old illustration and these brief descriptions. It was collected near Port-au-Prince, Haiti, on the Cul-de-sac, by Dr. W. L. Abbott and Mr. E. C. Leonard, April 1920 (No. 3500); also at the same locality by Mr. H. M. Pilkington, December 1920; also a single branch by Dr. Paul Bartsch at Thomazeau in 1917 (No. 221). The Abbott and Leonard material consists of wood-sections and herbarium specimens of branches, flowers, fruit, and seeds, supplemented by living specimens and by fruit and flowers in formalin, together with several habit photographs.

In habit it resembles *Dendrocereus*, its branches resemble *Acanthocereus*, and the small limb of the flower resembles *Leptocereus*; but the plant differs from all of these in bearing several flowers at the ends of terminal branches and in developing a kind of cephalium. In the last respect it approaches *Neoraimondia*, near which we would place it in our present classification.

Illustrations: Smiths. Misc. Coll. **72**[9]: pl. 1 to 4; pl. 2, f. 1, 2; Bull. Amer. Mus. Nat. Hist. **33**: 31. f. 11.

Figures 256 and 257 show the flower and fruit; figure 258 shows the top of a tree; figures 259 and 260 show the plant in its natural surroundings; figure 223*a*, page 248, is a reproduction of Plumier's plate.

On page 187, vol. II, under *Hylocereus undatus*, add to illustrations: De Laet, Cat. Gén. f. 31; Tribune Hort. **4**: pl. 140; Blanc, Cacti 37. No. 346; Ann. Inst. Roy. Hort.

Fromont **2**: pl. **1**, f. D; Gartenwelt **11**: 101; Watson, Cact. Cult. ed. **3**. pl. opp. 29; Rev. Hort. Belg. **40**: after 184; Meehans' Monthly **6**: 5; West Amer. Sci. **13**: 5; Gartenflora **55**: f. 2, as *Cereus triangularis*; De Tussac, Fl. Antill. **4**: pl. 26, as *Cactus triangularis*; Stand. Cycl. Hort. Bailey **3**: pl. 57, as *Hylocereus tricostatus*; Cañizares, Jard. Bot. Inst. Habana 98, as *H. triangularis*.

On page 189, vol. II, under *Hylocereus lemairei*, add to illustrations: Blühende Kakteen **3**: pl. 173, as *Cereus lemairei*.

On page 191, vol. II, under *Hylocereus napoleonis*, add to illustrations: Hartinger, Parad. **2**: 1, as *Cereus napoleonis*.

FIGS. 259 and 260.—Neoabbottia paniculata.

On page 192, vol. II, under *Hylocereus triangularis*, insert: The name *Cactus anizogonus* of English gardens is given as a synonym of *Cereus triangularis* by Rümpler (Förster, Handb. Cact. ed. 2. 764. 1885).

On page 192, vol. II, under *Hylocereus trigonus* insert: *Cereus triqueter* Haworth (Syn. Pl. Succ. 181. 1812) is some species of *Hylocereus* near *H. trigonus*. If really from South America, as stated by Haworth, it may be the same as *H. lemairei*.

On page 194, vol. II, under *Hylocereus* sp., insert after first paragraph: This species of *Hylocereus* from the Guianas should be studied in connection with *Cereus scandens* Salm-Dyck (Cact. Hort. Dyck. 1849. 219. 1850), which is said to have come from Guiana. The variety *C. scandens minor* Boerhaave (Monatsschr. Kakteenk. **1**: 82. 1891) is only mentioned.

After page proof had been read, some fine specimens of a *Hylocereus* were received from Surinam through Gerold Stahel, which we describe as follows:

Stems much elongated, 3-angled, 2 to 6 cm. broad, bluish or whitened, somewhat glaucous; ribs often thin; margins of ribs not horny, nearly straight, areoles distant, sometimes 6 cm. apart; spines brown, 2 or 3, very short, much swollen at base.

In the shape, number, and size of spines this specimen resembles *H. lemairei*, but differs from it in the whitened stems. We do not know its flowers.

On page 197, vol. II, under *Selenicereus grandiflorus*, insert: *Cereus haitiensis* Hortus is cited by Schelle (Handb. Kakteenk. 89. 1907) as a synonym of *C. grandiflorus*.

Cereus grandiflorus flemingii Rümpler (Förster, Handb. Cact. ed. 2. 751. 1885; *C. flemingii*, Monatsschr. Kakteenk. **3**: 109. 1893) is said to be a hybrid between *C. grandiflorus* and *C. speciosissimus*.

Add to illustrations: Fl. Serr. **3**: pl. 1-2, as *Cereus grandifloro-speciosissimus*; Balt. Cact. Journ. **1**: 56, as queen of the night; Remark, Kakteenfreund 8; Gartenflora **42**: 541. f. 110; **64**: 90. f. 22; Gartenwelt **16**: 613; **19**: 18; Gard. Chron. III. **14**: 187. f. 36; Thomas, Zimmerkultur Kakteen 15; Tribune Hort. **4**: pl. 139; Blanc, Cacti 32; De Laet, Cat. Gén. f. 29; Fl. Serr. **3**: 233-234; Weinberg, Cacti 8; Knippel, Kakteen pl. 1; Goebel, Pflanz. Schild. **1**: pl. 2, f. 5; Möllers Deutsche Gärt. Zeit. **14**: 340 to 343; **20**: 561, as *Cereus grandiflorus;* Cañizares, Jard. Bot. Inst. Habana 100.

On page 198, vol. II, under *Selenicereus urbanianus*, add to illustrations: Gartenwelt **12**: 255, as *Cereus urbanianus*; Roig, Cact. Fl. Cub. pl. [3,] f. 2; pl. [4], as *Selenicereus maxonii*.

On page 199, vol. II, under *Selenicereus coniflorus*, insert: Dr. J. K. Small finds this plant naturalized in pinelands near the Everglades, west of Halenville, Florida.

On page 200, vol. II, under *Selenicereus pteranthus*, also add to illustrations: Garden **13**: 291; Monatsschr. Kakteenk. **31**: 71; Watson, Cact. Cult. 63. f. 15; ed. 3. f. 10; Gartenflora **41**: f. 23, 24, as *Cereus nycticalus*.

Add the synonym: *Cereus nycticalus peanii* Beguin in Riccoboni, Boll. R. Ort. Bot. Giard. Col. Palermo **8**: 252. 1909.

On page 202, vol. II, under *Selencereus boeckmannii*, add to illustration: Blühende Kakteen **3**: pl. 175, 176, as *Cereus boeckmannii*.

On page 202, vol. II, under *Selenicereus macdonaldiae*, add the synonym: *Cereus grandiflorus macdonaldiae* Blanc, Cacti 34.

Also insert: *Cereus kewensis* Worsley (Journ. Roy. Hort. Soc. **39**: 92. 1913) is said to be a "garden hybrid between *C. macdonaldiae and* probably *C. nycticalus.*"

Also add to illustrations: Blanc, Cacti 34. No. 206, as *Cereus grandiflorus macdonaldiae*; Monatsschr. Kakteenk. **30**: 107; Gartenwelt **16**: 537; Möllers Deutsche Gärt. Zeit. **25**: 488. f. 22, No. 6, as *Cereus macdonaldiae*, Blühende Kakteen **3**: pl. 166, 167, as *Cereus grusonianus*.

Insert: *Cereus rothii* Weingart (Monatsschr. Kakteenk. **32**: 146. 1922) is of this relationship. It is a new name for the plant from South America called *Cereus macdonaldiae* by Spegazzini; we have not seen it.

On page 204, vol. II, under *Selenicereus hamatus*, add to illustrations: Tribune Hort. **4**: pl. 140; Floralia **42**: 371, as *Cereus rostratus*.

On page 209, vol. II, insert the following:

17. Selenicereus nelsonii (Weingart).

Cereus nelsonii Weingart, Zeitschrift Sukkulentenkunde **1**: 33. 1823.

A slender, much branched vine, 1 to 1.5 cm. in diameter, giving off occasional aërial roots; ribs 6 or 7, low, somewhat tubercled; areoles small, circular, about 1 cm. apart; spines about 12, acicular, white to yellowish, 5 to 7 mm. long; length of flower including ovary and closed perianth about 20 cm.; outer perianth-segments linear, pointed, reddish brown, the inner perianth-segments narrowly lanceolate, 7 cm. long, 12 to 15 mm. broad, acute; filaments numerous, weak, white; style long and slender, exserted beyond the withering perianth; stigma-lobes slender, white, entire; scales on the ovary and flower-tube minute, 1 to 1.5 mm. long, reddish brown, bearing white felt and white bristles in their axils; fruit crowned by the withering perianth, globular, 2 to 2.5 cm. in diameter, reddish, bearing numerous, small, circular areoles, these with clusters of acicular spines sometimes 1 cm. long.

Type locality: Southern Mexico.

Distribution: Mexico, but range not known.

We have had this plant under observation since 1914 when cuttings were sent us by C. Z. Nelson, an enthusiastic grower of cacti at Galesburg, Illinois, who obtained it from southern Mexico from Dr. J. L. Slater. This plant made two flowers during the week of May 17, 1922; the fruit ripens very slowly and did not mature until October 10, 1922.

According to Wilhelm Weingart, the same species has long been grown by Frantz de Laet at Contich, Belgium, also from Mexican material.

Illustration: Zeitschrift Sukkulentenk. 1: 33, as *Cereus nelsonii.*

Figure 261 shows a branch bearing a newly matured fruit.

FIG. 261.—Selenicereus nelsonii.

On page 216, vol. II, under *Werckleocereus tonduzii*, add: *Illustration:* Monatsschr. Kakteenk. **31**: 85, as *Cereus tonduzii.*

On page 218, vol. II, under *Aporocactus leptophis*, add to illustrations: Ann. Fl. Pom. **1839**: pl. 43, as *Cereus leptophis.*

On page 219, vol. II, under *Aporocactus flagelliformis*, add to illustrations: Tribune Hort. **1**: pl. 4, as *Cereus serpentinus*; Fl. Antill. **1**: pl. 67, as cierge queue de souris; Cact. Journ. **1**: 82; Rother, Praktischer Leitfaden Kakteen 57; Remark, Kakteenfreund 6; Watson, Cact. Cult. ed. 3. f. 11; Floralia **42**: 371; Gartenwelt **15**: 637; Blanc, Cacti 27. No. 104; Amer. Gard. **11**: 527 as *Cereus flagelliformis*; Cact. Journ. **1**: 125; **2**: 34; **2**: 153, as *C. flagelliformis cristatus.*

Insert: *Cereus smithianus* is a hybrid listed by Sweet (Hort. Brit. ed. 2. 237. 1830) which the Index Kewensis refers to *C. smithii*, a generic hybrid already referred to.

On page 221, vol. II, under *Aporocactus martianus*, add to the illustrations: Thomas, Zimmerkultur Kakteen 14, as *Cereus martianus.*

Corrections and Additions to Volume III.

On page 6, vol. III, under *Echinocereus scheeri*, add to illustrations: Thomas, Zimmerkultur Kakteen 27.

On page 7, vol. III, under *Echinocereus salm-dyckianus*, add to illustrations: Thomas, Zimmerkultur Kakteen 25.

On page 11, vol. III, under *Echinocereus polyacanthus*, add to illustrations: Förster, Handb. Cact. ed. 2. 212. f. 19, as *Cereus polyacanthus*; Floralia **42**: 376, as *Echinocereus polyacanthus* var.; Thomas, Zimmerkultur Kakteen 29; De Laet, Cat. Gén. f. 34, 37, 38; Möllers Deutsche Gärt. Zeit. **36**: 145. f. III; Succulenta **5**: 74.

On page 12, vol. III, under *Echinocereus acifer*, add to illustrations: Thomas, Zimmerkultur Kakteen 26, as *Echinocereus acifer trichacanthus*; Blühende Kakteen **3**: pl. 179, as *E. durangensis*.

On page 14, vol. III, under *Echinocereus coccineus*, add to illustrations: Pac. R. Rep. **4**: pl. 4, f. 1 to 3, as *Cereus phoeniceus*.

On page 17, vol. III, under *Echinocereus viridiflorus*, add the synonym: *Cereus viridiflorus minor* Engelmann, Proc. Amer. Acad. **3**: 278. 1856.

On page 21, vol. III, under *Echinocereus blanckii*, add to illustrations: Watson, Cact. Cult. 70. f. 18, as *Cereus blankii*; Watson, Cact. Cult. 68. f. 17; ed. 3. f. 13, as *C. berlandieri*.

On page 22, vol. III, under *Echinocereus pentalophus*, add to illustrations: Watson, Cact. Cult. 78. f. 23, as *Cereus leptacanthus*; Watson, Cact. Cult. 83. f. 27; ed. 3. f. 18, as *C. procumbens*; Balt. Cact. Journ. **2**: 218, as *Echinocereus procumbens*.

Also add the synonym: *Cereus propinquus subarticulatus* Pfeiffer in Förster, Handb. Cact. 373. 1846.

On page 23, vol. III, under *Echinocereus cinerascens*, insert: *Echinocactus deppii* Link and Otto (Steudel, Nom. ed. 2. **1**: 536. 1840) was given in error for *Echinocereus deppei*.

On page 25, vol. III, under *Echinocereus reichenbachii*, insert: Watson, Cact. Cult. ed. 3. f. 14, as *Cereus caespitosus*; West Amer. Sci. **7**: 237; **13**: 14; Gartenflora **23**: pl. 813, as *C. pectinatus*; Remark, Kakteenfreund 17; Balt. Cact. Journ. **2**: 218, as *Echinocereus caespitosus*.

On page 37, vol. III, under *Echinocereus enneacanthus*, add to illustrations: Bull. Univ. Tex. **60**: pl. 11, f. 1, as *Cereus longispinus*; Watson, Cact. Cult. 75. f. 21; ed. 3. f. 15, as *C. enneacanthus*.

Also insert: *Echinocereus saltillensis* is offered for sale by Haage and Schmidt, 1920, page 75.

Mr. C. R. Orcutt has called our attention to the following varieties which have been omitted: *Cereus englemannii* var. *albispinus* Cels, var. *caespitosus*, var. *fulvispinus* Cels, var. *pfersdorffii* Heiden, all of which are listed by him (Orcutt, Rev. Cact. **1**: 13. 1897),

On page 45, vol. III, insert: The name *Cactus bertini* was given for this plant when awarded a silver medal soon after its discovery (Hort. Franc. II. **5**: 222).

On page 45, vol. III, *Rebutia minuscula*, add to illustrations: Succulenta **3**: 96; Thomas, Zimmerkultur Kakteen 34; Kaktusy 25, as *Echinocactus minusculus*.

On page 48, vol. III, under *Chamaecereus silvestrii*, add to illustrations: Blühende Kakteen **3**: pl. 168, as *Cereus silvestrii*.

On page 48, vol. III, substitute for *Echinopsis deminuta:*

5. **Rebutia deminuta** (Weber).

> *Echinopsis deminuta* Weber, Bull. Mus. Hist. Nat. Paris **10**: 386. 1904.

Through the kindness of J. J. Verbeek Wolthuys, we have been able to examine a flower of this plant which shows that it belongs to the genus *Rebutia*.

On page 54, vol. III, under *Lobivia pentlandii*, add to illustrations: Watson, Cact. Cult. ed. 3. f. 32, as *Echinopsis pentlandii*.

On page 59, vol. III, insert the following:

21. Lobivia famatimensis (Spegazzini).

Echinocactus famatimensis Spegazzini, Anal. Soc. Cient. Argentina 92: 44. 1921.

Solitary or in clusters, short-cylindric, 3 to 3.5 cm. high, 2.5 to 2.8 cm. in diameter, strongly umbilicate at apex; ribs 24, low, obtuse, somewhat tuberculate; areoles approximate; spines small, appressed, whitish; flowers solitary, from the side near the middle, about 3 cm. long.

Type locality: Near Famatima, Argentina, altitude 2,000 to 3,000 meters.

Distribution: Province of La Rioja, Argentina.

Illustration: Anal. Soc. Cient. Argentina 92: f. 9, as *Echinocactus famatimensis*.

On page 64, vol. II, under *Echinopsis multiplex*, add to illustrations: Watson, Cact. Cult. ed. 3. f. 16, as *Cereus multiplex*; Rev. Hort. 48: 13. f. 1, as *Echinocactus multiplex*; Rev. Hort. 48: 13. f. 2, as *E. multiplex cristata*; Gard. Chron. III. 56: 145. f. 60.

On page 65, vol. III, under *Echinopsis oxygona*, add to illustrations: Thomas, Zimmerkultur Kakteen 23, as *Echinopsis oxygona inermis*; Succulenta 5: 85.

On page 66, vol. III, under *Echinopsis eyriesii*, add to illustrations: Rother, Praktischer Leitfaden Kakteen 47, as *Echinopsis triumphans*; Remark, Kakteenfreund 9; Rother, Praktischer Leitfaden Kakteen 45, 106.

On page 67, vol. III, under *Echinopsis turbinata*, add to illustrations: Watson, Cact. Cult. 131. f. 50; ed. 3. f. 30; Gard. Chron. III. 16: 625. f. 79, as *Echinopsis decaisneana*; Floralia 42: 374, as *E. gemmata*.

On page 67, vol. III, under *Echinopsis tubiflora*, add to illustrations: Thomas, Zimmerkultur Kakteen 20, as *Echinopsis tubiflora rohlandii*; Rother, Praktischer Leitfaden Kakteen 44, as *E. zuccariniana*.

On page 72, vol. III, under *Echinopsis leucantha*, add to illustrations: Thomas, Zimmerkultur Kakteen 21.

On page 74, vol. III, under *Echinopsis bridgesii*, insert: *Illustration:* Möllers Deutsche Gärt. Zeit. 25: 475. f. 7, No. 18, as *Echinopsis salmiana*.

On page 75, vol. III, under *Echinopsis formosa*, add to illustrations: Monatsschr. Kakteenk. 32: 149.

On page 76, vol. III, *Echinopsis formosissima*, insert: *Cereus formosissimus* Weber (Dict. Hort. Bois 471. 1896) was cited by Weber as a synonym of this species.

On page 80, vol. III, under *Ariocarpus retusus*, add the synonyms: *Cactus areolosus* Kuntze, Rev. Gen. Pl. 1: 260. 1891; *Cactus pulvilliger* Kuntze, Rev. Gen. Pl. 1: 260. 1891. Add to illustrations: Monatsschr. Kakteenk. 23: 66, 67, as *Ariocarpus trigonus*.

On page 82, vol. III, under *Ariocarpus kotschoubeyanus* insert: *Anhalonium kotschubeyi* Lemaire (Salm-Dyck, Cact. Hort. Dyck. 1849. 5. 1850), given as a synonym of *A. sulcatum,* is to be referred here.

On page 83, vol. III, under *Ariocarpus fissuratus*, add to illustrations: Bull Univ. Texas 60: pl. 11, f. 2, as *Ariocarpus fissuratus*; Gard. Chron. III. 12: 789. f. 130; Watson, Cact. Cult. 161. f. 61, as *Mammillaria fissurata*; Watson, Cact. Cult. ed. 3. f. 6, as *Anhalonium engelmannii*; Remark, Kakteenfreund 10; Balt. Cact. Journ. 1: 27; 2: 247, as *Anhalonium fissuratum*; Rother, Praktischer Leitfaden Kakteen 35.

On page 85, vol. III, under *Lophophora williamsii*, add to illustrations: Sci. Amer. 124: 492, as mescal button; Rother, Praktischer Leitfaden Kakteen 36; Remark, Kakteenfreund 11; Karsten and Schenck, Vegetationsbilder 2: pl. 20, B; Thomas, Zimmerkultur Kakteen 30; Monatsschr. Kakteenk. 31: 187, as *Echinocactus williamsii*; Balt. Cact. Journ. 1: 71; 2: 247; Watson, Cact. Cult. ed. 2. 243. f. 91; ed. 3. f. 7, as *Anhalonium williamsii*; Succulenta 2: 3; 4: 7.

On page 91, vol. III, under *Pediocactus simpsonii*, add to illustrations: Wiener Obst. Zeit. 2: 90. f. 13; Remark, Kakteenfreund 13, as *Echinocactus simpsonii*.

On page 106, vol. III, under *Hamatocactus setispinus*, add to illustrations: Schulz, 500 Wild Fl. San Antonio pl. 12, as *Echinocactus setispinus*.

On page 107, vol. III, under *Strombocactus disciformis*, add to illustrations: Remark, Kakteenfreund 12, as *Echinocactus turbiniformis*.

On page 123, vol. III, at end of *Echinofossulocactus*, insert the following:

Echinocactus tetracentrus Lemaire (Cact. Gen. Nov. Sp. 31. 1839) has not been identified. It is to be referred to one of the species of *Echinofossulocactus*.

Echinocactus barcelona (Cact. Journ. **2**: 79, 175, 191) was offered for sale by F. A. Walton.

On page 124, vol. III, under *Ferocactus stainesii*, insert: *Echinocactus pilosus canescens* Scheidweiler is listed in Index Bibliographique (286. 1887).

On page 129, vol. III, under *Ferocactus lecontei*, insert: *Echinocactus leopoldii* (Belg. Hort. **25**: 132. 1876) was only briefly described when awarded a prize in Belgium. Schumann referred it as a synonym of *E. cylindraceus*, while De Laet, who saw the plant, thought it was a form of *Ferocactus lecontei*. It is misspelled in Volume III.

Add to illustrations: Wiener Ill. Gart. Zeit. **11**: pl. 3, in part; Thomas, Zimmerkultur Kakteen 32; Watson, Cact. Cult. ed. 3. f. 26, as *Echinocactus lecontei*.

On page 130, vol. III, under *Ferocactus acanthodes*, add to illustrations: Thomas, Zimmerkultur Kakteen 40, as *Echinocactus cylindraceus*.

On page 132, vol. III, insert:

11a. Ferocactus johnstonianus Britton and Rose, sp. nov.

Plants simple, short-cylindric, 6 dm. high or less, up to 3.5 dm. in diameter; ribs 24 to 31, with margins undulate; areoles elliptic, rather closely set; spines 20 or more, subulate, very much alike, none hooked, slightly spreading and more or less outwardly recurved, 7 cm. long or less, yellow to brownish yellow, annulate; flowers including ovary 5 cm. long; perianth-segments narrow, yellowish, or the outer ones tinged with red, short-acuminate, the margins slightly erose; filaments yellowish below, becoming reddish above; stigma-lobes 8 to 13, flesh-colored; scales on the ovary orbicular; fruit small, 2.5 cm. in diameter, the seed dehiscing by a large pore at the base; seeds angled, black, pitted, 2 mm. long; hilum small, circular, depressed, white.

Collected by Ivan M. Johnston at Angel de la Guardia Island, Lower California May 2, 1921 (Nos. 3394, type, and 3395).

This species is perhaps nearest *Ferocactus diguetii* but is much smaller and has fewer ribs, many more spines in a cluster, and yellow flowers.

On page 140, vol. III, under *Ferocactus viridescens*, add to illustrations: Blühende Kakteen **3**: pl. 177, as *Echinocactus viridescens*.

On page 143, vol. III, under *Ferocactus latispinus*, add to illustrations: Remark, Kakteenfreund 13, as *Echinocactus cornigerus flavispinus*.

Also add the note: *Cactus cornigereus* Mociño and Sessé (De Candolle, Prodr. **3**: 461. 1828) is given as a synonym of *Echinocactus cornigerus*.

On page 144, vol. III, under *Ferocactus hamatacanthus*, add to illustrations: Rother, Praktischer Leitfaden Kakteen 106, as *Echinocactus longihamatus*.

On page 148, vol. III, under *Echinomastus erectocentrus*, insert: Mr. C. R. Orcutt has called our attention to the fact that in *Echinomastus erectocentrus* the fruit opens by splitting down one side, and in this respect differs from our generic description. This character in the description was drawn from a study of the fruits of *E. intertextus*, the only species in this genus of which we know much about the fruit. He also states that the fruit of *Astrophytum myriostigma* splits open on one side, an observation we had not recorded.

On page 150, vol. III, under *Echinomastus unguispinus*, add to illustrations: Thomas, Zimmerkultur Kakteen 31; Schelle, Handb. Kakteenk. 200. f. 132, as *Echinocactus unguispinus*.

On page 155, vol. III, under *Gymnocalycium denudatum*, insert: *Echinocactus denudatus multiflorus* (Monatsschr. Kakteenk. **14**: 178) is only a name.

Also add to illustrations: Thomas, Zimmerkultur Kakteen 41, as *Echinocactus denudatus*.

On page 157, vol. III, under *Gymnocalycium saglione*, add to illustrations: Thomas, Zimmerkultur 37, as *Echinocactus saglionis*.

On page 158, vol. III, under *Gynocalycium gibbosum* add to illustrations: Van Géel, Sert. Bot. **1**: 113, as *Cactus gibbosus*.

On page 159, vol. III, under *Gymnocalycium brachyanthum*, insert: *Illustration:* Möllers Deutsche Gärt. Zeit. **36**: 145. No. 11, as *Echinocactus brachyanthus*.

On page 161, vol. III, under *Gymnocalycium monvillei*, add to illustrations: Möllers Deutsche Gärt. Zeit. **36**: 145. f. 1, as *Echinocactus monvillei*.

On page 168, vol. III, under *Echinocactus grusonii*, add to illustrations: Watson, Cact. Cult. ed. **3**. f. 23; Rother, Praktischer Leitfaden Kakteen 30; Deutsche Garten-Zeitung 28. f. 6; Zeitschrift Sukkulentk. **1**: 15.

On page 168, vol. III, under *Echinocactus ingens*, add to illustrations: Remark, Kakteenfreund 14.

On page 171, vol. III, under *Echinocactus visnaga*, add to illustrations: Balt. Cact. Journ. **2**: 181.

On page 181, vol. III, insert the following paragraphs:

Echinocactus acutispinus Hildmann (Deutsche Garten-Zeitung **1886**: 116. f. 27. 1886) was described and figured, but the plant is a small, barren one which we have not been able to associate with any described species. It came from Mexico and may be one of the species of *Echinocactus*.

Echinocactus cylindricus Hortus (Forbes, Hort. Tour Germ. 152) was described as cylindrical, with 12 or 13 ribs, the radial spines white and the central ones light brown. It was introduced from Mexico in 1836. It can not be identified.

On page 182, vol. III, under *Homalocephala texensis*, add to illustrations: Schulz, 500 Wild Flowers of San Antonio, pl. 13 in part as *Echinocactus texensis*.

Echinocactus darrahii Schumann (Monatsschr. Kakteenk. **12**: 21. 1902) is only mentioned.

Echinocactus dicracanthus Hortus (Forbes, Journ. Hort. Tour Germ. 160) is only a name.

Echinocactus inflatus Gillies (Steudel, Nom. ed. 2. **1**: 536. 1840) seems never to have been published. Steudel simply states that it was from Chile.

Echinocactus praegnacanthus Förster (Handb. Gartenz. **17**: 160. 1861) is a plant from Chile which has never been identified.

Echinocactus purpureus (Monatsschr. Kakteenk. **5**: 106. 1895) is listed by Schumann as in Gruson's Garden.

Echinocactus rhodanthus (Forbes, Journ. Hort. Tour Germ. 151) is only a name.

On page 182, vol. III, under *Astrophytum myriostigma*, add to illustrations: Remark, Kakteenfreund 11; Balt. Cact. Journ. **1**: 82.

On page 185, vol. III, under *Astrophytum capricorne*, add to illustrations: Remark, Kakteenfreund 12, as *Echinocactus capricornis*.

On page 188, vol. III, under *Malacocarpus tephracanthus*, add the synonym: *Echinocactus sellowii tetracanthus* Lemaire in Schumann, Monatsschr. Kakteenk. **18**: 150. 1908.

Insert: *Echinocactus buchheimianus* Haage in Quehl (Monatsschr. Kakteenk. **9**: 74. 1899) has been described briefly but its flower and fruit are unknown. It is said to resemble *E. sellowii*.

Add to illustrations: Schumann, Gesamtb. Kakteen f. 15, as *Cereus tephracanthus.*

On page 193, vol. III, under *Malacocarpus concinnus*, add to illustrations: Succulenta **3**: 22, 48, as *Echinocactus concinnus.*

On page 193, vol. III, under *Malacocarpus scopa*, add to illustrations: Rother, Praktischer Leitfaden Kakteen 107, as *Echinocactus scopa cristatus*; Kaktusy 26, as *Echinocactus scopa*; Kaktusy 27, as *E. scopa cristata*; Kaktusy 28, as *E. scopa rubra.*

On page 195, vol. III, under *Malacocarpus linkii*, insert: *Echinocactus ottonis linkii* Hortus (Förster, Handb. Cact. ed. 2. 554. 1885) is given as a synonym of *E. linkii.*

On page 196, vol. III, under *Malacocarpus ottonis*, add to illustrations: Succulenta **3**: 56, as *Echinocactus ottonis tenuispinus*; Rother, Praktischer Leitfaden Kakteen 34, as *E. ottonis.*

On page 198, vol. III, under *Malacocarpus erinaceus*, add to illustrations: Watson, Cact. Cult. 98. f. 32, as *Echinocactus corynodes.*

On page 200, vol. III, under *Malacocarpus mammulosus*, add to illustrations: Rother, Praktischer Leitfaden Kakteen 30, as *Echinocactus submammulosus.*

On page 202, vol. III, under *Malacocarpus haselbergii*, add to illustrations: Succulenta **3**: 31, as *Echinocactus haselbergii.*

On page 205, vol. III, under *Malacocarpus leninghausii*, add to illustrations: Succulenta **3**: 39, as *Echinocactus leninghausii.*

On page 207, vol. III, under *Hickenia microsperma*, add to illustrations: Succulenta **3**: 71; **5**: pl. 1; Thomas, Zimmerkultur Kakteen 35, as *Echinocactus microspermus.*

On page 237, vol. III, insert the following:

19. Cactus oaxacensis sp. nov.

Globular to ovoid, 12 to 15 cm. thick, with a small, low crown only 2 to 3 cm. high and 3 to 4 cm. broad; ribs 11 to 15, prominent, usually rounded; radial spines 8 to 12, subulate, more or less recurved at first, reddish brown but grayish in age, 2 cm. long or less; central spines 1 or sometimes 2, erect or porrect; flowers slender, about 2 cm. long, dark rose; filaments and style light yellow; fruit thick-clavate, 2 to 4.5 cm. long, scarlet, shiny; seeds small, black.

This plant was illustrated and mentioned in the place here cited (Cactaceae **3**: 237. f. 249) but was not given a specific name. Since then C. R. Orcutt reports finding it at Salina Cruz and Dr. B. P. Reko sends us a photograph and flowers obtained by him in 1923, while Dr. J. A. Purpus re-collected it in 1923 (type) and has sent us living plants.

Fig. 262.—Cactus oaxacensis.

Illustration: Cactaceae **3**: 236. f. 249, as *Cactus* sp.

Figure 262 is from a photograph of the plant sent us by Dr. Reko.

On page 238, vol. III, insert: *Melocactus ellemeetii* Miquel (Nederl. Kruidk. Arch. **4**: 336. 1858) and *M. pachycentrus* Suringar (Verh. Akad. Wettensch. Amst. II. **8**: 28. 1901) have not been identified.

On page 238, vol. III, at end of *Cactus* add: *Cactus aculeatissimus* is listed by Stendel (Nom. 131. 1821) credited to Zeyher and cited by the Index Kewensis, but it has never been identified.

Cactus tuna major is used by Roxburgh (Hort. Beng. 37. 1814).

Cactus reptans Willdenow (Ann. Hort. Berol. Suppl. 33. 1813) was taken up in this work by mistake as *Cereus reptans*.

Cactus neglectus Dehnhardt (Rivist. Napol. 1. **3**: 166.), according to the Index Kewensis, is a species of *Pereskia*.

Cereus erinaceus is credited to Haworth by Steudel (Nom. ed. 2. 1: 334. 1840) and said to come from the West Indies. Steudel must have had in mind *Cactus erinaceus* Haworth; if so, the plant is from South America.

Cereus torrellianus (Monatsschr. Kakteenk. **20**: 42. 1910) is probably a misspelling for *C. tonelianus*.

Figure 263, shown below, gives a typical Arizona landscape in which *Carnegiea gigantea* is the dominant plant.

INDEX.

This index covers the four volumes. The large capitals occurring before the figures indicate the particular volume. Pages of principal entries are in heavy-faced type.

291